PRINCIPLES OF
Virology
4TH EDITION

VOLUME II *Pathogenesis and Control*

PRINCIPLES OF
Virology
4TH EDITION

Jane Flint
Department of Molecular Biology
Princeton University
Princeton, New Jersey

Vincent R. Racaniello
Department of Microbiology & Immunology
College of Physicians and Surgeons
Columbia University
New York, New York

Glenn F. Rall
Fox Chase Cancer Center
Philadelphia, Pennsylvania

Anna Marie Skalka
Fox Chase Cancer Center
Philadelphia, Pennsylvania

with
Lynn W. Enquist
Department of Molecular Biology
Princeton University
Princeton, New Jersey

ASM
PRESS
WASHINGTON, DC

Library of Congress Cataloging-in-Publication Data

Flint, S. Jane, author.
Principles of virology / Jane Flint, Department of Molecular Biology, Princeton University, Princeton, New Jersey; Vincent R. Racaniello, Department of Microbiology, College of Physicians and Surgeons, Columbia University, New York, New York; Glenn F. Rall, Fox Chase Cancer Center, Philadelphia, Pennsylvania; Anna Marie Skalka, Fox Chase Cancer Center, Philadelphia, Pennsylvania; with Lynn W. Enquist, Department of Molecular Biology, Princeton University, Princeton, New Jersey.—4th edition.
pages cm
Revision of: Principles of virology / S.J. Flint ... [et al.]. 3rd ed.
Includes bibliographical references and index.
ISBN 978-1-55581-933-0 (v. 1 pbk.)—ISBN 978-1-55581-934-7 (v. 2 pbk.)—ISBN 978-1-55581-951-4 (set pbk.)—ISBN 978-1-55581-952-1 (set ebook) 1. Virology. I. Racaniello, V. R. (Vincent R.), author. II. Rall, Glenn F., author. III. Skalka, Anna M., author. IV. Enquist, L. W. (Lynn W.), author. V. Title.
QR360.P697 2015
616.9'101--dc23
2015026213
doi:10.1128/9781555818951 (Volume I)
doi:10.1128/9781555818968 (Volume II)
doi:10.1128/9781555819521 (e-bundle)

10 9 8 7 6 5 4 3 2 1

Address editorial correspondence to ASM Press, 1752 N St., N.W., Washington, DC 20036-2904, USA

Send orders to ASM Press, P.O. Box 605, Herndon, VA 20172, USA
Phone: 800-546-2416; 703-661-1593
Fax: 703-661-1501
E-mail: books@asmusa.org
Online: http://www.asmscience.org

Illustrations and illustration concepting: Patrick Lane, ScEYEnce Studios
Cover and interior design: Susan Brown Schmidler
Cover image: Courtesy of Jason A. Roberts (Victorian Infectious Diseases Reference Laboratory, Doherty Institute, Melbourne, Australia)
Back cover photos: Peter Kurilla Photography

We dedicate this book to the students, current and future scientists, physicians, and all those with an interest in the field of virology, for whom it was written.
We kept them ever in mind.

We also dedicate it to our families:
Jonn, Gethyn, and Amy Leedham
Doris, Aidan, Devin, and Nadia
Eileen, Kelsey, and Abigail
Rudy, Jeanne, and Chris
And
Kathy and Brian

Oh, be wiser thou!
Instructed that true knowledge leads to love.
WILLIAM WORDSWORTH
Lines left upon a Seat in a Yew-tree
1888

Contents

Preface   xvii
Acknowledgments   xxi
About the Authors   xxiii

1   Infections of Populations: History and Epidemiology   2

Introduction to Viral Pathogenesis   3

A Brief History of Viral Pathogenesis   4

The Relationships between Microbes and the Diseases They Cause   4
The First Human Viruses Identified and the Role of Serendipity   4
New Techniques Led to the Study of Viruses as Causes of Disease   7

Viral Epidemics in History   8

Epidemics Shaped History: the 1793 Yellow Fever Epidemic in Philadelphia   8
Tracking Epidemics by Sequencing: West Nile Virus Spread to the Western
    Hemisphere   9
The Economic Toll of Viral Epidemics in Agricultural Organisms   9
Population Density and World Travel as Accelerators of Viral Transmission   10
Zoonotic Infections and Viral Epidemics   10

Epidemiology   10

Fundamental Concepts   11
Tools of Epidemiology   13
Surveillance   15

**Parameters That Govern the Ability of
a Virus To Infect a Population   16**

Environment   16
Host Factors   19

Perspectives   22

References   23

2   Barriers to Infection   24

Introduction   25

An Overview of Infection and Immunity   25

A Game of Chess Played by Age-Old Masters   25

Initiating an Infection   27

Successful Infections Must Modulate or Bypass Host Defenses   29

Skin   29

Respiratory Tract   31

Alimentary Tract   33

Urogenital Tract   35

Eyes   35

Viral Tropism   36

Accessibility of Viral Receptors   36

Host Cell Proteins That Regulate the Infectious Cycle   36

Spread throughout the Host   39

Hematogenous Spread   40

Neural Spread   42

Organ Invasion   45

Entry into Organs with Sinusoids   45

Entry into Organs That Lack Sinusoids   46

Organs with Dense Basement Membranes   46

Skin   47

The Fetus   47

Shedding of Virus Particles   47

Respiratory Secretions   48

Saliva   48

Feces   49

Blood   49

Urine   49

Semen   49

Milk   49

Skin Lesions   49

Perspectives   50

References   50

3   The Early Host Response: Cell-Autonomous and Innate Immunity   52

Introduction   53

The First Critical Moments of Infection: How Do Individual Cells Detect a Virus Infection?   54

Cell Signaling Induced by Receptor Engagement   55

Receptor-Mediated Recognition of Microbe-Associated Molecular Patterns   55

Cellular Changes That Occur Following Viral Infection   60

Intrinsic Responses to Infection   62
Apoptosis (Programmed Cell Death)   62
Other Intrinsic Immune Defenses   67
The Continuum between Intrinsic and Innate Immunity   74

Soluble Immune Mediators of the Innate Immune Response   74
Overview of Cytokine Functions   74
Interferons, Cytokines of Early Warning and Action   76
Chemokines   86

The Innate Immune Response   87
Complement   87
Natural Killer Cells   90
Other Innate Immune Cells of Relevance to Viral Infections   92

Perspectives   93
References   94

4 **Adaptive Immunity and the Establishment of Memory   98**

Introduction   99
Attributes of the Host Response   99
Speed   99
Diversity and Specificity   100
Memory   100
Self-Control   100

Lymphocyte Development, Diversity, and Activation   100
All Blood Cells Derive from a Common Hematopoietic Stem Cell   100
The Two Arms of Adaptive Immunity   101
The Major Effectors of the Adaptive Response: B Cells and T Cells   101
Diverse Receptors Impart Antigen Specificity to B and T Cells   107

Events at the Site of Infection Set the Stage for the Adaptive Response   108
Acquisition of Viral Proteins by Professional Antigen-Presenting Cells Enables Production of Proinflammatory Cytokines and Establishment of Inflammation   108
Antigen-Presenting Cells Leave the Site of Infection and Migrate to Lymph Nodes   111

Antigen Processing and Presentation   114
Professional Antigen-Presenting Cells Induce Activation via Costimulation   114
Presentation of Antigens by Class I and Class II MHC Proteins   115
Lymphocyte Activation Triggers Massive Cell Proliferation   119

The Cell-Mediated Response   119
CTLs Lyse Virus-Infected Cells   119
Control of CTL Proliferation   122
Noncytolytic Control of Infection by T Cells   122
Rashes and Poxes   122

The Humoral (Antibody) Response   122
Antibodies Are Made by Plasma Cells   122
Types and Functions of Antibodies   125
Virus Neutralization by Antibodies   125
Antibody-Dependent Cell-Mediated Cytotoxicity:
    Specific Killing by Nonspecific Cells   127
Immunological Memory   128
Perspectives   130
References   130

5   **Mechanisms of Pathogenesis   134**
Introduction   135
Animal Models of Human Diseases   135
Patterns of Infection   136
Incubation Periods   137
Mathematics of Growth Correlate with Patterns of Infection   138
Acute Infections   139
Persistent Infections   143
Latent Infections   150
"Slow" Infections   157
Abortive Infections   157
Transforming Infections   157
Viral Virulence   158
Measuring Viral Virulence   158
Alteration of Viral Virulence   159
Viral Virulence Genes   160
Pathogenesis   164
Infected Cell Lysis   164
Immunopathology   164
Immunosuppression Induced by Viral Infection   168
Oncogenesis   169
Molecular Mimicry   170
Perspectives   172
References   172

6   **Cellular Transformation and Oncogenesis   174**
Introduction   175
Properties of Transformed Cells   175
Control of Cell Proliferation   178
Oncogenic Viruses   182
Discovery of Oncogenic Viruses   182
Viral Genetic Information in Transformed Cells   187
The Origin and Nature of Viral Transforming Genes   189
Functions of Viral Transforming Proteins   192

Activation of Cellular Signal Transduction Pathways by Viral Transforming Proteins 192
Viral Signaling Molecules Acquired from the Cell 192
Alteration of the Production or Activity of
Cellular Signal Transduction Proteins 195

Disruption of Cell Cycle Control Pathways by Viral Transforming Proteins 201
Abrogation of Restriction Point Control Exerted by the Rb Protein 201
Production of Virus-Specific Cyclins 204
Inactivation of Cyclin-Dependent Kinase Inhibitors 204

Transformed Cells Must Grow and Survive 206
Mechanisms That Permit Survival of Transformed Cells 206

Tumorigenesis Requires Additional Changes in the Properties of Transformed Cells 209
Inhibition of Immune Defenses 210

Other Mechanisms of Transformation and Oncogenesis by Human Tumor Viruses 210
Nontransducing Oncogenic Retroviruses: Tumorigenesis with
Very Long Latency 210
Oncogenesis by Hepatitis Viruses 213

Perspectives 214

References 214

7 **Human Immunodeficiency Virus Pathogenesis 218**

Introduction 219
Worldwide Impact of AIDS 219

HIV Is a Lentivirus 219
Discovery and Characterization 219
Distinctive Features of the HIV Reproduction Cycle and the Functions of
Auxiliary Proteins 222
The Viral Capsid Counters Intrinsic Defense Mechanisms 230

Cellular Targets 230

Routes of Transmission 231
Modes of Transmission 231
Mechanics of Spread 233

The Course of Infection 234
The Acute Phase 234
The Asymptomatic Phase 235
The Symptomatic Phase and AIDS 236
Variability of Response to Infection 236

Origins of Cellular Immune Dysfunction 237
CD4$^+$ T Lymphocytes 237
Cytotoxic T Lymphocytes 238

Monocytes and Macrophages   238
B Cells   238
Natural Killer Cells   238
Autoimmunity   238

Immune Responses to HIV   238
Innate Response   238
The Cell-Mediated Response   238
Humoral Responses   239
Summary: the Critical Balance   240

Dynamics of HIV-1 Reproduction in AIDS Patients   241
Effects of HIV on Different Tissues and Organ Systems   242
Lymphoid Organs   242
The Nervous System   244
The Gastrointestinal System   245
Other Organs and Tissues   246

HIV and Cancer   246
Kaposi's Sarcoma   246
B Cell Lymphomas   248
Anogenital Carcinomas   248

Prospects for Treatment and Prevention   248
Antiviral Drugs   248
Confronting the Problems of Persistence and Latency   249
Gene Therapy Approaches   249
Immune System-Based Therapies   250
Antiviral Drug Prophylaxis   250

Perspectives   250
References   252

8   **Vaccines**   254

Introduction   255
The Origins of Vaccination   255
Smallpox: a Historical Perspective   255
Large-Scale Vaccination Programs Can Be Dramatically Effective   257

Vaccine Basics   260
Immunization Can Be Active or Passive   260
Active Vaccination Strategies Stimulate Immune Memory   261
The Fundamental Challenge   264

The Science and Art of Making Vaccines   265
Inactivated or "Killed" Virus Vaccines   265
Attenuated Virus Vaccines   269
Subunit Vaccines   270
Recombinant DNA Approaches to Subunit Vaccines   272
Virus-Like Particles   272
DNA Vaccines   273
Attenuated Viral Vectors and Foreign Gene Expression   274

Vaccine Technology: Delivery and Improving Antigenicity   275

Adjuvants Stimulate an Immune Response   275

Delivery and Formulation   276

Immunotherapy   276

The Quest for an AIDS Vaccine   277

Formidable Challenges and Promising Leads   277

Perspectives   279

References   279

9   **Antiviral Drugs**   282

Introduction   283

Historical Perspective   283

Discovering Antiviral Compounds   284

The Lexicon of Antiviral Discovery   284

Screening for Antiviral Compounds   285

Computational Approaches to Drug Discovery   287

The Difference between "R" and "D"   288

Examples of Some Antiviral Drugs   293

Approved Inhibitors of Viral Nucleic Acid Synthesis   293

Approved Drugs That Are Not Inhibitors of Nucleic Acid Synthesis   298

Expanding Target Options for Antiviral Drug Development   300

Entry and Uncoating Inhibitors   300

Viral Regulatory Proteins   300

Regulatory RNA Molecules   300

Proteases and Nucleic Acid Synthesis and Processing Enzymes   301

Two Success Stories: Human Immunodeficiency and Hepatitis C Viruses   301

Inhibitors of Human Immunodeficiency Virus and Hepatitis C Virus Polymerases   303

Human Immunodeficiency Virus and Hepatitis C Virus Protease Inhibitors   306

Human Immunodeficiency Virus Integrase Inhibitors   306

Hepatitis C Virus Multifunctional Protein NS5A   308

Inhibitors of Human Immunodeficiency Virus Fusion and Entry   309

Drug Resistance   309

Combination Therapy   310

Challenges Remaining   312

Perspectives   312

References   314

10   **Evolution**   316

Virus Evolution   317

Classic Theory of Host-Parasite Interactions   317

How Do Virus Populations Evolve?   318
Two General Survival Strategies Can Be Distinguished   319
Large Numbers of Viral Progeny and Mutants Are Produced in Infected Cells   319
The Quasispecies Concept   321
Sequence Conservation in Changing Genomes   321
Genetic Shift and Genetic Drift   324
Fundamental Properties of Viruses That Constrain and Drive Evolution   326

The Origin of Viruses   327

Host-Virus Relationships Drive Evolution   333
DNA Virus Relationships   333
RNA Virus Relationships   333
The Protovirus Hypothesis for Retroviruses   335

Lessons from Paleovirology   335
Endogenous Retroviruses   336
DNA Fossils Derived from Other RNA Viral Genomes   337
Endogenous Sequences from DNA Viruses   337
The Host-Virus "Arms Race"   337

Perspectives   339
References   340

11 Emergence   342

The Spectrum of Host-Virus Interactions   343
Stable Interactions   344
The Evolving Host-Virus Interaction   345
The Dead-End Interaction   345
Common Sources of Animal-to-Human Transmission   347
The Resistant Host   348

Encountering New Hosts: Ecological Parameters   348
Successful Encounters Require Access to Susceptible and Permissive Cells   349
Population Density, Age, and Health Are Important Factors   350
Experimental Analysis of Host-Virus Interactions   350
Learning from Accidental Infections   351

Expanding Viral Niches: Some Well-Documented Examples   351
Poliomyelitis: Unexpected Consequences of Modern Sanitation   351
Smallpox and Measles: Exploration and Colonization   352

Notable Zoonoses   352
Hantavirus Pulmonary Syndrome: Changing Climate and Animal Populations   352
Severe Acute and Middle East Respiratory Syndromes (SARS and MERS): Two New Zoonotic Coronavirus Infections   352
Acquired Immunodeficiency Syndrome (AIDS): Pandemic from a Zoonotic Infection   353

Host Range Can Be Expanded by Mutation, Recombination, or Reassortment   354
Canine Parvoviruses: Cat-to-Dog Host Range Change by Two Mutations   354
Influenza Epidemics and Pandemics: Escaping the Immune Response by Reassortment   354

New Technologies Uncover Hitherto Unrecognized Viruses   355

Hepatitis Viruses in the Human Blood Supply   357

A Revolution in Virus Discovery   357

Perceptions and Possibilities   358

Virus Names Can Be Misleading   359

All Viruses Are Important   359

What Next?   359

Can We Predict the Next Viral Pandemic?   359

Emerging Viral Infections Illuminate Immediate Problems and Issues   360

Humans Constantly Provide New Venues for Infection   360

Preventing Emerging Virus Infections   361

Perspectives   361

References   362

12   **Unusual Infectious Agents   364**

Introduction   365

Viroids   365

Replication   365

Sequence Diversity   366

Movement   366

Pathogenesis   368

Satellites   368

Replication   369

Pathogenesis   369

Virophages or Satellites?   369

Hepatitis Delta Satellite Virus   370

Prions and Transmissible Spongiform Encephalopathies   371

Scrapie   371

Physical Nature of the Scrapie Agent   371

Human TSEs   371

Hallmarks of TSE Pathogenesis   372

Prions and the *prnp* Gene   372

Prion Strains   374

Bovine Spongiform Encephalopathy   374

Chronic Wasting Disease   376

Treatment of Prion Diseases   377

Perspectives   377

References   378

APPENDIX Diseases, Epidemiology, and Disease Mechanisms of Selected Animal Viruses Discussed in This Book   379

Glossary   407

Index   413

Preface

*The enduring goal of scientific endeavor, as of all human enterprise, I imagine, is to
achieve an intelligible view of the universe. One of the great discoveries of modern science
is that its goal cannot be achieved piecemeal, certainly not by the accumulation of facts.
To understand a phenomenon is to understand a category of phenomena or it is nothing.
Understanding is reached through creative acts.*

A. D. HERSHEY
Carnegie Institution Yearbook 65

All four editions of this textbook have been written according to the authors' philosophy
that the best approach to teaching introductory virology is by emphasizing shared principles. Studying the phases of the viral reproductive cycle, illustrated with a set of representative viruses, provides an overview of the steps required to maintain these infectious agents in
nature. Such knowledge cannot be acquired by learning a collection of facts about individual
viruses. Consequently, the major goal of this book is to define and illustrate the basic principles of animal virus biology.

In this information-rich age, the quantity of data describing any given virus can be overwhelming, if not indigestible, for student and expert alike. The urge to write more and more
about less and less is the curse of reductionist science and the bane of those who write textbooks meant to be used by students. In the fourth edition, we continue to distill information with the intent of extracting essential principles, while providing descriptions of how
the information was acquired. Boxes are used to emphasize major principles and to provide
supplementary material of relevance, from explanations of terminology to descriptions of
trail-blazing experiments. Our goal is to illuminate process and strategy as opposed to listing
facts and figures. In an effort to make the book readable, rather than comprehensive, we are
selective in our choice of viruses and examples. The encyclopedic *Fields Virology* (2013) is recommended as a resource for detailed reviews of specific virus families.

What's New

This edition is marked by a change in the author team. Our new member, Glenn Rall, has
brought expertise in viral immunology and pathogenesis, pedagogical clarity, and down-to-
earth humor to our work. Although no longer a coauthor, our colleague Lynn Enquist has
continued to provide insight, advice, and comments on the chapters.

Each of the two volumes of the fourth edition has a unique appendix and a general glossary.
Links to Internet resources such as websites, podcasts, blog posts, and movies are provided;
the digital edition provides one-click access to these materials.

A major new feature of the fourth edition is the incorporation of in-depth video interviews with scientists who have made a major contribution to the subject of each chapter. Students will be interested in these conversations, which also explore the factors that motivated the scientists' interest in the field and the personal stories associated with their contributions.

Volume I covers the molecular biology of viral reproduction, and Volume II focuses on viral pathogenesis, control of virus infections, and virus evolution. The organization into two volumes follows a natural break in pedagogy and provides considerable flexibility and utility for students and teachers alike. The volumes can be used for two courses, or as two parts of a one-semester course. The two volumes differ in content but are integrated in style and presentation. In addition to updating the chapters and Appendices for both volumes, we have organized the material more efficiently and new chapters have been added.

As in our previous editions, we have tested ideas for inclusion in the text in our own classes. We have also received constructive comments and suggestions from other virology instructors and their students. Feedback from students was particularly useful in finding typographical errors, clarifying confusing or complicated illustrations, and pointing out inconsistencies in content.

For purposes of readability, references are generally omitted from the text, but each chapter ends with an updated list of relevant books, review articles, and selected research papers for readers who wish to pursue specific topics. In general, if an experiment is featured in a chapter, one or more references are listed to provide more detailed information.

Principles Taught in Two Distinct, but Integrated Volumes

These two volumes outline and illustrate the strategies by which all viruses reproduce, how infections spread within a host, and how they are maintained in populations. The principles of viral reproduction established in Volume I are essential for understanding the topics of viral disease, its control, and the evolution of viruses that are covered in Volume II.

Volume I The Science of Virology and the Molecular Biology of Viruses

This volume examines the molecular processes that take place in an infected host cell. It begins with a general introduction and historical perspectives, and includes descriptions of the unique properties of viruses (Chapter 1). The unifying principles that are the foundations of virology, including the concept of a common strategy for viral propagation, are then described. An introduction to cell biology, the principles of the infectious cycle, descriptions of the basic techniques for cultivating and assaying viruses, and the concept of the single-step growth cycle are presented in Chapter 2.

The fundamentals of viral genomes and genetics, and an overview of the surprisingly limited repertoire of viral strategies for genome replication and mRNA synthesis, are topics of Chapter 3. The architecture of extracellular virus particles in the context of providing both protection and delivery of the viral genome in a single vehicle are considered in Chapter 4. Chapters 5 through 13 address the broad spectrum of molecular processes that characterize the common steps of the reproductive cycle of viruses in a single cell, from decoding genetic information to genome replication and production of progeny virions. We describe how these common steps are accomplished in cells infected by diverse but representative viruses, while emphasizing common principles. Volume I concludes with a new chapter, "The Infected Cell," which presents an integrated description of cellular responses to illustrate the marked, and generally, irreversible, impact of virus infection on the host cell.

The appendix in Volume I provides concise illustrations of viral life cycles for members of the main virus families discussed in the text; five new families have been added in the fourth edition. It is intended to be a reference resource when reading individual chapters and a convenient visual means by which specific topics may be related to the overall infectious cycles of the selected viruses.

Volume II Pathogenesis, Control, and Evolution

This volume addresses the interplay between viruses and their host organisms. The first five chapters have been reorganized and rewritten to reflect our growing appreciation of the host immune response and how viruses cause disease. In Chapter 1 we introduce the discipline of epidemiology, provide historical examples of epidemics in history, and consider basic aspects that govern how the susceptibility of a population is controlled and measured. With an understanding of how viruses affect human populations, subsequent chapters focus on the impact of viral infections on hosts, tissues and individual cells. Physiological barriers to virus infections, and how viruses spread in a host, invade organs, and spread to other hosts are the topics of Chapter 2. The early host response to infection, comprising cell autonomous (intrinsic) and innate immune responses, are the topics of Chapter 3, while the next chapter considers adaptive immune defenses, that are tailored to the pathogen, and immune memory. Chapter 5 focuses on the classic patterns of virus infection within cells and hosts, the myriad ways that viruses cause illness, and the value of animal models in uncovering new principles of viral pathogenesis. In Chapter 6, we discuss virus infections that transform cells in culture and promote oncogenesis (the formation of tumors) in animals. Chapter 7 is devoted entirely to the AIDS virus, not only because it is the causative agent of the most serious current worldwide epidemic, but also because of its unique and informative interactions with the human immune defenses.

Next, we consider the principles involved in treatment and control of infection. Chapter 8 focuses on vaccines, and Chapter 9 discusses the approaches and challenges of antiviral drug discovery. The topics of viral evolution and emergence have now been divided into two chapters. The origin of viruses, the drivers of viral evolution, and host-virus conflicts are the subjects of Chapter 10. The principles of emerging virus infections, and humankind's experiences with epidemic and pandemic viral infections, are considered in Chapter 11. Volume II ends with a new chapter on unusual infectious agents, viroids, satellites, and prions.

The Appendix of Volume II provides snapshots of the pathogenesis of common human viruses. This information is presented in four illustrated panels that summarize the viruses and diseases, epidemiology, disease mechanisms, and human infections.

Reference

Knipe DM, Howley PM (ed). 2013. *Fields Virology,* 6th ed. Lippincott Williams & Wilkins, Philadelphia, PA.

For some behind-the-scenes information about how the authors created the fourth edition of *Principles of Virology*, see: http://bit.ly/Virology_MakingOf

Acknowledgments

These two volumes of *Principles* could not have been composed and revised without help and contributions from many individuals. We are most grateful for the continuing encouragement from our colleagues in virology and the students who use the text. Our sincere thanks also go to colleagues (listed in the Acknowledgments for the third edition) who have taken considerable time and effort to review the text in its evolving manifestations. Their expert knowledge and advice on issues ranging from teaching virology to organization of individual chapters and style were invaluable, and are inextricably woven into the final form of the book.

We also are grateful to those who gave so generously of their time to serve as expert reviewers of individual chapters or specific topics in these two volumes: Siddharth Balachandran (Fox Chase Cancer Center), Patrick Moore (University of Pittsburgh), Duane Grandgenett (St. Louis University), Frederick Hughson (Princeton University), Bernard Moss (Laboratory of Viral Diseases, National Institutes of Health), Christoph Seeger (Fox Chase Cancer Center), and Thomas Shenk (Princeton University). Their rapid responses to our requests for details and checks on accuracy, as well as their assistance in simplifying complex concepts, were invaluable. All remaining errors or inconsistencies are entirely ours.

Since the inception of this work, our belief has been that the illustrations must complement and enrich the text. Execution of this plan would not have been possible without the support of Christine Charlip (Director, ASM Press), and the technical expertise and craft of our illustrator. The illustrations are an integral part of the text, and credit for their execution goes to the knowledge, insight, and artistic talent of Patrick Lane of ScEYEnce Studios. We also are indebted to Jason Roberts (Victorian Infectious Diseases Reference Laboratory, Doherty Institute, Melbourne, Australia) for the computational expertise and time he devoted to producing the beautiful renditions of poliovirus particles on our new covers. As noted in the figure legends, many could not have been completed without the help and generosity of numerous colleagues who provided original images. Special thanks go to those who crafted figures or videos tailored specifically to our needs, or provided multiple pieces: Chantal Abergel (CNRS, Aix-Marseille Université, France), Mark Andrake (Fox Chase Cancer Center), Timothy Baker (University of California), Bruce Banfield (The University of Colorado), Christopher Basler and Peter Palese (Mount Sinai School of Medicine), Ralf Bartenschlager (University of Heidelberg, Germany), Eileen Bridge (Miami University, Ohio), Richard Compans (Emory University), Kartik Chandran (Albert Einstein College of Medicine), Paul Duprex (Boston University School of Medicine), Ramón González (Universidad Autónoma del Estado de Morelos), Urs Greber (University of Zurich), Reuben Harris (University of Minnesota), Hidesaburo Hanafusa (deceased), Ari Helenius (University of Zurich), David Knipe (Harvard Medical School), J. Krijnse-Locker (University of Heidelberg, Germany), Petr G. Leiman (École Polytechnique Fédérale de Lausanne), Stuart Le Grice (National Cancer Institute, Frederick MD), Hongrong Liu (Hunan Normal University), David McDonald (Ohio State University), Thomas Mettenleiter (Federal Institute for Animal Diseases, Insel Reims, Germany), Bernard Moss (Laboratory of Viral Diseases, National Institutes of Health), Norm Olson (University of California), B. V. Venkataram Prasad (Baylor College of Medicine), Andrew Rambaut (University of Edinburgh), Jason Roberts (Victorian Infectious Diseases Reference Laboratory, Doherty Institute, Melbourne, Australia), Felix Rey (Institut Pasteur, Paris, France), Michael Rossmann (Purdue University), Anne Simon (University of Maryland), Erik Snijder (Leiden University Medical Center), Alasdair Steven (National Institutes of Health), Paul Spearman (Emory University), Wesley Sundquist (University of Utah), Livia Varstag (Castleton State College, Vermont), Jiri Vondrasek (Institute of organic Chemistry and Biochemistry, Czech Republic), Matthew Weitzman (University of Pennsylvania), Sandra Weller (University of Connecticut Health Sciences Center, Connecticut), Tim Yen (Fox Chase Cancer Center), and Z. Hong Zhou (University of California, Los Angeles).

The collaborative work undertaken to prepare the fourth edition was facilitated greatly by several authors' retreats. ASM Press generously provided financial support for these retreats as well as for our many other meetings.

We thank all those who guided and assisted in the preparation and production of the book: Christine Charlip (Director, ASM Press) for steering us through the complexities inherent in a team effort, Megan Angelini and John Bell (Production Managers, ASM Press) for keeping us on track during production, and Susan Schmidler for her elegant and creative designs for the layout and cover. We are also grateful for the expert secretarial and administrative support from Ellen Brindle-Clark (Princeton University) that facilitated preparation of this text. Special thanks go to Ellen for obtaining many of the permissions required for the figures.

There is little doubt in undertaking such a massive effort that inaccuracies still remain, despite our best efforts to resolve or prevent them. We hope that the readership of this edition will draw our attention to them, so that these errors can be eliminated from future editions of this text.

This often-consuming enterprise was made possible by the emotional, intellectual, and logistical support of our families, to whom the two volumes are dedicated.

About the Authors

Jane Flint is a Professor of Molecular Biology at Princeton University. Dr. Flint's research focuses on investigation of the molecular mechanisms by which viral gene products modulate host cell pathways and antiviral defenses to allow efficient reproduction in normal human cells of adenoviruses, viruses that are widely used in such therapeutic applications as gene transfer and cancer treatment. Her service to the scientific community includes membership of various editorial boards and several NIH study sections and other review panels. Dr. Flint is currently a member of the Biosafety Working Group of the NIH Recombinant DNA Advisory Committee.

Vincent Racaniello is Higgins Professor of Microbiology & Immunology at Columbia University Medical Center. Dr. Racaniello has been studying viruses for over 35 years, including poliovirus, rhinovirus, enteroviruses, and hepatitis C virus. He teaches virology to graduate, medical, dental, and nursing students and uses social media to communicate the subject outside of the classroom. His Columbia University undergraduate virology lectures have been viewed by thousands at iTunes University, Coursera, and on YouTube. Vincent blogs about viruses at virology.ws and is host of the popular science program This Week in Virology.

Glenn Rall is a Professor and the Co-Program Leader of the Blood Cell Development and Function Program at the Fox Chase Cancer Center in Philadelphia. At Fox Chase, Dr. Rall is also the Associate Chief Academic Officer and Director of the Postdoctoral Program. He is an Adjunct Professor in the Microbiology and Immunology departments at the University of Pennsylvania, Thomas Jefferson, Drexel, and Temple Universities. Dr. Rall's laboratory studies viral infections of the brain and the immune responses to those infections, with the goal of defining how viruses contribute to disease in humans. His service to the scientific community includes membership on the Autism Speaks Scientific Advisory Board, Opinions Editor of *PLoS Pathogens*, chairing the Education and Career Development Committee of the American Society for Virology, and membership on multiple NIH grant review panels.

Anna Marie Skalka is a Professor and the W.W. Smith Chair in Cancer Research at Fox Chase Cancer Center in Philadelphia and an Adjunct Professor at the University of Pennsylvania. Dr. Skalka's major research interests are the molecular aspects of the replication of retroviruses. Dr. Skalka is internationally recognized for her contributions to the understanding of the biochemical mechanisms by which such viruses (including the AIDS virus) replicate and insert their genetic material into the host genome. Both an administrator and researcher, she has been deeply involved in state, national, and international advisory groups concerned with the broader, societal implications of scientific research, including the NJ Commission on Cancer Research and the U.S. Defense Science Board. Dr. Skalka has served on the editorial boards of peer-reviewed scientific journals and has been a member of scientific advisory boards including the National Cancer Institute Board of Scientific Counselors, the General Motors Cancer Research Foundation Awards Assembly, the Board of Governors of the American Academy of Microbiology, and the National Advisory Committee for the Pew Biomedical Scholars Program.

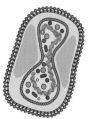

1

Infections of Populations: History and Epidemiology

Introduction to Viral Pathogenesis

A Brief History of Viral Pathogenesis

The Relationships Between Microbes and the Diseases They Cause

The First Human Viruses Identified and the Role of Serendipity

New Techniques Led to the Study of Viruses as Causes of Disease

Viral Epidemics in History

Epidemics Shaped History: the 1793 Yellow Fever Epidemic in Philadelphia

Tracking Epidemics by Sequencing: West Nile Virus Spread to the Western Hemisphere

The Economic Toll of Viral Epidemics in Agricultural Organisms

Population Density and World Travel as Accelerators of Viral Transmission

Zoonotic Infections and Viral Epidemics

Epidemiology

Fundamental Concepts

Tools of Epidemiology

Surveillance

Parameters That Govern the Ability of a Virus to Infect a Population

Environment

Host Factors

Perspectives

References

LINKS FOR CHAPTER 1

▶▶ *Video: Interview with Dr. W. Thomas London*
http://bit.ly/Virology_London

▶▶ *Epidemiology causes conclusions (p<0.05)*
http://bit.ly/Virology_Twiv169

▶▶ *Slow motion sneezing*
http://bit.ly/Virology_1-23-13

Swords, lances, arrows, machine guns, and even high explosives have had far less power over the fates of nations than the typhus louse, the plague flea, and the yellow-fever mosquito.

HANS ZINSSER
Rats, Lice and History 1934

Introduction to Viral Pathogenesis

While the title of Zinsser's classic volume *Rats, Lice and History* may trigger a wry smile, the ideas proposed in this classic volume about pathogens and the diseases they cause remain as relevant today as when they were published in 1934. As Zinsser argued, the global impact of pathogens, including viruses, has shaped human history as much as any war, natural disaster, or invention. This view may seem an exaggeration to today's student of virology, who probably perceives most viral infections as annoyances that cause unpleasant side effects and may result in a few missed classes or days of work. But in the context of history, epidemics of smallpox, yellow fever, human immunodeficiency virus, and influenza have resulted in an incalculable loss of life and have changed entire societies. Smallpox alone has killed over 300 million people, more than twice the number of deaths from all the wars in the 20th century. Huge empires fell to a relatively small number of invaders, in part because the conquerors inadvertently introduced viruses that crippled the empires' defense forces.

Although vaccines and antivirals have reduced, and even eliminated, some of these scourges, a recent influenza pandemic, the alarming number of human cases of Ebola virus in Africa, the lack of success in developing a human immunodeficiency virus vaccine, the resurgence of vaccine-preventable infections, and the emergence of "new" human viral pathogens, such as the coronavirus that causes Middle East Respiratory Syndrome, remind us of the challenges we still face. Of equal importance, while populations in resource-rich countries may be generally protected from some former viral foes, infections with vaccine-preventable viruses, including measles, polio, and hepatitis B virus, remain prevalent in countries that lack the money or infrastructure to ensure widespread vaccination.

The ways by which viruses cause diseases in their hosts, the tug-of-war among viruses and the host's defenses, and the impact that viral epidemics have had on human and animal populations are therefore not just interesting academic pursuits but rather life-and-death issues for all organisms. That said, it is important to bear in mind this critical fact: pathogenesis (the basis of disease) is often an unintended outcome of the parasitic lifestyle of viruses. As is true for humans, selective pressures that control viral evolution act only on the ability to survive and reproduce. From this perspective, one could argue that the most successful viruses are those that cause no apparent disease in their natural host.

In the first chapter of Volume I, we recounted an abbreviated history of virology and described milestones that established the foundation for our current understanding of viral reproduction. In this chapter, we return to history, focusing on watershed events that catalyzed the fields of viral epidemiology and pathogenesis. Subsequent chapters in this volume will consider the impact of viral infections on individual hosts, tissues, and cells. Our goal is to build on the principles of viral reproduction that were established in Volume I to provide a comprehensive and integrated view of how viruses cause disease in single cells, discrete hosts, and large populations.

PRINCIPLES   *Introduction to viral pathogenesis*

- Koch's postulates helped to identify causal relationships between a microbe and the disease it caused in the host, though these Postulates may not always be applicable to virus infections.

- Major insights in viral pathogenesis have come from exploitation of technical advances in the fields of molecular biology and immunology.

- The increased mobility of human and animal populations on the planet has accelerated the emergence of epidemics.

- Many viruses that can infect multiple species establish a reservoir in an animal host in which the virus causes negligible disease. Spread into new human hosts, called zoonoses, are usually dead-end infections.

- Epidemiology, the study of infections in populations, is the cornerstone of public health research.

- Social interactions, individual differences among prospective hosts, group dynamics and behaviors, geography, and weather all influence how efficiently a virus can establish infection within a population.

- National and international agencies charged with monitoring outbreaks, implementing surveillance plans, disseminating vaccines and antivirals, and educating the public, must coordinate data obtained from a vast network of researchers, clinics, and physicians in the field.

- The regional occurrence of viral infections may be due to the restriction of a vector or animal reservoir to a limited geographical area.

- Seasonal differences in the appearances of some viruses may be due to variations in viral particle stability at various temperatures or humidity, changes in the integrity of host barriers (such as the skin or mucosa), or seasonal changes in the life cycles of viral vectors, such as mosquitoes.

- Susceptibility to infection and susceptibility to disease are independent.

A Brief History of Viral Pathogenesis

The Relationships between Microbes and the Diseases They Cause

Long before any disease-causing microbes were identified, poisonous air (miasma) was generally presumed to cause epidemics of contagious diseases. The association of particular microorganisms, initially bacteria, with specific diseases can be attributed to the ideas of the German physician Robert Koch. With his colleague Friedrich Loeffler, Koch developed four criteria that, if met, would prove a causal relationship between a given microbe and a particular disease. These criteria, **Koch's postulates**, were first published in 1884 and are still used today as a standard by which pathogens are identified. The postulates are as follows:

- the microorganism must be associated regularly with the disease and its characteristic lesions but should not be found in healthy individuals;
- the microorganism must be isolated from the diseased host and grown in culture;
- the disease should be reproduced when a pure preparation of the microorganism is introduced into a healthy, susceptible host; and
- the same microorganism must be reisolated from the experimentally infected host.

Guided by these postulates and the methods developed by Pasteur for the sterile culture and isolation of purified preparations of bacteria, researchers identified and classified many pathogenic bacteria (as well as yeasts and fungi) during the latter part of the 19th century. Identifying a cause-and-effect relationship between a microbe and a pathogenic outcome set the stage for transformative therapeutic advances, including the development of antibiotics.

During the last decade of the 19th century, however, it became clear that not all epidemic diseases could be attributed to bacterial or fungal agents. This breakdown of the paradigm led to the identification of a new class of infectious agents: submicroscopic particles that came to be called viruses (see Volume I, Chapter 1). Koch's postulates can often be applied to viruses, but not all virus-disease relationships meet these criteria. While compliance with Koch's principles **will** establish that a particular virus is the causative agent of a specific disease, failure to comply does not rule out a possible cause-and-effect relationship (Box 1.1).

The First Human Viruses Identified and the Role of Serendipity

The first human virus that was identified was the agent responsible for causing yellow fever. The story of its identification in 1901 is instructive, as it highlights the contributions of creative thinking, collaboration, serendipitous timing, and even heroism in identifying new pathogens.

Yellow fever, widespread in tropical countries since the 15th century, was responsible for devastating epidemics associated with extraordinary rates of mortality (for example, over a quarter of infected individuals died in the New Orleans epidemic of 1853). While the disease can be relatively mild, with transient symptoms that include fever and nausea, more-severe cases result in major organ failure. Destruction of the liver causes yellowing of the skin (jaundice), the symptom from which the disease name is derived. Despite its impact, little was known about how yellow fever was spread, although it was clear that the disease was not transferred directly from person to person. This property prompted speculation that the source of the infection was present in the atmosphere and led to desperate efforts to "purify" the air, including burning barrels of tar and firing cannons. Others believed that the pathogen was carried on **fomites**, such as bedding or clothing, although this hypothesis was disproved when volunteers remained healthy after sleeping in the nightwear of yellow fever victims.

The first real advance in establishing the origin, or **etiology**, of yellow fever came in 1880, when the Cuban physician Carlos Juan Finlay proposed that a bloodsucking insect, most likely a mosquito, played a part in the transmission of the disease. A commission to study the basis of yellow fever was

Figure 1.1 Conquerors of yellow fever. This painting by Dean Cornwell (1939) depicts the experimental exposure of James Carroll with infected mosquitoes. Walter Reed, in white, stands at the head of the table, while Jesse Lazear applies the infected mosquitoes to Carroll's arm. Also depicted in this painting is Carlos Finlay, in a dark suit. Despite the care that Cornwell took to ensure accuracy of his portrayal of the participants and their uniforms, the event documented in this painting never took place; rather, artistic license was used to place all the major players in one depiction of a watershed moment in medical history. Photo courtesy of Wyeth Pharmaceuticals.

BOX 1.1

DISCUSSION
Why viruses may not fulfill Koch's postulates

Although Koch's postulates provided a framework to identify a pathogen unambiguously as an agent of a particular disease, some infectious agents, including viruses, cause disease but do not adhere to all of the postulates. In fact, it has been argued that the rigid application of these criteria to viral agents may have impeded early progress in the field of virology. Koch himself became aware of the limitations of his postulates upon discovery that *Vibrio cholerae*, the agent of cholera, could be isolated from both sick and healthy individuals.

Application of these criteria to viruses can be particularly problematic. For example, the first postulate, which states that the microorganism must be "regularly associated" with the disease, does not hold true for many animal reservoirs, such as bats, in which the virus actively reproduces but causes no disease. Similarly, arthropod vectors, such as mosquitoes, support reproduction of a variety of hemorrhagic viruses but do not themselves die of such infections. As another example of the problem of unilaterally applying Koch's postulates to viruses, the second postulate states that the microorganism must be grown in culture. However, many viruses, including papillomaviruses that cause warts and cervical cancer and hepatitis B virus that causes liver cirrhosis and cancer, cannot be cultured, or require complex culture conditions that must mimic the

tissue complexity found in the infected host. Consequently, it is generally accepted that the postulates are a guide, not an invariant set of requirements to fulfill.

More recently, detection methods based on nucleic acid sequence have rendered Koch's original postulates even less relevant. Such approaches alleviate the requirement to culture the suspected agent and are sufficiently sensitive to detect the presence of vanishingly small quantities of viral nucleic acid in an apparently healthy individual. As such, a revised set of Koch's postulates that takes into consideration new technical capabilities has been proposed (see Volume I, Box 1.4).

Assiduously applying the postulates has been particularly problematic for identifying

viruses that cause human tumors. As noted in a review by Moore and Chang, Koch's postulates "are a brilliant example of precision in scientific thinking, but they hold little practical value for 21st-century tumor virology since they cannot prove nor disprove most candidate tumor viruses to cause cancers." Whether Koch's postulates will continue to be a useful standard to identify pathogens or will become an historical footnote remains to be seen.

Fredericks, DN, Relman, DA. 1996. Sequence-based identification of microbial pathogens: a reconsideration of Koch's postulates. *Clin Microbiol Rev* **9:**18–33.
Moore PS, Chang Y. 2014. The conundrum of causality in tumor virology: the cases of KSHV and MCV. *Semin Cancer Biol* **26:**4–12.

More-sensitive technologies, including DNA sequencing, have triggered a reconsideration of Koch's postulates.

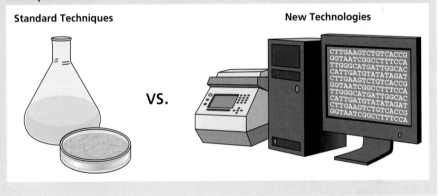

established in 1899 in Cuba by the U.S. Army under Colonel Walter Reed. This commission was formed in part because of the high **incidence** of the disease among soldiers who were occupying Cuba. Jesse Lazear, a member of Reed's commission, confirmed Finlay's hypothesis when he allowed himself to be bitten by a yellow fever virus-infected mosquito. "I rather think I am on the track of the real germ," wrote Lazear to his wife, sadly just days before he died of yellow fever himself. The results of the Reed Commission's study proved conclusively that mosquitoes are the **vectors** for this disease. In retrospect, a mosquito-borne mode of transmission made sense, as the disease was predominately found in warm and humid regions of the world (e.g., Cuba, New Orleans) where mosquitoes were, and remain, abundant. The members of this courageous

team, perhaps the first true epidemiologists, are depicted in a dramatic 1939 painting (Fig. 1.1).

The nature of the pathogen was established in 1901, when Reed and James Carroll injected diluted, filtered serum from the blood of a yellow fever patient into three healthy individuals. Two of the volunteers developed yellow fever, causing Reed and Carroll to conclude that a "filterable agent," which we now know as yellow fever virus, was the cause of the disease. In the same year, Juan Guiteras, a professor of pathology and tropical medicine at the University of Havana, attempted to produce immunity by exposing volunteers to mosquitoes that were allowed to take a blood meal from an individual who showed signs of yellow fever. Of 19 volunteers, 8 contracted the disease and 3 died. One of the deceased was Clara Louise Maass, a U.S. Army nurse. Maass' story is of interest, as she had volunteered to be inoculated by infected mosquitoes some time before, developed only mild symptoms, and survived. Her agreement to be infected a second time was to test if her earlier exposure provided protection from a subsequent challenge. This was a prescient idea, because at that time, virtually nothing was known about immune memory. Maass' death prompted a public outcry and helped to end yellow fever experiments in human volunteers.

Yellow fever had been **endemic** in Havana for 150 years, but the conclusions of Reed and his colleagues about the nature of the pathogen and the vector that transmitted it led to rapid implementation of effective mosquito control measures that dramatically reduced the incidence of disease within a year. To this day, mosquito control remains an important method for preventing yellow fever, as well as other viral diseases transmitted by arthropod vectors.

Other human viruses were identified during the early decades of the 20th century (Fig. 1.2). However, the pace of discovery was slow, in great part because of the dangers and difficulties associated with experimental manipulation of human viruses so vividly illustrated by the experience with yellow fever virus. Consequently, agents of some important human diseases were not identified for many years and only then with some good luck.

A classic example is the identification of the virus responsible for influenza, a name derived in the mid-1700s from the Italian language because of the belief that the disease resulted from the "influence" of contaminated air and adverse astrological signs. Worldwide epidemics (**pandemics**) of influenza had been documented in humans for well over 100 years. Such pandemics were typically associated with mortality among the very young and the very old, but the 1918-1919 pandemic following the end of World War I was especially devastating. It is estimated that one-fifth of the world's population was infected, resulting in more than 50 million deaths, far more than were killed in the preceding war. Unlike in previous epidemics that affected the elderly and the very young, healthy young adults were often victims (Fig. 1.3).

Despite many efforts, a human influenza virus was not isolated until 1933, when Wilson Smith, Christopher Andrewes, and Patrick Laidlaw serendipitously found that the virus could be propagated in an unusual host. Laidlaw and his colleagues at Mill Hill in England were using ferrets in studies of canine distemper virus, a paramyxovirus unrelated to influenza. Despite efforts to keep these ferrets isolated from both the environment and other pathogens (for example, all ferrets were housed separately, and all laboratory personnel had to disinfect themselves before and after entering a room), it is thought that a lab worker infected with influenza transmitted the virus to a ferret. This ferret then developed a disease very similar to influenza in humans. Realizing the implications of their observation, Laidlaw and colleagues then infected naive ferrets with throat washings from sick individuals and isolated the virus now known as

Figure 1.2 Pace of discovery of new infectious agents. Koch's introduction of efficient bacteriological techniques spawned an explosion of new discoveries of bacterial agents in the early 1880s. Similarly, the discovery of filterable agents launched the field of virology in the early 1900s. Despite an early surge of virus discovery, only 19 distinct human viruses had been reported by 1935. Adapted from K. L. Burdon, *Medical Microbiology* (MacMillan Co., New York, NY, 1939, with permission.)

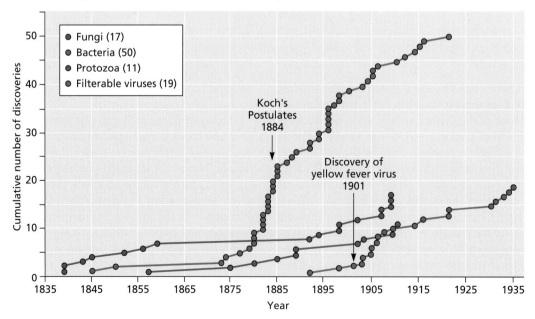

A

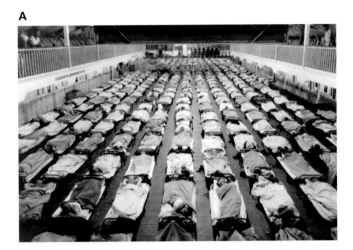

B

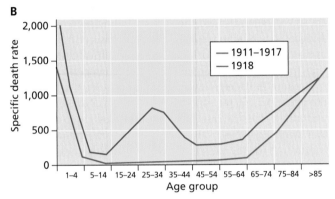

Figure 1.3 1918 flu consequences. (A) The 1918-1919 influenza pandemic infected a staggering number of people, resulting in the hasty establishment of cavernous quarantines in college gymnasia and large halls, filled with rows and rows of infected patients. Photo courtesy of the Naval History and Heritage Command. **(B)** Of particular concern, this epidemic had a high death rate among young, otherwise healthy, individuals compared to those of previous flu seasons. Adapted from R. Ahmed et al., *Nat. Immunol.* **8:**1188–1193, 2007, with permission.

influenza A virus. (Note the effective use of Koch's postulates in this study!) Subsequently, influenza A virus was shown to also infect adult mice and chicken embryos. The latter proved to be an especially valuable host system, as vast quantities of the virus are produced in the allantoic sac. Chicken eggs are still used today to produce influenza vaccines.

New Techniques Led to the Study of Viruses as Causes of Disease

Technological developments propelled advances in our understanding of how viruses are reproduced (Volume I, Chapter 1) and also paved the way for early insights into **viral pathogenesis**, the study of how viruses cause disease. The period from approximately 1950 to 1975 was marked by remarkable creativity and productivity, and many experimental procedures developed then are still in use today. With these techniques

in hand, scientists performed pioneering studies that revealed how viruses, including mousepox virus, rabies virus, poliovirus, and lymphocytic choriomeningitis virus, caused illness in susceptible hosts.

Revolutionary developments in molecular biology from the mid-1970s to the end of the 20th century further accelerated the study of viral pathogenesis. Recombinant DNA technology enabled the cloning, sequencing, and manipulation of host and viral genomes. Among other benefits, these techniques allowed investigators to mutate particular viral genes and to determine how specific viral proteins influence cell pathology. The polymerase chain reaction (PCR) was first among the many new offshoots of recombinant DNA technology that transformed the field of virology. PCR is used to amplify extremely small quantities of viral nucleic acid from infected samples. Once sufficient viral DNA has been obtained and the sequence determined, the virus can be more easily identified and studied. The ability to sequence and manipulate DNA also led to major advances in the related field of immunology and, consequently, had an important impact on the investigation of viral pathogenesis. While many of the early studies in immunology focused on immune cell development, others began to address how immune cells recognized and responded to pathogens. The Nobel Prizes of the 1980s and 1990s highlight the importance of these new technologies; they include awards for the establishment of transgenic animals, gene targeting, immune cell recognition of virus-infected cells, and RNA interference. These discoveries, and the ways that they helped to shape our current view of viral disease, will be discussed in later chapters.

The emergence of molecular biology and cell biology as distinct fields marked a transition from a descriptive era to one that focused on the mechanisms by which particular viral processes were controlled, among other advances. Genomes were isolated, proteins were identified, functions were deduced by application of genetic and biochemical methods, and new animal models of disease were developed. These approaches not only defined basic steps in the viral life cycle and functions of virus-encoded proteins but also ushered in practical applications, including the development of diagnostic tests, antiviral drugs, and vaccines. As the 20th century came to a close, another paradigm shift was occurring in virology, as many scientists realized the power of a more holistic strategy to study virus-host relationships. These scientists embraced the concept of **systems biology**, the notion that all the molecules or reactions that govern a biological process could be identified and monitored during an infection, allowing discovery of new processes that were missed by the more reductionist, one-gene-at-a-time approaches. These ideas were initially developed using microarray technology, which enabled a global and unbiased snapshot of the quantities of both host and viral mRNAs under defined conditions.

New tools continue to expand our capabilities, and methods once considered cutting edge are eclipsed by more-powerful, faster, or cheaper alternatives. Parallel developments in information technology and computer analyses (often called "data mining") have been critical to infer meaningful conclusions from the massive data sets now commonly collected. Computer-aided approaches have enabled scientists to define cellular pathways that are triggered during viral infection, to identify common features among seemingly diverse viruses, and to make structural predictions about small-molecule inhibitors that could prevent infection. While these new tools are exciting and powerful, it is likely that traditional approaches will still be required to validate and advance the hypotheses that are emerging from systems biology. New technological developments should be viewed as adding to, rather than replacing, experimental strategies from the past.

While the methods that virologists employ may be ever-changing, the fundamental question asked by early pioneers is still with us: how do viruses cause disease? The remainder of this chapter focuses on the impact of viral infections in large populations and how outbreaks and epidemics begin.

Viral Epidemics in History

In the popular movies *Outbreak* (1995) and *Contagion* (2011), fictional epidemics were depicted following introduction of a virus into a naïve human population. Each movie included a pivotal scene in which an epidemiologist ominously described the devastating consequences of uncontrolled, exponential viral spread through a population. These movies were terrifying, exciting, and ultimately comforting, as humans, with improbable speed, developed a vaccine and gained the upper hand. But how realistic is this Hollywood vision? One could argue that proof of our triumph over viral pathogens can be found in the eradication of smallpox and the development of vaccines to prevent infection by many viruses that historically resulted in much loss of life, but there is a risk in becoming complacent. We may ignore how quickly a virus can spread in a susceptible population and forget the fear and feeling of helplessness that accompany viral epidemics. The four stories that follow highlight the financial toll, loss of life, and historical ramifications of viral epidemics and underscore a new reality: the increased mobility of human and animal populations on the planet will almost certainly accelerate the emergence of epidemics.

Epidemics Shaped History: the 1793 Yellow Fever Epidemic in Philadelphia

One powerful example of a deadly viral epidemic that influenced American history and changed how cities are managed is the yellow fever outbreak in Philadelphia, Pennsylvania. In 1793, when this epidemic occurred (and a full century before

Walter Reed's commission), nothing was known about this disease or how it was spread. No one at the time knew that viruses existed, so the seemingly random way that individuals became sick compounded the confusion. Furthermore, this epidemic struck at a pivotal time for the fledgling union. At that time, Philadelphia was the new nation's temporary capital and was a city of active commerce and trade. One can easily imagine the panic in Philadelphia when scores of individuals became ill and died of this mysterious disease within a very short time frame. In the 101 days between August 1 and November 9, some 5,000 people died in a city of about 45,000, making this one of the most severe epidemics in the history of the United States (Fig. 1.4). There were few families that did not lose a relative to this disease, and many entire families were lost. Those who could flee the city did so, including the new President, George Washington, and his cabinet. Others stayed behind to aid the sick, including men of the Free African Society, who volunteered on the basis of the incorrect notion of Benjamin Rush, a prominent Philadelphia physician, that people of color were immune to infection.

Because Philadelphia was a major port city, it is likely that the agent, which we now know was the yellow fever virus, was transported by infected mosquitoes on cargo ships and that standing water in the city provided a hospitable breeding ground for the insects. Credit goes to Rush, who noticed identical symptoms in many victims and who recommended that individuals either leave the city or quarantine themselves, practices that helped to curtail the epidemic. Rush's belief that the scourge arose from a pile of rotting coffee beans left on a dock, and his treatment regimen of purging and bloodletting is less worthy of praise.

Figure 1.4 Deaths caused by the yellow fever epidemic in Philadelphia, 1793. This map records the locations of deaths due to yellow fever, with red and orange streets marking those with highest mortality. Yellow fever was most deadly near the northern wharves, where poorer people lived and where Hell Town was located. Both areas furnished breeding places for *Aedes aegypti*, the type of mosquito that transmits the disease. Adapted from Paul Sivitz and Billy G. Smith, with permission.

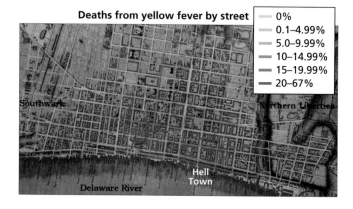

Deaths from yellow fever by street

— 0%
— 0.1–4.99%
— 5.0–9.99%
— 10–14.99%
— 15–19.99%
— 20–67%

Southwark

Northern Liberties

Hell Town

Delaware River

The city of Philadelphia was transformed after the epidemic. The outbreak, believed by many to be due to contaminated water (which was, in part, true), incentivized the local government to establish a municipal water system, the first major city in the world to do so. Infirmaries to tend to the sick (and isolate them from the healthy) were developed. Finally, the epidemic spurred a city-supported effort to keep streets free of trash, leading to the development of a sanitation program that would be a model for similar programs elsewhere.

Tracking Epidemics by Sequencing: West Nile Virus Spread to the Western Hemisphere

It took a full century to determine the cause of the Philadelphia epidemic, but technological advances have greatly accelerated our ability to understand the natural histories of some modern-day outbreaks. While the sudden appearance of West Nile virus in the Western Hemisphere in 1999 fortunately did not result in massive loss of life, this epidemic is notable for the role that viral genome sequencing played in defining its origin in the Middle East.

Prior to the summer of 1999, West Nile virus infections were restricted to Africa and the Mediterranean basin. Upon introduction to the United States, West Nile virus spread with remarkable speed; in 3 years, the incidence of infection expanded from eight cases in Queens (a borough of New York City) to virtually all of the United States and much of Canada, where it is now endemic (Fig. 1.5). The eight cases first identified in Queens held the key for major epidemiologic efforts to identify the source of this new infection. All victims had been healthy, and many had engaged in outdoor activities soon before showing signs of sickness. At about the same time, a high proportion of dead birds were found in and around New York City, prompting epidemiologists to consider that the same virus had infected both hosts. PCR and genome sequencing were used to confirm that West Nile virus was the cause of the bird deaths and the human illnesses.

How West Nile virus arrived in North America will never be known conclusively, but many think that the culprit was an infected mosquito (the natural **reservoir** for West Nile virus) that arrived as a stowaway on a flight from Israel to New York. This scenario was deduced from the remarkable similarity between genome sequences of a virus isolated in New York and an isolate obtained from an infected goose in Israel. It is sobering to contemplate that a virus that can now be found in virtually all states and provinces of North America may have begun with a single mosquito, perhaps trapped in a suitcase or purse: an invisible passenger on a trans-Atlantic flight.

The Economic Toll of Viral Epidemics in Agricultural Organisms

Epidemics can affect animals as well, especially those in dense farming populations. The outbreak of foot-and-mouth disease

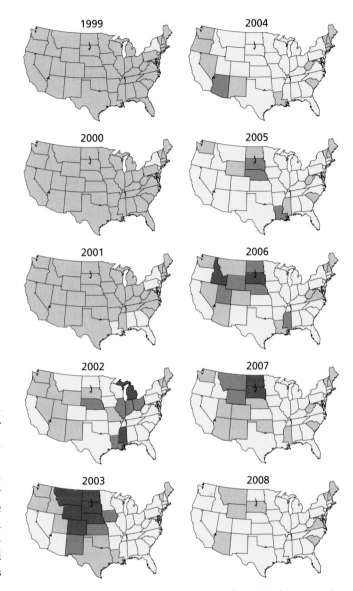

Figure 1.5 Spread of West Nile virus in the United States. The maps show West Nile virus incidence per 100,000 inhabitants in each state of the United States from 1999 to 2008. Credit: Centers for Disease Control and Prevention, with permission.

in the United Kingdom in 2001 caused an agricultural crisis of historical proportions; over 10 million sheep and cattle were killed (an average of 10,000 to 13,000 per day) in an attempt to stop the infection from spreading. In this instance, the original infected animal (the "**index case**") could be traced to one pig on a specific farm in Northumberland. Unfortunately, the owner did not inform the authorities of the appearance of foot-and-mouth disease, which is relatively easy to identify, on his farm. The epidemic spread rapidly, accelerated by the use of trucks by both contaminated and noncontaminated farms to transport animals to slaughterhouses. While this outbreak did not affect humans directly, the indirect financial impact on farming and tourism was enormous; it is estimated that this crisis cost the United Kingdom over $16 billion and almost brought down the government. A vaccine for foot-and-mouth disease virus exists and was available during the epidemic. However, at that time, vaccine use was rejected because farmers feared that they would then not be able to ship their meat to other countries, as vaccinated animals cannot be serologically distinguished from infected animals. A positive consequence of this preventable infection is that all farm animals in the United Kingdom are now vaccinated for foot-and-mouth disease virus. Other viruses of ruminants, however, including Bluetongue virus, remain threats to this day.

Population Density and World Travel as Accelerators of Viral Transmission

While the thought of an ocean cruise may evoke images of endless buffets and piña coladas by the pool, to viral epidemiologists, such pleasure ships appear as prime breeding grounds for viral epidemics. Norwalk virus, a member of the norovirus family, is most often associated with cruise ship outbreaks of gastroenteritis, although other viruses can also cause these nautical nightmares. Moreover, Norwalk virus is not restricted to ships; hot spots include any place in which many people from various locations are in close proximity for an extended period. Other high-density environments include prisons, long flights, day care facilities, dormitories, and elderly care communities. The risk of transmission is enhanced by the fact that noroviruses are quite hardy and can be transmitted either person to person or via contaminated food or surfaces, resulting in the need to decontaminate all shared surfaces with chlorine-containing solutions following an outbreak. While the gastrointestinal effects of a noroviral infection (nausea, vomiting, and diarrhea) are unpleasant, the disease is short-lived, and patients usually recover quickly. However, the frequency with which these outbreaks strike is a chilling reminder that, despite improved tools to characterize viral epidemics and reduce their spread, the ease and prevalence of world travel greatly facilitate the encounter between viruses and new hosts.

Zoonotic Infections and Viral Epidemics

Viral epidemics often appear seemingly without warning, raising questions about their origins. Some viral epidemics begin with a zoonotic infection, discussed in detail in Chapters 10 and 11. **Zoonoses** are infections transmitted between species, usually to humans from other animals. Many viruses that can infect multiple species establish a reservoir in a host in which the virus causes no disease or only nonlethal disease. When a new host is in proximity to an infected reservoir animal, a species jump may occur. While zoonotic transmission may cause disease in the new host, transspecies

infection is usually a dead end for the virus. Consequently, zoonotic infections rarely spread from human to human, as is the case for rabies virus, West Nile virus, and avian influenza. Although relatively rare, zoonotic infections are a concern to epidemiologists, because the new host will not have immunity and the disease that occurs in the new host may be different (often more severe) than that in the reservoir host. The transspecies spread of a human immunodeficiency virus-like ancestor from monkeys to humans is a prime example (Box 1.2).

As increased contact between species is the predominant risk factor for zoonotic infection, one can envision how changes in the environment or ecosystems of some animals may increase the risk for contact among different species. This is of particular concern when humans invade wilderness areas. For example, it is thought that Nipah virus, a paramyxovirus for which bats are the natural reservoir, underwent species-to-species transmission in 1999 in Malaysia, when pig farming began in the habitat occupied by infected fruit bats. The virus moved from bats to pigs and ultimately to the farmers themselves. However, it is not necessary to invoke an exotic locale or complex combination of animals for zoonosis to occur; petting zoos, open markets, and state fairs provide sufficient human-animal contact to allow a virus to jump species.

Epidemiology

The study of viruses can be likened to a set of concentric circles. The most basic studies in virology comprise the detailed analyses of the genome and the structures of viral particles and proteins, which are crucial to understanding the biochemical consequences of the interaction of viral with host cell proteins. How infection of individual cells affects the tissue in which the infected cells reside and how that infected tissue disturbs the biology of the host define the landscape of the field of viral pathogenesis (discussed in the next four chapters). But if a viral population is to survive, transmission must occur from an infected host to susceptible, uninfected hosts. The study of infections of populations is the discipline of epidemiology, the cornerstone of public health research.

An epidemiologist investigates outbreaks by undertaking careful data collection in the field (that is, where the infections occur) and performing statistical analyses. Often, questions such as "how might the symptoms observed in an infected individual implicate one mode of viral transmission over another?" or "can a timeline be established to trace back the origins of an epidemic to a single event?" are asked. The answers help epidemiologists learn more about the pathogen that caused the epidemic. Social interactions, individual differences among prospective hosts, group dynamics and behaviors, geography, and weather, all influence how efficiently a virus can establish infection within a population. Epidemiologists lack the luxury of performing controlled experiments in which only one variable is manipulated. Consequently, they must consider many parameters

BOX 1.2

TRAILBLAZER
Zoonotic transmission of human immunodeficiency virus to humans

Acquired immunodeficiency syndrome (AIDS) of humans is caused by one of two lentiviruses, human immunodeficiency viruses types 1 and 2 (HIV-1 and HIV-2). Both HIV types arose as a result of multiple cross-species transmissions of simian immunodeficiency viruses (SIVs), which naturally infect African primates (see the figure). While most species-to-species transfers resulted in viruses that spread to humans to only a limited extent, a transmission of SIV from chimpanzees (SIVcpz) in southeastern Cameroon gave rise to HIV-1 group M, the principal cause of the AIDS pandemic. An AIDS-like disease likely afflicted chimpanzees for a long period before the recognition of human immunodeficiency virus as a human pathogen. Tracing the genetic changes that occurred as SIVs crossed from monkeys to apes and from apes to humans provides a new framework to examine the requirements of successful host switches and to gauge future zoonotic risk.

Sharp PM, Rayner JC, Hahn BH. 2013. Great apes and zoonoses. *Science* 340:284–286.

Sharp PM, Hahn BH. 2011. Origins of HIV and the AIDS pandemic. *Cold Spring Harb Perspect Med* 1:a006841.

Evolution of the human immunodeficiency virus-1 from monkey and primate hosts. Adapted from P. M. Sharp and B. H. Hahn, *Cold Spring Harb. Perspect. Med.* **1**:a006841, 2011, with permission.

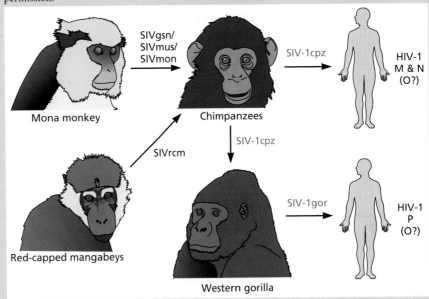

simultaneously to identify the source and transmission potential of a viral pathogen within a host community. Many of these variables are captured faithfully in various video games and apps that simulate outbreaks (Box 1.3). In the next section, we identify some crucial terms and concepts used in this field (For commentary and a personal account related to the topic, see the interview with Dr. Thomas London: http://bit.ly/Virology_London).

Fundamental Concepts

Incidence versus Prevalence

Determining the number of infected individuals is a primary goal of epidemiological studies. This information is required to establish both the incidence and the **prevalence** of infection. Incidence is defined as the number of new cases within a population in a specified period. Some epidemiologists use this term to determine the number of new cases in a community during a particular period of time, while others use incidence to indicate the number of new disease cases per unit of population per period. For example, the incidence of influenza can be stated as the number of reported cases in New York City per year or the number of new cases/1,000 people/year. Disease prevalence, on the other hand, is a measure of the number of infected individuals at one moment in time

divided by an appropriate measure of the population. A highly infectious and lethal disease (such as the 1793 epidemic of yellow fever in Philadelphia) may have a high incidence but a low prevalence because many of the infected individuals either died or cleared the infection. In contrast, a virus that can persist in a host for decades is likely to have high prevalence. An example of high prevalence is provided by hepatitis B virus; of the 300 to 400 million people infected globally, one-third live in China, with 130 million carriers. For this reason, incidence is an informative measure for acute infections, whereas prevalence is often used to describe persistent infections in which disease onset is not easily determined.

Prospective and Retrospective Studies

Infections of natural populations obviously differ from those under controlled conditions in the laboratory. Nevertheless, it is possible to determine if one or more variables affect disease incidence and viral transmission in nature. Two general experimental approaches are used: **prospective** (also called cohort or longitudinal) and **retrospective** (or case-controlled) studies. In prospective studies, a population is randomly divided into two groups (cohorts). One group then gets the "treatment of interest," such as a vaccine or a

DISCUSSION
Video games model infectious-disease epidemics

The hugely popular online video game *World of Warcraft* became a model for the transmission of virus infections. In this game, players interact in a fantasy world populated by humans, elves, orcs, and other exotic beasts. In late 2005, a dungeon was added in which players could confront and kill a powerful creature called Hakkar. In his death throes, Hakkar hit foes with "corrupted blood," infected with a virus that caused a fatal infection. The infection was meant to affect only those in the immediate vicinity of Hakkar's corpse, but the virus spread as players and their virtual pets traveled to other cities in the game. Within hours after the software update that installed the new dungeon, a full-blown virtual epidemic ensued as millions of characters became infected.

Although such games are meant only for entertainment, they do model disease spread in a mostly realistic manner. For example, the spread of the virus in Hakkar's blood depended on the ease of travel within the game, interspecies transmission by pets, and transmission via asymptomatic carriers. While computer models of epidemics have been developed, they lack the variability and unexpected outcomes found in real-world epidemics. Massive multiplayer online role-playing games have a large number of participants (at one point, over 10 million for *World of Warcraft*), creating an

excellent community for experimental study of infectious diseases. While entertainment is the central focus of such games, the players are serious and devoted, and their responses to dangerous situations approximate real-world reactions. For example, during the "corrupted-blood" epidemic, players with healing ability were the first to rush to the aid of infected players. This action probably affected the dynamics of the epidemic because infected players survived longer and were able to travel and spread the infection. Multiplayer video games provide a nontraditional but powerful opportunity to examine the consequences of human actions within a statistically significant and danger-free computer simulation.

A more reality-based smart phone app called *Plague Inc.* asks: "Can you infect the world?" and gives players the opportunity to choose a pathogen and influence its evolution. Players compete against the clock, trying to destroy humanity before the world can develop a cure. Like *World of Warcraft*, *Plague Inc.* is immensely popular and has been downloaded over 10 million times. Despite their macabre objective, such pursuits provide an education in epidemiology. Successful players learn to integrate multiple variables simultaneously, including environment, time, and population density, and these applications demonstrate how the reproductive cycle of a virus may

change over the course of an epidemic. However, the parallels to real-world epidemiology end there; a player can begin again with the click of a button or the flick of a finger. Alas, real life does not come with "do-overs."

Lofgren ET, Fefferman NH. 2007. The untapped potential of virtual game worlds to shed light on real world epidemics. *Lancet Infect Dis* 7:625–629.

---

drug, and the other does not. The negative-control population often receives a placebo. Whether a person belongs to the treatment or placebo cohort is not known to either the recipient or the investigator until the data are collected and the code is broken ("**double blind**"). This strategy removes potential investigator bias and patient expectations that may otherwise skew the data. Once the data are collected, the code is broken, and the incidence of disease or side effect is determined for each cohort and compared. Prospective studies require a large number of subjects, who often are followed for months or years. The number of subjects and time required depend on the incidence of the disease or side effect under consideration and the statistical **power** required to draw conclusions.

In contrast, retrospective studies are not encumbered by the need for large numbers of subjects and long study times. Instead, some number of subjects with the disease or side effect under investigation is selected, as is an equal number of subjects who do not have the disease. The presence of the variable under study is then determined for each group. For

example, in one retrospective study of measles vaccine safety and childhood autism, a cohort of vaccinated children and an equivalent cohort of age-matched unvaccinated children were chosen randomly. The proportion of children with or without autism was then calculated for each group to determine if the rate of occurrence of autism in the vaccinated group was higher, lower, or the same as in the unvaccinated group. The incidence of the side effect in each group is then calculated; the ratio of these values is the relative risk associated with vaccination. In this example, the rate of autism was not found to be different in the two groups, showing that vaccination is not a risk factor for the development of this disorder (see Chapter 8).

Mortality, Morbidity, and Case Fatality Ratios
Three other measures used in epidemiology can cause confusion because of the similarity of their definitions: **mortality**, **morbidity**, and **case fatality ratios** (Box 1.4). Mortality is expressed as a percentage of deaths in a known population

TERMINOLOGY

Morbidity, mortality, incidence, and case fatality

The terminology used to calculate the number of people who are infected and/or who become ill following a viral outbreak can be confusing. The following fictional example will be used to clarify these definitions.

Imagine that, in a city of 100,000 residents, a virus causes infection of 10,000 persons (as determined by serology). Of these 10,000,

7,000 develop signs of illness and 500 die of the infection.

- The **incidence** of this infection is the number of people infected divided by the population (10,000/100,000, or 10%).
- **Morbidity** is the number of individuals who became ill divided by

the number of infected individuals (7,000/10,000, or 70%).
- **Mortality** is the number of deaths divided by the number infected (500/10,000, or 5%).
- The **case-fatality ratio** is the number of deaths divided by the number of individuals with illness (500/7,000, or 7.1%).

of infected individuals. Thus, 40 deaths in a population of 2,000 infected individuals would be expressed as 2% mortality (40/2,000). The morbidity rate is similar but refers to the number of infected individuals in a given population that show symptoms of infection per unit of time. The morbidity percentage will always be higher than the mortality percentage, of course, because not all sick individuals will die of the infection.

In contrast, a case fatality ratio is a measure of the number of deaths among clinical cases of the disease, expressed as a percentage. As an example, if 200 people are diagnosed with a respiratory tract infection and 16 of them die, the case fatality ratio would be 16/200, or 8%. In a technical sense, the use of the word "ratio" is incorrect; a case fatality ratio is actually more a measure of relative risk than a ratio between two numbers.

While statistics are crucial to all studies in virology, they are of particular value in viral epidemiology, in which outcomes and causes are rarely black or white. An understanding of terms in statistics and some essential principles concerning the use of statistics in virology are provided in Box 1.5.

Tools of Epidemiology

We have considered some of the terms that epidemiologists use, but how do these scientists monitor and develop strategies to control the spread of viruses in populations?

An investigation begins at the site of an outbreak, where as much descriptive data as possible about the infected cases and the environment are gathered. In cases of viral infections in humans, information on recent travel, lifestyle, and preexisting health conditions is considered, along with the medical records of infected individuals to generate a testable hypothesis about the origin of the outbreak. The word "descriptive" can have a negative connotation in virology and is often used to mean the opposite of "mechanistic." However, in epidemiology, descriptive studies are essential to establish or exclude particular hypotheses about the origins of an outbreak. Indeed, descriptive epidemiology was the cornerstone for the discovery of human immunodeficiency virus during the AIDS epidemic in the 1980s (Box 1.6). Following the descriptive phase, analytical epidemiological methods are used to test hypotheses using control populations in either retrospectively or prospectively focused studies. Clinical epidemiology focuses on the collection of biospecimens, such as blood, sputum, urine, and feces, to search for viral agents or other pathogens and to help determine the potential route of transmission. Once specimens are collected, nucleic acid sequencing is often performed on the samples. In addition, such studies may include serological analyses, in which antibodies in the blood that implicate previous infection are identified. A timeline of the discovery of the H1N1 strain of influenza virus in 2009 illustrates the speed and coordination of epidemiological

BOX 1.5

METHODS
The use of statistics in virology

When studying viral infections in hosts, scientists do not always obtain results that are so clear and obvious that everyone agrees with the conclusions. Often the effects are subtle, or the data are highly varied from sample to sample or from study to study (sometimes referred to as "noise"). This ambiguity is particularly true in epidemiological studies, given the large number of parameters and potential outcomes. How do you know if the data that you generated (or that you are reading about in a paper) are significant?

Statistical methods, properly employed, provide the common language of critical analysis to determine whether differences observed between or among groups are significant (**Table 1.1**). Unfortunately, surveys of articles published in scientific journals indicate that statistical errors are common, making it even more difficult for the reader to interpret results. In fact, the term "significant difference" may be one of the most misused phrases in scientific papers, because the actual statistical support for the statement is often absent or incorrectly derived. While a detailed presentation of basic statistical considerations for virology experiments is beyond the scope of this text, some guiding principles are offered.

It is essential to consider experimental design carefully before going to the bench or to the field. A fundamental challenge in study design is to predict correctly the number of observations required to detect a significant difference. The significance level is defined as the probability of mistakenly saying that a difference is meaningful; typically, this probability is set at 0.05. Scientists do not usually refer to things as "true" or "false" but rather use quantitative approaches to provide a sense of the significance between two data sets (e.g., experimental versus control). An important concept is power, the probability of detecting a difference that truly is significant. In the simplest case, power can be increased by having a larger sample size (**Table 1.2**). Even when results seem black and white, having too few animals (or replicates) is insufficient for drawing a statistically meaningful conclusion.

It is critical to include a detailed description of how statistical analyses were performed in all communications linked with the data. The tests that were used to determine significance are just as important as the description of methods used to generate the data. Benjamin Disraeli, a 19th-century British Prime Minister, once said, "There are three kinds of lies: lies, damned lies, and statistics." Indeed, a gullible reader may be persuaded that a certain set of data is significant, but this conclusion depends on the stringency and appropriateness of the tests that were used, as well as the data points included in the analysis.

While this text cannot define what tests are applicable for which assays, we can make a few strong suggestions. Statistics should not be considered an afterthought or a painful process that one does in retrospect when putting data together for a publication. Reliable studies that stand the test of time have considered statistics throughout the scientific process, and good statistics are as critical as good study design. Do not be fearful of statistics. While it is true that the field can become quite complex, most of the tests used by virologists are reasonably straightforward. Computer programs such as Excel and GraphPad have made the calculations easy, but you need to know which tests to apply. Fortunately, there are excellent books available that make statistics logical and accessible (e.g., *Intuitive Biostatistics*, by Harvey Motulsky). For more-complex data, study design issues, and analyses, one may require consultation with a statistician.

Motulsky H. 2013. *Intuitive Biostatistics: a Nonmathematical Guide to Statistical Thinking*, 3rd ed. Oxford University Press, Oxford, United Kingdom.

Table 1.1 Statistical terms[a]

Term	Definition
Alternative hypothesis	Hypothesis that contradicts the null hypothesis
Binary data	Data that consist of only two values (e.g., positive and negative)
Cardinal data	Data that are on a scale in which common arithmetic is meaningful
Confidence interval	Likely range of the true value of a parameter of interest
Hypothesis testing	Use of statistical testing to objectively assess whether results seen in experiments are real or due to random chance
Nonparametric test	Statistical test that requires no assumptions regarding the underlying distribution of the data
Normally distributed data	Data which, when plotted in a histogram, look approximately like a bell-shaped curve
Null hypothesis	Hypothesis which presumes that there are no differences between treated and untreated groups; if hypothesis testing results in a statistically significant difference, the null hypothesis is rejected
P value	Probability of getting a result as extreme as or more extreme than the value obtained in one's sample, given that the null hypothesis is true
Parametric test	Statistical test that assumes that the data follow a particular distribution (e.g., normal)
Power	Probability of detecting a statistically significant difference that truly exists
Sample size	Number of experimental units in a study
Significance level	Probability of falsely finding a statistically significant difference

[a]Reprinted from B. A. Richardson and J. Overbaugh, *J. Virol.* **79:**669–676, 2005, with permission.

Table 1.2 *P* values for the differences in infection rates between experimental and control groups[a]

No. of animals per group	*P* value for indicated group[b]		
	All control animals infected and no experimental animals infected	All control animals and one experimental animal infected or one control animal infected and no experimental animal infected	One control animal infected and one experimental animal infected
3	0.1	0.4	1.0
4	0.03	0.1	0.5
5	0.008	0.05	0.2
6	0.002	0.02	0.08
7	<0.001	0.005	0.03
8	<0.001	0.001	0.01

[a]Reprinted from B. A. Richardson and J. Overbaugh, *J. Virol.* **79:**669–676, 2005, with permission.

[b]Determined by Fisher's exact test, using a two-sided hypothesis test with the significance level fixed at 0.05. Fisher's exact test is used because it is appropriate for experiments with small numbers of observations.

DISCUSSION
Descriptive epidemiology and the discovery of human immunodeficiency virus

Acquired immunodeficiency syndrome (AIDS) was first recognized as a new disease in the United States by physicians in New York, Los Angeles, and San Francisco, who independently noticed that some of the young homosexual male patients in their practices had developed unusual diseases, such as *Pneumocystis carinii* pneumonia (PCP) and Kaposi's sarcoma, which were typically associated with immunosuppressed patients. The first report in the medical literature that described this apparently new syndrome appeared in June 1981 and described five young homosexual men in Los Angeles with PCP. Other reports of a similar syndrome in individuals who injected drugs soon followed. While these "descriptive" observations raised many questions and incited much anxiety, they laid the foundation for the subsequent mechanistically focused work that identified the human immunodeficiency virus as a new human pathogen.

Centers for Disease Control and Prevention. 1981. Pneumocystis pneumonia—Los Angeles. *MMWR Morb Mortal Wkly Rep* **30**:250–252.

efforts to identify and thwart widespread dissemination of this virus (Fig. 1.6).

Surveillance

A final function of epidemiology is the establishment of vigilant surveillance procedures that can shorten the period between the beginning of an epidemic and its detection. One could argue that the development of worldwide surveillance programs and information sharing have had as profound an impact on limiting viral infections as antiviral medications and vaccines. The U.S. Centers for Disease Control and Prevention (CDC) was established in 1946 after World War II, with a primary mission to prevent malaria from spreading across the country. The scope of the CDC quickly expanded, and this institution is now a central repository for information and biospecimens available to epidemiologists; it also offers educational tools to foster awareness and ensure the safety of

the public. The World Health Organization (WHO), founded in 1948 as an international agency of the United Nations, is charged with establishing priorities and guidelines for the worldwide eradication of viral agents. The WHO provides support to countries that may not have the resources to combat infectious diseases and coordinates results from a global network of participating laboratories. While the WHO provides coordination, the experimental work is performed in hundreds of laboratories throughout the world, often in remote locations which process samples and relay information back to the WHO. These WHO-certified laboratories adhere to stringent standards to ensure consistency of methods and interpretations. The laboratories conduct field surveillance using wild and sentinel animals and perform periodic blood screening for signs of infection or immunity. Sentinel animals ("canaries in the coal mine") allow rapid identification of new pathogens that may have entered a particular ecosystem. The chief successes of such global-surveillance efforts to date include the eradications of smallpox virus and Rinderpest virus, a relative of measles virus that causes disease in animals used in agriculture, such as cattle and sheep.

The Internet is a powerful tool for data sharing and public education. Publications and websites help to distribute consistent and timely information to health care workers across the globe. The weekly *Morbidity and Mortality Weekly Report*, published by the CDC, provides a central clearinghouse for health care providers in the United States to communicate individual cases of infectious diseases or to report unusual observations. ProMED (Program for Monitoring Emerging Diseases), sponsored by the International Society for Infectious Diseases, is a worldwide effort to promote communication among members of "the international infectious disease community, including scientists, physicians, epidemiologists, public health professionals, and others interested in infectious diseases on a global scale" (http://www.promedmail.org/aboutus/). Reporting of individual cases, when considered by epidemiologists in the aggregate, may catch an epidemic in its earliest days, when intervention is most effective. Use of

Figure 1.6 Discovery of the H1N1 strain of influenza virus (swine flu). Less than one month transpired between the first case (in San Diego County, CA) and the first press conference announcing the new strain. Adapted from ECDC Technical Emergency Team, *Eurosurveillance* **14**(18):pii=19204, 2009, with permission.

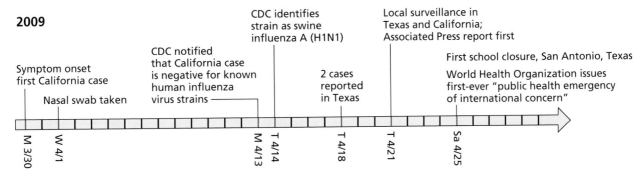

real-time data-gathering tools, such as Google Flu Trends, a Web-based application that surveys search queries from over 25 countries to predict influenza epidemics, have also emerged recently. While the predictions made from this application have been generally consistent with more-traditional surveillance data-gathering approaches, its accuracy and practical utility have not yet been proven. Nevertheless, the innovative use of keyword collection to monitor viral outbreaks underscores how collaboration between distinct fields (e.g., epidemiology and search engine design) can lead to creative ways to detect incipient outbreaks.

Parameters That Govern the Ability of a Virus To Infect a Population

One often hears that a virus is "going around," and such comments usually correlate with particular times of year (i.e., flu season). The seasonal appearance of some viruses, especially those that cause respiratory and gastrointestinal disease, raises the question of what parameters facilitate such spread in a population. This question is relevant both to viruses that cause widespread epidemics and to local, seasonal infections, such as the common cold. Identifying the variables associated with increased risk in a population has obvious value in clinical efforts to prevent outbreaks. As described below, multiple aspects of both host and environment contribute to maintaining a virus in a community.

Environment

Geography and Population Density

Some viruses are found only in specific geographical locations. The regional occurrence of viral infections may be due to the restriction of a vector or animal reservoir to a limited area. For example, most insect vectors are restricted to a specific region or ecosystem; unless this vector "escapes" its natural habitat, the viruses that it harbors will also be geographically constrained. Changes in migration routes or territory of a reservoir species may therefore influence the distribution of a virus and lead to new interactions with other species, increasing the risk of zoonotic transmission. A striking example of how a vector can change where a virus is found is provided by the global spread of the once-rare Chikungunya virus (Box 1.7).

BOX 1.7

DISCUSSION
An exotic virus on the move

Chikungunya virus is a togavirus in the alphavirus genus. The virus is spread by mosquitoes (primarily the notorious *Aedes aegypti*). The viral disease has been known for more than 50 years in the tropics and savannahs of Asia and Africa but had never been a problem of the developed countries in Europe or the United States. The disease is uncomfortable (rashes and joint pains) but not fatal. In the last 5 years, however, something changed dramatically and brought this once exotic disease into the forefront of public concern.

In 2004, outbreaks of Chikungunya disease spread rapidly from Kenya to islands in the Indian Ocean and then to India, where it had not been reported in over 30 years. In some of the Indian Ocean islands, more than 40% of the population fell ill. In 2007, there was an outbreak in Italy, the first ever in Europe. What had happened to change the pattern of infection?

An alarming finding was that the Asian tiger mosquito (*Aedes albopictus*) became an efficient new vector for the virus. A point mutation in the viral genome appears to be the cause of the vector expansion and, perhaps, for the epidemic spread of the disease in areas where it had been unknown.

A. albopictus, which has a greater geographical range than *A. aegypti*, is spreading across the globe from eastern Asia and is now found in mainland Europe and the United States. This mosquito is a maintenance (occasionally epidemic) vector of dengue viruses in parts of Asia and is a competent vector of several other viral diseases. Since its discovery in the United States, five arboviruses (Eastern equine encephalitis, Keystone, Tensaw, Cache Valley, and Potosi viruses) have been isolated from *A. albopictus*.

Enserink M. 2007. Chikungunya: no longer a Third World disease. *Science* **318**:1860–1861.

Projected distribution of *Aedes albopictus* in Europe, based on climate change models. Projections from two emission scenarios from the Intergovernmental Panel on Climate Change indicate that the habitat of *Aedes albopictus* will increase dramatically over the next century. From D. Fischer et al., *Int. J. Health Geogr.* **12**:51, 2013, with permission.

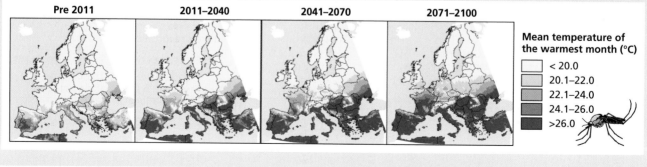

| Pre 2011 | 2011–2040 | 2041–2070 | 2071–2100 |

Mean temperature of the warmest month (°C)
< 20.0
20.1–22.0
22.1–24.0
24.1–26.0
>26.0

Host population density is a critical parameter for some virus populations to be sustained. Person-to-person transmission of some acute viral infections occurs only if the host population is large and interactive. For example, measles virus can be maintained only in human populations that exceed ~200,000, most likely because there is no animal reservoir, and infected individuals develop complete and long-lasting immunity. These infections are rarely found in isolated groups that might populate small islands or areas with extreme climates. Before global travel was possible, isolated host populations were the norm, and the distribution of viruses was far more limited. Now, however, as illustrated by the rapid colonization of the Western Hemisphere by West Nile virus, viruses are transported routinely and efficiently around the globe. In fact, epidemiologists have begun to think about the potential for epidemics in terms of the "effective distances" between airports, arguing that London is actually closer to New York than to other British towns, based upon air traffic densities; the larger the number of people that travel between airports and the cities that they serve, the smaller the effective distance.

Climate

In contrast to cultured cells that grow under conditions of invariant temperature and humidity or laboratory animals that live in strictly controlled enclosures, humans and other animals exist in ever-changing environments that directly influence viral biology. These changes include normal seasonal variations as well as progressive changes, such as global warming (Box 1.8).

Climate, including temperature and humidity, can have a profound influence on viral infections of populations. Indeed, there is a striking seasonal variation in the incidence of most acute viral diseases (Fig. 1.7). Respiratory virus infections occur more frequently in winter months, whereas infections of the gastrointestinal tract predominate in the summer. Seasonal differences in diseases caused by arthropod-borne viruses are clearly a consequence of the life cycle of the insect vector; when there are fewer mosquitos, there is a parallel reduction in the prevalence of the viruses that they harbor. However, the basis for the seasonal nature of infections by viruses that are not transmitted by arthropods is less obvious. It has been suggested that the seasonality of some infections is attributable to temperature- or humidity-based differences in the stability of virus particles. For example, poliomyelitis was known as a summertime disease in New England but not in Hawaii. The prevailing view is that poliovirus is inactivated during winter months when humidity is low, unlike other viruses, such as influenza virus, which remain infectious through the drier winter months.

A widely held belief is that large changes in temperature will increase a host's susceptibility to infection. In fact, as a parent likely warned you, transmission of "the flu" (specifically, influenza A virus particles) is more efficient at low temperature and humidity, and this property could contribute to increased rates of influenza in the winter months (Box 1.9). However, epidemiological studies with rhinoviruses that are also anecdotally associated with cold temperatures have failed to support any relationship between the cold and getting a cold; whether the "urban legends" associated with respiratory viral infections are true thus appears to depend on the virus in question (Box 1.10).

Climate-based variations in viral disease may also be caused by bodily changes in the host that influence its susceptibility. Such changes might be linked to **circadian rhythms** or be governed by alterations in the thicknesses of mucosal surfaces, production of virus receptors, or immune fitness.

BOX 1.8

DISCUSSION
How human behaviors and activities increase the risk of zoonoses

In his book *Spillover: Animal Infections and the Next Human Pandemic*, science writer David Quammen argues that the increase in zoonoses that we have seen in the past decades can be directly linked to human behavior and the ways in which we are irrevocably altering the world's ecosystems. The list of zoonotic infections that have impacted humans is impressive: Bolivian hemorrhagic fever caused by the Machupo virus (1961), Marburg hemorrhagic fever (1967), Lassa fever (1969), Ebola hemorrhagic fever (1976), HIV-1 diseases (inferred in 1981, first isolated in 1983), HIV-2 diseases (1986), hantavirus cardiopulmonary syndrome caused by Sin Nombre virus (1993), Hendra virus infection (1994), avian flu (1997), Nipah virus infection (1998), West Nile virus infection (1999), severe acute respiratory syndrome (SARS) (2003), and MERS. As Quammen notes,

What we're doing is interacting with wild animals and disrupting the ecosystems that they inhabit—all to an unprecedented degree. Of course, humans have always killed wildlife and disrupted ecosystems, clearing and fragmenting forests, converting habitat into cropland and settlement, adding livestock to the landscape, driving native species toward extinction, introducing exotics. But now that there are seven billion of us on the planet, with greater tools, greater hungers, greater mobility, we're pressing into the wild places like never before, and one of the things that we're finding there is... new infections. And once we've acquired a new infection, the chance of spreading it globally is also greater than ever.

Quammen D. 2012. *Spillover: Animal Infections and the Next Human Pandemic*. W. W. Norton and Company, New York, NY.

A Rubella, 1963–1968

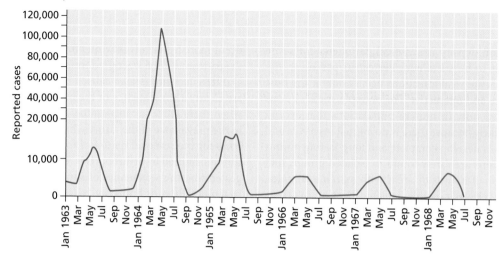

B Influenza, 1994–1999

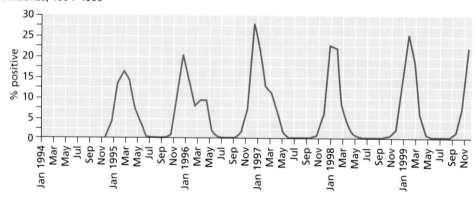

C Poliomyelitis, 1956–1957

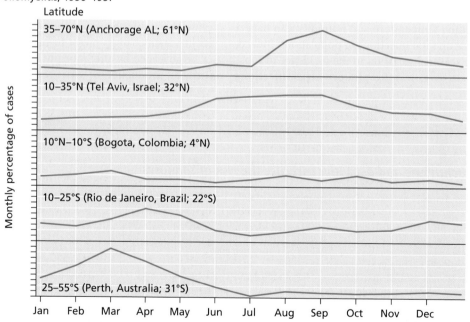

Figure 1.7 Seasonal variation in disease caused by three human pathogens in the United States.
(A) Annual cycles of rubella between larger epidemics, which occurred every 6 to 9 years (1963 to 1968). **(B)** Annual cycles of influenza virus infection (1994 to 1999). Note the strong seasonal prevalence. **(C)** Monthly incidence of poliomyelitis at different latitudes, with representative cities denoted. Adapted from S. F. Dowell, *Emerg. Infect. Dis.* **7:**369–374, 2001, with permission.

BOX 1.9

EXPERIMENTS
Seasonal factors that influence the transmission of influenza virus

Seasonality is a familiar feature of influenza: in temperate climates, the infection occurs largely from November to March in the Northern Hemisphere and from May to September in the Southern Hemisphere. There have been many hypotheses to explain this seasonality, but none had been supported by experimental data until recently. A guinea pig model was used to show that spread of the virus in aerosols is dependent upon both temperature and relative humidity.

Transmission experiments were conducted by housing infected and uninfected guinea pigs together in an environmental chamber. Transmission of infection was most effective at humidities of 20 to 35% and blocked at a humidity of 80%. In addition, transmission occurred with greater frequency when guinea pigs were housed at 5°C than at 20°C. The authors conclude that low temperature and humidity, conditions found during winter, favored influenza virus spread. The dependence of influenza virus transmission on low humidity might be related to the size of the droplets produced by coughing and sneezing (see the figure).

Model for the effect of humidity on the transmission of influenza virus. Transmission efficiency at 20°C (dashed line) or 5°C (solid line) is shown as a function of percent humidity. At 20°C, transmission is highest at low humidity, conditions which favor conversion of exhaled droplets into droplet nuclei (defined as droplets less than 5 mm in diameter and which remain airborne). Reduced particle stability at intermediate humidity is the cause of poor transmission. At high humidity, the conversion from droplets to droplet nuclei is inhibited, and the heavier droplets fall from the air, reducing transmission. At 5°C, transmission is more efficient than at 20°C, but there is a gradual loss of transmission with increasing humidity, presumably also as a consequence of the reduced formation of droplet nuclei. Adapted from A. C. Lowen et al., *PLoS Pathog.* **3:**1470–1476, 2007, with permission.

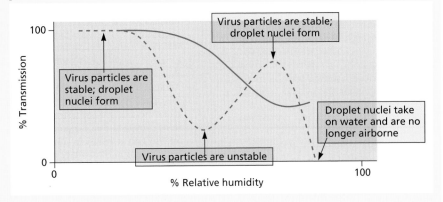

For example, if the mucosa is thinner in the winter or the skin is drier and cracked, the protective barriers that normally block viral entry into a host can become compromised.

Host Factors
The severity of disease following exposure to a viral pathogen can differ considerably from one person to another within a community in which the environmental variables discussed above are presumably the same. Some people become infected; others do not. Furthermore, among those who become infected, the severity of symptoms can vary widely, from mild to more serious, with long-lasting consequences. Susceptibility to infection and susceptibility to disease are independent; simply because someone is more susceptible to an infection does **not** mean that they will suffer more-severe illness. While we cannot yet predict who will become infected or how serious the infection will be, it is clear that both genetic and nongenetic (e.g., age) parameters determine how individuals in a population respond to viral infections (see Chapter 5).

Genetic and Immune Parameters
The basis of individual resistance or susceptibility in humans remains largely a mystery; we do not as yet understand why you always seem to catch a cold but your brother does not. However, in inbred organisms, such as laboratory mice and many plant species, proteins that make some members of a species more resistant than others have been identified. For example, the interferon-inducible myxovirus (influenza) resistance gene, *Mx1*, which is present in some strains of mice, encodes a guanosine triphosphatase (GTPase) that directly inhibits influenza virus replication. Similarly, resistance of some strains of mice to flavivirus disease has been mapped to the *Flv* locus. While virtually all mice strains lacking this locus die following infection, only ∼15% of mice with this gene succumb. Flavivirus titers in resistant mice are 1,000- to 10,000-fold lower than in susceptible animals, and the infection is cleared before disease symptoms develop. The product of this gene is 2′-5′-oligo(A) synthetase, an interferon-induced enzyme that activates ribonuclease L (RNase L), leading to degradation of host and viral mRNAs (see Chapter 3). While humans possess *MX* and *FLV* genes, whether they have parallel functions has not yet been determined. These genes could have different functional roles in humans, or other host genes in outbred humans could obscure the protective effects afforded by the products of these genes.

There are few human proteins known to influence susceptibility to viral infections but not to other pathogen encounters. Those that have been identified (and that differ among susceptible and nonsusceptible humans) enable a critical step in the infectious cycle, such as viral entry. Hosts who encode

---

BACKGROUND

Quiz: The origins and veracity of urban legends about infections

Which of these statements about colds and the flu are true, and which are myths?

1. You can catch the flu from a flu shot.
2. Stress increases your chances of getting sick.
3. Wearing a hat will help protect you from a cold.
4. Flying on an airplane will increase your risk of getting sick.
5. Pregnant or breastfeeding women should not get vaccinated.
6. Increasing how much you sweat (e.g., using lots of blankets) will speed up how quickly you resolve an infection.
7. "Feed a cold; starve a fever."
8. Your grandmother's chicken soup can help.
9. Eating garlic can help to prevent you from getting sick.
10. Over-the-counter cold "prevention" tablets or drinks are effective.

Answers

1. **Myth**: As discussed in Chapter 8, the injected flu vaccine is an inactivated ("dead") virus; it is therefore impossible to get the flu from the shot itself. The nasal flu mist contains a live, but drastically weakened, virus, and it is highly unlikely that someone will get influenza from the nasal vaccine.
2. Not yet proven, but probably **fact**: Stress alters hormones, hormones affect immunity, and immunity controls your response to viral infections, so it is quite possible that stress can affect your ability to respond to an infection.
3. **Myth**: Wearing a hat will keep your head warm, but that's it.
4. **Fact**: Recirculated air combined with a large number of people in close quarters is a perfect recipe for transmission of respiratory infections from person to person.
5. **Myth**: Flu symptoms are generally worse in pregnant women than in nonpregnant women, so it is of added importance that pregnant women be vaccinated. Many studies have shown that there are no adverse consequences of maternal vaccination to the fetus or the nursing neonate.
6. **Myth**: While piling on the blankets may make you feel better, it will not make the cold go away faster; the only thing proven to alter the duration of an infection is the use of antivirals within a short (1- to 2-day) window after symptoms appear.
7. **Myth**: How much you eat, or what you eat, will not influence how quickly you will resolve an infection. However, drinking lots of fluids **will** help, as staying hydrated, especially if you have a fever, will keep the mucus in respiratory passages loose. Moreover, colds and flu tend to cause a transient lack of appetite, so choosing food wisely (e.g., protein-rich) when recovering will hasten feeling better.
8. **Fact!**: It is a fact that warm liquids open up nasal passages and keep the mucus moving (a good thing), and chicken soup has also been proposed to mobilize neutrophils, important virus-fighting immune cells.
9. **Fact**: Garlic has powerful antioxidant activity, which boosts immunity. Eating **lots** of garlic will also keep away potentially infected friends and colleagues.
10. **Myth**: The small bottles that often appear at checkout lines in supermarkets and that promise protection from catching a cold are primarily just a large dose (usually 1,000 mg) of vitamin C. While it remains controversial whether vitamin C is beneficial, daily multivitamins (or, better still, a healthy diet) can provide as much of the key ingredient and for less money.

---

a viral receptor are permissive; those who do not are nonpermissive. For example, most, but not all, humans synthesize the chemokine receptor CCr5, a cofactor for entry of human immunodeficiency virus type 1 into cells. Those rare few individuals with CCr5 genes that do not encode a functional protein as a result of mutation appear to be resistant to HIV-1 infection. The same cellular gene product may affect different viruses in different ways; a mutation in the gene encoding CCr5 that protects against infection with human immunodeficiency virus type 1 actually **increases** susceptibility to lethal encephalitis caused by West Nile virus.

Recently, the genes that encode Toll-like receptor 3 (Tlr3) and Unc-93B were shown to protect humans from herpes simplex virus encephalitis, as mutations in these genes were associated with elevated risk. Both gene products govern the production of **interferons**. As the importance of interferons in host defense might suggest, mutations in these genes result in broad sensitivity to many viral pathogens, as is true experimentally in mice. However, as far as can be determined, humans with Tlr3 or Unc-93B mutations showed increased susceptibility to only herpes simplex virus and not other microbial pathogens. The implication is that certain immunological defenses may be exquisitely specific for a single pathogen or reaction in a pathogen's reproductive cycle. How such precision may have evolved and is controlled is discussed in Chapters 4 and 10.

Immunity also governs the susceptibility of a host population. Viral epidemiologists divide populations into two groups, susceptible and immune (or resistant). Humans or other animals that have been infected in the past are likely to be immune. Consequently, they cannot be infected again with the same virus and cannot transmit infection to others. Susceptible individuals are targets for infection and can develop disease and spread the virus to others. The persistence of a virus in a population depends on the presence of a sufficient number of those who are vulnerable. Immunization against viral infection by natural infection or vaccination reduces the number of potential hosts and therefore limits the foothold that a virus can establish in a community. For example, epidemics of polio were self-limiting, because the asymptomatic spread of the virus immunized the population. The competence of the immune response also determines the speed and efficiency with which an infection is spread and resolved and the severity of symptoms.

In addition to immune memory, inherent differences in an individual's ability to respond to pathogens may contribute to selective susceptibility. The class I and class II major histocompatibility complex (MHC) proteins bind to small viral peptides within an infected cell and present them on the infected cell surface to T cells (Chapter 4). Different MHC proteins bind to different peptides, and the ability of MHC proteins to interact with diverse epitopes determines, in part, the potency and quality of the antiviral response. These properties, in turn, dictate how efficiently the infection is cleared. The diversity among MHC genes increases the likelihood that there will be a suitable molecule to present peptides for any given infectious agent. An implication of this process is that some individuals may be intrinsically more or less able to mount an effective immune response to particular viral pathogens because of randomly encoded variations in the kinds of peptides that their MHC molecules can bind. This hypothesis is supported by the observations that individuals from isolated populations (e.g., dwellers of small islands) exhibit less polymorphism in these genes, a property that may account for their greater susceptibility to certain infections.

Nongenetic Risk Factors

Nongenetic risk factors include age, health, lifestyle, and occupation. In a classic story of virology history, Edward Jenner noticed that milkmaids appeared immune to smallpox, a trait that was later found to be due to "vaccination" by a relative of smallpox that infected cows. A present-day example of how one's profession influences the kinds of viruses one encounters is the careful attention paid to the health of poultry workers in China to monitor the emergence of the H5N1 or H7N9 influenza ("bird flu") strains.

The age of the host also plays an important role in determining the result of viral infections. Very young children and the elderly are generally more susceptible to disease. The increased susceptibility of these individuals can be explained by the immaturity or progressive decline, respectively, of their immune responses. However, when a virus causes **immunopathology**, infection of newborns and the elderly is

less severe than in immunocompetent adults. For example, intracerebral inoculation of lymphocytic choriomeningitis virus in adult mice is lethal because recruitment of T cells into the brain leads to swelling and death. In contrast, infant mice survive this challenge because of their weaker response (Box 1.11). Some viral infections, including those caused by poliovirus, mumps virus, and measles virus, are less severe in children than in adults, perhaps because the robust adult response contributes to disease. The 1918-1919 influenza pandemic was particularly lethal, not only for the very young and the very old but unexpectedly also for young adults, those 18 to 30 years of age (Fig. 1.3). It has been suggested that the increased death rate in young adults was caused by an overly aggressive immune response that flooded their systems with cytokines, sometimes referred to as a "**cytokine storm**."

BOX 1.11

DISCUSSION

Congenital brain infection: the lymphocytic choriomeningitis virus model

During development, the fetal brain is among the most vulnerable organs; viral infections of the fetus often result in severe brain injury. Unfortunately, many animal models of congenital brain infections do not mimic human disease for a variety of poorly understood reasons.

In contrast, the neonatal rat model for congenital lymphocytic choriomeningitis virus infection reproduces virtually all the neuropathological changes observed in congenitally infected humans. Within the developing rat brain, the virus selectively infects mitotically active neuronal precursors, a fact that explains the variation in pathology with time of infection during gestation. Lymphocytic choriomeningitis virus infection results in delayed-onset neuronal loss after the virus has been cleared by the immune system. Accordingly, many researchers think that this model can be used to study neurodegenerative or psychiatric diseases associated with loss of neurons or their function.

Bonthius D, Perlman, S. 2007. Congenital viral infection of the brain: lessons learned from lymphocytic choriomeningitis virus in the neonatal rat. *PLoS Pathog* 3:1541–1550.

Physiological differences other than immune fitness can also explain age-dependent variation in susceptibility. Infection of human infants with enteric coronaviruses is severe, because the alimentary canal is not fully active and presents a particularly hospitable niche for infection. In infants, the gastric pH tends to be less acidic, and digestive enzymes are less abundant. As animals age, the alveoli in the lungs become less elastic, the respiratory muscles weaken, and the cough reflex is diminished. These changes may explain, in part, why elderly people are at greater risk for acquiring respiratory tract infections. Other age-related variables that may be important include age-dependent changes in the tissues that a virus can infect. Respiratory syncytial virus causes severe lower respiratory tract infections in infants but only mild upper respiratory tract infections in adults. This is probably the result of both differences in the potencies of the host immune responses of children and adults and age-dependent variations in the susceptibilities of upper versus lower respiratory epithelia to viral infection.

Human males are more susceptible to viral infections than females, but the difference is slight and the reasons are not understood. Hormonal differences, which can alter the efficacy of the immune system, may be partly responsible. Pregnant women are more susceptible to infectious disease than nonpregnant women, probably for similar reasons. Moreover, the severity of disease caused by some viruses, including hepatitis A, B, and E and poliovirus, is exacerbated in pregnant women.

Malnutrition increases susceptibility to infection because the physical barriers, as well as immune fitness, are compromised. An example is the increased susceptibility to measles in children with protein deficiency. For this reason, measles is 300 times more lethal in poor countries. When children are malnourished, the small red spots inside the mouth that are hallmarks of a measles infection (called **Koplik's spots**) become massive ulcers, the skin rash is much more severe, and lethality may approach 10 to 50% (Chapter 5). Such severe measles infections are observed in children in tropical Africa and in aboriginal children in Australia.

As virologists begin to view individuals as products of their histories, genetics, environments, and life choices, rather than as masses of permissive cells and tissues, a more complete picture of susceptibility to an infection will emerge. This perspective has already yielded some insights. Corticosteroid hormones are known to affect susceptibility, because they are essential for the body's response to the stress of infection. These hormones have an anti-inflammatory effect, which is thought to limit tissue damage. Similarly, cigarette smoking increases susceptibility to respiratory infections as a result of the decreased capacity of the tar-coated lungs and airways to self-clean. Increased susceptibility can also occur in stressful life situations. We often refer to "our defenses being down," but what this probably means is that the balance of hormones that maintain homeostasis is altered, creating opportunities for a viral infection.

Perspectives

A fundamental principle of virology is that for a virus to be maintained in a host population, virus particles must be released from one infected host to infect another. This process of serial infection, while simple in principle, is difficult to study in natural systems given the mind-boggling number of host, viral, and environmental variables. Nevertheless, epidemiology, the study of this process, is evolving rapidly as new ways to track and identify infectious agents are developed. To thwart a potential epidemic, viral epidemiologists must possess the skills of a private investigator, sociologist, conductor, and chef at a popular restaurant. To track the origins of infection, epidemiologists must consider simultaneously multiple variables and clues, some of which are false leads. These investigators must understand the dynamics of the animal or human populations at risk and how aspects of behavior might increase the potential for infection. They must then integrate these diverse pieces of information, and because the investigation often begins only once an epidemic is under way and victims have been identified, epidemiologists must be able to work under great pressure, within a constrained time frame, and often under intense media scrutiny.

At the writing of this text, an Ebola outbreak in Africa has begun, killing thousands of people. Doctors cannot document all of the new cases, and many individuals who are ill are staying in their homes, fearful of the government response. Many believe that going to a clinic for care will isolate them from their families. Containing an epidemic under such pressures is a monumental challenge, especially when no certain therapy exists that can be offered to patients. The goal of epidemiology in this setting is to define the basis of the outbreak and to limit further transmission. Fortunately, Ebola virus is only transmitted via bodily fluids, such as blood and sputum; if this highly lethal virus could be acquired by means of aerosol droplets, the epidemiological challenge would obviously be far greater.

Our current understanding of the fundamental principles of viral pathogenesis comes largely from studies with animal models. For example, use of the large number of genetically identical, available mouse strains has led to the identification of many genes that confer susceptibility to particular viral infections. It is noteworthy that most of the gene products identified thus far impact a single virus or virus family, and many target a particular step in virus reproduction (Chapter 3). At one time, it seemed impossible to do similar studies with humans, because genetic techniques were not sufficiently powerful to detect rare susceptibility mutations in genetically different humans. However, several human genes that encode products that participate in intrinsic or innate defense for particular viral infections have recently been identified. Such investigations have heralded a new approach to answering fundamental questions in viral pathogenesis, the focus of the next four chapters.

References

Books

Burdon KL. 1939. *Medical Microbiology*. MacMillan Co, New York, NY.

Crosby AW. 2003. *America's Forgotten Pandemic: the Influenza of 1918*, 2nd ed. Cambridge University Press, Cambridge, England.

Crosby M. 2006. *The American Plague: the Untold Story of Yellow Fever, the Epidemic That Shaped Our History*. Berkley Books, New York, NY.

Diamond J. 1997. *Guns, Germs and Steel: the Fates of Human Societies*. W W Norton, New York, NY.

Katze MG. 2013. *Systems Biology. Current Topics in Microbiology and Immunology 363*. Springer, Berlin, Germany.

Krauss H, Weber A, Appel M, Enders B, Isenberg HD, Schiefer HG, Slenczka W, von Graevenitz A, Zahner H. 2003. *Zoonoses: Infectious Diseases Transmissible from Animals to Humans*, 3rd ed. ASM Press, Washington, DC.

McNeill WH. 1977. *Plagues and Peoples*. Anchor Press, Garden City, NY.

Mims CA, Nash A, Stephen J. 2001. *Mims' Pathogenesis of Infectious Disease*, 5th ed. Academic Press, Orlando, FL.

Murphy K, Travers P, Walport M. 2008. *Janeway's Immunobiology*, 7th ed. Garland Science, New York, NY.

Nathanson N (ed). 2007. *Viral Pathogenesis and Immunity*, 2nd ed. Academic Press, London, United Kingdom.

Notkins AL, Oldstone MBA (ed). 1984. *Concepts in Viral Pathogenesis*. Springer-Verlag, New York, NY.

Oldstone MBA. 1998. *Viruses, Plagues and History*. Oxford University Press, New York, NY.

Parham P. 2009. *The Immune System*, 3rd ed. Garland Science, New York, NY.

Richman DD, Whitley RJ, Hayden FG (ed). 2009. *Clinical Virology*, 3rd ed. ASM Press, Washington, DC.

Zinnser H. 1935. *Rats, Lice and History*. Transaction Publishers, Piscataway, NJ.

Review Articles

Ahmed R, Oldstone MBA, Palese P. 2007. Protective immunity and susceptibility to infectious diseases: lessons from the 1918 influenza pandemic. *Nat Immunol* **8:**1188–1193.

Biondi M, Zannino L. 1997. Psychological stress neuroimmunomodulation and susceptibility to infectious diseases in animals and man: a review. *Psychother Psychosom* **66:**3–26.

Casanova J, Abel L. 2007. Primary immunodeficiencies: a field in its infancy. *Science* **317:**617–619.

Collins PL, Graham BS. 2008. Viral and host factors in human respiratory syncytial virus pathogenesis. *J Virol* **82:**2040–2055.

Keesing F, Belden LK, Daszak P, Dobson A, Harvell CD, Holt RD, Hudson P, Jolles A, Jones KE, Mitchell CE, Myers SS, Bogich T, Ostfeld RS. 2010. Impacts of biodiversity on the emergence and transmission of infectious diseases. *Nature* **468:**647–652.

Korth MJ, Tchitchek N, Benecke AG, Katze MG. 2013. Systems approaches to influenza virus-host interactions and the pathogenesis of highly virulent and pandemic viruses. *Semin Immunol* **25:**228–239.

Moore PS, Chang Y. 2013. The conundrum of causality in tumor virology: the cases of KSHV and MCV. *Semin Cancer Biol* **26:**4–12.

Rall GF, Lawrence DMP, Patterson CE. 2000. The application of transgenic and knockout mouse technology for the study of viral pathogenesis. *Virology* **271:**220–226.

Sancho-Shimizu V, Zhang SY, Abel L, Tardieu M, Rozenberg F, Jouanguy E, Casanova JL. 2007. Genetic susceptibility to herpes simplex encephalitis in mice and humans. *Curr Opin Allergy Clin Immunol* **7:**495–505.

Sejvar JJ. 2003. West Nile virus: an historical overview. *Ochsner J* **5:**6–10.

Sharp PM, Hahn BH. 2011. Origins of HIV and the AIDS pandemic. *Cold Spring Harb Perspect Med* **1:**a006841.

Takada A, Kawaoka Y. 2002. The pathogenesis of Ebola hemorrhagic fever. *Trends Microbiol* **9:**506–511.

Virgin HW. 2007. *In vivo veritas*: pathogenesis of infection as it actually happens. *Nat Immunol* **8:**1143–1147.

Papers of Special Interest

Bodian D. 1955. Emerging concept of poliomyelitis infection. *Science* **122:**105–108.

Brockmann D, Helbing D. 2013. The hidden geometry of complex, network-driven contagion phenomena. *Science* **342:**1337–1342.

Casrouge A, Zhang S, Eidenschenk C, Jouanguy E, Puel A, Yang K, Alcais A, Picard C, Mahfoufi N, Nicolas N, Lorenzo L, Plancoulaine S, Sénéchal B, Geissmann F, Tabeta K, Hoebe K, Du X, Miller RL, Héron B, Mignot C, de Villemeur TB, Lebon P, Dulac O, Rotenberg F, Beutler B, Tardieu M, Abel L, Casanova JL. 2006. Herpes simplex virus encephalitis in human UNC-93B deficiency. *Science* **314:**308–312.

Cheung C, Poon L, Ng I, Luk W, Sia S-F, Wu M, Chan K-H, Yuen K-Y, Gordon S, Guan Y, Peiris J. 2005. Cytokine responses in severe acute respiratory syndrome coronavirus-infected macrophages in vitro: possible relevance to pathogenesis. *J Virol* **79:**7819–7826.

Evans A. 1978. Causation and disease: a chronological journey. *Am J Epidemiol* **108:**249–258.

Gibbs SE, Wimberly MC, Madden M, Masour J, Yabsley MJ, Stallknecht DE. 2006. Factors affecting the geographic distribution of West Nile virus in Georgia, USA: 2002–2004. *Vector Borne Zoonotic Dis* **6:**73–82.

Glass WG, McDermott DH, Lim JK, Lekhong S, Yu SF, Frank WA, Pape J, Cheshier RC, Murphy PM. 2006. CCR5 deficiency increases risk of symptomatic West Nile virus infection. *J Exp Med* **203:**35–40.

Goodman LB, Loregian A, Perkins GA, Nugent J, Buckles EL, Mercorelli B, Kydd JH, Palù G, Smith KC, Osterrieder N, Davis-Poynter N. 2007. A point mutation in a herpesvirus polymerase determines neuropathogenicity. *PLoS Pathog* **3:**1583–1592.

Leib DA, Machalek MA, Williams BR, Silverman RH, Virgin HW. 2000. Specific phenotypic restoration of an attenuated virus by knockout of a host resistance gene. *Proc Natl Acad Sci U S A* **97:**6097–6101.

Lindesmith LC, Donaldson EF, Lobue AD, Cannon JL, Zheng DP, Vinje J, Baric RS. 2008. Mechanisms of GII.4 norovirus persistence in human populations. *PLoS Med* **5:**269–290.

Lowen AC, Mubareka S, Steel J, Palese P. 2007. Influenza virus transmission is dependent on relative humidity and temperature. *PLoS Pathog* **3:**1470–1476.

McLean AR. 2013. Coming to an airport near you. *Science* **342:**1330–1331.

Perelygin AA, Scherbik SV, Zhulin IB, Stockman BM, Li Y, Brinton MA. 2002. Positional cloning of the murine flavivirus resistance gene. *Proc Natl Acad Sci U S A* **99:**9322–9327.

Rivers T. 1937. Viruses and Koch's postulates. *J Bacteriol* **33:**1–12.

Samuel M, Diamond M. 2006. Pathogenesis of West Nile virus infection: a balance between virulence, innate and adaptive immunity, and viral evasion. *J Virol* **80:**9349–9360.

Sangster MY, Heliams DB, MacKenzie JS, Shellam GR. 1993. Genetic studies of flavivirus resistance in inbred strains derived from wild mice: evidence for a new resistance allele at the flavivirus resistance locus. *J Virol* **67:**340–347.

Smith W, Andrewes CH, Laidlaw PP. 1933. A virus obtained from influenza patients. *Lancet* **222:**66–68.

Souza M, Azevedo M, Jung K, Cheetham S, Saif L. 2008. Pathogenesis and immune responses in gnotobiotic calves after infection with the genogroup II.4-HS66 strain of human norovirus. *J Virol* **82:**1777–1786.

Zhang S, Jouanguy E, Ugolini S, Smahi A, Elain G, Romero P, Segal D, Sancho-Shimizu V, Lorenzo L, Puel A, Picard C, Chapgier A, Plancoulaine S, Titeux M, Cognet C, von Bernuth H, Ku CL, Casrouge A, Zhang XX, Barreiro L, Leonard J, Hamilton C, Lebon P, Héron B, Vallée L, Quintana-Murci L, Hovanian A, Rozenberg F, Vivier E, Geissmann F, Tardieu M, Abel L, Casanova JL. 2007. TLR3 deficiency in patients with herpes simplex encephalitis. *Science* **317:**1522–1527.

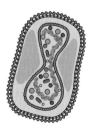

2 Barriers to Infection

Introduction

An Overview of Infection and Immunity

A Game of Chess Played by Age-Old Masters

Initiating an Infection

Successful Infections Must Modulate or Bypass Host Defenses

Skin

Respiratory Tract

Alimentary Tract

Urogenital Tract

Eyes

Viral Tropism

Accessibility of Viral Receptors

Host Cell Proteins That Regulate the Infectious Cycle

Spread throughout the Host

Hematogenous Spread

Neural Spread

Organ Invasion

Entry into Organs with Sinusoids

Entry into Organs That Lack Sinusoids

Organs with Dense Basement Membranes

Skin

The Fetus

Shedding of Virus Particles

Respiratory Secretions

Saliva

Feces

Blood

Urine

Semen

Milk

Skin Lesions

Perspectives

References

LINKS FOR CHAPTER 2

▶▶◄ *Video: Interview with Dr. Neal Nathanson*
http://bit.ly/Virology_Nathanson

▶▶◄ *Wookie viruses*
http://bit.ly/Virology_Twiv250

This earth of majesty, this seat of Mars
This other Eden, demi-paradise
This fortress built by Nature for herself
Against infection and the hand of war.
WILLIAM SHAKESPEARE,
RICHARD II (ACT 2, SCENE 1)

Introduction

Microbes are everywhere. They are on our hands, in our food, on the lips of those we kiss, on the ground and in the oceans, filling the air we breathe. For young children who play in dirt, scrape their knees, and pick their noses, interactions with potential pathogens are even more frequent and diverse. As we begin a series of chapters dedicated to immune responses and viral diseases, perhaps the right question to ask is not "What makes us sick?" but rather, "How can we possibly manage to stay healthy?"

If students of immunology are asked to list components of the host response to infection, typical responses will include mention of professional antigen-presenting cells, antibodies, cytotoxic T lymphocytes, and interferon gamma. These answers are not incorrect *per se*, but to focus only on attributes of the immune system misses the bigger picture: by the time a virus or other microbe has been engulfed by phagocytes or stimulated a T cell response, it has already bypassed an impressive fortress of defenses. These defenses, such as skin, mucus, and stomach acid, might seem much more primitive than the elegantly coordinated innate and adaptive immune responses. Nevertheless, they block the overwhelming majority of infections.

Such sentries, while effective, are imperfect despite millions of years of evolution in the presence of microbes. When viruses breach these initial barriers, infections of host cells and attendant disease can occur. The genomes of successful viruses encode gene products that modify, redirect, or block these, as well as other, host defenses. For every host defense, there will be a viral offense. It is remarkable that the genome of every known virus on the planet today encodes countermeasures to modulate the defenses of its host, even though some viral genomes are very small. As we shall see, many of

these "anti-host response" strategies (Box 2.1) are targeted at the body's physical barriers to infection. For comments and a personal account related to the chapter topic, see the interview with Dr. Neal Nathanson: http://bit.ly/Virology_Nathanson.

An Overview of Infection and Immunity
A Game of Chess Played by Age-Old Masters

Infection by viruses is often described in terms associated with warfare. There are opposing forces, each equipped with weapons to defeat the other. Once the battle ensues, each side fights with maximum force until a winner emerges. A more fitting metaphor to define the events pursuant to a viral infection would be a game of chess played by two masters. For each action, there follows a counteraction. Powerful tactics, such as induction of the adaptive immune response, may take many "moves" to be put into action. As one thinks about infection and immunity, it is imperative to bear in mind that we have coevolved with many of the viruses that infect us today. Such coevolution implies that, at a population level, **both** host and virus will survive. On an individual level, however, the consequence of infection is dictated by the host species, immune fitness, dose and strain of virus, and numerous environmental factors.

The pathogenesis of ectromelia virus, the agent of mousepox, highlights how the result of infection is affected by these variables (Fig. 2.1). Ectromelia virus is shed in the feces of its natural mouse host and gains access to naïve hosts via small abrasions in the footpad (or, in a laboratory setting, by injection into the footpad). Therefore, the first hurdle to be overcome is penetration of the dead skin of the footpad, which serves as an inhospitable barrier against infection. There is no guarantee that a mouse in a cage with infected feces will become infected. Virus particles must come in physical contact with permissive and susceptible cells for infection to occur, necessitating a break in the skin to allow access of the virus to live cells. Once the virus has gained entry, local reproduction in the epidermis and dermis of the footpad takes place. Within a day after exposure, the virus moves

---

PRINCIPLES **Barriers to infection**

- Three requirements must be met to ensure successful infection in an individual host: a sufficient number of infectious virus particles, access of the virus to susceptible and permissive cells, and absent or quiescent local antiviral defenses.

- Common sites of virus entry include the respiratory, alimentary, and urogenital tracts, the outer surface of the eyes (conjunctival membranes or cornea), and the skin.

- Each of these portals is equipped with anatomical or chemical features that limit viral entry and infection.

- Spread beyond the initial site of infection depends on the initial viral dose, the presence of viral receptors

on other cells, and the relative rates of immune induction and release of infectious virus particles.

- Disseminated infections typically occur by transport through the bloodstream, though some viruses can be transported by the peripheral nervous system.

- Effective transmission of virus particles from one host to another depends on the concentration of released particles and the mechanisms by which the virus particles are introduced into the next host.

- Viral transmission to a new host usually occurs through body fluids, including respiratory secretions, blood, saliva, semen, urine, and milk.

BOX 2.1

TERMINOLOGY
Is it evasion or modulation?

From the online *Merriam-Webster Dictionary*:
> **Evade**: to elude by dexterity or stratagem

> **Modulate**: to adjust to or keep in proper measure or proportion

The phrase "immune evasion" pervades the virology literature. It is intended to describe the viral mechanisms that thwart host immune defense systems. However, this phrase is imprecise and even misleading. The term "evasion" implies that host defenses are ineffective, similar to a bank robber evading capture by a hapless police force. In reality, a virus does not necessarily need to be invisible to the host response throughout its life cycle; it simply needs to delay or defer detection for a time sufficient to reproduce. If viruses really could evade the immune system, we might not be here discussing such semantic issues.

Perhaps a more accurate term to describe viral gene products that delay or frustrate host defenses is "immune modulators." The principle here is that, given the speed of viral reproduction, an infection can be successful even if host defenses are only suppressed transiently.

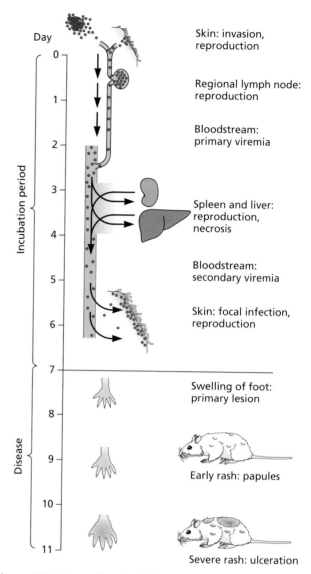

Figure 2.1 Ectromelia virus infection of mice. Infection begins through a break in the skin, allowing local viral reproduction and dissemination via the lymphatics within 1 to 2 days of exposure. Primary viremia occurs when the virus is released into the bloodstream, permitting infection of the spleen, liver, and other organs. Secondary viremia occurs due to release of virus from organs, resulting in infection of distal sites of the skin. The foot (the site of primary infection) swells due to the inflammatory response. In certain strains and wild mice, a severe rash may develop. Adapted from F. Fenner, et al., *The Biology of Animal Viruses* (Academic Press, New York, NY, 1974), with permission.

to draining lymph nodes, enters the bloodstream, and can be found in the spleen and liver by 3 days after infection. Thereafter, the virus continues to spread throughout the host, causing massive inflammation and severe skin lesions by 10 to 11 days of exposure.

From the moment of ectromelia virus entry, the host mounts a response to counteract the virus. The impact of such countermeasures is revealed by the effects of specific immune deficiencies, which lead to different kinds of disease. If the mouse lacks CD8+ T lymphocytes, it will die of extensive liver destruction by 4 to 5 days after infection. If instead the host lacks the critical cytokine interferon gamma, the virus may be controlled in the liver, even though death will occur by 10 to 12 days after infection as a consequence of uncontrolled viral reproduction in the skin. Even in immunocompetent mice, viral movement from tissue to tissue means that the immune response is continually playing catch-up as infection is controlled in the liver, infection of the skin appears. Moreover, while mice of a certain strain can contain the infection, immunocompetent mice of a different strain cannot, underscoring the critical involvement of more-subtle genetic regulators of immune control.

Just as ectromelia advances through various permissive tissues of the host, the host defenses are deployed in a coordinated, stepwise manner (Fig. 2.2). All surfaces of the mammalian body where pathogens may enter are protected by defensive layers provided by fur, skin, and mucus, or are protected by acidic environments. Once these barriers are

CONTINUOUS	IMMEDIATE	MINUTES/HOURS	HOURS/DAYS
Time Post-Exposure →			
Physical Barriers	**Intrinsic**	**Innate**	**Adaptive**
Mucus	Interferons	Natural killer cells	T cells
Saliva	Autophagy	Complement	B cells
Stomach acid	Apoptosis	Antigen-presenting cells	
Tears	MicroRNAs	Neutrophils	
Skin	CRISPRs		
Scabs			
Defensins			

Figure 2.2 The coordinated host response to infection. In healthy individuals, anatomical and chemical barriers are in place to prevent or repel infection by microbes. When viruses successfully bypass these defenses, intrinsic responses are engaged. These responses encompass those that already exist in the infected cell or host, poised to respond without the need for new transcription or translation. Within hours following exposure, cellular components of the innate immune response migrate to the site of infection, including professional antigen-presenting cells, neutrophils, and natural killer cells. These cells elaborate chemokines and cytokines that serve as a beacon for the subsequent recruitment of the adaptive immune response and synthesis of interferon-stimulated genes that induce an antiviral state. Within days following infection, antigen-specific T and B cells will be activated and undergo massive proliferation and migration to the site of infection, where, in most cases, resolution of the infection occurs. CRISPRs, clustered regularly interspaced short palindromic repeats.

crossed and cells become infected, **instrinsic cellular defenses** including cell-autonomous responses, such as interferon production, autophagy, editing, and, as a last resort, cell suicide, or **apoptosis**, are engaged. Because the virus may reproduce faster than an infected cell can control it, the "professional" immune response is also induced, beginning with the **early innate response** (Box 2.2). Finally, virus-specific cells of the **adaptive response** arrive at the site of infection, targeting infected cells and extracellular virus particles for destruction or elimination.

While this text generally avoids imparting actions to viruses, the impression one may have from the ectromelia virus example is that viruses are on a seemingly preordained, step-by-step path to gain access to their target cells of choice (for example, hepatitis viruses in hepatocytes, measles virus in epithelia, or human immunodeficiency virus in CD4$^+$ T cells). Likewise, one might think that the immune response is deployed in a synchronized and choreographed manner, much like actors performing a play night after night. These impressions would be wrong. As every game of chess is constrained by the same rules, but each game differs in execution and outcome, so too are viral infections and host immunity influenced by random, or stochastic, events. For example, tissues and the immune system may impose bottlenecks on the dissemination of a virus population. The diversity of viral populations enables some particles to pass through the bottleneck, while others are lost as the virus spreads (Chapter 10). Such bottlenecks include not only access to tissues but also immune restriction (Fig. 2.3). The stochastic view does not reject the idea that infection is a series of defined steps but rather adds random elements to the consequence of each.

Initiating an Infection

Three requirements must be met to ensure successful infection in an individual host: a sufficient number of infectious virus particles must be available to initiate infection; the cells at the site of infection must be physically accessible to the virus, **susceptible** (bear receptors for entry), and **permissive** (contain

---

BOX 2.2

TERMINOLOGY
Innate or intrinsic?

The line that distinguishes intrinsic immunity from innate immunity is a blurry one, and even the authors of this text disagree on some assignations. Many in the field use these terms interchangeably, adding to the confusion. For the purposes of clarity, we will define intrinsic responses to be those that preexist in the host or target cell and function without the need for new protein synthesis in response to pathogen detection.

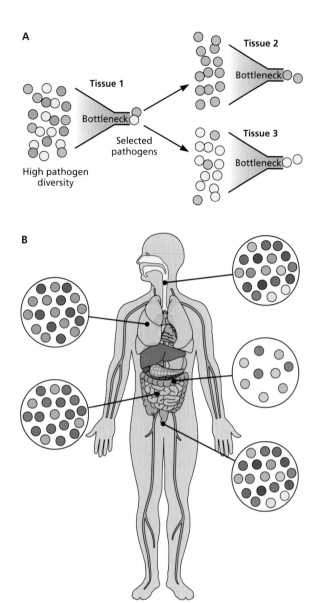

Figure 2.3 Infection seen as a series of stochastic events.
(A) Selection of viruses that can pass through bottlenecks in cells, organs, or hosts. In this case, the viral population enters the host as a diverse quasispecies with sufficient titer to establish infection. After entry, the population may encounter a host barrier that limits diversity. Pool members with the ability to overcome this barrier may then reproduce and restore diversity. The subsequent viral population in this tissue may have high diversity but differ in overall consensus sequence from the initial infecting population. Certain tissues may be highly permissive for viral infection with limited host or viral pressures. The viral population in these tissues may become dominated by more-fit variants with enhanced replicative capacity, resulting in high titer and low diversity. **(B)** Alternatively, the viral population may enter a highly restrictive environment where host and viral factors limit productive reproduction. The resulting population may have decreased diversity and titer or be eliminated from this tissue entirely. A productive infection in a host should allow efficient reproduction in one or more tissues, facilitating spread of a viral population at high titer and high diversity to optimize spread. Adapted from H. W. Virgin, *Nat. Immunol.* **8:**1143–1147, 2007, with permission.

intracellular gene products needed for viral reproduction); and local antiviral defenses must be absent or, at least initially, quiescent.

The first requirement imposes a substantial barrier to any infection and represents a significant limitation in the transmission of virus from host to host. Free virus particles face both a harsh environment and rapid dilution that can reduce their concentration. To remain infectious, viruses that are spread in contaminated water and sewage must remain stable in the presence of osmotic shock, pH changes, proteases, and sunlight. Aerosol-dispersed virus particles must remain hydrated and highly concentrated to infect the next host. These requirements account for why respiratory viruses spread most successfully in populations in which individuals are in close contact; the time that a virus particle is outside a host is minimized. In contrast, viruses that are spread by biting insects, contact with mucosal surfaces, or other means of direct contact, including contaminated needles, have virtually no environmental exposure; the virus is transmitted directly, for example, from mosquito to human.

Even after transmission from one host to another, infection may fail simply because the concentration of infectious virus particles is too low. In principle, a single West Nile virion should be able to initiate an infection, but host physical and immune defenses, coupled with the complexity of the infection process itself, usually require the presence of many particles. Even those particles that have remained intact may not encounter a target cell following entry into a host. One can envision many paths to failure: the virus particle may adhere to a dead or dying cell, become attached to nonsusceptible cells by nonspecific protein-protein interactions, be swept away in the bloodstream, get stuck in mucus, or be delivered to a lysosome upon entry into a target cell.

In addition, populations of viruses often contain particles that are not capable of completing an infectious cycle. Defective particles can be produced by incorporation of errors during virus genome replication or by interactions with inhibitory compounds in the environment. In the laboratory, a quantitative measure of the proportion of infectious viruses is the particle-to-plaque forming unit (PFU) ratio. The number of particles in a given preparation can be counted, usually with an electron microscope, and compared with the number of infectious units per unit volume (Volume I, Chapter 2). This ratio is a useful indicator of the quality of a virus preparation, as it should be relatively constant for a given virus. Some viruses, such as Semliki Forest virus, have a very low ratio (that is, virtually all particles are infectious), while other particle-to-PFU ratios, including those for poliovirus and some papillomaviruses, exceed 1,000 or 10,000. Why these ratios differ so drastically is not known, but the main point should be clear: not every virus particle that binds to a susceptible and permissive cell can induce all the steps needed to produce progeny virus particles, and even those that can may be thwarted at any step of the viral reproductive cycle.

Successful Infections Must Modulate or Bypass Host Defenses

In most mammals, common sites of virus entry include the mucosal linings of the respiratory, alimentary, and urogenital tracts, the outer surfaces of the eyes (conjunctival membranes or cornea), and the skin (Fig. 2.4). Each of these portals is equipped with anatomical or chemical features that limit viral entry and infection.

Skin

The skin is the largest organ of the body, weighing more than 5 kg in an average adult. It serves obvious protective functions but is also required for thermoregulation, control of hydration and evaporation, and integration of sensory information. The external surface of the skin, or epidermis, is composed of several layers, including a basal germinal layer of proliferating cells, a granular layer of dying cells, and an outer layer of dead, keratinized cells (Fig. 2.5). This

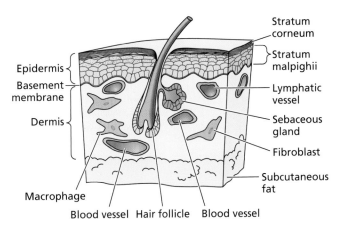

Figure 2.5 Schematic diagram of the skin. The epidermis consists of a layer of dead, keratinized cells (stratum corneum) over the stratum malpighii. The latter may have two layers of cells with increasing numbers of keratin granules and a basal layer of dividing epidermal cells. Below this is the basement membrane. The dermis contains blood vessels, lymphatic vessels, fibroblasts, nerve endings, and macrophages. A hair follicle and a sebaceous gland are shown. Adapted from F. Fenner et al., *The Biology of Animal Viruses* (Academic Press, New York, NY, 1974), with permission.

Figure 2.4 Sites of viral entry into the host. Sites of virus entry and shedding are indicated. The body is covered with skin, which has a relatively impermeable (dead) outer layer of keratinocytes covering a live layer of epithelial cells rich in capillaries. Moreover, other portals in the host, present to absorb food, exchange gases, and release urine, may allow access of viruses to host tissues.

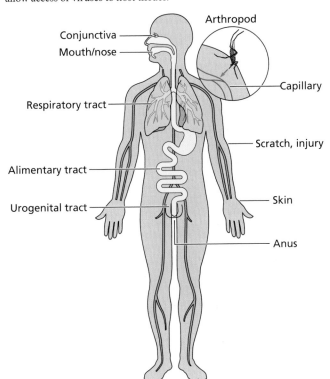

outermost layer is a rather literal coat of armor against viral infection: many virus particles that land on intact skin are inactivated by desiccation, acids, or other inhibitors secreted by commensal microorganisms or are simply removed from the body when dead cells slough off. Particles may also be washed away by soap and water. However, when the integrity of the dead cell layer is compromised by cuts, abrasions, or punctures (e.g., insect bites and needle sticks), virus particles can access blood vessels, epithelial cells, endothelial cells, and neuronal processes.

Examples of viruses that can gain entry via the skin are some human papillomaviruses, certain poxviruses (e.g., myxoma virus), and all tick- or mosquito-borne viruses that are transmitted by arthropod injection below the dead cell layer (Table 2.1). Even deeper inoculation into the tissue and muscle below the dermis can occur by hypodermic needle punctures, body piercing, tattooing, or sexual contact when body fluids are mingled as a result of skin abrasions or ulcerations. Animal bites can introduce rabies virus into tissue and muscle rich with nerve endings, through which virus particles can invade motor neurons. In contrast to the strictly localized reproduction in the epidermis (e.g., papillomaviruses that cause warts), viruses that initiate infection in dermal or subdermal tissues can reach nearby blood vessels, lymphatic tissues, and cells of the nervous system. As a consequence, they may spread to other sites in the body (Box 2.3).

The body's response to a breach in the critical barrier formed by the skin is to make rapidly a hard, water-resistant

Table 2.1 Different routes of viral entry into the host

Location	Virus(es)
Skin	
Arthropod bite	Bunyavirus, flavivirus, poxvirus, reovirus, togavirus
Needle puncture, sexual contact	Hepatitis C and D viruses, cytomegalovirus, Epstein-Barr virus, hepatitis B virus, human immunodeficiency virus, papillomavirus (localized)
Animal bite	Rabies virus
Respiratory tract	
Localized upper tract	Rhinovirus; coxsackievirus; coronavirus; arenaviruses; hantavirus; parainfluenza virus types 1–4; respiratory syncytial virus; influenza A and B viruses; human adenovirus types 1–7, 14, 21
Localized lower tract	Respiratory syncytial virus; parainfluenza virus types 1–3; influenza A and B viruses; human adenovirus types 1–7, 14, 21; severe acute respiratory syndrome coronavirus
Entry via respiratory tract followed by systemic spread	Rubella virus, arenaviruses, hantavirus, mumps virus, measles virus, varicella-zoster virus, poxviruses
Alimentary tract	
Systemic	Enterovirus, reovirus, adenovirus types 40 and 41
Localized	Coronavirus, rotavirus
Urogenital tract	
Systemic	Human immunodeficiency virus type 1, hepatitis B virus, herpes simplex virus
Localized	Papillomavirus
Eyes	
Systemic	Enterovirus 70, herpes simplex virus
Localized	Adenovirus types 8, 22

BOX 2.3

E X P E R I M E N T S
Dermal damage increases immunity and host survival

When it was still in use, the smallpox vaccine was delivered by a bifurcated needle (Volume I, Box 1.3), in a process referred to as scarification, that results in local damage to the skin and a subsequent and quickly resolved reaction or lesion in most individuals. Until recently, it was not appreciated that the scarification process itself was an important component of the vaccine's efficacy. Experiments using the smallpox-related virus, vaccinia virus, showed that intradermal inoculation of the virus into rabbits resulted in lethal disease by 8 days after infection, whereas delivery of the virus by scarification led to a protective host response. Scarified rabbits also responded more than a day before those inoculated by the intradermal route. Moreover, scarification in the absence of virus followed immediately by a same site intradermal challenge with virus resulted in significant protection to the infected rabbits. This dramatic difference can be attributed to the rapid induction of a non-specific host response. The act of scarification damages skin cells and the underlying epidermis, inducing the release of cytokines and chemokines that help to direct the host response to the site of infection and restrict the dissemination of the virus throughout the host.

Rice AD, Adams MM, Lindsey SF, Swetnam DM, Manning BR, Smith AJ, Burrage AM, Wallace G, MacNeill AL, Moyer RW. 2014. Protective properties of vaccinia virus-based vaccines: skin scarification promotes a nonspecific immune response that protects against orthopoxvirus disease. *J Virol* **88**: 7753–7763.

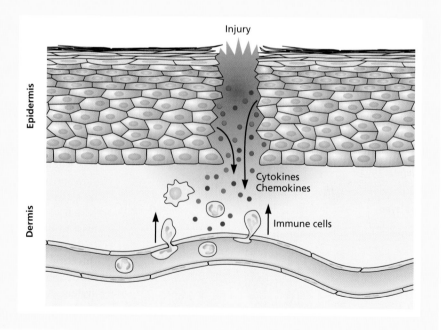

shell over the wound, called a scab. Scabs are more than just the dermis below the site of injury drying and hardening; neutrophils and macrophages are recruited in large numbers to a wound, primarily to engulf bacteria and other pathogens that may benefit from this breach in the skin to infect the host. Recruitment of immune cells before pathogen exposure is one way by which the host prepares for possible pathogen invasion. In addition to their anti-pathogen functions, macrophages recruited to the wound further aid the healing process by the production of growth factors that promote cell proliferation.

Respiratory Tract

Surfaces exposed to the environment but not covered by skin are lined by living cells and are at risk for infection despite the continuous actions of self-cleansing mechanisms. The most common route of viral entry is through the respiratory tract. In a human lung, there are about 300 million terminal sacs, called alveoli, which function in gaseous exchange between inspired air and the blood. Each sac is in close contact with capillary and lymphatic vessels. The combined surface area of the human lung is ~30 to 50 m², approximately the size of a studio apartment. At rest, humans inspire ~6 liters of air per minute. Together, the impressive surface area and large volumes of "miasma" that one inhales each minute imply that foreign particles, such as bacteria, allergens, and viruses, are introduced into the lungs with every breath.

Mechanical barriers play a significant role in antiviral defense in the respiratory tract. The tract is lined with a mucociliary blanket consisting of ciliated cells, mucus-secreting goblet cells, and subepithelial mucus-secreting glands (Fig. 2.6). Foreign particles

Figure 2.6 Sites of viral entry in the respiratory tract. (Left) A detailed view of the respiratory epithelium. A layer of mucus, produced by goblet cells, is a formidable barrier to virus particle attachment. Virus particles that traverse this layer may reproduce in ciliated cells or pass between them, reaching another physical barrier, the basement membrane. Beyond this extracellular matrix are tissue fluids, from which particles may be taken into lymphatic capillaries and reach the blood. Local macrophages patrol the tissue fluids in search of foreign particles. Adapted from C. A. Mims et al., *Mims' Pathogenesis of Infectious Disease* (Academic Press, Orlando, FL, 1995), with permission. **(Right)** Viruses that reproduce at different levels of the respiratory tract, with the associated clinical syndromes. SARS, severe acute respiratory syndrome; MERS, Middle East respiratory syndrome.

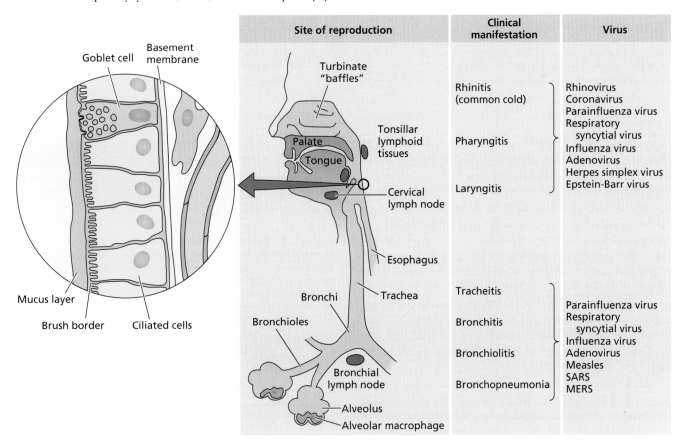

DISCUSSION
In praise of mucus

When you have a cold or sinus infection, it can be disconcerting to take a peek in your tissue after a sneeze. However, the mucus that accompanies many such infections actually serves a very important purpose. Mucus-producing cells line the mouth, nose, sinuses, throat, lungs, vagina, and entire gastrointestinal tract. In addition to its lubricant function, mucus acts as a protective blanket over these surfaces, preventing the tissue underneath from dehydrating. Mucus also acts as a sort of pathogen flypaper, trapping viruses and bacteria. More than being just a sticky goo, mucus contains antibodies, enzymes that kill the invaders it traps, and a variety of immune cells poised to respond to pathogens that attach to it.

It is a common misconception that discolored mucus is directly due to bacterial or viral presence, but the yellow or green mucus hue observed during infection is not due to bacteria or virus particles. When an individual acquires a respiratory tract infection, neutrophils, a key element of the host innate response, rush to the infected site. These cells contain an enzyme, myeloperoxidase, which is critical for the ability of neutrophils to eliminate pathogens, as individuals with a genetic loss of myeloperoxidase are immunocompromised, especially for respiratory tract infections. Myeloperoxidase is stored in azurophilic granules prior to release; these granules are naturally green or tan. Thus, when neutrophils are present in large numbers, the mucus appears green. One may indeed assume that discolored mucus is a sign of infection, as recruitment of neutrophils often accompanies infection.

One final thought that you may wish you did not know: while it is not a very socially acceptable practice, eating one's own nasal secretions (mucophagy), a habit of many young children, may have some evolutionary benefit. Some scientists argue that mucophagy provides benefits to the immune system, especially the underdeveloped host responses of children. As noted above, mucus destroys most of the pathogens that it tethers, so nasal secretions themselves are unlikely to be laden with infectious virus particles. Reintroducing these crippled microorganisms into the gut, where antigen-presenting cells are abundant, may be a form of "low-tech" vaccination or immune memory booster.

Bellows A. 2009. *Alien Hand Syndrome and Other Too-Weird-Not-To-Be-True Stories,* p 28–30. Workman Publishing, New York, NY.

deposited in the nasal cavity or upper respiratory tract are often trapped in mucus, carried to the back of the throat, swallowed, and destroyed in the low-pH environment of the gut (Box 2.4). In the lower respiratory tract, particles trapped in mucus are brought up from the lungs to the throat by ciliary action (Fig. 2.7). Cold temperatures, cigarette smoke, and very low humidity cause the cilia to stop functioning, likely accounting for the association of these environmental conditions with increased illness. When coughing occurs, both the host and the virus benefit; the host expels virus-laden mucus with each productive cough, and the virus is carried out of the host, perhaps to infect another nearby. The lowest portions of the tract, the alveoli, lack cilia or mucus, but macrophages lining the alveoli ingest and destroy virus particles.

Many viruses enter the respiratory tract in the form of aerosolized droplets expelled by an infected individual by coughing or sneezing (Table 2.1; Fig. 2.8). Infection can also spread through contact with respiratory secretions or saliva from an infected individual. Larger virus-containing droplets are deposited in the nose, while smaller droplets can penetrate deeper into the airways or the alveoli. To infect the respiratory tract successfully, virus particles must not be captured or swept away by mucus, neutralized by antibody, or destroyed by alveolar macrophages.

Figure 2.7 Cilia help to move debris trapped in the mucus of the respiratory tract out of the body. Cells in the cell membrane under the mucus have tiny hair-like projections called cilia. Usually, the mucus traps incoming particles. In coordinating waves, the cilia sweep the mucus either up to the nasal passages or back into the throat, where it is swallowed rather than inhaled into the lungs. The acid of the stomach destroys most pathogens not inactivated by the mucus.

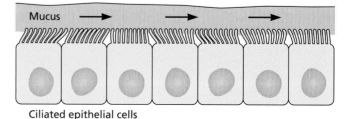

Mucus

Ciliated epithelial cells

Figure 2.8 A picture is worth a thousand words. A group of applied mathematicians evaluated the distance and "hang time" of various sized droplets produced after a sneeze, using the same strategies as ballistics experts studying gunfire. As many as 40,000 droplets can be released in a single sneeze, some traveling at over 200 miles an hour. Heavier droplets (seen in the photo) succumb to gravity and fall quickly, while smaller droplets (less than 50 μm in diameter) can stay in the air until the droplet dehydrates. Courtesy of CDC/ Brian Judd, CDC-PHIL ID#11161.

Alimentary Tract

The alimentary tract is another major site of viral invasion and dissemination (Table 2.1). Eating, drinking, kissing, and sexual contact routinely place viruses in the gut. Virus particles that infect by the intestinal route must, at a minimum, be resistant to extremes of pH, proteases, and bile detergents. Many enveloped viruses do not initiate infection in the alimentary tract, because viral envelopes are susceptible to dissociation by detergents, such as bile salts.

As depicted in Fig. 2.4, the lumen of the alimentary tract, from mouth to anus, is "outside" of our bodies, and thus the anatomy of the alimentary tube possesses many of the features of the skin. Like the skin, the gut has numerous physical, chemical, and protein-based barriers that collectively limit viral survival and infection: the stomach is acidic, the intestine is alkaline, and proteases and bile detergents are present at high concentrations. In addition, mucus lines the entire tract, and the luminal surfaces of the intestines contain antibodies and phagocytic cells. Moreover, the small and large intestines are coated in a thick (50-μm) paste of symbiotic bacteria that not only aids in digestion and homeostasis but also imposes a formidable physical barrier for virus particles to access the cells beneath (Box 2.5).

BOX 2.5

EXPERIMENTS
Commensal bacteria aid enteric virus infection

On a per-cell basis, humans are more bacterial than mammalian; we are "metaorganisms." Our gastrointestinal tract teems with bacteria, most of which aid in food digestion and promote good health. Consequently, both our eukaryotic defenses and the commensal bacteria that occupy the small intestine can be barriers to viral infection.

In many cases, however, commensal bacteria actually facilitate viral infection of the host. For example, when the intestinal microbiota of mice was depleted with antibiotics before inoculation with poliovirus, an enteric virus, the animals were found to be less susceptible to disease. Further investigation showed that poliovirus binds lipopolysaccharide, the major outer component of Gram-negative bacteria, and exposure of poliovirus to bacteria enhanced host cell association and infection. Furthermore, three other unrelated enteric viruses, reovirus, mouse mammary tumor virus, and murine norovirus, also have enhanced infection in the presence of intestinal bacteria. These results indicate that interactions with intestinal microbes promote enteric virus infection.

Baldridge MT, Nice TJ, McCune BT, Yokoyama CC, Kambal A, Wheadon M, Diamond MS, Ivanova Y, Artyomov M, Virgin HW. 2015. Commensal microbes and interferon-lambda determine persistence of enteric murine norovirus infection. *Science* 347:266-269.

Jones MK, Watanabe M, Zhu S, Graves CL, Keyes LR, Grau KR, Gonzalez-Hernandez MB, Iovine NM, Wobus C, Vinje J, Tibbetts SA, Wallet SM, Karst SM. 2014. Enteric bacteria promote human and murine norovirus infection of B cells. *Science* 346:755-759.

Kane M, Case L K, Kopaskie K, KozlovaA, MacDearmid C, Chervonsky AV, Golovkina TV. 2011. Successful transmission of a retrovirus depends on microbiota. *Science* 334:245-249.

Kuss SK, Best GT, Etheredge CA, Pruijssers AJ, Frierson JM, Hooper LV, Dermody TS, Pfeiffer JK. 2011. Intestinal microbiota promote enteric virus replication and systemic pathogenesis. *Science* 334:249–252.

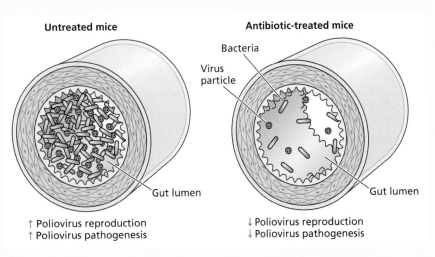

Untreated mice

Antibiotic-treated mice

Bacteria

Virus particle

Gut lumen

Gut lumen

↑ Poliovirus reproduction
↑ Poliovirus pathogenesis

↓ Poliovirus reproduction
↓ Poliovirus pathogenesis

Saliva in the mouth presents an initial obstacle to virus entry. While saliva is mostly water, it does contain lysozymes and other enzymes which aid in the breakdown of food but also can destabilize viral particles. One type of antibody found in saliva, secretory IgA (Chapter 4), may directly bind and inactivate incoming viral particles. A protein known as salivary agglutinin can directly interfere with influenza virus and human immunodeficiency virus, although ingestion is not the traditional route of infection by these viruses.

While passage from the mouth to the stomach is generally considered a quick trip following a swallow, cells in the oropharynx (for example, the tonsils and the back of the throat) appear to be permissive for human papillomaviruses, which can cause oropharyngeal squamous cell carcinoma. Papillomaviruses, traditionally thought to be restricted to the genitourinary or urogenital tract, are likely delivered to the throat during oral sex and can affect both men and women, as both semen and vaginal secretions can carry infectious papillomavirus particles.

Once in the stomach, a virus particle must endure stomach acid, which typically has a pH of 1.5 to 3.0, sufficiently low to denature most proteins of incoming food and many opportunistic viruses. Mucus is also abundant in the stomach, where it coats the lining and helps to prevent the highly corrosive gastric acid from attacking the stomach itself. Mucus also serves as a trap for virus particles, much as in the respiratory tract.

Nearly the entire small intestinal surface is covered with columnar villous epithelial cells with apical surfaces that are densely packed with microvilli (Fig. 2.9). This brush border, together with a surface coat of glycoproteins and glycolipids and the overlying mucus layer, is permeable to electrolytes and nutrients but presents a barrier to microorganisms. Once in the small intestine, pathogens can be attacked by small antimicrobial peptides called **defensins**, which are secreted by Paneth cells. These cells, which lie at the base of the microvillus crypt, secrete large granules filled with enteric alpha-defensins, also called cryptdins. These small (~30-amino-acid) peptides serve primarily to inactivate bacteria by destabilizing the bacterial cell wall or by interfering with bacterial metabolism. Recently, a role of these small peptides in antiviral defense has also been demonstrated. While a widely held view is that defensins exert their antimicrobial functions by disrupting lipid membranes, studies with viruses, including nonenveloped viruses without a lipid coat, reveal more-diverse functions of these peptides, including negative effects on viral entry and movement to the nucleus. Defensins actually **promote** the infection of some viruses, such as human immunodeficiency virus type 1 and human adenovirus, likely by increasing attachment of the virus particles to their cellular receptors.

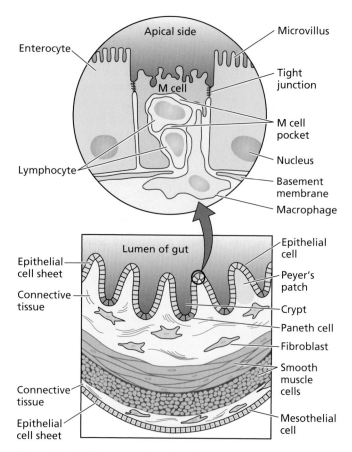

Figure 2.9 Cellular organization of the small intestine. A simplified view of the cellular composition of the small intestine is shown, with Paneth cells lining the base of the intestinal crypts and M cells providing the thin barrier between the intestinal lumen and the Peyer's patches beneath. Schematic drawing of the intestinal wall. This organ is made up of epithelial, connective, and muscle tissues. Each is formed by different cell types that are organized by cell-cell adhesion within an extracellular matrix. A section of the epithelium has been enlarged, and a typical M cell is shown surrounded by two enterocytes. Lymphocytes and macrophages move in and out of invaginations on the basolateral side of the M cell. Adapted from A. Siebers and B. B. Finlay, *Trends Microbiol.* **4:**22–28, 1996, and B. Alberts et al., *Molecular Biology of the Cell* (Garland Publishing, New York, NY, 1994), with permission.

Despite the formidable barriers, some viruses reproduce extensively in intestinal epithelial cells. Scattered throughout the intestinal mucosa are lymphoid follicles that are covered on the luminal side with a specialized follicle-associated epithelium consisting mainly of columnar absorptive cells and M (membranous epithelial) cells. The M cell cytoplasm is very thin, resulting in a membrane-like bridge that separates the intestinal lumen from the subepithelial space. M cells deliver antigens to the underlying lymphoid tissue (termed Peyer's patches) by **transcytosis**. In this process, material taken up on the luminal side of the M cell traverses the cytoplasm virtually intact and is delivered to the underlying basal membranes and extracellular space (Fig. 2.9). It is thought that M cell transcytosis is the mechanism by which some enteric viruses gain access to deeper tissues of the host. After crossing the mucosal epithelium, a virus particle could enter lymphatic vessels and capillaries of the circulatory system, facilitating spread within the host. A particularly well-studied example is transcytosis of reovirus. After attaching to the M cell surface, reovirus subviral particles are transported to cells underlying the lymphoid follicle, where the virus is reproduced and then spreads to other tissues. Some viruses actively reproduce only within M cells and not underlying tissues. For example, infection by human rotavirus and the coronavirus, transmissible gastroenteritis virus, destroys M cells, resulting in mucosal inflammation and diarrhea, but is not spread beyond the gut.

In some cases, the hostile environment of the alimentary tract actually **facilitates** infection. For example, reovirus particles are converted by host proteases in the intestinal lumen into infectious subviral particles, the form that subsequently infects intestinal cells.

While most viruses that can infect via the alimentary tract gain access via the mouth, it is possible for virus particles to enter the body through the lower gastrointestinal tract without passing through the upper tract and its defensive barriers. Human immunodeficiency virus can be introduced efficiently as a result of anal intercourse. Anal sex can cause abrasions within the rectum, stripping away the protective mucus and damaging the epithelial lining, resulting in broken capillaries. Human immunodeficiency virus particles can pass through such damaged epithelia to gain access to the blood for efficient transport to lymph nodes, where infection and reproduction can ensue. Moreover, as M cells are abundant in the lower colon, they are likely to provide a portal of entry for this virus into susceptible lymphocytes in the underlying lymphoid follicles. Once in the follicle, the virus can infect migratory lymphoid cells and spread throughout the body.

Urogenital Tract

Some viruses enter the urogenital tract, most typically as a result of sexual practices (Table 2.1). Like the alimentary tract, the urogenital tract is well protected by mucus and low pH. The vagina maintains a pH that is typically between 3.4 and 4.5; when the pH increases toward neutrality (as a result of antibiotic use or natural changes in the menstrual cycle, for example), many pathogens, including bacteria and yeast, can flourish. Sex can result in tears or abrasions in the vaginal epithelium or the urethra, allowing virus particles to enter. Some viruses infect the epithelium and produce local lesions (for example, human papillomaviruses, which cause genital warts). Others penetrate deeper, gaining access to cells in the underlying tissues and infecting cells of the immune system (human immunodeficiency virus type 1) or the peripheral nervous system (herpes simplex virus type 2). Infection by the latter

two viruses invariably spreads from the initial urogenital site to other tissues in the host, thereby establishing lifelong infections.

Viruses that gain entry by the urogenital tract are extremely common. Approximately one in six people between 15 and 50 years of age has genital herpes, and as this is a lifelong infection, the risk of transmission to future sex partners is high. Herpesvirus infection is often asymptomatic, although the virus can still be shed and infect others. Infections by these viruses pose a particular risk to the developing fetus and can result in miscarriage, early delivery, or lifelong infection that begins in the neonate. These dangers can be mitigated by Caesarian delivery. It is sobering to note that individuals may be affected by multiple sexually transmitted pathogens, and a preexisting infection with one may predispose to infection with another. For example, a genital herpes lesion provides an excellent portal for human immunodeficiency virus.

Eyes

The epithelia that cover the exposed part of the sclera (the outer fibrocollagenous coat of the eyeball) and form the inner surfaces of the eyelids (conjunctivae) provide the route of entry for several viruses. Every few seconds, the eyelid closes over the sclera, bathing it in secretions that wash away foreign particles. Like the saliva, tears that are routinely produced to keep the eye hydrated also contain small quantities of antibodies and lysozymes, which can destroy the peptidoglycan layer of some bacteria. Of interest, the chemical composition of tears differs, depending on whether they are "basal" tears produced constantly in the healthy eye, "psychic tears" produced in response to emotion or stress, or "reflex tears" produced in response to noxious irritants, such as tear gas or onion vapor. The concentration of antimicrobial molecules increases in reflex tears, but not psychic tears, underscoring the fact that host defenses are finely calibrated to respond to changes in the environment.

The primary function of tears is to wash away dust particles, viruses, and other microbes that land on the eye or under the eyelid. There is usually little opportunity for viral infection of the eye, unless it is injured by abrasion. Direct inoculation into the eye may occur during ophthalmologic procedures or from environmental contamination, such as improperly sanitized swimming pools and hot tubs. In most cases, viral reproduction is localized and results in inflammation of the conjunctiva, a condition called conjunctivitis or "pink eye." Systemic spread of the virus from the eye is rare, although it does occur; paralytic illness after enterovirus 70 conjunctivitis is one example. Herpesviruses, in particular herpes simplex virus type 1, can also infect the cornea, mainly at the site of a scratch or other injury, and immunocompromised individuals are at greater risk of retinal infection with cytomegalovirus. Such infections may lead to immune destruction of the cornea or the retina and eventual blindness. Inevitably, herpes simplex virus infection of the cornea is followed by spread of the virus to sensory neurons and then to neuronal cell bodies in the sensory ganglia, where a latent infection is established. Injury to the eye that allows for viral entry need not be a major trauma: small dust particles or rubbing one's eyes too aggressively may be sufficient to damage the protective layer and form a doorway for virus particles to access permissive cells.

While one may not normally think of eyelashes and eyebrows as key components of host defenses, these well-placed patches of hair help to capture fomites that might invade the eye. An intriguing thought is that, as evolution progressed from apes to humans, dense hair was lost from all except a few parts of the body: on top of the head, in the pubic region, and around the eye. It is tempting to speculate that individuals who retained these patches of hair may have had an evolutionary advantage because they were more resistant to certain infections (Box 2.6).

BOX 2.6

D I S C U S S I O N
Is intuition a host defense?

As sentient humans (and animals), we are constantly surveying our environment for dangers and opportunities. Such senses help all organisms to evade predation and to locate food sources but may also be useful in avoiding infections. For example, a rather typical human behavior upon locating some food toward the back of the refrigerator is to smell it to see if it is still "good," and people are usually quite adept at knowing when food is no longer acceptable to eat. In principle, food surveillance is a kind of quality control to ensure that the foods we eat do not carry dangerous microbes. Similarly, avoiding a murky hot tub or declining the advances of a dubious sexual partner could also be considered finely honed skills that may help to avoid contact with pathogens.

Viral Tropism

Before we describe how viruses move throughout a host, it will be useful to discuss briefly viral **tropism**: the cellular and anatomical parameters that define the cells in which a virus can reproduce *in vivo*. Most viruses do not infect all the cells of a host but are restricted to specific cell types in certain organs. For example, an enterotropic and a neurotropic virus reproduce in cells of the gut and nervous system, respectively. Some viruses are **pantropic**, infecting many cell types and tissues.

Tropism is governed by at least four parameters. It can be determined by the distribution of receptors for entry (**susceptibility**) or by a requirement for differentially produced intracellular gene products required to complete the infectious cycle (**permissivity**). However, even if the cell is susceptible and permissive, infection may not occur because virus particles are physically prevented from interacting with the tissue (**accessibility**). Finally, an infection may not occur, even when the tissue is accessible and the cells are susceptible and permissive, because of intrinsic and innate immune defenses.

Tropism influences the pattern of infection, pathogenesis, and long-term virus survival. Human herpes simplex virus is considered neurotropic because of its ability to infect, and be reactivated from, the nervous system, but in fact, this virus is pantropic and reproduces in many cells and tissues in the host. By infecting neurons, it may establish a stable latent infection, but because it is pantropic, infection may spread to other tissues. Consequently, if an infection is not contained by host defenses at the site of inoculation, the virus may cause disseminated disease, as can occur when herpes simplex virus infects infants and immunocompromised adults (Fig. 2.10). On rare occasions, this virus can enter the central nervous system and cause fatal encephalitis.

Accessibility of Viral Receptors

A cell may be susceptible to infection if the viral receptor(s) is present and functional. However, the receptor may not be accessible to the virus. If the cellular receptor is present only on the basal cell membrane of polarized epithelial cells, a virus cannot infect cells unless it first reaches that location by some means. Alternatively, if the viral receptor is located between adjacent cells (at the tight junctions), another cell surface protein may be necessary to ferry the virus particle from the apical membrane to the site of the viral receptor (Volume I, Chapter 5). Nonsusceptible (non-receptor-producing) cells can still be infected by alternative routes; for example, virus particles bound to antibodies can be taken up by Fc receptors (see "Immunopathological lesions caused by B cells" in Chapter 4).

Host Cell Proteins That Regulate the Infectious Cycle

Sequences in viral genomes that control transcription of viral genes, such as enhancers, may be determinants of viral tropism.

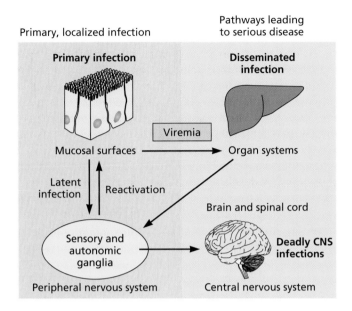

Figure 2.10 Outline of the spread of alphaherpesviruses and relationship to disease. CNS, central nervous system.

In the brain, JC polyomavirus reproduces only in oligodendrocytes (Box 2.7), because the JC virus enhancer is active only in this cell type. Other examples include the liver-specific enhancers of hepatitis B virus, the keratinocyte-specific enhancer of human papillomavirus type 11, and the enhancers in the long terminal repeat of human immunodeficiency virus type 1 that are active in cells of the immune system.

Cellular proteases are often required to cleave viral proteins to form the mature infectious virus particle (Volume I, Chapter 13). For example, a cellular protease cleaves the influenza virus HA0 precursor into two subunits so that fusion of the viral envelope and cell membrane can proceed. In mammals, the reproduction of influenza virus is restricted to epithelial cells of the upper and lower respiratory tracts. The tropism of this virus is thought to be influenced by the limited production of the protease that processes HA0. This serine protease, called tryptase, is secreted by nonciliated club cells of the bronchial and bronchiolar epithelia (Fig. 2.11). The purified enzyme can cleave and activate HA0 in virus particles *in vitro*. Alteration of the hemagglutinin (HA) cleavage site so

---

BOX 2.7

BACKGROUND
JC virus, a ubiquitous human polyomavirus

JC virus is widespread in the human population: 70 to 90% of adults are infected. Most humans experience inapparent childhood infections, but the virus then persists for life in the kidneys, brain, and gut tissue. If the immune system is compromised by pregnancy or chemotherapy, virus often reactivates from kidney tissues, and infectious virus particles can be found in the urine, approaching titers of greater than 100,000 particles/ml. On rare occasions, JC virus reactivates in the brain, causing the serious disease progressive multifocal leukoencephalopathy (PML) that affects the myelin-producing oligodendrocytes. This disease is often seen in patients with acquired immunodeficiency syndrome and after immunosuppressive therapy for organ transplants. The genome of JC virus found in the central nervous system of PML patients generally contains differences in the promoter sequence compared to those in healthy individuals. It is thought that such differences in promoter sequence contribute to the fitness of the virus in the central nervous system and thus to the development of PML. Given the rarity of the disease and the lack of suitable animal models, it has been difficult to determine how the genomes of these viruses are maintained and replication reactivated.

Progression of PML in an AIDS patient. Over the course of 4 months, the severity of white matter damage, seen in these magnetic resonance images as bright patches, increases significantly. Reprinted from A.K. Bag et al., *AJNR Am J Neuroradiol* **31:**1564-1576, 2010, with permission.

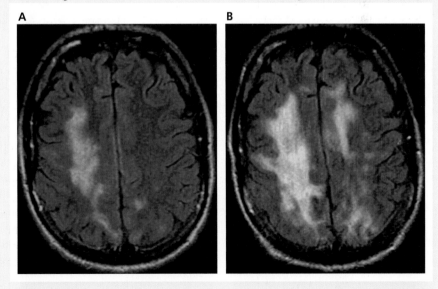

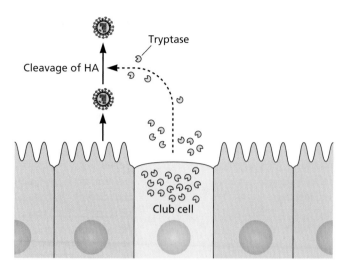

Figure 2.11 Cleavage of influenza virus HA0 by club cell tryptase. Influenza viruses reproduce in respiratory epithelial cells in humans. These virus particles contain the uncleaved form of HA (HA0) and are noninfectious. Club cells (once called Clara cells) secrete a protease, tryptase, which cleaves the HA0 of extracellular particles, thereby rendering the viral particles infectious. Adapted from M. Tashiro and R. Rott, *Semin. Virol.* **7**:237–243, 1996, with permission. Note: In previous editions of this text, club cells were referred to as "Clara cells," named after the German scientist who discovered them. Because Clara was an active member of the Nazi party, in 2013, the lung physiology community elected to change the name of these cells to "club cells." We have adopted this convention.

BOX 2.8

DISCUSSION
A mechanism for expanding the tropism of influenza virus is revealed by analyzing infections that occurred in 1940

Until the isolation of the H5N1 virus from 16 individuals in Hong Kong, viruses with the HA0 cleavage site mutation that permits cleavage by ubiquitous furin proteases had not been found in humans. However, the WSN/33 strain of influenza virus, produced in 1940 by passage of a human isolate in mouse brain, is pantropic in mice. Unlike most human influenza virus strains, WSN/33 can reproduce in cells in culture in the absence of added trypsin, because its HA can be cleaved by serum plasmin. Surprisingly, it was found that the NA of WSN/33 is necessary for HA0 cleavage by serum plasmin. This altered NA protein can bind plasminogen, sequestering it on the cell surface, where it is converted to the active form, plasmin (see figure, panel A). Plasmin then cleaves HA0 into HA1 and HA2. Therefore, a change in NA, not in HA, allowed cleavage of HA by a ubiquitous cellular protease. This property may, in part, explain the pantropic nature of WSN/33.

Proposed mechanism for activation of plasminogen and cleavage of HA. (A) Plasminogen binds to NA, which has a lysine at the carboxyl terminus. A cellular protein converts plasminogen to the active form, plasmin. Plasmin then cleaves HA0 into HA1 and HA2. **(B)** When NA does not contain a lysine at the carboxyl terminus, plasminogen cannot interact with NA and is not activated to plasmin. Therefore, HA is not cleaved. Adapted from H. Goto and Y. Kawaoka, *Proc. Natl. Acad. Sci. U. S. A.* **95**:10224–10228, 1998, with permission.

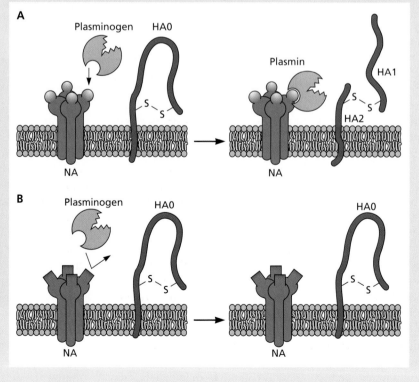

Goto H, Kawaoka Y. 1998. A novel mechanism for the acquisition of virulence by a human influenza A virus. *Proc Natl Acad Sci U S A* **95**:10224–10228.

Taubenberger JK. 1998. Influenza virus hemagglutinin cleavage into HA1 and HA2: no laughing matter. *Proc Natl Acad Sci U S A* **95**:9713–9715.

that it can be recognized by other cellular proteases changes the tropism of the virus and its pathogenicity dramatically; some highly virulent avian influenza virus strains contain an insertion of multiple basic amino acids at the cleavage site of HA0. This new sequence permits processing by ubiquitous intracellular proteases, such as furins. As a result, these variant viruses are released in active form and are able to infect many organs of birds, including the spleen, liver, lungs, kidneys, and brain. Naturally occurring mutants of this type cause high mortality in poultry farms. Avian influenza viruses isolated from 16 people in Hong Kong contained similar amino acid substitutions at the HA cleavage site. Indeed, many of these individuals had gastrointestinal, hepatic, and renal symptoms as well as respiratory disease. A virus with such an HA site alteration had not been previously identified in humans, and its isolation led to fears that an influenza pandemic was imminent. To prevent the virus from spreading, all chickens in Hong Kong were slaughtered. Changes in other viral proteins can influence HA cleavage indirectly (Box 2.8).

Spread throughout the Host

Following reproduction at the site of entry, virus particles can remain localized or can spread to other tissues (Table 2.1). Spread beyond the initial site of infection depends on multiple parameters, including the initial viral dose, the presence of viral receptors on other cells, and the relative rates of immune induction and release of infectious virus particles. Localized infections in the epithelium are usually limited by the physical constraints of the tissue and are brought under control by the intrinsic and innate defenses discussed in Chapter 3. An infection that spreads beyond the primary site of infection is said to be **disseminated**. If many organs are viral targets, the infection is described as **systemic.** Spread beyond the primary site requires that the host's physical barriers be breached. For example, virus particles may be able to cross the basement membrane when the integrity of that structure is compromised by inflammation and epithelial cell destruction. Below the basement membrane are subepithelial tissues, where virus particles encounter tissue fluids, the lymphatic system, and phagocytes. All three make substantial contributions in clearing foreign particles but may also allow infectious virus particles to be disseminated beyond the primary site of infection.

One important mechanism for avoiding local host defenses and facilitating spread within the body is the directional release of virus particles from polarized cells at a mucosal surface (Volume I, Chapter 12). Virus particles can be released from the apical surface, from the basolateral surface, or from both (Fig. 2.12). After reproduction, particles released from the apical surface are back where they started, that is, "outside" the host. Such directional release facilitates the dispersal of many newly synthesized enteric viruses in the feces (e.g., poliovirus)

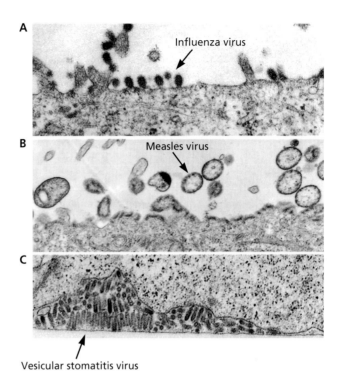

Figure 2.12 Polarized release of viruses from cultured epithelial cells visualized by electron microscopy. (A) Influenza virus released by budding from the apical surface of canine kidney cells. **(B)** Budding of measles virus on the apical surface of human colon carcinoma cells. **(C)** Release of vesicular stomatitis virus at the basal surface of canine kidney cells. Arrows indicate virus particles. Magnification, ×324,000. Reprinted from D. M. Blau and R. W. Compans, *Semin. Virol.* **7:**245–253, 1996, with permission. Courtesy of D. M. Blau and R. W. Compans, Emory University School of Medicine, Atlanta, GA.

or the respiratory tract (e.g., rhinoviruses). In general, virus particles released at apical membranes establish a localized or limited infection and do not penetrate deeply beyond the primary site of infection. In this case, local lateral spread from cell to cell may occur in the infected epithelium, but the underlying lymphatic and circulatory vessels are rarely infected. In contrast, virus particles released from the basolateral surfaces of polarized epithelial cells can access underlying tissues, facilitating systemic spread. The consequences of directional release are striking. Sendai virus, which is normally released from the apical surfaces of polarized epithelial cells, causes only a localized infection of the respiratory tract. In stark contrast, a mutant strain of this virus which is released from both apical and basal surfaces is disseminated, and the infected animals suffer higher morbidity and mortality.

When spread occurs by neural pathways, innervation at the primary site of inoculation determines which neuronal circuits will be infected. The only areas in the brain or spinal cord that are targets for herpes simplex virus infection are those that contain neurons with axon terminals or dendrites connected to common sites of inoculation in the body. Reactivated herpes simplex virus uses the same neural circuits to return to those sites, where it causes lesions (for example, cold sores in the mouth). After peripheral infection, poliovirus never reaches certain areas of the spinal cord and brain. However, reproduction occurs in these locations if the virus is inoculated directly into the brain or neural pathways.

There are two primary ways to gain access to tissues distal to the site of the inoculation: via the blood and via the nervous system.

Hematogenous Spread

Disseminated infections typically occur by transport through the bloodstream (hematogenous spread). Entry may occur through broken blood vessels (human immunodeficiency virus), through direct inoculation (for example, from the proboscis of an infected arthropod vector or the bite of a dog), or by basolateral release of virus particles from infected capillary endothelial cells. Because every mammalian tissue is nourished by a web of blood vessels, virus particles in the blood have access to all host organs, provided that susceptible cells exist in other tissues (Fig. 2.13).

Figure 2.13 Entry, dissemination, and shedding of blood-borne viruses. Shown are the target organs for some viruses that enter at epithelial surfaces and spread via the blood. The sites of virus shedding (red arrows), which may lead to transmission to other hosts, are shown. Adapted from N. Nathanson (ed), *Viral Pathogenesis* (Lippincott-Raven Publishers, Philadelphia, PA, 1997), with permission.

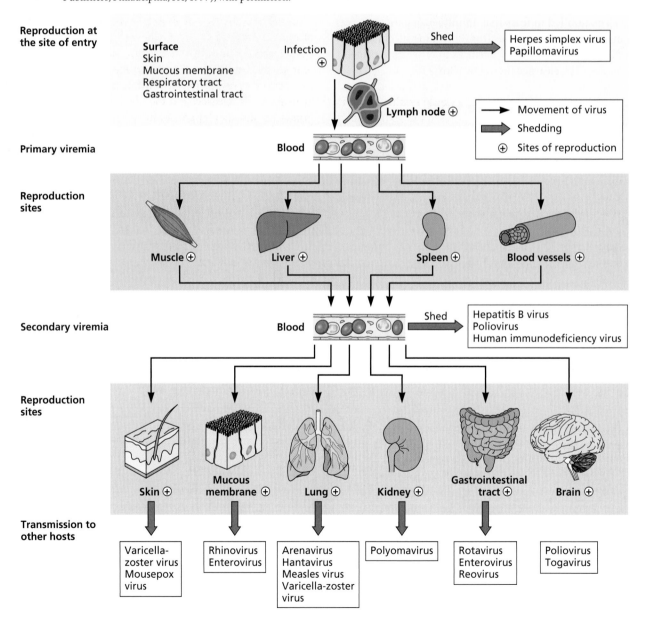

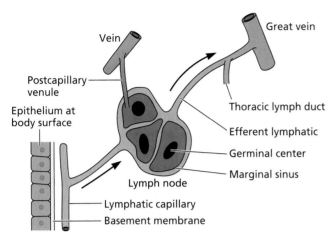

Figure 2.14 The lymphatic system. Lymphocytes flow from the blood into the lymph node through postcapillary venules. Adapted from C. A. Mims et al., *Mims' Pathogenesis of Infectious Disease* (Academic Press, Orlando, FL, 1995), with permission.

Hematogenous spread begins when newly synthesized particles produced at the entry site are released into extracellular fluids and are taken up by the local lymphatic vascular system (Fig. 2.14). Lymphatic capillaries are considerably more permeable than circulatory system capillaries, facilitating virus entry. As lymphatic vessels ultimately drain into the circulatory system, virus particles in lymph have free access to the bloodstream. In the lymphatics, virus particles pass through lymph nodes, where they encounter lymphocytes and monocytes. When viruses reproduce in cells of the immune system, such as human immunodeficiency virus and measles virus, once virus particles reach the lymph node, where susceptible cells are in abundance, they initiate a robust phase of viral reproduction.

The migratory nature of many immune cells allows some viruses to move throughout the host. Because the viral genome is inside a cell during transport, it is effectively shielded from antibody recognition. Traversing the blood-brain barrier poses a particular challenge, as the capillaries that make up this unique barrier limit the access of serum proteins to the brain. However, activated macrophages can pass through, freely delivering viruses such as human immunodeficiency virus into the brain tissue. This process is often referred to as the Trojan Horse approach, because of its similarity to the legend of how the Greeks invaded and captured the protected fortress of Troy (Box 2.9).

The term **viremia** describes the presence of infectious virus particles in the blood. Active viremia is a consequence of reproduction in the host, whereas passive viremia results when particles are introduced into the blood without viral reproduction at the site of entry (as when an infected mosquito inoculates a susceptible host with West Nile virus). Progeny virus particles released into the blood after initial reproduction at the site of entry constitute the **primary viremia** phase. The concentration of particles during this early stage of infection is usually low. However, subsequent dissemination of the virus to other sites results in the release of considerably more virus particles. The delayed appearance of a high concentration of infectious virus in the blood is termed **secondary viremia** (Fig. 2.15). The two phases of viremia were first described in classic studies of mousepox (Fig. 2.1).

The concentration of virus particles in blood is determined by the rate of their synthesis in permissive tissues and by how

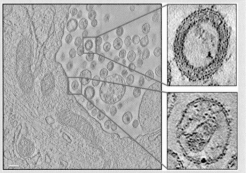

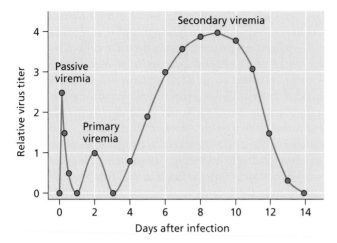

Figure 2.15 Characteristics of viremia. The graph was produced using data from different viral infections. For passive viremia, La Crosse virus, a bunyavirus, was injected into weanling mice, and virus titers in plasma, brain, and muscle were determined at different times thereafter. No virus can be detected in the blood after 1 day. For primary viremia, ectromelia virus was inoculated into the footpads of mice; after local multiplication, the virus enters the blood. For secondary viremia, viral progeny produced by reproduction of ectromelia virus in the target organs were counted. Virus reaches these organs during primary viremia. Adapted from N. Nathanson (ed.), *Viral Pathogenesis and Immunity* (Academic Press, London, United Kingdom, 2007), with permission.

quickly they are released into, and removed from, the blood. Circulating particles are engulfed and destroyed by phagocytic cells of the **reticuloendothelial system** in the liver, lungs, spleen, and lymph nodes. When serum antibodies appear, virus particles in the blood may be bound by them and be neutralized (Chapter 4). Formation of a complex of antibodies and virus particles facilitates uptake by Fc receptors carried by macrophages lining the circulatory vessels. These virus-antibody complexes can be sequestered in significant quantities in the kidneys, spleen, and liver, prior to elimination from the host via urine or feces. The average duration of an individual virus particle in the blood varies from 1 to 60 min, depending on parameters such as the physiology of the host (e.g., age and health) and the size and structural integrity of virus particles. Some viral infections are noteworthy for the long-lasting presence of infectious particles in the blood. Hosts infected with hepatitis B and C viruses or mice infected with lymphocytic choriomeningitis virus may have active viremia that persists for months to years.

Viremia is of diagnostic value to monitor the course of infection in an individual over time, and epidemiologists use the detection of viremia to identify infected individuals within a population. Frequently, it may be difficult, or technically impossible, to quantify infectious particles in the blood, as is the case for hepatitis B virus. In these situations, the presence of characteristic viral proteins, such as the reverse transcriptase for human immunodeficiency virus, provides surrogate markers for viremia.

However, the presence of infectious virus particles in the blood also presents practical problems. Infections can be spread inadvertently in the population when pooled blood from thousands of individuals is used for therapeutic purposes (transfusions) or as a source of therapeutic proteins (gamma globulin or blood-clotting factors). We have learned from unfortunate experience that blood-borne viruses, such as the hepatitis viruses and human immunodeficiency virus, can be spread by contaminated blood and blood products. The World Health Organization estimates that, as of 2000, inadequate blood screening resulted in 1 million new human immunodeficiency virus infections worldwide. Careful screening for these viruses in blood supplies before it is transfused into patients is now standard procedure. However, sensitive detection methods and stringent purification protocols are useful only when we know what we are looking for; as-yet-undiscovered viruses may still be transmitted through the blood supply.

Neural Spread

Some viruses spread from the primary site of infection by entering local nerve endings. In some cases, neuronal spread is the definitive characteristic of pathogenesis, notably by rabies virus and alphaherpesviruses, which cause infections that primarily impact neuronal function or survival. In other cases, invasion of the nervous system is a rare, typically dead-end, diversion from their normal site of reproduction (e.g., poliovirus and reovirus). Mumps virus, human immunodeficiency virus, and measles virus reproduce in the brain but access the central nervous system by the hematogenous route. The molecular mechanisms that dictate spread by neural or hematogenous pathways are not well understood. While viruses that infect the nervous system are often said to be neurotropic (Box 2.10), they are generally capable of infecting a variety of cell types. Viral reproduction usually occurs first in nonneuronal cells, with virus particles subsequently spreading into afferent (e.g., sensory) or efferent (e.g., motor) nerve fibers that innervate the infected tissue (Fig. 2.16).

Neurons are polarized cells with structurally and functionally distinct processes (axons and dendrites) that can be separated by enormous distances. For example, in adult humans, the axon terminals of motor neurons that control stomach muscles can be 50 centimeters away from the cell bodies and dendrites in the brain stem. While our understanding of how viral particles move in and among neurons of the nervous system is incomplete, what is certain is that neurotropic viruses do not traverse these great distances by Brownian (random) motion. Rather, the neuronal cytoskeleton, including microtubules and actin, provides the "train tracks" that enable movement of mitochondria, synaptic vesicles, and virus particles to and from the synapse. Molecular motor proteins, such as dynein and kinesin, are the "engines" that move along these

TERMINOLOGY
Infection of the nervous system: definitions and distinctions

A **neuroinvasive virus** can enter the central nervous system (spinal cord and brain) after infection of a peripheral site.

A **neurotropic virus** can infect neurons; infection may occur by neural or hematogenous spread from a peripheral site.

A **neurovirulent virus** can cause disease of nervous tissue, manifested by neurological symptoms and often death.

Examples:

Herpes simplex virus has low neuroinvasiveness but high neurovirulence. It always enters the peripheral nervous system but rarely gains access to the central nervous system. When it does, the consequences are severe, often fatal.

Mumps virus has high neuroinvasiveness but low neurovirulence. Most infections lead to invasion of the central nervous system, but neurological disease is mild.

Rabies virus has high neuroinvasiveness and high neurovirulence. It readily infects the peripheral nervous system and spreads to the central nervous system with 100% lethality, unless postinfection vaccination is given.

Primary mouse hippocampal neurons expressing a measles virus receptor, CD46, and infected with measles virus for 48 h. Virus-infected cells are stained brown. Original magnification = ×200.

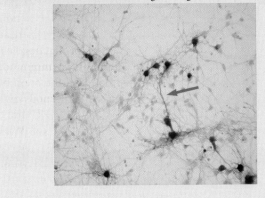

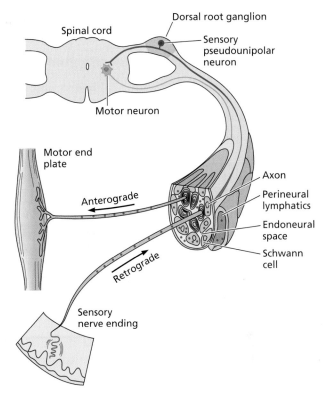

Figure 2.16 Possible pathways for the spread of infection in nerves. Virus particles may enter sensory or motor neuron endings. They may be transported within axons, in which case viruses taken up at sensory endings reach dorsal root ganglion cells. Those taken up at motor endings reach motor neurons. Viruses may also travel in the endoneural space, perineural lymphatics, or infected Schwann cells. Directional transport of virus particles inside the sensory neuron is defined as anterograde [movement from the (−) to the (+) ends of microtubules] or retrograde [movement from the (+) to the (−) ends of microtubules]. Adapted from R. T. Johnson, *Viral Infections of the Nervous System* (Raven Press, New York, NY, 1982), with permission.

thoroughfares (Box 2.11). Drugs, such as colchicine, that disrupt microtubules efficiently block the spread of many neurotropic viruses from the site of peripheral inoculation to the central nervous system (Volume I, Chapter 12).

With few exceptions (Box 2.12), cells of the peripheral nervous system are the first to be infected. These neurons represent the first cells in circuits connecting the innervated peripheral tissue with the spinal cord and brain. Once in the peripheral nervous system, alphaherpesviruses and some rhabdoviruses (e.g., rabies virus), flaviviruses (e.g., West Nile virus), and paramyxoviruses (e.g., measles and canine distemper virus) can spread among neurons connected by synapses (Box 2.13). Virus spread by this mode can continue through chains of connected neurons of the peripheral nervous system

and may eventually reach the spinal cord and brain, often with devastating results (Fig. 2.10). Nonneuronal support cells and satellite cells in ganglia may also be infected.

Movement of virus particles and their release from infected cells are important features of neuronal infections. As is true for polarized epithelial cells discussed earlier, directional release of virions from neurons affects the outcome of infection. Alphaherpesviruses become latent in peripheral neurons that innervate the site of infection. Reactivation from the latent state results in viral reproduction in the primary neuron and subsequent transport of progeny virus particles from the neuron cell body back to the innervated peripheral tissue where the infection originated. Alternatively, virus particles can spread from the peripheral to the central nervous system (Fig. 2.10). The direction taken is the difference between a minor local infection (a cold sore) and a life-threatening viral encephalitis. Luckily, spread back to the peripheral site is by far more common.

BOX 2.11

TERMINOLOGY
Which direction: anterograde or retrograde?

Those who study virus spread in the nervous system often use the words **retrograde** and **anterograde** to describe direction. Unfortunately, confusion arises because the terms can be used to describe directional movement of virus particles inside a cell, as well as spread between synaptically connected neurons. Spread from the primary neuron to the second-order neuron in the direction of the nerve impulse is called anterograde spread (see figure). Spread in the opposite direction is termed retrograde. Spread inside a neuron is defined by microtubule polarity. Transport on microtubules from $(-)$ to $(+)$ ends is anterograde, while transport on microtubules from $(+)$ to $(-)$ ends is retrograde.

Retrograde and anterograde spread of virus in nerves. (A) Retrograde spread of infection. Virus invades at axon terminals and spreads to the cell body, where reproduction occurs. Progeny virus particles spread to a neuron at sites of synaptic contact. Particles enter the axon terminal of the second neuron to initiate a second cycle of replication and spread. **(B)** Anterograde spread of infection. Virus invades at dendrites or cell bodies and reproduces. Virus particles then spread to axon terminals, where virus particles cross synaptic contacts to invade dendrites or cell bodies of the second neuron.

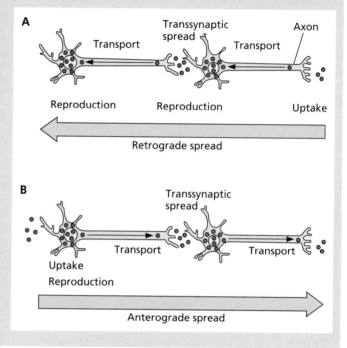

BOX 2.12

BACKGROUND
The path rarely taken: direct entry into the central nervous system by olfactory routes

Olfactory neurons are unusual in that their cell bodies are present in the olfactory epithelia and their axon termini are in synaptic contact with olfactory bulb neurons. These conduits to the brain project from cells that are in direct contact with the environment. The olfactory nerve fiber passes through the skull via an opening called the arachnoid. Remarkably, few viruses enter the brain by the olfactory route, despite significant reproduction of many in the nasopharyngeal cavity. Adapted from R. T. Johnson, *Viral Infections of the Nervous System* (Raven Press, New York, NY, 1982), with permission.

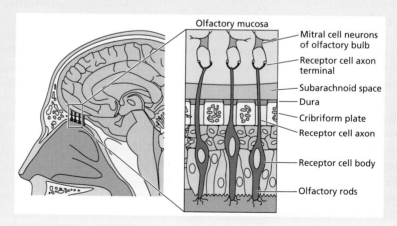

DISCUSSION
Tracing neuronal connections in the nervous system with viruses

The identification and characterization of synaptically linked multineuronal pathways in the brain are important to understanding the functional organization of neuronal circuits. Conventional tracing methodologies have relied on the use of markers, such as wheat germ agglutinin-horseradish peroxidase or fluorochrome dyes. The main limitations of these tracers are their low specificity and sensitivity. During experimental manipulation, it is difficult to restrict the diffusion of tracers to a particular cell group or nucleus, which can lead to false-positive labeling of a circuit. Furthermore, neurons located one or more synapses away from the injection site receive progressively less tracer, which is diluted at each stage of transneuronal transfer.

Some alphaherpesviruses and rhabdoviruses have considerable promise for use as self-amplifying tracers of synaptically connected neurons. Under proper conditions, second- and third-order neurons will show the same labeling intensity as those infected initially. Moreover, the specific pattern of infected neurons observed in such tracing studies is consistent with transsynaptic passage of virus rather than lytic spread through the extracellular space.

Viruses are typically detected by light microscopy using immunohistochemical staining to localize viral antigens. More recently, reporter genes, such as that encoding the green fluorescent protein (GFP) of *Aequorea victoria*, have been introduced into the genomes of neurotropic viruses for direct visualization of viral infection.

Ekstrand M, Pomeranz L, Enquist LW. 2008. The alpha-herpesviruses: molecular pathfinders in nervous system circuits. *Trends Mol Med* **14**:134–140.

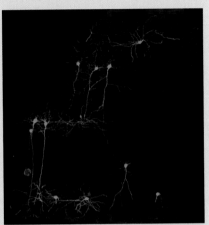

Identification of a possible microcircuit in the rodent visual cortex (V2) after injection of a GFP-expressing strain of pseudorabies virus into the synaptically connected, but distant, V1 region. Infection spread via V1 axons (V1 cell bodies are located far out of the field of view) in a retrograde manner to a subset of V2 cell bodies seen here. Confocal microscopy and image reconstruction by Botond Roska, Friedrich Miescher Institute, Basel, Switzerland.

Organ Invasion

Once virus particles enter the blood or neurons and are dispersed from the primary site, any subsequent reproduction requires invasion of other cells. We have already discussed viral movement into and among neurons to access the brain and spinal cord and will return to this issue in Chapter 5 when we discuss neuropathogenesis resulting from viral infections.

There are three main types of blood vessel-tissue junctions that serve as portals for tissue invasion (Fig. 2.17). In some tissues, the endothelial cells are continuous with a dense basement membrane. At other sites, the endothelium contains gaps, and at still others, there may be **sinusoids**, in which macrophages form part of the blood-tissue junction. Viruses can traverse all three types of junctions.

Entry into Organs with Sinusoids

Organs such as the liver, spleen, bone marrow, and adrenal glands are characterized by the presence of sinusoids lined with macrophages. Such macrophages, known somewhat misleadingly as the reticuloendothelial system (these macrophages are neither endothelial nor a "system"), function to filter the blood and remove foreign particles, similar to a HEPA filter purifying incoming air. The macrophages often provide the portal

Figure 2.17 Blood-tissue junction in a capillary, venule, and sinusoid. (Left) Continuous endothelium and basement membrane found in the central nervous system, connective tissue, skeletal and cardiac muscle, skin, and lungs. **(Center)** fenestrated endothelium found in the choroid plexus, villi of the intestine, renal glomerulus, pancreas, and endocrine glands. **(Right)** sinusoid, lined with macrophages of the reticuloendothelial system, as found in the adrenal glands, liver, spleen, and bone marrow. Adapted from C. A. Mims et al., *Mims' Pathogenesis of Infectious Disease* (Academic Press, Orlando, FL, 1995), with permission.

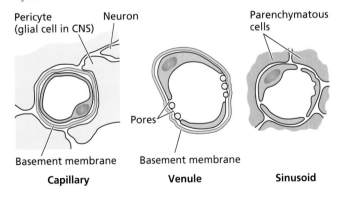

Pericyte (glial cell in CNS)   Neuron   Parenchymatous cells

Pores

Basement membrane   Basement membrane

Capillary   **Venule**   **Sinusoid**

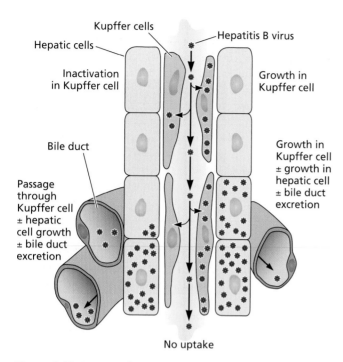

Figure 2.18 Routes of viral entry into the liver. Two layers of hepatocytes are shown, with the sinusoid at the center lined with Kupffer cells. Endothelial cells are not shown. Adapted from C. A. Mims et al., *Mims' Pathogenesis of Infectious Disease* (Academic Press, Orlando, FL, 1995), with permission.

for entry of viral particles into tissues. For example, the hepatitis viruses that infect the liver, which is the major filtering and detoxifying organ of the body, usually enter from the blood. The presence of virus particles in the blood invariably leads to the infection of **Kupffer cells**, the macrophages that line liver sinusoids (Fig. 2.18). Virus particles may be transcytosed across Kupffer and endothelial cells without reproduction to reach the underlying hepatic cells. Alternatively, viruses may multiply in these cells and then infect underlying hepatocytes. Either mechanism may induce inflammation and necrosis of liver tissue, a condition termed hepatitis.

Entry into Organs That Lack Sinusoids

To enter tissues that lack sinusoids (Fig. 2.17), virus particles must first adhere to the endothelial cells lining capillaries or venules, where the blood flow is slowest and the walls are thinnest. To increase the chances of adhesion, virus particles must be present in a high concentration and circulate for a sufficient period. Clearly there is a "race" between adhesion and removal of virus particles by macrophages. Once blood-borne virus particles have adhered to the vessel wall, they can readily invade the renal glomerulus, pancreas, ileum, or colon, because the endothelial cells that make up the capillaries of these tissues are fenestrated, permitting virus particles or

virus-infected cells to cross. Some viruses traverse the endothelium while being carried by infected monocytes or lymphocytes, a process called **diapedesis**.

Organs with Dense Basement Membranes

In the central nervous system, connective tissue, and skeletal and cardiac muscle, capillary endothelial cells are supported by a dense basement membrane, which raises an additional barrier to viral passage into the tissue (Fig. 2.17 and 2.19). In the central nervous system, the basement membrane, formed in part by astrocyte processes that align with the basolateral surface of the capillary endothelium, is the foundation of the blood-brain barrier (Fig. 2.20).

In several well-defined parts of the brain, the capillary epithelium is fenestrated (with "windows" between cells, loosely joined together), and the basement membrane is sparse, affording an easier passage for some neurotropic viruses. These highly vascularized sites include the choroid plexus, a sheet of tissue that lies within the brain ventricles and that produces more than 70% of the cerebrospinal fluid that bathes the spinal cord. Some viruses (mumps virus and certain togaviruses) pass through the capillary endothelium and enter the stroma of the choroid plexus, where they may cross the epithelium into the cerebrospinal fluid either by transcytosis or by directed release following production of progeny virus particles. Once in the cerebrospinal fluid, infection spreads to the ependymal cells lining the ventricles and the underlying brain tissue (Fig. 2.20). Other viruses (picornaviruses) may infect directly, or be transported across, the capillary endothelium.

Figure 2.19 How viruses travel from blood to tissues. Schematic of a capillary illustrating different pathways by which viruses may leave the blood and enter underlying tissues. Adapted from N. Nathanson (ed.), *Viral Pathogenesis and Immunity* (Academic Press, London, United Kingdom, 2007), with permission.

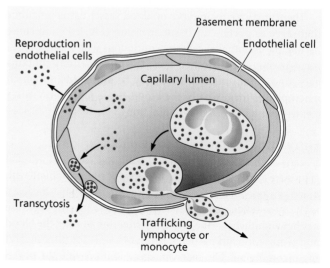

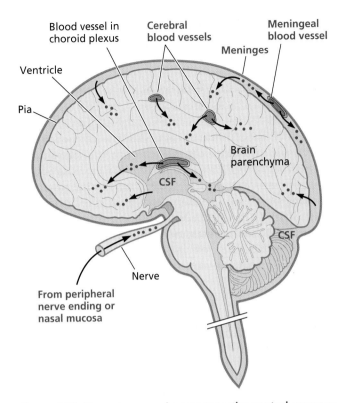

Figure 2.20 How viruses gain access to the central nervous system. A summary of the mechanisms by which viruses can enter the brain is shown. CSF, cerebrospinal fluid. Adapted from C. A. Mims et al., *Mims' Pathogenesis of Infectious Disease* (Academic Press, Orlando, FL, 1995), with permission.

Some viruses (human immunodeficiency virus and measles virus) cross the endothelium within infected monocytes or lymphocytes (the Trojan Horse approach, described earlier). Increased local permeability of the capillary endothelium, caused, for example, by certain hormones, may also facilitate virus entry into the brain and spinal cord.

Skin

In a number of systemic viral infections, rashes are produced when virus particles leave blood vessels. Viruses that cause rashes include measles virus, rubella virus (German measles), varicella-zoster virus (chicken pox and shingles), some parvoviruses (fifth disease), poxviruses (smallpox), and coxsackieviruses (hand, foot, and mouth disease). Skin lesions are distinguished by size, color, frequency, and elevation (an indication of inflammation), may appear coincident with or subsequent to an infection, and are sufficiently distinct in appearance to be easily associated with a particular virus. Destruction of cells by virus reproduction and the host immune system are the primary causes of these skin lesions.

Rashes are not restricted to the skin. Lesions may also occur in mucosal tissues, such as those in the mouth and throat.

Because these surfaces are wet, vesicles break down more rapidly than on the skin. During measles infection, vesicles in the mouth become ulcers before the appearance of skin lesions. Such Koplik spots are diagnostic for measles virus infection and appear 2 to 4 days before the skin rash. After virus particles leave the subepithelial capillaries in the respiratory tract, only a single layer of cells must be traversed before particles reach the exterior. Hence, during infections with measles virus and varicella-zoster virus, particles appear in respiratory tract secretions a few days before the skin rash appears. By the time that the infection is recognized from the skin rash, viral transmission to other individuals may already have occurred.

The Fetus

Basement membranes are less well developed in the fetus, and infection can occur by invasion of the placental tissues and subsequent entry into fetal tissue. Infected circulating cells, such as monocytes, may enter the fetal bloodstream directly. In a pregnant female, viremia may result in infection of the developing fetus. Transplacental infections are distinct from perinatal infections, in which the virus is acquired via contact with maternal blood as the baby is delivered through the birth canal. Many pregnant women who are positive for human immunodeficiency virus deliver their children by C-section, but before this became a widespread practice, perinatal infection with human immunodeficiency virus was a major cause of infant morbidity, responsible for an estimated 4 million deaths since the start of the pandemic.

While perinatal infections can be avoided by Caesarian delivery, transplacental infections cannot. Historically, the primary transplacental infections of concern were rubella, cytomegalovirus, and herpes simplex. These viruses, along with the parasite *Toxoplasma*, comprise the four pathogens defined by the acronym TORCH. These pathogens pose a substantial threat to the fetus. The risk of fetal infection in infants whose mothers were infected with rubella virus during the first trimester is approximately 80%. Similarly, intrauterine transmission of human cytomegalovirus occurs in approximately 40% of pregnant women with primary infection. We now know that transplacental transfer of other viruses, including some parvoviruses, measles virus, human immunodeficiency virus, and varicella-zoster virus, can also occur.

Shedding of Virus Particles

As viruses that cannot spread from host to host face extinction, viruses must exit one host to infect another. The release of virus particles from an infected host is called **shedding**. While most transmission events are attributable to such release, there are some exceptions. These exceptions include the direct transmission to host progeny of viral genomes in the host germ line and viruses transmitted via blood transfusions or organ transplantation, such as the human immunodeficiency viruses or hepatitis viruses.

During localized infections in or near one of the body openings, shedding can occur from the primary site of virus reproduction. The papillomaviruses cause genital warts; these viruses reproduce locally in the genital epithelium and are transmitted to naïve hosts via sexual contact. In contrast, release of virus particles that cause disseminated infections can occur from many sites. Effective transmission of virus particles from one host to another depends on the concentration of released particles and the mechanisms by which the virus particles are introduced into the next host. The shedding of small quantities of virus particles may be insufficient to cause new infections, while the shedding of high concentrations may facilitate transmission via minute quantities of tissue or body fluid. For example, the concentration of hepatitis B virus particles in blood can be so high that a few microliters is sufficient to initiate an infection. The stability of virus particles in the environment also influences the efficiency of transmission.

Respiratory Secretions

Respiratory transmission depends on the incorporation of airborne particles in aerosols. Aerosols are produced during speaking and normal breathing, while coughing produces even more forceful expulsion. Transmission from the nasal cavity is facilitated by sneezing and is much more effective if infection induces the production of nasal secretions. A sneeze produces up to 20,000 droplets (in contrast to several hundred expelled by coughing), and all may contain rhinovirus if the individual has a common cold. As noted when we discussed viral entry, the size of a droplet affects its "hang time": large droplets fall to the ground, but smaller droplets (1 to 4 μm in diameter) may remain suspended in the air indefinitely. Such particles may not only come in contact with a naïve host but may be able to reach the lower respiratory tract. Nasal secretions also frequently contaminate hands or tissues. The infection may be transmitted when these objects contact another person's fingers and that person in turn touches his or her nose or conjunctiva. In today's crowded society, the physical proximity of people may select for viruses that spread efficiently by this route. Sneezing may be the body's way of trying to eliminate an irritant in the respiratory tract. Some have speculated that viruses may have been selected that induce sneezing in their hosts to ensure transmission to new hosts (Box 2.14).

Saliva

Some viruses that reproduce in the lungs, nasal mucosa, or salivary glands are shed into the oral cavity. Transmission may occur through aerosols, as discussed above, via contaminated fingers, or by kissing or spitting. Animals that lick, nibble, and groom may also transmit infections in saliva. Perhaps the best-known human virus that is transmitted via saliva is Epstein-Barr virus, which results in mononucleosis, or "kissing disease." Remarkably, the incubation time for this virus

---

BOX 2.14

DISCUSSION
A ferret model of influenza virus infection ignites irrational fears

Ferrets, which are carnivorous mammals, are excellent models for the study of influenza virus infection, pathogenesis, and transmission. Human and avian influenza viruses reproduce in the ferret airway, and infected animals develop many characteristic signs of the flu, including fever and sneezing. The release of infectious virus through nasal discharges allows for ferret-to-ferret transmission of influenza, an observation first reported in 1933.

In 2011, influenza virus experiments using this well-established model came under intense media scrutiny when two research groups genetically engineered an H5N1 strain of influenza virus that was suspected to be a possible origin of "pandemic" strain in humans. These investigators showed that the engineered viruses were transmissible in ferrets, raising concern that, if the viruses or an infected ferret escaped or was otherwise released, this could trigger a new influenza pandemic. The debate, which continues to impact the scientific and lay communities, centers around "dual-use" experiments: studies that have both a potential public health benefit but that could also be used for bioterrorism or could endanger humans. Some scientists have contended that work such as this should never have been done, given the risks. Others, including many virologists, counter with multiple points. First, that a virus can be transmitted in ferrets does not indicate that it will also spread in humans. However, knowing the genetic changes that affect transmission would have great benefit should an H5N1 virus infection occur in humans by enabling better monitoring of infections and more-rapid development of antivirals. Finally, high-level biocontainment facilities and procedures for such experiments have been mandated to prevent accidental release of nefarious viruses

(or infected animals). While it may seem like the basis of a terrific thriller, work with these agents has been strictly controlled.

Belser JA, Katz JM, Tumpey TM. 2011. The ferret as a model organism to study influenza A virus infection. *Dis Model Mech* **4:**575–579.

(that is, the time between infection and disease) is 4 to 7 weeks. Consequently, an Epstein-Barr virus-infected Lothario has lots of time to transmit far more than a radiant smile and magnetic personality before being "found out." Human cytomegalovirus, mumps virus, and some retroviruses can also be transmitted through saliva via an oral route.

Feces

Particles of enteric and hepatic viruses that are shed into the intestine and transmitted via a fecal route are generally more resistant to inactivation by environmental conditions than those released at other sites. An important exception is hepatitis B virus, which is shed in bile that is released into the intestine but is inactivated as a consequence and therefore not transmitted in feces. Viruses transmitted by fecal spread usually survive dilution in water, as well as drying.

Inefficient sewage treatment or its absence, contaminated irrigation systems, and the use of animal manures are prime sources of fecal contamination of food, water supplies, and living areas. Any one of these conditions provides an effective mode for continual reentry of these viruses into the alimentary canals of new hosts. Two hundred years ago, such contamination was inevitable in most of the world, as disposal of human feces in the streets was a common practice. Communities downstream of sites of defecation and waste removal used contaminated water for cooking and drinking. With modern sanitation, the fecal-oral cycle has been largely interrupted in developed countries but remains a major cause of viral spread throughout the rest of the world.

Blood

Viremia is a common feature of many viral infections, and exposure to viremic blood is a primary mode of virus transmission. Arthropods acquire virus particles when they bite viremic hosts and may transmit them to subsequent hosts upon the next blood meal. Hepatitis viruses and human immunodeficiency virus can be transmitted by virus-laden blood during transfusions and injections. Virus particles may also be transmitted from blood during sex or childbirth, and consumption of raw meat may place contaminated blood in contact with the alimentary and respiratory tracts. Health care professionals, emergency rescue workers, and dentists are exposed routinely to blood from individuals who may harbor infections. Indeed, for many of the viruses that cause fatal hemorrhagic fevers (such as members of the *Bunyaviridae* and *Filoviridae*), the only mode of transmission to humans is via contaminated blood and body fluids. Consequently, health care workers often are the first people to become infected and show symptoms in an outbreak of such viral diseases.

Urine

Virus-containing urine is a common contaminant of food and water supplies. The presence of virus particles in the urine is called **viruria**. Hantaviruses and arenaviruses that infect rodents cause persistent viruria. Consequently, humans may be infected by exposure to dust that contains dried urine from infected rodents. A few human viruses, including the polyomaviruses JC and BK, reproduce in the kidneys and are shed in urine.

Semen

Some retroviruses, including human immunodeficiency virus type 1, herpesviruses, and hepatitis B virus, are shed in semen and transmitted during sex. Herpesviruses that infect the genital mucosa are shed from lesions and transmitted by genital secretions, as are papillomaviruses.

Recently, it was shown that human immunodeficiency virus in semen is different from the virus found in blood of the same patients. While sequences of genomes of viruses isolated from the blood are heterogeneous, sequences of the viruses in the semen were much more homogeneous. Two mechanisms were proposed to account for this difference: clonal amplification and compartmentalization. In the first mechanism, one to several viruses are proposed to reproduce in T cells in the seminal tract over a short period of time, such that the population detected in semen is relatively homogeneous (compared to the complex population in the blood). In the second mechanism, the virus is proposed to reproduce in these same cells but over a period of time that is long enough to allow a population genetically distinct from the virus in the blood to be selected.

Milk

Mouse mammary tumor virus is transmitted to offspring primarily via mother's milk into which the virus is shed, as are some tick-borne encephalitis viruses. Mumps virus and cytomegalovirus are shed into human milk but are probably not often transmitted by this route.

Skin Lesions

Many viruses reproduce in the skin, and the lesions that form from such infections contain virions that can be transmitted to other hosts. In these cases, the virus is usually transmitted by direct body contact. For example, herpes simplex virus causes a common rash in wrestlers, known as herpes gladiatorum. Warts caused by certain poxviruses and papillomavirus may also be transmitted by direct, skin-to-skin contact.

Varicella-zoster virus, the agent of chicken pox, is released from the skin in a particularly effective manner. The lesions that form during an acute chicken pox infection are small, lymph-filled blisters that erupt, leaving a crusty scab. Virus concentrations in this fluid are high. Despite the availability of an effective vaccine, acute infections still occur in unvaccinated individuals. Alarmingly, some parents have elected to allow their children to become infected by encouraging close exposure to acutely infected peers (Box 2.15).

DISCUSSION
Chicken pox parties

Prior to the widespread use of the varicella-zoster virus vaccine, some parents who wanted to control when their child would get chicken pox (often considered a childhood rite of passage) would host chicken pox parties, in which uninfected children would share lollipops licked by infected children. Given the presence of the virus in the oral mucosa, this ensured that the lollipop contained a high dose of the virus and virtually guaranteed infection. Moreover, because the incubation period for varicella is quite precise (about 14 days following exposure), parents could preplan days off of work to be with their sick child. Even today, there is a "black market" of virus-laced items (such as lollipops) available through the Web. Such practices are an almost inconceivably bad idea; infections by these viruses can be quite severe, and effective, safe vaccines do exist. Moreover, infected children pose risks to immunocompromised adults, such as the elderly and cancer patients receiving immunosuppressive chemotherapy.

Perspectives

Despite the complexity and diversity of viral infection cycles, at a minimum, all viruses must get in, and they must get out. This is true not only for the infected cell (a major theme of Volume I) but also for the infected host. In this chapter, we discussed the many ways by which an organism may acquire pathogens. It is not hyperbole to note that pathogens, including viruses, bacteria, parasites, and fungi, are truly everywhere, and because they have coevolved with their hosts, viruses have coopted our most intimate behaviors to ensure host-to-host transmission.

Fortunately, our counterdefenses pose formidable obstacles. Viruses are trapped in mucus, repelled by dead layers of skin, brushed away by cilia, and destroyed by stomach acid, but simply capturing a rook and a bishop does not end this age-old game of chess. Some viruses can bypass these defenses to reach target cells deep within organs. When viruses breach one of these walls, it is up to the elite forces of the host immune system, the precise strategies of the intrinsic, innate, and adaptive responses, to either end the game in checkmate or suffer the fateful capture of the King.

References

Books

Brock T (ed). 1961. *Milestones in Microbiology*. American Society for Microbiology, Washington, DC.

Johnson RT. 1982. *Viral Infections of the Nervous System*. Raven Press, New York, NY.

Mims CA, Nash A, Stephen J. 2001. *Mims' Pathogenesis of Infectious Disease*, 5th ed. Academic Press, Orlando, FL.

Nathanson N (ed). 2007. *Viral Pathogenesis and Immunity*, 2nd ed. Academic Press, London, United Kingdom.

Notkins AL, Oldstone MBA (ed). 1984. *Concepts in Viral Pathogenesis*. Springer-Verlag, New York, NY.

Richman DD, Whitley RJ, Hayden FG (ed). 2009. *Clinical Virology*, 3rd ed. ASM Press, Washington, DC.

Review Articles

Bieniasz PD. 2004. Intrinsic immunity: a front-line defense against viral attack. *Nat Immunol* **5**:1109–1115.

Collins P, Graham B. 2008. Viral and host factors in human respiratory syncytial virus pathogenesis. *J Virol* **82**:2040–2055.

Diefenbach R, Miranda-Saksena M, Douglas M, Cunningham A. 2008. Transport and egress of herpes simplex virus in neurons. *Rev Med Virol* **18**:35–51.

Esteban DJ, Buller RM. 2005. Ectromelia virus: the causative agent of mousepox. *J Gen Virol* **86**:2645–2659.

Kennedy PGE. 1992. Molecular studies of viral pathogenesis in the central nervous system: the Linacre lecture 1991. *J R Coll Physicians Lond* **26**:204–214.

Lancaster KZ, Pfieffer JK. 2012. Viral population dynamics and virulence thresholds. *Curr Opin Microbiol* **15**:1–6.

Malamud D, Abrams WR, Barber CA, Weissman D, Rehtanz M, Golub E. 2011. Antiviral activities in human saliva. *Adv Dent Res* **23**:34–37.

Pereira L, Maidji E, McDonagh S, Tabata T. 2005. Insights into viral transmission at the uterine-placental interface. *Trends Microbiol* **13**:164-174.

Tellinghuisen T, Evans M, von Hahn T, You S, Rice C. 2007. Studying hepatitis C virus: making the best of a bad virus. *J Virol* **81**:8853–8867.

Virgin HW. 2007. *In vivo veritas*: pathogenesis of infection as it actually happens. *Nat Immunol* **8**:1143–1147.

Wiens ME, Wilson SS, Lucero CM, Smith JG. 2014. Defensins and viral infection: dispelling common misconceptions. *PLoS Pathol* **10**:e1004186.

Yan N, Chen ZJ. 2014. Intrinsic antiviral immunity. *Nat Immunol* **13**:214–222.

Zampieri C, Sullivan N, Nabel G. 2007. Immunopathology of highly virulent pathogens: insights from Ebola virus. *Nat Immunol* **8**:1159–1164.

Papers of Special Interest

Bodian D. 1955. Emerging concept of poliomyelitis infection. *Science* **122**:105–108.

Crawford S, Patel D, Cheng E, Berkova Z, Hyser J, Ciarlet M, Finegold M, Conner M, Estes M. 2006. Rotavirus viremia and extraintestinal viral infection in the neonatal rat model. *J Virol* **80:**4820–4832.

Croxford J, Olson J, Anger H, Miller S. 2005. Initiation and exacerbation of autoimmune demyelination of the central nervous system via virus-induced molecular mimicry: implications for the pathogenesis of multiple sclerosis. *J Virol* **79:**8581–8590.

Hamelin M-E, Yim K, Kuhn K, Cragin R, Boukhvalova M, Blanco J, Prince G, Boivin G. 2005. Pathogenesis of human metapneumovirus lung infection in BALB/c mice and cotton rats. *J Virol* **79:**8894–8903.

Ida-Hosonuma M, Iwasaki T, Yoshikawa T, Nagata N, Sato Y, Sata T, Yoneyama M, Fujita T, Taya C, Yonekawa H, Koike S. 2005. The alpha/beta interferon response controls tissue tropism and pathogenicity of poliovirus. *J Virol* **79:**4460–4469.

Ku C-C, Besser J, Abendroth A, Grose C, Arvin A. 2005. Varicella-zoster virus pathogenesis and immunobiology: new concepts emerging from investigations with the SCIDhu mouse model. *J Virol* **79:**2651–2658.

Maidji E, Genbacev O, Chang H-T, Pereira L. 2007. Developmental regulation of human cytomegalovirus receptors in cytotrophoblasts correlates with distinct replication sites in the placenta. *J Virol* **81:**4701–4712.

Morrison LA, Fields BN. 1991. Parallel mechanisms in neuropathogenesis of enteric virus infections. *J Virol* **65:**2767–2772.

Publicover J, Ramsburg E, Robek M, Rose J. 2006. Rapid pathogenesis induced by a vesicular stomatitis virus matrix protein mutant: viral pathogenesis is linked to induction of tumor necrosis factor alpha. *J Virol* **80:** 7028–7036.

Rambaut A, Pybus O, Nelson M, Viboud C, Taubenberger J, Holmes E. 2008. The genomic and epidemiological dynamics of human influenza A virus. *Nature* **453:**615–619.

Rivers T. 1937. Viruses and Koch's postulates. *J Bacteriol* **33:**1–12.

Sacher T, Podlech J, Mohr C, Jordan S, Ruzsics Z, Reddehase M, Koszinowski U. 2008. The major virus-producing cell type during murine cytomegalovirus infection, the hepatocyte, is not the source of virus dissemination in the host. *Cell Host Microbe* **3:**263–272.

Samuel M, Diamond M. 2006. Pathogenesis of West Nile virus infection: a balance between virulence, innate and adaptive immunity, and viral evasion. *J Virol* **80:**9349–9360.

Smith GA, Pomeranz L, Gross SP, Enquist LW. 2004. Local modulation of plus-end transport targets herpesvirus entry and egress in sensory neurons. *Proc Natl Acad Sci U S A* **101:**16034–16039.

Souza M, Azevedo M, Jung K, Cheetham S, Saif L. 2008. Pathogenesis and immune responses in gnotobiotic calves after infection with the genogroup II.4-HS66 strain of human norovirus. *J Virol* **82:**1777–1786.

Tyler K, McPhee D, Fields B. 1986. Distinct pathways of virus spread in the host determined by reovirus S1 gene segment. *Science* **233:**770–774.

Tyler KL, Virgin HW, IV, Bassel-Duby R, Fields BN. 1989. Antibody inhibits defined stages in the pathogenesis of reovirus serotype-3 infection of the central nervous system. *J Exp Med* **170:**887–900.

Wallace GD, Buller RM. 1985. Kinetics of ectromelia virus (mousepox) transmission and clinical response in C57BL/6j, BALB/cByj and AKR/J inbred mice. *Lab Anim Sci* **35:**41–46.

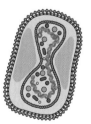

3

The Early Host Response: Cell-Autonomous and Innate Immunity

Introduction

The First Critical Moments of Infection: How Do Individual Cells Detect a Virus Infection?

Cell Signaling Induced by Receptor Engagement

Receptor-Mediated Recognition of Microbe-Associated Molecular Patterns

Cellular Changes That Occur Following Viral Infection

Intrinsic Responses to Infection

Apoptosis (Programmed Cell Death)

Other Intrinsic Immune Defenses

The Continuum between Intrinsic and Innate Immunity

Soluble Immune Mediators of the Innate Immune Response

Overview of Cytokine Functions

Interferons, Cytokines of Early Warning and Action

Chemokines

The Innate Immune Response

Complement

Natural Killer Cells

Other Innate Immune Cells of Relevance to Viral Infections

Perspectives

References

LINKS FOR CHAPTER 3

▶▶ *Video: Interview with Dr. George Stark*
http://bit.ly/Virology_Stark

▶▶ *Brought to you by the letters H, P, and eye*
http://bit.ly/Virology_Twiv336

▶▶ *Jumpin' Jack Flash, it's a GAS GAS GAS*
http://bit.ly/Virology_Twiv222

Introduction

In the previous chapter, anatomical and chemical barriers that repel the vast majority of potential pathogens were described. When viruses breach these barriers and cells become infected, a distinct and more complicated panoply of reactions is triggered. Some of these responses are deployed moments after pathogen encounter, while others are induced hours to days following infection. Collectively, this constellation of events is defined as the "immune response," but this designation is too broad to describe with any precision what actually happens in the body after viral infection. Consequently, it is helpful to consider antiviral immune responses in temporal stages. When a cell is infected, cell-intrinsic defensive actions are initiated almost immediately. These defensive actions confer antiviral protection, activating and recruiting components of the innate and adaptive immune response (Fig. 2.2). Following pathogen clearance, a memory state is established; memory T and B cells are rapidly reactivated if that same pathogen is encountered again. A corollary to immune activation and viral resolution is the importance of suppressing the host response. Moderating the swift, coordinated, and aggressive assault on a viral infection is crucial for host survival, as antiviral effector actions usually include the destruction of infected cells and production of toxic inflammatory mediators. When improperly stimulated or regulated, the immune response can lead to massive cellular damage, organ failure, chronic illness, or host death. How antiviral host defenses are induced, utilized, and integrated will be the focus of this chapter and Chapter 4.

In Chapter 5, we will turn our focus to the mechanisms by which viruses cause disease, and how improper activation of the immune cascade can lead to tissue damage. Before embarking on these thematically linked chapters, we offer four introductory points to help frame the discussion.

Conveying immunological complexity is challenging. The field of immunology is bewildering to many, even to those who work in closely related disciplines. Contemplating the sheer number and diversity of cell types, soluble proteins, signal transduction pathways, and anatomical locations can be overwhelming. Everything appears to interact with everything else, and there is no evident foundation on which subsequent layers of information can be built. In efforts to convey this complexity, textbooks and reviews often revert to comparisons either with warfare, and the notion that different aspects of the military possess different skills, or with the medical profession, in which immediate immune responders are equated with emergency medical technicians and the later adaptive response compared to specialized surgeons. While initially appealing, these metaphors fail, not simply because they anthropomorphize the immune response, but because they are inaccurate: the immune response is not merely an assortment of different cells with distinct functions, nor does it function in a prescribed sequence of events in which A leads to B and then to C, independently of the pathogen.

As metaphors can be helpful, we propose that the immune response is much like an orchestra playing a symphony: many instruments contribute at discrete times and with unique sounds to create the final piece. The bassoon may appear in both the first movement and the third, the violins may be active throughout but carrying different tunes and played at different volumes, and the cymbals may be silent until the final climactic measures. While listening to an individual part will give you an appreciation of that particular instrument's

---

PRINCIPLES   *The early host response: cell-autonomous and innate immunity*

- The immediate response to an infection is based on two coupled processes: detection and alarm.

- Microbes contain unique components, including certain carbohydrates, nucleic acids, and proteins, that are recognized by cellular pattern recognition receptors present either on the cell surface or in the cytoplasm.

- Binding of a particular ligand to a pattern recognition receptor initiates a signal transduction cascade that results in activation of cytoplasmic transcription regulatory proteins such as Nf-κb and interferon regulatory factors.

- Apoptosis is a normal biological process that can be induced by the biochemical alterations initiated by virus infection.

- Most cells synthesize interferon when infected, and the released interferon inhibits reproduction of a wide spectrum of viruses.

- Phagocytes gather information and initiate the host immune response by taking up cellular debris and extracellular proteins released from dying or apoptosing cells.

- Mechanisms to limit viral reproduction that do not result in the death of the infected cell include autophagy, epigenetic silencing, RNA interference, cytosine deamination, and Trim protein interference.

- Infected cells, sentinel phagocytes, and cellular components of the innate and adaptive immune response secrete many different proteins that can result in activation and recruitment of immune cells, induction of signaling pathways, tissue damage, and fever.

- The innate immune response is crucial in antiviral defense because it can be activated quickly, functioning within minutes to hours of infection.

role in the symphony, only hearing them played simultaneously will allow you to appreciate what the composer hoped to achieve. Moreover, the same set of instruments can be used to play many symphonies. And so it is with immunology. As we delve into interferon γ's signaling pathways or the manner in which antibodies bind to virus particles, do not lose sight of the purpose of the host response: to eliminate a foreign invader quickly with minimal damage to host cells and tissues.

Critical elements are still unknown. The challenge of clearly explaining how the dynamic defensive response is coordinated is further compounded by the fact that we still do not know all the parameters that govern the timing of the host response. That is, if we revert to our symphony metaphor, we know of no specific conductor who leads the entire process. The field of molecular immunology is still quite new, and we have discovered critical players, including Toll-like receptors, T regulatory cells, and T_h17 cells, only within the past couple of decades. Our incomplete understanding of the cast of characters means that many important questions cannot yet be answered. Consequently, we must constantly reevaluate what we thought we understood as new principles and players are identified. Progress is rapid: since the last edition of this textbook, many critical questions have been answered or clarified.

Descriptors sometimes fail us. This text describes the immune response from a temporal viewpoint; in this chapter, we discuss the events that occur immediately after infection (the intrinsic response) through ~2 to 4 days postchallenge (the innate response). We use terms such as "intrinsic" and "innate" because they aid in telling the story, but, of course, the immune response knows no such distinctions. It will therefore be useful to focus on the larger continuum of host immunity.

Virology illuminates immunology. The coevolution of viruses with their hosts has resulted in the selection of viruses that can survive, despite host defenses. The genomes of successful pathogens, which can evolve far faster than the hosts they infect, encode proteins that modify, redirect, or block each step of host defense. Indeed, for every host defense, there will be a viral counter-offense, even for those viruses with genomes that encode a small number of proteins. Consequently, the exploration of how viruses reproduce in their hosts led to the discovery of crucial immunological principles, as we shall see throughout this chapter and the next.

The First Critical Moments of Infection: How Do Individual Cells Detect a Virus Infection?

A viral infection in a host can begin only once physical and chemical barriers are breached and virions encounter living cells that are both susceptible and permissive (Chapter 2). But once these hurdles are overcome, the host's awareness of a foreign invader must transition from a single cell sending out an "I'm infected" alarm to a full-scale, whole-body response, and these processes must be exquisitely coordinated and timed. The importance of an early and appropriate response cannot be overstated: if the response is delayed or weak, the host may die from the consequences of unrestricted infection; if it is too aggressive, the host may suffer from damage by its own immune cells and proteins.

All cells have the capacity to react defensively to various stresses, such as starvation, temperature extremes, irradiation, and infection. Some of these safeguards maintain cellular homeostasis, while others have evolved to detect cellular invaders rapidly. These cell-autonomous (that is, can be accomplished by a single cell in isolation), protective programs, which are inherent in all cells of the body, are termed **intrinsic cellular defenses** to distinguish them from the specialized defenses possessed by "professional" cells of the innate and adaptive arms of the immune system. As a quick response is key, most intrinsic defenses are "ready-to-go"; that is, they do not require transcription and translation, but rather are present in the cell, ready to act, or awaiting a signal to become activated immediately. Intrinsic defenses are among the most conserved processes in all of life, shared by humans, fruit flies, plants, and bacteria. In contrast, specialized immune cells and effector proteins appeared much later in evolution, during the emergence of multicellular organisms.

How the intrinsic defenses are induced in the first cell to be infected within a host, or in an adjacent phagocytic cell, such as a macrophage or dendritic cell, is quite similar to our own experience when something in our environment changes: we perceive a difference only in the context of what we recall as normal. As we can distinguish "familiar" and "different," the immune response distinguishes "self" from "nonself." The immediate response to an infection is based on two coupled processes: detection and alarm. First, the microbe must be recognized by the infected cell. This achievement, alone, is fascinating: in addition to the specific function(s) of any individual cell in the body, virtually **all** mammalian cells are equipped with a detection system that is triggered when a pathogen has engaged or crossed the plasma membrane. Furthermore, the cell is able to identify aspects of the intruder, including whether it is DNA based or RNA based; cytoplasmic or nuclear; an intracellular bacterium, virus, or parasite. Specific protein detectors recognize structures that are unique to microbes or their genomes.

Once a microbe has been detected, the infected cell must then sound the alarm to initiate the series of events that lead to an appropriate defense. The virus infection may be halted at any step along this continuum (Fig. 3.1). The molecular coupling between the pathogen detectors and

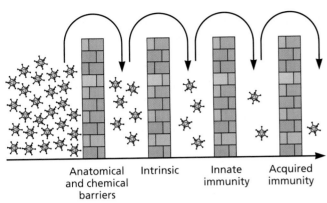

Figure 3.1 Integration of intrinsic defense with the innate and adaptive immune response. The sequential nature of host defenses is depicted as the breaching of successive barriers by viral infection. Most infections are prevented by anatomical or chemical barriers (Chapter 2). When these barriers are penetrated, additional host defenses, including intrinsic and innate defenses, come into play to contain the infection. Activation of acquired immune defenses (also called adaptive immunity) is usually sufficient to contain and clear any infections that escape intrinsic and innate defense. In rare instances, host defenses may be absent or inefficient, and severe or lethal tissue damage and host sickness or death can result. Adapted from D. T. Fearon and R. M. Locksley, *Science* **272:**50–54, 1996, with permission.

the response effectors is understood only in a broad sense, although this is an intense area of research. Tailoring the immune response to the pathogen continues as the infection proceeds, and at each critical juncture of immune defense. There are at least three ways by which a cell can become alerted to infection.

Cell Signaling Induced by Receptor Engagement

As soon as virus particles engage their receptors, cellular signal transduction pathways are activated. Remember that no cell surface protein is only a viral receptor: viruses have been selected to co-opt cellular proteins for viral entry (Volume I, Chapter 5). Many cell surface proteins are linked to intracellular molecules, such that binding of the normal ligand to the receptor triggers pathways that enable the cell to respond to changes in its environment. For example, when it is bound by a ligand, the CD46 receptor activates signaling molecules that impact cell proliferation, cellular polarity, and gene expression, including the synthesis of type I **interferons** (IFNs). Because this protein is the receptor for measles virus and some adenoviruses and herpesviruses, binding of these virus particles to CD46 can elicit the same signals, with the same consequences. Thus, even before the virus enters the cell, the dynamics of ion flow, membrane permeability, protein modification and localization, and host gene transcription may change. Note that noninfectious virus particles, which can bind to receptors but cannot reproduce, may also induce these signals.

Receptor-Mediated Recognition of Microbe-Associated Molecular Patterns

A second way that cells respond to infection is by interaction of intracellular components with microbial proteins or nucleic acids. Microbes contain unique components, including bacterial and fungal carbohydrates (e.g., lipopolysaccharide [LPS]), nucleic acids (such as single-stranded [ss] DNA or double-stranded [ds] RNA), polypeptides (such as flagellin), and lipoteichoic acids from Gram-positive bacteria. Such molecules, initially called **pathogen-associated molecular patterns (PAMPs)**, are detected by cellular **pattern recognition receptors**, which can be present either on the cell surface or in the cytoplasm. Pattern recognition receptors have been selected to be highly pathogen specific, and some also detect host components that are released upon cellular damage, including uric acid (termed **damage-associated molecular patterns [DAMPs]**). Recognition by pattern recognition receptors rapidly triggers the synthesis of antimicrobial products, including inflammatory cytokines, chemokines, and type I IFNs. While the term "PAMP" is relatively new, **all** microbes, including those that are not pathogenic, possess these molecules, prompting some to reconsider them as **microbe**-associated molecular patterns (MAMPs). This is not merely a semantic distinction, as this discovery implies that even nonthreatening entities could sound a cellular alarm.

Pattern recognition receptors were first identified in plants, which exhibit the simplest detection-to-alarm process: certain plant proteins are both the detector **and** the signal transducer that drives cell behavior. In contrast, most mammalian pattern recognition receptors transmit signals by engaging with multiple cytoplasmic adapter molecules that eventually provoke a cellular response, usually by influencing gene expression. Perhaps the use of multiple adapter proteins in animals allowed for diversification of the alarm to a common detection signal, or conversely, for the integration of diverse signals to a common node, such as nuclear factor-κb (Nf-κb). Our first insights into the immunological nature of these pathogen receptors came from *Drosophila* developmental genetics (Box 3.1). We now understand that all intrinsic and innate defense systems arose early in the evolution of multicellular organisms, and remain absolutely essential for survival of mature organisms in a microbe-filled world.

Members of different families of receptors detect specific motifs that are characteristic of invading microbes in single cells (Table 3.1 and Fig. 3.2). The Toll-like receptor (Tlr) family consists of 10 members in humans (12 in mice) that are present either on the cell surface or within lysosomes, where entering viruses first appear (Box 3.2). Tlrs 1, 2, 4, 5, and 6 are found on the extracellular membrane and recognize primarily extracellular microbes, such as bacteria, fungi,

BOX 3.1

TRAILBLAZER
Toll receptors: the fruit fly connection

- The Toll signaling pathway was defined initially as being essential for the establishment of the dorsal-ventral axis in *Drosophila* embryos. Eric Wieschaus and Christiane Nüsslein-Volhard discovered the first Toll mutants. When Wieschaus showed the unusual mutant *Drosophila* embryos to Nüsslein-Volhard, she exclaimed, "Toll!" (a German slang term comparable to "Cool!" or "Awesome!").
- Toll signaling also initiates the response of larval and adult *Drosophila* to microbial infections.
- Toll-like receptors in both flies and mammals bind to a variety of microbe-specific components and trigger a defensive reaction via signal transduction pathways and activation of new gene expression.
- Insect Toll receptors are activated by an endogenous protein ligand produced indirectly by exposure to microbes. Vertebrate Toll receptors bind microbial ligands directly.

Anderson KV. 2000. Toll signaling pathways in the innate immune response. *Curr Opin Immunol* **12**:13–19.
Gay NJ, Gangloff M. 2007. Structure and function of Toll receptors and their ligands. *Annu Rev Biochem* **76**:141–165.
Mushegian A, Medzhitov R. 2001. Evolutionary perspective on innate immune recognition. *J Cell Biol* **155**:705–710.

and protozoa, although the coat proteins of some viruses, such as measles virus, can be recognized by these cell surface molecules. Tlrs 3, 7, 8, and 9 are present within endocytic compartments of the cell and recognize primarily nucleic acid MAMPs derived from viruses or intracellular bacteria, including unmethylated DNA and dsRNA. The specificity of these molecules can be even more precise: Tlr9 binds to DNA-binding proteins, Tlr3 binds to dsRNA, and Tlr7 binds to ssRNA.

While Tlrs are the prototype pattern recognition receptors, their synthesis is generally restricted to macrophages and dendritic cells that engulf pathogens at points of entry into the body, including the skin and mucous membranes. Other receptors are produced more ubiquitously, including RNA helicases (e.g., Rig-I and Mda5) that detect foreign RNA in the cytoplasm (Box 3.3). These RNA helicases recognize chemical features of viral RNAs that do not

appear on cellular RNA, such as 5′ triphosphate groups and replication intermediates with extensive tracts of dsRNA.

Binding of a particular ligand to a pattern recognition receptor initiates a signal transduction cascade that results in activation of transcription regulatory proteins such as Nf-κb and interferon regulatory factors (Irfs). These regulatory proteins, in turn, stimulate expression of cytokine genes including those that encode IFN-α and IFN-β, as well as other proinflammatory cytokines. As we shall see later in this chapter, IFN-α and IFN-β play important roles in the recruitment of innate immune cells to the site of an infection, and the amplification of intrinsic cellular defenses by binding to IFN receptors on the infected cell surface or on the surface of adjacent, uninfected cells. For example, Rig-I and Mda5 result in the synthesis of IFNs, which in turn amplify Rig-I and Mda5 gene expression.

Table 3.1 Intracellular detectors of viral infection[a]

Receptor	Cellular compartment	Ligand(s) detected	Virus infection(s) detected
Rig-I	Cytoplasm	dsRNA; ssRNA with 5′ phosphate	Influenza virus
Mda5	Cytoplasm	dsRNA	Encephalomyocarditis virus, measles virus
Tlr2	Plasma and endosomal membranes	Measles virus HA protein	Human cytomegalovirus
Tlr4	Plasma and endosomal membranes	Mouse mammary tumor virus envelope protein	Respiratory syncytial virus
Tlr3	Plasma and endosomal membranes	dsRNA	Murine cytomegalovirus, reovirus, West Nile virus
Tlr7 and Tlr8	Plasma and endosomal membranes	ssRNA	Human immunodeficiency virus, influenza virus
Tlr9	Plasma and endosomal membranes	dsDNA; synthetic, unmethylated CpG DNA	Herpes simplex virus 1 and 2

[a]Data from T. Saito and M. Gale, Jr., *Curr Opin Immunol* **19**:17–23, 2007; G. Trinchieri and A. Sher, *Nat Rev Immunol* **7**:179–190, 2007; and N. J. Gay and M. Gangloff, *Annu Rev Biochem* **76**:141–165, 2007.

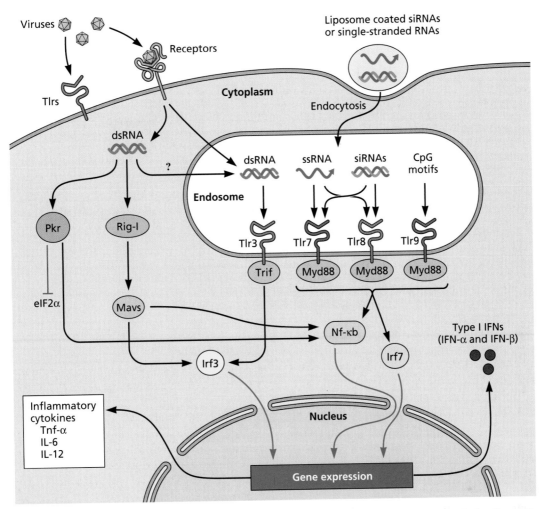

Figure 3.2 Recognition of foreign nucleic acids in mammalian cells. The Tlrs, Rig-I, and Pkr all contribute to detection of microbe-associated molecular patterns including ssRNA, dsRNA, RNA nucleotides, siRNAs, and unmethylated CpG-containing oligonucleotides. As the receptor's cognate nucleic acid is bound on the cell surface, in the cytoplasm, or in the lumen of endosomes, signal transduction leads to activation of Nf-κb, Irf3, or Irf7 to induce expression of inflammatory cytokines and IFN-α/β. Important cytoplasmic proteins in the signal transduction cascade, including Trif and Myd88, bind the cytoplasmic tails of endosomal Tlr proteins after they have engaged their cognate ligand. Viral RNA and DNA may be exposed in the lumen of endosomes after degradation or uncoating events. Pkr is autophosphorylated when dsRNA is bound, leading to phosphorylation of its substrates. One such substrate is the α subunit of eukaryotic translation initiation factor 2 (eIF2α). Phosphorylation of this protein blocks protein synthesis.

New pattern recognition receptors that recognize novel microbial products continue to be identified. For example, dsDNA is rarely found in the cytoplasm, and its presence could indicate either cellular damage or infection. Sensors of dsDNA, including cGAS, interferon-inducible protein 16 (Ifi16), and others, activate the cellular signaling molecule Sting to induce type I IFN synthesis (Fig. 3.3). The distribution and concentration of these receptors vary among different cell types and tissues.

While cytoplasmic sensors of infection are reasonably well characterized, less is known about nuclear detectors of foreign nucleic acids. It is known that a DNA damage response may ensue in the nucleus as a result of detection of single-stranded nucleic acid (e.g., from a single-stranded nuclear virus, such as influenza) or exposed double-strand ends of DNA. For example, Ifi16 binds herpesvirus DNA in the nucleus, triggering expression of antiviral cytokines.

Notwithstanding this impressive array of cell-based virus detectors, every virus that exists must reproduce to some extent in the face of these cellular responses. Clearly, viral gene products can bypass or modulate the intracellular detectors

BOX 3.2

DISCUSSION
The Toll-like receptors

Toll-like receptors (Tlrs) are the prototypical pattern recognition molecules. They are synthesized predominantly by dendritic cells and macrophages, but can be found in other cell types. Tlrs are type I transmembrane proteins that are conserved from sea urchins to humans, and they can recognize intracellular as well as extracellular microbial ligands. Endocytosed proteins and virus particles end up in lysosomal compartments of dendritic cells, where they can be digested. Some Tlrs, including Tlr3 and Tlr9, are also located in endosomes and lysosomes, perfectly placed to bind these unusual viral products. Ligands that might characterize viral infections include CpG-containing DNA, dsRNA, and unique features of ssRNA. Unmethylated CpG tracts are present in bacterial and most viral DNA

genomes, while dsRNA and ssRNA are commonly found in virus-infected cells.

After ligand binding, the Tlrs, like many receptors, aggregate in the plasma or lysosomal membrane, an event that stimulates binding of adapter proteins (see the figure). Many downstream signaling steps in pattern recognition and inflammation are mediated by common components. For example, when ligands bind Tlrs, a domain on their cytoplasmic tails, called the Toll/interleukin-1 receptor (Tir) domain, binds the adapter protein myeloid differentiation primary response protein 88 (Myd88), which then in turn binds to the IL-1 receptor-associated kinase (Irak) through common death domain motifs. With the exception of Tlr3, all Tlrs engage Myd88. Irak then activates conserved downstream

pathways. Nf-κb activation leads to transcription of genes encoding inflammatory cytokines and T cell costimulatory molecules, whereas extracellular signal-regulated kinase (Erk) and Jun N-terminal protein kinase (Jnk) signaling affect cytoskeletal organization, cell survival, and proliferation. The p38 pathway can lead to stabilization of short-lived mRNAs and increased production of various cytokines.

Tlrs are critical players in antiviral defense. Respiratory syncytial virus persists longer in the lungs of infected *tlr4*-null mice than in wild-type mice. Moreover, two vaccinia virus proteins, A46R and A52R, are similar in sequence to segments in the cytoplasmic domain of Tlrs and IL-1 receptors. These two viral proteins can inhibit IL-1- and Tlr4-mediated signal transduction, respectively. Vaccinia virus may modulate host immune responses by competing with this domain-dependent intracellular signaling.

Meylan E, Tschopp J, Karin M. 2006. Intracellular pattern recognition receptors in the host response. *Nature* **442**:39–44.

Quicke KM, Suthar MS. 2013. The innate immune playbook for restricting West Nile virus. *Viruses* **5**: 2643–2658.

Saito T, Gale M, Jr. 2007. Principles of intracellular viral recognition. *Curr Opin Immunol* **19**:17–23.

Signaling pathways and consequences pursuant to Tlr engagement. The cytoplasmic signaling pathways following Tlr4 binding (used as an example) are shown. These pathways lead to transcription of specific genes, including the interferons, as well as stabilization of RNA molecules.

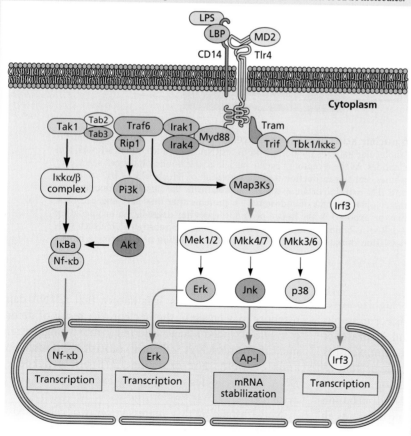

Toll-like receptors recognize microbial macromolecular patterns[a]

Toll-like receptor	Pattern recognized
Tlr1	Bacterial lipoproteins
Tlr2	Lipoproteins, viral glycoproteins, Gram-positive peptidoglycan
Tlr3	dsRNA
Tlr4	LPS, viral glycoproteins
Tlr5	Bacterial flagellin
Tlr6	Bacterial lipoproteins
Tlr7	ssRNA
Tlr8	ssRNA
Tlr9	CpG DNA, unmethylated CpG oligonucleotides
Tlr10	Unknown
Tlr11	Profilin

[a]Data from G. Barton and R. Medzhitov, *Curr Opin Biol* **14**:380–383, 2002; and S. Uematsu and S. Akira, *Handb Exp Pharmacol* **183**:1–20, 2008.

BOX 3.3

DISCUSSION
Detecting viral invaders within the cell

A fundamental problem solved over eons of evolution is the detection by individual cells of invading viral RNA or DNA. A challenge for cells in detecting viruses is that these obligate intracellular parasites are made of the same basic building materials as the cell itself. This property imposes a significant challenge for their detection. How an invading nucleic acid is distinguished from cellular RNA and DNA remained an enigma for decades. While their names seem disconnected from their now known functions, the retinoic acid-inducible protein I (Rig-I) and melanoma differentiation-associated protein (Mda5) represent a class of virus detectors that recognize RNA in the cytoplasm. These RNA helicases of the DEXD/H box family share tandem caspase activation and recruitment domains (CARDs). After binding their ligands, Rig-I and Mda5 signal via interaction of these domains with an adapter protein. This adapter is an outer mitochondrial membrane protein (Mavs) that, when bound to the CARD domain of either Rig-I or Mda5, activates Irf3 and Nf-κb by the pathways shown in the figure. (Note that the protein we refer to as Mavs actually has three other names, all still in use: Ips1, Cardif, and Visa. Apparently, the field felt that immunology was not sufficiently complex!) Mavs also localizes to peroxisomes, and both peroxisomal and mitochondrial Mavs are required for robust antiviral responses.

Mavs binds to Traf6 and induces its polyubiquitination by E1 ligase and Ubc12/Uve1A. In turn, Tak1 kinase and Tab2 adapter protein bind and activate Jnk kinases and the Iκk complex, phosphorylating the inhibitor of Nf-κb, IκB, and causing it to be released from the complex. Free Nf-κb can enter the nucleus and induce gene expression. The common adapter Mavs integrates two different interactions with a common output. The coordination of these three signal transduction pathways leads to the assembly of a multiprotein enhancer complex in the nucleus, which drives expression of the IFN-β gene.

A crucial question is how Rig-I and Mda5 distinguish viral from cellular RNA. It has been clear for some time that the two receptors have different specificities and actions *in vivo*. Use of mice that lacked either of these two RNA detectors showed that Rig-I is required for the flaviviruses and influenzaviruses. The Mda5 detector seems to be essential for the antiviral response to encephalomyocarditis virus and measles virus. Rig-I binds to RNAs with a 5′ phosphate group, which is common in viral RNAs and distinguishable from the 5′ cap structures present on all cellular mRNAs. While certainly an

important finding, the 5′ phosphate recognition must be just the tip of the iceberg, as the abundant human 7SL RNA has a 5′ triphosphate group yet does not activate an IFN response.

Picornaviral RNAs do not have cap structures **or** triphosphates on their 5′ ends, and cannot be detected by Rig-I, but they can be detected by Mda5. Mda5 also recognizes the synthetic dsRNA analog poly(I:C). Interestingly, short poly(I:C) is preferentially recognized by Rig-I, rather than by Mda5, suggesting that Rig-I and Mda5 recognize different lengths of dsRNA, which may be the basis of their apparently preferential recognition of viruses. Furthermore, some viruses, such as dengue virus and West Nile virus, require recognition by **both** Rig-I and Mda5 to generate a robust innate immune response.

Pichlmair A, Schulz O, Tan CP, Näslund TI, Liljeström P, Weber F, Reis e Sousa C. 2006. RIG-I-mediated antiviral responses to single-stranded RNA bearing 5′-phosphates. *Science* **314**:997–1001.
Saito T, Gale M, Jr. 2007. Principles of intracellular viral recognition. *Curr Opin Immunol* **19**:17–23.
Yoneyama M, Onomoto K, Jogi M, Akaboshi T, Fujita T. 2015. Viral RNA detection by RIG-I-like receptors. *Curr Opin Immunol* **32**:48–53.

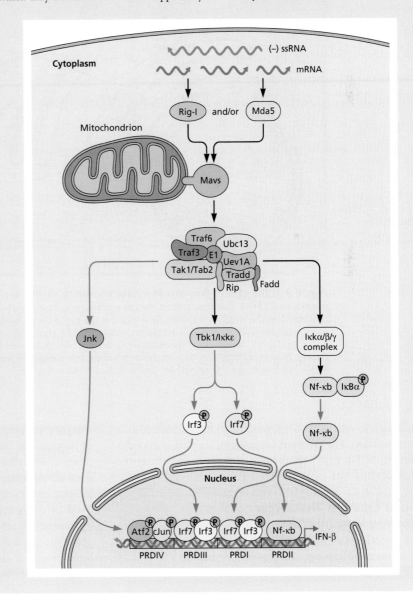

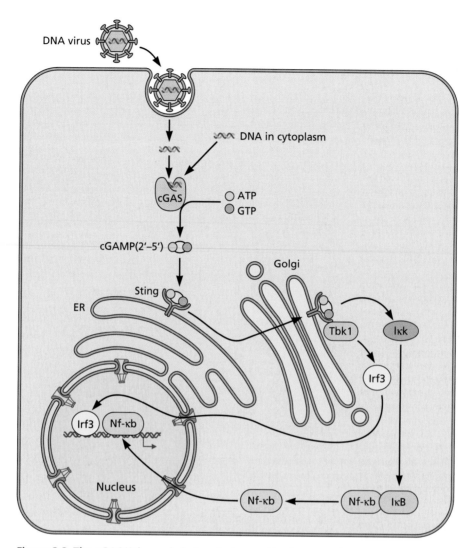

Figure 3.3 The cGAS/Sting axis in innate immunity. dsDNA in the cytoplasm (either from microbes with DNA genomes or from the cell nucleus) is detected by cyclic GMP-AMP (cGAMP) synthase (cGAS), which is activated to synthesize the cyclic dinucleotide (CDN) cGAMP(2′-5′) as its second messenger molecule (using the substrates ATP and GTP). cGAMP(2′-5′) then binds and activates the endoplasmic reticulum (ER)-resident receptor stimulator of interferon genes (Sting). Activated Sting then translocates to a perinuclear Golgi compartment, where it binds to TANK-binding kinase 1 to activate Irf3 and to induce Nf-κb activation.

of infection to enable the viral genome to be maintained in a host population. For example, herpes simplex virus 1 encodes a protein, ICP0, that suppresses Ifi16-dependent immune responses (Box 3.4).

Cellular Changes That Occur Following Viral Infection

Viral infection can also trigger the target cell to initiate a host response following viral protein production (Table 3.2). For example, viral proteins may induce alterations in cell

integrity, collectively called **cytopathic effect**, causing release of intracellular components that may be engulfed by adjacent phagocytic cells (Fig. 3.4). The phagocytic cells then produce inflammatory cytokines and present the engulfed cellular and viral components as antigens to naïve T and B cells. Viral proteins may also block essential host processes such as translation, DNA and RNA synthesis, and vesicular transport. The most dramatic cellular response to infection when homeostasis is altered or when signaling cytokines bind to their receptors is that the cell undergoes suicide, a process

BOX 3.4

EXPERIMENTS
Viral proteins that block pattern receptor recognition function

Viruses have been selected to interfere with multiple steps of the Tlr signaling cascade. For example, the V proteins of paramyxoviruses bind to Mda5, thereby inhibiting the activation of the IFN-β promoter, whereas the paramyxovirus P protein suppresses Tlr signaling through induction of the ubiquitin-modifying enzyme A20. The adenoviral E3 protein 14.7K inhibits antiviral immunity and inflammation by blocking the activity of Nf-κb. The hepatitis C virus NS3/4A protein cleaves Ips1, releasing this protein from the mitochondria and blunting subsequent pattern recognition receptor signaling.

Large DNA viruses, including vaccinia virus, encode immunomodulatory proteins that antagonize important components of the intrinsic immune response, as well. The vaccinia virus protein A52R blocks the activation of Nf-κb induced by multiple Tlrs by associating with Irak2 and Traf6 and inhibiting their activity. Likewise, the viral protein A46R targets multiple Tlr adapter molecules, like Myd88 and Trif, thereby contributing to virulence. A recombinant vaccinia virus lacking A52R is attenuated in a murine intranasal model, demonstrating the importance of this protein to viral reproduction.

A recent publication underscores the effective ways by which virus-encoded proteins can blunt this critical early step in induction of immune responses. Epstein-Barr virus is a human herpesvirus that persistently infects >90% of adults worldwide. Epstein-Barr virus encodes enzymes, called deubiquitinases, that remove ubiquitin tags from substrate proteins. The activation of many proteins in the Tlr signaling cascade is regulated by the addition of ubiquitin tags. A virus-encoded deubiquitinase, Bplf1, deubiquitinylates components of the Tlr signaling pathway during Epstein-Barr virus production and is packaged into newly produced virus particles (see the figure).

van Gent M, Braem SG, de Jong A, Delagic N, Peeters JG, Boer IG, Moynagh PN, Kremmer E, Wiertz EJ, Ovaa H, Griffin BD, Ressing ME. 2014. Epstein-Barr virus large tegument protein BPLF1 contributes to innate immune evasion through interference with Toll-like receptor signaling. *PLoS Pathog* **10**:e1003960. doi:10.1371/journal.ppat.1003960.

Xagorari A, Chlichlia K. 2008. Toll-like receptors and viruses: induction of innate antiviral immune responses. *Open Microbiol J* **2**:49–59.

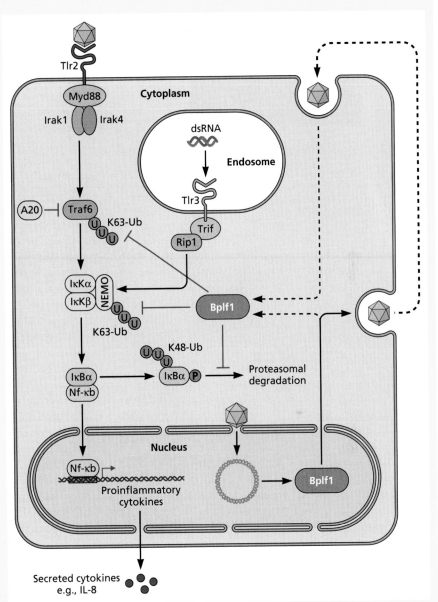

Schematic model of Bplf1-mediated Tlr evasion during Epstein-Barr virus infection. The Epstein Barr virus-encoded deubiquitinase Bplf1 counteracts Tlr-mediated Nf-κb activation and blocks ubiquitination of crucial signaling intermediates. Bplf1 is expressed as a full-length protein during the late phase of productive infection, is incorporated into the tegument of viral particles, and can subsequently be released into newly infected cells.

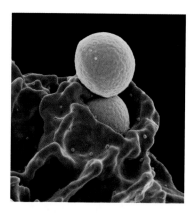

Figure 3.4 Phagocytosis. A scanning electron micrograph image of methicillin-resistant *Staphylococcus aureus* ingestion by a neutrophil. Bacteria pictured in yellow; neutrophil in purple. Courtesy of NIAID/National Institutes of Health/Science Photo Library, with permission.

called **apoptosis**. This response is remarkably effective in curtailing virus propagation, especially when it occurs early in the infection: the cellular factory required to make more infectious particles is destroyed and fragments of such dead or dying cells are taken up by tissue-resident dendritic cells, which then amplify the host response.

Even an approximation of the various ways by which the intrinsic antiviral response is induced is overwhelming (Fig. 3.5). To complicate matters further, we are beginning to appreciate that different cell populations use particular pathways selectively, and that substantial cross-talk likely exists among these various sensors and their effector responses within a responding cell. While daunting, such knowledge reminds us of the numerous mechanisms that are in place to protect the host from rampant infection. Even so, viral evolution in the face of these responses has selected for viral gene products that counter, modulate, or even bypass intrinsic cell defenses.

Intrinsic Responses to Infection
Apoptosis (Programmed Cell Death)

Apoptosis is a normal biological process used chiefly used to eliminate particular cells during development and differentiation and to maintain organ size (Box 3.5). For example, the separation of fingers and toes during fetal development is a result of apoptosis of the cells between the digits. Such programmed cell death continues throughout life: every day, in an adult human, >50 billion cells die by apoptosis.

Apoptosis can also be induced by the biochemical alterations initiated by virus infection. Cell death is the result of a cascade of reactions that ultimately leads to nuclear membrane breakdown, chromatin condensation, loss of membrane integrity (called "blebbing," in which bubbles of cytoplasm appear on the cell surface [Fig. 3.6]), and eventually DNA degradation. When a cell undergoes apoptosis, cellular debris is taken up by macrophages and dendritic cells, which then migrate to local lymph nodes and produce cytokines to amplify the host response.

Regardless of the nature of the initiation signal that triggers the apoptotic response, all converge on common effectors, the **caspases**. Caspases are members of a family of cysteine proteases that cleave after aspartate residues. These proteases are first synthesized as precursors with little or no activity, and exist in normal cells in this inactive state. A mature caspase with full activity is produced after cleavage by a protease, often another caspase. Increasing the concentration of some caspase precursors results in cleavage-independent activation. These protease cascades are not unlike the many sequential steps that result in blood clotting or the complement cascade (discussed below). This property underscores the "ready-to-go" nature of the intrinsic response: induction of apoptosis requires successive activation of proteins that are already present in the cytoplasm (a rapid event), in contrast to a more time-consuming response that includes new or enhanced gene transcription and mRNA translation.

Table 3.2 Host alterations are early signals of infection

Alteration	Virus	Viral protein	Target
Inhibition of transcription	Poliovirus	3C	Tbp-TfiIIc complex
Blocking accumulation of host mRNA in cytoplasm	Adenovirus	E1B-55K E4-34K	Cellular protein involved in mRNA transport?
Inhibition of 5′-end-dependent translation	Poliovirus	2A^pro	eIF4G
Alteration of MAP4	Poliovirus	3C	MAP4
Increased plasma membrane permeability	Sindbis virus	?	Na,K-ATPase
Fusion of cell membranes, syncytium formation	Paramyxovirus	F protein	Plasma membrane
Inhibition of transport and processing of host RNA	Herpes simplex virus	ICP27	SR splicing proteins
Depolymerization of cytoskeleton	Many viruses	?	Actin filaments, microtubules, intermediate filaments

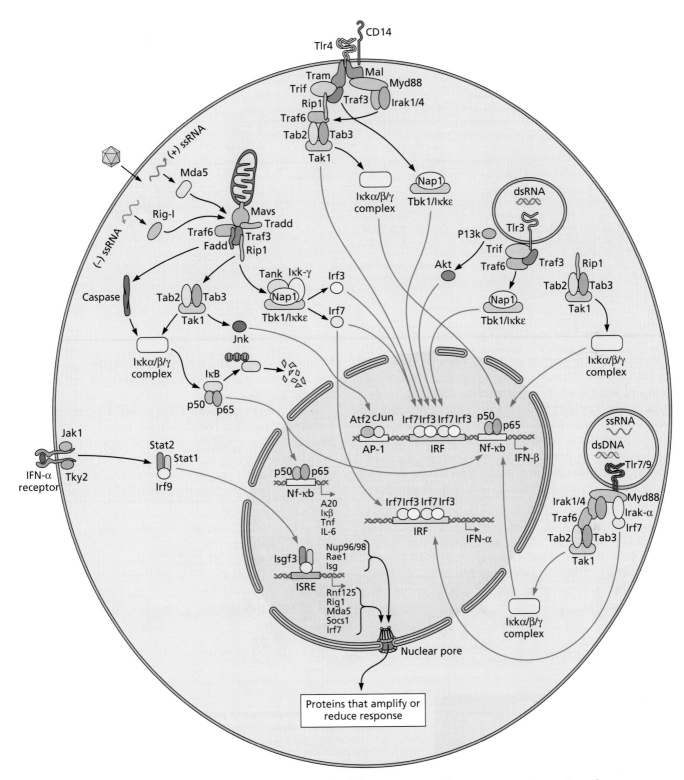

Figure 3.5 Summary of some established intrinsic defense responses. Every cell has receptors on the surface and inside that bind microbial proteins and nucleic acids. A generic cell is indicated here, but it should be clear that not all cells have the same constellation of responses. Upon binding their cognate ligand, pattern recognition receptors initiate reactions leading to production of potent cytokines. Cytokines, such as IFN, are the primary response mediators that emanate from a single infected cell.

TERMINOLOGY
What to do with the second "p"?

How does one pronounce the word "apoptosis"? The most straightforward approach, and the one that implies its actual function, is to pronounce the second "p": a-POP-tosis. But many in the field find this to be an amateur mistake, and prefer to keep the second "p" silent: a-po-tosis.

In Greek, "apoptosis" translates to the "dropping off" of petals or leaves from plants or trees. Hippocrates co-opted the term for medical use to describe "the falling off of the bones." In English, the *p* of the Greek *pt* consonant cluster is typically silent at the beginning of a word (as in "pterodactyl" or "Ptolemy"), but usually pronounced when preceded by a vowel (as in "helicopter" or "chapter"), but not always (as in "receipt"). What to do?

In the original *British Journal of Cancer* paper by Kerr, Wyllie, and Currie, there is a footnote regarding the pronunciation: "We are most grateful to Professor James Cormack of the Department of Greek, University of Aberdeen, for suggesting this term. To show the derivation clearly, we propose that the stress should be on the penultimate syllable, the second half of the word being pronounced like 'ptosis' (with the 'p' silent), which comes from the same root 'to fall' . . ."

A-po-tosis, it is.

Kerr JF, Wyllie AH, Currie AR. 1972. Apoptosis: a basic biological phenomenon with wide-ranging implications in tissue kinetics. *Br J Cancer* **26**:239–257.

Two convergent caspase activation cascades are known: the extrinsic and intrinsic pathways (Fig. 3.7). The **extrinsic pathway** begins when a cell surface receptor binds a proapoptotic ligand (e.g., the cytokine tumor necrosis factor α [Tnf-α]). Ligand binding trimerizes the receptor so that death-inducing signaling proteins are recruited to the receptor's now clustered cytoplasmic domain (Fig. 3.7A). The latter proteins in turn attract procaspase-8, which becomes activated following receptor engagement.

Caspase-8 cleaves and activates procaspase-3 to produce caspase-3, the final effector common for both extrinsic and intrinsic pathways.

The **intrinsic pathway**, often called the mitochondrial pathway, integrates cellular stress responses as well as internal developmental cues. Common intracellular initiators include DNA damage and ribonucleotide depletion. In these situations, the cell cycle regulatory protein p53 is activated (Chapter 6) and apoptosis typically follows in virus-infected

A

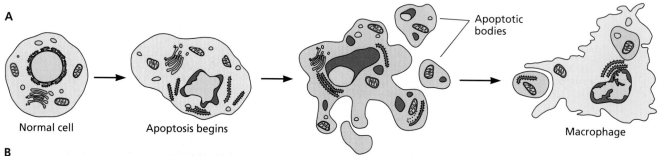

Normal cell   Apoptosis begins   Apoptotic bodies   Macrophage

B

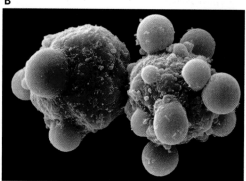

Figure 3.6 Apoptosis, the process of programmed cell death. (A) Apoptosis can be recognized by several distinct changes in cell structure. A normal cell is shown at the left. When programmed cell death is initiated, the first visible event is the compaction and segregation of chromatin into sharply delineated masses that accumulate at the nuclear envelope (dark blue shading around periphery of nucleus). The cytoplasm also condenses, resulting in shrinkage of the cell and nuclear membranes. The process can be rapid: within minutes, the nucleus fragments and the cell surface convolutes, giving rise to the characteristic "blebs" and stalked protuberances illustrated. These blebs then separate from the dying cell and are called apoptotic bodies. Macrophages (the cell at the right) engulf and destroy these apoptotic bodies. Adapted from J. A. Levy, *HIV and the Pathogenesis of AIDS*, 2nd ed. (ASM Press, Washington, DC, 1998). **(B)** A liver cell undergoing programmed cell death, with characteristic blebbing of the plasma membrane. Credit: David McCarthy/Science Photo Library, with permission.

A Extrinsic death receptor pathway

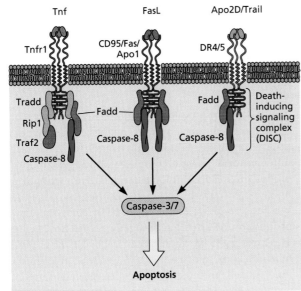

B Intrinsic death receptor pathway

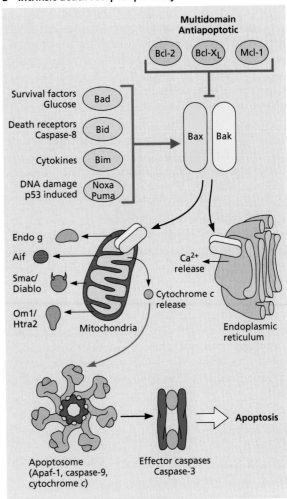

Figure 3.7 Pathways to apoptosis. (A) The extrinsic death receptors and their death-inducing signaling proteins. Three receptors found on the surfaces of cells can initiate the apoptosis pathway: the Tnf receptor, the Fas ligand (FasL) receptor (CD95), and the Apo2/Trail receptor (Dr4/5). When these receptors engage their respective ligands, the cytoplasmic domains of each protein complex form a scaffold for assembly of the death-inducing signaling complex (DISC). Important cytoplasmic proteins in this complex are shown. Caspase-8 is activated when it binds these complexes, leading to activation of caspase-3, the main effector of apoptosis. Adapted from N. N. Danial and S. J. Korsmeyer, *Cell* **116:** 205–219, 2004, with permission. **(B)** The process of intrinsic apoptosis is controlled by the Bcl-2 family of proteins. The critical antiapoptotic regulators are Bcl-2, Bcl-X$_L$, and Mcl-1. These proteins keep Bax and Bak (proapoptotic proteins) from assembling at the mitochondrial or endoplasmic reticulum membranes and causing release of cytochrome *c* and calcium, respectively. When not sequestered in this way, these proteins permeabilize mitochondrial membranes, and internal stores of cytochrome *c* are released. Cytochrome *c* in the cytoplasm binds to a cellular protein (Apaf-1), which oligomerizes in the presence of deoxyadenosine 5′ triphosphate (dATP) or ATP. The oligomeric assembly, called the apoptosome, then binds and cleaves procaspase-9, which in turn activates procaspase-3, the effector caspase that produces the characteristic events of controlled cell suicide. Four other classes of Bcl-2 regulatory proteins that also bind to different subsets of the Bcl-2 proteins are indicated at the top left (blue ellipses). These four classes act under conditions in which survival is threatened (Bad), when the extrinsic pathway is stimulated (Bid), when certain cytokines are produced (Bim), and when DNA damage is detected and p53 is induced (Noxa and Puma).

cells, with formation of the apoptosome and consequent activation of caspase-9. Members of a single family of proteins, the Bcl-2 family (Fig. 3.7B), are the master regulators that inhibit intrinsic apoptosis. Their chief function is to restrict **pro**apoptotic proteins, which, curiously, are also Bcl-2 family members. The differential binding of other antiapoptotic Bcl-2

proteins, including Bcl-X$_L$, enables tissue-specific regulation and stress-specific responses.

The extrinsic and intrinsic signaling pathways can converge in other ways. For example, if the extrinsic pathway is activated, mature caspase-8 may cleave the proapoptotic protein Bid, which then translocates to the mitochondria to trigger the intrinsic pathway. There is ample evidence to suggest that the intrinsic pathway can amplify the extrinsic pathway. Consequently, these pathways are not discrete roads to a common effector, but rather may intersect throughout the response to cellular stress.

Apoptosis Is a Defense against Viral Infection

Because virus particles engage cell receptors, and because viral reproduction requires engagement of many or all of the host's metabolic pathways (Volume I, Chapter 14), a variety of signals can activate the extrinsic and intrinsic pathways (Box 3.6). In many infections, the target cell is quiescent and hence unable to provide the enzymes and other proteins needed by the infecting virus to reproduce. Consequently, viral proteins, including the adenoviral E1A proteins or simian virus 40 large T protein, may cause the cell to leave the resting state and enter the cell cycle. Cell cycle checkpoint proteins may then respond to this unscheduled event by inducing apoptosis, which limits the production of infectious progeny (a form of antiviral defense by suicide). Perhaps not surprisingly, to ensure that infected cells survive long enough to produce some progeny, viral genomes encode counter-response proteins that block cell-initiated apoptosis.

Viral Gene Products That Modulate Apoptosis

The discovery of viral proteins that modulate the apoptotic pathway proved exceedingly valuable in dissecting the complex interactions and regulatory circuits that control this pathway in normal cells. Apoptosis is normally held in check by viral regulatory proteins called inhibitors of apoptosis. The prototype gene in this group was described in baculovirus genomes by the late Lois Miller and colleagues in 1993. This seminal work led to the discovery of cellular orthologs in yeasts, worms, flies, and humans. Mutant viruses unable to inhibit apoptosis were detected originally because the host DNA of infected cells was unstable; the cells lysed rapidly upon infection; and, as a consequence, viral yields were reduced, resulting in small plaques. Since then, we have discovered many viral gene products that modulate apoptosis (Table 3.3).

Human cytomegalovirus encodes an abundant, 2.7-kb noncoding RNA ($\beta 2.7$) that binds to and inhibits the mitochondrial protein complex that triggers apoptosis. Not only is this response blocked, but the mitochondrial membrane potential is maintained, enabling continued ATP biosynthesis, resulting in prolonged cell viability to allow for the production of infectious progeny before host cell collapse. This strategy is particularly favorable for those viruses, like human cytomegalovirus, with long infectious cycles. Alternatively, in cells infected by some viruses, including poliovirus, apoptosis provides a way for progeny virus particles to get out of the cell.

Many viral genomes that encode mimics of cellular proteins that hold apoptosis in check can contribute to transformation. For example, the human adenovirus E1B 19-kDa protein, one of the first viral homologs of cellular antiapoptotic proteins to be identified, allows survival and transformation of rodent cells that also contain the viral growth proliferation-promoting E1A protein. Similarly, human herpesvirus 8 encodes the antiapoptotic protein vFLIP, which has been associated with the development of Kaposi's sarcoma.

BOX 3.6

BACKGROUND
The many ways by which virus infections activate apoptotic pathways

At the Cell Surface
Production of apoptosis-inducing cytokines after virus particles bind their receptors
Alteration of membrane integrity or composition via membrane fusion or virus particle passage into the cytoplasm via receptor-mediated endocytosis.

In the Cytoplasm
Production of inhibitors (e.g., arrest of host translation)
Modification of cytoskeleton (e.g., disruption of actin microfilaments)
Disruption of signal transduction pathways (e.g., death domain proteins and kinase- and phosphatase-binding proteins)

In the Nucleus
Degradation of, and damage to, DNA
Alteration of gene expression (e.g., increased expression of heat shock genes)
Disruption of the cell cycle (e.g., inactivation of p53 or phosphoretinoblastoma [pRb])

Hay S, Kannourakis G. 2002. A time to kill: viral manipulation of the cell death program. *J Gen Virol* **83:**1547–1564.
Miller LK, White E. 1998. Apoptosis in viral infections. *Semin Virol* **8:**443–523.

Table 3.3 Some viral regulators of apoptosis[a]

Cellular Target	Virus	Gene	Function
Bcl-2	Adenovirus	E1B 19K	Bcl-2 homolog
	Epstein-Barr virus	LMP-1	Increases synthesis of Bcl-2; mimics CD40/Tnf receptor signaling
Caspases	Adenovirus	14.7K	Inactivates caspase-8
Cell cycle	Hepatitis B virus	pX	Blocks p53-mediated apoptosis
	Human papillomavirus	E6	Targets p53 degradation
	Simian virus 40	Large T	Binds and inactivates p53
Fas/Tnf receptors	Adenovirus	E3 10.4/14.5K	Internalizes Fas
	Cowpox	CrmB	Neutralizes Tnf and LT-α
	Myxoma virus	MT-2	Secreted Tnf receptor homolog
vFLIPs; DED box-containing proteins	Human herpesvirus 8	K13	Blocks activation of caspases by death receptors
Oxidative stress	Molluscum contagiosum virus	MC066L	Inhibits UV- and peroxide-induced apoptosis; homologous to human glutathione peroxidase
Transcription	Human cytomegalovirus	IE1, IE2	Inhibits Tnf-α but not UV-induced apoptosis

[a]Data from D. Tortorella et al., *Annu Rev Immunol* 18:861–926, 2000; S. Redpath et al., *Annu Rev Microbiol* 55:531–560, 2001; and S. Hay and G. Kannourakis, *J Gen Virol* 83: 1547–1564, 2002.

Apoptosis Is Monitored by Sentinel Cells

Specialized phagocytes, dendritic cells and macrophages, reside in areas of the body where microbes are most likely to invade, including skin and mucous membranes. These remarkable cells, sometimes referred to as "sentinels" because they keep watch for invading microbes, are critical players in early defense as well as in activating a more global immune response. Phagocytes gather information (as packets of proteins) by taking up cellular debris and extracellular proteins released from dying or apoptosing cells. This process activates the sentinel cell to migrate to local lymph nodes, where it presents its collected peptide cargo to lymphocytes of the adaptive immune system. This cell-cell communication, the interface between the innate and adaptive arms of host immunity, informs T cells about the nature of the insult that is killing cells in peripheral tissues, and specific T cells become activated accordingly. Moreover, at the site of viral invasion, the sentinel cells, as well as the damaged and dying cells, produce cytokines, such as Tnf-α, that can induce apoptosis in nearby infected cells and contain the infection until the cells of the adaptive immune response arrive. A core principle that this process exemplifies should now be quite familiar: a modest initial signal that occurs in a single infected cell (induction of an intrinsic apoptotic pathway) leads to a cascade of subsequent events that vastly amplify the host response.

In a curious adoption of a cellular process by a virus, the vaccinia virus envelope is derived from membranes of a dying cell. These membranes have cellular markers of apoptosis, including phosphatidylserine, which normally bind to receptors that are present on the surface of phagocytic cells and that initiate endocytosis of debris. When vaccinia virus particles, with their envelopes marked by these "eat me"

phospholipids, bind to the cell surface of a susceptible cell, they trigger engulfment, normally appropriate for apoptotic debris, allowing the virus to gain access to target cells.

Programmed Necrosis (Necroptosis)

As viruses can block the apoptotic cascade, alternative mechanisms are in place to ensure cell death in many cells following infection. One recently identified pathway is through programmed necrosis (necroptosis). This form of cell death is mediated by two kinases, Rip1 and Rip3. Their activation following infection leads to phosphorylation, trimerization, and activation of a pseudokinase, Mlkl, which inserts into cell membranes, including the plasma membrane, where it punches holes and leads to osmotic imbalance and necrotic death of the cell. The importance of this pathway has been shown in Rip3 knockout mice, which have a higher susceptibility to various viruses, because the host cannot clear these infections.

Other Intrinsic Immune Defenses

While killing an infected cell is a good way to contain an infection, other processes that limit infection but do not result in the death of the infected cell have also been identified. These conserved cellular processes include autophagy, epigenetic silencing, RNA interference, cytosine deamination, and Trim protein interference, described below.

Autophagy

Cells can degrade cytoplasmic contents by formation of specialized membrane compartments related to lysosomes. This process, **autophagy** (from the Greek, "to eat oneself"), allows for recycling of cellular components, but is also an effective way to target incoming viruses to the lysosomal pathway. Autophagy is evoked by stressors, such as nutrient starvation or viral infection. In contrast to apoptosis, which results in cell death, autophagy is a cellular effort to consolidate resources and "weather the storm."

Infection by many viruses induces a state of metabolic stress that normally triggers intrinsic defenses. These include stress-induced alterations in translation that are modulated in part by eIF2α kinases. Phosphorylated eIF2α can trigger autophagy, a process that, in turn, leads to engulfment and digestion of virus particles or other viral components. Capturing virus particles and targeting them for degradation in lysosomes is called **xenophagy**. Virus reproduction can be blocked when xenophagy is induced, and some viruses encode gene products that block this process, including the herpes simplex virus 1 ICP34.5 protein and the Nef protein of human immunodeficiency virus type 1.

Autophagy can also exert antiviral functions by delivering viral genetic material to endosomes where engagement of the resident ssRNA-sensing Tlr7 leads to production of type I IFNs, or by targeting viral peptides to major histocompatibility complex (MHC) class II compartments to enhance CD4+ T cell responses. These processes integrate virus-induced stress responses with the molecular detectors of viral nucleic acid (e.g., cytoplasmic Rig-I/Mda5 and endosomal Tlrs).

For some RNA viruses, including hepatitis C virus, poliovirus, and rotavirus, the autophagy machinery may be co-opted to create scaffolds that aid in viral genome replication or morphogenesis. Poliovirus may also use these structures to promote the noncytolytic release of new virus particles from cells. All of these viruses presumably have means to block the maturation of these structures into destructive autolysosomes.

Epigenetic Silencing

Epigenetic silencing is another normal cellular process that specifies transcriptionally repressed regions of host chromatin. Epigenetic changes do not affect the nucleotide sequence, but rather add or remove modifications to the chromosomal histones that make it either more condensed (and therefore generally less accessible to the transcriptional machinery) or less compacted and more available to transcription factors. Chromatin packing is maintained, in part, by histone acetylation and deacetylation, reactions that are catalyzed by histone acetyltransferase and histone deacetylase activity, respectively. Acetylation removes the positive charge on the histones, thereby converting acetylated regions of chromatin into a more relaxed structure that is associated with easier access to the DNA and greater degrees of transcription. This relaxation can be reversed by histone deacetylase activity (as well as other histone modifications), resulting in condensed chromatin. DNA methylation at CpG sites, which is catalyzed by cellular DNA methyltransferases, is another mechanism for gene silencing (Fig. 3.8).

The genomes of viruses that reproduce in the nucleus, including those of many DNA viruses, can be susceptible to these modifications. Upon entering the nucleus, foreign DNA molecules can be quickly organized into transcriptionally silent chromatin, an intrinsic effort that limits viral replication but may also allow the virus to establish a long-term infection in the host cell. Histone deacetylation can maintain the viral genome in a quiescent state for long periods, often over many cell divisions. Organized collections of proteins in the nucleus, called **promyelocytic leukemia (Pml) bodies**, may be nuclear sites where such repression occurs. These structures are implicated in antiviral defense for many reasons, including the fact that IFN stimulates synthesis of some of the proteins that comprise them, increasing the number and size of the Pml bodies (Fig. 3.9).

As might be predicted, viral proteins that counter epigenetic silencing have been identified. The human cytomegalovirus protein pp71 binds to Daxx, a cellular protein that interacts with histone deacetylases and DNA methyltransferases to maintain transcriptional repression. By engaging Daxx, and marking it for degradation, silencing is avoided and transcription of the cytomegalovirus genome can proceed. Other DNA virus-encoded proteins can function in this manner as well: the global repression

Figure 3.8 Epigenetic silencing of DNA. Histone acetylation and deacetylation impact host chromatin condensation and the access of transcriptional regulators to cellular genes. Generally, though not always, condensed chromatin and silenced genes are associated with nonacetylated histones, whereas acetylated histones are associated with open chromatin in which DNA is transcriptionally active.

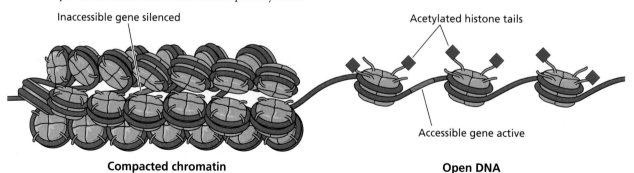

Inaccessible gene silenced

Acetylated histone tails

Accessible gene active

Compacted chromatin   **Open DNA**

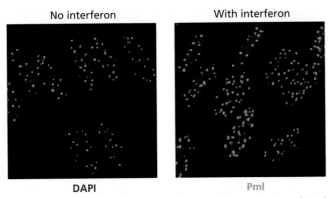

No interferon   With interferon

DAPI   Pml

Figure 3.9 Interferon increases the number and size of Pml bodies. Human foreskin fibroblasts were treated with 500 U of type I IFN and imaged by immunofluorescence 24 h thereafter. The images are confocal z-stack projections. DAPI, 4′,6-diamidino-2-phenylindole. From J. S. Chahal et al., *PLoS Pathog* **8:**e1002853, 2012.

of Pml-bound DNA can be relieved by viral proteins such as the ICP0 protein of herpes simplex virus 1. This protein accumulates in Pml bodies and induces the proteasome-mediated degradation of several of their protein components. The human cytomegalovirus IE1 proteins, the Epstein-Barr virus nuclear antigen Ebna5 protein and the adenovirus E4 Orf3 protein all affect Pml protein localization or synthesis.

Viruses that cause long-lasting, persistent or latent infections may benefit from heritable epigenetic changes that produce an environment that is conducive to long-term infection of a host cell. For example, Epstein-Barr virus, which establishes a latent infection and is associated with Burkitt's lymphoma, nasopharyngeal carcinoma, and gastric cancers, induces hypermethylation of DNA in host cells. Methylation of distinct sets of gene promoters in Epstein-Barr viral DNA enables this virus to establish persistent infection and temporarily escape immune detection, and may also silence host tumor suppressor genes.

Epigenetic silencing manifests in many ways; those studying gene transfer with retrovirus vectors are often frustrated to find that expression of their favorite gene is low or completely turned off in the infected cell. We now understand that integrated retroviral DNA is subject to reversible epigenetic silencing, a prominent process in embryonic or adult stem cells (Volume I, Chapter 8).

RNA Interference

RNA silencing is a mechanism of sequence-specific inhibition of gene expression that operates in diverse plants and animals. It is likely to have arisen early in the evolution of eukaryotes to detect and destroy foreign nucleic acids. RNA silencing is related to a process called RNA interference (RNAi) that was identified first in petunias, and subsequently in many eukaryotes (Volume I, Chapter 10).

RNAi is mediated by microRNAs (miRNAs) or small interfering RNAs (siRNAs). While these short RNA molecules are derived from distinct RNA sources, both RNAi types can bind to mRNA molecules and decrease gene expression, for example, by blocking translation of the mRNA. Both pathways are also controlled by the enzyme Dicer, which cleaves long dsRNAs into short fragments of 20 to 25 nucleotides in length. Each siRNA is unwound into ssRNAs, and then, as part of a larger assembly of proteins called the RNA-induced silencing complex (Risc), binds to a complementary sequence on an mRNA to influence its translation or stability. RNAi is a valuable research tool because synthetic dsRNA introduced into cells can selectively suppress expression of genes of interest (Box 3.7).

When a ssRNA virus infects a cell, among the first steps is the synthesis of the viral RNA-dependent RNA polymerase, which makes a complementary strand to the incoming genome. This strand is used as a template to generate progeny genomes, but with each round of replication, the viral genome becomes transiently double stranded. This double-stranded intermediate may be targeted by the RNAi machinery, which chops dsRNA into siRNA fragments.

miRNAs are single-stranded, noncoding host RNAs of 19 to 22 nucleotides that regulate gene expression. While the contributions of cellular miRNAs to the antiviral response in plants and invertebrates have been known for some time, a role in antiviral action in mammals has been demonstrated only recently. In one study, IFN-β treatment of the human hepatoma cell line Huh7, as well as freshly isolated primary murine hepatocytes, resulted in an induction of numerous cellular miRNAs. Eight of these miRNAs targeted hepatitis C virus genomic RNA: treatment of infected cells with synthetic miRNAs of the same sequence blocked virus reproduction. IFN-β treatment also reduces expression of liver-specific miR-122, an RNA essential for hepatitis C virus reproduction (Volume I, Chapter 10). In another example, host miRNAs are required to shut down human immunodeficiency virus type 1 transcription in blood mononuclear cells from infected donors. However, in other cell types, viral infection suppresses the host miRNAs that would normally repress proviral gene expression.

We are learning more about RNAi as an antiviral defense, as many viral genomes, including those of plant, insect, fish, and human pathogens, encode suppressors of this process. Many of these suppressors are RNA-binding proteins without a preference for siRNAs. While we have known for some time that simpler organisms use RNAi to fend off invading viruses, until recently it was unclear if mammals use this strategy for fighting viruses. While RNAi may be active in embryonic stem cells and other undifferentiated cells, most evidence suggests that RNAi is not a primary antiviral response in mammalian somatic cells. Thus, whether virus-directed RNA silencing is important in human viral infections remains controversial.

BOX 3.7

EXPERIMENTS
Use of viral vectors encoding siRNAs

Viruses can be used as delivery vehicles to shut down gene expression in cells. In 2003, a method to generate a large number of transgenic mice in which expression of specific genes could be downregulated was described. The technology built on the use of lentiviral vectors, combined with the use of RNAi. As proof of principle, a lentiviral vector capable of producing siRNA specific for green fluorescent protein (GFP) after transduction of 293T-GFP cell lines showed no GFP fluorescence. Furthermore, no GFP-specific RNA could be detected.

Tiscornia G, Singer O, Ikawa M, Verma IM. 2003. A general method for gene knockdown in mice by using lentiviral vectors expressing small interfering RNA. *Proc Natl Acad Sci U S A* **100**:1844–1848.

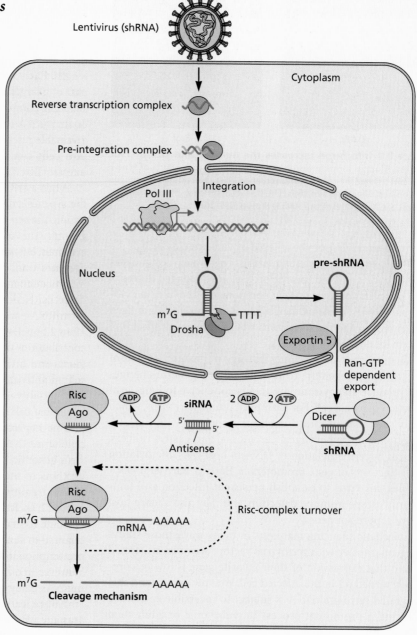

Following integration of a short hairpin RNA (shRNA)-containing lentivirus, the cellular machinery that generates shRNA is activated. This results in the cleavage of the shRNA by Drosha, its nuclear export, and the formation of the cytoplasmic Risc complex that targets a designated cellular gene.

Editing and Cytosine Deamination

While genomic editing (the process of modifying specific nucleotides) was considered initially to be restricted to a small number of viruses with RNA genomes, a growing literature has shown that this is a commonly employed device in the intrinsic immunity toolkit. A number of studies have established the importance of adenosine deaminase acting on RNA (Adar) -specific enzymes that primarily lead to A-to-I editing during viral infections (Table 3.4). Somewhat surprisingly, the effect of these enzymes on the virus-host interaction can be either antiviral or proviral, dependent upon the specific virus and mammalian host cell combination and the amount of Adar protein synthesized. Our current understanding suggests that Adars may act either directly by editing a viral RNA in a manner that influences the outcome of the infection, or indirectly by editing a cellular RNA and hence a cellular protein that participates in the antiviral response. It is also possible that Adars may function in an editing-independent manner, by altering protein- or nucleic acid-binding interactions that subsequently affect the outcome of the viral infection. In the case of viruses that reproduce in the nucleus and that display less common A-to-G and U-to-C substitutions in viral sequences, either Adar1 or Adar2 could be responsible for the editing events, as these nuclear proteins are enzymatically active deaminases. Editing is not limited to viruses with a nuclear stage of their life cycle, however. Cytoplasmic viruses may be edited by the p150 isoform of Adar1, as this isoform is the only known cytoplasmic Adar in mammalian cells.

Reproduction of some viruses may be enhanced by these cellular Adars. Adar1 promotes production of human immunodeficiency virus type 1 particles by an editing-independent mechanism that does not require its deaminase activity. This effect appears to be the result of Adar1 inhibition by RNA-activated protein kinase (Pkr). The latter enzyme, which is synthesized and activated in response to IFN, phosphorylates the translation initiation protein eIF2B to inhibit translation (see Volume I, Chapter 11). It is well established that inhibition of Pkr by Adar1 increases the efficiency of reproduction of several viruses with $(-)$ strand RNA genomes.

Table 3.4 Some viral gene products that suppress RNA interference

Virus	Gene product	Mechanism
Human adenovirus type 5	VA-RNA I and VA-RNA II[a]	Competition for binding to exportin-5 and Dicer
Ebola virus	VP35 protein	Binding to dsRNA
Influenza A virus	NS1 protein	Binding to dsRNA
Vaccinia virus	E3L protein	Binding to dsRNA
Human immunodeficiency virus type 1	Tat	Inhibition of Dicer?

[a]Both RNAs are cleaved by Dicer, and the products are incorporated into Riscs.

In contrast to the editing enzymes that can modulate infection by a broad range of viruses described above, Trim and Apobec proteins appear to function only against retroviruses. Like the cellular components already discussed, these proteins are constitutively produced, and therefore are part of the intrinsic host response.

Trim proteins. Trim (tripartite motif family) proteins prevent some cell types from being infected by certain retroviruses (see Chapter 7). The hypothesis is that *Trim* genes evolved independently in various species to protect against endemic retroviruses. We know by the presence of large numbers of retroviral proviruses in vertebrate genomes that retroviruses have existed for millions of years. We also know that many cell types are resistant to infection by some retroviruses, despite carrying functional receptors. For example, human immunodeficiency virus type 1 infections were found to be restricted in some, but not all, cell types, prompting a massive effort to identify the constitutive inhibitor, and perhaps develop it as a new antiviral. Despite its ability to enter, human immunodeficiency virus type 1 is unable to reproduce in the cells of Old World monkeys. Infection is blocked soon after entry, but before reverse transcription, resulting in an aborted infection. Introduction of a rhesus macaque cDNA library into permissive cells led to the identification of a dominant gene that blocked reproduction, tripartite interaction motif 5α (Trim5α). While humans do encode a Trim5α homolog, the human allele does not restrict human immunodeficiency virus type 1 reproduction. One could speculate that if humans had the rhesus macaque *Trim5α* gene, the AIDS pandemic might not have occurred (Chapter 10).

Rhesus macaque Trim5α targets the human immunodeficiency virus type 1 capsid protein, but not the capsid proteins of other retroviruses. When synthesis of Trim5α was reduced experimentally using siRNA molecules, the block of human immunodeficiency virus infection in monkey cells was relieved, but this treatment had no effect on a different retrovirus, murine leukemia virus. Trim5α binds to motifs within the capsid proteins and interferes with the uncoating process, preventing reverse transcription and nuclear entry of the retroviral genome. Trim5α is a ubiquitin ligase, promoting ubiquitinylation of capsids and their subsequent degradation by the proteasome. While Trim5α is made constitutively, IFN treatment increases its synthesis in both human and monkey cells.

The idea that a retroviral capsid can be a target for intrinsic defense is not new. Host restriction (or exclusion) of mouse retrovirus infection has been known for more than 30 years. The prototypical host gene blocking early retroviral replication events was identified using the Friend strain of murine leukemia virus. The locus is called Fv1 (Friend virus susceptibility). The Fv1 protein blocks reproduction of some strains of murine leukemia virus soon after reverse transcription. Surprisingly, the mouse *Fv1* gene encodes sequences related to that of the capsid of an endogenous retrovirus resident in the mouse genome! Recent studies with wild mice have revealed that *Fv1* from various strains can recognize and restrict a wide range of retroviruses, including examples from the gammaretrovirus, lentivirus, and foamy virus genera. How Fv1 acts remains unclear, but restriction depends on a specific interaction with the capsid protein of the incoming virus. This finding is a remarkable demonstration of selection, turning endogenous retroviral gene-related expression against potential infection by other retroviruses.

Apobec proteins. Apobec (apolipoprotein B mRNA-editing enzyme, catalytic polypeptide-like) proteins have various functions, including RNA editing of host genes (Fig. 7.6). Several members of the Apobec3 family are induced by IFN and are intrinsic antiretroviral proteins packaged into virus particles. After infection, some of these cellular enzymes, including Apobec3g, inhibit reverse transcription and introduce point mutations. When the viral reverse transcriptase begins to copy viral RNA into DNA, Apobec deaminates ssDNA, specifically the nascent (−) strand, which is synthesized first. The enzyme converts C's to U's with the consequence that when the deaminated (−) DNA strand is copied, the U pairs with A, producing a G-to-A transition. Consequently, the new proviral genome is mutated in a very characteristic pattern (many GC pairs become AT pairs), and some codons may be converted into stop codons. One retrovirologist called Apobec a "WMD"—a weapon of mass deamination.

The action of Apobec should be lethal for retroviruses that incorporate this enzyme into virus particles. However, human immunodeficiency virus type 1 Vif protein counters this potential lethal defense by binding to Apobecs, as well as to a host ubiquitin ligase, and promoting the ubiquitinylation and subsequent degradation of the editing enzymes by the proteasome. Interestingly, mouse Apobec3g cannot be targeted by human immunodeficiency virus type 1 Vif, a main reason why we cannot model this infection in mice.

ISGylation

Interferon-stimulated gene 15 (Isg15) is expressed primarily in monocytes and lymphocytes, and acts much like a cytokine. A 17-kDa antiviral protein encoded in the *Isg15* gene in humans shares properties with ubiquitin-like molecules that mark cellular proteins for degradation. However, in contrast to ubiquitin-like molecules, Isg15 has not been identified in yeasts, invertebrates, or plants.

The first identified viral target of Isg15 was the NS1 protein of influenza A virus (NS1A), a multifunctional protein made in influenza A virus-infected cells but not incorporated into virus particles. It was observed that multiple lysines in the Ns1A proteins of several influenza A virus strains were

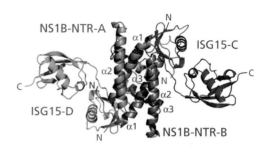

Figure 3.10 Influenza NS1B-Isg15 interactions. A heterotetramer, comprising two NS1B N-terminal region (Ntr) molecules (NS1B-Ntr-A, green; NS1B-Ntr-B, brown), which form an interwoven dimer, binds to two Isg15 molecules (Isg15-C, magenta; Isg15-D, cyan). Reprinted from R. Guan et al., *Proc Natl Acad Sci U S A* **108:**13468–13473, 2011, with permission.

modified by addition of Isg15; such changes resulted in a loss of Ns1A protein binding to critical cellular components, including importin α, and concomitant loss of infectivity. Isg15 conjugation targets newly synthesized proteins. As viral proteins constitute the majority of those newly synthesized molecules in infected cells, this property may explain the specificity with which Isg15 acts. It was subsequently shown that this protein inhibits the reproduction of a large number of other viruses, including Sindbis virus, human immunodeficiency virus type 1, and herpes simplex virus 1.

Less than 5% of the total target protein population is modified at any one time during virus infection. Some have conjectured that modification of a limited number of viral targets may confer a dominant-negative effect on the function of unconjugated potential targets, especially if the ISGylated viral protein functions as an oligomer in infected cells.

The genomes of influenza B and vaccinia viruses encode viral proteins that bind to Isg15 and prevent its conjugation with target proteins. The NS1B protein of influenza virus engages with human and nonhuman primate Isg15 (Fig. 3.10), but not with the mouse homolog, perhaps accounting for why influenza B virus does not reproduce in mice. Another countermeasure against Isg15 conjugation, encoded by both nairovirus [(−) strand RNA viruses] and arterivirus [(+) strand RNA viruses] genomes, is a viral protease that removes Isg15 from conjugated protein targets.

CRISPRs

The intrinsic antiviral defenses described above are found in mammals, and some are unique for nonhuman primates and humans. However, bacteria and archaea also mount defenses to impede or prevent invasion by foreign DNAs, including viral genomes (Box 3.8). A prokaryotic mechanism to silence exogenous DNA, also known as clustered regularly interspaced short palindromic repeats (CRISPRs), consists

BACKGROUND
Ancient mechanisms of intrinsic immunity

Antiviral defense systems in prokaryotes act at virtually all stages of the bacteriophage life cycle (see the figure). An effective and simple means of acquiring phage resistance is to block adsorption to the host cell, either by masking the receptor (for example, as a result of the expression of polysaccharides on the cell surface) or by downregulating or altering receptor molecules. Obstruction of the entry of the viral genome into the host's cytoplasm is a second line of defense. The proteins that block DNA injection are usually localized in association with or in close proximity to the membrane/cell wall and can be encoded by a plasmid or a prophage. If bacteriophage adsorption and DNA injection are not prevented, intracellular defense systems may act directly on the viral DNA. The restriction-modification (R-M) system is a broad-range prokaryotic immune system that targets DNA. A typical R-M system consists of a DNA methyltransferase, which modifies specific DNA sequences; and a restriction endonuclease, which cleaves the same sequences when unmodified. The general principle of these systems is that the host's genomic DNA is methylated and protected against cleavage, whereas exogenous DNA is unmodified and subject to degradation. Altruistic cell suicide, similar to apoptosis in eukaryotic cells, may also be triggered; the Abi system is one such pathway that can result in prokaryotic death.

An unusual structure of repetitive DNA downstream from the *Escherichia coli iap* gene consisting of invariant direct repeats (29 nucleotides) and variable spacer sequences (32 nucleotides) was discovered in 1987. This unique arrangement is now called the CRISPR system. All CRISPR/Cas systems operate in three stages: adaptation, expression, and interference. During the adaptation stage, resistance is acquired by integration of a new spacer sequence, corresponding to a section of the genome of the invading phage, into the CRISPR array. During the expression stage, *cas* genes are transcribed and translated;

in addition, CRISPRs are transcribed into precursor CRISPR RNAs (pre-CrRNAs), which are subsequently cleaved by a Cas6 homolog. Mature CrRNAs contain only a single spacer sequence, and hence can recognize only a single target. During the interference stage, the CrRNA guides one or more Cas proteins to cleave the genome of an invading bacteriophage genome at the sequence that is complementary to the spacer.

As expected, viral genomes also encode CRISPR systems of their own to antagonize those of the host. For example, it was recently shown that a bacteriophage-encoded CRISPR/Cas system can counteract a phage-inhibitory chromosomal island of the bacterial host, enabling a successful lytic infection.

Ishino Y, Shinagawa H, Makino K, Amemura M, Nakata A. 1987. Nucleotide sequence of the *iap* gene, responsible for alkaline phosphatase isozyme conversion in *Escherichia coli*, and identification of the gene product. *J Bacteriol* **169**:5429–5433.

Seed KD, Lazinski DW, Calderwood SB, Camilli A. 2013. A bacteriophage encodes its own CRISPR/Cas adaptive response to evade host innate immunity. *Nature* **494**:489–491.

Westra ER, Swarts DC, Staals RH, Jore MM, Brouns SJ, van der Oost J. 2012. The CRISPRs, they are a-changin': how prokaryotes generate adaptive immunity. *Annu Rev Genet* **46**:311–339.

Overview of bacterial defense systems. Bacterial cells possess several mechanisms to defend against bacteriophage infection. Such strategies include blocking of phage adsorption or DNA injection or those that act directly on the phage DNA, such as R-M and CRISPR/Cas.

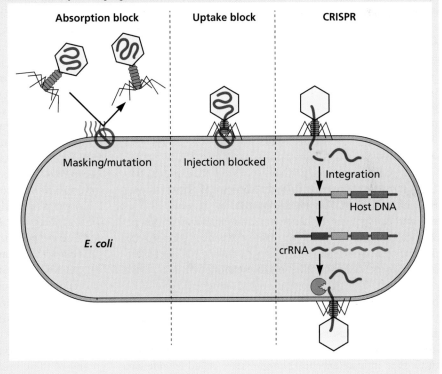

of short repetitions of nucleotide sequences in about half of eubacterial genomes and virtually all archaeal genomes. CRISPRs are associated with **cas genes**, which code for proteins related to CRISPRs. The CRISPR/Cas system confers prokaryotic resistance to foreign genetic elements such as bacteriophages. In a manner analogous to RNAi, CRISPR/Cas systems include an RNA-guided DNA endonuclease, Cas9, to generate double-strand breaks in invasive DNA during the bacterial immune response. The ability to program Cas9 for DNA cleavage at specific sites defined by guide RNAs has led to its adoption as a versatile platform for genome engineering and gene regulation. By delivering the Cas9 protein and appropriate guide RNAs into a cell, the organism's genome can be cut at any desired location.

The Continuum between Intrinsic and Innate Immunity

At the beginning of this section, we noted that intrinsic immune responses are "ready-to-go" when a virus infection occurs and do not require new protein synthesis. The attentive reader may have noted that some of the systems we described above depend on IFN stimulation, and do require transcription and translation to produce the antiviral molecules.

As is discussed in the following section, IFNs released from infected cells can bind to receptors on adjacent, uninfected cells as a warning of a pending possible infection. Binding of IFN to uninfected cells stimulates the production of antiviral defenses in advance of viral infection, such that if the virus **does** infect the cell, these defenses are ready to deploy. One could argue that IFN-stimulated genes, such as *Isg15*, encode molecules that are "innate" in already-infected cells, but "intrinsic" in those that have yet to be infected.

A warning at the beginning of this chapter noted that distinctions such as these are often imprecise, and that, despite our efforts to impose order on the host immune response to make it easier to explain, the reality is far more complex. This is especially true when considering the soluble proteins of the host immune response.

Soluble Immune Mediators of the Innate Immune Response

Infected cells, sentinel phagocytes, and cellular components of the innate and adaptive immune response secrete many different proteins that can result in activation and recruitment of immune cells, induction of signaling pathways, tissue damage, and fever. The presence of cytokines in the blood is one of the first indications that a host has been infected and that immune defenses have been called into play. Traditionally, soluble mediators that chiefly influence the migration of immune cells to sites of infection are called **chemokines**, those that activate antiviral programs are called **interferons** (for their ability to interfere with viral infection), and the diverse collection of other soluble molecules are traditionally referred to as **cytokines**, although members of this latter category share few properties, either in sequence, structure, or function.

For purposes of clarity, we will adhere to these traditions, though a few introductory comments are warranted. With the potential exception of type I IFNs, these soluble products rarely, if ever, operate alone: the inflammatory response at the site of a viral infection comprises a heterogeneous mix of many different cell types and cytokines, and it is possible that many cytokines can bind to their receptors on a single cell, transducing distinct (perhaps conflicting) signals. Scientists often refer to the cytokine response as a "storm," though most laboratory-based experiments assess the function of only one cytokine at a time for simplicity. Whether cytokine A behaves the same when it is used alone in a controlled laboratory experiment as when it is part of a storm of cytokines A to Z in an infected host is not yet known.

In addition, while immunologists refer to soluble effector proteins as "cytokines," neurobiologists call them "neurotransmitters" and endocrinologists refer to them as "hormones." We are beginning to appreciate that these distinctions are artificial, made in an effort by humans to comprehend a complex topic. But neurotransmitters have been shown to possess immune cell-activating properties; hormones can alter neuronal behavior; and cytokines, such as interleukin-1β (IL-1β), can act on the central nervous system. An interested student would find a fascinating literature developing in the area of "cross-disciplinary" soluble molecules.

Overview of Cytokine Functions

IFN production is the initial response when a single cell detects a foreign nucleic acid or when a Tlr is engaged (Fig. 3.2). In turn, locally released IFN activates a more global innate immune response, should viral reproduction continue unabated. Most of these pattern recognition receptor-stimulated pathways converge on two critical cellular transcriptional activators, Irfs and Nf-κb (Fig. 3.11). Secreted IFNs engage receptors on sentinel dendritic cells, macrophages, and adjacent uninfected cells, which then synthesize a new burst of cytokines, amplifying the initial response. The first cytokines to appear in high concentrations are IFN-α and IFN-β, followed by Tnf-α, IL-6, IL-12, and IFN-γ. While their individual functions vary, all are potent molecules, capable of inducing a response at nanograms-per-milliliter concentrations. More than 100 cytokines are known, and the list continues to grow.

Consequences of massive cytokine induction include many of the clinical symptoms we typically associate with viral infection (Fig. 3.12). Some cytokines act directly on cells of the nervous system to produce fever, sleepiness, lethargy, muscle pain (myalgia), appetite suppression, and nausea. Proinflammatory cytokines stimulate the liver to synthesize acute-phase proteins, many of which are required to repair tissue damage and to clear the infection. Members of the colony-stimulating factor class of cytokines, which are made in the bone marrow after an inflammatory response,

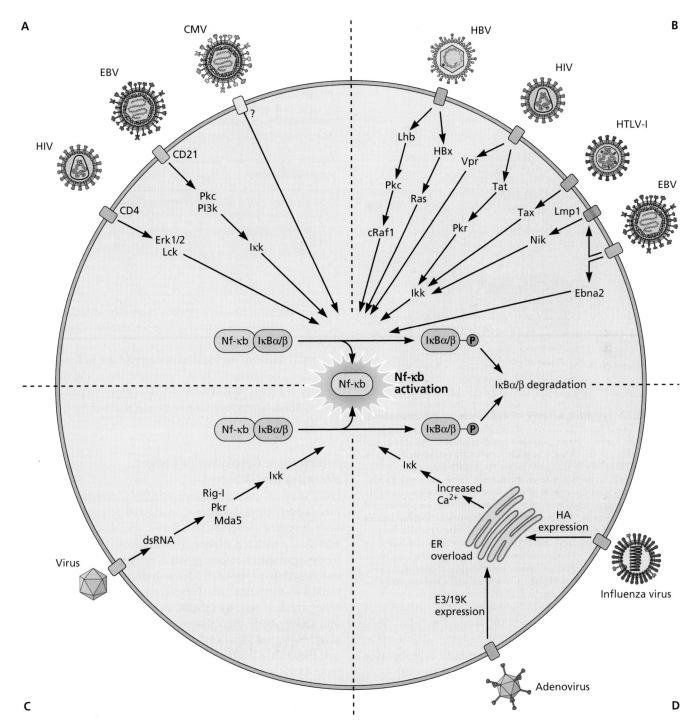

Figure 3.11 Activation of the transcription regulator Nf-kB by viral infection. Nf-κB is a transcription control protein important in the initial response to viral infection. Four diverse mechanisms that result in activation of Nf-κB are illustrated. In all cases, activation of these pathways leads to destruction of the inhibitor of Nf-κB, IκB, freeing Nf-κB to transit to the nucleus and initiate transcription of IFN genes. **(A)** Signal transduction pathways are activated upon binding of a virus particle to its receptor; **(B)** viral proteins synthesized in the infected cell directly engage signal transduction pathways that culminate in Nf-κB activation; **(C)** Pkr binds double-stranded viral RNA, or Rig-I/Mda5 bind single-stranded viral RNA, leading to activation of Nf-κB; and **(D)** overproduction of viral proteins in the endoplasmic reticulum (ER) leads to calcium release, which, in turn, activates Nf-κB. HIV, human immunodeficiency virus; EBV, Epstein-Barr virus; CMV, cytomegalovirus; HBV, hepatitis B virus; HTLV, human T cell lymphotropic virus.

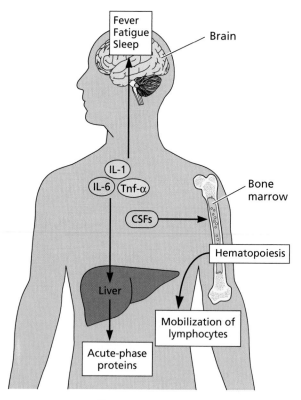

Figure 3.12 Systemic effects of cytokines in inflammation.
A localized viral infection often produces global effects, including fever
and lethargy, lymphocyte mobilization (swollen glands), and appearance
of new proteins in the blood. The proinflammatory cytokines IL-1, IL-6,
and Tnf all act on the brain (particularly the hypothalamus) to produce
a variety of effects, including fever and fatigue. These cytokines also act
in the liver to cause the release of iron, zinc, and acute-phase proteins,
including mannose-binding protein, fibrinogen, C-reactive protein, and
serum amyloid protein. These acute-phase proteins have innate immune
defense capabilities: e.g., C-reactive protein binds phosphorylcholine on
microbial surfaces and activates complement. The colony-stimulating
factors (CSFs) activated by an inflammatory response have long-range
effects in the bone marrow on hematopoiesis and lymphocyte mobilization. Adapted from A. S. Hamblin, *Cytokines and Cytokine Receptors*
(IRL Press, Oxford, United Kingdom, 1993), with permission.

control the proliferation and maturation of lymphocytes
and other cells essential in antiviral defense (Box 3.9).

Predictably, many viral gene products can mimic or modulate cytokine responses. The former proteins have been
called **virokines** if they mimic host cytokines, or **viroceptors**
if they mimic host cytokine receptors. The arsenal includes
remarkable proteins such as soluble IL-1 receptors, a variety
of chemokine antagonists, and functional homologs of IL-10
and IL-17. Viral DNA genomes encode most of the well-
known virokines and viroceptors, but smaller viral RNA
genomes contain some surprises. For example, the envelope
protein of respiratory syncytial virus is a mimic of fractalkine,
the only known chemokine that is a non-secreted membrane

protein. The viral envelope protein competes with the binding
of cellular fractalkine to its receptor, which also functions as a
receptor for the virus.

In this section, we focus on two cytokine groups, type I
IFNs and chemokines. Other cytokines, including the interleukins and the type II IFN, IFN-γ, are primarily produced by
cells of the adaptive immune response, and will be discussed
in Chapter 4.

Interferons, Cytokines of Early Warning and Action

IFNs are synthesized by mammals, birds, reptiles, amphib-
ians, and fish, and are critical signaling proteins of the
host frontline defense (Box 3.10). The discovery of IFN
was first reported almost simultaneously in the 1950s
by two groups of investigators. One group observed that
chicken cells exposed to inactivated influenza virus contained a substance that interfered with the infection of
other chicken cells by infectious influenza virus. The second group made their discovery using vaccinia virus. We
now know that most cells synthesize IFN when infected,
and the released IFN inhibits reproduction of a wide spectrum of viruses.

There are three types of IFN: types I, II, and III. All type
I IFNs bind to a specific cell surface receptor known as the
IFN-α receptor (IFNar) that consists of IFNar1 and IFNar2
chains (Fig. 3.13). The type I IFNs present in humans are
IFN-α, IFN-β, IFN-ε, IFN-κ, and IFN-ω, though this text
will primarily focus on IFN-α and IFN-β. The sole type
II IFN, IFN-γ, binds to the heterodimeric IFN-γ receptor. Less is known about the third type of IFN, which
includes IFN-λ1, -2, and -3. (Rather unhelpfully, these also
have interleukin designations: IL-29, IL-28A, and IL-28B,
respectively.)

EXPERIMENTS
The interferon system is crucial for antiviral defense

Many steps in a viral infectious cycle can be inhibited by IFN, depending on the virus family and cell type. Binding of type I IFN to its receptor leads to increased transcription of nearly 2,000 genes. A website, called interferome (http://interferome.its.monash.edu.au/interferome/home.jspx), maintains a searchable index of the known IFN gene targets. The proteins produced inhibit viral penetration and uncoating, synthesis of viral mRNAs or viral proteins, replication of the viral genome, and assembly and release of progeny virus particles. Multiple steps in virus reproduction can be inhibited, providing a strong cumulative effect.

The contribution of IFN can be demonstrated in animal models in which the antiviral response is reduced by treatment with anti-IFN antibodies, or in mice harboring mutations that delete or inactivate IFN genes, IFN receptor genes, genes that regulate the IFN response, or genes that are induced by IFNs. Animals with a defective IFN response typically exhibit a reduced ability to contain viral infections, and often show an increased incidence of illness or death. When the IFN-α/β response is impaired, there is a global increase in susceptibility to most viruses (see the figure). These observations indicate that IFN-α/β is crucial as a general antiviral defense.

Huang S, Hendriks W, Althage A, Hemmi S, Bluethmann H, Kamijo R, Vilcek J, Zinkernagel RM, Aguet M. 1993. Immune response in mice that lack the interferon-gamma receptor. *Science* **259:**1742–1745.

Ida-Hosonuma M, Iwasaki T, Yoshikawa T, Nagata N, Sato Y, Sata T, Yoneyama M, Fujita T, Taya C, Yonekawa H, Koike S. 2005. The alpha/beta interferon response controls tissue tropism and pathogenicity of poliovirus. *J Virol* **79:**4460–4469.

Stojdl DF, Abraham N, Knowles S, Marius R, Brasey A, Lichty BD, Brown EG, Sonenberg N, Bell JC. 2000. The murine double-stranded RNA-dependent protein kinase PKR is required for resistance to vesicular stomatitis virus. *J Virol* **74:**9580–9585.

Zhou A, Paranjape JM, Der SD, Williams BR, Silverman RH. 1999. Interferon action in triply deficient mice reveals the existence of alternative antiviral pathways. *Virology* **258:**435–440.

Poliovirus receptor transgenic mice

Poliovirus receptor transgenic mice x interferon alpha beta receptor knockout mice

Immunohistochemical detection of poliovirus proteins in infected poliovirus receptor (Pvr) transgenic mice (top panels) and Pvr mice crossed to IFN-α/β receptor knockout mice (bottom panels). (A/B) Representative liver sections, day 1 postinfection. **(C/D)** Representative spleen sections, day 1 postinfection. **(E/F)** Representative pancreas sections, day 3 postinfection. Bar, 125 μm. From M. Ida-Hosonuma et al. *J Virol* **79:**4460–4469, with permission.

Liver
1 day post-infection

Spleen
1 day post-infection

Pancreas
3 days post-infection

Type I IFN Synthesis

While type I IFNs are induced principally following detection of infection and signaling by pattern recognition receptors and their downstream partners, other signals can also lead to IFN production. Structural proteins of some viruses stimulate IFN synthesis upon binding of virus particles to cells. For example, engagement of CD46, which is an entry receptor for vaccine strains of measles virus, induces potent IFN responses. In other cases, virus-induced degradation of the inhibitor of Nf-κb (Iκbα) leads to transcription of the genes encoding IFNs (Fig. 3.14).

Virus-infected cells usually produce IFN, but uninfected macrophages and dendritic cells that patrol tissues also make this cytokine when their Tlrs bind products released from infected cells (by apoptosis, for example). Such products include viral proteins, viral nucleic acids, and cellular stress proteins (e.g., heat shock proteins). If the infection is not contained at this stage and spreads to more cells, large quantities of IFN may be synthesized by specialized dendritic cell precursors in the blood called plasmacytoid dendritic cells. Systemic circulation of IFN leads to many of the general symptoms we associate with feeling sick following viral infection.

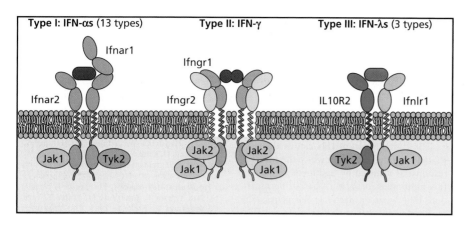

Figure 3.13 Interferon receptors. Type I IFNs interact with IFN-α receptors 1 and 2 (Ifnar1 and Ifnar2); type II IFN with IFN-γ receptors 1 and 2 (Ifngr1 and Ifngr2); and type III IFNs with IFN-λ receptor 1 (Ifnlr1) and IL-10 receptor 2 (IL10R2). Important cytoplasmic proteins that bind to the intracellular domains of each of these receptors are indicated.

Figure 3.14 Type I interferon synthesis, secretion, receptor binding, and signal transduction. Viruses or viral components are bound by Tlrs that trigger downstream signaling cascades leading to the production of type I IFNs (α and β) and Nf-κB-regulated genes. The type I IFNs are released from the cell, and can then bind to IFN receptors on the surfaces of adjacent cells to stimulate synthesis of IFN-responsive genes.

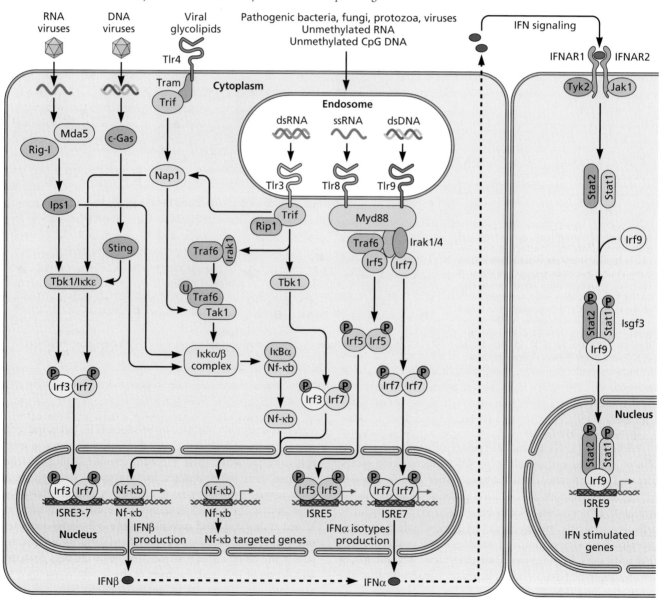

We tend to describe the two main type I IFNs, α and β, in similar terms, but they are quite different. There is only ~50% homology between α and β proteins, and while there is only one type I IFN-β gene, there are at least 20 IFN-α genes in the mouse and 13 in humans. Although these different type I IFN-α genes appear to be redundant (they are >80% identical to each other), they can be expressed differentially upon infection. The selective pressures that led to the diversity of type I IFNs remain to be discovered. In addition, transcription of the human *IFN-β* gene precedes expression of the IFN-α genes. The *IFN-β* enhancer possesses several remarkable properties that allow precise temporal control of transcription (Box 3.11).

The production of IFN by infected cells and uninfected, immature dendritic cells at the site of infection is rapid and robust, but transient; it occurs within hours of infection and generally declines by 10 h postexposure. Furthermore, the quantity of IFN released from cells infected by different isolates of a particular virus is astonishingly variable. In the case of vesicular stomatitis virus infection, the released IFN concentration can vary over a 10,000-fold range, depending on the serotype of the virus. As discussed later, many viral proteins affect the quantity of IFN made, as well as its action.

IFN Affects Only Cells with IFN Receptors

IFN functions only when it occupies its receptor on the surfaces of cells. Cells that produce IFN may also respond to IFN, but it must first be secreted, and then bind to cellular receptors (Fig. 3.14). A cell without IFN receptors may synthesize IFN, but cannot be affected by this cytokine. Binding of IFN to its receptor initiates a signal transduction cascade that culminates in increased transcription of many genes. A simplified outline of this signaling pathway is shown in Fig. 3.15.

IFNs affect gene expression by signaling via the Jak/Stat pathway. Members of this signal transduction family can also respond to IL-6 and other cytokines. There are four known Jak kinases and seven structurally and functionally related Stat proteins. Their targeted disruption of the respective genes in mice has revealed much about their functions. For example, a mouse in which the *stat1* gene has been deleted has no innate response to viral or bacterial infection, whereas deletion of *stat4* and *stat6* leads to inhibition of specific functions of the adaptive response. *Stat* gene homologs are encoded in the genomes of *Drosophila melanogaster* and *Dictyostelium discoideum*, underscoring the ancient evolutionary origin of this pathway. Signaling via Jak/Stat activates transcription dependent on specific promoter sequences. These sequences are found in the promoters of ~300 well-characterized IFN-activated genes, though more sensitive methods to detect gene expression changes have shown that 1,000 to 2,000 genes may be affected by this critical early cytokine.

IFN Action Produces an Antiviral State

As the name aptly indicates, IFN interferes with the reproduction of a wide variety of viruses in cells in culture and animals. Shortly after infection of the host, newly made IFN released from infected cells and local immature dendritic cells can be found circulating in the body, but its concentration is highest at the site of infection, where it is bound by any cell with the appropriate receptor. Cells that bind and respond to IFN do not support propagation of many different viruses; they are said to be in an **antiviral state**.

Many genes are induced by IFN signaling, but their mix and concentrations vary according to cell type and specific IFN. Which subset of the hundreds of IFN-inducible proteins establishes the antiviral state in any given cell remains unknown. Many of the products of IFN-inducible genes possess potent, broad-spectrum antiviral activities, but the relevant molecular mechanisms of only a few are understood. IFN not only induces death of the infected cells, but also ensures that uninfected cells in the vicinity are prepared to kill themselves should they become infected. Such a local cauterizing response has led some to characterize IFN action as a molecular firebreak to infection: IFNs define the boundary of infection by inducing either the death of infected cells or the antiviral state to prevent the virus from spreading beyond the local region of infection (Fig. 3.16).

IFN-induced proteins are functionally diverse and participate in signal transduction, chemokine action, antigen presentation, regulation of transcription, the stress response, and control of apoptosis. Some of these proteins are induced by other stimuli, including dsRNA, bacterial LPSs, Tnf-α, or IL-1. Because IFN induces the synthesis of many cytotoxic gene products, a common outcome of IFN signaling is cell death. We next describe some of the better-characterized IFN-stimulated gene products and their specific contributions to the antiviral state.

Some IFN-Induced Gene Products and Their Antiviral Actions

dsRNA-activated protein kinase. Viral and cellular protein synthesis in infected cells is often stopped abruptly. In many cases, this phenomenon, mediated by a cellular dsRNA-activated protein kinase (Pkr), is lethal to both the virus and the infected cell (also described in Volume I, Chapter 11). Establishment of the Pkr-mediated antiviral state is a two-step process, in which IFN promotes the increased production and accumulation of an inactive protein that can become activated only when it encounters double-stranded viral RNA.

All mammalian cells contain low concentrations of inactive Pkr, a serine/threonine kinase with antiviral, antiproliferative, and antitumor activities. The signal transduction cascade initiated by IFN binding to its receptor leads to a dramatic increase in the concentration of inactive Pkr. Metaphorically, this

BOX 3.11

DISCUSSION
Switching IFN-β transcription on and off

The regulation of the process that controls transcriptional activation and cessation of the *IFN-β* gene is the result of the coordinated action of many cellular proteins. It is therefore a significant scientific achievement that we have been able to understand this process.

Viral infection activates transcription of the human *IFN-β* gene, but only for a short period. This on-off response is controlled by an enhancer located immediately upstream of the core promoter. Like other enhancers, this regulatory sequence contains binding sites for multiple transcriptional activators, including Nf-κb and members of the Ap-1 and Atf families. However, the *IFN-β* enhancer possesses several remarkable properties that allow precise temporal control of transcription.

- The enhancer contains four binding sites for the architectural protein Hmg1(Y), which alters DNA conformation to direct the assembly of a precisely organized nucleoprotein complex on the enhancer.
- In contrast to typical modular enhancers, all binding sites **and** their natural arrangement are essential for activation of *IFN-β* transcription.
- Formation of the complex takes place in stages, and is not complete until several hours after infection.
- Activation of transcription requires sequential recruitment of the histone acetylase general control nonderepressible 5 (Gcn5), the coactivator cellular cAMP response element binding protein (Creb)-binding protein (Cbp), RNA polymerase II, and the chromatin-remodeling complex Swi/Snf.
- In addition to modifying nucleosomes, Gcn5 acetylates one of the Hmg proteins (A1) at Lys71. This modification stabilizes the complex.
- The Hmg(A1) protein is also acetylated by Cbp at another residue, Lys65. However, **this** modification impairs DNA-binding activity and results in disruption of the complex and cessation of *IFN-β* transcription.
- Remarkably, this inhibitory modification by Cbp is blocked for several hours by prior Gcn5 acetylation of Hmg(A1). As a result, the "off" switch is delayed for a sufficient period to allow a burst of *IFN-β* transcription.

Munshi N, Agalioti T, Lomvardas S, Merika M, Chen G, Thanos D. 2001. Coordination of a transcriptional switch by HMG1(Y) acetylation. *Science* **293:** 1133–1136.

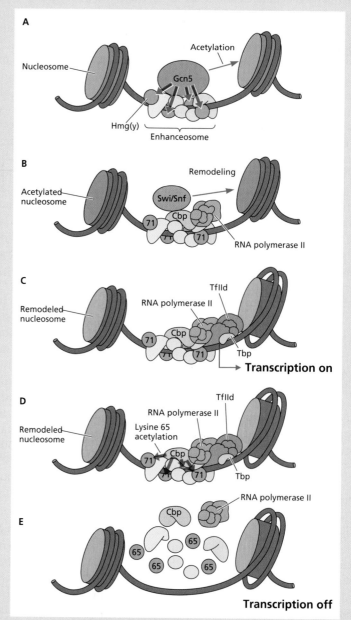

(A) Viral infection of human cells leads to assembly of multiple proteins on the IFN-β enhancer, which lies in a nucleosome-free region of the gene. The signals that direct binding of transcriptional activators (blue, yellow, and tan, collectively called the enhanceosome in the figure) and Hmg(A1) (green) are not fully understood. The precisely organized surface of the complex allows binding of Gcn5, which acetylates both histones in nearby nucleosomes and Lys71 of bound Hmg(A1) molecules (green arrows). The latter modification stabilizes the enhanceosome. **(B)** A complex of Cbp, RNA polymerase II, and the chromatin-remodeling protein Swi/Snf binds sequentially to the stabilized complex. The Swi/Snf complex alters the adjacent nucleosome that contains the core promoter DNA (green arrow). **(C)** Such alteration allows binding of TfIId and activation of transcription. Because Lys71 of Hmg(A1) is acetylated, Cbp cannot acetylate Lys65. **(D)** Eventually, Cbp does acetylate Lys65 of Hmg(A1) (red arrows), but how the inhibition induced by Lys71 acetylation is overcome is not yet clear. **(E)** Hmg(A1) modification by Cbp disrupts the complex and switches off transcription. Adapted from K. Struhl, *Science* **293:** 1054–1055, 2001, with permission.

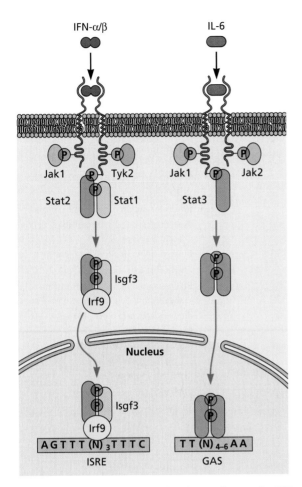

Figure 3.15 Common signal transduction pathways for IFN-α/β and IL-6. IFN signals via the Jak/Stat pathway, characterized by a family of tyrosine kinases given the acronym Jak (Janus kinases; Janus, a Roman god, guardian of gates and doorways, is represented with two faces and therefore faces in two directions at once) and a set of transcription proteins named Stat (signal transduction and activators of transcription). The receptors for IFN-α/β and IL-6 are different, but all affect components of the Jak/Stat signal transduction pathway. Type I IFNs and IL-6 bind to their receptors with high affinity (equilibrium dissociation constant [K_d] of $\sim 10^{-10}$ M). Binding of IFN or IL-6 to the appropriate receptor leads to the phosphorylation of tyrosine in tyrosine kinases as well as in the receptor itself. These modifications are followed by phosphorylation of tyrosine in the Stat proteins. In mammals there are seven *Stat* genes. The phosphorylated Stat proteins then form a variety of dimers that enter the nucleus. Within that organelle, Stat dimers bind, in some cases in conjunction with other proteins (e.g., Irf9), to specific transcriptional control sequences of IFN-α/β- and IL-6-inducible genes called interferon-stimulated response elements (ISREs) and IFN-gamma activated sequence (GAS) elements, respectively. Later in the transcriptional response to IFN, a second transcriptional activator called Irf1 replaces Isgf3.

means that the cell may go from 5 hand grenades to 50, but the pin is still in place, and the hand grenades are therefore not dangerous. If the cell is infected, this enzyme becomes activated by binding viral dsRNA; the pin is pulled. Active Pkr phosphorylates the α subunit of the eIF2 translation initiation

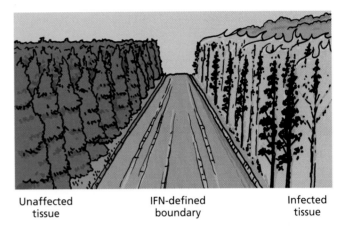

Figure 3.16 The interferon-induced firebreak that restricts viral spread beyond the site of infection.

protein (eIF2α), rendering it incapable of supporting protein synthesis in the cell (see Volume I, Chapter 11). Phosphorylated eIF2α does not invariably lead to cell death, as this modified protein can also trigger autophagy.

Many viral genomes encode proteins that can block the lethal actions of Pkr (Table 3.5). For example, a herpes simplex virus 1 protein (ICP34.5) redirects the cellular protein phosphatase 1 to dephosphorylate eIF2α after it has been phosphorylated and inactivated by Pkr. While wild-type virus is fully virulent in mice, ICP34.5-null mutants are markedly attenuated, particularly in brain infections. Significantly, this mutant regains wild-type virulence in mice lacking the *pkr* gene. This observation provides convincing evidence that Pkr mediates defense against herpes simplex virus infection in mice.

RNase L and 2′-5′-oligo(A) synthetase. Another well-studied antiviral response induced by IFN is mediated by two enzymes and dsRNA. RNase L can degrade most cellular and viral RNA species. Its concentration increases 10- to 1,000-fold after IFN treatment, but the protein remains inactive unless a second enzyme is synthesized. This enzyme, 2′-5′-oligo(A) synthetase, makes oligomers of adenylic acid, but only when triggered by dsRNA. These unusual nucleotide oligomers then activate RNase L, which in turn begins to degrade all host and viral mRNA. We now know from studies of mouse mutants defective in RNase L that this enzyme is important for the IFN-β response to viral infection. RNA fragments produced by RNAse L have double-stranded regions that allow them to be identified by RNAse L and Mda5, enhancing the production of IFN-β.

Mx proteins. Unlike the broad-spectrum antiviral effects of Pkr and RNase L, at least one IFN-induced mouse protein and two related human proteins appear to be directed against specific viruses. Mouse strains that have an IFN-inducible

Table 3.5 Some viral modulators of the interferon response[a]

Type of modulation	Representative viruses	Viral protein, if known	Mechanism of action
Inhibition of IFN synthesis	Epstein-Barr virus	Bcrf1	IL-10 homolog, inhibits production of IFN-γ
	Vaccinia virus	A18R	Regulates dsRNA production
	Foot-and-mouth disease virus	L	Host protein synthesis block
IFN receptor decoys	Vaccinia virus	B18R	Soluble IFN-α/β decoy receptor
Inhibition of IFN signaling	Adenovirus	E1A	Decreases quantity of Stat1 and P48, blocks Isgf3 formation, interferes with Stat1 and Cbp/p300 interactions
	Vaccinia virus	VH1	Viral phosphatase reverses Stat1 activation
	Human papillomavirus 16	E7	Binds p48
	Hepatitis C virus	NS5a	Blocks formation of Isgf3 and Stat dimers
	Nipah virus	V protein	Prevents Stat1 and Stat2 activation and nuclear accumulation
Block function of IFN-induced proteins	Adenovirus	VA-RNA I	Binds dsRNA, blocks Pkr
	Herpes simplex virus 1	US11	Blocks Pkr activation
		ICP34.5	Redirects protein phosphatase 1α to dephosphorylate eIF2α, reverses Pkr action
	Vaccinia virus	E3L	Binds dsRNA and blocks Pkr
		K3L	Pkr pseudosubstrate, decoy
	Human immunodeficiency virus type 1	TAR RNA	Blocks activation of Pkr
		Tat	Pkr decoy
	Hepatitis B virus	Capsid protein	Inhibits MxA
	Influenza virus	NS1	Binds dsRNA and Pkr, blocks action of Isg15
	Reovirus	σ3	Binds dsRNA, inhibits Pkr and 2′-5′-oligo(A) synthase

[a]For further examples and details, see B. B. Finlay and G. McFadden, *Cell* **124:**767–782, 2006.

gene called *mx1* are completely resistant to influenza virus infection. (The name, Mx1, is from the former name for influenza, myxovirus.) The Mx1 protein is part of a small family of IFN-inducible guanosine triphosphatases (GTPases) with potent activities against various (−) strand RNA viruses. After IFN induction, this protein accumulates in the nucleus and inhibits the unusual influenza virus "cap-snatching" mechanism (Volume I, Chapter 6). It is likely that the Mx1 protein interferes with the function of the viral polymerase subunit PB2, as overproduction of this viral protein overcomes the antiviral effect of Mx1. The significance of the *mx1* gene in the biology of influenza virus or of mice is not clear, as influenza virus does not circulate among wild mice. Moreover, about one-quarter of the mouse population (including many inbred strains used in laboratory research) lacks a functional *mx1* gene, with no obvious deleterious consequences.

The two human genes related to the murine *mx1* gene are termed *mxA* and *mxB*. Expression of these genes is also induced by IFN, but unlike the murine protein, the human proteins reside in the cytoplasm. MxA, but not MxB, blocks reproduction of influenza virus. In contrast to murine Mx1, which inhibits only influenza virus, the human MxA protein also prevents reproduction of vesicular stomatitis virus, measles virus, and other (−) strand RNA viruses. These human Mx proteins are related to members of the dynamin superfamily of GTPases, which regulate endocytosis and vesicle transport, but how this property relates to their antiviral activities is unknown.

Promyelocytic leukemia proteins. The promyelocytic leukemia (Pml) proteins are present in both the nucleoplasm and discrete multiprotein complexes known as nuclear bodies or Pml bodies (discussed earlier, and in Volume I, Chapter 9). These structures are important in the intrinsic cellular response to infection because their components bind foreign DNA that enters the nucleus. Pml and other proteins present in the complexes are thought to exert their antiviral effects by transcriptional repression and nucleosome remodeling. Many viral infections promote the dismantling of Pml bodies, in part as a measure to override global repression.

Ubiquitin-proteasome pathway components. The proteasome is a large, multisubunit protease that degrades cytoplasmic and nuclear proteins targeted for proteolysis by polyubiquitylation. Such degradation is important for the removal of abnormal or damaged proteins, the turnover of short-lived regulatory proteins, and the production of peptides for assembly of MHC class I proteins that are critical for induction of adaptive immunity. All IFNs induce transcription of a number of genes that encode proteins of the ubiquitin-proteasome pathway. In fact, many IFN-stimulated genes encode ubiquitin ligases. Increased protein degradation may contribute to the antiviral response to some viruses. For example, proteasome inhibitors block the anti-hepatitis B virus action of type I IFN. In this case, activation of the proteasome may be **the** major antiviral effect, because the results of other experiments demonstrate that the Pkr and RNase L systems are completely ineffective.

Tetherin/Bst2. Tetherin, or bone marrow stromal antigen 2 (Bst2), is a lipid raft-associated protein encoded by the *Bst2* gene in humans. This protein is constitutively made in some cells of the immune system, but can be induced by IFN in many others. Most of what is known about this protein relates to its antiviral properties, though recently its role has been identified in uninfected cells (Chapter 7). Tetherin blocks many enveloped viruses from budding from the infected cell surface by tethering the budding viral membranes to each other and to the plasma membrane. Tetherin protein spans the plasma membrane at one end and is attached to membranes by a glycosylphosphatidylinositol anchor at the other end. This unique topology and the tendency of tetherin to form dimers aid in retention of enveloped viral particles at the plasma membrane and prevent their separation (Fig. 3.17).

The human immunodeficiency virus type 1 protein Vpu can overcome this restriction by ubiquitinylating tetherin, leading to its degradation. Consistent with its importance in preventing human immunodeficiency virus type 1 transmission, tetherin gene variants are associated with disease progression.

IFN regulatory proteins. Members of the Irf protein family are required for sustained transcription of the IFN genes after induction. Mice lacking the *irf1* gene are incapable of mounting an effective IFN response to viral infection. Other members of this gene family (*irf2* to *irf9*) were discovered because their protein products bound to the interferon-stimulated response element (ISRE) in promoters of IFN-regulated genes. Irf4 is synthesized only in T and B cells, and Irf8 is made only in cells of the macrophage lineage. Mice defective for *irf8* gene expression are markedly more susceptible to infection and cannot synthesize proinflammatory cytokines. The protein Irf9 is the DNA-binding component of the transcriptional regulator IFN-stimulated gene factor 3 (Isgf3) (Fig. 3.15). Several viral Irf-like proteins that block IFN action have been identified (Table 3.5).

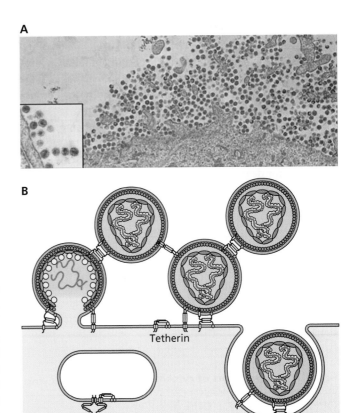

Figure 3.17 Tetherin prevents budding of enveloped viruses. Human immunodeficiency virus type 1 (HIV-1) virus particles lacking a functional Vpu protein are trapped at the surface of a tetherin-expressing cell by apparent particle-to-particle, as well as particle-to-cell, tethering. Panel A: from S. J. Neil et al., *Nature* **451**:425–430, 2008. **(B)** Tetherin anchors enveloped virus particles to the cell membrane, and to each other, preventing release. Human immunodeficiency virus type 1 protein Vpu antagonizes this intrinsic restriction mechanism.

Other IFN-induced proteins. Other proteins with antiviral effects surely remain to be discovered among the many IFN-induced genes that have been identified. For example, the IFN response is required to clear human cytomegalovirus infections, but Pkr, Mx, and RNase L proteins are not. Similarly, uncharacterized IFN-induced proteins block penetration and uncoating of simian virus 40 and some retroviruses. Others impair the maturation, assembly, and release of vesicular stomatitis virus, herpes simplex virus, and some retroviruses by unknown mechanisms. Specific combinations of the products of Isgs are probably needed to control viral reproduction, and this likely depends on the nature of the infecting virus.

Regulators of the IFN Response

As many of the gene products characteristic of the antiviral state are highly cytotoxic, it is imperative to suppress the response once viral reproduction has been controlled. Such containment is accomplished by the action of members of the suppressor of cytokine signaling (Socs) protein family, which act in a classical negative-feedback loop to attenuate cytokine signal transduction (Fig. 3.18). The SH2 domains of Socs proteins interact with activated cytokine receptors, including the Jaks, blocking their ability to activate Stat molecules. The Socs proteins combine specific inhibitory interactions with a general mechanism of targeting associated signaling molecules for degradation.

Gene-knockout studies have shown that Socs proteins are indispensable regulators of important physiological systems: Socs1 is an essential homeostatic regulator of IFN signaling that is crucial to allow the beneficial immunological effects of IFN without the damaging pathological responses. Mice that lack Socs1 die early in life, even in the absence of viral infections; they have liver disease, inflammatory lesions, lymphopenia, apoptosis of lymphoid organs, and anomalous T cell activation, all probably the result of unrestricted IFN signaling.

Viral Gene Products That Counter the IFN Response

The term "antiviral state" implies that the IFN response confers complete resistance to virus infection. However, viruses vary considerably in their sensitivity to the effects of this cytokine. The reproduction of some viruses, such as vesicular stomatitis virus, is so sensitive to IFN that this property is used to titrate the cytokine. Other viruses can be more resistant to IFN. We now know that numerous viral mechanisms confound IFN production or action.

Many viral genomes encode dsRNA-binding proteins that interfere with detection by pattern recognition receptors and IFN induction. The reovirus σ3 protein, the multifunctional influenza virus NS1 protein, and the hepatitis B virus core antigen are all well-characterized dsRNA-binding proteins with anti-IFN effects. The vaccinia virus E3L protein and the herpes simplex virus 1 US11 protein also have binding properties that correlate with inhibition of IFN induction. Adenovirus VA-RNA I acts as a dsRNA decoy and blocks the activation of Pkr by binding to this enzyme directly.

Figure 3.18 Suppressors of cytokine signaling. In unstimulated cells, Socs genes are not expressed. However, when IFN is present, Socs proteins are among the genes induced, which then act in a negative-feedback loop to block signal transduction. Socs1 interacts directly with Jaks and Socs3 inhibits Jaks after gaining access by receptor binding. In addition, Socs proteins interact with the cellular ubiquitination machinery through the Socs box and might direct associated proteins, such as Jaks or receptors, for ubiquitin-mediated proteasomal degradation.

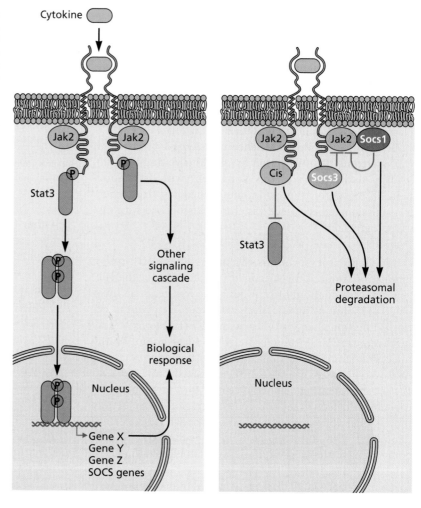

An inescapable inference from the various counter-measures encoded by the genomes of diverse viruses is that IFN is an essential host defense component (Fig. 3.19). But numerous questions remain. For example, infections by some viruses (e.g., Newcastle disease virus) are inhibited only by IFN-α, while others (e.g., herpes simplex virus 1) are inhibited primarily by IFN-β. IFN synthesis is induced after infection by vaccine strains of measles virus, while little IFN is made after wild-type virus infection. Perhaps most fascinating, the IFN response varies depending on the route of infection when animals are inoculated experimentally (Box 2.3). Despite much progress, and elucidation of many details of

Figure 3.19 Virus-mediated modulation of interferon production and action. Viral gene products modulate most steps in the IFN response from the infected cell to the responding cell. Such modulation affects the dynamics of cytokine production and action in ways that are not fully understood. For example, dendritic cells detect viral infection or products of viral infection and produce type I IFN and IFN-γ. However, viral infection may lead to reduction of IFN production in these primary defense cells (red line, IFN antagonist). The IFN produced by dendritic cells can bind to receptors on innate immune cells (e.g., NK cells) or T cells, leading to production of IFNs and other IFN-inducible gene products (indicated by the question mark). The combination of NK cell and T cell action should produce soluble antiviral effectors, leading to destruction of other infected host cells (e.g., epithelial cells). However, viral gene products produced in these infected target cells can impair IFN signaling or block recognition of the infected cell by NK cells or T cells. As a result, virus-infected cells are exposed to a rapidly changing cytokine array, not only by the infected cell, but also by innate and adaptive immune cells reacting to the infection. Adapted from A. García-Sastre and C. A. Biron, *Science* **312**:879–882, 2006, with permission.

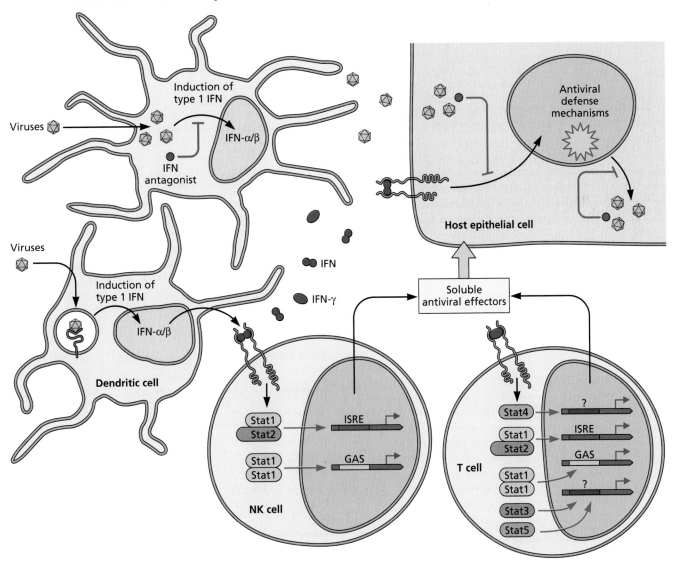

IFN biology (for example, how the *IFN-β* gene is turned on and off), it is likely that major principles of IFN synthesis, activity, and regulation remain to be discovered.

Chemokines

Imagine that, somewhere in a large U.S. city, a person has just started a fire while cooking dinner. His efforts to put out the fire, perhaps dousing it with water, may limit the blaze, but these localized efforts on the part of the chef may not be completely successful; the professionals, who offer tools, expertise, and experience, must be called in. But how, in this vast city, are the firefighters alerted? In this example, one would call 911. For the host response, chemokines are the emergency alert, attracting circulating immune cells to the specific site of damage.

The ability of the immune system to respond to the presence of foreign antigens, tissue damage, and other physiological insults depends on chemokine gradients to recruit lymphocytes to the right place and to activate these cells at the right time. Chemokines also coordinate cellular movement in normal processes, including lymphocyte and neural development and new blood cell formation. Chemokines, secreted by local macrophages and some infected cells, bind to G protein-coupled receptors on circulating lymphocytes, inducing signaling pathways involved in cell movement and activation.

A representation of the process by which chemokines aid the migration of a white blood cell from the blood, across the endothelium, and into an affected tissue during an inflammatory response is shown in Fig. 3.20. First, selectins on the endothelium interact with mucin receptors on the leukocyte, causing it to roll along the cell surface, slowing its transit, and enabling it to migrate through the blood vessel. Chemokines bind glucoasminoglycans on the endothelial cell surface to induce production of additional adhesion molecules, including integrins, which further retain lymphocytes near the site of viral infection. The cells then pass from the blood into tissue, squeezing between endothelial cells that comprise blood vessels, following chemokine gradients.

Approximately 50 human chemokines and 20 receptors have been discovered (Table 3.6). Early in the chemokine literature, these molecules and their receptors were given names based on their presumed functions. This quickly became confusing, because the names reflected only some of the actual functions of these molecules. Since 2000, chemokines are named based on the number and location of conserved cysteine residues. There are four families: CXC, CC, C, and CX3C (in which "X/X3" represents one or three noncysteine amino acids). Chemokine ligands are denoted with an "l" (as in CCl2), and their receptors are designated with an "r" (as in CCr2).

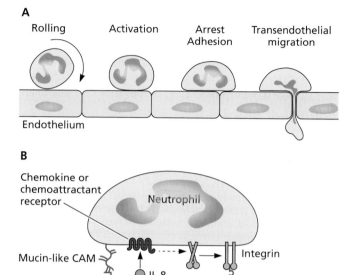

Figure 3.20 Steps in immune cell extravasation into tissues, and the role of chemokines. (A) The sequential steps of lymphocyte migration from the blood into a tissue parenchyma. **(B)** The critical cell adhesion molecules that result in anchoring of a blood cell (here, a neutrophil) to the endothelium. CAM: cell adhesion molecule.

Although chemokines were selected to benefit the host, inappropriate regulation or utilization of these proteins can contribute to or cause many diseases, including autoimmune disorders (e.g., psoriasis, rheumatoid arthritis, and multiple sclerosis), pulmonary diseases (asthma and chronic obstructive pulmonary disease), cancer, and vascular disease, presumably by disrupting cell mobility within the host. In addition, the CXCr4 and CCr5 receptors serve as coreceptors by human immunodeficiency virus type 1 for entry into cells.

Table 3.6 Some chemokine receptors and their ligands[a]

Receptor[b]	Old chemokine ligand name	New chemokine ligand name
CCr1	Mip1α, Rantes	CCl3
CCr2	Mcp1	CCl2
CCr5	Rantes	CCl5
CXCr2	IL-8	CXCl8
CXCr3	Ip-10	CXCl10

[a]Data from C. R. Mackay, *Curr Biol* 7:R384–R386, 1997.

[b]The four families of chemokine receptors are distinguished by the pattern of cysteine residues near the amino terminus and are abbreviated CXC, CC, C, and CX3C. Only two types are listed in this table. The CXC family has an amino acid between two cysteines; the CC family has none; the C family has only one cysteine; and the CX3C family has three amino acids between two cysteines. Subfamilies of these major four groups also exist.

The Innate Immune Response

The staggering number and diversity of local mechanisms to contain or eliminate viruses underscores the importance of a powerful frontline defense. While this text focuses on viruses, almost all of these mechanisms also operate against other types of microbial challenges.

When intrinsic cell defenses are unable to stop the spread of infection, the combination of cell death, local increasing concentrations of cytokines, and release of other stress-related molecules around the area of infection leads to activation of the next phase of host defense, the innate immune response. We have already introduced some of the critical players in this response: the local **sentinel cells** (dendritic cells and macrophages), which bring peptides derived from viral proteins to the lymph node to induce the adaptive response, and which synthesize soluble, antiviral mediators. These mediators, chemokines and type I IFNs, dampen viral spread, forewarn uninfected cells that are adjacent to infected areas, and are a beacon for the subsequent recruitment of components of the innate and adaptive responses. In addition to these effectors, the innate response also incorporates a large collection of serum proteins termed **complement**, and cytolytic lymphocytes called **natural killer cells (NK cells)**. Neutrophils and other granulocytic white blood cells are also important in innate defense in response to the initial burst of cytokines from dendritic cells, macrophages, and infected cells.

The innate immune response is crucial in antiviral defense because it can be activated quickly, functioning within minutes to hours of infection. Such rapid action contrasts with the activation of the adaptive response, which is far slower than the infectious cycles of some viruses. It takes days to weeks to orchestrate the effective response of antibodies and activated lymphocytes specifically tailored to the infecting virus. While the speed and potency of innate immunity is important, this response must also be transient, because its continued activity is damaging to the host.

Complement

The **complement system** was identified in 1890 as a heat-labile serum component that lysed bacteria in the presence of antibody. The name "complement" derived from the ability of this blood component to cooperate with antibodies and phagocytic cells to clear an infection (Box 3.12). We now know that the complement system comprises many proteins that function in a complex cascade, in which inactive precursors are sequentially triggered, leading to the massive amplification of the response and the activation of the membrane attack complex. More than 30 proteins and protein fragments make up the three distinct pathways in the complement system: the **classical pathway**, **alternative pathway**, and **lectin pathway**. Unfortunately, the nomenclature of the complement proteins can be confusing, as they were named in order of their discovery.

BOX 3.12

DISCUSSION

The complement cascade has four major biological functions

Lysis
Membrane disruption and lysis occur when specific activated complement components (C6, C7, C8, and C9) polymerize on a foreign cell or enveloped virus, forming pores or holes that disrupt the lipid bilayer and compromise its function. The cell or virus is disrupted by osmotic effects.

Activation of Inflammation
Inflammation is stimulated by several peptide products of complement proteins produced during the complement cascade. These peptides (C3a, C4a, and C5a) bind to vascular endothelial cells and various classes of lymphocytes to stimulate the inflammatory response and to enhance responses to foreign antigens.

Opsonization
Complement proteins (typically C3b and C1q) can bind to virus particles so that phagocytic cells carrying appropriate receptors can then engulf the coated viruses and destroy them; this process is called opsonization. Complement receptors such as Cr1 present on phagocyte surfaces bind C3b-coated particles and initiate their endocytosis.

Solubilization of Immune Complexes
Noncytopathic viral infections commonly result in pathological accumulations of antigen-antibody complexes in lymphoid organs and kidneys. Complement proteins can disrupt these complexes, by binding to both antigen and antibody, and facilitate their clearance from the circulatory system.

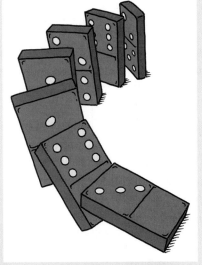

One push triggers an inevitable cascade.

Complement proteins are present in the blood and in various tissues in uncleaved, inactivated forms. Complement action can be initiated by direct recognition of a microbial invader by C1q (a component of C1) in the classical pathway, or by recognition of cleaved C3b proteins in the alternative pathway (Fig. 3.21). The mannan-binding lectin pathway triggers complement action upon binding of a lectin similar to C1q to mannose-containing carbohydrates on bacteria or viruses. Importantly, complement can also function as an effector of the adaptive defense system by the binding of C1q to antigen-antibody complexes on the surface of a microbe or infected cell (the classical pathway).

The Complement Cascade

In all three pathways, a protease cascade leads to the activation of two critical proteases called **C3 convertase** and

Figure 3.21 Activation and regulation of the complement system. The complement system can be activated through three pathways: classical, lectin, and alternative. Complement component 1 (C1) comprises C1q (a pattern recognition protein), C1r, and C1s. The complement cascade is activated when C1 binds an antigen-antibody complex on the surface of an infected cell or a virus particle; C1 also links the classical and lectin activation pathways by interacting with the mannose-binding lectin (MBL)-associated serine protease (Masp). These complexes contain proteases that cleave complement proteins C2 and C4, which then form the C3 and C5 convertases for the classical and lectin pathways. The alternative pathway activates complement without going through the C1-C2-C4 complex. For the alternative pathway, factor B is the C2 equivalent. Factor B is cleaved by factor D. Factor P (properdin) stabilizes the alternative pathway convertases. All three pathways culminate in the formation of the C3 and C5 convertases (orange box), which produce the three primary actions of activated complement: inflammation, cell lysis, and coating of foreign antigens so that they can be taken up by phagocytes (opsonization). The C3a and C5a proteins are potent stimulators of the inflammatory response (also called anaphylatoxins). The membrane attack complex is formed by the complement proteins C5b, C6, C7, C8 and C9 and forms a hole in membranes, leading to lysis of cells. The C3b (opsonin) coats bacteria and virions and also amplifies the alternative pathway. See C. Kemper and J. P. Atkinson, *Nat Rev Immunol* 7:9–18, 2007.

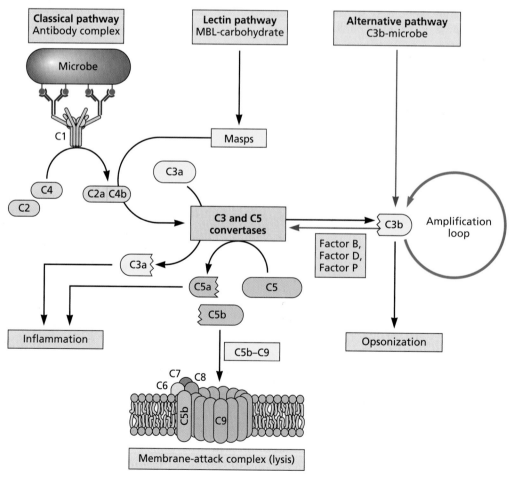

C5 convertase (note that the three pathways yield the same enzyme activity, but the proteins comprising each convertase are different). A crucial property of C3 and C5 convertase enzymes is that they are bound covalently to the surface of the pathogen or the infected cell. The action of surface-bound C3 convertase on its substrate yields C3b, the primary effector of all three complement pathways, and C3a, a potent soluble mediator of inflammation. C3b remains on the pathogen surface, where it binds more complement components to stimulate a protease cascade that produces other bioactive proteins. The protease cleavage products stimulate inflammation, attract lymphocytes, potentiate the adaptive response, and kill infected cells. C3b also stimulates phagocytic cells to take up the C3b-coated complex.

More than 90% of plasma complement components are made in the liver. Other sites of synthesis include the major portals of pathogen entry. For example, the initiator complex C1 is synthesized mainly in the gut epithelium, and mannan-binding lectin is found in the respiratory tract. In addition, monocytes, macrophages, lymphocytes, fibroblasts, endothelial cells, and cells lining kidney glomeruli or synovial cavities all make most proteins of the complement system.

One important consequence of complement cascade activation is the initiation of a local, broad-spectrum defense. Complement components released locally aid in recruitment of monocytes and neutrophils to the site of infection, stimulate their activities, and increase vascular permeability (Table 3.7). The antiviral effects of complement are both direct and indirect. The membrane attack complex lyses infected cells and inactivates enveloped viruses, while phagocytes engulf and destroy virus particles coated with C3b protein. Complement components stimulate a local inflammatory response that can limit infection, and aid in presenting signals of the invader to the adaptive immune system. The activated complement system "instructs" the humoral and T cell responses much as activated dendritic cells communicate with T cells, and is one of the bridges between frontline early defenses and adaptive immunity.

"Natural Antibody" Protects against Infection

The classical complement pathway of humans and higher primates can be activated by a particular collection of antibodies present in serum prior to viral infection (historically called "natural antibody"). Synthesis of some of these antibodies is triggered by the antigen galactose α(1,3)-galactose (α-Gal) found as a terminal sugar on glycosylated cell surface proteins. Lower primates, most other animals, and bacteria synthesize the enzyme galactosyltransferase, which attaches α-Gal to membrane proteins, but humans and higher primates lack the enzyme and do not make this antigen. Because of constant exposure to bacteria producing α-Gal in the gut, human serum contains high levels of antibodies specific for

Table 3.7 Biological activities of proteins and peptides released during the complement cascade

Substance	Biological activity
C5b, C6, C7, C8, and C9	Lytic membrane attack complex
C3a	Peptide mediator of inflammation, smooth-muscle contraction; vascular permeability increase; degranulation of mast cells, eosinophils, and basophils; histamine release; platelet aggregation
C3b	Opsonization of particles and solubilization of immune complexes; facilitation of phagocytosis
C3c	Neutrophil release from bone marrow; leukocyte lysis
C3dg	Molecular adjuvant; profound influence on adaptive response
C4a	Smooth-muscle contraction; vascular permeability increase
C4b	Opsonin for phagocytosis, processing, and clearance of antibody-antigen immune complexes
C5a	Peptide mediator of inflammation, smooth-muscle contraction; vascular permeability increase; degranulation of mast cells, basophils, and eosinophils; histamine release; platelet aggregation; chemotaxis of basophils, eosinophils, neutrophils, and monocytes; hydrolytic enzyme release from neutrophils
Bb	Inhibition of migration and induction of monocyte and macrophage spreading
C1q	Opsonin for phagocytosis, clearance of apoptotic cells, and processing and clearance of antibody-antigen immune complexes

this antigen. Indeed, >2% of the IgM and IgG populations is directed against this sugar. It is this antibody that triggers the complement cascade and subsequent lysis of foreign cells and enveloped viruses bearing α-Gal antigens. The anti-α-Gal antibody-complement reaction is probably the primary reason why humans and higher primates are resistant to infection by enveloped viruses of other animals, despite the ability of many of these viruses to infect human cells efficiently in culture. Consistent with this view, when such viruses are grown in nonhuman cells, they are sensitive to inactivation by human serum. Anti-α-Gal antibodies provide a mechanism for cooperation of the adaptive immune system and the innate complement cascade to provide immediate, "uninstructed" action.

Regulation of the Complement Cascade

Any amplified antiviral defense system as lethal as the complement cascade must be regulated with precision.

Spontaneous activation of any one of the three pathways must be blocked, and triggering by minor infections, nonpathogenic microbes, or noninfectious proteins avoided. Some regulation is intrinsic to the complement proteins themselves. For example, many are large and cannot leave blood vessels to attack infected tissues unless there is localized tissue damage and capillary wall breakdown that exposes cells directly to blood. Consequently, minor infections do not activate a substantial complement response. Moreover, many cascade intermediates do not exist long enough to diffuse far from the site of infection: they are short-lived, with millisecond half-lives. Further control is maintained by complement-inhibitory proteins present in the serum and on the surface of many cells (e.g., the complement receptor type 1 protein [Cr1], decay-accelerating protein [Daf, or CD55], protectin [CD59], and membrane cofactor protein [CD46]). These proteins are regulators that can limit the alternative-pathway cascade by binding to complement components such as C3b and C4b and preventing the accidental deposition of these cascade triggers on host cells. Some viruses have co-opted these molecules to protect themselves from complement-mediated lysis. Human immunodeficiency virus type 1 and the extracellular form of vaccinia virus incorporate CD46, CD55, and CD59 in their envelopes, providing protection from complement-mediated lysis.

Many viral genomes encode proteins that interfere with the complement cascade. For example, alphaherpesvirus glycoprotein C binds the C3b component, and several poxvirus proteins bind C3b and C4. The variola Spice protein (smallpox inhibitor of complement enzymes) inactivates human C3b and C4b and is a major contributor to the high mortality caused by this virus.

Several viral receptors, including those for measles virus and certain picornaviruses, are complement control proteins. Epstein-Barr virus particles bind to CD21 (the Cr2 complement receptor), with profound consequences for the host and virus: this interaction activates the Nf-κb pathway, which then allows transcription from an important viral promoter. Epstein-Barr virus binding to the complement receptor enables viral reproduction in resting B cells that would otherwise be incapable of supporting viral transcription. Moreover, binding of measles virus to the complement regulatory protein CD46 induces an IFN response. As only vaccine strains bind to this receptor, this interaction may be the reason why these measles strains induce a protective response, rather than initiate a pathogenic infection.

Pattern Recognition by C1q, the Collectins, and the Defensins

The action of the complement initiator protein C1q exemplifies a definitive property of intrinsic and innate defense: C1q can recognize molecular patterns characteristic of pathogens, much like the Tlrs. C1q is a calcium-dependent, sugar-binding protein (a **lectin**) in the collectin family of proteins. These proteins bind to polysaccharides on a wide variety of microbes and act as opsonins or activators of the complement cascade. Defensins represent another class of antimicrobial lectins. They are small (29- to 51-residue), cysteine-rich, cationic proteins produced by lymphocytes and epithelial cells that are active against bacteria, fungi, and enveloped viruses. Collectins and defensins bind the glycoproteins of a number of enveloped viruses, including human immunodeficiency virus type 1, herpes simplex viruses, Sindbis virus, and influenza virus. The antiviral activity of some collectins and defensins can be observed with cells in culture. The basis for such activity appears to be inhibition of membrane fusion. An attractive hypothesis is that they function by cross-linking viral glycoproteins and blocking displacement of other proteins from the fusion site. While these lectins display antiviral activity in the laboratory, their physiological contributions have not been well studied. Some have been modified for testing as antiviral compounds to be delivered systemically or topically.

It is useful to reiterate the multiple parts that certain immune proteins play in host defense. Complement can be considered a component of the intrinsic, innate, or adaptive response. Some complement proteins are pattern recognition receptors, others need to be activated by upstream signals, and still others can distinguish self from nonself.

Natural Killer Cells

Natural killer cells are at the front line of innate defense: they are ready to recognize and kill some virus-infected cells, and do not need selection or stimulation to do so. Like T cells, NK cells can distinguish infected cells amidst vast numbers of uninfected cells. However, the mechanism of recognition is completely different: NK cells recognize "missing self." We previously introduced the concept that humans recognize that something is different in their environment based on their recollection of what was there before. For NK cells, "different" means an **absence** of something familiar.

NK cells are abundant, representing ~2% of circulating lymphocytes. They are large, granular cells, distinguished by the absence of the antigen receptors found on B and T cells (Chapter 4). When an NK cell binds an infected target cell, it releases a mix of cytokines (notably IFN-γ and Tnf-α) that contribute to a local inflammatory response and alert cells of the adaptive immune system. They also can produce prodigious quantities of IL-4 and IL-13, the major cytokines that stimulate antibody production. In addition, NK cells participate later in adaptive defense by binding to infected cells coated with IgG antibody and inducing antibody-dependent cell-mediated cytotoxicity.

The number of NK cells increases quickly after viral infection and then declines as the adaptive immune response is educated and amplified. NK cells are stimulated to divide whenever infected cells and sentinel dendritic cells make IFN. The NK cells kill after contact with the target by releasing perforins and granzymes that perforate membranes and trigger caspase-mediated cell death, respectively, in a process identical to how cytotoxic T lymphocytes kill their targets. In humans, NK cells are particularly important in controlling primary infection by many herpesviruses, as patients with NK cell deficiencies suffer from severe infections with varicella-zoster virus, human cytomegalovirus, and herpes simplex viruses. While a role for direct NK cell-mediated killing in antiviral defense is difficult to establish experimentally, NK cell production of IFN-γ clearly provides significant antiviral action.

NK Cell Recognition of Infected Cells: Detection of "Missing Self" or "Altered Self" Signals

A collection of cell surface proteins called the MHC proteins are important receptors in the adaptive immune response (Chapter 4). MHC class I proteins are found on the surfaces of most cells of the body and "present" microbial peptides to T cells. The MHC class I molecules are the self antigens that, when missing, cause the NK cell to kill the target cell. A mechanism for detection of missing self is illustrated in Fig. 3.22. At least two receptor-binding interactions that cooperate to send either a "go" or "stop" signal to the prearmed NK cell are required for such discrimination.

Figure 3.22 NK cells distinguish normal, healthy target cells by a two-receptor mechanism. Both positive (stimulating) and negative (inhibiting) signals may be received when an NK cell contacts a target cell. The converging signal transduction cascades from the two classes of receptor regulate NK cell cytotoxicity and release of cytokines. The inhibitory receptors dominate all interactions with normal, healthy cells. Their ligands are the MHC class I proteins. When NK cells contact MHC class I molecules on the surface of the target cell, signal transduction blocks the response of activating receptors.

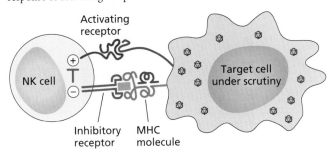

The "go" signal is delivered when an NK activating receptor binds a pathogen-specific ligand (e.g., virus-infected cells may present new glycoproteins on their surface). As a consequence, a signal transduction cascade is initiated and the NK cell is stimulated to secrete cytokines and kill the cell. However, a dominant-negative regulatory signal is produced when an inhibitory receptor on the NK cell engages MHC class I molecules on the surface of the same target cell. Because many infected cells carry fewer MHC class I molecules on their surfaces (Chapter 5), they are prime NK cell targets. In essence, NK cells serve a counter-counter-response to those viruses that downregulate MHC molecules on the infected cell surface to escape T cell detection. The two-receptor recognition system employed by NK cells ensures that normal cells that synthesize MHC class I proteins, even those that may be virus infected, are not killed by NK cells.

MHC Class I Receptors on NK Cells Produce Inhibitory Signals

Human NK cells synthesize two inhibitory MHC class I receptors of either the C-type lectin family or the immunoglobulin family (called killer cell immunoglobulin-like inhibitory receptors, or Kirs). NK cells also can recognize and spare target cells carrying HLA-E, an unusual MHC class I protein that binds peptides derived from the signal sequences of other MHC class I molecules. The presence of HLA-E protein bound to signal peptide informs the NK cell that MHC class I synthesis is normal. An intriguing finding is that infection by human cytomegalovirus induces synthesis of HLA-E protein, thereby escaping potential NK cell recognition and lysis.

Viral Proteins Modulate NK Cell Actions

Many viral genomes encode proteins that block or confound NK cell recognition and killing (Fig. 3.23). At least five distinct categories of modulation can be described. NK modulators have been identified in the genomes of several virus families including *Flaviviridae*, *Papillomaviridae*, *Herpesviridae*, *Retroviridae*, and *Poxviridae*. Some viral genomes encode more than one distinct NK modulator. For example, human cytomegalovirus encodes at least seven such gene products that modulate the NK cell response. One striking example of viral interference with NK cell activity is provided by the hepatitis C virus E2 envelope protein, which binds to CD81, a protein on the surface of NK cells, and blocks activation signals. As a result, the NK cell no longer recognizes infected cells.

NK Cell Memory

NK cells may have a "memory" state, a property normally thought to be unique to cells of the adaptive immune

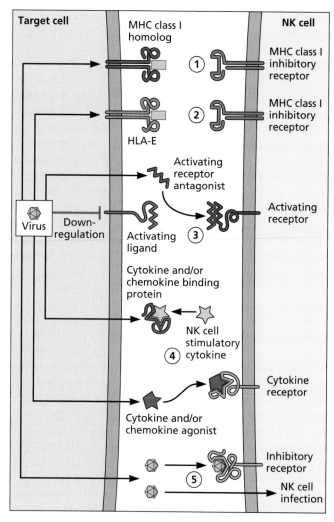

liver, where this memory pool can access the circulation and maintain immune surveillance at a low but constant level. When the same antigen is encountered in the periphery, antigen-specific NK cells then accumulate at the site of challenge, where they orchestrate local effector responses. Moreover, most NK cells can acquire certain memory-like properties even without exposure to a specific antigen, similar to the cytokine-driven, antigen-independent "bystander" response of CD8+ T cells (Chapter 4).

Other Innate Immune Cells of Relevance to Viral Infections

While the main cellular actors that govern the outcome of a viral infection are T cells, B cells, and NK cells, the heterogeneity of blood cells and the presence of multiple "minor players" at sites of infection indicate that other blood-borne cells likely also contribute to antiviral immunity.

Neutrophils

Neutrophils are, by far, the most abundant type of cell in the blood, comprising >50% of the circulating white blood cells. These cells also produce soluble mediators, such as cytokines, reactive oxygen species, and perforating granules. Neutrophils participate primarily in the resolution of bacterial infections, in part as a result of the release of nets that capture extracellular bacteria, much like a spiderweb. These neutrophil extracellular networks (NETs), comprising DNA decorated with cellular histones, are highly charged, making them "sticky" (Fig. 3.24). But such innovative strategies

Figure 3.23 Virus-encoded mechanisms for modulation of NK cell activity. (Left) An infected target cell. (Right) An NK cell. The infected target cell should be lysed by an activated NK cell. However, five categories of NK cell-modulating strategies are illustrated (circled numbers). Viral proteins produced in the infected cell are labeled in red. (1) Inhibition by a viral protein with homology to cellular MHC class I proteins. (2) Inhibition of expression or cell surface localization of host HLA-A or HLA-B (human MHC class I homologs) resulting in an increase in the amount of host HLA-E (or HLA-C) on the target cell surface. (3) Release of virus-encoded cytokine-binding proteins that block the action of NK cell-activating cytokines (also, viral proteins can reduce the amount of the activating ligand on the surface of the infected cell so that the NK cell is not activated). (4) Inhibition of action of NK cell-stimulating cytokines by binding these cytokines or by producing a chemokine antagonist. (5) Effect of newly produced virus particles, which can engage the NK cell, block an inhibitory NK cell receptor, or infect the NK cell itself to disrupt various effector functions or even kill the cell.

Figure 3.24 Neutrophils produce a "net" to capture extracellular pathogens. In this image, a *Klebsiella pneumoniae* bacterium is captured in the extracellular chromatin net produced by neutrophils within the lung. Credit: Science Photo Library.

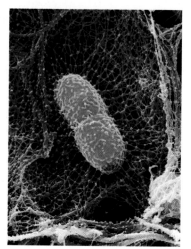

response. Exposure to activating cytokines, such as IL-12 and IL-18, elicits a form of memory in splenic NK cells whereby the primed cells mediate enhanced IFN-γ responses after restimulation by cytokines or by antibody-mediated ligation of activating receptors. Memory NK cells are found in the would presumably not work for viruses, which spend much of their life cycle within cells, or would likely be too small to be ensnared by these mesh-like DNA structures derived from decondensed chromatin DNA within the cell. It had therefore been a puzzle why neutrophils are found at sites of viral infections.

A study using the poxvirus vaccinia virus showed that these same structures may, in fact, have antiviral properties. Following infection, NETs within the liver microvasculature were found to significantly reduce the number of infected host cells: a direct role for the NETs was shown using DNase treatment (destroying the NETs, but not the neutrophils), which abrogated their protective effect. Other studies indicate that such sticky solutions may be operative for other viruses as well, including human immunodeficiency virus type 1 and influenza virus. At present, we understand little about this remarkable defense mechanism.

NKT Cells

NKT cells are so named because they share features of both NK and T cells. They make an αβ T cell receptor, but also possess cell surface markers that are found primarily on NK cells. These cells are rare, constituting only 0.1% of peripheral blood T cells. This T cell receptor recognizes the nonpolymorphic molecule CD1d, an antigen-presenting molecule that binds to lipids and glycolipids. This interaction may be important for recognizing infections by bacteria that have glycolipids on their surfaces, but a contribution of these cells in viral resolution has yet to be defined.

γδ Cells

As we will discuss in Chapter 4, conventional T cells are characterized by the dimeric T cell receptor, which comprises α and β chains. γδ cells possess many of the same T cell markers, but have a distinct receptor, called γδ to distinguish them from their more abundant and well-characterized cousins. These T cells are highly prevalent in the gut mucosa: a clue that they may be critical for early recognition of invading microbes. Conventional T cells recognize peptides in the context of class I MHC determinants, but γδ T cells do not, although some recognize class Ib MHC molecules. It is thought that these cells are particularly suited to bind to lipid (as opposed to protein-based) antigens.

These unconventional cells lie at the intersection of the innate and adaptive response. They can be considered adaptive, in that they rearrange their T cell receptors and establish memory, but much like NK cells, they do not recognize processed antigen nor need extensive education and amplification to be functionally active. Their specific contributions to antiviral immunity have not yet been explored.

Perspectives

This chapter began with some warnings: the immune response is elaborate, and defies our efforts to order neatly where each effector process "belongs" in the overall response. From an evolutionary perspective, one must marvel at the number and diversity of ways in which our body's defenses continually try to keep us safe from pathogens. Intrinsic and innate defenses are always on high alert: unlike the cells of the adaptive response, which sit patiently in the spleen or lymph node waiting for their cognate antigen to appear, these defenses are constitutively surveying all possible portals of microbial invasion.

To illuminate important principles in this chapter, it is useful to consider a hypothetical acute viral infection that is cleared by the host response (Fig. 3.25). To initiate the primary infection, physical barriers are breached and virus particles enter permissive cells. Almost immediately, viral proteins and viral nucleic acids are bound by pattern recognition receptors. Signal transduction cascades then result in the activation of transcriptional regulators that drive the production of cytokines, such as IFN. As new viral proteins are produced, the cell initiates other intrinsic defenses, such as apoptosis or autophagy. Local sentinel cells (the immature dendritic cells and macrophages) respond to the locally released cytokines and internalize viral proteins produced by infected cells. The first response of the immature dendritic cell is to produce massive quantities of IFN and other cytokines. If viral anti-IFN or antiapoptotic gene products are made, progeny virus particles are released. If the newly infected cells have already bound IFN, protein synthesis is inhibited when viral nucleic acid is produced. Soon thereafter, NK cells recognize the infected cells because of new surface antigens and a low or aberrant display of MHC class I proteins. The IFN produced by infected cells stimulates the NK cells to intensify their activities, which include target cell destruction and synthesis of IFN-γ. In some cases, serum complement can be activated to destroy enveloped viruses and infected cells. In general, the intrinsic and innate defenses bring most viral infections to an uneventful close before the adaptive response is required. Even if all these responses prove insufficient, the immune response still has one powerful trick up its sleeve, as we shall see in the next chapter.

One cannot help but be impressed by such a swift, diverse, coordinated, integrated response. But bear in mind this central fact: no matter how adept our host defense may be at detecting and neutralizing viruses, all successful viruses encode gene products that frustrate their host's defenses, and many cause disease (Chapter 5). In these instances, the struggle has barely begun.

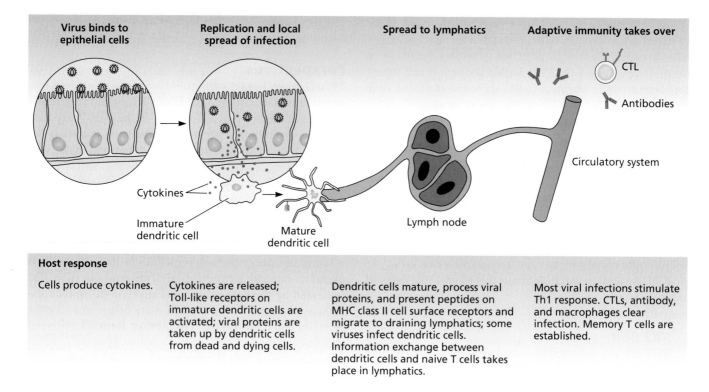

Virus binds to epithelial cells	Replication and local spread of infection	Spread to lymphatics	Adaptive immunity takes over

Cytokines

Immature dendritic cell

Mature dendritic cell

Lymph node

CTL

Antibodies

Circulatory system

Host response

Cells produce cytokines.	Cytokines are released; Toll-like receptors on immature dendritic cells are activated; viral proteins are taken up by dendritic cells from dead and dying cells.	Dendritic cells mature, process viral proteins, and present peptides on MHC class II cell surface receptors and migrate to draining lymphatics; some viruses infect dendritic cells. Information exchange between dendritic cells and naive T cells takes place in lymphatics.	Most viral infections stimulate Th1 response. CTLs, antibody, and macrophages clear infection. Memory T cells are established.

Figure 3.25 Critical events in a hypothetical acute virus infection. CTL, cytotoxic T lymphocyte.

References

Books

Murphy K, Travers P, Walport M. 2007. *Janeway's Immunobiology*, 7th ed. Garland Science, Garland Publishing Inc, New York, NY.

Nathanson N (ed). 2007. *Viral Pathogenesis and Immunity*, 2nd ed. Academic Press, London, United Kingdom.

Parham P. 2009. *The Immune System*, 3rd ed. Garland Science, Garland Publishing Inc, New York, NY.

Reviews

Aaronson DS, Horvath CM. 2002. A road map for those who don't know JAK-STAT. *Science* **296:**1653–1655.

Alsharifi M, Müllbacher A, Regner M. 2008. Interferon type I responses in primary and secondary infections. *Immunol Cell Biol* **86:**239–245.

Bieniasz PD. 2004. Intrinsic immunity: a front-line defense against viral attack. *Nat Immunol* **5:**1109–1115.

Bowie AG. 2008. TRIM-ing down Tolls. *Nat Immunol* **9:**348–350.

Carroll MC. 2004. The complement system in regulation of adaptive immunity. *Nat Immunol* **5:**981–986.

Colonna M, Trinchieri G, Liu YJ. 2004. Plasmacytoid dendritic cells in immunity. *Nat Immunol* **5:**1219–1226.

Coscoy L. 2007. Immune evasion by Kaposi's sarcoma-associated herpesvirus. *Nat Rev Immunol* **7:**391–401.

Cullen BR. 2007. Immunology. Outwitted by viral RNAs. *Science* **317:**329–330.

Cullen BR, Cherry S, tenOever BR. 2013. Is RNA interference a physiologically relevant innate antiviral immune response in mammals? *Cell Host Microbe* **14:**374–378.

Finlay BB, McFadden G. 2006. Anti-immunology: evasion of the host immune system by bacterial and viral pathogens. *Cell* **124:**767–782.

Fire A. 2005. Nucleic acid structure and intracellular immunity: some recent ideas from the world of RNAi. *Q Rev Biophys* **38:**303–309.

García-Sastre A, Biron CA. 2006. Type 1 interferons and the virus-host relationship: a lesson in détente. *Science* **312:**879–882.

Gay NJ, Gangloff M. 2007. Structure and function of Toll receptors and their ligands. *Annu Rev Biochem* **76:**141–165.

Grove J, Marsh M. 2011. The cell biology of receptor-mediated virus entry. *J Cell Biol* **195:**1071–1082.

Hornung V, Hartmann R, Ablasser A, Hopfner KP. 2014. OAS proteins and cGAS: unifying concepts in sensing and responding to cytosolic nucleic acids. *Nat Rev Immunol* **14:**521–528.

Kudchodkar SB, Levine B. 2009. Viruses and autophagy. *Rev Med Virol* **19:**359–378.

Kumar H, Kawai T, Akira S. 2011. Pathogen recognition by the innate immune system. *Int Rev Immunol* **30:**16–34.

Li F, Ding SW. 2006. Virus counterdefense: diverse strategies for evading the RNA-silencing immunity. *Annu Rev Microbiol* **60:**503–531.

Lodoen MB, Lanier LL. 2005. Viral modulation of NK cell immunity. *Nat Rev Microbiol* **3:**59–69.

Luban J. 2007. Cyclophilin A, TRIM5, and resistance to human immunodeficiency virus type 1 infection. *J Virol* **81:**1054–1061.

Medzhitov R. 2007. Recognition of microorganisms and activation of the immune response. *Nature* **449:**819–826.

Mercer J, Helenius A. 2010. Apoptotic mimicry: phosphatidylserine-mediated macropinocytosis of vaccinia virus. *Ann N Y Acad Sci* **1209:**49–55.

Meylan E, Tschopp J, Karin M. 2006. Intracellular pattern recognition receptors in the host response. *Nature* **442:**39–44.

Mossman KL, Ashkar AA. 2005. Herpesviruses and the innate immune response. *Viral Immunol* **18:**267–281.

Orange JS, Fassett MS, Koopman LA, Boyson JE, Strominger JL. 2002. Viral evasion of natural killer cells. *Nat Immunol* **3:**1006–1012.

Saito T, Gale M, Jr. 2007. Principles of intracellular viral recognition. *Curr Opin Immunol* **19:**17–23.

Sarnow P, Jopling CL, Norman KL, Schütz S, Wehner KA. 2006. MicroRNAs: expression, avoidance and subversion by vertebrate viruses. *Nat Rev Microbiol* **4:**651–659.

Selsted ME, and Ouellette AJ. 2005. Mammalian defensins in the antimicrobial immune response. *Nat Immunol* **6:**551–557.

Sen GC. 2001. Viruses and interferons. *Annu Rev Microbiol* **55:**255–281.

Silverman RH. 2007. Viral encounters with 2′-5′-oligoadenylate synthetase and RNase L during the interferon antiviral response. *J Virol* **81:**12720–12729.

Steinman RM. 2000. DC-SIGN: a guide to some mysteries of dendritic cells. *Cell* **100:**491–494.

Towers GJ. 2007. The control of viral infection by tripartite motif proteins and cyclophilin A. *Retrovirology* **4:**40. doi:10.1186/1742-4690-4-40.

Trinchieri G, Sher A. 2007. Cooperation of Toll-like receptor signals in innate immune defence. *Nat Rev Immunol* **7:**179–190.

Classic Papers

Fire A, Xu S, Montgomery MK, Kostas SA, Driver SE, Mello CC. 1998. Potent and specific genetic interference by double-stranded RNA in *Caenorhabditis elegans. Nature* **391:**806–811.

Isaacs A, Lindenmann J. 1957. Virus interference. I. The interferon. *Proc R Soc Lond B Biol Sci* **147:**258–267.

Isaacs A, Lindenmann J, Valentine RC. 1957. Virus interference. II. Some properties of interferon. *Proc R Soc Lond B Biol Sci* **147:**268–273.

Janeway CA, Jr. 1989. Approaching the asymptote? Evolution and revolution in immunology. *Cold Spring Harb Symp Quant Biol* **54:**1–13.

Marcus PI, Sekellick MJ. 1976. Cell killing by viruses. III. The interferon system and inhibition of cell killing by vesicular stomatitis virus. *Virology* **69:**378–393.

Matzinger P. 1994. Tolerance, danger, and the extended family. *Annu Rev Immunol* **12:**991–1045.

Nagano Y, Kojima Y. 1958. Inhibition of vaccinia infection by a liquid factor in tissues infected by homologous virus. *C R Seances Soc Biol Fil* **152:**1627–1629. (In French.)

Seth RB, Sun L, Ea CK, Chen ZJ. 2005. Identification and characterization of MAVS, a mitochondrial antiviral signaling protein that activates NF-κB and IRF 3. *Cell* **122:**669–682.

Stremlau M, Owens CM, Perron MJ, Kiessling M, Autissier P, Sodroski J. 2004. The cytoplasmic body component TRIM5α restricts HIV-1 infection in Old World monkeys. *Nature* **427:**848–853.

Vanderplasschen A, Mathew E, Hollinshead M, Sim RB, Smith GL. 1998. Extracellular enveloped vaccinia virus is resistant to complement because of incorporation of host complement control proteins into its envelope. *Proc Natl Acad Sci U S A* **95:**7544–7549.

Selected Papers

Intrinsic Immunity

Balachandran S, Roberts PC, Brown LE, Truong H, Pattnaik AK, Archer DR, Barber GN. 2000. Essential role of the dsRNA-dependent protein kinase PKR in innate immunity to viral infection. *Immunity* **13:**129–141.

Birdwell CE, Queen KJ, Kilgore PC, Rollyson P, Trutschl M, Cvek U, Scott RS. 2014. Genome-wide DNA methylation as an epigenetic consequence of Epstein-Barr virus infection of immortalized keratinocytes. *J Virol* **88:**11442–11458.

Chahal JS, Qi J, Flint SJ. 2012. The human adenovirus type 5 E1B 55 kDa protein obstructs inhibition of viral replication by type I interferon in normal human cells. *PLoS Pathog* **8:**e1002853. doi:10.1371/journal.ppat.1002853.

Diner BA, Lum KK, Javitt A, Cristea IM. 18 February 2015. Interactions of the antiviral factor IFI16 mediate immune signaling and herpes simplex virus-1 immunosuppression. *Mol Cell Proteomics* doi:10.1074/mcp.M114.047068.

Gitlin L, Barchet W, Gilfillan S, Cella M, Beutler B, Flavell RA, Diamond MS, Colonna M. 2006. Essential role of mda-5 in type I IFN responses to polyriboinosinic:polyribocytidylic acid and encephalomyocarditis picornavirus. *Proc Natl Acad Sci U S A* **103:**8459–8464.

Kaiser SM, Malik HS, Emerman M. 2007. Restriction of an extinct retrovirus by the human TRIM5α antiviral protein. *Science* **316:**1756–1758.

Karikó K, Buckstein M, Ni H, Weissman D. 2005. Suppression of RNA recognition by Toll-like receptors: the impact of nucleoside modification and the evolutionary origin of RNA. *Immunity* **23:**165–175.

Kato H, Takeuchi O, Sato S, Yoneyama M, Yamamoto M, Matsui K, Uematsu S, Jung A, Kawai T, Ishii KJ, Yamaguchi O, Otsu K, Tsujimura T, Koh CS, Reis e Sousa C, Matsuura Y, Fujita T, Akira S. 2006. Differential roles of MDA5 and RIG-I helicases in the recognition of RNA viruses. *Nature* **441:**101–105.

Katz RA, Jack-Scott E, Narezkina A, Palagin I, Boimel P, Kulkosky J, Nicolas E, Greger JG, Skalka AM. 2007. High-frequency epigenetic repression and silencing of retroviruses can be antagonized by histone deacetylase inhibitors and transcriptional activators, but uniform reactivation in cell clones is restricted by additional mechanisms. *J Virol* **281:**2592–2604.

Köck J, Blum HE. 2008. Hypermutation of hepatitis B virus genomes by APOBEC3G, APOBEC3C and APOBEC3H. *J Gen Virol* **89:**1184–1191.

Lee HK, Lund JM, Ramanathan B, Mizushima N, Iwasaki A. 2007. Autophagy-dependent viral recognition by plasmacytoid dendritic cells. *Science* **315:**1398–1401.

Mercer J, Helenius A. 2008. Vaccinia virus uses macropinocytosis and apoptotic mimicry to enter host cells. *Science* **320:**531–535.

Meylan E, Curran J, Hofmann K, Moradpour D, Binder M, Bartenschlager R, Tschopp J. 2005. Cardif is an adaptor protein in the RIG-I antiviral pathway and is targeted by hepatitis C virus. *Nature* **437:**1167–1172.

Okeoma CM, Lovsin N, Peterlin BM, Ross SR. 2007. APOBEC3 inhibits mouse mammary tumour virus replication *in vivo. Nature* **445:**927–930.

Orvedahl A, Alexander D, Tallóczy Z, Sun Q, Wei Y, Zhang W, Burns D, Leib DA, Levine B. 2007. HSV-1 ICP34.5 confers neurovirulence by targeting the Beclin 1 autophagy protein. *Cell Host Microbe* **1:**23–35.

Pedersen IM, Cheng G, Wieland S, Volinia S, Croce CM, Chisari FV, David M. 2007. Interferon modulation of cellular microRNAs as an antiviral mechanism. *Nature* **449:**919–922.

Pichlmair A, Schulz O, Tan CP, Näslund TI, Liljeström P, Weber F, Reis e Sousa C. 2006. RIG-I-mediated antiviral responses to single-stranded RNA bearing 5′-phosphates. *Science* **314:**997–1001.

Reeves MB, Davies AA, McSharry BP, Wilkinson GW, Sinclair JH. 2007. Complex I binding by a virally encoded RNA regulates mitochondrial-induced cell death. *Science* **316:**1345–1348.

Saffert RT, Kalejta RF. 2006. Inactivating a cellular intrinsic immune defense mediated by Daxx is the mechanism through which the human cytomegalovirus pp71 protein stimulates viral immediate-early gene expression. *J Virol* **80:**3863–3871.

Sun L, Wu J, Du F, Chen X, Chen ZJ. 2013. Cyclic GMP-AMP synthase is a cytosolic DNA sensor that activates the type I interferon pathway. *Science* **339:**786–791.

Takaoka A, Wang Z, Choi MK, Yanai H, Negishi H, Ban T, Lu Y, Miyagishi M, Kodama T, Honda K, Ohba Y, Taniguchi T. 2007. DAI (DLM-1/ZBP1) is a cytosolic DNA sensor and an activator of innate immune response. *Nature* **448:**501–505.

Thome M, Schneider P, Hofmann K, Fickenscher H, Meinl E, Neipel F, Mattmann C, Burns K, Bodmer JL, Schröter M, Scaffidi C, Krammer PH, Peter ME, Tschopp J. 1997. Viral FLICE-inhibitory proteins (FLIPs) prevent apoptosis induced by death receptors. *Nature* **386:**517–521.

Vartanian JP, Guétard D, Henry M, Wain-Hobson S. 2008. Evidence for editing of human papillomavirus DNA by APOBEC3 in benign and precancerous lesions. *Science* **320:**230–233.

Wu J, Sun L, Chen X, Du F, Shi H, Chen C, Chen ZJ. 2013. Cyclic GMP-AMP is an endogenous second messenger in innate immune signaling by cytosolic DNA. *Science* **339:**826–830.

Yoneyama M, Kikuchi M, Natsukawa T, Shinobu N, Imaizumi T, Miyagishi M, Taira K, Akira S, Fujita T. 2004. The RNA helicase RIG-I has an essential function in double-stranded RNA-induced innate antiviral responses. *Nat Immunol* **5:**730–737.

Innate Immunity

Brown MG, Dokun AO, Heusel JW, Smith HR, Beckman DL, Blattenberger EA, Dubbelde CE, Stone LR, Scalzo AA, Yokoyama WM. 2001. Vital involvement of a natural killer cell activation receptor in resistance to viral infection. *Science* **292:**934–937.

Collins SE, Noyce RS, Mossman KL. 2004. Innate cellular response to virus particle entry requires IRF3 but not virus replication. *J Virol* **78:**1706–1717.

Dokun AO, Kim S, Smith HR, Kang HS, Chu DT, Yokoyama WM. 2001. Specific and nonspecific NK cell activation during virus infection. *Nat Immunol* **2:**951–956.

Lubinski J, Wang L, Mastellos D, Sahu A, Lambris JD, Friedman HM. 1999. *In vivo* role of complement-interacting domains of herpes simplex virus type 1 glycoprotein gC. *J Exp Med* **190:**1637–1646.

4 Adaptive Immunity and the Establishment of Memory

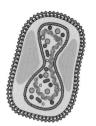

Introduction

Attributes of the Host Response

Speed

Diversity and Specificity

Memory

Self-Control

Lymphocyte Development, Diversity, and Activation

All Blood Cells Derive from a Common Hematopoietic Stem Cell

The Two Arms of Adaptive Immunity

The Major Effectors of the Adaptive Response: B Cells and T Cells

Diverse Receptors Impart Antigen Specificity to B and T Cells

Events at the Site of Infection Set the Stage for the Adaptive Response

Acquisition of Viral Proteins by Professional Antigen-Presenting Cells Enables Production of Proinflammatory Cytokines and Establishment of Inflammation

Antigen-Presenting Cells Leave the Site of Infection and Migrate to Lymph Nodes

Antigen Processing and Presentation

Professional Antigen-Presenting Cells Induce Activation via Costimulation

Presentation of Antigens by Class I and Class II MHC Proteins

Lymphocyte Activation Triggers Massive Cell Proliferation

The Cell-Mediated Response

CTLs Lyse Virus-Infected Cells

Control of CTL Proliferation

Noncytolytic Control of Infection by T Cells

Rashes and Poxes

The Humoral (Antibody) Response

Antibodies Are Made by Plasma Cells

Types and Functions of Antibodies

Virus Neutralization by Antibodies

Antibody-Dependent Cell-Mediated Cytotoxicity: Specific Killing by Nonspecific Cells

Immunological Memory

Perspectives

References

LINKS FOR CHAPTER 4

▶▶❙ *Video: Interview with Dr. Peter Doherty*
http://bit.ly/Virology_Doherty

▶▶❙ *More than one way to skin a virus*
http://bit.ly/Virology_Twiv175

▶▶❙ *Concerto in B*
http://bit.ly/Virology_Twiv161

▶▶❙ *How ZMapp antibodies bind to Ebola virus*
http://bit.ly/Virology_11-25-14

▶▶❙ *Viruses might provide mucosal immunity*
http://bit.ly/Virology_7-2-13

Introduction

There are some who find the study of immunology confounding. It can be dizzying to try to comprehend the many types of T cells, markers, cytokines, and signaling pathways, especially for a student new to the field. For example, immunologists often identify particular immune cell subtypes on the basis of the presence or absence of a panel of proteins. A regulatory T cell may therefore be referred to as "$CD4^+/CD8^-/CD25^+/Foxp3^+$." Such lists of protein markers can be perplexing, but they are simply a means to distinguish one cell population from another, much as humans are distinguished by traits such as hair color, height, and voice. Compounding the challenge of mastering the jargon of immunology are these additional attributes: the immune response is dynamic and dependent on diverse cell-cell interactions, the tissues and cells that produce the host response are scattered throughout the body, and many lymphocytes can morph from one functional state to another during their lifetimes. But this complexity is also one of the most fascinating aspects of immunology: that so many diverse, potent, interacting, and overlapping strategies are in place to thwart pathogenic encounters underscores the importance of a formidable host defense. As many students of immunology appreciate, the more one knows, the more amazing the immune system seems.

In this text, we have divided our discussion of the antiviral response into distinct chapters: the physical barriers to infection (Chapter 2), the cell-intrinsic and innate immunity (Chapter 3), and, in this chapter, adaptive immunity.

These distinctions, however, are not meant to imply that the immune response happens in discrete temporal phases. Indeed, interferons are present and functional throughout the host response, memory T and B cells may be called into play soon after exposure to a pathogen, and neutrophils are prevalent and active at the sites of infection during the peak of the adaptive response. Nevertheless, the ways by which the innate and adaptive arms of host defense recognize and control a virus infection are quite distinct. The innate immune system, which recognizes pathogen-specific properties, provides critical frontline control to limit viral spread, but it is the adaptive response that executes the highly specific assault on virus particles and infected cells. This system is called "adaptive" because it not only differentiates between infected and uninfected cells, but also is tailored to the particular microbe or antigen. Understanding how such precision is achieved was among the most important advances in immunology, and defining which viral proteins are important for eliciting this response remains an essential step in modern vaccine design.

Attributes of the Host Response

Speed

The interval between viral infection and immune-mediated resolution defines the window during which disease may occur. (Exceptions are those virus infections that result in immunopathology, in which case the host response itself is the cause of tissue damage; see Chapter 5). The consequences of infection and the development of immunity are often described as a race against the clock: the virus replication rate, yield, and distribution in the host are pitted against the efficiency of detection and clearance by host defenses. While elements of

---

PRINCIPLES *Adaptive immunity and the establishment of memory*

- The adaptive response is characterized by speed, antigen specificity, memory, and self control.

- The degradation of "foreign" proteins (e.g., viral proteins) by professional antigen presenting cells such as dendritic cells, and their presentation to naive lymphocytes, are the critical steps that bridge the innate and the adaptive responses.

- Activation of tissue-resident dendritic cells causes them to leave the site of infection and migrate to lymphoid tissues, where naïve T and B cells are found.

- The cell-mediated response (chiefly, T cells) facilitates recovery from a viral infection primarily because it eliminates virus-infected cells without damaging uninfected cells.

- The humoral immune response (chiefly, antibodies produced by B cells) contributes to antiviral defense by binding to, and causing the elimination of, free virus particles.

- Viral peptides can be presented on the cell surface in the groove of either a class I or a class II major histocompatibility complex protein. Class I proteins present internally synthesized antigens; class II MHC proteins present antigens that were phagocytosed.

- Two primary types of T cells exist: $CD4^+$ T cells and $CD8^+$ T cells.

- $CD4^+$ T cells interact with MHC class II-expressing cells (including professional antigen-presenting cells and B cells), and synthesize cytokines and growth factors that stimulate ("help") the specific classes of lymphocytes with which they interact.

- $CD8^+$ T cells, also called cytotoxic T lymphocytes (CTLs), interact with antigen presented in the context of MHC class I proteins; when productively engaged, CTLs can destroy the cell presenting the peptides.

- Once a specific adaptive response has been established and the viral infection is resolved, the individual is immune to subsequent infection by the same pathogen; this is the core principle of vaccination.

intrinsic and innate immunity usually keep the virus in check during the critical early days following infection, the subsequent massive clonal expansion of antigen-specific T and B cells is often the fatal blow to a virus infection: individuals with mutations that affect T or B cell function fare poorly following most infections. The conversion of a small number of quiescent, naive lymphocytes into a mob of activated, cytokine-producing effector cells over such a short period is a marvel of cell biology. For example, CD8$^+$ T cells specific for the lymphocytic choriomeningitis virus divide as many as 20 times following infection of the host, resulting in up to a 50,000-fold increase in total number in just a few days.

Rapid induction of the adaptive response is also facilitated by the presence of lymph nodes throughout the body. These "immunological chat rooms" are the sites at which antigen-presenting cells, transported in the lymph, encounter naive lymphocytes that circulate in the blood. Lymph nodes are located strategically near areas of the body that are sites of virus entry, including the respiratory and gastrointestinal tracts, thereby minimizing the distance an antigen-presenting cell must travel from the tissue. The intersection between lymphatic circulation and the blood that occurs in the lymph node increases the probability that a dendritic cell that presents viral antigens will find the appropriate naive T lymphocytes, and, as a consequence, accelerates the activation and amplification of antigen-specific T and B cells.

Diversity and Specificity

Naive T cells specific for every possible pathogen circulate in all humans: while your chances of contracting Ebola virus are vanishingly small, rest assured that there are naive T cells capable of recognizing Ebola virus antigens circulating in you now, at the ready. Most of these naive cells will never encounter their cognate antigen, and consequently their numbers will never be increased. But the process of generating this astounding diversity is one of the more interesting properties of immune cell development. T and B cells possess receptors on their surface that can recognize small portions of a viral protein (or three-dimensional facets of a protein) termed **epitopes**. Receptor diversity accounts for the capacity of these cells to recognize and respond to virtually any pathogen: for example, it has been estimated that there are >20 million distinct T cell specificities. As T and B cell receptors are encoded by host genes (of which humans have only ~30,000), it is not possible that each of the millions of T and B cell receptors are encoded by a discrete gene. Rather, random, somatic rearrangements of a limited number of segments of lymphocyte receptor genes create many putative receptors that then pass through a process of quality control before release into the circulation. An interesting consequence of this stochastic process is that everyone's T and B cell repertoire is distinct, even among closely related individuals. The abundance and diversity of the immune repertoire may explain why otherwise healthy individuals may respond differently to an encounter with the same pathogen.

Memory

Once T and B cells have become activated by interaction with a cognate antigen, a small number are retained as memory cells, equivalent to an idling car stopped at a red light. While these cells do not produce the effector functions of the majority of the activated cells, they are poised to reenter the cell cycle rapidly and to expand clonally immediately upon reexposure to their cognate antigen (the "green light"). At the next encounter with the virus, preservation of memory cells dramatically skews the immune response race in favor of the host: while the initial conversion of naive to activated cells requires multiple steps and a few days, the amplification of memory cells begins almost immediately. Consequently, subsequent infections with the same agent are met with a robust and highly specific defense that usually stops the infection as soon as it starts, with minimal reliance on the innate response. As discussed later in this chapter, and in Chapter 8, this property is the basis of vaccination.

Self-Control

Mounting an immune defense results in the production of large quantities of cytokines and expansion of immune cells that can make the host quite ill: most of the unpleasant symptoms of infections (fever, muscle aches) result from the host response (e.g., fever-inducing cytokines) rather than from the virus itself. Therefore, once the pathogen has been vanquished, this response must be blunted quickly to avoid further risk to the host. Processes that are intrinsic to activated lymphocytes ensure their demise within a short period after activation: the life of an activated immune cell is exciting, but brief. The need to dampen an activated immune response has been explored in detail only recently, but it is clear from these studies that turning off the antiviral cascade is as important as turning it on.

Lymphocyte Development, Diversity, and Activation

Some knowledge of developmental immunology will be useful as we explore how lymphocytes recognize antigens presented by infected cells or by professional antigen-presenting cells (Box 4.1). In this section, we introduce the developmental origins of the cells in the blood, describe how lymphocyte receptor diversity is generated, and introduce the two major players in the adaptive response: T cells and B cells.

All Blood Cells Derive from a Common Hematopoietic Stem Cell

All cells in the blood are derived from a common lineage, originating with a multipotential hematopoietic stem cell

BOX 4.1

TERMINOLOGY
Pathogens, antigens, and epitopes

T and B lymphocytes do not recognize complete virus particles, but rather small, linear pieces of a viral protein (generally true for T cells) or three-dimensional facets of folded viral proteins (generally true for B cells and antibodies). However, the terminology to distinguish what these lymphocytes recognize can be confusing. A **pathogen** is a microbe, such as a virus particle, that can cause disease (hence, "pathology"). Not all microbes are pathogens: the normal gut flora includes many types of bacteria, but in most cases, these microbes do not make the host sick or they contribute positively to the host's welfare. Proteins made by pathogens that are capable of inducing a host immune response are called **antigens**. The term was originally derived from "antibody-generating" proteins, although antigens can also be bound by T cells. Moreover, antigens can be DNA, polysaccharides, or lipids. An **epitope** is the portion of the antigen that is bound by an antibody or that is recognized by a T cell receptor. Consequently, a protein antigen (for example, the measles virus hemagglutinin protein) may have multiple epitopes to which different T cells and antibodies can bind.

---

that differentiates into two discrete progenitor populations, the common myeloid stem cell and the common lymphoid stem cell. These two precursors subsequently give rise to all blood cell types (Fig. 4.1). The parental hematopoietic stem cells reside in the bone marrow and are self-renewing. Lymphocyte differentiation is marked by an orchestrated loss of stem cell-specific proteins and a concomitant acquisition of those that are characteristic of fully differentiated leukocytes (Box 4.2). Although new immune cells are generated throughout life, the rate of production declines with age, a property that is generally considered a major contributor to the greater vulnerability of the elderly to infection. Given the common ancestry of all blood cells, the process of distinguishing among them was a notable challenge for early immunologists. The advent of flow cytometry, combined with the development of fluorescent antibody reagents specific for leukocyte signature proteins, were crucial technical accomplishments that enabled immunologists to define the functional contributions of the diverse family of blood cells.

The Two Arms of Adaptive Immunity

The adaptive response comprises two complementary actions, the **humoral response** (B cells and the antibodies they produce) and the **cell-mediated response** (helper and effector T cells) (Fig. 4.2). As we discuss the features of these lymphocytes and the processes that characterize each component, it is important to understand that **both** are essential in antiviral defense, and they function in concert. In general, antibodies bind and inactivate virus particles in the bloodstream and at mucosal surfaces, whereas T cells recognize and kill infected cells, the "factories" that generate new virus particles. The relative contribution of each in any given infection varies with the nature of the virus, as well as with host parameters including age, organs infected, and previous immunological exposures.

The Major Effectors of the Adaptive Response: B Cells and T Cells

B Cells

B cells develop in the bone marrow. As they mature, each synthesizes an antigen receptor, which is a membrane-bound antibody. Antigen binding in a specific manner to a membrane-bound antibody on an immature B cell initiates a signal transduction cascade. As a consequence, new gene products are made and the cell begins to divide rapidly. The daughter cells produced by each division differentiate into effector **plasma cells** and a small number of memory B cells.

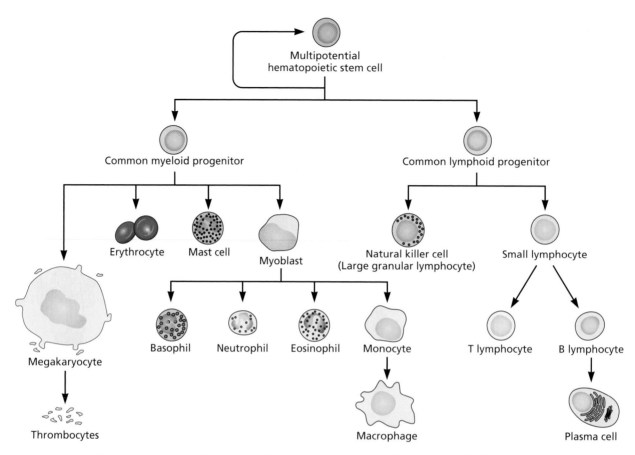

Figure 4.1 Development of leukocytes from a common stem cell precursor. All cells within the blood are derived from a common stem cell precursor, termed the multipotential hematopoietic stem cell. This self-renewing cell population, which exists in the bone marrow, generates two additional precursors. The common myeloid progenitor can differentiate further into red blood cells (erythrocytes), mast cells, and myeloblasts, which give rise to basophils, neutrophils, eosinophils, and monocytes. The common lymphoid progenitor differentiates into natural killer cells, T lymphocytes, and B lymphocytes.

BOX 4.2

TERMINOLOGY
Leukocytes and lymphocytes

While these names sound similar, lymphocytes and leukocytes are not synonymous. "Leukocyte" is a general term for a white blood cell, and includes lymphocytes, neutrophils, eosinophils, and macrophages. Lymphocytes are a subset of leukocytes, specifically T and B cells and NK cells, that possess variable antigen-detecting cell surface receptors (the T cell receptor and the B cell receptor).

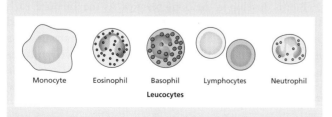

As their name implies, memory B cells, and their clonal progeny, are long-lived and continue to produce the parental, membrane-bound antibody receptor. In contrast, plasma cells live for only a few days and no longer make membrane-bound antibody, but instead synthesize the same antibody in secreted form. These antibodies can bind and inactivate extracellular pathogens, including virus particles. A single plasma cell can secrete >2,000 antibody molecules per second. Like dendritic cells, the B cell is an antigen-presenting cell that uses the major histocompatibility complex (MHC) class II system and exogenous antigen processing.

T Cells

T cell precursors are also produced in the bone marrow, but in contrast to those of B cells, a T cell precursor must migrate to the thymus gland to mature; hence the "T" in "T cell." The thymus gland is located in the thoracic cavity, above the heart. Subsets of T cells have distinct functions.

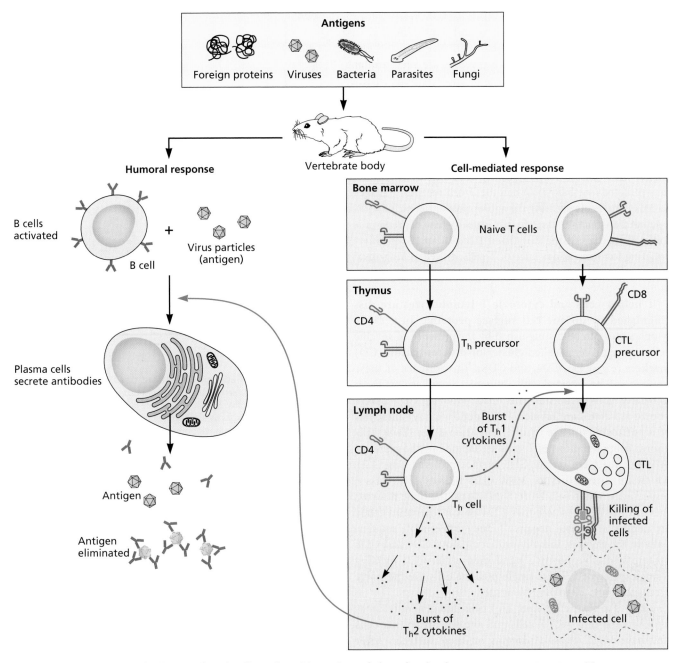

Figure 4.2 The humoral and cell-mediated branches of the adaptive immune system. A variety of foreign proteins and particles (antigens) may stimulate adaptive immune responses after recognition by intrinsic and innate defense systems. (Left) The humoral branch comprises lymphocytes of the B cell lineage, which produce antibodies, the important effector molecules of this response. The process begins with the interaction of a specific receptor on precursor B lymphocytes with an antigen. Binding of antigen promotes differentiation into antibody-secreting cells (plasma cells). (Right) The cell-mediated branch comprises lymphocytes of the T cell lineage that arise in the bone marrow and are differentiated further in the thymus. The activation process is initiated in lymph nodes when the T cell receptor on the surface of naive T lymphocytes binds viral peptides that are in a complex with MHC class II protein on the surface of dendritic cells or B cells. Two subpopulations of naive T cells are illustrated: the T helper (T_h)-cell precursor and the cytotoxic T lymphocyte (CTL) precursor. The T_h cell recognizes antigens bound to MHC class II molecules and produces cytokines that "help" activated B cells to differentiate into antibody-producing plasma cells (T_h2 cytokines), or CTL precursors (T_h1 cytokines) to differentiate into CTLs capable of recognizing and killing virus-infected cells. The T_h1 or T_h2 cytokines are produced by different subsets of T_h cells and promote or inhibit cell division and gene activity of B cell or CTL precursors.

The T cell receptor is a disulfide-linked heterodimer composed of either α and β or γ and δ protein chains. The peptide-binding site of the T cell receptor and the epitope-binding site of the B cell receptor are very similar structures, formed by the folding of three regions in the amino-terminal domains of the proteins that participate in epitope recognition (the so-called hypervariable regions). However, unlike the B cell receptor, which can recognize the epitope as part of an intact folded protein, the T cell receptor can recognize **only** a peptide fragment produced by proteolysis. Furthermore, the peptide must be bound to MHC cell surface proteins (see below). When the T cell receptor engages an MHC molecule carrying the appropriate antigenic peptide, a signal transduction cascade that leads to gene expression is initiated. As a result, the stimulated T cell is capable of differentiating to form various effector T cells, as well as long-lived memory cells.

T helper cells and cytotoxic T lymphocytes are distinguished by unique cell surface proteins. In general, lymphocytes can be distinguished by the presence of specific cell surface proteins called **cluster-of-differentiation (CD) markers** (e.g., CD3, CD4, and CD8). The presence of these proteins can be detected with antibodies raised against them in heterologous organisms. The >350 individual CD markers known (to date!) are invaluable in identifying lymphocytes of a particular lineage or differentiation stage. Two well-known subpopulations of T cells are defined by the presence of either the CD4 or the CD8 surface proteins (Fig. 4.3), which are coreceptors for MHC class II and MHC class I, respectively. When immature T cells leave the bone marrow, they do not synthesize either CD4 or CD8 proteins (they are said to be "double-negative"). They differentiate sequentially in the thymus, initially producing both CD4 and CD8 proteins ("double-positive") and then either CD8 or CD4 ("single-positive"). These single-positive cells are the naive T cells that migrate to peripheral sites.

CD4$^+$ T cells, or T helper (T$_h$) cells, are generally capable of interacting with B cells and antigen-presenting cells that have MHC class II proteins on their surfaces. After such interactions, CD4$^+$ T$_h$ cells mature into T$_h$1 or T$_h$2 cells (see below). T$_h$ cells synthesize cytokines and growth factors that stimulate ("help") the specific classes of lymphocytes with which they interact. **CD8$^+$ T cells** differentiate into cytotoxic T lymphocytes (CTLs) that can interact with MHC class I proteins, found on almost all cells of the body. CTLs recognize foreign peptides bound to MHC class I proteins and, when productively engaged, destroy the cell presenting the peptides. Mature cytotoxic T cells play important roles in eliminating virus-infected cells from the body by cell lysis and by production of interferon γ (IFN-γ) and tumor necrosis factor α (Tnf-α).

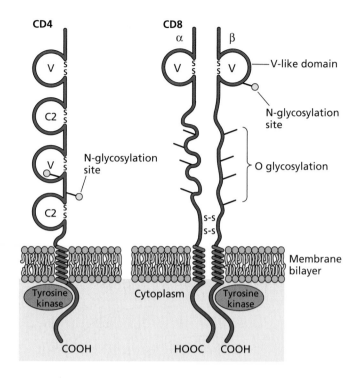

Figure 4.3 Simplified representations of CD4 and CD8 coreceptor molecules. These two molecules associate with the T cell receptor on the surface of T cells. The CD4 molecule is a glycosylated type 1 membrane protein and exists as a monomer in membranes of T cells. It has four characteristic immunoglobulin-like domains labeled V and C2. The V domains in the tertiary structure are similar to the variable domain of immunoglobulin. The first two domains form a binding site for a region on MHC class II proteins. The cytoplasmic domain interacts with specific tyrosine kinases, endowing CD4 with signal transduction properties. The CD8 molecule is a type 1 membrane protein with both N- and O-glycosylation sites. It is a heterodimer of an α chain and a β chain covalently linked by disulfide bonds, and interacts with a region on MHC class I proteins. The two polypeptides are quite similar in sequence, each having an immunoglobulin-like V domain thought to exist in an extended conformation. Tyrosine kinases also associate with the CD8 cytoplasmic domain and participate in signal transduction reactions.

T$_h$1 and T$_h$2 cells. Recall that tissue-resident antigen-presenting cells, usually dendritic cells, engulf antigens at the site of pathogen entry and then move to local lymph nodes to present these antigens to naive T cells. When naive T$_h$ cells engage mature dendritic cells in lymphoid tissue, cytokines and receptor-ligand interactions stimulate the T cell to differentiate into one of two T$_h$ cell types, called T$_h$1 and T$_h$2 (Fig. 4.4). These two cell types can be distinguished by the cytokines they produce and the processes they invoke. T$_h$1 cells are important for controlling most, but not all, viral infections; such cells promote the cell-mediated response by stimulating the maturation of cytotoxic-T cell precursors. T$_h$1 cells accomplish this, in part, by producing interleukin-2 (IL-2) and IFN-γ, cytokines that stimulate inflammation (the

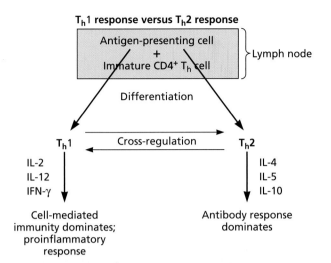

Figure 4.4 The T$_h$1 versus the T$_h$2 response. Immature CD4$^+$ T$_h$ cells differentiate into two general subtypes called T$_h$1 and T$_h$2, defined functionally according to the cytokines they secrete. T$_h$1 cells produce cytokines that promote the inflammatory response and activity of cytotoxic T cells, and T$_h$2 cells synthesize cytokines that stimulate the antibody response. The cytokines made by one class of T$_h$ cell tend to suppress production of those of the other class.

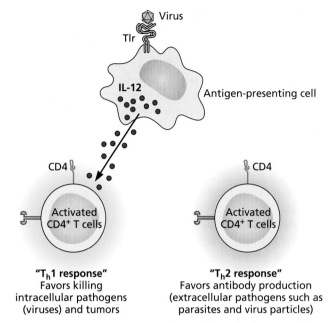

Figure 4.5 Interleukin-12 skews the T cell response toward a T$_h$1 profile. Engagement of Toll-like receptors (Tlrs) on the antigen-presenting cell surface drive the expression of IL-12, which promotes a T$_h$1 T cell response.

proinflammatory response). In addition, T$_h$1 cells provide stimulating cytokines to the antigen-presenting dendritic cell so that it can communicate with naive CD8$^+$ T cells. If IL-12 is secreted by antigen-presenting cells at the time of antigen recognition, immature T$_h$ cells differentiate into T$_h$1 cells. IL-12 also stimulates natural killer (NK) and T$_h$1 cells to secrete IFN-γ, thereby increasing the activity of macrophages at sites of inflammation (Fig. 4.5).

In the presence of IL-4, perhaps secreted by innate immune cells such as NKT cells (Chapter 3), immature T$_h$ cells differentiate into T$_h$2 cells, which stimulate the antibody response rather than the cell-mediated, proinflammatory response. T$_h$2 cells promote the antibody response by inducing maturation of immature B cells and resting macrophages. They also reduce the inflammatory response by producing IL-4, IL-6, and IL-10, but not IL-2 or IFN-γ. T$_h$2 cells are more active after invasion by extracellular bacteria or multicellular parasites. Nevertheless, the T$_h$2 response is critical for controlling infections that result in accumulation of large quantities of virus particles in the blood.

In general, T$_h$1 and T$_h$2 responses coexist in a carefully orchestrated balance: as one increases, the other decreases (Fig. 4.4). While IFN-γ turns up the T$_h$1 response, it also inhibits the synthesis of IL-4 and IL-5 by T$_h$2 cells. On the other hand, production of T$_h$2 cytokines is an important mechanism to shut off the proinflammatory and potentially dangerous T$_h$1 response. An added complexity is that these two cell populations are distinguished only by the types of cytokines they secrete, not by particular cell surface receptors. Many immunologists believe that an individual T cell's subtype may vary based on the tissue and cytokine environment in which it exists.

How a particular type of pathogen triggers synthesis of interleukins that skew the T helper response toward either a T$_h$1 or T$_h$2 profile remains unknown, but one idea is that mature dendritic cells produce proinflammatory cytokines (e.g., IFN-γ) as their default pathway, poised to activate a T$_h$1 response unless appropriate T$_h$2 signals are provided. An alternative view is that when dendritic cells detect CpG sequences, single-stranded RNA, or double-stranded RNA via their Toll-like receptors, nuclear factor-κb (Nf-κb) is activated and T$_h$1 cytokine genes are transcribed.

We know that many viral proteins modulate the T$_h$1-T$_h$2 balance in interesting ways. For example, infection of B cells by Epstein-Barr virus and equine herpesvirus 2 should stimulate an active T$_h$1 response. However, both viral genomes encode proteins homologous to IL-10, a regulatory cytokine that represses the T_h1 response. Viral IL-10 foils the T_h1 antiviral defense that would kill infected B cells, while promoting their differentiation into memory B cells that are important for long-term survival of the viral genome. Measles virus, which can infect antigen-presenting cells such as macrophages, may blunt production of IL-12, a crucial driver of a T_h1 response (Chapter 5). These properties identified with cells in culture are borne out in human studies: measles virus-infected patients have large quantities of IL-10 in their serum, indicative of a skewed T_h1-T_h2 balance. This shift from a (protective) T_h1 response to a less appropriate T_h2 response may, in part, account for the transient immunosuppression associated with measles virus infection and mortality.

For most viral infections, a given T_h response represents a spectrum of some T_h1 and some T_h2 cells, and consequently a mixture of cytokines. Establishment of the proper repertoire of T_h cells is therefore an important early event in host defense; an inappropriate response has far-reaching consequences. For example, synthesis of the T_h2 cytokine IL-4 by an attenuated strain of mousepox virus resulted in lethal, uncontained spread of this virus in vaccinated animals. As the design of potent and effective vaccines depends on stimulating the appropriate spectrum of response, understanding how this balance of cytokines is achieved has direct therapeutic implications.

T_h17 cells. In the past decade, a new class of CD4⁺ helper cells that plays a central role in control of the inflammatory response was identified (Fig. 4.6). These cells are found in the skin and in the lining of the gastrointestinal tract and at other interfaces between the external and internal environments. When dendritic cells present antigens to them in the presence of transforming growth factor β (Tgf-β) and IL-6, these T cells secrete IL-17 and IL-21. In addition, the stimulated T_h17 cells now produce the receptor for IL-23, which leads to their massive proliferation. Such activated cells stimulate a strong inflammatory response, secrete defensins, and recruit neutrophils to the site of activation. T_h17 cells are probably important in the control of bacterial infections, as hosts that lack these cells are susceptible to opportunistic infections. However, because these cells are potent inducers of inflammation, they can exacerbate autoimmune diseases that lead to chronic inflammation, including psoriasis, Crohn's disease, multiple sclerosis, and rheumatoid arthritis. Their importance in controlling viral infections is only now being understood. For example, individuals with large numbers of T_h17 cells in their gut mucosa appear to be able to control human immunodeficiency virus type 1 infections much better than individuals with reduced numbers of these lymphocytes. T_h17 cells are also important in the fatal central nervous system infection caused by arboviruses.

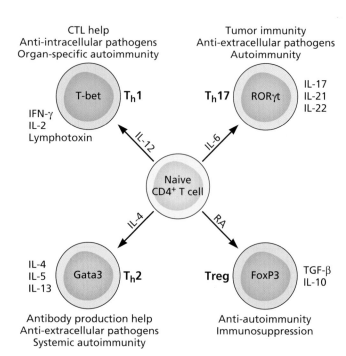

Figure 4.6 Differentiation of T helper subsets. T cell subset differentiation is modulated by cytokines released from dendritic cells and other immune cells. T_h1 cells are induced in the presence of IL-12 via induction of the transcriptional activator T-bet, and aid the resolution of intracellular pathogens. T_h1 cells are also the primary effectors of autoimmunity. T_h2 cells promote production of antibodies, contributing to the clearance of extracellular pathogens and systemic autoimmunity, and are induced by IL-4 and the transcription factor Gata3. T_h1 and T_h2 differentiation is inhibited by Tgf-β. Both Treg and T_h17 cells are induced by Tgf-β. In addition, Tregs require RA, and T_h17 cells require IL-6. Treg cells make the transcription activator forkhead box protein 3 (Foxp3) and secrete the anti-inflammatory cytokines Tgf-β and IL-10, which can suppress autoimmunity and immune responses to pathogens. T_h17 cells synthesize the transcriptional regulator Rorc2/Rorγt (humans/mice) and contribute to defense against extracellular pathogens, tumor immunity, and autoimmunity. RA, retinoic acid.

Regulatory T cells. The regulatory T cell (Treg) subset of T cells (once called suppressor T cells) has been recognized for some time, but their importance in controlling antiviral immunity has become a subject of intense study only recently. Tregs are pivotal players in the end-stage immune response to most, if not all, infectious agents. Their primary function is to terminate the response and return the immune system to a quiescent state (Fig. 4.7). As noted in the discussion of critical attributes of the adaptive response, curtailing an aggressive antiviral response is needed to minimize immunopathology.

Treg cells are important for immune suppression, self-tolerance, and control of the inflammatory response. These cells serve to maintain a balance between protection and immune pathology, but in some cases, Treg action also may limit the effectiveness of vaccines because they shut down the immune response.

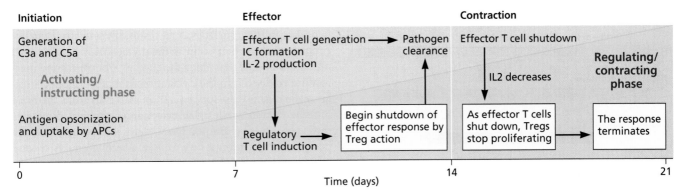

Figure 4.7 Regulation of the T cell response by complement and regulatory T cells. The magnitude and duration of the adaptive immune response is controlled in part by regulatory T cells (Treg cells). This model of an acute viral infection provides a view of how complement (part of the innate immune response; Chapter 3) may regulate the three phases of a T cell-mediated response. (Initiation) Soon after infection, antigen-presenting cells (APCs) take up viral proteins and make their way to local lymph nodes, where the T cell response is initiated. The complement cascade stimulated at the site of infection produces a variety of effector proteins, including C3a, C5a, and C3b. The C3b opsonin facilitates the uptake of C3b-coated antigens by APCs, while C3a and C5a stimulate their maturation. (Effector) Mature APCs then engage potential effector T cells in lymph nodes, resulting in production of IL-2 and activation of Treg cells. Ligands for CD46 include C3b-opsonized immune complexes (IC formation). Effector CTLs and T_h1 cells are stimulated by the APCs and leave the lymph node to resolve the infection at sites of virus reproduction. The balance between activated CTLs and Treg cells determines the extent of CTL action as well as the degree of immunopathology. Too many CTL cells can cause damage, but too few cannot clear the viral infection; conversely, too many Treg cells shut down the effector response prematurely, while too few Treg cells promote continued CTL action and potential immunopathology. (Contraction) The relative dynamics of CTL and Treg cell proliferation promote the controlled contraction of the CTL response. Both CTLs and Treg cells rapidly decline in numbers at this stage. The contraction occurs in part because CD46-stimulated Treg cells divide more quickly than CTLs, and through the action of Treg cytokines, the CTL response shuts down. Because the activated CTLs and T_h cells produce IL-2 necessary for Treg cell replication, the pool of Treg cells then diminishes as the system returns to its unstimulated state. During this phase, memory CTL and memory Treg cells are also produced. Adapted from C. Kemper and J. P. Atkinson, *Nat Rev Immunol* 7:9–18, 2007, with permission.

Diverse Receptors Impart Antigen Specificity to B and T Cells

Like the innate response, the adaptive response must distinguish infected from uninfected cells. However, this feat is accomplished in a markedly different fashion than in the innate immune system. Highly specific molecular recognition is mediated by two antigen receptors. Membrane-bound antibody on B cells and the T cell receptor on T lymphocytes both bind foreign antigens, but they do so in different ways. The B cell receptor engages discrete epitopes in intact proteins. In contrast, the T cell receptor binds short, linear peptides derived from proteolytically processed proteins, presented in the context of a class I MHC protein. The binding to an epitope has profound effects on the lymphocyte bearing that receptor: the T or B lymphocyte may respond by producing cytokines, entering a period of rapid cell division, killing the cell that bears the foreign protein or peptide, or synthesizing antibodies. The events catalyzed by epitope binding comprise the adaptive immune response.

The diversity of the B and T cell receptors is generated during the process of differentiation into mature naive cells in the bone marrow (for B cells) or the thymus (for T cells).

Much is known about how such receptor diversity is generated. In brief, each transmembrane cell surface receptor possesses a constant (C) region that transduces critical signals following antigen engagement and a variable (V) region that engages the epitope for which it is specific. The diversity of these receptors enables the lymphocyte to distinguish among an extraordinary number of potential epitopes.

During the development of the individual T and B cell, DNA rearrangements generate this diversity. The genetic locus of the variable domain of T and B cell receptors comprises three main protein-coding regions, variable (V), diversity (D), and joining (J), each of which comprises many small modules. As T and B cells develop, DNA rearrangements occur in this region of the genome, in which a given allele from each of the V, D, and J regions is randomly selected, and the selected alleles for that particular lymphocyte are then spliced together during DNA recombination (Fig. 4.8). Because the DNA splicing event is inherently imprecise, additional nucleotides may be added or removed from each splice junction. From a limited number of modules, it is possible to generate extraordinary diversity: for example, the estimated total number of antibody specificities in a human is 10^{11}.

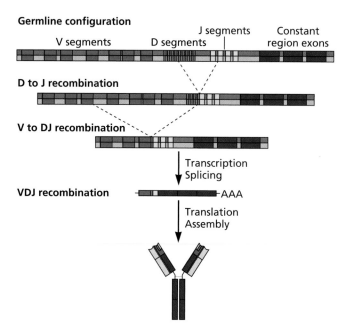

Germline configuration

V segments   D segments   J segments   Constant region exons

D to J recombination

V to DJ recombination

Transcription
Splicing

VDJ recombination   AAA

Translation
Assembly

Figure 4.8 Generation of receptor diversity. The T and B cell receptor alleles in developing lymphocytes comprise small modules, called segments, that are clustered into three regions: variable (V), diversity (D), and joining (J). During development, the DNA of each lymphocyte undergoes recombination in which a module from each region is selected and spliced with a selected module from another region. D-to-J splicing occurs first, followed by V-to-DJ splicing. Incorporation of random nucleotides at these splice junctions further contributes to receptor diversity.

A large number of enzymes that mediate DNA cleavage, nucleotide additions, and ligation participate in this process, including the recombinase-activating genes, or RAGs. Many laboratories that study viral pathogenesis use RAG knockout mice to determine the contributions of the adaptive immune response, because mice lacking RAG genes cannot recombine their T and B cell receptors. As a result, lymphocyte development is blocked, and the mice are unable to mount defenses to many viral challenges, despite possessing intact intrinsic and innate immune responses.

Because the process of shuffling and recombining these sequence modules is random (in terms of which alleles are chosen, and the extent of the interjunctional nucleotide additions or deletions), many of the pre-T cells and B cells that are generated are flawed because they encode proteins that comprise rearrangements that do not form functional receptors or, worse, that form receptors that can bind to **host**-encoded epitopes. Such host-specific lymphocytes could lead to autoimmune diseases. Consequently, two concurrent quality control processes, **positive selection** and **negative selection**, weed out T cells and B cells with dysfunctional receptors. Positive selection allows the survival of those T cells that can bind appropriate surface molecules via the rearranged T cell receptor to survive, whereas negative selection efficiently kills those

T cells that recognize target cells displaying host (or "self") peptides on their surfaces. As a result of these strict quality control processes that occur in the thymus, only 1 to 2% of all immature T cells that enter the thymus from the bone marrow are released into the circulation as mature T cells. Similar mechanisms exist for generation of B cells from the bone marrow. Of note, while a dual selection mechanism eliminates most autoreactive lymphocytes, many still escape into the peripheral circulation. As we will see, a second quality control checkpoint, costimulation, reduces the risk that such circulating, naive lymphocytes will become activated against host tissues.

Events at the Site of Infection Set the Stage for the Adaptive Response

At the conclusion of Chapter 3, we summarized the preexisting intrinsic processes that are deployed following infection, the critical contributions of interferons in restricting viral reproduction, and the influx of innate immune cells to the site of viral reproduction that limit viral spread. But at this stage, all the action is occurring at the site of the infection; rather literally, "naive" T cells wait in lymph nodes or circulate in the blood, unaware that a virus has entered the host. How, then, are T cells in the blood and lymph tissues alerted to an infection so that adaptive immunity can be initiated? Bridging this divide is one of the critical jobs of the antigen-presenting cell.

Acquisition of Viral Proteins by Professional Antigen-Presenting Cells Enables Production of Proinflammatory Cytokines and Establishment of Inflammation

A consequence of local innate defense is that tissue-resident myeloid cells, which include dendritic cells, engulf remnants of dying cells and virus particles, a process called **phagocytosis** ("phago-": "to devour"). Phagocytosis is a specific form of endocytosis that leads to the vesicular internalization of cellular debris, bacteria, and nutrients. This process is initiated when dendritic cells are activated by Toll-like receptor attachment to microbe-associated molecular patterns on infected cells (Chapter 3), although nonspecific activation may occur following tissue damage or local inflammation. Dendritic cell maturation triggers Nf-κb activation (Fig. 3.9), which, among other functions, results in the induction of the actin-myosin contractile system that is required for phagocytosis (Fig. 4.9). When a dendritic cell ingests a virus particle or a portion of a dying cell that contains the virus, the pathogen becomes trapped in an intracellular phagosome, which then fuses with a lysosome. Enzymes within these specialized vesicles digest viral proteins. Immature B cells and cells of the monocyte lineage (e.g., macrophages) are also considered to be professional antigen-presenting cells. Importantly, phagocytosis by these cells is about more than cleaning up debris: the degradation

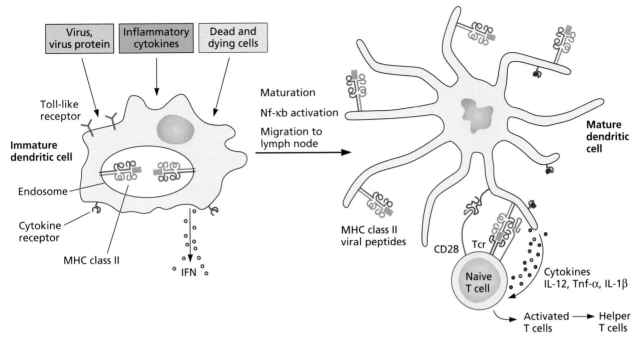

Figure 4.9 Dendritic cells provide cytokine signals and packets of protein information to naive T cells.
Immature dendritic cells actively take up extracellular proteins by endocytosis and store the proteins internally. They do not present MHC class II proteins on their surfaces. Binding of ligands to the Toll-like receptors or cytokine receptors induces differentiation into mature dendritic cells. These cells no longer have the capacity for endocytosis of proteins and display a new repertoire of cell surface receptors. Some of these are chemokine receptors that enable the dendritic cell to migrate to local lymph nodes. The proteins ingested by the immature cell are now processed into peptides and loaded onto MHC class II proteins for subsequent transport to the cell surface. Mature cells extend long dendritic processes to increase the surface area for binding of naive T cells in the lymph node. Mature dendritic cells release proinflammatory cytokines to stimulate T cell differentiation. Naive but antigen-specific T cells bind to the MHC class II-peptide assemblies via recognition of specific peptides by their T cell receptors. The interaction is strengthened by the presence of increased concentrations of costimulatory ligands (e.g., CD28) on the mature dendritic cell. The T cell is activated, begins the maturation process into its final effector state, and moves out of the lymph node into the circulation.

of viral proteins and their presentation to naive lymphocytes is the critical bridge between the innate and the adaptive responses.

The Inflammasome and Cytokine Release

In addition to phagocytosis, dendritic cell activation triggers two additional reactions: cytokine release and migration to a local lymph node. Initiation of the inflammatory response is held under strict control by many regulatory proteins, as an overexuberant or inappropriately triggered proinflammatory response may set in motion a cascade of cytopathic events. Important checkpoints that require multiple, independent switches to be engaged to result in a "go" signal help to guard against faulty or premature induction of these powerful pathways.

One such checkpoint is a cytoplasmic protein complex in antigen-presenting cells called the "inflammasome." This multiprotein assembly of >700 kDa links the sensing of microbial products with the activation of proinflammatory cytokines (Fig. 4.10). When a microbe-associated molecular pattern is engaged by a Toll-like receptor, a signal transduction cascade results in the activation of Nf-κb, and the synthesis of two precursor cytokine molecules, pro-IL-1β and pro-IL-18, is triggered. These precursor proteins must be cleaved for release from the dendritic cell as functional cytokines. Cleavage is achieved via a second signal, the inflammasome, which results from pathogen-nonspecific stimuli such as potassium release or elevated intracellular reactive oxygen species. The induction of the inflammasome leads to the synthesis of caspases that cleave the pro- forms of IL-1β and IL-18 to create secreted and functional cytokines. Secretion of these potent **interleukins** (so named because they enable communication among leukocytes) catalyzes production of a number of proinflammatory cytokines and chemokines, resulting in the further recruitment of immune cells to the site of infection.

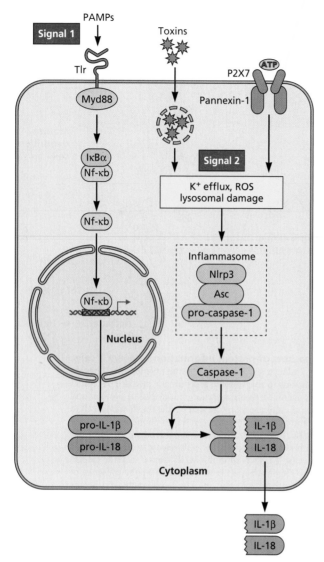

Figure 4.10 The inflammasome. The best characterized inflammasome is the Nlrp3 inflammasome which comprises the Nlrp3 protein (<u>N</u>acht, <u>L</u>rr, and <u>P</u>YD domain-containing protein 3), the adapter Asc (apoptosis-associated speck like protein containing a CARD domain), and procaspase-1. Maturation and release of IL-1β requires two distinct signals: the first signal leads to synthesis of pro-IL-1β and other components of the inflammasome, such as Nlrp3 itself; and the second signal results in the assembly of the Nlrp3 inflammasome, caspase-1 activation, and IL-1β secretion.

Inflammation

The rapid release of cytokines and the appearance of soluble mediators of the complement cascade at the site of infection have far-reaching consequences. Multifunctional Tnf-α, one of the signals of early warning, is produced by activated antigen-presenting cells. Tnf-α changes the local permeability of capillaries that attract, and facilitate entry of, circulating leukocytes to the site of infection. Tnf-α also can

induce an antiviral response directly when it binds to receptors on infected cells. Within seconds, the combination of infection and binding of Tnf-α to its receptor initiates a signal transduction cascade that activates caspases, resulting in apoptotic cell death. Viral proteins that modulate the function of Tnf-α have been identified, including the core (capsid) protein of hepatitis C virus. Many DNA viruses that encode homologs of a cellular protein, cellular FLICE-like inhibitory protein (cFLIP), inhibit Tnf-α-mediated apoptosis. These viral counterparts, called vFLIPs, inhibit caspase activation in infected cells (Chapter 3). However, even when the caspase-dependent cell death pathway is blocked by viral proteins, infected cells can still induce their own demise via caspase-independent, programmed necrosis. This thrust-and-parry relationship between the altruistic suicide of the host cell and the virus-encoded proteins that keep the cell alive to prolong virus reproduction underscores the "chess match between masters" tension between host and virus.

One visible response to Tnf-α is **inflammation**. The four classical signs of inflammation are redness, heat, swelling, and pain. Such symptoms result from increased blood flow and capillary permeability, influx of phagocytic cells, and tissue damage. While unpleasant, these signs of infection are important to localize the host response to the site of damage. Local inflammation causes blood vessels to constrict, resulting in swelling of the capillary network in the area of infection. This response produces redness (erythema) and an increase in local tissue temperature. Capillary permeability increases, further facilitating an efflux of fluid and cells from the engorged capillaries into the surrounding tissue. The cells that migrate into the damaged area are largely mononuclear phagocytes. These circulating antigen-presenting cells are attracted by chemokines synthesized by virus-infected cells; cytokines elaborated by local defensive systems; and secondary reactions that facilitate adherence of phagocytic cells and other cells of the innate response, including neutrophils, to capillary walls near sites of damage (Fig. 4.11). Infiltrating monocytes are also important in the healing reactions that take place after the infection is resolved.

In the past, investigators experimentally induced inflammation by the use of adjuvants, such as Freund's adjuvant (killed mycobacterial cells in oil emulsion) or aluminum hydroxide gels, at sites of antigen administration. The adjuvant-induced inflammation mimics an infection to cultivate an environment conducive for the induction of a strong immune response to injected viral proteins in vaccines. Indeed, most vaccines would not work without adjuvants to stimulate the adaptive immune response (Chapter 8).

The nature and extent of the inflammatory response to viral infection depend on the tissue that is infected, as well as whether the virus is cytopathic; in general, noncytopathic viruses do not induce a strong inflammatory response. The early reactions at the site of infection dictate the type

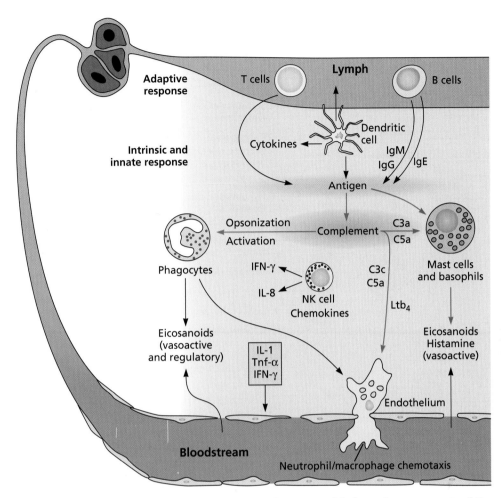

Figure 4.11 Inflammation provides integration and synergy with the main components of the immune system. Viral infection at entry sites in the body often triggers an inflammatory response. A stylized section of infected tissue served by the lymphoid system (top, green) and the circulatory system (bottom, red) is shown. Inflammation reactions can be initiated in several ways, for example, by cytokines such as IFN that are released by immature dendritic cells as they detect infection; by the classical or alternative pathway of complement activation; or by mast cells that migrate to sites of cell damage in response to cytokine release, where they can be activated by IgE antibody and antigen. C3a, C3c, and C5a are protease digestion products of the complement cascade that stimulate the inflammatory response. C3a increases vascular permeability and activates mast cells and basophils; C3c stimulates neutrophil release; and C5a increases vascular permeability and chemotaxis of basophils, eosinophils, neutrophils, and monocytes and stimulates neutrophils. The cytokines IL-1, IFN-γ, and Tnf-α act on the local capillary endothelium to enhance leukocyte adhesion and migration. IL-8 and other chemokines promote lymphocyte and monocyte chemotaxis. IL-1 and Tnf-α bind to receptors on epithelial and mesenchymal cells to induce division and collagen synthesis and to stimulate prostaglandin and leukotriene synthesis. Ltb$_4$ is a particularly active leukotriene that is vasoactive and chemotactic. The activities of cells that enter an infected site where inflammation reactions are occurring are controlled by locally produced cytokines, particularly Tnf-α, IL-1, and IFN-γ. Adapted from D. Male et al., *Advanced Immunology*, 3rd ed. (Mosby, St. Louis, MO, 1996), with permission.

of adaptive response that will predominate, and hence can influence the outcome of a viral infection. Tissues that have reduced access to the circulatory system (e.g., the brain and the interior of the eye) are less accessible to mediators of inflammation. As a result, the kinetics, extent, and final outcome of viral infections of these tissues are often markedly different from those of more-vascularized tissues.

Antigen-Presenting Cells Leave the Site of Infection and Migrate to Lymph Nodes

Activation of tissue-resident dendritic cells as a result of cytokine production and local inflammation causes them to leave the site of infection and migrate to lymphoid tissues, which include the extensive network of lymph nodes strategically located throughout the body (Fig. 4.12).

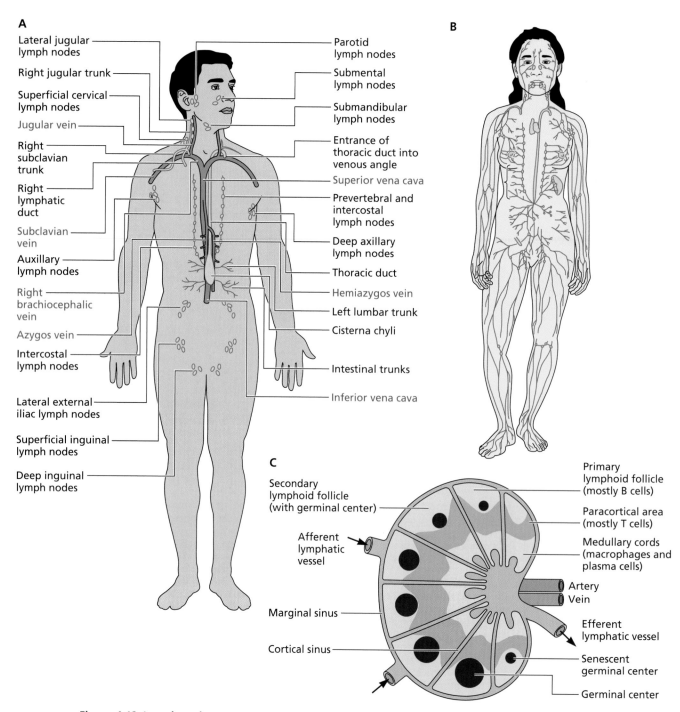

A

Lateral jugular lymph nodes

Right jugular trunk

Superficial cervical lymph nodes

Jugular vein

Right subclavian trunk

Right lymphatic duct

Subclavian vein

Auxillary lymph nodes

Right brachiocephalic vein

Azygos vein

Intercostal lymph nodes

Lateral external iliac lymph nodes

Superficial inguinal lymph nodes

Deep inguinal lymph nodes

Parotid lymph nodes

Submental lymph nodes

Submandibular lymph nodes

Entrance of thoracic duct into venous angle

Superior vena cava

Prevertebral and intercostal lymph nodes

Deep axillary lymph nodes

Thoracic duct

Hemiazygos vein

Left lumbar trunk

Cisterna chyli

Intestinal trunks

Inferior vena cava

B

C

Secondary lymphoid follicle (with germinal center)

Afferent lymphatic vessel

Marginal sinus

Cortical sinus

Primary lymphoid follicle (mostly B cells)

Paracortical area (mostly T cells)

Medullary cords (macrophages and plasma cells)

Artery
Vein

Efferent lymphatic vessel

Senescent germinal center

Germinal center

Figure 4.12 Lymph node anatomy. Lymph from the extracellular space carries antigens and antigen-presenting cells such as dendritic cells and macrophages from the tissues to the lymph nodes. The lymph enters the node at several points along the lymphatic system through afferent lymphatic vessels. The blood supply enters and leaves the lymph node at the hilum via small arteries that create a capillary network within the node. Lymphocytes in the blood can then enter the lymph node across the walls of postcapillary venules, which are also known as high endothelial venules (HEVs). These HEVs merge into small veins, which then carry blood away from the node. **(A)** Major lymph nodes in humans and their relation to major blood vessels; **(B)** the lymphatic system; **(C)** the anatomy of a lymph node.

Mucosal immunity is usually the first adaptive defense to be engaged. The lymphoid tissues below the mucosa of the respiratory and gastrointestinal tracts (mucosa-associated lymphoid tissue [MALT] and gut-associated lymphoid tissue [GALT]) (Fig. 4.13A) are vital in antiviral defense. These clusters of lymphoid cells include the tonsils in the pharynx, the submucosal follicles of the upper airways, Peyer's patches in the lamina propria of the small intestine, and the appendix. A specialized epithelial cell of mucosal surfaces is the **M cell** (microfold or membranous epithelial cell), which samples and delivers antigens to the underlying lymphoid tissue by transcytosis (Chapter 2). M cells have invaginations (pockets) of their

Figure 4.13 Components of the human lymphatic and mucosal immune systems. (A) Cellular components of the mucosal immune system in the gut (gut-associated lymphoid tissue). The lumen of the small intestine is depicted at the top of the figure. The mucosal epithelial cells are shown with their basal surface oriented toward the lamina propria. Cross sections of a lymphatic vessel and a capillary are shown, illustrating their juxtaposition to cells of the mucosal immune system. M cells have large intraepithelial pockets filled with B and CD4$^+$ T lymphocytes, macrophages, and dendritic cells. M cells and intraepithelial lymphocytes are important in the transfer of antigen from the intestinal lumen to the lymphoid tissue in Peyer's patches, where an immune response can be initiated. **(B)** The cutaneous immune system (skin-associated lymphoid tissue) comprises three cell types: keratinocytes, Langerhans cells, and T cells. Keratinocytes secrete various cytokines, including Tnf-α, IL-1, and IL-6, and have phagocytic activity. They also synthesize both MHC class I and MHC class II proteins and can present antigens to T and B cells if stimulated by IFN-γ. Langerhans cells are migratory dendritic cells and are the major antigen-presenting cells in the epidermis. When products of viral infections in the skin are detected, Langerhans cells secrete IFN and undergo maturation. Mature dendritic cells migrate to the local draining lymph node, where they present viral peptides on both MHC class I and MHC class II proteins to antigen-specific T cells. Dedicated skin-tropic T cells can cross the endothelium to enter the epidermis, where they can mature into T_h1 or T_h2 cells depending on the antigen and cytokine milieu. Activated T cells synthesize cytokines, including IFN-γ, that increase antigen presentation via MHC expression from keratinocytes and Langerhans cells. Tcr, T cell receptor. Adapted from A. K. Abbas et al., *Cellular and Molecular Immunology* (The W. B. Saunders Co., Philadelphia, PA, 1994), with permission.

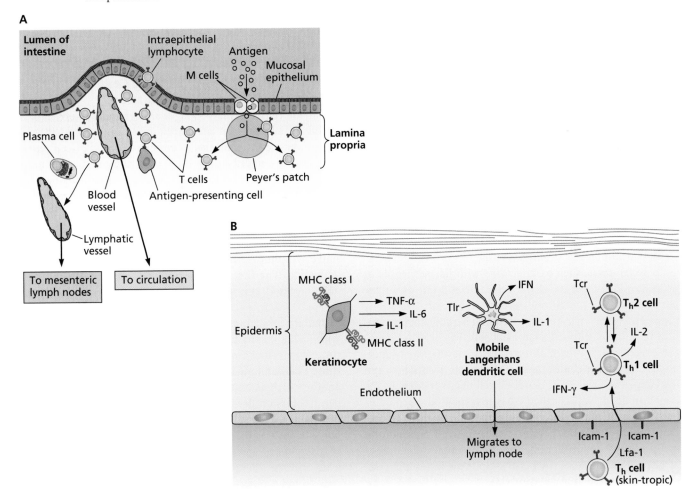

membranes that harbor immature dendritic cells, B and CD4⁺ T lymphocytes, and macrophages. The secreted antibody IgA (important in antiviral defense at mucosal surfaces) is made by B cells that accumulate at adhesion sites in these M-cell membrane pockets. After viruses or viral components transit through M cells, they emerge to come into intimate contact with all the appropriate immune cells. This process represents an essential step for the development of mucosal immune responses.

The skin, the largest organ of the body, possesses its own diverse community of organized immune cells. Lymphocytes and Langerhans cells comprise the **cutaneous immune system** (also called skin-associated lymphoid tissue) (Fig. 4.13B). These cells are important in the initial response and resolution of viral infections of the skin. In particular, Langerhans cells, the predominant scavenger antigen-presenting cells of the epidermis, function as the sentinels or outposts of early warning and reaction. These abundant, mobile cells sample antigens and migrate to regional lymph nodes to transfer information to T cells, and to activate B lymphocytes directly. Certain T cells in the circulation have tropism for the skin and, after binding to the vascular endothelium, enter the epidermis to interact with Langerhans cells and keratinocytes. These skin-tropic T cells are important for the production of the virus-specific skin rashes and poxes characteristic of measles virus and varicella-zoster virus infections.

Virus particles at the primary site of infection can interact with lymphoid cells that are associated with the mucosal and cutaneous immune systems to suppress their responses by inducing lysis or misregulation of such cells. These interactions can govern the outcome of the primary infection and often establish the pattern of infection that is characteristic of a given virus. The M cells in the mucosal epithelium have been implicated in the spread from the pharynx and the gut to the lymphoid system of a variety of viruses, including poliovirus, enteric adenoviruses, human immunodeficiency virus type 1, and reovirus. M cells have also been suggested to be sites of the persistent or latent infection of a number of other viruses, including herpes simplex virus.

In some cases, such as infections with herpesviruses or influenza virus, the dendritic cells that migrate from the periphery to the lymph node are not those that eventually present antigen to naive lymphocytes. In herpes simplex virus infection, for example, Langerhans cells in the skin phagocytose debris from dying infected cells and migrate to local nodes. Thereafter, some transfer of antigen occurs between the Langerhans cell and a dendritic cell that is resident in the lymph node, a process called cross-presentation. Such complexity may be important following infection by those viruses that can enter and kill dendritic cells: antigen presentation by uninfected dendritic cells, and the eventual induction of adaptive immunity, can still occur, despite the loss of the original antigen-presenting cell (Box 4.3).

BOX 4.3

BACKGROUND
Infection of the sentinels: dysfunctional immune modulation

When viruses infect immature or mature dendritic cells, the immune system's first command-and-control link is compromised. Some of the many possible consequences of such infection, any one of which could lead to suppression of the immune response locally or systemically, include
- destruction of immature dendritic cells
- interference with their maturation
- impairment of antigen uptake or processing
- inhibition of migration of dendritic cells to lymphoid tissue
- prevention of their activation of T cells

López CB, Yount JS, Moran TM. 2006. Toll-like receptor-independent triggering of dendritic cell maturation by viruses. *J Virol* **80:**3128–3134.

Mellman I, Steinman RM. 2001. Dendritic cells: specialized and regulated antigen processing machines. *Cell* **106:**255–258.

Finally, some viruses take advantage of the migration of dendritic cells to lymph nodes. Human immunodeficiency virus type 1 binds to a protein on the antigen-presenting cell surface, called Dc-Sign, to be ferried to the lymph node, where millions of naive T cells, the target cell for this virus, await.

Antigen Processing and Presentation
Professional Antigen-Presenting Cells Induce Activation via Costimulation

Dendritic cells and macrophages are generally referred to as "professional" antigen-presenting cells, implying that they have enhanced capabilities as compared to other cell types. Like virtually all cells in the body, dendritic cells present antigens in the groove of an MHC molecule, but they alone are able to activate naive lymphocytes, a process called **priming**. The ability to activate naive cells is mediated by cell surface proteins that enable costimulation: two protein-protein interactions between an antigen-presenting cell and a quiescent, naive lymphocyte must happen simultaneously to induce activation. The first of these signals is antigen specific (the MHC class II-epitope complex interacting with the T cell receptor), while the second signal is mediated via interactions of molecules such as CD80 and CD86 on the dendritic cell with CD28 on the naive T cell (Fig. 4.14). This two-trigger process is called costimulation. Other interactions include those that facilitate adhesion of T cell and target cell to one another, such as binding of CD2 or leukocyte function antigen 1 (Lfa1) on the CTL with Lfa-3 or intercellular adhesion molecule 1 (Icam1), respectively, on the infected cell. This fail-safe ensures that naive T cells are not activated in error: engagement of naive T cells in the absence of costimulation leads to T cell anergy or tolerance, a response in which these cells do not proliferate further. B cells, which can also present

A

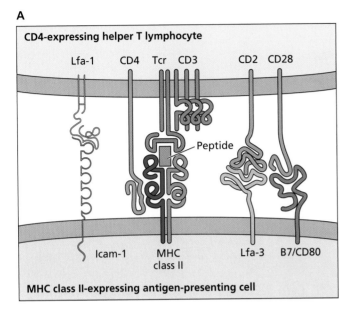

B

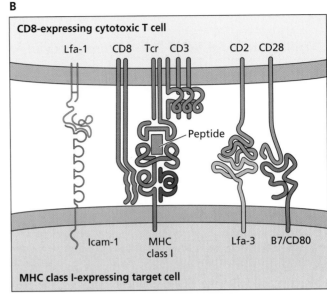

Figure 4.14 T cell surface molecules and ligands. The interactions of these receptors and ligands are important for antigen recognition and initiation of signal transduction and other T cell responses. **(A)** Interaction of a CD4 coreceptor-expressing T$_h$ cell with an antigen-presenting cell. The antigen-presenting cell exhibits an MHC class II-peptide complex in addition to Icam-1, Lfa-3, and CD80 (B7) membrane proteins. These cell surface proteins all are capable of binding cognate receptors on the T$_h$ cell as illustrated. **(B)** Interaction of a CTL bearing the CD8 coreceptor with its target cell. The target cell exhibits an MHC class I-peptide complex in addition to Icam-1, Lfa-3, and CD80 (B7) membrane proteins. Tcr, T cell receptor.

antigen, are costimulated in a similar way during engagement with T cells via a CD40-CD40L interaction.

Presentation of Antigens by Class I and Class II MHC Proteins

Earlier in this chapter, the generation of T and B cell receptor diversity was introduced, a process that enables receptor-bearing lymphocytes to "recognize" their cognate antigen. Antigens on infected cells or professional antigen-presenting cells can be displayed by one of two related but distinct protein assemblies: class I and class II MHC proteins. MHC class I proteins display protein fragments on the surface of almost all cells, whereas MHC class II proteins are generally found only on the surfaces of mature dendritic cells, macrophages, and B lymphocytes (the professional antigen-presenting cells). How antigens are presented by these proteins differs, but they have one major feature in common: fragments of proteins are presented in an outward-facing groove of the MHC molecule, similar to a hot dog in a hot dog bun. In this way, the T or B cell receptor detects the presence of "altered self": a familiar "self" protein (the MHC molecule) that is modified by the virus-specific epitope.

MHC class I proteins present antigens from intracellular pathogens, whereas MHC II molecules display antigens from extracellular pathogens or those that have been engulfed by phagocytosis. Corresponding to the two classes of MHC molecules, there are two types of effector T lymphocytes. CD8$^+$ CTLs defend against intracellular infections (e.g., viruses) and chiefly recognize class I-presented epitopes. CD4$^+$ T$_h$ lymphocytes recognize class II-presented peptides and afford protection against extracellular pathogens (e.g., many bacteria and parasites) as a result of the activation of B cells and the production of antibodies. Elucidation of the basis of MHC recognition systems, defined by immunologists as MHC restriction, was a major step forward, not only explaining how T cells recognize their targets, but also providing broader insights into how cells communicate with one other (Box 4.4).

Cytotoxic T Cells Recognize Infected Cells by Engaging MHC Class I Receptors

An imperative of the host response is to destroy virus-infected while ignoring uninfected cells. The former are identified, in part, because they display small viral peptides in association with MHC class I proteins on their surfaces. These viral (and cellular) peptides are produced and displayed via a pathway called **endogenous antigen presentation** (Fig. 4.15).

In all uninfected and infected cells, a fraction of most newly synthesized proteins is degraded by the proteasome. The targeted protein is marked for destruction by the covalent attachment of multiple copies of the small protein

BOX 4.4

TRAILBLAZER
Virology provides Nobel Prize-winning insight: MHC restriction

In 1974, Rolf Zinkernagel and Peter Doherty performed a classic experiment that provided insight into how CTLs recognize virus-infected cells. Initially, they teamed up to determine the mechanism of the lethal brain destruction observed when mice are infected with lymphocytic choriomeningitis virus, a noncytolytic arenavirus. They anticipated that the brain damage was due to CTLs responding to replication of the noncytopathic virus in the brain.

When they infected mice of a particular MHC type with the virus and then isolated T cells, these cells lysed virus-infected target cells *in vitro* **only** when the target cells and the T cells were of identical MHC type. Uninfected target cells were not lysed, even when they shared identical MHC alleles. This requirement for MHC matching was called MHC restriction.

The Nobel Prize-winning insight was that a CTL must recognize two determinants present on a virus-infected cell: one specific for the virus and one specific for the MHC of the host. We now know that CTLs recognize a short peptide derived from viral proteins and only engage the peptide when it is bound to MHC class I proteins present on the surface of target cells. (For an interview with Dr. Peter Doherty, see: http://bit.ly/Virology_Doherty)

Zinkernagel RM, Doherty PC. 1974. Restriction of *in vitro* T cell-mediated cytotoxicity in lymphocytic choriomeningitis within a syngeneic or semiallogeneic system. *Nature* **248**:701–702.

ubiquitin, and following ATP-dependent unfolding, the protein is broken down in the inner chamber of the proteasome. The peptide products are released and transported into the endoplasmic reticulum (ER) by proteins that span the ER membrane, called transport-associated proteins (Taps). Within the ER, binding of peptides to newly synthesized MHC class I proteins allows these MHC molecules to adopt their native conformation and to be transported to the cell surface via the secretory pathway. In this manner, MHC class I proteins constitutively "present," or display, a representative sampling of intracellular epitopes on their surface; when a cell is infected, the sampling includes epitopes from the virus. Binding of a viral peptide-MHC class I complex on the surface of an infected cell by the T cell receptor triggers a series of reactions that activate the CTL for killing (see below). Presentation of cellular proteins is usually ignored, because such autoreactive T cells were either deleted during development or were never appropriately costimulated.

MHC class I proteins are found on the surfaces of nearly all nucleated cells. These proteins comprise two subunits called the α chain (often called the heavy chain) and β_2-microglobulin (the light chain). Lymphocytes possess the highest concentration, with about 5×10^5 molecules per cell. In contrast, fibroblasts, muscle cells, hepatocytes, and neurons carry much smaller quantities, sometimes 100 or fewer molecules per cell. There are three MHC class I loci in humans (*A*, *B*, and *C*) and two in mice (*K* and *D*). Because there are many allelic forms of these genes in outbred populations, they are said to be **polymorphic genes**. For example, at the human MHC class I locus *hla-B*, >149 alleles with pairwise differences ranging from 1 to 49 amino acids have been identified. When cells bind cytokines such as type 1 IFN and IFN-γ, transcription of the MHC class I α chains, β_2-microglobulin, and the linked proteasome and peptide transporter genes is markedly increased.

T_h Cells Recognize Professional Antigen-Presenting Cells by Engaging MHC Class II Receptors

Both antibody and CTL responses are controlled by cytokines produced by T_h cells, which are activated upon antigen recognition via MHC class II proteins presented on the surfaces of professional antigen-presenting cells. As dendritic cells mature, MHC class II glycoproteins loaded with peptides produced from their stores of endocytosed antigens appear on their surfaces. Mature antigen-presenting cells also carry high surface concentrations of costimulatory T cell adhesion molecules that bind receptors on naive T_h cells in the lymphoid tissue.

The process by which viral proteins are taken up from the outside of the cell and digested, and by which resulting peptides are loaded onto MHC class II molecules, is called **exogenous antigen presentation** (Fig. 4.16). Unlike proteins processed by the endogenous pathway, viral proteins presented by MHC class II are not produced inside the cell, as the cell is not infected. Rather, phagocytosed viral particles are broken down and their proteins digested in endosomes rather than in the proteasome. Viral peptides and MHC class II molecules are then brought together by vesicular fusion. Similar to peptide-bound MHC class I presentation, the assembly is then transported to the surface of the antigen-presenting cell, where it is available to interact with appropriate T cells in the lymph node. Interaction of T cell receptors on the naive CD4$^+$ T_h cell with the MHC class II-peptide induces concerted changes in the T_h cell, leading to its activation and differentiation.

T_h cells activated in this fashion produce IL-2, as well as a high-affinity receptor for this cytokine. The secreted IL-2 binds to the newly synthesized receptors to induce autostimulation and proliferation of T_h cells. Such clonal expansion of specific T_h1 or T_h2 cells then promotes the activation of CTLs and B lymphocytes (Fig. 4.2).

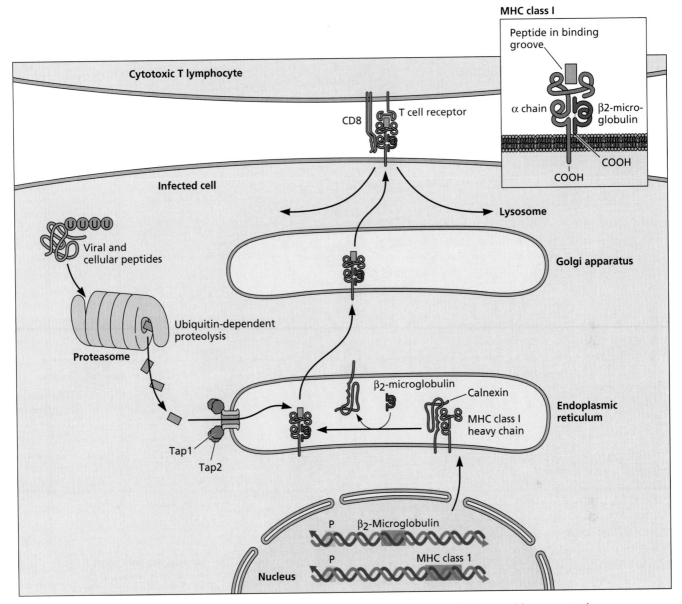

Figure 4.15 Endogenous antigen processing: the pathway for MHC class I peptide presentation.
Intracellular proteins derived from both the host cell and the virus are degraded in the cytoplasm. Proteins are marked for destruction by polyubiquitinylation and are then taken up and degraded by the proteasome. The resulting short peptides are transported into the lumen of the ER by the Tap1-Tap2 heterodimeric transporter in a reaction requiring ATP. Once in the ER lumen, the peptides associate with newly synthesized MHC class I molecules that bind weakly to the Tap transporter. Assembly of the α chain and β_2-microglobulin (β_2m) of the MHC class I molecule is facilitated by the ER chaperone calnexin, but formation of the final native structure requires peptide loading. The MHC class I complex loaded with peptide is released from the ER to be transported via the Golgi compartments to the cell surface, where it is available for interaction with the T cell receptor of a cytotoxic T cell carrying the CD8 coreceptor. (Inset) The MHC class I molecule is a heterodimer of the membrane-spanning type I glycoprotein α chain (43 kDa) and β_2-microglobulin (12 kDa) that does not span the membrane. The α chain folds into three domains, 1, 2, and 3. Domains 2 and 3 fold together to form the groove where peptide binds, and domain 1 folds into an immunoglobulin-like structure. Adapted from D. Male et al., *Advanced Immunology*, 3rd ed. (Mosby, St. Louis, MO, 1996), with permission.

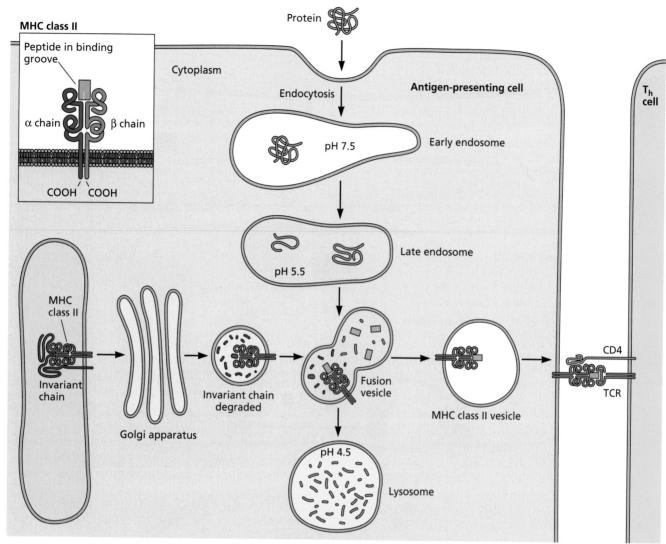

Figure 4.16 Exogenous antigen processing in the antigen-presenting cell: the pathway for MHC class II peptide presentation. Peptides in the ER lumen of the antigen-presenting cell are prevented from binding to the MHC class II peptide groove by association of MHC class II molecules with a protein called the invariant chain. The complex is transported through the Golgi compartments to a post-Golgi vesicle, where the invariant chain is removed by proteolysis. This reaction frees MHC class II molecules to accept peptides. The peptides are derived not from endogenous proteins but from extracellular proteins that enter the antigen-presenting cell. In some cells, the proteins enter by endocytosis (top) and are internalized to early endosomes with neutral luminal pH. Endocytotic vesicles traveling to the lysosome via this pathway are characterized by a decrease in pH as they "mature" into late endosomes. The lower pH activates vesicle proteases that degrade the exogenous protein into peptides. Internalized endosomes with their peptides fuse at some point with the vesicles containing activated MHC class II. The newly formed peptide-MHC class II assembly then becomes competent for transport to the cell surface, where it is available for interaction with the T cell receptor (TCR) of a T_h cell carrying the CD4 coreceptor. (Inset) The MHC class II molecule is a heterodimer of the membrane-spanning type I α-chain (34-kDa) and β-chain (29-kDa) glycoproteins. Each chain folds into two domains, 1 and 2, and together the α and β chains fold into a structure similar to that of MHC class I. The two amino-terminal domains from α and β chains form the groove in which peptide binds. Unlike the closed MHC class I peptide groove, the MHC class II peptide-binding groove is open at both ends. The second domain of each chain folds into an immunoglobulin-like structure. Human genomes contain three MHC class II loci (*DR*, *DP*, and *DQ*), and mouse genomes have two (*IA* and *IE*). The three sets of human genes give rise to four types of MHC class II molecules. Adapted from D. Male et al., *Advanced Immunology*, 3rd ed. (Mosby, St. Louis, MO, 1996), with permission.

Although synthesis of MHC class II proteins occurs primarily in professional antigen-presenting cells, other cell types, including fibroblasts, pancreatic β cells, endothelial cells, and astrocytes, can synthesize MHC class II molecules, but only after exposure to IFN-γ. As with MHC class I, there are many alleles of MHC class II genes.

Both classes of MHC protein have a peptide-binding cleft that is sufficiently flexible to accommodate many epitopes. Even so, not all possible peptides are bound. The ability of MHC molecules to bind and display particular epitopes on the cell surface varies from individual to individual as a result of the many MHC alleles that exist within the human population. Such allelic diversity plays an important role in an individual's capacity to respond to various infections: the greater the diversity, the wider the capacity to respond. This fact has dramatic consequences for the spread of viral diseases. For example, individuals in inbred populations lose MHC diversity over time and have a concomitantly limited capacity to respond to infections. Protective immunity is difficult to establish in such populations, increasing the risk of epidemics.

Lymphocyte Activation Triggers Massive Cell Proliferation

Following a productive interaction between an antigen-presenting cell and a naive T cell, a massive expansion of the naive population ensues. Only a few cells in this population, whether in lymphoid tissues or elsewhere in the body, participate in the initial encounter with any foreign epitope. For example, the frequency of B or T lymphocytes that recognize infected cells on first exposure is as few as 1 in 10,000 to 1 in 100,000. Following activation, this precursor population is amplified substantially during the next 1 or 2 weeks: the number of virus-specific lymphocytes increases >1,000-fold, in some cases by as much as 50,000-fold. Because much of this expansion occurs in lymph nodes, individuals suffering from virus infection often note swelling in the neck or the groin, areas of high lymph node density. Activation is accompanied by differentiation such that each daughter cell has the same specific immune reactivity as the original parent (often called a clonal response).

Before discussing how activated lymphocytes resolve a virus infection, we pause here to note that many of the fundamental discoveries described above were accelerated by the development of powerful methodologies. As examples, restriction enzymes and DNA sequencing enabled precise determination of the genetic origins of the T and B cell receptors, flow cytometry allowed isolation of lymphocytes at distinct stages of maturation, and recombinant DNA approaches made possible the generation of transgenic and knockout mice with genetically altered immune systems. T_h17 and Treg cells are fairly new discoveries in immunology, but their identification prompted a reevaluation of what

we thought we understood when CD4$^+$ T cells were "simply" either T_h1 or T_h2. As new technologies come along, it is exciting, and a bit intimidating, to consider how our current understanding of the host response will change based on the discoveries that lie ahead.

The Cell-Mediated Response

The cell-mediated response facilitates recovery from a viral infection primarily because it eliminates virus-infected cells without damaging uninfected cells. While the T_h2-promoted antibody response is important for some infections in which virus particles spread in the blood, antibody alone is often unable to contain and clear an infection. Indeed, antibodies have little or no effect in many natural infections that spread by cell-to-cell contact, including those caused by many neurotropic viruses that spread transsynaptically, or by viruses that establish long-term or noncytolytic infections, such as the hepatitis viruses. These infections can be stopped only by CTL-produced antiviral cytokines and direct killing of infected cells.

CTLs Lyse Virus-Infected Cells

CTLs are superbly equipped to kill virus-infected cells, and following lysis of an infected cell, they can detach and kill again. Signaling from the T cell receptor pursuant to engagement of the peptide antigen-MHC complex requires clustering (aggregation) of a number of T cell receptors and reorganization of the T cell cytoskeleton to form a specialized structure called the **immunological synapse** (Fig. 4.17; see also Fig. 7.14). Only after these reactions have taken place can the CTL lyse an infected cell.

The term "immunological synapse" was coined because the proteins that mediate target and T cell recognition show an unexpected degree of spatial organization at the site of T cell–target cell contact, not unlike a neuronal synapse. The synapse structure contributes to stabilizing signal transduction by the T cell receptor for the prolonged periods required for activation of gene expression. In addition, membrane proteins in the synapse engage the underlying cytoskeleton and polarize the secretion apparatus so that a high local concentration of effector molecules is attained. Small numbers of peptide ligands bound to MHC class I molecules apparently can stimulate a T cell because they serially engage many T cell receptors on the opposing cell surface within the synapse. Unengaged T cell receptors subsequently entering this zone have an increased likelihood of binding a specific ligand and amplifying a signal.

Given the central roles of the T cell receptor and formation of the immunological synapse in adaptive immune defense, it should come as no surprise that viral gene products can alter the structure, function, and localization of the T cell receptor and the various coreceptors. Indeed, infection by

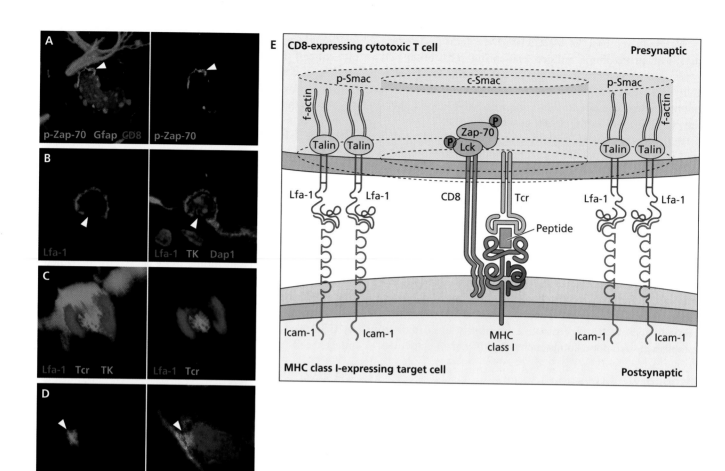

Figure 4.17 The immunological synapse. (A–D) The morphological characteristics of an *in vivo* immunological synapse between CD8+ CTLs and adenovirus-infected astrocytes is illustrated. The striatum of rats was injected with a recombinant adenovirus vector carrying an expression cassette for the herpes simplex virus thymidine kinase (TK) gene. **(A)** Interaction between a CD8+ CTL (red) and an infected astrocyte (Gfap, magenta [which marks astrocytes]) stimulates T cell receptor (Tcr) signaling, resulting in phosphorylation and polarization of tyrosine kinases such as Zap70 (green) toward the interface with the infected cell. The white arrow indicates polarized pZap70; blue stain (Dap1) indicates nuclei. **(B)** Adhesion molecules such as Lfa-1 (red) aggregate to form a peripheral ring (p-SMAC [peripheral supramolecular activation cluster]) at the junction formed by the immunological synapse. The postsynaptic astrocyte process can be identified by staining with antibody to TK, a marker of adenovirus infection (green). Note the characteristic absence of Lfa-1 at the central portion of the immunological synapse between the T cell and the infected astrocyte (white arrow). **(C)** A rotated image from a three-dimensional reconstruction demonstrates the typical central polarization (c-SMAC) of Tcr molecules (green) toward the infected astrocyte (TK, white) and the peripheral distribution of Lfa-1 in the p-SMAC (red). **(D)** The effector molecule IFN-γ (green) within a Tcr+ (red) CTL is directed toward the site of close contact with an infected target cell (TK, white); the white arrow indicates the T-target cell contact zone. The diameter of a CTL is ~10 μm. **(E)** Schematic cross section of an immunological synapse showing the characteristic polarized arrangement of the cytoskeleton (actin and talin proteins indicated) and organization of the adhesion molecule Lfa-1 toward the p-SMAC. The Tcr molecules are directed toward the c-SMAC. The phosphorylated TKs (Zap70 and Lck) and effector IFN-γ molecules (not shown) are in the center of the immunological synapse. See C. Barcia et al., *J Exp Med* **203**:2095–2107, 2006. Figure coutesy of Pedro Lowenstein, Kurt Kroeger, and Maria Castro.

several members of the *Retroviridae* and *Herpesviridae* leads to reduction of T cell receptor function by inhibiting the synthesis of one or more of the T cell receptor protein subunits. Viral infection can also modulate the abundance of various accessory molecules on cell surfaces and therefore alter CTL recognition and subsequent effector function.

CTLs kill by two primary mechanisms: transfer of cytoplasmic granules from the CTL to the target cell, and induction of apoptosis. These killing systems develop during cellular differentiation. The maturing CTL fills with cytoplasmic granules that contain macromolecules required for lysis of target cells, such as perforin, a membrane pore-forming protein; and granzymes, members of a family of serine proteases. Granules are released by CTLs in a directed fashion when in membrane contact with the target cell, and are taken up by that cell via receptor-mediated endocytosis. Perforin, as its name implies, punctures holes in the plasma membrane, allowing access of granzymes that induce apoptosis in the infected cell (Fig. 4.18). CTL killing by the perforin pathway is rapid, occurring within minutes after contact and recognition. Activated CTLs can also induce apoptotic cell death via binding of the Fas ligand on their surfaces to the Fas receptor on target cells, although this pathway is much slower than perforin-mediated killing. Many activated CTLs also secrete IFN-γ, which, as we have seen, is a potent inducer of both the antiviral state in neighboring cells and synthesis of MHC class I and II proteins (Box 4.5). Activated CTLs also secrete powerful cytokines, including IL-16, and chemokines such as CCl5. Their release by virus-specific CTLs following recognition of an infected target cell may assist in coordination of the antiviral response.

Typically, following infection by a cytopathic virus, CTL activity appears within 3 to 5 days, peaks in about a week,

Figure 4.18 CTL lysis. Granzymes participate in target cell apoptosis in association with perforin. Granzymes diffuse into the target cell cytoplasm upon perforin "puncturing," where they activate caspases, leading to apoptotic death. Pores can either allow release of cytoplasmic contents **(A)** or injection of granzymes that result in cleavage of procaspases, leading to cell death **(B)**.

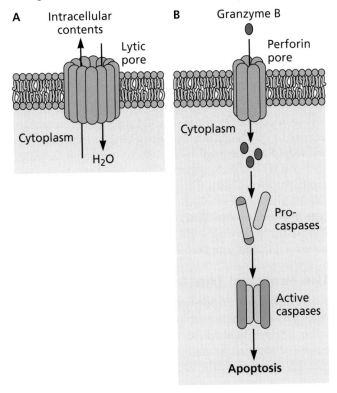

BOX 4.5

BACKGROUND
Interferon γ signaling

IFN-γ is the only type II member of the interferon family, which also comprises type I (IFN-α/β) and type III IFNs. Unlike type I IFNs, which are expressed by most cells soon after infection, IFN-γ is produced chiefly by activated cells of the immune system such as NK cells and T cells. IFN-γ initiates a cellular response by binding to the IFN-γ receptor, Ifngr (a heterotetramer of the Ifngr1 and Ifngr2 subunits). Binding triggers activation of receptor-associated Janus kinases (Jaks) 1/2 and subsequent phosphorylation of tyrosine in the cytoplasmic tail of the Ifngr1 subunits. Signal transducer and activator of transcription (Stat) 1 is recruited to the phosphorylated Ifngr1 subunit, where Stat1 becomes phosphorylated, homodimerizes, and translocates to the nucleus. The phosphorylated Stat1 homodimer binds to IFN-γ-activated site (GAS) elements within IFN-γ-responsive genes (Isgs) to initiate transcription. More than 250 genes are induced in this manner to inhibit viral spread. While Stat1 is required for a classical IFN-γ response, a substantial number of studies have also demonstrated the presence of IFN-γ-dependent, Stat1-independent pathways.

The classical pathways of type I and type II IFN signaling are shown for comparison. Irf9, IFN regulatory factor 9; Isgf3, IFN-stimulated gene factor 3; ISRE, IFN-stimulated response element; P, phosphate.

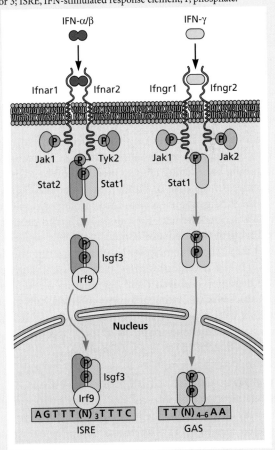

and declines thereafter. The magnitude of the CTL response depends on such variables as viral titer, route of infection, and age of the host. The critical contribution of CTLs to antiviral defense is demonstrated by **adoptive transfer** experiments in which virus-specific CTLs from an infected animal can confer protection to nonimmunized recipients following virus inoculation. However, CTLs can also cause direct harm by large-scale cell killing. Such immunopathology often follows infection by noncytopathic viruses, when cells can be infected yet still function. For example, the liver damage, cirrhosis, and hepatocellular cancer associated with infection by hepatitis viruses is actually caused by CTL killing of persistently infected liver cells and the consequent necessity for their constant regeneration (Chapter 5).

Control of CTL Proliferation

Massive CTL precursor expansions after acute primary infections by viruses such as lymphocytic choriomeningitis virus and Epstein-Barr virus have been identified (Box 4.6). For example, >50% of CTLs from the spleen of a lymphocytic choriomeningitis virus-infected mouse were specific for a **single** viral peptide. The response reached a maximum 8 days after infection, but up to 10% of virus-specific T cells were still detectable after a year postchallenge. Such results are in contrast to those for hepatitis B virus or human immunodeficiency virus infection, in which <1% of the CTLs from spleens of infected patients are specific for a single viral peptide. The basis for this large range in CTL epitope recognition is not understood.

Viral proteins can blunt the CTL response, with far-reaching effects, ranging from rapid onset of severe symptoms and death of the host to long-lived, persistent infections (Chapter 5). Many of these proteins confound CTL recognition by disguising or reducing antigen presentation by MHC class I molecules. In the case of human immunodeficiency virus type 1, the viral genome encodes three proteins that interfere with CTL action: Nef and Tat induce Fas ligand production and subsequent Fas-mediated apoptosis of CTLs, whereas Env engages the CXCr4 chemokine receptor, triggering the death of the CTL. Nef also reduces MHC class I presentation by cell surface depletion and lysosomal degradation of the protein (Fig. 7.10). Human cytomegalovirus-infected cells contain at least six viral proteins that interfere with the MHC class I pathway and also evoke apoptosis of virus-specific CTLs by increasing synthesis of Fas ligand (Chapter 5).

Noncytolytic Control of Infection by T Cells

Complete clearance of intracellular viruses by the adaptive immune system does not depend solely on the destruction of infected cells by CTLs. The production of cytokines, such as IFN-γ and Tnf-α, by these cells can lead to purging of viruses from infected cells without cell lysis. This mechanism requires that the infected cell retain the ability to activate antiviral pathways induced by binding of these cytokines to their receptors, and that viral reproduction is sensitive to the resulting antiviral response.

In certain circumstances, such as infection of the liver by hepatitis B and C viruses, there are orders of magnitude more infected cells than there are virus-specific CTLs. Furthermore, when nonrenewable cell populations, such as neurons, are infected, CTL killing would do more harm to the host than good (Box 4.7). In such circumstances, cytokine-mediated viral clearance represents an optimal strategy, and the results can be highly effective: when hepatitis B virus-specific CTLs are experimentally transferred to another animal (via adoptive transfer), the IFN-γ and Tnf-α produced appear to clear the infection from thousands of cells without their destruction.

The resolution of infections by noncytolytic viruses via CTL-produced cytokines such as IFN-γ and Tnf-α has been documented for many DNA and RNA viruses. Indeed, CTL production of powerful, secreted, antiviral cytokines provides a simple explanation for how CTLs are able to control massive numbers of infected cells. Additional cytokines produced by a variety of immune system cells are likely to participate in such viral clearance. CD4$^+$ T cells can also clear some infections with noncytolytic viruses with little involvement of CTLs. Such cases include infections by vaccinia virus, vesicular stomatitis virus, and Semliki Forest virus.

Rashes and Poxes

Many infections, including those caused by measles virus, smallpox virus, and varicella-zoster virus, produce a characteristic rash or lesion over extensive areas of the body (Fig. 4.19), even though the primary infection began at a distant mucosal surface. This phenomenon results when the primary infection escapes the local defenses and virus particles or infected cells spread in the circulatory system to initiate foci of infected cells in the skin. T_h1 cells and macrophages activated by the initial infection home to these secondary sites and respond by aggressive synthesis of cytokines, including IL-2 and IFN-γ. Such cytokines then act locally to increase capillary permeability, which is partially responsible for a characteristic local immune response referred to as **delayed-type hypersensitivity**. This response, which usually requires 2 to 3 days to develop, is the basis of many virus-promoted rashes and lesions with fluid-filled vesicles.

The Humoral (Antibody) Response
Antibodies Are Made by Plasma Cells

After a B cell emerges from the bone marrow into the circulation, it travels to lymph and lymphoid organs and differentiates and synthesizes antibody only if its surface antibody receptor is bound to the cognate antigen. Antigen binding causes clustering of receptors on the B cell surface, which then

BOX 4.6

METHODS
Measuring the antiviral cellular immune response

The Classic Assays: Limiting Dilution and Chromium Release

For more than 50 years, virologists have assessed the presence of CTLs in blood, spleen, or lymphoid tissues of immune animals by using the limiting-dilution assay and radioactive chromium release from lysed target cells. This assay measures CD8⁺ CTL precursors, or memory CTLs, based on two attributes: (i) the ability of foreign proteins and peptides to stimulate CTL proliferation and (ii) the capacity of activated CTLs to lyse target cells.

Lymphocytes are obtained from an animal that has survived virus infection and are cultured for 1 to 2 weeks in the presence of whole inactivated virus, its proteins, or synthetic viral peptides. Under these conditions, virus-specific CTL precursors are induced to divide (clonal expansion). The expanded CTL population is then tested for its ability to destroy target cells that display viral peptides on MHC class I molecules. The target cells loaded with the viral antigen in question are then exposed to chromium-51, a radioactive isotope that binds to most intracellular proteins (Panel A). After being washed to remove external isotope, cultured CTLs are incubated with the target cells, and lysis is measured by the release of chromium-51 into the supernatant. Within the last decade, nonradioactive substrates have become commercially available; these alternatives function in the same manner and offer many of the same attributes, but are much safer than ⁵¹Cr.

Serial dilution before assay provides an estimate of the number of CD8⁺ CTL precursors in the original cell suspension (providing the name of the assay: "limiting dilution"). This assay provides a quantitative measure of cellular immunity, but it is time-consuming, technically demanding, and expensive.

Identifying and Counting Virus-Specific T Cells

The limiting-dilution assay defines T cells by function, but not by their peptide specificity. Until recently, investigators have tried with little success to identify individual T cells based on their peptide recognition properties. This failure has been attributed to the low affinity and high "off" rates of the MHC–peptide–T cell receptor interaction. Without this information, it was difficult, if not impossible, to measure antigen-specific T cell responses.

An important advance is the use of an **artificial MHC tetramer** as an epitope-specific, T cell-staining reagent (Panel B). The extracellular domains of MHC class I proteins are produced in *Escherichia coli*. These engineered MHC class I molecules have an artificial C-terminal 13-amino-acid sequence that enables them to be biotinylated. The truncated MHC class I proteins are folded *in vitro* with a synthetic peptide that will be recognized by a specific T cell. Biotinylated tetrameric complexes are purified and mixed with isolated T cell pools from virus-infected animals. Individual T cells that bind the biotinylated MHC class I-peptide complex are detected by a variety of immunohistochemical techniques. Cell-sorting techniques can be used, and stained cells are viable.

The **intracellular cytokine assay** is a relatively rapid method to count specific CTLs. Fresh lymphoid cells are treated with brefeldin A. This fungal metabolite blocks the secretory pathway and prevents the secretion of cytokines. The cells are then fixed with a mild cross-linking chemical that preserves protein antigenicity, such as glutaraldehyde. Treated cells are permeabilized so that a specific antibody for a given cytokine can react with the intracellular cytokines. Cells that react with the antibody can be quantified by fluorescence-activated cell sorting. With appropriate software and calibration, the staining intensity corresponds to cytokine concentration, and the number of cells responding to a particular epitope and MHC class I molecule can be determined.

Measuring the Antiviral Antibody Response

Antibodies are the primary effector molecules of the humoral response. There are many methods to detect antibodies. However, a standard method in virology is the **neutralization assay**: viral infectivity is determined in the presence and absence of antibody. Two variations on this general theme include the **plaque reduction assay** and the **neutralization index**. In the former, a known number of PFU are exposed to serial dilutions of the antibody or serum in question. The highest dilution that will reduce the plaque count by 50% is taken as the plaque reduction titer of the serum or antibody. The index is calculated as the difference in viral titers in the presence and absence of test antibody or serum. Obvious requirements for these assays are that the virus in question can be propagated in cultured cells and that a measure of virus infectivity such as plaque formation is available.

Klenerman P, Cerundolo V, Dunbar PR. 2002. Tracking T cells with tetramers: new tales from new tools. *Nat Rev Immunol* 2:263–272.

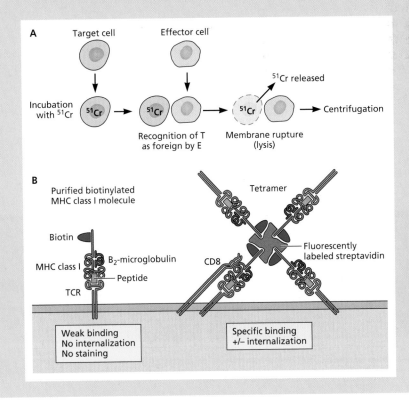

BOX 4.7

DISCUSSION
The immune system and the brain

Cells of the central nervous system (CNS; brain and spinal cord) can initiate a robust and transient intrinsic defense (for example, by the production of type I IFNs), but, surprisingly, cannot mount an adaptive response. A primary reason for this deficit is that the CNS lacks lymphoid tissue and dendritic cells. In addition, the CNS of vertebrates is separated from many cells and proteins of the bloodstream by tight endothelial cell junctions that comprise the so-called blood-brain barrier (see the figure). As a consequence of these features, viral infections of the CNS can have unexpected outcomes. For example, if virus particles are injected directly into the ventricles or membranes covering the brain, which are in contact with the bloodstream, the innate immune system is activated and a strong inflammatory and an adaptive response occur. In contrast, if virus particles are injected directly into brain tissue, avoiding the blood vessels and ventricles, only a transient inflammatory response is produced. The adaptive response is not activated; antibodies and antigen-activated T cells are not produced.

Although the CNS is unable to initiate an adaptive immune response, it is not isolated from the immune system, and hence an earlier concept of the brain being "immune privileged" has fallen out of favor. Indeed, the blood-brain barrier is open to entry of activated immune cells circulating in the periphery. Antigen-specific T cells regularly enter and travel through the brain, performing immune surveillance. Moreover, at least

two glial cell types, astrocytes and microglia, have a variety of cell surface receptors that can engage the T cells. Astrocytes, the most numerous cell type in the central nervous system, respond to a variety of cytokines made by immune system cells. All natural brain infections begin in peripheral tissue, and any infection that begins outside the CNS activates the adaptive immune response. However, if the infection spreads to the brain, the resulting immune attack on this organ can be devastating.

Virus particles can be injected experimentally into the brain directly without infecting peripheral tissues; in such cases, an adaptive response is completely avoided. On the other hand, if the animal is first immunized by injecting virus particles into a peripheral tissue, such that the adaptive response is activated, subsequent injection of the same virus into the brain of the immunized animal elicits massive immune attack on that organ. Any virus-infected target in the brain is recognized and destroyed by the peripherally activated T cells. In both natural and experimental infections, the inflammatory response is not transient but sustained, resulting in capillary leakage, swelling, and cell death. Uncontrolled inflammation in the closed confinement of the skull has many deleterious consequences, and when it is coupled with bleeding and cell death, the results are disastrous. In such extreme cases, the swollen brain has nowhere to go but to be extruded out the foramen at the base of the skull.

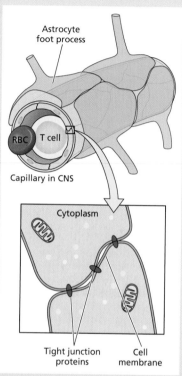

Cross section of a brain capillary. Capillaries within the central nervous system are comprised of tightly packed endothelial cells and astrocyte processes on the "brain" side of the capillary. This barrier prevents free access of blood-borne proteins and cells from the blood to the brain (termed the "blood-brain barrier"), though this barrier is permeable to activated lymphocytes.

Figure 4.19 Virus-induced rashes and poxes.

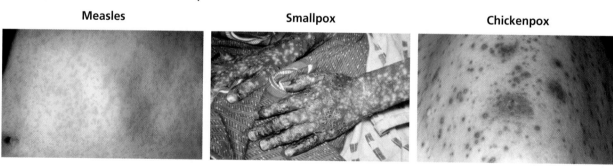

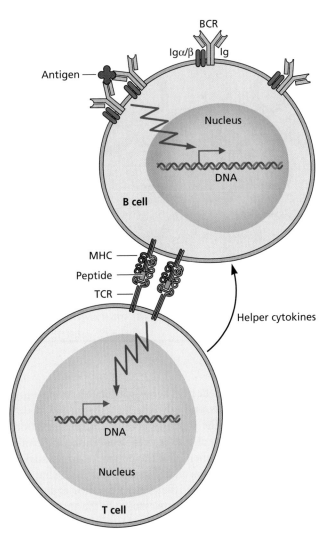

BCR

Igα/β   Ig

Antigen ——

Nucleus

DNA

B cell

MHC ——

Peptide ——

TCR ——

Helper cytokines

DNA

Nucleus

T cell

Figure 4.20 Activation of B cells to produce antibodies. When antigen binds and causes clustering of B cell receptors (BCRs), a signal transduction cascade is triggered that leads to the activation of the B cell and production of soluble antibodies. Peptides derived from the antigen bound to MHC class II are transported to the cell surface, where they are bound by T_h-cell receptors, leading to T cell activation. In turn, the activated T cell provides cytokine "help."

activates signaling via Src family tyrosine kinases that associate with the cytoplasmic domains of the closely opposed receptors. B cell coreceptors, such as CD19, CD21, and CD8, enhance signaling by recruiting tyrosine kinases to clustered antigen receptors and coreceptors (Fig. 4.20).

Binding of antigen to the B cell receptor is only part of the activation process: cytokines from T_h cells are also required. When the T cell receptor of T_h2 cells recognizes MHC class II-peptide complexes present on the B cell surface, the T_h2 cells produce a locally high concentration of stimulatory cytokines, as well as CD40 ligand (a protein homologous to Tnf). The engagement of CD40 ligand with its B cell receptor

facilitates a local exchange of cytokines that further stimulates proliferation of the activated B cell and promotes its differentiation. Fully differentiated, antibody-producing plasma cells make huge quantities of specific antibodies: the rate of synthesis of IgG can be as high as 30 mg/kg of body weight/day. For a human of ~50 kg, this value equates to >1 g of antibody made every 24 h.

Types and Functions of Antibodies

All antibodies (immunoglobulins) have common structural features, illustrated in Fig. 4.21. Five classes of immunoglobulin (IgA, IgD, IgE, IgG, and IgM) are defined by their distinctive heavy chains (α, δ, ϵ, γ, and μ, respectively). Their properties are summarized in Table 4.1. IgG, IgA, and IgM are commonly produced after viral infection. During B cell differentiation, "switching" of the constant region of heavy-chain genes occurs by somatic recombination and is regulated in part by specific cytokines, including IL-4 and IL-5, which bind to their receptors on the target B cell. Importantly, while the Fc region of these antibodies changes, the antigen-binding region (the Fab portion of the antibody) does not, and hence the specificity of the antigen remains the same. Class switching allows for changes in the effector functions of the antibodies and increases their functional diversity.

During the **primary antibody response**, which follows initial contact with antigen, the production of antibodies follows a characteristic sequence. The IgM antibody appears first, followed by IgA on mucosal surfaces or IgG in the serum. The IgG antibody is the major antibody of the response and is remarkably stable, with a half-life of 7 to 21 days. Specific IgG molecules remain detectable for years, because of the presence (and occasional reactivation) of memory B cells. A subsequent challenge with the same antigen or viral infection promotes a rapid secondary antibody response (Fig. 4.22).

Virus Neutralization by Antibodies

Antibodies contribute to antiviral defense chiefly by binding to, and causing the elimination of, free virus particles. Viruses that infect mucosal surfaces or that circulate in the blood will be exposed to IgA antibodies and IgG and IgM antibody molecules, respectively. For example, immunodeficient animals can be protected from some lethal viral infections by injection with virus-specific antiserum or purified monoclonal antibodies (also called passive immunization). A therapy based on this principle was used in the recent Ebola outbreak in 2014. The experimental therapeutic, ZMapp, comprises a cocktail of three Ebola virus-specific monoclonal antibodies that was shown to have efficacy in preventing Ebola virus disease in macaques experimentally infected with Ebola virus. At the time of the outbreak, it had not been tested in humans. Because no other therapeutic options were available, ZMapp was used to treat

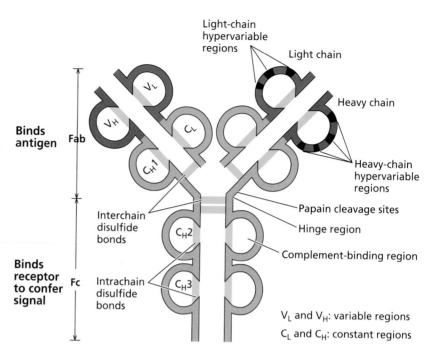

Figure 4.21 The structure of an antibody molecule. This is a schematic representation of an IgG molecule delineating the subunit and domain structures. The light and heavy chains are held together by disulfide bonds (yellow bars). The variable regions of the heavy (V_H) and light (V_L) chains, as well as the constant regions of the heavy (C_H) and light (C_L) chains, are indicated at the left part of the molecule. The hypervariable regions and invariable regions of the antigen-binding domain (Fab) are emphasized. The constant region (Fc) performs many important functions, including complement binding (activation of the classical pathway) and binding to Fc receptors found on macrophages and other cells. Clusters of papain protease cleavage sites are indicated, as this enzyme is used to separate the Fab and Fc domains.

seven Ebola virus-infected Americans; of these, five individuals survived. While these data are not sufficiently robust to show efficacy, many believe that postexposure treatment with such preparations or derivatives will be an effective method to prevent disease in Ebola virus-exposed individuals.

Perhaps the best example of the importance of antibodies in antiviral defense is the success of the poliovirus vaccine in preventing poliomyelitis, as the type of antibody produced can significantly influence the outcome of a poliovirus infection. Poliovirus infection stimulates strong IgM and IgG responses in the blood, but it is mucosal IgA that is vital in defense. This antibody isotype can neutralize poliovirus directly in the gut, the site of primary infection. The live attenuated Sabin

poliovirus vaccine is effective because it elicits a strong mucosal IgA response. This antibody type is synthesized by plasma cells that underlie the mucosal epithelium. This antibody is secreted as dimers of two conventional immunoglobulin subunits. The dimers then bind the polymeric immunoglobulin receptor on the basolateral surface of epithelial cells (Fig. 4.23). This complex is then internalized by endocytosis and moved across the cell (**transcytosis**) to the apical surface. Protease cleavage of the receptor releases dimeric IgA into mucosal secretions, where it can interact with incoming virus particles.

IgA may also block viral reproduction inside infected mucosal epithelial cells (Fig. 4.23). Because IgA molecules

Table 4.1   The five classes of immunoglobulins

Property	IgA	IgD	IgE	IgG	IgM
Function	Mucosal; secretory	Surface of B cell	Allergy; anaphylaxis; epithelial surfaces	Major systemic immunity; memory responses	Major systemic immunity; primary response; agglutination
Subclasses	2	1	1	4	1
Light chain	κ, λ	κ, λ	κ, λ	κ, λ	κ, λ
Heavy chain	α	δ	ε	γ	μ
Concn in serum (mg/ml)	3.5	0.03	0.00005	13	1.5
Half-life (days)	6	2.8	2	25	5
Complement activation					
Classical	−	−	−	+	++
Alternative	−/+	−	+	−	−

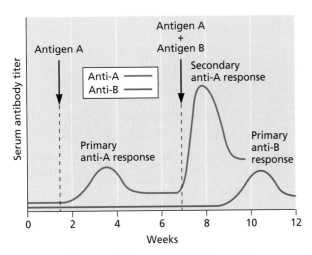

Figure 4.22 The specificity, self-limitation, and memory of the adaptive immune response. This general profile of a typical adaptive antibody response demonstrates the relative concentration of serum antibodies after time (weeks) of exposure to antigen A or a mixture of antigens A and B. The antibodies that recognize antigens A and B are indicated by the red and blue lines, respectively. The primary response to antigen A takes about 3 to 4 weeks to reach a maximum. When the animal is injected with a mixture of both antigens A and B at 7 weeks, the secondary response to antigen A is more rapid and more robust than the primary response. However, the primary response to antigen B again takes about 3 to 4 weeks. These properties demonstrate immunological memory. Antibody levels (also termed titers) decline with time after each immunization. This property is called self-limitation or resolution. Reprinted from A. K. Abbas et al., *Cellular and Molecular Immunology* (The W. B. Saunders Co., Philadelphia, PA, 1994), with permission.

must pass through such a cell en route to secretion from the apical surface, they are available during transit for interaction with viral proteins produced within the cell. The antigen-binding domain of intracellular IgA lies in the lumen of the ER, the Golgi compartment, and any transport vesicles of the secretory pathway. It can therefore bind to the external domain of any type I viral membrane protein that has the cognate epitope of that IgA molecule. Such interactions have been demonstrated with Sendai virus and influenza virus proteins during infection of cells in culture. In these experiments, antibodies colocalized with viral antigen only when the IgA could bind to the particular viral envelope protein. These studies suggest that clearing viral infection from mucosal surfaces need not be limited to the lymphoid cells of the adaptive immune system.

It is widely assumed that the primary mechanism of antibody-mediated neutralization of viruses is via steric blocking of virus particle-receptor interaction (Fig. 4.24). While some antibodies do prevent virus particles from attaching to cell receptors, the vast majority of virus-specific antibodies are likely to interfere with the concerted structural changes that are required for entry. Antibodies can also promote aggregation of virus particles, thereby reducing the effective concentration of viruses that can initiate infection. Many enveloped viruses can be destroyed *in vitro* when antiviral antibodies and serum complement disrupt membranes (the classical complement activation pathway).

Much of what we know about antibody neutralization comes from the isolation and characterization of "antibody escape" mutants or **monoclonal antibody-resistant mutants**. These mutants are selected by propagating virus in the presence of neutralizing antibody. The analysis of the mutant viruses allows a precise molecular definition, not only of antibody-binding sites but also of parts of viral proteins important for entry. Antigenic drift (see Chapters 5 and 10) is a consequence of selection and establishment of antibody escape mutants in viral populations.

Antibodies can provoke other remarkable responses in virus-infected cells. For example, in a process analogous to CTL lysis, antibodies that bind to the surface proteins of many enveloped viruses can clear the particles from persistently infected cells. This process is noncytolytic and complement independent. In this case, antibodies act synergistically with IFN and other cytokines. Virus-specific antibodies bound to surfaces of infected cells can inhibit virus budding at the plasma membrane and also reduce surface expression of viral membrane proteins by inducing endocytosis.

Nonneutralizing antibodies are also prevalent after infection: they bind specifically to virus particles, but do not interfere with infectivity. In some cases, such antibodies can even enhance infectivity: antibody bound to virus particles is recognized by Fc receptors on macrophages, and the entire complex is brought into the cell by endocytosis. This process, antibody-dependent enhancement, is the basis of disease following a secondary exposure to dengue virus (Chapter 5).

Antibody-Dependent Cell-Mediated Cytotoxicity: Specific Killing by Nonspecific Cells

T_h1 stimulation results in production of a particular isotype of IgG in B cells that can bind to antibody receptors on macrophages and some NK cells. These receptors are specific for the more conserved Fc region of an antibody molecule (Fig. 4.21). If an antiviral antibody is bound in this manner, the amino-terminal antigen-binding site is still free to bind viral antigen on the surface of the infected cell. In this way, the antiviral antibody targets the infected cell for elimination by macrophages or NK cells. This process is called **antibody-dependent cell-mediated cytotoxicity** (often referred to as ADCC). The antibody provides the specificity for killing by a less discriminating NK cell. Efforts are under way to harness the power of this process in the development of a universal influenza virus vaccine.

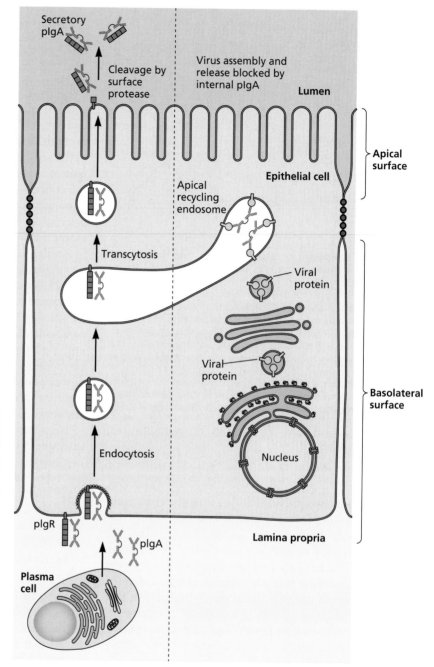

Figure 4.23 Secretory antibody, IgA, is critical for antiviral defense at mucosal surfaces. A single polarized epithelial cell is illustrated. The apical surface is shown at the top, and the basal surface is shown at the bottom. (Left) Antibody-producing B cells (plasma cells) in the lamina propria of a mucous membrane secrete the IgA antibody (also called polymeric IgA [pIgA]). pIgA is a dimer, joined at its Fc ends. For IgA to be effective in defense, it must be moved to the surface of the epithelial cells that line body cavities. This process is called transcytosis. (Right) A virus particle infecting an epithelial cell potentially can be bound by internal IgA if virus components intersect with the IgA in the lumen of vesicles during transcytosis. This process is likely to occur for enveloped viruses, as their membrane proteins are processed in many of the same compartments as those mediating transcytosis. Adapted from M. E. Lamm, *Annu Rev Microbiol* 51:311–340, 1997, with permission.

Immunological Memory

Once a specific response has been established and the viral infection is subdued, the individual is immune to subsequent infection by the same pathogen. Immunological memory of previous infections is one of the most powerful features of the adaptive immune system, and makes vaccines possible. While the primary adaptive response takes days to reach its antiviral potential, a subsequent encounter with the same pathogen awakens a massive reaction that occurs within hours of pathogen entry. This process occurs because a subset of B and T lymphocytes, called **memory cells**, is maintained after each encounter with a foreign antigen. These cells survive for years and are ready to respond immediately to any subsequent encounter by rapid proliferation and synthesis of their antiviral effector functions. Because such a secondary response is usually stronger than the primary one, childhood infection protects adults, and immunity conferred by vaccination can last for years, sometimes for a lifetime. An important

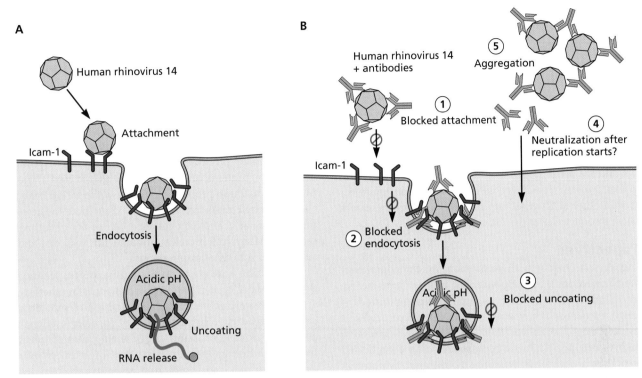

Figure 4.24 Interactions of neutralizing antibodies with human rhinovirus 14. (A) The normal route of infection. The virus attaches to the Icam-1 receptor and enters by endocytosis. As the internal pH of the endosome decreases, the particle uncoats and releases its RNA genome into the cytoplasm. **(B)** Possible mechanisms of neutralization of human rhinovirus 14 by antibodies. With well-characterized monoclonal antibodies, at least five modes of neutralization have been proposed and are illustrated: (1) blocked attachment—binding of antibody molecules to virus results in steric interference with virus-receptor binding; (2) blocked endocytosis—antibody molecules binding to the capsid can alter the capsid structure, affecting the process of endocytosis; (3) blocked uncoating—antibodies bound to the particle fix the capsid in a stable conformation so that pH-dependent uncoating is not possible; (4) blocked uncoating, inside cell—antibodies themselves may be taken up by endocytosis and interact with virions inside the cell after infection starts; (5) aggregation—because all antibodies are divalent, they can aggregate virus particles, facilitating their destruction by phagocytes. (A) Adapted from T. J. Smith et al., *Semin Virol* 6:233–242, 1995, with permission.

concept is that a memory response does not protect against reinfection, but rather against the symptoms that arise following unrestricted infection. Consequently, an individual may be exposed repeatedly to particular pathogens and never be aware of it, because the memory response eliminates the virus before illness appears.

The events that give rise to a memory response are quite similar to those that induce a primary response. The primary differences are that memory T and B cells are more abundant and more easily activated. Moreover, memory B cells produce more-effective antibodies (e.g., IgG) than the low-affinity IgM made at the beginning of a primary infection. This property, which is unique to the humoral arm of the adaptive response, is the result of somatic hypermutation of the genes encoding the virus-specific antibodies that refine and focus the antibody response after each successive exposure to the pathogen. As a result, repeated exposures strengthen B cell memory and antibody affinity.

Unlike B cells, which undergo isotype switching that distinguishes them from the primary response, the expression of the T cell receptor on memory T cells does not change. Although a comparison of the proteins that are differentially synthesized by effector and memory T cells does reveal some subtle differences in their gene expression profiles, the differences between these populations was poorly understood until recently. A major current focus in immunology is to define different types of memory T cells and to ascertain how each contributes to long-term protection of the host. Two sets of memory T cells, with distinct activation requirements, have been identified: **effector memory T cells** and **central memory T cells**. On encountering a specific viral antigen, effector memory cells quickly produce cytokines of either a T_h1 or T_h2 response. These cells are generally found in the circulatory system, and have higher concentrations of particular adhesion molecules that equip them to readily enter peripheral tissues. Central memory T cells, by contrast, are more abundant in lymph nodes and other lymphoid organs, and have the capacity for self-renewal. Their restricted localization may ensure that a depot of memory cells is preserved in the lymph node "library" for future exposures.

Both T and B cell memory are maintained without the need for persisting antigen. As discussed in Chapter 8, it was once thought that the host may preserve a small "reminder" of previous virus encounters to restimulate memory, but this has since been shown not to be the case. While most memory cells are found in a quiescent state in an uninfected host, a small proportion are dividing. Some immunologists have hypothesized that cytokines, produced constitutively or during infections with other pathogens, may help to maintain these cells by causing some to enter the cell cycle.

Perspectives

We began this chapter with a comment on the complexity of the adaptive immune response. But as Confucius noted, "Life is simple, but we insist on making it complicated"—although the details of how virus particles are recognized and eliminated, how T cells "see" their cognate antigen, and how memory is retained for many decades are surely complicated concepts, the host defense serves to do one thing only: protect the host from pathogens. One of the features that makes the study of viruses so fascinating is how well these microbes subvert host immunity: for almost every host defense, there is a virus counter-response. In fact, much of what we know about host immunity was gleaned from the careful observation of viruses: if a viral protein that thwarts host immunity is made, it must point to an essential element of the host response.

As this third chapter focused on the host response ends, we return to the hypothetical virus infection introduced at the end of Chapter 3 (Fig. 3.25). In general, the intrinsic and innate defenses bring most viral infections to an uneventful close before the adaptive response is required. However, if viral reproduction outpaces the innate defense, a critical threshold is reached: increased IFN production by circulating immature dendritic cells elicits a more global host response, and flu-like symptoms are experienced by the infected host. As viral reproduction continues, viral antigens are delivered by mature dendritic cells to the local lymph nodes or spleen to establish sites of information exchange with T cells. T cell recirculation is shut down because of the massive recruitment of lymphocytes into lymphoid tissue. The swelling of lymph nodes that is so often characteristic of infection is a sign of this stage of immune action.

Within days, T_h cells and CTLs appear; these cells are the first signs of activation of the adaptive immune response. T_h cells produce cytokines that begin to direct the amplification of this response. The synthesis of antibodies, first of IgM and then of other isotypes, follows quickly. The relative concentrations of antibody isotypes are governed by the route of infection and the pattern of cytokines produced by the T_h cells.

As the immune response is amplified, CTLs kill infected cells or purge virus from them, and antibodies bind to, and inactivate, virus particles. Specific antibody-virus complexes can be recognized by macrophages and NK cells, which induce antibody-dependent cell-mediated cytotoxicity and can also activate the classical complement pathway. Both of these processes lead to the directed killing of infected cells and virus particles by macrophages and NK cells.

An inflammatory response often occurs as infected cells die and innate and adaptive responses develop. Cytokines, chemotactic proteins, and vasodilators are released at the site of infection. These proteins, invading white blood cells, and various complement components all contribute to the swelling, redness, heat, and pain characteristic of the inflammatory response. Many viral proteins modulate this response and the subsequent activity of immune cells.

If infection spreads from the primary site, second and third rounds of virus reproduction can occur in other organs. T cells that were activated at the initial site of infection can cause delayed-type hypersensitivity (usually evident as a characteristic rash or lesion) at these later sites of infection. Immunopathology, particularly after infection by noncytopathic viruses, can result from an overly exuberant host response. Most infections are resolved: the combination of innate and adaptive responses clears the infection, and the host becomes immune because of the presence of memory T and B cells and antibodies. The high concentrations of lymphocytes drop dramatically as these cells die by apoptosis, and the system eventually returns to its preinfection state. The adaptive response can be avoided completely, or in part, when organs or tissues that have poor or nonexistent immune responses are infected, when new viral variants are produced rapidly because of high mutation rates, or when progeny virus particles spread directly from cell to cell.

Despite these formidable defenses, viruses can make us sick, and, in infections caused by human immunodeficiency virus, Ebola virus, influenza virus, and others, can result in substantial loss of life. The various ways by which viruses contribute to human disease and, in some cases, fatality are considered in Chapter 5.

References

Books

Murphy K, Travers P, Walport M. 2007. *Janeway's Immunobiology*, 7th ed. Garland Science, Garland Publishing Inc, New York, NY.

Nathanson N (ed). 2007. *Viral Pathogenesis and Immunity*, 2nd ed. Academic Press, London, United Kingdom.

Reviews

Belkaid Y, Rouse BT. 2005. Natural regulatory T cells in infectious disease. *Nat Immunol* **6:**353–360.

Chang JT, Wherry EJ, Goldrath AW. 2014. Molecular regulation of effector and memory T cell differentiation. *Nat Immunol* **15:**1104–1115.

Clark RA. 2015. Resident memory T cells in human health and disease. *Sci Transl Med* **7:**269rv1. doi:10.1126/scitranslmed.3010641.

Colonna M, Trinchieri G, Liu YJ. 2004. Plasmacytoid dendritic cells in immunity. *Nat Immunol* **5:**1219–1226.

Coscoy L. 2007. Immune evasion by Kaposi's sarcoma-associated herpesvirus. *Nat Rev Immunol* **7:**391–401.

Delon J, Germain RN. 2000. Information transfer at the immunological synapse. *Curr Biol* **10:**R923–R933.

DiVico AL, Gallo RC. 2004. Control of HIV-1 infection by soluble factors of the immune response. *Nat Rev Microbiol* **2:**401–413.

Doherty PC, Christensen JP. 2000. Accessing complexity: the dynamics of virus-specific T cell responses. *Annu Rev Immunol* **18:**561–592.

Fauci AS, Mavilio D, Kottilil S. 2005. NK cells in HIV infection: paradigm for protection or targets for ambush. *Nat Rev Immunol* **5:**835–844.

Finlay BB, McFadden G. 2006. Anti-immunology: evasion of the host immune system by bacterial and viral pathogens. *Cell* **124:**767–782.

Guidotti LG, Chisari FV. 2001. Noncytolytic control of viral infections by the innate and adaptive immune response. *Annu Rev Immunol* **19:**65–91.

Hickey WF. 2001. Basic principles of immunological surveillance of the normal central nervous system. *Glia* **36:**118–124.

Iannacone M, Sitia G, Guidotti L. 2006. Pathogenetic and antiviral immune responses against hepatitis B virus. *Future Virol* **1:**189–196.

Jegaskanda S, Reading PC, Kent SJ. 2014. Influenza-specific antibody-dependent cellular cytotoxicity: toward a universal influenza vaccine. *J Immunol* **193:**469–475.

Kaech SM, Wherry EJ. 2007. Heterogeneity and cell-fate decisions in effector and memory CD8$^+$ T cell differentiation during viral infection. *Immunity* **27:**393–405.

Kahan SM, Wherry EJ, Zajac AJ. 2015. T cell exhaustion during persistent viral infections. *Virology* **479–480C:**180–193.

Kim H, Ray R. 2014. Evasion of TNF-α-mediated apoptosis by hepatitis C virus. *Methods Mol Biol* **1155:**125–132.

Kurosaki T, Kometani K, Ise W. 2015. Memory B cells. *Nat Rev Immunol* **15:**149–159.

Lane TE, Buchmeier MJ. 1997. Murine coronavirus infection: a paradigm for virus-induced demyelinating disease. *Trends Microbiol* **5:**9–14.

Latz E, Xiao TS, Stutz A. 2013. Activation and regulation of the inflammasomes. *Nat Rev Immunol* **13:**397–411.

Lodoen MB, Lanier LL. 2005. Viral modulation of NK cell immunity. *Nat Rev Microbiol* **3:**59–69.

López CB, Yount JS, Moran TM. 2006. Toll-like receptor-independent triggering of dendritic cell maturation by viruses. *J Virol* **80:**3128–3134.

Lowenstein PR. 2002. Immunology of viral-vector-mediated gene transfer into the brain: an evolutionary and developmental perspective. *Trends Immunol* **23:**23–30.

Marcus A, Raulet DH. 2013. Evidence for natural killer cell memory. *Curr Biol* **23:**R817–R820.

Martinon F, Mayor A, Tschopp J. 2009. The inflammasomes: guardians of the body. *Annu Rev Immunol* **27:**229–265.

Mossman KL, Ashkar AA. 2005. Herpesviruses and the innate immune response. *Viral Immunol* **18:**267–281.

Novak N, Peng WM. 2005. Dancing with the enemy: the interplay of herpes simplex virus with dendritic cells. *Clin Exp Immunol* **142:**405–410.

Orange JS, Fassett MS, Koopman LA, Boyson JE, Strominger JL. 2002. Viral evasion of natural killer cells. *Nat Immunol* **3:**1006–1012.

Park SH, Bendelac A. 2000. CD1-restricted T cell responses and microbial infection. *Nature* **406:**788–792.

Redpath S, Angulo A, Gascoigne NR, Ghazal P. 2001. Immune checkpoints in viral latency. *Annu Rev Microbiol* **55:**531–560.

Rother RP, Squinto SP. 1996. The α-galactosyl epitope: a sugar coating that makes viruses and cells unpalatable. *Cell* **86:**185–188.

Rouse BT. 1996. Virus-induced immunopathology. *Adv Virus Res* **47:**353–376.

Sallusto F, Geginat J, Lanzavecchia A. 2004. Central memory and effector memory T cell subsets: function, generation, and maintenance. *Annu Rev Immunol* **22:**745–763.

Selin LK, Welsh RM. 2004. Plasticity of T cell memory responses to viruses. *Immunity* **20:**5–16.

Simas JP, Efstathiou S. 1998. Murine gammaherpesvirus 68: a model for the study of gammaherpesvirus pathogenesis. *Trends Microbiol* **6:**276–282.

Streilein JW, Dana MR, Ksander BR. 1997. Immunity causing blindness: five different paths to herpes stromal keratitis. *Immunol Today* **18:**443–449.

Tortorella D, Gewurz BE, Furman MH, Schust DJ, Ploegh HL. 2000. Viral subversion of the immune system. *Annu Rev Immunol* **18:**861–926.

Trapani JA, Smyth MJ. 2002. Functional significance of the perforin/granzyme cell death pathway. *Nat Rev Immunol* **2:**735–747.

Zuniga E, Hahm B, Edelmann K, Oldstone M. 2005. Immunosuppressive viruses and dendritic cells: a multifront war. *ASM News* **71:**285–290.

Classic Papers

Janeway CA, Jr. 2001. How the immune system works to protect the host from infection: a personal view. *Proc Natl Acad Sci U S A* **98:**7461–7468.

Murali-Krishna K, Altman JD, Suresh M, Sourdive DJ, Zajac AJ, Miller JD, Slansky J, Ahmed R. 1998. Counting antigen-specific CD8 T cells: a reevaluation of bystander activation during viral infection. *Immunity* **8:**177–187.

Zinkernagel RM. 1996. Immunology taught by viruses. *Science* **271:**173–178.

Zinkernagel RM, Doherty PC. 1974. Restriction of *in vitro* T cell-mediated cytotoxicity in lymphocytic choriomeningitis within a syngeneic or semiallogeneic system. *Nature* **248:**701–702.

Selected Papers

Brown MG, Dokun AO, Heusel JW, Smith HR, Beckman DL, Blattenberger EA, Dubbelde CE, Stone LR, Scalzo AA, Yokoyama WM. 2001. Vital involvement of a natural killer cell activation receptor in resistance to viral infection. *Science* **292:**934–937.

Callan MF, Tan L, Annels N, Ogg GS, Wilson JD, O'Callaghan CA, Steven N, McMichael AJ, Rickinson AB. 1998. Direct visualization of antigen-specific CD8$^+$ T cells during the primary immune response to Epstein-Barr virus *in vivo*. *J Exp Med* **187:**1395–1402.

Cecchinato V, Trindade CJ, Laurence A, Heraud JM, Brenchley JM, Ferrari MG, Zaffiri L, Tryniszewska E, Tsai WP, Vaccari M, Parks RW, Venzon D, Douek DC, O'Shea JJ, Franchini G. 2008. Altered balance between Th17 and Th1 cells at mucosal sites predicts AIDS progression in simian immunodeficiency virus-infected macaques. *Mucosal Immunol* **1:**279–288.

Deshpande SP, Lee S, Zheng M, Song B, Knipe D, Kapp JA, Rouse BT. 2001. Herpes simplex virus-induced keratitis: evaluation of the role of molecular mimicry in lesion pathogenesis. *J Virol* **75:**3077–3088.

Evans CF, Horwitz MS, Hobbs MV, Oldstone MB. 1996. Viral infection of transgenic mice expressing a viral protein in oligodendrocytes leads to chronic central nervous system autoimmune disease. *J Exp Med* **184:**2371–2384.

Gebhard JR, Perry CM, Harkins S, Lane T, Mena I, Asensio VC, Campbell IL, Whitton JL. 1998. Coxsackievirus B3-induced myocarditis: perforin exacerbates disease, but plays no detectable role in virus clearance. *Am J Pathol* **153:**417–428.

Klimstra WB, Ryman KD, Bernard KA, Nguyen KB, Biron CA, Johnston RE. 1999. Infection of neonatal mice with Sindbis virus results in a systemic inflammatory response syndrome. *J Virol* **73:**10387–10398.

Kulcsar KA, Baxter VK, Greene IP, Griffin DE. 2014. Interleukin 10 modulation of pathogenic Th17 cells during fatal alphavirus encephalomyelitis. *Proc Natl Acad Sci U S A* **111:**16053–16058.

Leikina E, Delanoe-Ayari H, Melikov K, Cho MS, Chen A, Waring AJ, Wang W, Xie Y, Loo JA, Lehrer RI, Chernomordik LV. 2005. Carbohydrate-binding molecules inhibit viral fusion and entry by crosslinking membrane glycoproteins. *Nat Immunol* **6:**995–1001.

Raftery MJ, Schwab M, Eibert SM, Samstag Y, Walczak H, Schönrich G. 2001. Targeting the function of mature dendritic cells by human cytomegalovirus: a multilayered viral defense strategy. *Immunity* **15:**997–1009.

Sigal LJ, Crotty S, Andino R, Rock KL. 1999. Cytotoxic T-cell immunity to virus-infected non-haematopoietic cells requires presentation of exogenous antigen. *Nature* **398**:77–80.

Stevenson PG, Austyn JM, Hawke S. 2002. Uncoupling of virus-induced inflammation and anti-viral immunity in the brain parenchyma. *J Gen Virol* **83**:1735–1743.

Stevenson PG, Freeman S, Bangham CR, Hawke S. 1997. Virus dissemination through the brain parenchyma without immunologic control. *J Immunol* **159**:1876–1884.

Thomas CE, Schiedner G, Kochanek S, Castro MG, Löwenstein PR. 2000. Peripheral infection with adenovirus induces unexpected long-term brain inflammation in animals injected intracranially with first-generation, but not with high-capacity, adenovirus vectors: toward realistic long-term neurological gene therapy for chronic diseases. *Proc Natl Acad Sci U S A* **97**:7482–7487.

Trevejo JM, Marino MW, Philpott N, Josien R, Richards EC, Elkon KB, Falck-Pedersen E. 2001. TNF-α-dependent maturation of local dendritic cells is critical for activating the adaptive immune response to virus infection. *Proc Natl Acad Sci U S A* **98**:12162–12167.

5

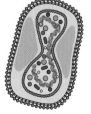

Mechanisms of Pathogenesis

Introduction

Animal Models of Human Diseases

Patterns of Infection

Incubation Periods

Mathematics of Growth Correlate with Patterns of Infection

Acute Infections

Persistent Infections

Latent Infections

"Slow" Infections

Abortive Infections

Transforming Infections

Viral Virulence

Measuring Viral Virulence

Alteration of Viral Virulence

Viral Virulence Genes

Pathogenesis

Infected Cell Lysis

Immunopathology

Immunosuppression Induced by Viral Infection

Oncogenesis

Molecular Mimicry

Perspectives

References

LINKS FOR CHAPTER 5

▶▶ *Video: Interview with Dr. Rafi Ahmed*
http://bit.ly/Virology_Ahmed

▶▶ *The enemy of my enemy is not my friend*
http://bit.ly/Virology_Twiv316

▶▶ *How influenza virus infection might lead to gastrointestinal symptoms*
http://bit.ly/Virology_12-10-14

▶▶ *The running mad professor*
http://bit.ly/Virology_Twiv308

▶▶ *Transmission of Ebola virus*
http://bit.ly/Virology_9-27-14

▶▶ *Why do viruses cause disease?*
http://bit.ly/Virology_2-7-14

Introduction

The study of viruses has been instrumental in a multitude of fundamental discoveries in science, including revelations in basic cell biology, structural biology, the origins of cancer, and the function of immune responses. We have described the amazing diversity of viruses in terms of their structures, reproduction strategies, and methods of counteracting host defenses. Furthermore, we have emphasized that not all viruses are "bad": indeed, viruses can be used in gene therapy, or for the precise delivery of toxic payloads to cancer cells.

But for most people, the value of learning more about viruses is based on a somewhat less academic viewpoint: viruses scare us. Smallpox has killed 1 in every 20 people that have ever lived, scientists warn of the global impact of the next influenza pandemic, human immunodeficiency virus continues to be a modern plague, and Ebola virus epidemics result in high mortality and global anxiety. Virus infections of animals and crops have led to billions of dollars in lost products, and the vaccine industry has invested equally impressive resources in the development of vaccines and antivirals. Some of our fears are justified, of course, as viruses **can** kill their hosts. But it is worth reiterating that causing disease is not the purpose of any viral reproduction strategy. For example, in animal infections, cell lysis is a common mechanism for exit of virus particles from infected cells; that the loss of this particular cell may have deleterious consequences for the host is generally irrelevant for viral propagation. In some cases, virus-triggered disease is more a result of changes in the host immune response rather than the virus infection itself. Immunosuppressive viruses prolong the period in which they can reproduce unchecked: that the host may now be more vulnerable to other infections is collateral damage. Similarly, an overly aggressive antiviral response can result in immunopathology or autoimmunity. Still other viruses cause disease as a consequence of interference with normal cellular processes. For example, some nonlytic viruses inhibit specific functions of differentiated cells, such as the ability of a neuron to synthesize a particular neurotransmitter. While this effect would have little bearing on the infected cell, the consequences for the host could be considerable.

Despite the distinct ways by which viruses cause disease, infection must be viewed from a holistic perspective to understand viral pathogenesis: cell culture systems are valuable for understanding basic aspects of viral reproduction, but cannot model the complexities of infections in the host that will include numerous cells and tissues that participate in the response to the infection. In this chapter, we focus on the classic patterns of virus infection within cells and hosts, the myriad ways that viruses cause illness, and the value of animal models in uncovering new principles of viral pathogenesis.

Animal Models of Human Diseases

Viral pathogenesis refers to the adverse physiological consequences that occur as a result of viral infection of a host organism: in essence, the study of the origins of viral disease. Pathogenesis following infection is determined by many parameters in addition to the impact on the infected cells themselves. The tissues in which those cells reside, the fitness of the host response, the age, gender, health, and immunological history of the host, the size of the host population, and the environment in which it resides all are contributing components. Conclusions about the nature of pathogenesis that are derived from reductionist approaches, such as focusing on the function of a viral receptor protein in cultured cells, are frequently called into question when tested in animals (Box 5.1).

PRINCIPLES *Mechanisms of pathogenesis*

- Cell culture systems cannot replicate the complexities of infections in the host that will include numerous cells and tissues that participate in the response to the infection.

- Viral pathogenesis refers to the adverse physiological consequences that occur as a result of viral infection of a host organism.

- The laboratory mouse has been particularly useful in viral pathogenesis studies, owing to its similar physiology to humans, and our ability to manipulate the mouse genome.

- Some virus infections kill the cell rapidly (cytopathic viruses), others result in the release of virus particles without causing immediate host cell death (noncytopathic viruses), and still others remain dormant in the host cell, neither killing it nor producing any progeny.

- Antigenic variation refers to changes in virus proteins in response to antibody selection, and can arise by two distinct processes: antigenic drift that results from selection of virus particles with slightly altered surface proteins, and antigenic shift, in which particles have a major change in the surface protein(s).

- Viruses have multiple strategies to establish persistent infections, including modulation of the host response and selective reproduction in tissues with limited immune surveillance.

- Latent infections are characterized by an intact, but transcriptionally quiescent, viral genome that results in poor recognition by the host immune response.

- Viruses can cause disease by direct cell death, immunopathology, immunosuppression, oncogenesis, or more recently recognized mechanisms including molecular mimicry.

- For many noncytolytic viruses, including the hepatitis viruses and some herpesviruses, immunopathology is the primary basis of disease.

BOX 5.1

EXPERIMENTS
Of mice and humans

The conclusion that human influenza virus strains are preferentially bound by sialic acids attached to galactose via an α(2,6) linkage was derived by studying the binding of virus particles to cells in culture and to purified sugars. As this sugar is the major sialic acid present on the surface of cells of the human respiratory epithelium, it was thought that it was the receptor bound by virus during infection of most animals. This hypothesis was tested using mice that lack the gene encoding the sialyltransferase, ST6Gal I, the enzyme used for linking α(2,6) sialic acid to glycoproteins. Such mice have no detectable α(2,6) sialic acid in the respiratory tract. Nevertheless, human influenza viruses replicated efficiently in the lung and trachea of these mice, indicating that α(2,6) sialic acid is **not** essential for influenza virus infection, at least in mice.

The lesson to be learned is clear: even when the results of experiments performed in tissue culture seem to have obvious relevance to infection in the host, such notions must always be validated *in vivo*.

Glaser L, Conenello G, Paulson J, Palese P. 2007. Effective replication of human influenza viruses in mice lacking a major α(2,6) sialyltransferase. *Virus Res* **126**:9–18.

Experiments that provide one answer in cell culture may result in quite different outcomes when done *in vivo*.

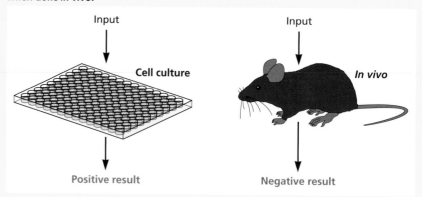

Studying pathogenesis in animals can be challenging, because so many variables come into play that it is often impossible to define precise mechanisms of disease. To quote a famous U.S. politician, there are probably many "unknown unknowns" when studying complex organisms such as humans, or even mice. Infection of inbred littermates in the same cage can lead to different pathogenic outcomes, despite identical histories, environment, and genetics: if there is such discrepancy in response even within inbred mouse populations, imagine how difficult it is to dissect the variables that result in different outcomes in humans. Consequently, viral pathogenesis has often been called a phenomenological discipline, in which observations are many, but mechanistic insights are few. In the past two decades, however, this view has been changing, thanks to the development of new experimental tools and more precise animal models, coupled with improved studies in humans and with human tissues.

Some viruses that infect humans have a broad host range and can infect other animals such as monkeys, ferrets, and guinea pigs. As we shall see later in this chapter, these various animal models have been invaluable for understanding viral pathogenesis. The laboratory mouse has been particularly useful: because the mouse genome can be manipulated readily, and the physiology of the mouse and the human are similar, it is possible to engineer mice to allow susceptibility to some human viruses (Box 5.2). Likewise, the ability to add, modify, or delete specific genes in mice enables the assessment of the function of individual proteins in pathogenesis.

For example, we have learned a lot about immunity from the use of mice with targeted deletions of specific immune cell populations or effector proteins. In some cases, insights into human disease are gleaned by studying relatives of the human viruses. An example is simian immunodeficiency virus infection of monkeys, a useful surrogate to study the pathogenesis of human immunodeficiency virus infections. Although the knowledge obtained from animal models is essential for understanding how viruses cause disease in humans, the results of such studies must be interpreted with caution. No human disease is completely reproduced in an animal model: what is true for a mouse is not always (perhaps even rarely) true for a human. Differences in size, metabolism, organ systems, aging, immune histories, and developmental program bear substantially on pathogenesis. Furthermore, as most mice used in viral pathogenesis studies are heavily inbred, they cannot provide much insight into the subtle effects of human genetic diversity. Nevertheless, the study of animal models of virus infections has yielded many principles and mechanistic insights that were corroborated in follow-up studies with human tissues.

Patterns of Infection

Studying the biology of an infected cell is a useful first step in understanding what kind of pathology the virus will cause in the host. Some virus infections kill the cell rapidly, producing a burst of new particles (**cytopathic viruses**), while other infections result in the release of virus particles without

BOX 5.2

METHODS

Transgenic and knockout mice for studying viral pathogenesis

Mice have always played an important role in the study of viral pathogenesis (see figure). Because it is possible to manipulate this animal genetically, much new information about how viruses cause disease has been collected. Introducing a gene into the mouse germ line to produce a transgenic mouse and ablating specific genes (gene knockouts) both have wide use in virology. The application of Cre-lox technology allows selective removal of a host gene within a particular cell type or at a chosen time in development.

Mice are not susceptible to infection by all viruses, however. In those cases in which mice lack virus receptors, it has been possible to engineer transgenic mice that produce the human receptor, with the goal of enabling a human virus to infect mouse cells. A model of disease can be established, assuming all the other necessary cellular proteins are present to allow reproduction of the human virus. In cases where viral receptors have not been identified, or are not sufficient for infection, an alternative approach is to express either the entire viral genome or a selected viral gene in mice. For example, transgenic mice that express the hepatitis B virus genome have been used to study interactions between the virus and the host immune response. Transgenic mice that express T cell receptor transgenes or genes encoding soluble immune mediators have also been produced. Such mice have been used to study the effect of immune cells on virus clearance, as well as the protective and deleterious effects of cytokines.

Mice lacking specific components of the immune response have proven invaluable for studying immunity and immunopathogenesis. For example, mice lacking the gene encoding perforin, a molecule essential for the ability of cytotoxic T lymphocytes to lyse target cells (Chapter 4), cannot clear infection by lymphocytic choriomeningitis virus, despite the presence of an otherwise intact immune response.

More recently, "humanized" mice have been developed to study viruses that are specific to humans. These mice carry human genes, cells, or tissues; when human cells or tissues are transplanted, the recipient mice are usually immunodeficient to prevent them from mounting an antigraft immune response. Such "Frankenstein" mice have been used for the study of human immunodeficiency virus and Epstein-Barr virus infections, among others.

Denton PW, Garcia JV. 2011. Humanized mouse models of HIV infection. *AIDS Rev* **13**:135–148.

Rall GF, Lawrence DMP, Patterson CE. 2000. The application of transgenic and knockout mouse technology for the study of viral pathogenesis. *Virology* **271**:220–226.

Various approaches to the use of transgenic and knockout mice in the study of viral reproduction and pathogenesis.

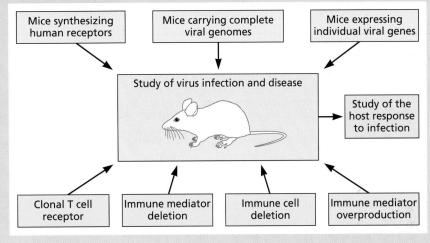

causing immediate host cell death (**noncytopathic viruses**). Alternatively, some infections neither kill the cell, nor produce any progeny, but rather remain dormant or become **abortive infections.**

Infections in host organisms can be categorized based on their duration: rapid and self-limiting (**acute infections**) or long-term (**persistent infections**). Variations and combinations of these two modes are common (Fig. 5.1). It can be argued that all virus-host encounters begin with an acute infection, and that differences in the subsequent management of that infection diversify the ultimate outcome. For example, most **latent infections**, in which no infectious particles are produced, begin as an acute infection. Conversion to a latent infection enables the viral genome to persist undetected, perhaps to be reactivated in the future. Intermediate patterns that lie between rapid viral growth and latent infec-

tion can be thought of as "smoldering infections" in which low-level viral reproduction occurs in the face of a strong immune response.

Incubation Periods

Once anatomical and chemical barriers to infection have been breached and an infection is established, a cascade of new defensive reactions occurs in the host (see Chapters 3 and 4). Symptoms and pathologies may or may not be obvious, depending upon the virus, the infected tissue and host, and the antiviral immune response. The period before the symptoms of a disease are obvious is called the **incubation period.** During this window, viral genomes are replicating and the local induction of the innate immune response produces cytokines such as interferon (IFN). Often, the initial symptoms (fever, malaise, aches, pains, and nausea) detected pursuant to infection are

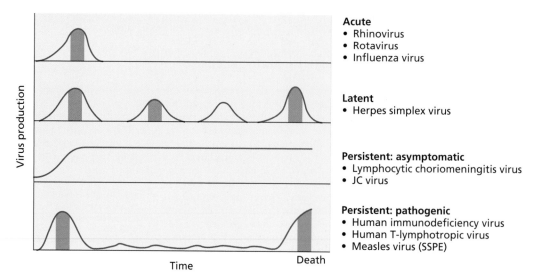

Figure 5.1 General patterns of infection. Relative virus particle production is plotted as a function of time after infection (blue line). The time during which symptoms are present is indicated by the orange shaded area. The top panel is the typical profile of an acute infection, in which virus particles are produced, symptoms appear, and the infection is cleared, often within 7 to 10 days. The second panel depicts a latent infection in which an initial acute infection is followed by a quiescent phase and repeated rounds of reactivation. Reactivation may or may not be accompanied by symptoms, but generally results in the production of infectious virus particles. The bottom two panels are variations of the profile of a persistent infection. The third panel is the typical profile of a persistent infection, in which virus production continues for the life of the host, often at low levels, or in tissues that immune cells do not routinely patrol. Symptoms may or may not be apparent, depending upon the virus. The fourth panel depicts a persistent infection, in which a period of years intervenes between a typical primary acute infection and the usually fatal appearance of symptoms. The production of infectious particles during the long period between primary infection and fatal outcome may be continuous (e.g., human immunodeficiency virus) or not detectable (e.g., measles virus, in the case of subacute sclerosing panencephalitis). Adapted from F. J. Fenner et al., *Veterinary Virology* (Academic Press, Inc., Orlando, FL, 1993), with permission.

consequences of type I IFN production. Remarkably, incubation periods can vary greatly (Table 5.1).

The intrinsic and innate responses limit and contain many acute infections. When these defenses are lacking or compromised, acute infections can be become disastrous, primarily because the infection becomes systemic and multiple organs can be damaged (provided that receptor-positive, susceptible cells are present in other tissues). If the infection spreads to multiple organs quickly, the host adaptive response may not be able to contain the infection in these various sites.

Mathematics of Growth Correlate with Patterns of Infection

Before discussing the various patterns of viral infection, it will be informative to consider the constraints of viral reproduction based on the populations that are infected. In ecology, this concept is often referred to as the r/K selection theory, in which the principle is quite simple: an organism can focus upon either increased number of offspring (with minimal attention to offspring quality), or reduced number of offspring with a corresponding increased parental investment. ***r*-selection** favors large numbers of offspring with low cost per individual, while ***K*-selected species** devote high cost in reproduction to produce

Table 5.1 Incubation periods of some common viral infections

Disease	Incubation period (days)[a]
Influenza virus	1–2
Rhinovirus	1–3
Ebola virus	2–21
Acute respiratory disease (adenoviruses)	5–7
Dengue	5–8
Herpes simplex	5–8
Coxsackievirus	6–12
Poliovirus	5–20
Human immunodeficiency virus	8–21
Measles	9–12
Smallpox	12–14
Varicella-zoster virus	13–17
Mumps	16–20
Rubella	17–20
Epstein-Barr virus	30–50
Hepatitis A	15–40
Hepatitis B and C	50–150
Rabies	30–100
Papilloma (warts)	50–150

[a]Until first appearance of prodromal symptoms.

BOX 5.3

METHODS
Mathematical approaches to understanding viral population dynamics

The changes in the size of a viral population can be described by a single, simple concept: the rate of increase in the size is the difference between the rate of reproduction and the rate of elimination. We can write this statement as

$$dN/dt = (b - d)N$$

where dN/dt is the rate of change of the population (N) with respect to time (t). The terms b and d are the average rates of birth and death, respectively (although, of course, virus particles are neither born nor die). The term $(b - d)$ is usually written as a constant r, the intrinsic rate of increase. Therefore, we obtain equation 5.1:

$$dN/dt = rN \qquad (5.1)$$

$$and \ln N = rt$$

This is the equation for exponential population growth. Plotting $\ln N$ versus t yields a straight line with slope r (Fig. 5.2A).

If b far exceeds d (as is the case for infections in cultured cells), progeny accumulate. When b equals d, the population maintains a stable size. If we assume a linear relationship for increase and decrease of the population, then the slope of the increase of reproduction rate is equal to k_b and the slope of death or removal rate is equal to k_d. The stability of the population N then can be written as follows:

$$b_0 - k_b N = d_0 + k_d N$$

or

$$N = (b_0 - d_0)/(k_d + k_b)$$

This description of N, the viral population, is called the **carrying capacity (K)** of the environment. The term "environment" can define a single cell, an individual, or the entire host population. For any value of N greater than K, the viral population will decrease, and for any value of N less than K, the viral population will increase. The carrying capacity K is of particular interest in virology, as it defines the upper boundary of the growing population and influences patterns of infection.

Therefore, by knowing that $r = (b_0 - d_0)$ and $K = (b_0 - d_0)/(k_b + k_d)$, we can substitute these values in equation 5.1 to obtain the basic equation for growth and regulation of a population, sometimes called the logistic growth equation (equation 5.2).

$$dN/dt = rN(K - N/K) \qquad (5.2)$$

Plotting $\ln N$ versus t yields the curve illustrated in Fig 5.2B. Here, K is easily seen to be the upper limit to growth, and the rate of increase is r.

a low number of offspring. One strategy is not necessarily better than the other: the environment determines which of them will predominate. For example, the average gestation period for a mouse is 21 days, an average litter is 7 to 12 pups, and the female can become impregnated again on the same day she delivers her litter. This example is an r-selection strategy, in which volume (and not quality) is evolutionarily favored, likely because mice have many natural predators. Compare this to humans, with a gestation period of 9 months, and an average litter size of one. Such a strategy may be favored in humans because of the relative paucity of predators.

How can this ecological principle be applied to viruses? Production of large numbers of viral progeny maintained by a steady, unbroken lineage of serial infections is consistent with the r-replication strategy. Such viral reproduction will never reach a limit as long as susceptible hosts are available (Box 5.3, equation 5.1; Fig. 5.2A). The alternative is the **K-replication strategy**, in which the host population is at or close to its saturation density (e.g., new susceptible hosts are rare or nonexistent, or for which rates of viral propagation may be slow or very low [Box 5.3, equation 5.2; Fig. 5.2B]).

r-replication strategies often manifest as acute infections characterized by short reproductive cycles with production of many progeny, and extensive viral spread. Acute infections following an r-replication strategy will "burn out" if the number of susceptible hosts becomes limiting. One can mimic an r-selection environment in cell culture by low multiplicity-of-infection (MOI) infections: in this case, permissive cells sustain multiple rounds of replication, but transmission stops when all the cells become infected. K-replication strategies often appear as persistent or latent infections. In this case, infected hosts survive for extended times and faster viral reproduction confers no selective advantage. Viruses and their hosts exist along a continuum of values for r and K.

The growth equations, as written in their simplest form, can be used to model replication in identical cells in culture. However, to describe accurately how a viral infection is propagated and maintained in a large population of host organisms, more variables must be considered. These additional parameters include the rate of shedding from infected individuals, the rate of transmission to other hosts, the probability that one infected individual will infect more than one other, and the number and density of susceptible individuals. Some of these parameters are discussed in greater detail in Chapter 10.

Acute Infections

The term "acute" refers to rapid onset of viral reproduction that may be accompanied by disease with a short, but occasionally severe, course. Hallmarks of an acute viral infection include the rapid production of large numbers of virus particles (hence: an r-replication strategy), followed by immune-mediated destruction of the particles and virus-infected cells. Acute infections are the typical, expected course for agents such as influenza virus, norovirus, and rhinovirus (Fig. 5.3). The disease symptoms tend to resolve over a period of days. Nevertheless, during the rapid reproduction

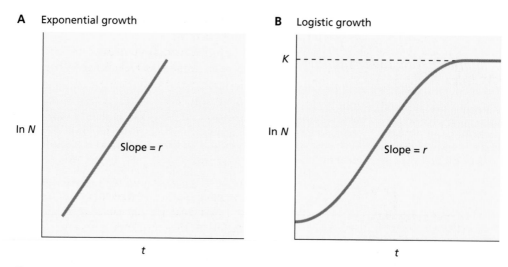

Figure 5.2 Two plots of standard growth equations. (A) A graph of simple exponential reproduction. **(B)** A graph of the pattern termed logistic growth illustrating *K*, the limit to reproduction. *r* is the slope in both plot types.

phase, some progeny are invariably shed and spread to other hosts before the infection is contained. If the initial infection modulates local immune defenses and virus spreads via hematogenous or neural routes to other parts of the body, several rounds of reproduction may occur in different tissues within the same animal, with new and distinctive symptoms. A classic example is varicella-zoster virus, an alphaherpesvirus that causes the childhood disease chickenpox, but that can recur later in life to cause a different (and far more painful) skin rash, shingles (Fig. 5.4).

An acute infection may result in limited or no obvious symptoms. Indeed, inapparent (or asymptomatic) acute infections are quite common, and can be major sources of transmission within populations. They are recognized by the presence of virus-specific antibodies with no reported history of disease. For example, over 95% of the unvaccinated population of the United States has antibody to varicella-zoster virus, but fewer than half of these individuals report that they have had chickenpox. In such infections, sufficient virus particles are made to maintain the virus in the host population, but the quantity is below the threshold required to induce symptoms. The usual way an inapparent infection is detected is by elevated antiviral antibody concentrations in an otherwise healthy individual. Well-adapted pathogens often cause asymptomatic infections, as demonstrated by poliovirus, in which more than 90% of infections are inapparent.

Figure 5.3 The course of a typical acute infection. Relative virus reproduction plotted as a function of time after infection. The concentration of virus particles increases with time, as indicated by the red line. During the establishment of infection, only intrinsic and innate defenses are at work. If the infection reaches a certain threshold (which is specific to the virus and host [purple]), adaptive immunity is initiated. After 4 to 5 days, effector cells and molecules of the adaptive response begin to clear infected tissues and virus particles (green shaded area). After this action, memory cells are produced, and the adaptive response is suppressed. Antibodies and memory cells provide lasting protection should the host be reinfected at a later date. Redrawn from C. A. Janeway, Jr., and P. Travers, *Immunobiology: the Immune System in Health and Disease* (Current Biology Ltd. and Garland Publishing, New York, NY, 1996), with permission.

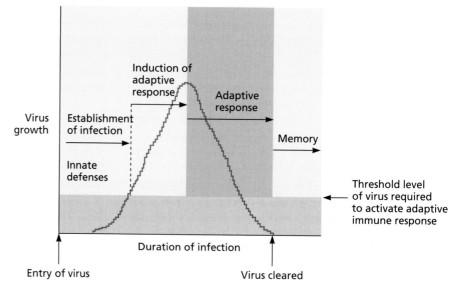

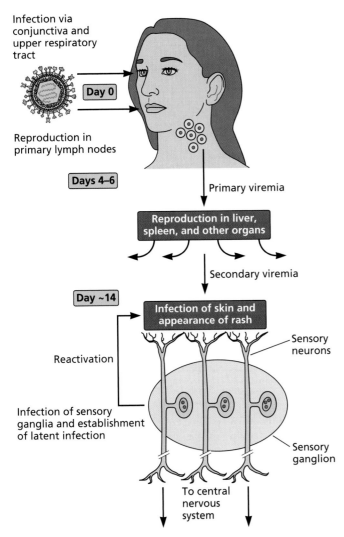

Infection via
conjunctiva and
upper respiratory
tract

Day 0

Reproduction in
primary lymph nodes

Days 4–6

Primary viremia

Reproduction in liver,
spleen, and other organs

Secondary viremia

Day ~14

Infection of skin and
appearance of rash

Reactivation

Sensory
neurons

Infection of sensory
ganglia and establishment
of latent infection

Sensory
ganglion

To central
nervous
system

Figure 5.4 Model of varicella-zoster virus infection and spread. Infection initiated on the conjunctiva or mucosa of the upper respiratory tract spreads to regional lymph nodes. After 4 to 6 days, infected T cells enter the bloodstream, causing a primary viremia. These infected cells subsequently invade the liver, spleen, and other organs, initiating a second round of infection. Virus particles and infected cells are then released into the bloodstream in a secondary viremia. Infected skin-homing T cells efficiently invade the skin and initiate a third round of infection about 2 weeks after the initial infection. The characteristic vesicular rash of chicken pox appears as a result of immune defensive action. Next, virus particles produced in the skin infect sensory nerve terminals and spread to dorsal root ganglia of the peripheral nervous system, where a latent infection is established. The latent infection is maintained by active immune surveillance. Later in life, perhaps as the robustness of the immune system wanes, viral reactivation can occur, in which another infectious cycle is initiated. In this case, virus particles leave the peripheral neurons to infect the skin. The characteristic recurrent disease, called shingles, is often accompanied by a long-lasting painful condition called postherpetic neuralgia. Normally, an infected individual experiences only one visible reactivation event, probably because reactivation stimulates the immune system. Such restimulation of the immune system is the rationale for administering the varicella-zoster live vaccine to adults to prevent reactivation and shingles.

Antigenic Variation Facilitates Repeated Acute Infections

If an individual survives a typical acute infection, he or she is often immune to rechallenge by the same virus. Nevertheless, some acute infections occur repeatedly, despite the host mounting a robust immune response to them. These recurring infections are possible because selection pressures during the initial acute infections lead to release of virus particles that are resistant to immune clearance. Mutations in the genome may affect the structural properties of the virus and the capacity of neutralizing antibodies to block infectivity, or of T cells to recognize particular viral epitopes (Chapter 4).

Viral particles that can tolerate many amino acid substitutions in their structural proteins and remain infectious are said to have **structural plasticity** (e.g., influenza virus and human immunodeficiency virus). Populations of virus particles can include antibody-resistant mutants that are selected in the presence of neutralizing antibody. Such altered particles can reinfect individuals, even when there is preexisting immunity to the original virus.

Other viruses cannot tolerate many amino acid changes in their structural proteins (e.g., those of poliovirus, measles virus, and yellow fever virus). In these cases, even if the mutation rate is high, antibody-resistant infectious particles have a low probability of being generated, but have been observed (Box 5.4). This property ensured that vaccines effective in the 1950s are just as potent in the 21st century.

In contrast, the structural plasticity of rhinoviruses is a manifestation of the circulation of over 100 serotypes in the human population at any one time. This property accounts for the fact that individuals may contract more than one common cold each year, and also explains why it is difficult to produce a vaccine to prevent this disease.

Similarly, enveloped influenza virus particles that are resistant to antibodies are readily selected, whereas the enveloped particles of measles virus and yellow fever virus exhibit little variation in membrane protein amino acid sequence, and antibody-resistant variants are rarely observed. Consequently, an influenza vaccine is required every year, while a single measles virus vaccination typically lasts a lifetime.

Antigenic variation refers to changes in virus proteins in response to antibody selection. In an immunocompetent host, antigenic variation arises by two distinct processes. **Antigenic drift** is the appearance of virus particles with slightly altered surface proteins (antigen) following passage in the natural host (Fig. 5.5). In contrast, **antigenic shift** denotes a major change in the surface protein(s) of a virus particle, as genes encoding novel surface proteins (or substantial variants of known proteins) are acquired (Chapter 10). Fortunately, most year-to-year changes in the circulating influenza strains are due to antigenic drift. Consequently, last year's vaccine generally confers some protection against this year's virus. More

BOX 5.4

DISCUSSION
Poliovirus escapes antibodies

(This Box is adapted from a blog post by Dr. Vincent Racaniello (www.virology.ws) on August 29, 2014.)

Antigenic variation is a hallmark of influenza virus that allows host defenses to be circumvented. Consequently, influenza vaccines need to be reformulated frequently to keep up with changing viruses. In contrast, antigenic variation is not a hallmark of poliovirus, and the same poliovirus vaccines have been used for nearly 60 years to control infections by this virus. An exception is poliovirus type I that caused a 2010 outbreak in the Republic of Congo.

The 2010 outbreak (445 paralytic cases) was unusual because the case fatality ratio of 47% was higher than typically observed (usually less than 10% of patients with confirmed disease die). The first clue that something was different in this outbreak was the finding that sera from some of the fatal cases failed to neutralize effectively infection of cells by the strain of poliovirus isolated during this outbreak (the strain is called PV-RC2010). The same sera effectively neutralized the three Sabin vaccine viruses as well as wild type 1 polioviruses isolated from previous outbreaks. Therefore, gaps in vaccination coverage were not solely responsible for this outbreak.

Examination of the nucleotide sequence of the genome of type I polioviruses isolated from 12 fatal cases identified two amino acid changes within a site on surface of the viral capsid that is bound by neutralizing antibodies (illustration). The sequence of this site, called 2a, was changed from **Ser**-Ala-**Ala**-Leu to **Pro**-Ala-**Asp**-Leu. This particular combination of amino acid substitutions has never been seen before in poliovirus. Virus PV-RC2010, which also contains these two amino acid mutations,

is completely resistant to neutralization with monoclonal antibodies that recognize antigenic site 2.

Poliovirus neutralization titers were determined using sera from Gabonese and German individuals who had been immunized with Sabin vaccine. These sera effectively neutralized the type I strain of Sabin poliovirus, as well as type 1 polioviruses isolated from recent outbreaks. However, the sera had substantially lower neutralization activity against PV-RC2010. Nucleotide sequence analysis of PV-RC2010 showed that it is related to a poliovirus strain isolated in Angola in 2009, the year before the Republic of Congo outbreak. The Angolan virus had just one of the two amino acid changes in antigenic site 2a found in PV-RC2010.

It is possible that the relative resistance of the polioviruses to antibody neutralization might have been an important contributor to the high virulence observed during the Republic of Congo outbreak. The reduced ability of serum antibodies to neutralize virus would have led to higher concentrations of virus particles in the blood and a greater chance of entering the central nervous system. Another factor could also be that many of the cases of poliomyelitis were in adults, in which the disease is known to be more severe.

An important question is whether poliovirus strains such as PV-RC2010 pose a global threat. Typically, the fitness of antigenically variant viruses is not the same as wild type, and therefore such viruses are not likely to spread in well immunized populations. The incomplete poliovirus immunization coverage in some parts of the world, together with the reduced circulation of wild-type polioviruses leads to reduced population immunity. Such a

situation could lead to the evolution of antigenic variants. This situation occurred in Finland in 1984, when an outbreak caused by type 3 poliovirus took place. The responsible strains were antigenic variants that evolved as a result of use of a suboptimal poliovirus vaccine in that country.

The poliovirus outbreaks in the Republic of Congo and Finland were stopped by immunization with poliovirus vaccines, which boosted the population immunity. These experiences show that poliovirus antigenic variants such as PV-RC2010 will not cause outbreaks as long as we continue extensive immunization with poliovirus vaccines, coupled with environmental and clinical testing for the presence of such viruses.

Reconstruction of a poliovirus particle bound by antibodies. Figure courtesy of Jason Roberts, Victorian Infectious Diseases Reference Laboratory, Doherty Institute, Melbourne, Australia.

rarely, antigenic shift occurs, but when it does, it is accompanied by a huge increase in the number of cases, as few individuals have existing immunity to the reassortant virus.

Acute Infections Pose Common Public Health Problems

An acute infection is most frequently associated with serious outbreaks or epidemics, affecting millions of individuals every year (e.g., influenza virus, norovirus). The nature of an acute infection presents serious obstacles for physicians, epidemiologists, drug companies, and public health officials: by the time people report symptoms, they will probably have transmitted the virus to a naïve host. Such infections can be

difficult to diagnose retrospectively, or to control in large populations or crowded environments (such as day care centers, military camps, college dormitories, and nursing homes). When distrust in the government and public health care are also present, as in the 2014 Ebola virus outbreak in Liberia, Sierra Leone, and Guinea, limiting the exposure of naïve individuals to those who are infected becomes a Herculean task. Effective antiviral drug therapy requires treatment early in the infection, often before symptoms are apparent. Antiviral drugs can be given in anticipation of an infection, but this strategy demands that the drugs be affordable, safe, and free of side effects. Moreover, our arsenal of antiviral drugs

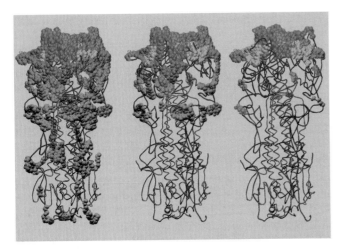

Figure 5.5 Antigenic drift: distribution of amino acid residue changes in hemagglutinins (HA) of influenza viruses isolated during the Hong Kong pandemic era (1968 to 1995). The space-filling models represent the virus-receptor binding site (yellow) and the substituted amino acids (green). **(Left)** All substitutions in HAs of virus particles isolated between 1968 and 1995; **(middle)** amino acid substitutions that were retained in subsequent years; **(right)** amino acid substitutions detected in monoclonal antibody-selected variants of A/Hong Kong/68 HA. The α-carbon tracings of the HA1 and HA2 chains are shown in blue and red, respectively. Adapted from T. Bizebard et al., *Curr. Top. Microbiol. Immunol.* **260:**55–64, 2001, with permission.

is modest (Chapter 9), and drugs effective for most common acute viral diseases do not exist.

Persistent Infections

Persistent infections occur when the primary infection is not cleared by the host immune response. Instead, virus particles, proteins, and genomes continue to be produced or persist for long periods, often for the life of the host. Virus particles may be produced continuously or intermittently for months or years, even in the presence of an ongoing immune response. In some instances, viral genomes remain after viral proteins can no longer be detected.

The persistent pattern is particularly common for noncytopathic viruses (Table 5.2). Some viruses, including the arenavirus, lymphocytic choriomeningitis virus, are inherently noncytopathic in their natural hosts and maintain a persistent infection if the host cannot mount a sufficient immune response. Other viruses toggle between a cytolytic phase and a noncytopathic phase. Epstein-Barr virus infections are typified by alternative transcription and replication programs that maintain the viral genome in some cell types with no production of viral particles. In other infections, such as those of some adenoviruses, circoviruses, polyomaviruses, and human herpesvirus 7, viral reproduction and shedding take place, but are uneventful in most individuals. What is clear from these examples is that no single mechanism is responsible for establishing a persistent viral infection. However, when viral cytopathic effects are minimized, and host defenses are suppressed, a persistent infection is likely.

Multiple Cellular Mechanisms Promote Viral Persistence

Whether viral infection leads to multiple rounds of reproduction or persistence may be based on the behavior of the infected cell. The alphavirus, Sindbis virus, provides a good example. Apoptosis is a common intrinsic cellular defense that can limit or expand viral reproduction and spread (Chapter 3). In some vertebrate cell lines, Sindbis virus infection is acute and cytopathic because apoptosis is induced.

Table 5.2 Some persistent viral infections of humans

Virus	Site(s) of persistence	Consequence(s)
Adenovirus	Adenoids, tonsils, lymphocytes	None known
Epstein-Barr virus	B cells, nasopharyngeal epithelia	Burkitt's lymphoma, Hodgkin's disease
Human cytomegalovirus	Kidneys, salivary gland, lymphocytes,[a] macrophages,[a] stem cells,[a] stromal cells[a]	Pneumonia, retinitis
Hepatitis B virus	Liver, lymphocytes	Cirrhosis, hepatocellular carcinoma
Hepatitis C virus	Liver	Cirrhosis, hepatocellular carcinoma
Human immunodeficiency virus	CD4[+] T cells, macrophages, microglia	AIDS
Herpes simplex virus types 1 and 2	Sensory and autonomic ganglia	Cold sore, genital herpes
Human T lymphotropic virus types 1 and 2	T cells	Leukemia, brain infections
Papillomavirus	Skin, epithelial cells	Papillomas, carcinomas
Polyomavirus BK	Kidneys	Hemorrhagic cystitis
Polyomavirus JC	Kidneys, central nervous system	Progressive multifocal leukoencephalopathy
Measles virus	Central nervous system	Subacute sclerosing panencephalitis, measles inclusion body encephalitis
Rubella virus	Central nervous system	Progressive rubella panencephalitis
Varicella-zoster virus	Sensory ganglia	Zoster (shingles), postherpetic neuralgia

[a]Proposed but not certain.

However, Sindbis virus causes a persistent infection of cultured postmitotic neurons because these cells synthesize Bcl2, a cellular protein that blocks apoptosis, and are therefore intrinsically resistant to virus-induced apoptosis. The *in vitro* studies are recapitulated in host animals. When Sindbis virus is injected into an adult mouse brain, a persistent noncytopathic infection is established. In contrast, when the same inoculum is injected into neonatal mouse brains, the infection is cytopathic and lethal, because neonatal neurons do not synthesize the gene products that block virus-induced apoptosis.

The intrinsic IFN response can also be important in determining patterns of infection. For example, bovine viral diarrhea virus, a pestivirus in the family *Flaviviridae*, establishes a lifelong persistent infection in a large proportion of cattle around the world. Persistently infected animals have no detectable antibody or T cell responses to viral antigens. Cytopathic and noncytopathic strains have been useful in understanding how persistence is established in the apparent absence of a host response. Infection of pregnant cattle by a cytopathic strain is contained quickly and eliminated. This phenotype depends on a rapid fetal IFN response that clears the infection. In contrast, infection of pregnant cattle with the noncytopathic virus during the first half of gestation results in birth of sickly, but viable, persistently infected calves. Noncytopathic infection of fetal tissue does not stimulate production of IFN, presumably because the virus is perceived as "self" during development, and does not invoke immunity. Consequently, the adaptive immune system is not activated, and because the virus does not kill cells, a persistent infection is established.

Modulation of the Adaptive Immune Response Can Perpetuate a Persistent Infection

Interference with Toll-like receptor detection and signaling. Viral infection triggers an early host response through activation of pattern recognition receptors, including Toll-like receptors (Tlr). Given the central role of Tlrs in the early immune response, it should not be surprising that these pathways are modulated following viral infection. Epstein-Barr virus activates Tlrs, including Tlr2, Tlr3, and Tlr9, but the expression of, and signaling by, Tlrs is attenuated during productive infection, probably as a result of the action of at least three viral deubiquitinases, including the tegument protein Bplf1, which suppresses Tlr-mediated activation of Nf-κB.

While DNA viruses that encode many "nonessential" genes have the greatest number and diversity of these immune-interfering proteins, small RNA viruses with more limited coding capacity can also block the host response. For example, hepatitis C virus, notorious for its ability to establish persistent liver infection, encodes the NS3/4A serine protease which degrades Trif, an adapter protein that is essential for signaling from Tlr3 to induce a multitude of antiviral defenses. Consequently, hepatitis C virus establishes a persistent infection in hepatocytes by interfering with a critical early step in the antiviral immune response, and by affecting the synthesis of numerous antiviral proteins.

Interference with production and function of MHC proteins. Cell lysis and production of inflammatory cytokines by cytotoxic T lymphocytes (CTL) are among the most powerful weapons in the antiviral arsenal (Chapter 4). CTLs make cytokines and cause cytolysis following engagement of the T cell receptor with viral peptides presented by major histocompatibility complex (MHC) class I proteins on the surface of the infected cell. Consequently, any mechanism that prevents viral peptides from binding to MHC class I molecules, even transiently, provides a potential selective advantage for the virus. Not surprisingly, then, the production of MHC class I proteins is modulated after many acute infections. Presumably such modulation prevents or delays elimination by CTLs so that sufficient progeny can be disseminated.

Many of the MHC-processing or regulatory steps were not known until the viral proteins that interfere with them were elucidated: several viral proteins block presentation of MHC class I molecules at the cell surface by interfering with various steps in the pathway (Fig. 5.6). Peptide presentation by MHC class I proteins can be reduced by lowering the expression of the MHC genes, blocking the production of peptides by the proteasome, or interfering with subsequent assembly and transport of the MHC-peptide complex to the cell surface.

Human cytomegalovirus deserves special mention, because MHC class I presentation of viral antigens is inhibited by multiple mechanisms during an acute infection. This betaherpesvirus causes a common childhood infection with inapparent to mild effects in healthy individuals. These infections are not cleared, and a persistent infection is established in salivary and mammary glands as well as the kidneys. Virus particles are secreted in saliva, milk, and urine. In addition, a latent infection is established in early precursor cells of the monocyte/macrophage lineage; no virus particles are produced from these cells. When latently infected individuals become immunosuppressed, cytomegalovirus reproduction resumes, often causing a life-threatening disease. The cytomegalovirus US6 protein inhibits translocation of viral peptides into the endoplasmic reticulum lumen, while the viral US3 protein binds to, and detains, MHC class I proteins in the endoplasmic reticulum, preventing their transport to the cell surface. Simultaneously, the US11 and US2 proteins eject MHC class I molecules from the endoplasmic reticulum lumen into the cytoplasm, where they are degraded by the proteasome. One possible reason that the genome of this virus encodes so many proteins to block antigen presentation is that multiple gene products act additively or synergistically to delay immune clearance until macrophage/monocyte precursors are infected, and a latent infection established.

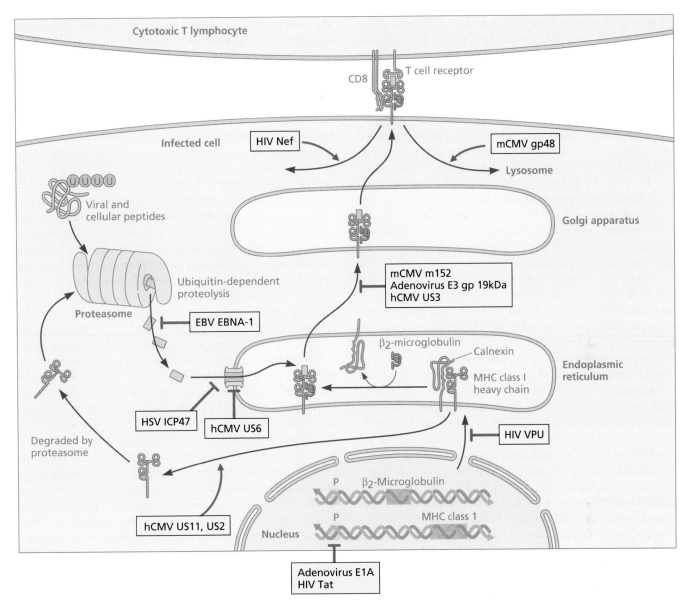

Figure 5.6 Viral proteins block cell surface antigen presentation by the MHC class I system. Specific gene products from diverse virus families block presentation of viral peptides with class I major histocompatibility complex molecules at almost every step of antigen processing and MHC assembly. Green arrows indicate stimulation; red bars indicate inhibition. HSV, herpes simplex virus; hCMV, human cytomegalovirus; mCMV, mouse cytomegalovirus; HIV, human immunodeficiency virus; EBV, Epstein-Barr virus.

Ubiquitinylation of proteins is an important regulatory mechanism that governs endocytosis, sorting, and degradation (Volume 1, Box 9.10). The genomes of many gammaherpesviruses and poxviruses encode a class of zinc-binding RING finger proteins with E3 ubiquitin ligase activity, which can interfere with class I MHC antigen presentation, stimulate viral replication, and inhibit apoptosis. The K3 and K5 genes of human herpesvirus 8 and the MK3 gene of murine gammaherpesvirus 68 encode such proteins. The K5 and MK3 proteins are related type III transmembrane proteins but,

unexpectedly, act at different steps in the MHC class I antigen presentation pathway: the K5 protein targets MHC class I proteins and costimulatory molecules for degradation by adding ubiquitin to the cytoplasmic domains of these integral membrane proteins and stimulating their endocytosis and destruction. In contrast, ubiquitinylation by the related MK3 protein promotes proteasomal destruction of MHC class I proteins soon after they appear in the endoplasmic reticulum. The genome of myxoma virus (a poxvirus) encodes a similar RING finger E3 ligase called MV-LAP that also directs proteasomal destruction by a mechanism analogous to that used by the K5 protein. Importantly, while the effects of K5 protein on human infections cannot be assessed, myxoma virus mutants that lack the MV-LAP gene are markedly attenuated in rabbits, the natural host.

Epstein-Barr virus, a member of the herpesvirus family, is among the most common human infections, responsible for infectious mononucleosis as well as certain cancers, including Burkitt's lymphoma, nasopharyngeal carcinoma, and Hodgkin's lymphoma. Early observations indicated that Epstein-Barr virus-infected individuals do not produce CTLs capable of recognizing the viral protein EBNA-1. This phosphoprotein is found in nuclei of latently infected cells, and is regularly detected in malignancies associated with the virus. T cells specific for other Epstein-Barr virus proteins are amplified, indicating that EBNA-1 must possess some intrinsic feature that allows escape from T cell detection. In fact, this protein contains an amino acid sequence that renders it invisible to the host proteasome. As a result, no EBNA-1 epitopes are produced. Remarkably, this inhibitory sequence can be fused to other proteins to inhibit their processing and the subsequent presentation of peptide antigens normally produced from them. The biological relevance of this mechanism is evident after acute infection of B cells: T cells kill all productively infected cells, sparing only those rare cells that produce EBNA-1, which harbor a latent viral genome.

MHC class II modulation after infection. In the exogenous pathway of antigen presentation, proteins are internalized and degraded, producing peptides that can bind to MHC class II molecules (Chapter 4). These complexes are transported to the cell surface, where they can be recognized by the CD4$^+$ T cell receptor. Activated CD4$^+$ T helper (T$_h$) cells stimulate the development of CTLs, and help coordinate an antiviral response to the pathogen. In many respects, they are the master regulators of the adaptive response. Consequently, any viral protein that modulates the MHC class II antigen presentation pathway would interfere with T$_h$ cell activation, and subsequent coordination of the adaptive immune response. Numerous mechanisms of viral interference in MHC class II presentation have been identified. The human cytomegalovirus US2 protein promotes proteasomal destruction of class II molecules. A protein encoded by a less pathogenic herpes simplex virus removes the MHC class II complex from the endocytic compartment.

Bypassing CTL lysis by mutation of immunodominant epitopes. Although many peptides are generated following proteolysis, T cells respond to very few viral peptides. These peptides are said to be **immunodominant**. An extreme example of a limited CTL response is observed after infection of C57BL/6 mice with herpes simplex virus type 1.

The virus-specific CTLs respond almost exclusively to a **single** peptide in the viral envelope protein gB.

A narrow repertoire of viral peptides to which immune cells can respond provides a ready opportunity for avoidance of T cell recognition, as a limited number of mutations in the coding sequence for the immunodominant peptides will render the infected cell virtually invisible to the T cell response. Viruses with these mutations, called CTL escape mutants, are thought to contribute to the accumulation of virus particles as a result of decreased immune efficacy. For example, CTL escape mutants, which are of central importance in the pathogenesis of human immunodeficiency virus type 1, arise as a consequence of error-prone genome replication and the selective pressure from constant exposure of the virus population to an activated immune response. In some cases, the sequence encoding the T cell epitope is completely deleted from the viral gene. Understanding how immunodominant peptides are selected, maintained, and bypassed is essential if effective vaccines against human immunodeficiency virus are to be developed. For example, a vaccine designed to target a dominant T cell peptide that is part of a critical structural motif in a viral protein may be useful, because CTL escape mutants (in which the critical motif is altered) will be less likely to be maintained and participate in subsequent spread of the virus.

Immunodominant epitopes and CTL escape mutants are crucial players in the common and dangerous infection caused by hepatitis C virus. The CTL response stimulated by acute infection is effective in fewer than 20 to 30% of individuals, and the majority of patients become persistently infected. After several years, this persistent infection can lead to serious liver damage and fatal hepatocellular carcinoma. Persistently infected chimpanzees harbor hepatitis C viruses with CTL escape mutations in their genomes. In contrast, the viral population isolated from animals that resolved the infection during the acute phase includes no such mutants. The principle derived from these observations is clear: if CTL escape mutations are present or arise early in the infection, a persistent infection is likely, but if CTLs clear the infection before escape mutants arise, persistent infection cannot occur.

The CTL epitope need not be deleted or radically altered to escape CTL recognition. Single nucleotide changes in the gene, which alter the protein coding sequence by only one amino acid, can be sufficient to evade detection by an activated T cell. This inherent vulnerability in the host response is particularly important for immune modulation by RNA viruses. Given their reduced coding capacity relative to the larger DNA viruses, RNA virus genomes rarely encode immune modifying proteins. However, as their RNA-dependent RNA polymerases lack the error correction mechanisms found in DNA-dependent RNA polymerases, they can survive by producing large numbers of viral mutants, some of which may be shielded from T cell recognition: in essence, a virological invisibility cloak.

Destruction of activated T cells. In some instances, when a CTL engages with an infected cell, the CTL dies instead of the infected target. This unexpected turn of events is another remarkable example of a viral counteroffense to the host defense. Activated T cells carry a membrane receptor called Fas on their surfaces, which is related to the Tnf family of membrane-associated cytokine receptors, and binds a membrane protein called Fas ligand (FasL). When Fas on activated T cells binds FasL on target cells, the receptor trimerizes, triggering a signal transduction cascade that results in apoptosis of the T cell. Consequently, if viral proteins increase the concentration of FasL on the cell surface, any T cell (CD4$^+$ or CD8$^+$) that engages it will undergo cell suicide. This mechanism has been proposed to explain the relatively high frequency of "spontaneous" T cell apoptosis that occurs in human immunodeficiency virus-infected patients. The viral Nef, Tat, and SU proteins, human T lymphotropic virus Tax protein, and the human cytomegalovirus IE2 protein have all been implicated in promoting increased synthesis of FasL within infected cells. Of note, this insidious mechanism for killing T cells is co-opted from an important host process to limit immunopathology: normal Fas-mediated CTL killing removes activated T cells when they are no longer needed after infection, or when their presence in a tissue may be detrimental. For example, certain irreplaceable organs or tissues, such as those in the eye, remain free of potentially destructive T cells by maintaining a high concentration of FasL on cell surfaces.

Persistent Infections May Be Established in Tissues with Reduced Immune Surveillance

Cells and organs of the body differ in how extensively they are patrolled by circulating immune cells. Those with less surveillance may be a fortuitous site for establishment of a persistent infection. Possibly the most extreme example of a virus family that escapes immune detection are the papillomaviruses that cause skin warts. Production of infectious particles occurs only in the outer, terminally differentiated skin layer, where an immune response is impossible because of the absence of capillaries at the skin surface. Moreover, dry skin continually flakes off, ensuring efficient spread of infection. The dust on your desk or the particles that catch the sunlight beaming through a window are most likely keratin from human skin.

Certain compartments of the body, such as the central nervous system and vitreous humor of the eye, lack initiators and effectors of the inflammatory response, because these tissues can be damaged by the fluid accumulation, swelling, and ionic imbalances that accompany inflammation. The brain, for example, is shielded by the skull that provides protection against blunt injury, but also constrains the tissue it surrounds. Consequently, even modest inflammation in the brain would be dangerous, as the brain has no "room" in which to expand. In addition, because most neurons do not regenerate, cytolytic immune defenses could be catastrophic. Because of these unique aspects of the central nervous system, the antiviral response to viral infections is notably distinct in the brain, favoring noncytolytic clearance via cytokine release. Persistent infections occur in such tissues more frequently than in those such as the lung and gastrointestinal tract with extensive immune surveillance (Table 5.2).

Persistent Infections May be Established in Cells of the Immune System

Some viruses such as measles virus, Epstein-Barr virus and human immunodeficiency virus, can infect cells of the immune system. Infected lymphocytes and monocytes rapidly disseminate throughout the host, providing efficient delivery of virus particles to new tissues. If infected immune cells die or become impaired during an acute infection, the host response could be rendered ineffective, and a persistent infection may be established. Systemic immunosuppression as a result of viral infection is discussed later in this chapter.

Human immunodeficiency virus type 1 infects not only CD4$^+$ Th cells, but also enters monocytes, dendritic cells, and macrophages, all of which can transport the virus to lymph nodes, the brain, and other organs. One might expect that the immune system would crash within a few days of the initial infection, but this does not happen, primarily because immune cells are continuously replenished. The new cells can be infected subsequently and die, but on average, the immune system remains functional for years following seroconversion: virus-triggered death is balanced by immune cell replacement. As a result, an untreated individual with human immunodeficiency virus continues to produce very large quantities of virus particles in the face of a highly activated immune system. It is only at the end stage of disease, when viral reproduction finally outpaces replenishment, that massive and fatal immune collapse occurs.

Two Viruses That Cause Persistent Infections

Measles virus infection in humans and lymphocytic choriomeningitis virus infection in mice are two well-studied examples that illustrate the establishment of persistence and the diseases associated with chronic virus reproduction.

Measles virus. This member of the family *Paramyxoviridae* is a common human pathogen with no known animal reservoir. Measles is one of the most contagious human viruses, and each year ~20 to 30 million infections occur worldwide, resulting in more than 100,000 deaths (predominantly of children). The incidence of measles varies widely in the world: most cases and fatalities occur in Southeast Asia and Africa, in contrast to Europe and the United States, where measles virus is generally well-controlled (Fig. 5.7A). Fortunately, aggressive vaccination campaigns over the past decade have reduced the global incidence of measles virus infection (Fig. 5.7B).

A

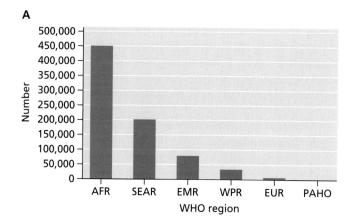

B

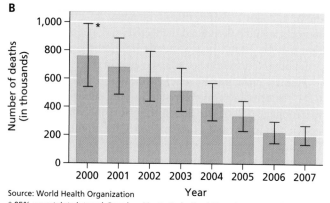

Source: World Health Organization

* 95% uncertainty interval. Based on Monte Carlo simulations that account for uncertainty in key input variables (i.e. vaccination coverage and case-fatality ratios).

Figure 5.7 Worldwide burden of measles virus. (A) Deaths due to measles virus based on geographical region. **(B)** Reduced mortality due to measles virus as a consequence of an effective worldwide vaccination campaign. AFR, African Region; SEAR, South Eastern Asian Region; EMR, Eastern Mediterranean Region; WPR, Western Pacific Region; EUR, European Region; PAHO, Pan American Health Organization. Data from CDC, *Morb Mortal Wkly Rep* **52:**471-475 and **57:**1303-1306.

After primary reproduction in the respiratory tract, measles virus infects resident monocytes and lymphoid cells that migrate to draining lymph nodes, where a small proportion enter the circulation. Infection of lymph tissues results in a secondary viremia that leads to epithelial cell infection in the lungs and skin. The course of an uncomplicated acute infection runs about two weeks, and is associated with cough, fever, the characteristic rash, and conjunctivitis (Fig. 5.8).

The vast majority of measles-infected individuals have an uneventful recovery, and lifelong immunity is established. However, during the course of acute infection, and for about two weeks after the infection is resolved, the host is transiently immunosuppressed. Consequently, secondary infections by other pathogens during this period may be uncontested by host defenses, and the results can be serious or fatal, if immediate intervention and care are not provided (see "Immunosuppression induced by viral infection" below).

In rare cases, severe, life-threatening diseases can occur when measles virus enters the brain, carried by infected lymphocytes that traverse the blood-brain barrier. The most common central nervous system complication is acute postinfectious encephalitis, which occurs in about 1 in 3,000 infections. The other is a very rare, but often lethal, brain infection called subacute sclerosing panencephalitis (SSPE). About one in 10 to 100,000 individuals with acute measles infection eventually develop SSPE, within a 6- to 8-year incubation period (Box 5.5). SSPE is most prevalent in children, especially those infected in their first year or two of life. Although the brains of SSPE victims are described histologically as "decorticated" because of massive cell loss, fully assembled particles cannot be detected in brains from autopsy specimens, perhaps because alterations in envelope proteins lead to ineffective particle assembly. Nevertheless, viral nucleoprotein complexes are produced, and infectious genomes probably spread between synaptically connected neurons. It is thought that the viral fusion protein, but not any of the known viral receptors, is necessary for spread of such complexes in the absence of assembled virus particles.

A long-standing mystery is the state of measles virus in the brain during the multiyear period between acute infection and the clinical appearance of SSPE. One possibility is that a true latency is established, in which no viral genomes are made. Alternatively, there may be a slow accumulation of progeny, with disease apparent only after a sufficient number of neurons are infected or a particular brain substructure is reached. In support of this latter hypothesis, a large number of brain samples taken from elderly individuals who had died of non-viral- and non-brain-related causes (for example, heart attacks), were positive for measles virus RNA, indicating that this virus may be able to establish life-long, central nervous system infections. It could therefore be surmised that some viral genome replication had to occur in order to sustain viral RNA for decades following acute challenge. If this were the case, the implication would be that not all viruses that enter the brain are necessarily pathogenic. However, more subtle long-term consequences of central nervous system infection by viruses has not been explored in any detail. Why, in some cases, measles infection of the brain leads to devastating diseases such as SSPE remains unknown, in part, because these diseases are so rare that they are difficult to study.

Lymphocytic choriomeningitis virus. This member of the family *Arenaviridae* was the first virus associated with aseptic meningitis in humans, although it has been most valuable as a model infection in mice. Use of this animal model has illuminated fundamental principles of immunology and viral pathogenesis, including insights into persistent infection, CTL recognition and MHC production, and immunopathogenesis. Early in the study of this virus, it was found

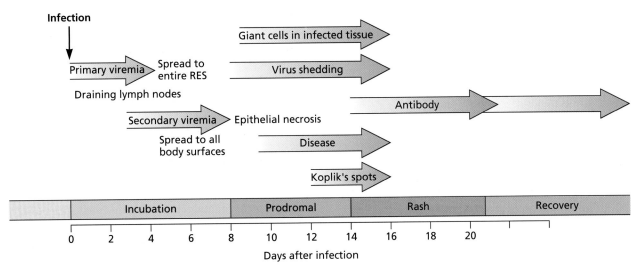

Figure 5.8 Infection by measles virus. Course of clinical measles infection and spread within the body. Four clinically defined temporal stages occur as infection proceeds (illustrated at the bottom). Characteristic symptoms appear as infection spreads by primary and secondary viremia from the lymph node to phagocytic cells, and finally to all body surfaces. The timing of typical reactions that correspond to the clinical stages is shown by the colored arrows. The telltale spots on the inside of the mouth (Koplik's spots) and the skin lesions of measles consist of pinhead-sized papules on a reddened, raised area. RES, reticuloendothelial system. Adapted from A. J. Zuckerman et al., *Principles and Practice of Clinical Virology*, 3rd ed. (John Wiley & Sons, Inc., New York, NY, 1994), with permission.

that the infection can spread zoonotically from rodents (the natural host) to humans, resulting in severe neurological and developmental damage. Infected rodent carriers excrete large quantities of virus particles in feces and urine throughout their lives without any apparent pathogenic consequence. The carrier state is established because the virus is not cytopathic and, if introduced to mice congenitally or immediately after birth (the main route of infection in the wild), viral peptides cannot be recognized as foreign ("non-self") by the developing immune response. In sharp contrast, if as few as 1 to 2 plaque-forming units are introduced intracerebrally into adult mice, the animals die of massive edema

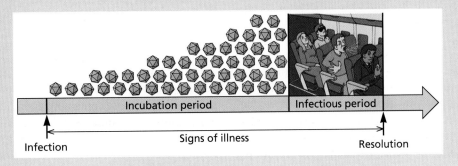

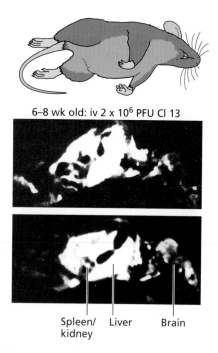

6–8 wk old: iv 2 x 10⁶ PFU Cl 13

Spleen/   Liver   Brain
kidney

Figure 5.9 Persistent infection with lymphocytic choriome-ningitis virus. Mice challenged with lymphocytic choriomeningitis virus as neonates were collected as adults, and whole body sections were made and subjected to RNA hybridization using radiolabeled probes. All "white" areas indicate active presence of viral nucleic acid. Adapted from M.B.A. Oldstone, *PLoS Pathog* **5**(7):e1000523 doi:10.1371/journal.ppat.1000523, 2009, with permission.

and encephalitis. The cerebral disease is immunopathological, as infection of adult mice lacking a functional immune response leads to persistence. In such persistently infected mice, virtually all tissues contain infectious virus, although most animals show no outward signs of illness (Fig. 5.9). More careful analyses have revealed that persistently infected mice are less "smart" than their uninfected peers, leading to the idea that persistent infections may cause nonlethal (that is, more subtle) forms of disease. The flexibility of this model system has paved the way for substantial insights into the long-term consequences of persistent infection within a host, and immune exhaustion resulting from chronic immune activation (Chapter 4).

Latent Infections

Latent infections are characterized by three general properties: viral gene products that promote virus reproduction are not made, or are synthesized in only small quantities; cells harboring the latent genome are poorly recognized by the immune system; and the viral genome persists intact so that a productive infection can be initiated at some later time to ensure the spread of viral progeny to new hosts (Fig. 5.1). The latent genome can be maintained as a nonreplicating chromosome in a nondividing cell (neuronal infection with herpes

simplex virus or varicella-zoster virus), as an autonomous, self-replicating chromosome in a dividing cell (Epstein-Barr virus infection in B cells or cytomegalovirus infection in salivary and mammary glands), or be integrated into a host chromosome, where it is replicated in concert with the host genome (adenovirus-associated virus).

There is no single mechanism to account for how all viruses can establish and maintain a latent infection. An emerging principle is that epigenetic alterations of viral genomes may facilitate the switch from productive reproduction to a latent state. Reactivation may be spontaneous (stochastic) or may follow trauma, stress, or other insults. While members of other virus families can establish latency, this property is a cardinal feature of the herpesviruses, and much is known about the establishment, maintenance, and reactivation of latency in this group of human pathogens. We therefore discuss the biology of herpes simplex type 1 and Epstein-Barr virus in some detail. How latency is established and reactivated following infection with these two herpesviruses, and the diseases associated with them, are remarkably distinct. These examples illustrate that, even within the same virus family, a common outcome can result from very different strategies.

Herpes Simplex Virus

The vast majority of adults in the United States have antibodies to herpes simplex virus type 1 or 2 and harbor latent viral genomes in their peripheral nervous systems. Approximately 40 million infected individuals will experience recurrent herpes disease as a result of virus reactivation at some point in their lifetimes. Many millions more carry latent viral genomes in their nervous systems, but never report reactivated infections. Why some people are more likely to suffer from the consequences of reactivation is poorly understood (Box 5.6). Although no animal reservoirs are known, several animals, including rats, mice, guinea pigs, and rabbits, can be infected experimentally. The alphaherpesviruses, of which herpes simplex virus type 1 is the type species, are unique in establishing latent infections predominantly in terminally differentiated, nondividing neurons of the peripheral nervous system.

The primary infection. Herpes simplex virus infections usually begin in epithelial cells at mucosal surfaces (Fig. 5.10). Virus particles are released from the basal surface in close proximity to sensory nerve endings. Because sensory terminals are abundant, they are easily infected, but autonomic nerve terminals may also be infected if deeper layers of the skin, including those containing endothelial cells of capillaries, are exposed to viral particles. If infection occurs in the eye, parasympathetic and cranial nerve endings may be invaded. Fusion of the viral envelope with any of these nerve endings releases the nucleocapsid with inner tegument proteins into

BOX 5.6

DISCUSSION

The hygiene hypothesis: why people vary in their response to herpes simplex virus infection

More than 80% of the adult population in the developed world harbor latent herpesviral genomes in their peripheral nervous system. Some individuals suffer from lesions after reactivation while others do not, although what accounts for the high infectivity yet marked diversity in host response to infection remains obscure.

It has been hypothesized that the capacity of the intrinsic and innate immune responses to stimulate appropriate adaptive immunity (T_h1 versus T_h2) is shaped by the individual's exposure to microbes early in life. A highly sanitized environment may lead to reduced stimulation of innate immunity during this critical period when the immune response learns to differentiate harmless substances (e.g., allergens) from those that can cause illness. Lack of immune education early in life may result in the reduced capacity to control infections later.

According to this hypothesis, the rising incidence of allergy and asthma, as well as of herpes simplex virus infections, in Western societies results from "hypersanitized" living conditions. Such conditions arise from use of sterilized baby food, excessive application of germicidal soaps,

antibiotics, and cleaners, and limited exposure of newborns to other individuals. Individuals who had limited exposure to microbes in early life will experience more reactivations of latent herpesvirus with severe symptoms because of their inability to mount an effective T_h1-dominated response. Instead, with inadequate early stimulation of innate immunity by microbial infections, subsequent exposure to foreign antigens may stimulate an inappropriate T_h2 response. Testing the hygiene hypothesis is not an easy matter; many observations that apparently support or refute the hypothesis are anecdotal or poorly controlled. Nevertheless, the

idea has stimulated considerable research and debate. Get out there and make some mud pies!

Camateros P, Moisan J, Henault J, De S, Skamene E, Radzioch D. 2006. Toll-like receptors, cytokines and the immunotherapeutics of asthma. *Curr Pharm Des* **12:**2365–2374.

Rouse BT, Gierynska M. 2001. Immunity to herpes simplex virus: a hypothesis. *Herpes* **8**(Suppl. 1)**:**2A–5A.

Strachan D. 1989. Hay fever, hygiene, and household size. *Br Med J* **299:**1259–1260.

Zock JB, Plana E, Jarvis D, Anto JM, Kromhout H, Kennedy SM, Kunzli N, Villani S, Olivieri M, Toren K, Radon K, Sunyer J, Dahlman-Hoglund A, Norback D, Kogevinas M. 2007. The use of household cleaning sprays and adult asthma: an international longitudinal study. *Am J Respir Crit Care Med* **176:**735–741.

A few not-so-representative examples of messy kids. (These examples are some of the rather untidy offspring of the authors).

the axoplasm. Dynein motors then move the internalized nucleocapsid on microtubules over long distances to the cell bodies of the neurons that innervate the infected peripheral tissue. A productive infection may be initiated in these neurons when the viral DNA enters the nucleus.

While it is commonplace to focus on neurons in this pattern of infection by herpes simplex virus, only 10% of the cells in a typical sensory ganglion are neurons; the remaining 90% are nonneuronal satellite cells and Schwann cells associated with a fibrocollagenous matrix. These nonneuronal cells are in intimate contact with neurons within ganglia. Some of the nonneuronal cells are infected during initial invasion of the ganglion, and may be the major source of infectious particles isolated from infected ganglia.

Establishment and maintenance of the latent infection. Soon after infection in neurons, the viral genome is coated with nucleosomes and may be silenced. In this case, transcription of viral genetic information is limited, and a quiescent, latent infection is established (Volume I, Chapter 8). As we will see, the establishment of this latent state is likely

to depend both on viral regulatory proteins and RNA, and the intrinsic and innate immune defenses that protect these tissues.

In general, most neurons neither replicate their genomes nor divide, and so once a silenced viral genome is established in the nucleus, no further viral reproduction is required for it to persist. Standard antiviral drugs and vaccines cannot cure a latent infection. Consequently, latency is sustained for the life of the host, or, as one herpesvirologist put it, "herpes is forever."

In several animal models and presumably in infected humans, peripheral ganglia support a robust acute infection with production of appreciable numbers of virus particles followed by a strong inflammatory response. Nevertheless, after 1 or 2 weeks, infectious particles can no longer be isolated from the ganglia, and establishment of the latent infection is inevitable. Inflammatory cells may persist in the latently infected ganglia for months or years, perhaps as a result of continuous or frequent low-level reactivation and production of viral proteins in latently infected tissue. Reactivated virus could come from either infected satellite cells or neurons.

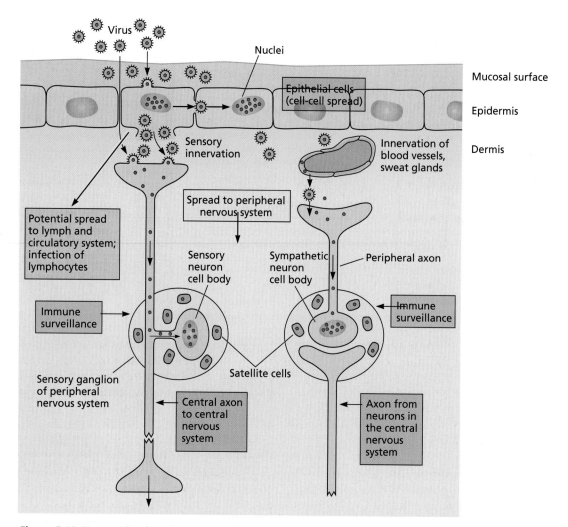

Figure 5.10 Herpes simplex virus primary infection of sensory and sympathetic ganglia. Viral reproduction occurs at the site of infection, usually in the mucosal epithelium. The infection may or may not manifest as a lesion. Host intrinsic and innate defenses, including type I IFN and other cytokines, normally limit the spread of infection at this stage. Virions may infect local immune effector cells, including dendritic cells and infiltrating natural killer cells. The infection also spreads locally between epithelial cells and may spread to deeper layers to engage fibroblasts, capillary endothelial cells, sweat glands, and other dermal cells such as those present in piloerector muscles around hair follicles. Particles that are released from basal surfaces infect nerve terminals in close contact. These axon terminals can derive from sensory neurons in dorsal root ganglia (**Left**) or from autonomic neurons in sympathetic ganglia (**Right**). Viral envelopes fuse with neuron axonal membranes, and the nucleocapsid with outer tegument proteins is transported within the axon to the neuronal cell body by microtubule-based systems (dynein motors), where viral DNA is delivered to the neuronal nucleus. Spread of productive infection to the central nervous system from these peripheral nervous system ganglia is rare. Unlike the brain and spinal cord, peripheral nervous system ganglia are in close contact with the bloodstream and are exposed to lymphocytes and humoral effectors of the immune system (immune surveillance). Consequently, infected ganglia become inflamed and populated with lymphocytes and macrophages. Infection of the ganglion is usually resolved within 7 to 14 days after primary infection, virus particles are cleared, and a latent infection of some neurons in the ganglion is established.

We do not understand why neurons are the favored site for a latent infection. Furthermore, it is difficult to explain how neurons in ganglia survive the primary infection by this markedly cytolytic virus. Most relevant to human disease, we do not understand why the infection stops in the first-order neurons of the peripheral nervous system and rarely spreads to the central nervous system, which is in direct synaptic contact with peripheral neurons. Were this not to be the case, devastating encephalitis would presumably be much more commonplace.

The latency-associated transcripts. Many latently infected neurons synthesize RNA molecules termed latency-associated transcripts (LATs) (discussed in Volume I, Chapter 8). Some researchers argue that all latently infected neurons synthesize LATs, while others report that only 5 to 30% do so. As in many studies of the latent state, the results depend on the animal model.

After infection of rabbits, viral mutants that do not synthesize LATs establish a latent infection, but spontaneous reactivation is markedly reduced. Despite this observation, which suggested that LATs contribute to reactivation, identifying molecular functions for the LATs continues to be a challenge. The major LAT contains two prominent open reading frames with potential to encode two proteins, but there is little evidence that these proteins are produced. Furthermore, disruption of these open reading frames has no effect on latency establishment or reactivation, and the sequences are not conserved in the closely related herpes simplex type 2 genome.

If the herpes simplex virus LATs are not translated, then the RNA molecules themselves may have biological activity. One hypothesis is that they are microRNA precursors that lead to degradation or reduced translation of host mRNAs. MicroRNAs may be a common feature of herpesvirus latency systems, as they now are suspected to be important for latent infections caused by the betaherpesviruses and gammaherpesviruses as well. Another proposal is that the herpes simplex virus type 1 LATs block apoptosis upon primary infection of neurons (or following reactivation). Some have contended that LATs maintain the latent state through antisense inhibition of the translation of ICP0 (a crucial viral transcription activator). Finally, it has been shown that herpes simplex virus type 1 LATs mediate the transition to latency by altering chromatin structure, perhaps by a process similar to mammalian X chromosome inactivation by the Xist RNA.

Reactivation. After reactivation of a latent infection in sensory ganglia, virus particles appear in the mucosal tissues innervated by that particular ganglion, an effective means of ensuring transmission of virus particles after reactivation, because mucosal contact is common among affectionate humans. However, in order to infect another person, sufficient virus particles must be produced in an individual who has already generated an antiviral response. Virus progeny can be produced in the face of existing immunity, in part, because the viral protein ICP47 blocks MHC class I presentation of viral antigens to T cells and facilitates spread of infection within epithelia. Such activity may provide sufficient time for virus reproduction to occur before the infected cell is eliminated by activated CTLs.

Murine models have been used to show efficient establishment of latency in neurons even in the presence of an antibody response in vaccinated animals, or in animals that receive passive immunization with virus-specific antibodies prior to infection. The immune response after reactivation is usually robust and clears the infected epithelial cells in a few days, but not before virus particles are shed. The typical "cold sore" lesion of herpes labialis is the result of the inflammatory immune response attacking the infected epithelial cells that were in contact with axon terminals of reactivating neurons. Some individuals with latent herpes simplex virus experience reactivation every 2 to 3 weeks, while others report rare (or no) episodes of reactivation. Indeed, reactivation may result in the shedding of infectious particles in the absence of obvious lesions or symptoms (Fig. 5.1). A final aspect of this reactivation phenomenon is that the virus can move directly from latently infected neurons to epithelial cells without the release of infectious progeny. Consequently, the host response would not be alerted until productive infection of epithelial cells occurs. This feature of herpesvirus reactivation presents extreme difficulties to those who strive to produce efficacious vaccines.

Reactivation from ganglia: not "all or none." The triggers that reactivate a latent infection include sunburn, stress, nerve damage, steroid use, heavy metals (the chemicals, not the music), and trauma, including dental surgery. Despite the apparent systemic nature of most reactivation stimuli, when reactivation does occur in animal models, only about 0.1% of neurons in a ganglion that contain the viral genome synthesize viral proteins and produce virus particles. The regulatory network in operation does not include an "on or off" circuit that affects all latently infected neurons, but rather may be sensitive to some nonuniformity within the latent population. Not only are different types of neurons infected in peripheral ganglia, but also the number of viral genomes in a given neuron varies dramatically (Box 5.7). Indeed, it is likely that one facet of competency for reactivation is the number of viral genomes within a given neuron: the more genomes, the more likely to reactivate.

Signaling pathways in reactivation. The diversity of potential reactivation signals may be surprising. However, it is likely that they all converge to stimulate production or action of specific cellular proteins needed for transcription of the herpes simplex virus immediate-early genes, and consequently activate the productive transcriptional program. Indeed, all of these exogenous signals have the capacity to induce the synthesis of cell cycle and transcriptional regulatory proteins that may render neurons permissive for viral reproduction. The synthesis of the viral immediate-early protein ICP0 is sufficient to reactivate a latent infection in model systems, and this viral protein has opposing functions to LATs in modulation of chromatin structure. In a single latently infected neuron, reactivation may be an all-or-none process requiring but a single reaction such as chromatin structural changes to "flip the switch" that triggers the cascade of gene expression of the

BOX 5.7

DISCUSSION

Neurons harboring latent herpes simplex virus often contain hundreds of viral genomes

The number of neurons in a ganglion that will ultimately harbor latent genomes following primary infection depends upon the host, the strain of virus, the concentration of infecting virus particles, and the conditions at the time of infection. A mouse trigeminal ganglion contains about 20,000 neurons. It is possible to infect as few as 1% to as many as 50% of the neurons in a ganglion. In controlled experiments with mice, the number of latently infected neurons increases with the dose.

Many infected neurons contain multiple copies of the latent viral genome, from fewer than 10 to more than 1,000; a small number have more than 10,000 copies. The significance of such variation in copy number has been enigmatic. Does it reflect multiple infections of

a single neuron, or is it the result of replication in a stimulated permissive neuron after infection by one particle? If it is the latter, how does the neuron recover from what should be an irreversible commitment to the productive cycle?

When virus reproduction is blocked by mutation or antiviral drugs, the number of latently infected neurons with multiple genomes is reduced significantly. Therefore, a single neuron may be infected by multiple virus particles, each of which participates in the latent infection.

Sawtell NM. 1997. Comprehensive quantification of herpes simplex virus latency at the single-cell level. *J Virol* **71:**5423–5431.

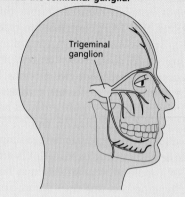

Localization of the trigeminal ganglia, also called the semilunar ganglia.

Trigeminal ganglion

---

productive pathway. Glucocorticoids are excellent examples of such activators, as they stimulate transcription rapidly and efficiently while inducing an immunosuppressive response. These properties explain the observation that clinical administration of glucocorticoids frequently results in reactivation of latent herpesviruses. While reactivation from a single neuron is likely all or none, the virus is reactivated in only a small number of neurons within an infected ganglia. Why so few of such neurons are affected, especially as glucocorticoids or trauma must impact all neurons within the ganglia, remains an important question in this field.

Epstein-Barr Virus

Epstein-Barr virus, also called human herpesvirus 4, is the type species of the gamma subfamily of herpesviruses, and is one of the most common viruses to infect humans (its only host). Indeed, in the United States, up to 95% of adults are seropositive and carry the viral genome in latently infected B cells. Two strains of Epstein-Barr virus that differ in their terminal repeats, as well as in production of nuclear antigens and small RNAs during latent infection, are recognized. Epstein-Barr virus 1 is about 10 times more prevalent in the United States and Europe than is Epstein-Barr virus 2, while both strains are equally represented in Africa. Most people are infected with the virus at an early age and have no symptoms, but some develop **infectious mononucleosis** ("mono") (Box 5.8).

Epstein-Barr virus establishes latent infections in B lymphocytes, and is one of the herpesviruses consistently associated with human cancers (Table 5.2; Appendix). As we will

learn in Chapter 6, B cell immortalization is a consequence of the mechanisms by which a latent infection is established, but the fact that the patient develops B cell lymphoma provides no selective advantage to the virus. In contrast to the nonpathogenic latent state of herpes simplex virus, the latent state of Epstein-Barr virus has been implicated in several serious diseases.

The primary infection. Epstein-Barr virus particles infect both epithelial cells and B cells. Some have proposed that distinct viral ligands engage different entry receptors on the surfaces of these two cell types, but this issue remains to be demonstrated. In any case, infection initiates in epithelial cells, usually those of the mucosal epithelia in the oropharyngeal cavity, which are sites for shedding of virus particles. Oral epithelium and tonsil tissue are rich in lymphoid cells and provide the perfect environment for the next stage of infection. After productive infection of epithelial cells, released particles can infect B lymphocytes, in which a modified transcriptional program can lead to establishment of a latent infection. The viral DNA genome exists as a circular, self-replicating episome in the B cell nucleus (Volume I, Chapter 9). This episome becomes associated with nucleosomes and undergoes progressive methylation at CpG residues. When latently infected B cells are in contact with epithelial cells, the virus may be reactivated, resulting in production of progeny particles that can infect epithelial cells. Infectious particles are shed predominantly in the saliva, but shedding from lung and cervical epithelia has also been reported.

BOX 5.8

DISCUSSION
Epstein-Barr virus, depression, and pregnancy

Reactivation of Epstein-Barr virus has been linked with depression both in late-term pregnant women, and in mothers soon after delivery. Up to 25% of women will experience depression either before or after delivery, and many of these individuals have a higher prevalence of Epstein-Barr virus reactivation. In addition to the challenges of the depression itself, the consequences for the developing fetus could be substantial: short-term depression in the pregnant female could irrevocably alter critical glucocorticoid signaling pathways.

Studies such as these, although provocative, can be interpreted in a number of ways. First, as Epstein-Barr virus is abundant within the human population, ascribing a direct cause-and-effect relationship is difficult; the

correlation could be purely circumstantial. Alternatively, a common trigger (such as stress) may cause both depression and virus reactivation, as opposed to virus reactivation being the cause of the depression. Regardless of the nature of the link, studies such as these serve as interesting reminders that viruses may be associated with symptoms or outcomes that we typically do not attribute to them.

Haeri S, Johnson N, Baker AM, Stuebe AM, Raines C, Barrow DA, Boggess KA. 2011. Maternal depression and Epstein-Barr virus reactivation in early pregnancy. *Obstet Gynecol* 117:862–866.

Zhu P, Chen Y-J, Hao J-H, Ge J-F, Huang K, Tao R-X, Jiang X-M, Tao F-B. 2013. Maternal depressive symptoms related to Epstein-Barr virus reactivation in late pregnancy. *Sci Rep* 3:3096.

Persistent infection. Both latently infected and productively infected B cells circulate among activated, virus-specific CTLs in the blood of infected individuals, and antibodies specific for viral proteins are abundant. How latency is maintained in B cells in the face of an active immune response is consequently a critical issue.

Children and teenagers are commonly infected, usually after oral contact (hence the name "kissing disease"). The acute infection requires expression of most viral genes and rapidly stimulates a potent immune response. Spread of infection to B cells in an individual with a competent immune system induces the infected cells to divide, augmenting immune and cytokine responses. The resulting disease is called infectious mononucleosis. The ensuing immune response destroys most infected cells, but approximately 1 in 100,000 survive. They persist as small, nonproliferating memory B cells that make **only** latent membrane protein 2A (LMP-2A) mRNA. They home to lymphoid organs and bone marrow, where they are maintained. These cells do not produce the B7 coactivator receptor, and therefore are not recognized or killed by CTLs (see Chapter 4).

When peripheral blood cells of an infected individual are cultured, growth factors in the medium stimulate proliferation of the rare, latently infected B cells, while uninfected B cells die. It is important to understand that these cultured immortal B cells (lymphoblasts) are not the same as latently infected cells that circulate *in vivo*. Nevertheless, because they can be propagated indefinitely, these cells comprise the best-understood model of Epstein-Barr virus latent infection. These immortal lymphoblasts synthesize a set of at least 10 viral proteins, including six nuclear proteins (termed EBNAs),

three viral membrane proteins (LMPs), small RNA molecules called EBER-1 and EBER-2, and at least 20 microRNAs. With the exception of LMP-1, the contributions of these viral products to transformation are unknown, as many are not synthesized in human cancers associated with Epstein-Barr virus infection. At least three distinct phenotypes or programs can be distinguished according to the viral gene products made in an infected B cell (called latency I to III). Synthesis of distinct sets of viral proteins and RNA in these latency types are also linked to particular Epstein-Barr virus associated diseases (Table 5.3).

The complicated collection of different B cell phenotypes is best understood in the context of normal B cell biology. To enter the resting state and become a memory cell, an uninfected B cell must have bound its cognate antigen and received appropriate signals from helper T cells in germinal centers of lymphoid tissue. During latent infection, the viral LMP-1 and LMP-2a proteins mimic all of these steps, such that the infected B cell can differentiate into a memory cell in the absence of external cues.

Table 5.3 Epstein-Barr latency programs

Latency program	Expressed viral genes	Disease(s)
0	None	None
I	LMP-2A/EBNA-1	Burkitt's Lymphoma
II	EBNA-1, LMP-1, LMP-2A, -2B	Hodgkin's disease, nasopharyngeal carcinoma
III	EBNA-1, -2, -3, -4, -5, -6, LMP-1, -2A, 2B	Infectious mononucleosis, AIDS-related immunoblastic B cell lymphoma

Although immunocompetent individuals maintain CTLs directed against many of the viral proteins synthesized in latently infected B cells, these cells are not eliminated. Some viral proteins, such as LMP-1, inhibit apoptosis or immune recognition of the latently infected cells. Moreover, EBNA-1 peptides are not presented to T cells, as discussed previously in this chapter. When the equilibrium between proliferation of latently infected B cells and the immune response that kills them is altered (e.g., upon immunosuppression), the immortalized B cells can form lymphomas (Fig. 5.11; see also Chapter 6).

Reactivation. The signals that reactivate latent Epstein-Barr virus reproduction in humans are not well understood,

Figure 5.11 Epstein-Barr virus primary and persistent infection. (Left) Primary infection. Epstein-Barr virus infects epithelial cells in the oropharynx (e.g., the tonsils). Virus particles can then infect resting B cells in the lymphoid tissue. Virus-infected B cells produce the full complement of latent viral proteins and RNAs (e.g., LMP-1 and LMP-2), and are stimulated to enter mitosis and proliferate. They produce antibody and function as B cell blasts. The latently infected B cells are attacked by natural killer cells and CTLs. **(Right)** Persistent infection. Most infected B cells are killed as a result of innate and immune defenses, but a few (approximately 1 in 100,000) persist in the blood as small, nonproliferating memory B cells that synthesize only LMP-2A mRNA. These memory B cells are presumably the long-term reservoir of Epstein-Barr virus *in vivo* and the source of infectious virus when peripheral blood cells are removed and cultured. A limited immune response to these infected B cells leads to self-limiting proliferation, infectious mononucleosis, or unlimited proliferation (polyclonal B cell lymphoma). When stimulated or propagated in culture, viral proteins needed to replicate and maintain the viral genome are again produced. Some latently infected B cells traffic to lymphoid tissues in close proximity to epithelial cells in the oropharynx. Here, the B cells are stimulated to produce particles capable of infecting and replicating in epithelial cells. Virus particles are produced and shed into the saliva for transmission to another host.

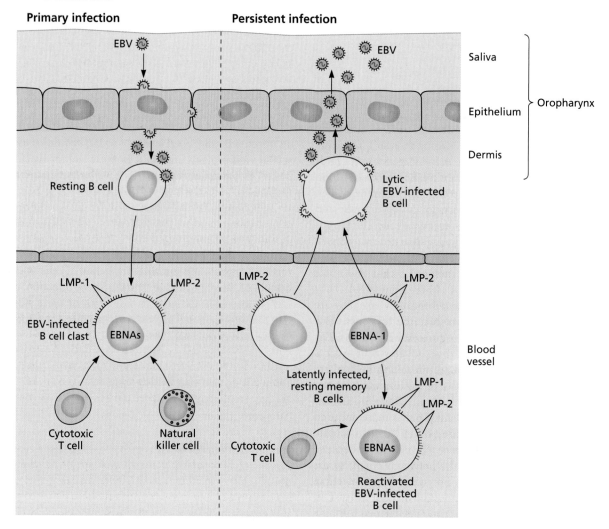

but considerable information has been obtained from studies of cultured cells. Certain signal transduction cascades that result in the production of the essential viral transcriptional activator, Zta (also called Z or zebra protein), reinitiates productive infection (Volume I, Chapter 8). However, Zta induces the full productive program only when specific promoters are methylated at CpG residues. Recall that in latently infected cells, the viral genome slowly acquires methylated cytosine residues, thereby facilitating reactivation when Zta is made. In essence, the very modifications that enable the transition to latency are also those that are critical for viral reactivation. Zta also represses the latency-associated promoters, and is responsible for recognition of the lytic origin of replication.

Many signal transduction pathways cooperate to reactivate Epstein-Barr virus from the latent state. Given this fact, it is surprising that latent infection is so stable. We now know that virus-encoded LMP-2A makes an important contribution to maintaining the latent infection by inhibiting tyrosine kinase signal transduction pathways. It is the first example of a viral protein that blocks reactivation of a latent infection. While the parameters that cause Epstein-Barr virus reactivation are less well defined than those that cause herpes simplex virus reactivation, many of the same conditions, including stress, have been implicated.

"Slow" Infections

Some fatal brain diseases of mammals are characterized by ataxia (movement disorders) or dementia (severe cognitive impairment), and likely to be a result of a variation of persistent infection, called slow infection (Fig. 5.1). In these instances, it may be years from the time of initial contact of the infectious agent with the host until the appearance of recognizable symptoms, but death then usually follows quickly.

Elucidating the molecular mechanisms responsible for an infectious-disease process of such long duration is a formidable challenge. Experimental analysis of these unusual diseases began in the 1930s, when a flock of Karakul sheep was imported from Germany to Iceland, where they infected the native sheep, causing a disease called maedi/visna. Thanks to careful work of Bjorn Sigurdsson, we now know that the maedi/visna syndrome is caused by a lentivirus. The striking feature that Sigurdsson discovered is the slow progression to disease after primary infection, often taking more than 10 years. He developed a framework of experimentation for studying these slow, progressive, and fatal brain infections that laid the foundation for later studies on human immunodeficiency virus, as well as prions and other "infectious" diseases of protein misfolding (Chapter 14).

Viruses such as measles virus, the polyomavirus JC virus, and retroviruses such as human immunodeficiency virus and human T lymphotropic virus can establish slow infections with severe nervous system pathogenesis at the end stage of disease. In many cases, the persistent infection is maintained in peripheral compartments with no apparent effect, and only enters the brain after many years. Consequently, whether "slow infection" is an accurate descriptor is of some debate: it may be that the virus is sequestered in a peripheral tissue in a latent state, reactivates, and transits to the brain where a more "typical" acute infection occurs. As mentioned earlier in this chapter, it is hard to imagine that the virus reproduces at such a slow pace that years are required for symptoms to be manifest. It is perhaps more meaningful to think of "slow" in relation to the incubation period rather than the actual rate of viral reproduction.

Abortive Infections

In an abortive infection, virus particles infect susceptible cells or hosts, but reproduction is not completed, usually because an essential viral or cellular gene is not expressed. Clearly, an abortive infection is nonproductive. Even so, it is not necessarily uneventful or benign for the infected host. Viral interactions at the cell surface and subsequent uncoating can initiate membrane damage, disrupt endosomes, or activate signaling pathways that cause apoptosis or cytokine production. In some instances, abortively infected cells may not be recognized by the immune system, and if they do not divide, the viral genome may persist as long as the cell survives. In some cases, an infection may proceed far enough that the infected cell is recognized by CTLs. Even though this scenario would not result in infectious progeny, an inflammatory response may nevertheless damage the host if sufficient cells participate.

With the advent of modern viral genetics, virologists can construct intentionally defective viral genomes, which initiate an abortive infection in the absence of a complementing gene product. One approach has been to use such defective genomes as vectors for gene therapy or as vaccines. For this idea to be effective, cytopathic genes of a prospective viral vector certainly must be eliminated. Many of the well-known, defective viral vectors lack essential genes and are designed to express only the therapeutic cloned gene. Given that intrinsic and innate defenses can be activated in response to replication-defective particles, prudence in assuming the safety of viral vectors is essential (Box 5.9). Cytotoxicity and inflammatory host responses are of particular concern if the therapeutic gene is to be delivered to a substantial number of cells, a process that requires administration of many virus particles.

Transforming Infections

A transforming infection is a special type of persistent infection. A cell infected by certain DNA viruses or retroviruses may exhibit altered growth properties and begin to proliferate faster than uninfected cells. In some cases, this change is

DISCUSSION
A viral vector leads to lethal immunopathology

In September 1999, an 18-year-old man, Jesse Gelsinger, participated in a clinical trial at the University of Pennsylvania to test the safety of a defective adenovirus designed as a gene delivery vector. It seemed like a routine procedure: normally, even replication-competent adenoviruses cause only mild respiratory disease. Most humans harbor adenoviruses as persistent colonizers of the respiratory tract and produce antibodies against the virus. The young man was injected with a very large dose of the viral vector. Four days after the injection, he died of multiple-organ failure. What

caused this devastating response to such an apparently benign virus?

Some relevant facts:

- A very large dose of virus was injected directly into his bloodstream.
- Natural adenovirus infection never occurs by direct introduction of virus particles into the circulation. Most infections occur at mucosal surfaces with rather small numbers of infecting virus particles.
- Most humans have antibodies to the adenoviral vectors used for gene therapy.

One compelling idea is that most of the infecting defective virus particles were bound by antibody present in the young man's blood. As a consequence, the innate immune system, primarily complement proteins, responded to the antibody-virus complex, resulting in massive activation of the complement cascade, and hence widespread inflammation in the vessel walls of the liver, lungs, and kidneys, and resulting in multiple-organ failure. This tragic end to a young man's life led to a large-scale reassessment of gene delivery methods and clinical protocols.

---

accompanied by integration of viral genetic information into the host genome. In others, viral genome replication occurs in concert with that of the cell. Virus particles may no longer be produced, but some or all of their genetic material generally persists. We characterize this pattern of persistent infection as transforming because of the change in cell behavior. Some transformed cells cause cancer in animals. This important infection pattern is discussed in detail in Chapter 6.

Viral Virulence

In the previous section, we discussed patterns of viral infection within individual cells and host organisms, and considered some of the diseases that may result from such infections. The manifestation of disease is an expression of viral **virulence**: a virulent virus causes disease, whereas an avirulent virus does not.

From the earliest days of experimental virology, it was recognized that viral strains often differ in virulence despite having similar reproduction rates. Virologists correctly hypothesized that the study of viruses with reduced virulence (**attenuated**), especially when compared with more virulent relatives, would provide insights into how viruses cause disease. This approach is still widely used in viral pathogenesis studies. We can experimentally alter viral genomes, and produce viruses of such limited virulence that they can be used as replication-competent vaccines (Chapter 8). Today, the methods of recombinant DNA technology allow us to mutate all genes in an unbiased way to accelerate the discovery of virulence genes.

Measuring Viral Virulence

Virulence can be quantified in a number of ways. One approach is to determine the quantity of virus that causes death or disease in 50% of the infected animals. This parameter is called the 50% lethal dose (LD_{50}), the 50% paralytic dose (PD_{50}), or the

50% infectious dose (ID_{50}), depending on the outcome that is measured (Box 5.10). Other measurements of virulence include time to death (Fig. 5.12A), the appearance of symptoms (such as a rash), the degree of fever, and weight loss. Virus-induced tissue damage can be measured directly by examining histological sections or blood (Fig. 5.12B). For example, the safety of replication-competent, attenuated poliovirus vaccine strains is determined by assessing the extent of pathological lesions in the central nervous system of experimentally inoculated animals. The reduction in the concentration of CD4+ lymphocytes in blood as a result of human immunodeficiency virus type 1 infection is another example. Indirect measures of virulence include assays for concentrations of liver enzymes (alanine or aspartate aminotransferases) that are released into the blood following infection with liver-tropic viruses such as the hepatitis viruses.

It is important to recognize that virulence is relative, and that the pathogenesis resulting from infection with a single virus strain may vary dramatically depending on the route of infection, as well as on the species, age, gender, and susceptibility of the host (Box 5.11). Consequently, the assays must be identical when comparing the virulence of two similar viruses. Furthermore, quantitative terms such as LD_{50} cannot be used to compare virulence among different viruses.

TERMINOLOGY
Measures of viral virulence

LD_{50}: Median Lethal Dose: the number of infectious particles that will kill 50% of the infected recipients.

ID_{50}: Median Infectious Dose: the number of infectious particles that will establish an infection in 50% of the challenged recipients.

A

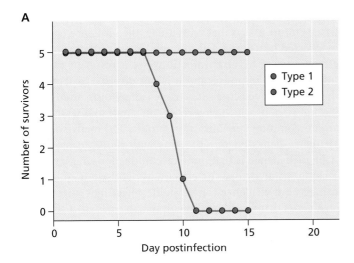

B

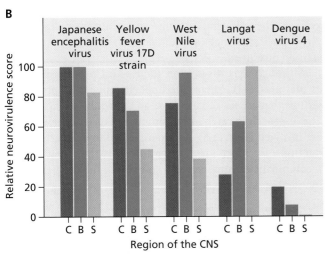

Figure 5.12 Two methods for measuring viral virulence. **(A)** Measurement of survival. Mice (5 per virus) were inoculated intracerebrally with either type 1 or type 2 poliovirus, and observed daily for survival. **(B)** Measurement of pathological lesions. Monkeys were inoculated intracerebrally with different viruses, and lesions in different areas of the central nervous system were assigned numerical values. C, cerebrum; B, brain stem; S, spinal cord. **(A)** Adapted from V. Racaniello, *Virus Res.* **1:**669–675, 1984, with permission. **(B)** Adapted from N. Nathanson (ed.), *Viral Pathogenesis and Immunity* (Academic Press, London, United Kingdom, 2007) with permission.

BOX 5.11

EXPERIMENTS
Viral virulence is dependent on multiple parameters

Lymphocytic choriomeningitis virus, a member of the arenavirus family, has been used extensively in studies of viral pathogenesis, in part owing to the distinct outcomes that occur following infection of mice via different routes. When adult immunocompetent mice are infected by a peripheral route (e.g., subcutaneously or intraperitoneally), virus reproduction is restricted to peripheral organs, the mouse mounts a robust immune response, and the virus is cleared with all mice surviving. Impressively, this outcome is independent of the original viral dose: mice can survive delivery of as many as 100,000 plaque-forming units. In sharp contrast, delivery of even 1 plaque-forming unit by an intracranial route kills all challenged mice. In these mice, the same robust immune response is made, but localized to infected cells within the brain, including the meninges, where it causes massive destruction and edema, leading to seizures that precede death (see Figure). Because disease is due solely to immunopathology, challenge of immunodeficient mice (such as recombinase-activating gene knockout mice) results in a third outcome: lifelong viral persistence throughout the mouse with no overt signs of illness.

Oldstone MBA. 2007. A suspenseful game of "hide and seek" between virus and host. *Nat Immunol* **8:**325–327.

Inoculation of mice by an intraperitoneal route with as many as 100,000 infectious units results in immunity and survival in most mice, whereas inoculation of as few as 1 infectious unit by an intracerebral route results in mortality in all challenged mice. LCMV, lymphocytic choriomeningitis virus.

Virus	LCMV-Armstrong	LCMV-Armstrong
Dose	100,000 PFU	1 PFU
Route	Intraperitoneal	Intracranial

Alteration of Viral Virulence

Before the era of modern virology, several approaches were used to identify viral virulence genes. Occasionally, avirulent viruses were isolated from clinical specimens. For example, although wild-type strains of poliovirus type 2 readily cause paralysis after intracerebral inoculation into monkeys, an isolate from the feces of healthy children was shown to be completely avirulent after inoculation by the same route. A second approach to isolate viruses with reduced virulence was to serially passage viruses either in animal hosts or in cell culture (Chapter 8).

Although these approaches were useful, their success was unpredictable. To overcome this limitation, viral genomes were often altered experimentally by exposing the viruses to mutagens (as described in Volume I, Chapter 2), and the mutant viruses were then assayed for virulence in animals. However, controlling the degree of mutagenesis was difficult, and multiple mutations were often introduced. Until the

A

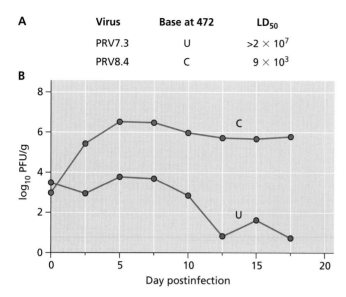

Virus	Base at 472	LD$_{50}$
PRV7.3	U	$>2 \times 10^7$
PRV8.4	C	9×10^3

Figure 5.13 Attenuation of viral virulence by a point mutation. Mice were inoculated intracerebrally with two strains of poliovirus that differ by a single base change at nucleotide 472. **(A)** The dose of virus causing death in 50% of the animals (LD$_{50}$) was determined. The change from C to U is accompanied by a large increase in LD$_{50}$. **(B)** Viral reproduction in mice was determined by plaque assay of spinal cord homogenates. The change from C to U decreases viral replication in the spinal cord. Adapted from N. La Monica, J. W. Almond, and V. R. Racaniello, *J. Virol.* **61:**2917–2920, 1987, with permission.

advent of recombinant DNA technology, the ability to identify precisely (or create) mutations in a candidate virulence gene was limited. With the ability to sequence entire viral genomes, amplify specific genomic segments by polymerase chain reaction, and perform site-directed mutagenesis, the progress in identifying candidate viral virulence genes and their products has been greatly accelerated (Fig. 5.13).

Viral Virulence Genes

Despite these powerful technological advances, the identification and analysis of virulence genes in a systematic way

has not been straightforward. Part of the problem is that no tissue culture assays that recapitulate the virulence observed in infected hosts exist. For example, many of the pathogenic effects caused by viral infections are a result of the host inflammatory response, and it is not possible to reproduce their complicated actions in a tissue culture dish. Additionally, it is not obvious which viral genes contribute to disease. There are no common "signatures" or motifs, and many so-called virulence genes encode proteins with multiple functions. A final challenge is that our expectations for viral virulence include major cytopathic effect and overt signs of cellular damage, but virulence can be subtle, affecting expression of host genes that would be difficult to appreciate in a standard cell culture-based analysis. Relevant animal models of disease are preferred for studying virulence and pathogenesis, but as noted earlier, such models are not always readily available. Nevertheless, considerable progress has been made in recent years. In the following sections, we discuss examples of viral virulence genes that can be placed in one of four general classes (Box 5.12).

Although this discussion focuses on producing viruses that are less virulent, the opposite approach, producing viruses that are **more** virulent than the wild type, is possible. The approach is rarely used, simply because unknown risks are involved (Box 11.5). An example of the inadvertent production of a more virulent pathogen is the isolation of a recombinant ectromelia virus containing the gene encoding interleukin-4 (IL-4) (Box 5.13).

Gene Products That Alter Virus Replication

Mutations in putative viral virulence genes can have one of two effects: some lead to poor reproduction of the virus, while others allow efficient reproduction, but reduced virulence (Fig. 5.14). Viral mutants that exhibit reduced or no reproduction in the animal host (or in culture) rarely cause disease, simply because they fail to produce sufficient viral progeny: this phenotype may be caused by mutations in virtually any viral gene. Some investigators mistake reduced reproduction for reduced virulence. Alternatively, some viruses exhibit

---

BOX 5.12

TERMINOLOGY
Four classes of viral virulence genes

The viral genes affecting virulence can be sorted into four general classes (and some may be included in more than one). The genes in these classes specify proteins that
- affect the ability of the virus to reproduce,
- modify the host's defense mechanisms,
- facilitate virus spread in and among hosts,
- are directly toxic.

As might be expected, mutations in these genes often have little or no effect on virus reproduction in cell culture and, as a consequence, are often called "nonessential genes," an exceedingly misleading appellation.

Virulence genes require careful definition, as exemplified by the first general class listed above (ability of the virus to reproduce).

Any defect that impairs virus reproduction or propagation often results in reduced virulence. In many cases, this observation is not particularly insightful or useful. The difficulty in distinguishing an indirect effect caused by inefficient reproduction from an effect directly relevant to disease has plagued the study of viral pathogenesis for years.

BOX 5.13

EXPERIMENTS
Inadvertent creation of a more virulent poxvirus

Australia had a wild mouse infestation, and scientists were attempting to attack this problem with a genetically engineered ectromelia virus, a member of the family *Poxviridae*. The idea was to introduce the gene for the mouse egg shell protein zona pellucida 3 into a recombinant ectromelia virus: when the virus infected mice, the animals would mount an antibody response that would destroy eggs in female mice. Unfortunately, the strategy did not work in all the mouse strains that were tested, and it was decided to incorporate the gene for IL-4 into the recombinant virus. This strategy was based on the previous observation that incorporation of this gene into vaccinia virus boosts antibody production in mice. The presence of IL-4 was therefore expected to increase the immune response against zona pellucida.

To the researchers' great surprise, rampant reproduction of the recombinant virus in inoculated mice destroyed their livers and killed them. Moreover, even mice that were vaccinated against ectromelia could not survive infection with the recombinant virus; half of them died. Essentially, they had shown that the common laboratory technique of recombinant DNA technology could be used to overcome the host immune response and create a more virulent poxvirus. Those who conducted this work debated whether to publish their findings, but eventually did so. Their results raised alarms about whether such technology could be used to produce biological weapons, and the incident was widely reported in the press.

Jackson RJ, Ramsay AJ, Christensen CD, Beaton S, Hall DF, Ramshaw IA. 2001. Expression of mouse interleukin-4 by a recombinant ectromelia virus suppresses cytolytic lymphocyte responses and overcomes genetic resistance to mousepox. *J Virol* **75:**1205–1210.
Müllbacher A, Lobigs M. 2001. Creation of killer poxvirus could have been predicted. *J Virol* **75:**8353–8355.

impaired virulence in animals, but show no defects in cells in culture (except perhaps in cells of the tissue in which disease develops). Such mutants should provide valuable insight into the basis of viral pathogenesis, because they allow identification of genes specifically required for disease.

A primary requirement for genome replication of DNA viruses is access to large pools of deoxyribonucleoside triphosphates. This need poses a significant obstacle for viruses that infect terminally differentiated, nondividing cells such as neurons or epithelial cells. The genomes of many small

Virus	Growth in cell culture	Effect on mice	Virulence phenotype
Wild type		Reproduction	Neurovirulent
Mutation leading to a general defect in reproduction		Poor reproduction	Attenuated
Mutation in a gene specifically required for virulence		Poor reproduction	Attenuated

Figure 5.14 Different types of virulence genes. Examples of virulence genes that affect viral reproduction, using intracerebral neurovirulence in adult mice as an example. Wild-type viruses reproduce well in cell culture, and after inoculation into the mouse brain, they reproduce and are virulent. Mutants with replication defects do not grow well in cultured cells, or in mouse brain, and are attenuated. Mutants with a defect in a gene specifically required for virulence reproduce well in certain cultured cells, but not in the mouse brain, and are attenuated. Adapted from N. Nathanson (ed.), *Viral Pathogenesis* (Lippincott-Raven Publishers, Philadelphia, PA, 1997), with permission.

DNA viruses encode proteins that alter the cell cycle; by forcing the cell to enter the cell cycle, substrates for DNA synthesis are produced. Another solution, exemplified by alphaherpesviruses, is to encode enzymes that function in nucleotide metabolism, such as thymidine kinase and ribonucleotide reductase. These virally encoded proteins help to increase the availability of nucleotides within infected cells. Mutations in these genes often reduce the neurovirulence of herpes simplex virus because the mutants cannot reproduce in neurons or in any other cell unable to complement the deficiency. However, because reproduction of these viruses is not affected in all cells, such genes can be classified as virulence genes.

Noncoding Sequences That Affect Virus Replication

The attenuated strains that comprise the live Sabin poliovirus vaccine are examples of viruses with mutations that are not in protein-coding sequences (Chapter 8). Each of the three serotypes in the vaccine contains a mutation in the 5′ noncoding region of the viral RNA that impairs virus reproduction in the brain (Fig. 5.15). These mutations also reduce translation of viral messenger RNA in cultured cells of neuronal origin, but not in other cell types. An interesting finding is that attenuated viruses

bearing these mutations apparently do not reproduce efficiently at the primary site of infection in the gut. Consequently, many fewer virus particles are available for hematogenous or neural spread to the brain. Mutations in the 5′ noncoding regions of other picornaviruses also affect virulence in animal models. For example, deletions in the long poly(C) tract within the 5′ noncoding region of mengovirus reduce disease in mice without affecting viral reproduction in cell culture.

Gene Products That Modify Host Defense Mechanisms

The study of viral virulence genes has identified a diverse array of viral proteins that sabotage the host's intrinsic, innate, and adaptive defenses. Some of these viral proteins are called **virokines** (secreted viral proteins that mimic cytokines and growth factors, but that do not transduce the same signals) or **viroceptors** (homologs of host receptors for these proteins). In most cases, these proteins are decoys that bind to cellular receptors, or that engage soluble immune mediators, and prevent them from performing their specific function, acting as a "sink" to delay the host immune response. Mutations in genes encoding either class of protein affect virulence, but these genes are **not** required for virus reproduction in cell culture.

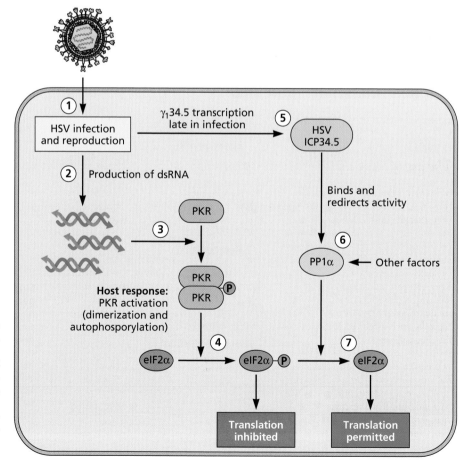

Figure 5.15 Summary of protein kinase R (PKR)-mediated protein shut-off and HSV ICP34.5 defense. Upon entry, herpes simplex virus type 1 produces double-stranded RNA molecules that are detected by the cellular PKR response. Activated PKR then phosphorylates eIF-2α, inhibiting protein translation. However, herpes simplex virus produces ICP34.5 which associates with cellular phosphatase PP1α, leading to the dephosphorylation of eIF2α and an induction of translation.

BOX 5.14

DISCUSSION
Poxviruses encode very efficient immune-modulating proteins that affect viral virulence

Variola virus, which causes the human disease smallpox, is the most virulent member of the *Orthopoxvirus* genus. The prototype poxvirus, vaccinia virus, does not cause disease in immunocompetent humans, and is used to vaccinate against smallpox. Both viral genomes encode inhibitors of the complement pathway. The vaccinia virus complement control protein is secreted from infected cells and functions as a cofactor for the serine protease factor I. The variola virus homolog, called smallpox inhibitor of complement, differs from the vaccinia virus protein by 11 amino acids. Because the variola virus protein had not been studied, it was produced by changing the 11 codons in DNA encoding the vaccinia

virus homolog. The variola virus protein produced in this way was found to be 100 times more potent than the vaccinia virus protein at inactivating human complement. These findings suggest that the virulence of variola virus, and the avirulence of vaccinia virus, might be controlled in part by complement inhibitors encoded in the viral genome. Furthermore, if smallpox should reemerge, the smallpox inhibitor of complement might be a useful target for intervention.

Poxviruses encode other immune-modifying proteins, including a decoy receptor for IFN type 1, the T1-IFN binding protein, which is essential for virulence. This protein attaches to uninfected cells surrounding infected foci

in the liver and the spleen, thereby impairing their ability to receive T1-IFN signaling, and facilitating virus spread. Remarkably, this process can be reversed; mousepox infection can be cured late in infection by treating with antibodies that block the biological function of the T1-IFNbp.

Rosengard AM, Liu Y, Zhiping N, Jimenez R. 2002. Variola virus immune evasion design: expression of a highly efficient inhibitor of human complement. *Proc Natl Acad Sci USA* **99:**8808–8813.

Xu RH, Rubio D, Roscoe F, Krouse TE, Truckenmiller ME, Norbury CC, Hudson PN, Damon IK, Alcami A, Sigal LJ. 2012. Antibody inhibition of a viral type 1 interferon decoy receptor cures a viral disease by restoring interferon signaling in the liver. *PLoS Pathog* **8:**e1002475.

Most virokines and viroceptors are encoded in the genomes of large DNA viruses (Box 5.14).

Other viral proteins interfere with the cellular intrinsic host response. Deletion of the herpes simplex virus gene encoding the ICP34.5 protein produces a mutant virus so dramatically attenuated that it is difficult to determine an LD_{50}, even when injected directly into the brain. Such mutants can reproduce in some cell types within the brain, but are

unable to grow in postmitotic neurons. ICP34.5 has multiple functions, including counteracting the activation of the IFNβ gene and opposing the innate antiviral activity of Pkr (Chapter 3). Viral ICP34.5 mutants are attenuated because they cannot prevent the cell from inducing the intrinsic responses of Pkr-dependent inhibition of translation (Fig. 5.15). Such cell type-selective attenuated viruses are under consideration as agents to selectively kill brain tumor cells (Box 5.15).

BOX 5.15

DISCUSSION
The use of attenuated herpes simplex viruses to clear human brain tumors

Malignant glioma, a prevalent type of brain tumor, is almost universally fatal within a year after diagnosis, despite clinical advances. Several groups have proposed the use of cell-specific replication mutants of herpes simplex virus to kill glioma cells *in situ*. One such virus under study carries a deletion of the ICP34.5 gene and the gene encoding the large subunit of ribonucleotide reductase. These mutant viruses reproduce efficiently in dividing cells, such as glioma cells, but not in nondividing cells, such as neurons. The hypothesis is that attenuated virus injected into the glioma will replicate and kill the dividing tumor cells, but will not impair nondividing neurons.

This idea works in principle: studies of mice have indicated that direct injection of this mutant virus into human gliomas transplanted into mice clears the tumor. The virus is attenuated and

safe: injection of 1 billion virus particles into the brain of a monkey that is highly sensitive to herpes simplex virus had no pathogenic effect. This degree of attenuation is remarkable.

Several human trials are in progress to test safety and dosage. In one study, up to 10^5 PFU was inoculated directly into the brain tumors of nine patients. No encephalitis, adverse clinical symptoms, or reactivation of latent herpes simplex virus were observed. Higher concentrations will be used until a therapeutic effect is attained.

Ning J, Wakimoto H. 2014. Oncolytic herpes simplex virus based strategies: toward a breakthrough in glioblastoma therapy. *Front Microbiol* **5:**303.

Rampling R, Cruickshank G, Papanastassiou V, Nicoll J, Hadley D, Brennan D, Petty R, MacLean A, Harland J, McKie E, Mabbs R, Brown M. 2000. Toxicity evaluation of replication-competent herpes simplex virus (ICP 34.5 null mutant 1716) in patients with recurrent malignant glioma. *Gene Ther* **7:**859–866.

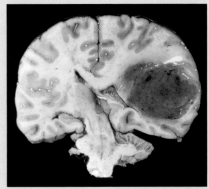

Gross anatomy of a glioblastoma. Credit: CNRI/Science Photo Library.

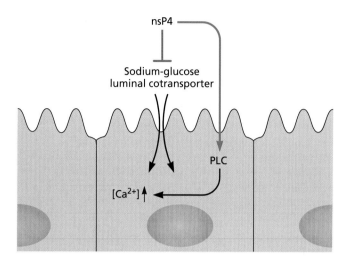

Figure 5.16 Model for rotavirus-induced diarrhea. nsP4, produced during rotavirus reproduction in intestinal epithelial cells, inhibits the sodium-glucose luminal cotransporter. Because this transporter is required for water reabsorption in the intestine, its inhibition by nsP4 could be one mechanism of diarrhea induction. nsP4 also induces a phospholipase C (PLC)-dependent calcium signaling pathway. The increase in the concentration of intracellular calcium could induce calcium-dependent chloride secretion. Adapted from M. Lorrot and M. Vasseur, *Virol. J.* **4:**31–36, 2007, with permission.

Gene Products That Enable the Virus to Spread in the Host

The mutation of some viral genes disrupts the spread from peripheral sites of inoculation to the organ in which disease is manifested (Fig. 5.16). For example, after intramuscular inoculation in mice, reovirus type 1 spreads to the central nervous system through the blood, while type 3 spreads by neural routes. Studies of recombinants between types 1 and 3 indicate that the gene encoding the viral outer capsid protein s1, which recognizes the cell receptor, determines the route of spread, and viruses with alterations in this protein are attenuated for neuroinvasion and neurovirulence.

Other viral membrane proteins have also been implicated in neuroinvasiveness. For instance, the change of a single amino acid in the gD glycoprotein of herpes simplex virus type 1 blocks neuron-to-neuron spread to the central nervous system following footpad inoculation. Similarly, studies of neuroinvasive and nonneuroinvasive strains of bunyaviruses have shown that the G1 glycoprotein is an important determinant of entry into the central nervous system from the periphery. Although it is tempting to speculate that these viral glycoproteins, which participate in entry into other cells, facilitate direct access of the virus to the nerve termini, the mechanisms by which they govern neuroinvasiveness are unknown.

Pathogenesis

We have discussed different patterns of viral infection, distinct ways by which viral virulence occurs, and the diversity of viral proteins associated with virulence. But how, ultimately, do viruses make us ill? The rapidly expanding field of viral pathogenesis attempts to integrate viral biology, cell biology, and host physiology (such as immunocompetence, age, and previous exposures) to elucidate the origins of viral disease. Some signs of virus-induced disease have been known for many centuries (as in the "dropped foot" consequence of poliovirus infection [Volume I, Chapter 1]). Other, more recently identified manifestations interfere with the health of the host in subtle ways, perhaps affecting synthesis or function of a small number of cellular proteins without overt cell destruction. We will first discuss the overt signs or symptoms of virus infection, and conclude this section with more recently identified—and more covert—ways by which viruses can cause illness in human hosts.

Infected Cell Lysis

Cell lysis is a common outcome of viral infection by most nonenveloped viruses, and some enveloped viruses. The destruction of the host cell membrane permits the release of viral progeny, and causes the death of the infected cell. There are multiple mechanisms whereby loss of membrane integrity can occur, including the production of **viroporins**.

Viroporins are hydrophobic, virus-encoded proteins that promote the release of virus particles from infected cells by associating with, and disrupting, the cellular plasma membrane. As their name implies, these proteins form pores in the membrane. Examples of viroporins include the influenza virus M2 protein, the picornavirus 2B protein, and the hepatitis C virus p7 protein. This growing family of virus proteins can accelerate the production of viral progeny either directly (by facilitating viral release) or indirectly, by destabilizing membrane polarity and integrity, allowing viruses to escape from the dying host cell.

Immunopathology

The clinical signs and symptoms of viral disease (e.g., fever, tissue damage, aches, pains, and nausea) result primarily from the host's immune response to infection (Table 5.4). This damage is called **immunopathology**, and it may be the price paid by the host to eliminate a viral infection. In fact, for noncytolytic viruses, including the hepatitis viruses and some herpesviruses, it is likely that immunopathology is the primary basis of disease. Most virus-triggered immunopathology is caused by activated T cells, but there are examples in which B cells, antibodies, or an excessive innate response are the source of disease. Because immunopathology is the result of an uncontrolled host reaction, the consequences can be severe, even life threatening (Box 5.9).

Immunopathological Lesions

Lesions caused by cytotoxic T lymphocytes. Infection of mice with lymphocytic choriomeningitis virus provides one of the most extensively characterized experimental examples

Table 5.4 Cells and mechanisms associated with immunopathology

Proposed mechanism	Virus
CD8+ T cell mediated	Coxsackievirus B
	Lymphocytic choriomeningitis virus
	Sin Nombre virus
	Human immunodeficiency virus type 1
	Hepatitis B virus
CD4+ T cell mediated	Theiler's virus
T_h1	Mouse coronavirus
	Semliki Forest virus
	Measles virus
	Visna virus
	Herpes simplex virus
T_h2	Respiratory syncytial virus
B cell mediated (antibody)	Dengue virus
	Feline infectious peritonitis virus

of cytotoxic T lymphocyte (CTL)-mediated immunopathology. The virus itself is noncytopathic and induces tissue damage only in immunocompetent animals. Experiments using adoptive transfer of T cell subtypes, depletion of cells, and gene knockout and transgenic mice, showed that the tissue damage following infection requires CTLs. Mice lacking CTLs, as well as perforin, the major cytolytic protein of CTLs, develop negligible disease after infection, whereas wild-type animals inoculated intracerebrally develop rapid rupturing of the cells that line the ventricles, resulting in massive edema, seizures, and death (choriomeningitis). The CTLs may also contribute to immunopathology indirectly, by releasing proteins that recruit inflammatory cells to the site of infection, which in turn elaborate proinflammatory cytokines.

Liver damage caused by hepatitis B virus also appears to depend on the action of CTLs. Production of the viral envelope proteins in transgenic mice has no effect. When the mice are injected with hepatitis B virus-specific CTLs, liver lesions that resemble those observed in acute human viral hepatitis develop. In this model, CTLs attach to the viral envelope protein-expressing hepatocytes and induce apoptosis. Cytokines released by these lymphocytes recruit neutrophils and monocytes, which exacerbate cell damage.

Lesions caused by CD4+ T cells. CD4+ T lymphocytes secrete larger quantities and a greater diversity of cytokines than do CTLs, resulting in the recruitment and activation of nonspecific effector cells. Such inflammatory reactions are usually called "delayed-type" hypersensitivity reactions, because of the longer period of time that must elapse for the reaction to occur, as compared to other, more immediate, hypersensitivity reactions. Most of the recruited cells are neutrophils and mononuclear cells, which can cause tissue damage as a result of release of proteolytic enzymes, reactive free radicals such as peroxide and nitric oxide (see below), and cytokines such as Tnf-α. For noncytopathic persisting viruses, the CD4+-mediated inflammatory reaction is largely immunopathological.

CD4+ T_h1 cells. The cytokines produced by CD4+ T_h1 cells facilitate the cell-mediated response, but not the antibody response. These cytokines include IL-2, IFN-β and tumor necrosis factor-alpha. CD4+ T_h1 cells are necessary for the demyelination seen in several rodent models of virally induced multiple sclerosis. When mice are inoculated with Theiler's murine encephalomyelitis virus (a picornavirus), proinflammatory cytokines produced by infected cells activate macrophages and microglial cells that mediate neuronal demyelination. It has been proposed that the activated phagocytic cells release superoxide and nitric oxide free radicals that, in combination with the T_h1 cell proinflammatory cytokines, destroys oligodendrocytes, which are the source of myelin. Similar observations have been made following infection with mouse hepatitis virus, a coronavirus. The similar demyelinating pathology caused by very different viruses suggests that these viruses may trigger a common immunopathological host reaction.

Herpes stromal keratitis is one of the most common causes of vision impairment in developed countries. The eye damage results almost entirely from immunopathology. In humans, herpes simplex virus infection of the eye induces lesions on the corneal epithelium, and repeated infections result in opacity and reduced vision. Studies of a mouse model for this disease demonstrated that CD4+ T_h1 cells contribute to immunopathology, but in an unusual manner. The surprise was that while viral reproduction occurs in the corneal epithelium, CD4+ T cell-mediated inflammation was restricted to the underlying, and uninfected, stromal cells. In fact, viral reproduction in the cornea had ceased by the time that CD4+ T cells attacked the stromal cells. It is thought that the damage to uninfected cells in the stroma is stimulated by secreted cytokines produced by infected cells in the corneal epithelium. Binding of the CD4+ T cells may be due to errant recognition of host proteins on the stromal cells by virus-specific T cells, a process called **molecular mimicry**, discussed in more detail later in this chapter.

CD4+ T_h2 cells. The cytokines produced by CD4+ T_h2 cells, including IL-4, IL-5, and IL-10, evoke strong antibody responses and eosinophil accumulation, typical responses to extracellular pathogens such as parasites and some bacteria. However, such cytokines have been implicated in some viral respiratory diseases. Respiratory syncytial virus is an important cause of lower respiratory tract disease in infants and the elderly. Models for this particular disease have been difficult to establish, but there has been some success using immunosuppressed mice. When these animals are infected, lesions of the respiratory tract are minor, but they become severe after adoptive transfer of viral antigen-specific, CD4$^+$ T$_h$2 cells. The respiratory tract lesions contain many eosinophils, which may be responsible for pathology.

The balance of T$_h$1 and T$_h$2 cells. T$_h$1 and T$_h$2 cytokine responses are not all or none, and both can be made following viral challenge. As a result, changes in the optimal balance of these powerful immune inducers can also result in immunopathology. For example, infection with respiratory syncytial virus induces a predominately T$_h$1 response in young children. However, when children were vaccinated with a formalin-inactivated whole-virus vaccine that elicited a primarily T$_h$2 response, they not only remained susceptible to infection but also developed an atypically severe disease, characterized by increased eosinophil infiltration into the lungs. This particular pathology had been predicted by adoptive transfer of CD4$^+$ T$_h$2 cells in mice.

Immunopathological lesions caused by B cells. Antibodies neutralize virus particles by binding and targeting them for elimination. Virus-antibody complexes accumulate to high concentrations when extensive viral reproduction occurs at sites that are inaccessible to the cellular immune system or continues in the presence of an inadequate immune response. Such complexes are not cleared efficiently by the reticuloendothelial system and continue to circulate in the blood. These large complexes can become deposited in small capillaries and cause lesions that are exacerbated when the complement system is activated (Fig. 5.17). Deposition of such immune complexes in blood vessels, kidneys, and brain may result in vasculitis, glomerulonephritis, and neuroinflammation, respectively. This type of immunopathology was first described in mice infected with lymphocytic choriomeningitis virus. Although immune complexes have been found in humans, viral antigens have been found in the complexes only in hepatitis B virus infections.

Antibodies may also enhance viral infection, as in dengue hemorrhagic fever. This disease is transmitted by mosquitoes and is endemic in the Caribbean, Central and South America, Africa, and Southeast Asia, where billions of people are at risk. Primary infection with dengue virus is usually asymptomatic, but in some cases, an acute febrile illness with severe headache, back and limb pain, and rash can develop. Although the infection is normally self-limiting, and patients recover in 7 to 10 days, the disease is referred to as "breakbone fever," owing to extraordinary muscle and join pain. There are four viral serotypes, and antibodies to any one serotype do not protect against infection by another. When an individual who has antibodies to one serotype is infected by a different serotype, nonprotective antibodies bind virus particles and facilitate their uptake into normally nonsusceptible peripheral blood monocytes via binding to Fc receptors. Consequently,

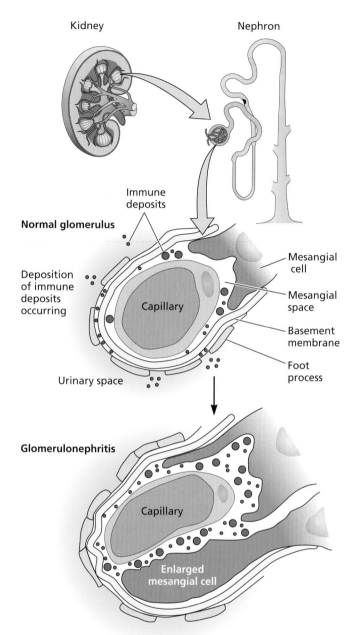

Figure 5.17 Deposition of immune complexes in the kidneys, leading to glomerulonephritis. (Top) Normal glomerulus and its location in the nephron and kidney. **(Middle)** Normal glomerulus. Red dots are immune complexes. The smaller complexes pass to the urine, and the larger ones are retained at the basement membrane. **(Bottom)** Glomerulonephritis. Complexes have been deposited in the mesangial space and around the endothelial cell. The function of the mesangial cell is to remove complexes from the kidney. In glomerulonephritis, the mesangial cells enlarge into the subepithelial space. This results in constriction of the glomerular capillary, and foot processes of the endothelial cells fuse. The basement membrane becomes leaky, filtering is blocked, and glomerular function becomes impaired, resulting in failure to produce urine. Adapted from C. A. Mims et al., *Mims' Pathogenesis of Infectious Disease* (Academic Press, Orlando, FL, 1995), with permission.

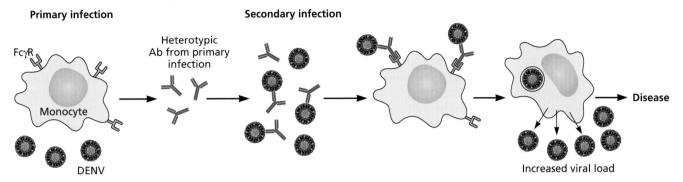

Primary infection   **Secondary infection**

FcγR

Monocyte

DENV

Heterotypic
Ab from primary
infection

Disease

Increased viral load

Figure 5.18 Model of antibody-dependent enhancement of dengue infection. Monocytes are not directly susceptible to dengue virus infection. However, when preexisting antibodies are present, a second exposure to dengue (for example, with a different serotype), allows for antibody-virus conjugates to bind to Fcγ receptors (FcγR) on circulating monocytes. Monocyte infection results in an increase in viral reproduction and a higher risk of severe dengue. DENV, dengue virus; Ab, antibody.

the infected monocytes contribute to an elevated viral load, and produce proinflammatory cytokines, which in turn stimulate T cells to produce more cytokines. This vicious cycle triggers the plasma leakage and hemorrhaging that are characteristic of dengue hemorrhagic fever (Fig. 5.18). In these instances, there may be so much internal bleeding that the often-fatal dengue shock syndrome results. Dengue hemorrhagic fever is generally rare, occurring in approximately 1 in 14,000 primary infections. However, after infection with a dengue virus of another serotype, the incidence of hemorrhagic fever increases dramatically to 1 in 90.

Systemic Inflammatory Response Syndrome (SIRS). An important tenet of immune defense is that virus reproduction induces a rapid, specific, and integrated host response to contain the infection. Typically, the scale of this response is appropriate to the pathogen, and once the pathogen is eliminated, inflammation is suppressed or limited. Precisely how this threshold is determined is not fully understood, but if it is breached too rapidly, or if the immune response is not proportional to the infection, the large-scale production and systemic release of inflammatory cytokines and stress mediators can overwhelm an infected host. Such a disastrous outcome can occur if the host is naive and has not coevolved with the invading virus (zoonotic infections; see Chapter 10), or if the host is very young, malnourished, or otherwise compromised. This type of pathogenesis is called the **systemic inflammatory response syndrome** (SIRS), and is sometimes referred to as a "cytokine storm." The lethal effects of the 1918 influenza virus have been attributed by some to this response, as well as those of human infections with Ebola and Marburg viruses. When this uncontrolled and systemic inflammation is induced by pathogens, it is referred to as **sepsis**, although noninfectious causes of SIRS, including trauma, burns, and anaphylaxis, also exist.

Heterologous T cell immunity. Much of Chapter 4 was dedicated to extolling the precise, antigen-driven induction of the adaptive immune response and resulting memory cells, but it turns out that memory cells are not always as specific as once thought. The first insight that this may be the case was the clinical observation that common infections can run surprisingly different courses in different individuals. Many variables may contribute to differential responses, but from experiments with genetically identical mice, it became clear that the history of previous encounters with pathogens can dictate the outcome of a new infection. The phenomenon is called **heterologous T cell immunity**: memory T cells specific for a particular virus epitope can be resurrected during infection with a completely unrelated virus. At first glance, it may seem advantageous for the host to quickly turn on potent immune effectors, but the consequences of activating T cells that are not tailored to the "new" pathogen may induce an inappropriate or poorly coordinated response. When mice are immunized against one of several viruses and then challenged with a panel of other viruses, the animals show partial, but not necessarily reciprocal, protection to the heterologous infection. Challenge with the arenavirus, lymphocytic choriomeningitis virus, provided substantial protection against the poxvirus, vaccinia virus, but not vice versa. For other virus pairs such as murine cytomegalovirus and vaccinia, the protection was partially reciprocal. The significance of these findings to human infections is emerging. For example, patients experiencing Epstein-Barr virus-induced mononucleosis may have a strong T cell response to a particular influenza virus epitope rather than the typical response to an immunodominant Epstein-Barr virus epitope. In these individuals, it appears that Epstein-Barr virus infection activated memory T cells that were produced by a previous exposure to influenza virus. These individuals had a different course of mononucleosis, often more severe, than did those with no previous exposure to influenza virus. Heterologous T cell immunity is a variation of a concept known as "original antigenic sin." In this scenario, a primary infection by a virus (for example), induces a protective host response against the immunodominant viral antigens, leading to resolution. If that same individual is challenged later in life with an altered virus in which that immunodominant epitope is replaced by a different epitope, the host will still make the primary response to the former (now subdominant) epitope. This weaker host response would lead to either inefficient or delayed clearance, and attendant pathogenesis.

T cell cross-reactivities among heterologous viruses are more frequent than commonly expected, but not yet well understood. Our limited knowledge about immune redundancy following pathogen exposure may be due to our heavy use of mouse models as surrogates for human infections. One of the limitations of working with mouse models is that, generally, mice are infected with a single virus, parameters of interest are examined, and the mice are killed. As a result, most experimental mice have no "immune histories" to other infections. The important principle that is emerging from more sophisticated polymicrobial studies in animal models, which are aimed to mimic more closely human virus encounters, is that prior infections can affect the defense against pathogens that have not yet been encountered, sometimes in dangerous ways.

Superantigens "Short Circuit" the Immune System

Some viral proteins are extremely powerful T cell mitogens known as **superantigens**. These proteins interact with the stalk of the Vβ chain of the T cell receptor, rather than with the antigen-binding pocket as in typical MHC-T cell receptor interactions. As approximately 2 to 20% of **all** T cells produce the particular Vβ chain that binds the superantigen, these viral proteins short circuit the interaction of MHC class II-peptide complex and the T cell. As a consequence, rather than activating a small, specific subset of T cells (only 0.001 to 0.01% of T cells usually respond to a given antigen), **all** subsets of T cells producing the Vβ chain to which the superantigen binds are activated and proliferate, regardless of specificity.

All known superantigens are microbial products, and many are produced after viral infection. The best-understood viral superantigen is encoded in the U3 region of the mouse mammary tumor virus long terminal repeat. This retrovirus is transmitted efficiently from mother to offspring via milk. However, the virus reproduces poorly in most tissues in the mother. Neonates can become infected upon ingestion of milk. When B cells in the neonatal small intestine epithelium are infected, the viral superantigen is produced and recognized by T cells carrying the appropriate T cell receptor Vβ chain. Consequently, extraordinarily large numbers of T cells are activated, producing growth factors and other

molecules that stimulate proliferation of the infected B cells. These cells then carry the virus to the mammary gland, enhancing transmission to the progeny of these mice, and increasing the risk for tumor formation (Fig. 5.19). Infection of mice with mutants harboring a deletion of the superantigen gene results in limited viral reproduction and minimal transmission to offspring via milk.

Mechanisms Mediated by Free Radicals

Two free radicals, superoxide (O_2^-) and nitric oxide, are produced during the inflammatory response and are probably critical effectors of virus-induced pathology. Superoxide is produced by the enzyme xanthine oxidase, present in phagocytes. The production of O_2^- is significantly increased in hypoxic cells and tissues, for example, in the lungs of mice infected with influenza virus or cytomegalovirus. Inhibition of xanthine oxidase protects mice from virus-induced death.

Nitric oxide is abundant in virus-infected tissues during inflammation as part of the innate immune response (Chapter 3). This compound inhibits the production of many viruses in cultured cells and in animal models. It acts within the cell to limit viral reproduction, but the molecular sites of action are not well understood. Nitric oxide is produced by three different IFN-inducible isoforms of nitric oxide synthase. Although low concentrations of nitric oxide have a protective effect, high concentrations or prolonged exposure can contribute to pathogenesis. While nitric oxide is relatively inert, it reacts rapidly with O_2^- to form peroxynitrite ($ONOO^-$), which is much more reactive than either molecule and may be responsible for cytotoxic effects on cells.

Immunosuppression Induced by Viral Infection

Virus-mediated suppression of immune defenses can range from a mild and rather specific attenuation to a marked global inhibition of the response. Immunosuppression by viral infection was first observed over 100 years ago when patients were unable to respond to a skin test for tuberculosis during and after measles infection. However, progress in understanding the phenomenon was slow until the human immunodeficiency virus epidemic was under way (Chapter 7).

While it is well known that human immunodeficiency virus is immunosuppressive, severe immune suppression can also result from infection by other human viruses such as rubella and measles. For example, the vast majority of the tens of thousands of measles virus-induced childhood deaths each year in Third World countries is due to opportunistic infections that arise during transient immunosuppression caused by this virus.

The ability of measles virus to cripple the entire immune response while infecting a very small proportion of cells of the immune system (only about 2% of total T cells are infected at the peak of viremia) was a long-standing mystery.

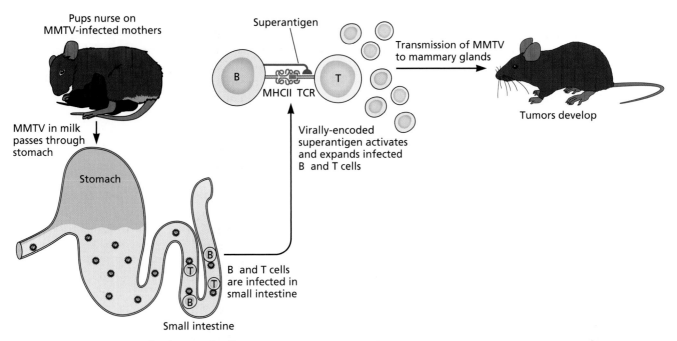

Figure 5.19 Infectious cycle of mouse mammary tumor virus. This retrovirus is produced in the mammary glands of infected female mice and is transmitted to newborn pups through the mother's milk. The ingested virus infects B and T cells in the gut-associated lymphoid tissues. The infected B cells express a superantigen that activates T cells nonspecifically, providing more target cells for infection. In the mammary gland, hormonal stimulation during pregnancy and lactation dramatically increases mouse mammary tumor virus (MMTV) reproduction, and can lead to insertional mutagenesis of proto-oncogenes and the development of mammary tumors.

It is now appreciated that measles virus infection of macrophages and dendritic cells may be critical. One of the first bits of evidence to suggest a role of dendritic cell impairment was the observation that infection of these cells resulted in reduced expression of the cytokine, IL-12. This cytokine is important for skewing the immune response toward a T_h1 profile that favors clearance of virus infections. Low IL-12 directs the T cell response toward a T_h2 profile, which promotes induction of the humoral (antibody) response. When IL-12 production by dendritic cells is reduced, the cytokine microenvironment is not conducive to a cytolytic response, and T cells cannot proliferate in response to interaction with infected dendritic cells (Fig. 5.20). In measles virus-infected macaques, decreased IL-12 and increased IL-4 (a marker of a T_h2 response) is observed, and the concentration of IL-12 is also greatly reduced in the blood of measles virus-infected humans.

In separate studies, it was shown that infection of cells with the related paramyxovirus, Hendra virus, can limit the induction of an innate response by restricting nuclear access of critical signal transducers. When interferons bind to cell receptors, Stat1 is rapidly phosphorylated and homodimerizes, exposing a nuclear localization signal that allows the protein to enter nuclei and bind to interferon response elements within promoters of interferon-inducible genes (Chapter 4).

Remarkably, infection with Hendra virus precludes nuclear localization of phosphorylated Stat1 (and Stat2), and thereby impedes the efficient induction of interferon genes (Fig. 5.21).

Other mechanisms that have been proposed to account for measles virus-induced transient immunosuppression include impaired development of infected dendritic cell precursors and decreased proliferation of infected T and B lymphocytes because of cell cycle arrest. These findings parallel clinical observations in humans, in which measles virus infection in immunosuppressed individuals is associated with profound reductions in the number of circulating white blood cells, and recovery from immunosuppression is directly correlated with the rate of synthesis of new cells from the bone marrow. This observation may explain why young children recover faster from infection than older children or adults. The diversity of the various mechanisms induced by this relatively simple virus (encoding only 9 proteins) underscores the evolutionary pressures that have been selected to frustrate host immunity.

Oncogenesis

Over 20% of all human cancers are associated with virus infection, and for some cancers, including liver cancer and cervical cancer, viruses are the major cause. Moreover, the history of cancer biology is, in great part, the history of

A

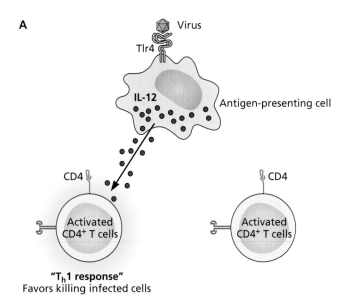

"T$_h$1 response"
Favors killing infected cells

B

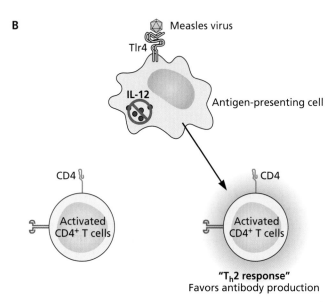

"T$_h$2 response"
Favors antibody production

Figure 5.20 Measles virus infection of antigen-presenting cells blocks IL-12 production. One of the proposed ways by which measles virus induces global suppression is that infection of a small proportion of monocytes blocks the synthesis of a critical cytokine, IL-12. Normally, IL-12 is made in response to viral infections and skews the resulting T cell response toward a primarily T$_h$1-like profile. When IL-12 is blocked by measles virus, a T$_h$2 cytokine profile predominates. Consequently, although the host is making an aggressive response, it is not the optimal response to eliminate an intracellular viral infection. Adapted from C.L. Karp et al. *Science* **273**:228–231, 1996, with permission.

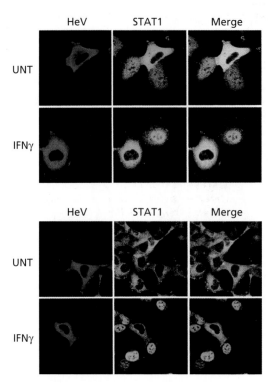

Figure 5.21 Hendra virus infection restricts nuclear localization of activated Stat molecules. An additional way by which paramyxoviruses can induce immunosuppression is by binding to signal transduction molecules that are needed to induce an antiviral response, and either retaining them in the cytoplasm or triggering their degradation.

Molecular Mimicry

Autoimmunity is caused by an immune response directed against host tissues (often described as "breaking immune tolerance"). One can envision multiple scenarios by which viral infections can trigger autoimmunity. Cytolytic virus reproduction leads to the release, and subsequent recognition, of self-antigens that are normally sequestered from the immune system. Additionally, cytokines, or virus-antibody complexes that modulate the activity of proteases in antigen-presenting cells, might cause the unmasking of self-antigens. For example, cytokines produced during infection may stimulate inappropriate surface accumulation of cellular membrane proteins that are recognized by host defenses. Another possibility is that, during virus particle assembly, host proteins that are not normally exposed to the immune system are packaged in particles that may be recognized as foreign upon entry into a different cell type. While these processes, in principle, could occur, to date none have been formally proven to be an etiological cause of human autoimmunity.

An additional hypothesis, termed "molecular mimicry," proposes cross-reactivity between a particular viral and host epitope, and is based on two observations. The first is that humans possess many putative autoreactive T cells that rarely cause disease because they are not appropriately activated via

virology: discovery of tumor suppressors and oncogenes occurred as a result of studying DNA and RNA tumor viruses. Consequently, an entire chapter is dedicated to the various mechanisms by which transformation and oncogenesis occurs pursuant to viral infection (Chapter 6).

BOX 5.16

EXPERIMENT
Viral infections promote or protect against autoimmune disease

Transgenic mice that synthesize proteins of lymphocytic choriomeningitis virus in β cells of the pancreas or oligodendrocytes within the brain have been developed. Synthesis of these viral proteins has no consequence; the mice are healthy. The viral transgene products are present in the mouse throughout development, and therefore are perceived by the host defense as a self-antigen.

Infection of these transgenic animals with lymphocytic choriomeningitis virus stimulates an immune response in which the self-antigen is recognized, leading to insulin-dependent diabetes mellitus (when the protein is made in the pancreas) or central nervous system demyelinating disease (when made in the brain).

Fujinami R. 2001. Viruses and autoimmune disease—two sides of the same coin? *Trends Microbiol* **9**:377–381.

von Herrath M, Oldstone M. 1996. Virus-induced autoimmune disease. *Curr Opin Immunol* **8**:878–885.

Holmes S, Friese MA, Siebold C, Jones EY, Bell J,. Fugger L. 2005. Multiple sclerosis: MHC associations and therapeutic implications. *Exp Rev Mol Med* **7**:1–17.

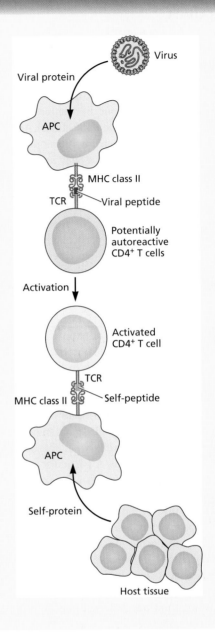

A potentially autoreactive T cell, possessing T cell receptors that recognize both a foreign (viral) peptide and a self-peptide, is activated by a virus-derived peptide. Abbreviations: APC, antigen-presenting cell; MHC, major histocompatibility complex; TCR, T cell receptor. Adapted from S. Holmes, M. A. Friese, C. Siebold, E. Y. Jones, J. Bell and L. Fugger. *Exp Rev Molec Med* **7**: 1-17, 2005, with permission.

costimulation (Chapter 4). The second observation is that viral and host proteins can share antigenic determinants. The hypothesis is that infection leads to the activation of immune cells specific for a viral epitope; if this epitope is presented on host cells (derived from a cellular protein, and thus occurring by chance), the activated immune response may target such cells, even if they are not infected. Although many peptide sequences are shared among viral and host proteins, direct evidence for this hypothesis has been difficult to obtain. One reason for the difficulty is the long interval between events

that trigger human autoimmune diseases and the onset of clinical symptoms. To circumvent this problem, transgenic mouse models, in which the products of foreign genes are expressed as self antigens, were established. Such model systems allowed proof-of-concept studies, which showed that this process can occur (Box 5.16). Although human parallels have yet to be ascribed definitively to molecular mimicry, some diseases, including multiple sclerosis and the neuropathology resulting from human immunodeficiency virus infection, have been proposed to be due to this process.

Perspectives

Patterns of viral infection are most likely established in the first minutes to hours after the initial inoculation. It appears that two distinct strategies of viral propagation have emerged during evolution: one produces large numbers of progeny (high reproductive output or *r*-replication; acute infections) and one results in a lower reproductive output, but better competition for resources (*K*-replication; persistent infections). Acute infections occur primarily because host defenses are modulated (passively or actively), at least for a short time. Such infections may progress beyond physical, intrinsic, and innate defenses only to be blocked and cleared by the adaptive immune response. Large numbers of new hosts are required to sustain the acute pattern of infection, as immune memory (or sometimes host death) limits the duration of a virus in a particular host.

Persistent infections result when essentially all defenses, including the adaptive immune system, are ineffective, often for long periods. Ineffective does not always mean nonfunctional. For example, in some persistent infections, such as those with hepatitis B virus, the low rate of viral reproduction is equal to the rate of immune elimination. This particular persistent infection pattern can be characterized as "smoldering," as it continues for very long periods in the face of active host defenses. While persisting viruses do not need to constantly hop from one infected individual to another, success is ensured only if there is a mechanism for periodic production of virus particles and their transmission to new hosts.

The existence of only two primary patterns of infection confront us with several questions. A particular pattern can be a defining characteristic of a virus family (e.g., influenza virus always produces an acute infection; herpes infections are forever). But this raises the fascinating philosophical question of why one particular pattern has been selected over another for these viruses. We know that acute and persistent infections are determined by properties of both host and virus. The patterns are not mutually exclusive, as some infections exhibit both acute and persistent phases. Some of the answers are discussed in Chapter 10, where we point out that viral populations emerge and prosper as a consequence of selection pressures. The selective advantages or disadvantages of rapid or limited reproduction manifest themselves quickly. Those host and viral genomes that can adapt survive to carry on the relationship for another day.

The role of pathogenesis as a selective force in the establishment and maintenance of viral infections is a subject of much research and debate. One hypothesis is that successful patterns result in symbiosis, neither helping nor harming the host. In this context, as suggested by Lewis Thomas, pathogenesis is an aberration of symbiosis, an overstepping of boundaries. Benign symbiosis is a recipe for stability, but many apparently successful viral infections are far from stable. Accordingly, another hypothesis posits that pathogenesis is a necessary survival feature of the viral population, and is selected during evolution of the relationship. Some individuals may be harmed in the short run to achieve long-term survival of the virus population. This discussion continues in Chapter 10.

Selection works in unexpected ways. In laboratory situations, adaptation to *r*-selection conditions (low MOI, rapid growth) yields viral populations that are less fit when exposed to *K*-selection conditions (high MOI, reduced growth) and *vice versa*. This fact may be obscured by the stability of one pattern compared to the other. Moreover, we are only beginning to appreciate how coincident infections may influence each other: infections with one virus can change the outcome of completely unrelated infections by systemic or local immunosuppression, by accidental induction of previous memory responses, or by recruitment of powerful immune responses to the "wrong" tissue. We have much to learn about how these viruses affect our lives, but remember that viruses are not "bad" *per se*. The pressure is for a virus to reproduce, not to debilitate its host: a powerful selective force.

References

Books

Arvin A, Campadelli-Fiume G, Mocarski E, Moore P, Roizman B, Whitley R, Yamanishi K. 2007. *Human Herpesviruses: Biology, Therapy and Immunoprophylaxis.* Cambridge University Press, Cambridge, United Kingdom.

Keeling M, Rohani P. 2007. *Modeling Infectious Diseases in Humans and Animals.* Princeton University Press, Princeton, NJ.

Mims CA, Nash A, Stephen J. 2001. *Mims' Pathogenesis of Infectious Disease,* 5th ed. Academic Press, San Diego, CA.

Richman DD, Whitley RJ, Hayden FG (ed). 2002. *Clinical Virology,* 2nd ed. ASM Press, Washington, DC.

Reviews

Anderson R, May R. 1979. Population biology of infectious diseases: part I. *Nature* **280:**361–367.

Borrow P, Oldstone MBA. 1995. Measles virus-mononuclear cell interactions. *Curr Top Microbiol Immunol* **191:**85–100.

Efstathiou S, Preston C. 2005. Towards an understanding of the molecular basis of herpes simplex virus latency. *Virus Res* **111:**108–119.

Hislop A, Taylor G, Sauce D, Rickinson A. 2007. Cellular responses to viral infection in humans: lessons from Epstein-Barr virus. *Annu Rev Immunol* **25:**587–617.

Kerdiles Y, Sellin C, Druelle J, Horvat B. 2006. Immunosuppression caused by measles virus: role of viral proteins. *Rev Med Virol* **16:**49–63.

Khanna KM, Lepisto AJ, Decman V, Hendricks RL. 2004. Immune control of herpes simplex virus during latency. *Curr Opin Immunol* **16:**463–469.

Knipe D, Cliffe A. 2008. Chromatin control of herpes simplex virus lytic and latent infection. *Nat. Rev Microbiol* **6:**211–221.

Koyuncu OO, Hogue IB, Enquist LW. 2013. Virus infections in the nervous system. *Cell Host Microbe* **13:**379–393.

Ku C, Besser J, Abendroth A, Grose C, Arvin A. 2005. Varicella-zoster virus pathogenesis and immunobiology: new concepts emerging from investigations with the SCIDhu mouse model. *J Virol* **79:**2651–2658.

Oldstone MBA. 2006. Viral persistence: parameters, mechanisms and future predictions. *Virology* **334:**111–118.

Peters C. 2006. Lymphocytic choriomeningitis virus—an old enemy up to new tricks. *N Engl J Med* **354:**2208–2211.

Rall G. 2003. Measles virus 1998–2002: progress and controversy. *Annu Rev Microbiol* **57**:343–367.

Redpath S, Angulo A, Gascoigne NRJ, Ghazal P. 2001. Immune checkpoints in viral latency. *Annu Rev Microbiol* **55**:531–560.

Rickinson A. 2002. Epstein-Barr virus. *Virus Res* **82**:109–113.

Rima B, Duprex W. 2005. Molecular mechanisms of measles virus persistence. *Virus Res* **111**:132–147.

Suttle C. 2007. Marine viruses—major players in the global ecosystem. *Nat Rev Microbiol* **5**:801–812.

Taubenberger JK, Morens DM. 2013. Influenza viruses: breaking all the rules. *mBio* **4**:e00365–13.

Thormar H. 2005. Maedi-visna virus and its relationship to human immunodeficiency virus. *AIDS Rev* **7**:233–245.

Tortorella D, Gewurz BE, Furman MH, Schust DJ, Ploegh HL. 2000. Viral subversion of the immune system. *Annu Rev Immunol* **18**:861–926.

Weiland S, Chisari F. 2005. Stealth and cunning: hepatitis B and hepatitis C viruses. *J Virol* **79**:9369–9380.

Weiss RA. 2002. Virulence and pathogenesis. *Trends Microbiol* **10**:314–317.

Wenner M. 2008. The battle within. *Nature* **451**:388–389.

Xu X, Screaton GR, McMichael AJ. 2001. Virus infections: escape, resistance, and counterattack. *Immunity* **15**:867–870.

Yanagi Y, Taleda M, Ohno S. 2006. Measles virus: cellular receptors, tropism, and pathogenesis. *J Gen Virol* **87**:2767–2779.

Yewdell JW, Hill AB. 2002. Viral interference with antigen presentation. *Nat Immunol* **3**:1019–1025.

Zinkernagel R. 2002. Lymphocytic choriomeningitis virus and immunology. *Curr Top Microbiol Immunol* **263**:1–5.

Papers of Special Interest

Barton ES, White DW, Cathelyn JS, Brett-McClellan KA, Engle M, Diamond MS, Miller VL, Virgin HW. 2007. Herpesvirus latency confers symbiotic protection from bacterial infection. *Nature* **447**:326–329.

Borderia A, Elena S. 2002. R-and K-selection in experimental populations of vesicular stomatitis virus. *Infect Genet Evol* **2**:137–143.

Bornkamm G, Behrends U, Mautner J. 2006. The infectious kiss: newly infected B cells deliver Epstein-Barr virus to epithelial cells. *Proc Natl Acad Sci USA* **103**:7201–7202.

Cohrs R, Gilden D. 2007. Prevalence and abundance of latently transcribed varicella-zoster virus genes in human ganglia. *J Virol* **81**:2950–2956.

Charleston B, Fray MD, Baigent S, Carr BV, Morrison WI. 2001. Establishment of persistent infection with non-cytopathic bovine viral diarrhoea virus in cattle is associated with a failure to induce type I interferon. *J Gen Virol* **82**:1893–1897.

Cui C, Griffiths A, Li G, Silva L, Kramer M, Gaasterland T, Wang X, Coen D. 2006. Prediction and identification of herpes simplex virus 1-encoded microRNA. *J Virol* **80**:5499–5508.

Decman V, Kinchington P, Harvey S, Hendricks R. 2005. Gamma interferon can block herpes simplex virus type 1 reactivation from latency, even in the presence of late-gene expression. *J Virol* **79**:10339–10347.

Epstein M, Achong B, Barr Y. 1964. Virus particles in cultured lymphoblasts from Burkitt's lymphoma. *Lancet* **i**:702–703.

Erickson AL, Kimura Y, Igarashi S, Eichelberger J, Houghton M, Sidney J, McKinney D, Sette A, Hughes AL, Walker CM. 2001. The outcome of hepatitis C virus infection is predicted by escape mutations in epitopes targeted by cytotoxic T lymphocytes. *Immunity* **15**:883–895.

Griffith TS, Brunner T, Fletcher SM, Green DR, Ferguson TA. 1995. Fas ligand-induced apoptosis as a mechanism of immune privilege. *Science* **270**:1189–1192.

Guidotti LG, Rochford R, Chung J, Shapiro M, Purcell R, Chisari FV. 1999. Viral clearance without destruction of infected cells during HBV infection. *Science* **284**:825–829.

Halford WP, Schaffer PA. 2001. ICP0 is required for efficient reactivation of herpes simplex virus type 1 from neuronal latency. *J Virol* **75**:4528–4537.

Karp CL, Wysocka M, Wahl LM, Ahearn JM, Duomo PJ, Sherry B, Trinchieri G, Griffin DE. 1996. Mechanism of suppression of cell-mediated immunity by measles virus. *Science* **273**:228–231.

Khanna K, Bonneau R, Kinchington P, Hendricks R. 2003. Herpes simplex virus specific memory CD8+ T cells are selectively activated and retained in latently infected sensory ganglia. *Immunity* **18**:593–603.

Korom M, Wylie KM, Wang H, Davis KL, Sangabathula MS, Delassus GS, Morrison LA. 2013. A proautophagic antiviral role for the cellular prion protein identified by infection with a herpes simplex virus 1 ICP34.5 mutant. *J Virol* **87**:5882–5894.

Knickelbein JE, Khanna KM, Yee MB, Baty CJ, Kinchington PR, Hendricks RL. 2008. Noncytotoxic lytic granule-mediated CD8+ T cell inhibition of HSV-1 reactivation from neuronal latency. *Science* **322**:268–271.

Kubat NJ, Amelio AL, Giordani NV, Bloom DC. 2004. The herpes simplex virus type 1 latency-associated transcript (LAT) enhancer/rcr is hyperacetylated during latency independently of LAT transcription. *J Virol* **78**:12508–12518.

Lawrence D, Patterson C, Gales T, D'Orazio J, Vaughn M, Rall G. 2000. Measles virus spread between neurons requires cell contact but not CD46 expression, syncytium formation, or extracellular virus production. *J Virol* **74**:1908–1918.

Lee A, Hertel L, Louie R, Burster T, Lacaille V, Pashine A, Abate D, Mocarski E, Mellins E. 2006. Human cytomegalovirus alters localization of MHC class II and dendrite morphology in mature Langerhans cells. *J Immunol* **177**:3960–3971.

Li K, Foy E, Ferreon JC, Nakamura M, Feereon ACM, Ikeda M, Ray SC, Gale M, Lemon SM. 2005. Immune evasion by hepatitis C virus NS3/4A protease-mediated cleavage of the Toll-like receptor 3 adaptor protein TRIF. *Proc Natl Acad Sci USA* **102**:2992–2997.

Lutley R, Petursson G, Aalsson P, Georgsson G, Klein J, Nathanson N. 1983. Antigenic drift in visna: virus variation during long-term infection of Icelandic sheep. *J Gen Virol* **64**:1433–1440.

Poiesz B, Ruscetti F, Reitz M, Kalyanaraman V, Gallo R. 1981. Isolation of a new type C retrovirus (HTLV) in primary uncultured cells of a patient with Sezary T-cell leukaemia. *Nature* **294**:268–271.

Reyburn HT, Mandelboim O, Vales-Gomez M, Davis DM, Pazmany L, Strominger JL. 1997. The class I MHC homologue of human cytomegalovirus inhibits attack by natural killer cells. *Nature* **386**:514–517.

Sawtell N, Thompson R. 2004. Comparison of herpes simplex virus reactivation in ganglia in vivo and in explants demonstrates quantitative and qualitative differences. *J Virol* **78**:7784–7794.

Sen J, Liu X, Roller R, Knipe DM. 2013. Herpes simplex virus US3 tegument protein inhibits Toll-like receptor 2 signaling at or before TRAF6 ubiquitination. *Virology* **439**:65–73.

Stevens J, Wagner K, Devi-Rao G, Cook M, Feldman L. 1987. RNA complementary to a herpes gene mRNA is prominent in latently infected neurons. *Science* **235**:1056–1059.

York IA, Roop C, Andrews DW, Riddell SR, Graham FL, Johnson DC. 1994. A cytosolic herpes simplex virus protein inhibits antigen presentation to CD81 T lymphocytes. *Cell* **77**:525–535.

van Gent M, Braem SG, DeJong A, Delagic N, Peters JG, Boer IG, Moynagh PN, Kremmer E, Wiertz EJ, Ovaa H, Griffin BD, Ressing ME. 2014. Epstein-Barr virus large tegument protein BPLF1 contributes to innate immune evasion through interference with toll-like receptor signaling. *PLoS Path* **10**:e1003960.

van Lint A, Kleinert L, Clarke S, Stock A, Heath W, Carbone F. 2005. Latent infection with herpes simplex virus is associated with ongoing CD8+ T-cell stimulation by parenchymal cells within sensory ganglia. *J Virol* **79**:14843–14851.

Zabolotny JM, Krummenacher C, Fraser NW. 1997. The herpes simplex virus type 1 2.0-kilobase latency-associated transcript is a stable intron which branches at a guanosine. *J Virol* **71**:4199–4208.

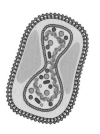

6

Cellular Transformation and Oncogenesis

Introduction
Properties of Transformed Cells
Control of Cell Proliferation

Oncogenic Viruses
Discovery of Oncogenic Viruses
Viral Genetic Information in Transformed Cells
The Origin and Nature of Viral Transforming Genes
Functions of Viral Transforming Proteins

Activation of Cellular Signal Transduction Pathways by Viral Transforming Proteins
Viral Signaling Molecules Acquired from the Cell
Alteration of the Production or Activity of Cellular Signal Transduction Proteins

Disruption of Cell Cycle Control Pathways by Viral Transforming Proteins
Abrogation of Restriction Point Control Exerted by the Rb Protein

Production of Virus-Specific Cyclins
Inactivation of Cyclin-Dependent Kinase Inhibitors

Transformed Cells Must Grow and Survive
Mechanisms That Permit Survival of Transformed Cells

Tumorigenesis Requires Additional Changes in the Properties of Transformed Cells
Inhibition of Immune Defenses

Other Mechanisms of Transformation and Oncogenesis by Human Tumor Viruses
Nontransducing Oncogenic Retroviruses: Tumorigenesis with Very Long Latency
Oncogenesis by Hepatitis Viruses

Perspectives

References

LINKS FOR CHAPTER 6

▶▶◀ *Video: Interview with Dr. Michael Bishop*
http://bit.ly/Virology_Bishop

▶▶◀ *Movie 6.1: Mitosis in HeLa cells*
http://bit.ly/Virology_V2_Movie6-1

▶▶◀ *Moore tumor viruses*
http://bit.ly/Virology_Twiv160

Introduction

Cancer is a leading cause of death in developed countries: about 8.2 million individuals succumb each year worldwide. Consequently, efforts to understand and control this deadly disease have long been high priorities for public health institutions. Our general understanding of the mechanisms of **oncogenesis**, the development of cancer, as well as of normal cell growth, has improved enormously since the latter part of the 20th century. This progress can be traced in large part to efforts to elucidate how members of several virus families cause cancer in animals. In fact, as we discuss in this chapter, study of oncogenic viruses has led to a detailed understanding of the molecular basis of this disease.

It is now clear that cancer (defined in Box 6.1) is a genetic disease: it results from the growth of successive populations of cells in which mutations and/or epigenetic modifications of genes and their associated nucleosomes have accumulated (Box 6.2). These changes affect various steps in the regulatory pathways that control cell communication, growth, and division, and lead to uncontrolled cell proliferation, increasing tissue disorganization, and ultimately cancer. One or more of these genetic changes may be inherited (Box 6.2), or they may arise as a consequence of endogenous DNA damage and exposure to environmental carcinogens or infectious agents, including viruses. It is estimated that viruses are a contributing factor in ~20% of all human cancers. For some, such as liver and cervical cancer, they are the major cause. However, it is important to understand that the induction of malignancy generally is **not** a requirement for the propagation of oncogenic viruses. A singular exception is described in a later section ("Discovery of Oncogenic Viruses"). In all other instances, this unfortunate outcome for the host is a side effect of either infection or the host's response to the presence of the virus. From this perspective, viruses can be thought of as cofactors, or unwitting initiators of oncogenesis.

Understanding the development of cancer ultimately depends on knowledge of how individual cells normally behave within an animal. As described in Chapters 1 to 4, analysis of viral pathogenesis must encompass a consideration of the organism as a whole, especially the body's immune defenses. However, elucidation of how members of several virus families cause cancer in animals began with studies of cultured cells in the laboratory. In particular, early investigators noticed that the growth properties and morphologies of some normal cells in culture could be changed upon infection with certain viruses. We therefore describe such cells as being **transformed**. The advantages of these cell culture systems are many: the molecular virologist can focus attention on particular cell types or specific viral genes and can readily distinguish effects specific to the virus. In many cases, cells transformed by viruses in culture can form tumors when implanted in animals. But it is important to realize that transformed cultures are **not** tumors. The major benefit of cell culture systems is that they allow researchers to study the molecular events that establish an oncogenic potential in virus-infected cells. Such studies were of great importance: they led to the identification of viral and cellular oncogenes and elucidation of the molecular circuits that control cell proliferation.

Properties of Transformed Cells

Cellular Transformation

The proliferation of cells in the body is a strictly regulated process. In a young animal, total cell multiplication exceeds cell death as the animal grows to maturity. In an adult, the processes of cell multiplication and death are carefully balanced. For some cells, high rates of proliferation are required to maintain this balance. For example, human intestinal cells and white blood cells have half-lives of only a few days and need to be replaced rapidly. On the other hand, red blood cells live for

---

PRINCIPLES *Cellular transformation and oncogenesis*

- Members of DNA and RNA virus families cause or contribute to ~20% of human cancers.

- Cancer is a disease of unregulated cell division, which can be the result of inherited mutations; exposure to environmental carcinogens; or infection with pathogens, including viruses.

- Immortalization, transformation, and oncogenesis are distinct states, but are part of a continuum.

- Transformed cells are distinguished from normal cells by their immortality, loss of contact inhibition, and often production of their own growth factors.

- With few exceptions, transformation is not required for viral reproduction.

- Retroviruses can either encode oncogenes (once derived from host genes) or integrate into the cellular genome and activate adjacent cellular proto-oncogenes.

- Small transforming DNA viruses encode proteins that bind to specific cellular proteins, notably the tumor suppressors Rb and p53, to promote cell cycle progression and block checkpoints.

- Proteins encoded by transforming viruses also can prevent cell death, block immune recognition, and promote blood vessel formation.

- Some viruses associated with human cancers do not transform cells directly, but rather induce a chronic immune response that, with time, results in tissue damage and the emergence of malignant cells.

Some cancer terms

Benign: An adjective used to describe a growth that does not infiltrate into surrounding tissues; opposite of malignant

Cancer: A malignant tumor; a growth that is not encapsulated and that infiltrates into normal tissues, replacing normal with abnormal cells; it is spread by the lymphatic vessels to other parts of the body; death is caused by destruction of organs to a degree incompatible with their function, by extreme debility and anemia, or by hemorrhage

Carcinogenesis: The multistage process by which a cancer develops

Carcinoma: A cancer of epithelial tissue

Endothelioma: A cancer of endothelial cells

Fibroblast: A cell derived from connective tissue

Fibropapilloma: A solid tumor of cells of the connective tissue

Hepatocellular carcinoma: A cancer of liver cells

Leukemia: A cancer of white blood cells

Lymphoma: A cancer of lymphoid tissue

Malignant: An adjective applied to any disease of a progressive and fatal nature; opposite of benign

Neoplasm: An abnormal new growth, i.e., a cancer

Oncogenic: Causing a tumor

Retinoblastoma: A cancer of cells of the retina

Sarcoma: A cancer of fibroblasts

Tumor: A swelling, caused by abnormal growth of tissue, not resulting from inflammation; may be benign or malignant

Genetic alterations associated with the development of colon carcinoma

Colorectal cancer is the third-most-common cancer worldwide and the leading cause of cancer-associated deaths. The clinical stages in the development of this cancer are particularly well defined. Furthermore, as shown in the figure, several genes that are frequently mutated to allow progression from one stage to the next have been identified. The early adenomas or polyps that form initially are benign lesions. Their conversion to malignant metastatic carcinomas correlates with the acquisition of additional, loss-of-function mutations in the *p53* and *dcc* (deleted in colon carcinoma) tumor suppressor genes. Inherited mutations can greatly increase the risk that an individual will develop colon carcinoma. For example, patients with familial adenomatous polyposis can inherit defects in the *apc* (adenomatous polyposis coli) gene that result in the development of hundreds of adenomatous polyps. The large increase in the **number** of these benign lesions increases the chance that some will progress to malignant carcinomas. In contrast, patients with hereditary nonpolyposis colorectal cancer develop polyps at the same rate as the general population. However, their polyps develop into carcinomas more frequently, because these patients inherit defects in mismatch repair genes, resulting in a higher mutation rate that promotes oncogenesis. Consequently, the likelihood that an individual polyp will develop into a malignant lesion increases from 5 to 70%.

The genes shown in the figure were identified by classical methods in human genetics. More recently, mutations associated with colorectal cancer and many other types of human tumors have been catalogued by high-throughput sequencing of tumor genomes and protein-coding sequences (exomes). Such studies have confirmed that most tumors are caused by the accumulation of sequential mutations over a long period. They also establish that these mutations alter the function or production of components of a limited number of signal transduction pathways that govern cell proliferation and survival, determination of cell fate, and maintenance of genome integrity. In the case of colorectal cancer, somatic mutations detected by high-throughput sequencing methods in at least 80% of the patients examined are in genes that encode components of the mitogen-activated protein kinase, the Wnt/Apc, and p53 signaling pathways. Additional inherited mutations associated with predisposition to colorectal cancer have also been identified, notably in the gene that encodes the proofreading exonuclease of DNA polymerase δ (described in Volume I, Chapter 9).

Kim TM, Lee SH, Chung YJ. 2013. Clinical applications of next-generation sequencing in colorectal cancers. *World J Gastroenterol* 19:6784–6793.

Vogelstein B, Papadopoulos N, Velculescu VE, Zhou S, Diaz LA, Jr, Kinzler KW. 2013. Cancer genome landscapes. *Science* 339:1546–1558.

Wheeler DA, Wang L. 2013. From human genome to cancer genome: the first decade. *Genome Res* 23:1054–1062.

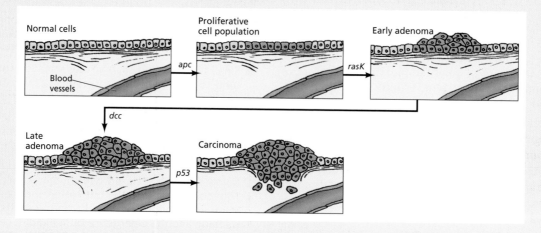

A Mouse cells

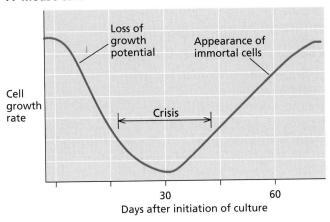

B Human cells

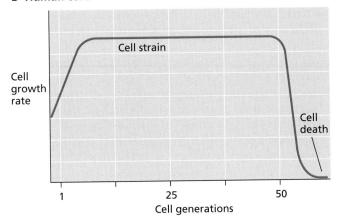

Figure 6.1 Stages in the establishment of a cell culture.
(A) Mouse or other rodent cells. When mouse embryo cells are placed in culture, most cells die before healthy growing cells emerge. As these cells are maintained in culture, they begin to lose growth potential and most cells die (the culture goes into crisis). Very rarely, cells do not die but continue growth and division until their progeny overgrow the culture. These cells constitute a cell line, which will grow indefinitely if it is appropriately diluted and fed with nutrients: the cells are immortal. **(B)** Human cells. When an initial explant is made (e.g., from foreskin), some cells die and others (mainly fibroblasts) start to grow; overall, the growth rate increases. If the surviving cells are diluted regularly, the cell strain proliferates at a constant rate for about 50 cell generations, after which growth and division begin to decrease. Eventually, all the cells die.

over 100 days, and healthy neuronal cells rarely die. Occasionally, this carefully regulated process breaks down, and a particular cell begins to grow and divide even though the body has sufficient numbers of its type; such a cell behaves as if it were immortal. Acquisition of **immortality** is generally acknowledged to be an early step in oncogenesis. An immortalized cell may acquire one or more additional genetic changes to give rise to a clone of cells that is able to expand, ultimately forming a mass called a **tumor**. Some tumors are **benign**; they do not

enter neighboring tissue and are not life-threatening. Other tumor cells grow and divide indefinitely to form invasive **malignant** cancers that damage and impair the normal function of organs and tissues. Some cells in a malignant tumor may acquire additional genetic changes that confer the ability to escape the boundary of the mass, to invade surrounding tissue, and to be disseminated to other parts of the body, where the cells take up residence. There they continue to grow and divide, giving rise to secondary tumors called **metastases**, which cause the most serious and life-threatening disease.

Many studies of the molecular biology of oncogenic animal viruses employed primary cultures of normal cells, for example, rat or mouse embryo fibroblasts. Such primary cells, like their normal counterparts in the animal, have a finite capacity to grow and divide in culture. Cells from some animal species, such as rodents, undergo a spontaneous transformation when maintained in culture. Immortalized cells appear after a "crisis" period in which the great majority of the cells die (Fig. 6.1A). As these surviving cells are otherwise normal, and do not induce tumors when introduced into animals, they can be used to identify viral gene products needed for steps in oncogenesis subsequent to immortalization. For reasons that are not fully understood, human and simian cells rarely undergo spontaneous transformation to immortality when passaged in culture (Fig. 6.1B). In fact, established lines of human cells generally can be derived only from tumors, or following exposure of primary cells to chemical carcinogens or to oncogenic RNA or DNA viruses (or their transforming genes). The realization that transformed cells share a number of common properties, regardless of how they were obtained, provided a major impetus for the investigation of viral transformation.

Properties That Distinguish Transformed from Normal Cells

The definitive characteristic of transformed cells is independence from the signals or conditions that normally control DNA replication and cell division. This property is illustrated by the list of growth parameters and behaviors provided in Table 6.1. As noted above, transformed cells are immortal: they can grow and divide indefinitely, provided that they are diluted regularly into fresh medium. Production

Table 6.1 Growth parameters and behavior of transformed cells

Immortal: can grow indefinitely
Reduced requirement for serum growth factors
Loss of capacity for growth arrest upon nutrient deprivation
Growth to high saturation densities
Loss of contact inhibition (can grow over one another or normal cells)
Altered morphology (appear rounded and refractile)
Anchorage independence (can grow in soft agar)
Tumorigenic

of **telomerase**, the enzyme that maintains telomeric DNA at the ends of chromosomes, is necessary for immortalization. In addition, transformed cells typically exhibit a reduced requirement for the growth factors present in serum. Some transformed cells actually produce their own growth factors and the cognate receptors, providing themselves **autocrine growth stimulation**. Normal cells cease to grow and enter a quiescent state (called G_0, described in "Control of Cell Proliferation" below) when essential nutrient concentrations drop below a threshold value. Transformed cells are deficient in this capacity, and some may even kill themselves by trying to continue to grow in an inadequate environment.

Transformed cells grow to high densities. This characteristic is manifested by the cells piling up and over each other. They also grow on top of untransformed cells, forming visually identifiable clumps called **foci** (Fig. 6.2). Transformed cells behave in this manner because they have lost **contact inhibition**, a response in which normal cells cease proliferation when they sense the presence of their neighbors. Unlike normal cells, many transformed cells have also lost the need for a surface on which to adhere, and we describe them as being **anchorage independent**. Some anchorage-independent cells form isolated colonies in semisolid media (e.g., 0.6% agar). This property correlates well with the ability to form tumors in animals and is often used as an experimental surrogate for malignancy. Transformed cells also **look** different from normal cells; they are more rounded, with fewer processes, and as a result many appear more refractile when observed under a microscope (Fig. 6.2).

There are other ways in which transformed cells can be distinguished from their normal counterparts. These properties include metabolic differences and characteristic changes in cell surface and cytoskeletal components. However, the list in Table 6.1 comprises the standard criteria used to judge whether cells have been transformed.

Control of Cell Proliferation

Sensing the Environment

Because proliferation of cells in an organism is strictly regulated to maintain tissue or organ integrity and normal physiology, normal cells possess elaborate pathways that receive and process growth-stimulatory or growth-inhibitory signals transmitted by other cells in the tissue or organism. Much of what we know about these pathways comes from study of the cellular genes transduced or activated by oncogenic retroviruses. Signaling often begins with the secretion of a growth factor by a specific type of cell. The growth factor may enter the circulatory system, as is the case for many hormones, or may simply diffuse through the spaces around cells in a tissue. Growth factors bind to the external portion of specific receptor molecules on the surface of the same or other types of cell. Alternatively, signaling can be initiated by binding of a receptor on one cell to a specific protein (or proteins) present on the surface of another cell or to components of the extracellular matrix (Volume I, Chapter 5). The binding of the ligand triggers a change, often via oligomerization of receptor molecules, which is transmitted to the cytoplasmic portion of the receptor. In the case illustrated in Fig. 6.3, the cytoplasmic domain of the receptor possesses protein tyrosine kinase activity, and interaction with the growth factor ligand triggers autophosphorylation. This modification sets off a **signal transduction cascade**, a chain of sequential physical interactions among, and biochemical modifications of,

Figure 6.2 Foci formed by avian cells transformed with two strains of Rous sarcoma virus. Differences in morphology are due to genetic differences in the transduced *src* oncogene. **(A)** A focus of infected cells with fusiform morphology shown on a background of flattened, contact-inhibited, uninfected cells. **(B)** Higher magnification of a fusiform focus showing lack of contact inhibition of the transformed cells. **(C)** A focus of highly refractile infected cells with rounded morphology and reduced adherence. **(D)** Higher magnification of rounded infected cells, showing tightly adherent normal cells in the background. Courtesy of P. Vogt, The Scripps Research Institute.

A   B   C   D

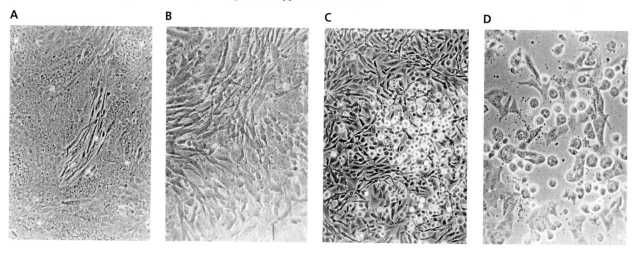

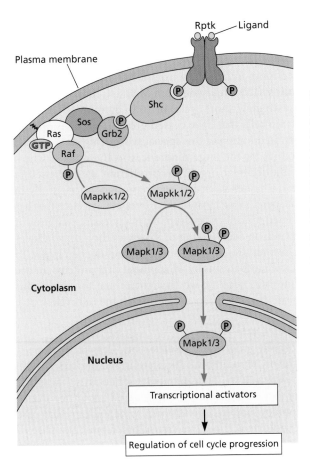

Figure 6.3 The mitogen-activated protein kinase signal transduction pathway. In the pathway shown, signal transduction is initiated by binding of a ligand to the extracellular domain of a receptor protein tyrosine kinase (Rptk), for example, the receptors for epidermal growth factor or platelet-derived growth factor. Binding of ligand (yellow circles) induces receptor dimerization and autophosphorylation of tyrosine residues in the cytoplasmic domain. Adapter proteins like Shc and the Grb2 component of the Grb2-Sos complex are recruited to the membrane by binding to these phosphotyrosine-containing sequences (or to a substrate phosphorylated by the activated receptor), along with Ras. Sos is the guanine nucleotide exchange protein for the small guanine nucleotide-binding protein Ras and stimulates exchange of GDP for GTP bound to Ras. The GTP-bound form of Ras binds to members of the Raf family of serine/threonine protein kinases. Raf then becomes autophosphorylated and initiates the mitogen-activated protein kinase (Mapk) cascade. The pathway shown contains dual-specificity Map kinase kinases (Mapkk1/2) and Mapk1/3. Phosphorylated Mapk1/3 molecules can enter the nucleus, where they modify and activate transcriptional regulators. These kinases can also regulate transcription indirectly, by effects on other protein kinases. Human cells contain multiple Mapkks and Mapks, and the pathway can be activated via plasma membrane receptors that respond to ligands such as inflammatory cytokines. Signal transduction cascades can also include enzymes that produce small molecules (e.g., cyclic AMP [cAMP] and certain lipids) that act as diffusible second messengers in the signal relay. Changes in ion flux across the plasma membrane, or in membranes of the endoplasmic reticulum, may also contribute to transmission of signals. Relay of the signal can terminate at cytoplasmic sites to alter metabolism or cell morphology and adhesion, but signaling to transcriptional regulators, as indicated, is common. To terminate signaling, ligand-bound receptor tyrosine kinases are internalized, GTPase-activating proteins induce hydrolysis of GTP bound to small G proteins like Ras, and protein phosphatases catalyze the hydrolysis of phosphate groups on signaling proteins.

membrane-bound and cytoplasmic proteins (Fig. 6.3). Ultimately, the behavior of the cell is altered.

Many signaling cascades culminate in the modification of transcriptional activators or repressors, and thereby alter the expression of specific cellular genes. The products of these genes either allow the cell to progress through another cell division cycle or cause the cell to stop growing, to differentiate, or to die, whichever response is appropriate to the situation. Errors in the signaling pathways that regulate these decisions can lead to transformation. The molecular features that transmit information are readily reversed, or short-lived, so that signal transduction pathways can be reset once the initiating cue is no longer present. Alterations that impair such mechanisms of termination of signal transmission can also contribute to transformation and oncogenesis.

Integration of Mitogenic and Growth-Promoting Signals

Prior to division, cells must increase in size and mass as they duplicate their components in preparation for the division that produces two daughter cells. Consequently, signals that induce cell proliferation also lead to the metabolic changes required to promote and sustain cell growth. Not surprisingly, the mechanisms that regulate growth of normal cells are

integrated with those that lead to cell proliferation in response to mitogenic signals. The small G protein Ras and the protein kinase Akt are important components of the networks that achieve such integration: their activation leads to not only increased production of proteins that drive progression through the cell cycle (e.g., D-type cyclins), but also stimulation of translation and regulation of the production or activity of many metabolic enzymes (Fig. 6.4).

Regulation of the Cell Cycle

The capacity of cells to grow and divide is controlled by a molecular timer. The timer comprises an assembly of proteins that integrate stimulatory and inhibitory signals received by, or produced within, the cell. Eukaryotic cells do not divide until all their chromosomes have been duplicated and are precisely organized for segregation into daughter cells. Nor are DNA synthesis and chromosome duplication initiated until the previous cell division is complete, or unless the extra- and intracellular environments are appropriate. Consequently, the molecular timer controls a tightly ordered **cell cycle** comprising intervals, or phases, devoted to specific processes.

The duration of the phases in the cell cycle shown in Fig. 6.5 is typical of those of many mammalian cells growing actively in culture. However, there is considerable variation

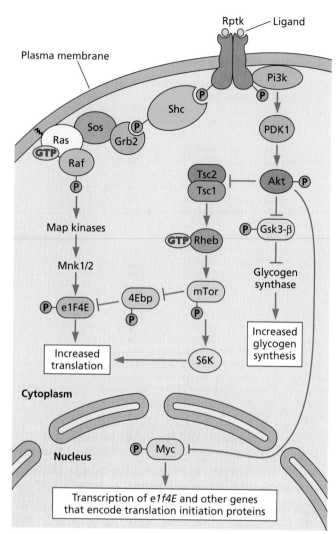

Figure 6.4 Some signaling pathways that promote cell growth. Upon activation, in this example by signaling initiated by binding of its ligand to a receptor protein tyrosine kinase (Rptk), signaling via Ras and the Map kinase cascade activates Map kinase-interacting serine/threonine protein kinases (Mnk1/2), which phosphorylate and activate the translation initiation protein eIF4E. The activity of this initiation protein is also increased when signaling from the Rptk via phosphatidylinositol 3-kinase (Pi3k) and 3-phosphoinositide-dependent protein kinase 1 (PDK1) stimulates the protein kinase Akt. The action of this kinase inhibits the tuberous sclerosis complex (Tsc1/2) and activates the small G protein Rheb (Ras homology enriched in brain) and mTor (mammalian target of rapamycin). Phosphorylation of the inhibitory eIF4E-binding protein (4Ebp) by mTor suppresses its ability to inactivate eIF4E. The transcription of the genes encoding eIF4E and other translation initiation proteins is stimulated via effects on the transcriptional activator Myc. Akt-dependent phosphorylation of ribosomal protein S6 kinase (S6K) increases the rate of translation elongation. These mechanisms increase the availability and activity of proteins crucial for protein synthesis and allow cells to provide proteins at a rate that sustains cells growth. Signaling from Akt also regulates metabolism via phosphorylation and inactivation of glycogen synthase kinase (Gsk3-β) and as a result of effects of activated mTor on lipid metabolism.

in the length of the cell cycle, largely because of differences in the **gap phases (G$_1$ and G$_2$)**. For example, early embryonic cells of animals dispense with G$_1$ and G$_2$, do not increase in mass, and move immediately from the **DNA synthesis phase (S)** to **mitosis (M)** and again from M to S. Consequently, they possess extremely short cycles of 10 to 60 min. At the other extreme are cells that have ceased growth and division. The variability in duration of this specialized **resting state**, termed **G$_0$**, accounts for the large differences in the rates at which cells in multicellular organisms proliferate. As discussed in Volume I, Chapter 9, viruses can reproduce successfully in cells that spend all or most of their lives in G$_0$, a state that has been likened to "cell cycle sleep." In many cases, synthesis of viral proteins in such resting or slowly cycling cells induces them to reenter the cell cycle and grow and divide rapidly. To describe the mechanisms by which these viral proteins induce such abnormal activity, we first introduce the molecular mechanisms that control passage through the cell cycle.

The Cell Cycle Engine

The orderly progression of eukaryotic cells through periods of growth, chromosome duplication, and nuclear and cell division is driven by intricate regulatory circuits. The elucidation of these circuits must be considered a tour de force of contemporary biology. The first experimental hint that cells contain proteins that control transitions from one phase of the cell cycle to another came more than 40 years ago. Nuclei of slime mold (*Physarum polycephalum*) cells in early G$_2$ were found to enter mitosis immediately following fusion with cells in late G$_2$ or M. This crucial observation led to the conclusion that the latter cells must contain a mitosis-promoting factor. Subsequently, similar experiments with mammalian cells in culture identified an analogous S-phase-promoting factor. The convergence of many observations eventually led to the identification of the highly conserved components of the cell cycle engine (Fig. 6.6A).

Mitosis-promoting factor proved to be an unusual protein kinase: its catalytic subunit is activated by the binding of an unstable regulatory subunit. Furthermore, the concentration of the regulatory subunit was found to oscillate reproducibly during each and every cell cycle. The regulatory subunit was therefore given the descriptive name **cyclin**, and the associated protein kinase was termed **cyclin-dependent kinase (Cdk)**. Similar proteins were implicated in cell cycle control in the yeast *Saccharomyces cerevisiae*, and it soon became clear that all eukaryotic cells contain multiple cyclins and Cdks, which operate in specific combinations to control progression through the cell cycle. The cyclins are related in sequence to one another, and they share such properties as activation of cyclin-dependent kinases and controlled destruction by the proteasome.

Various mammalian cyclin-Cdk complexes accumulate during the successive phases in the cell cycle (Fig. 6.6A).

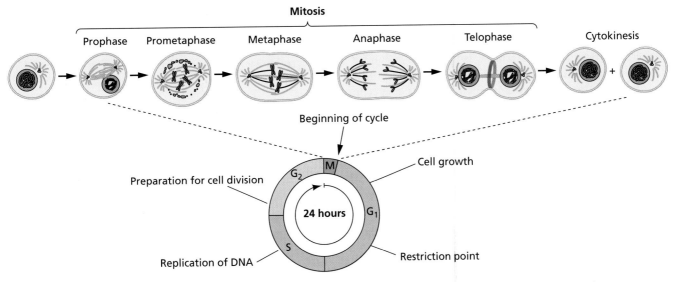

Figure 6.5 The phases of a eukaryotic cell cycle. The most obvious phase morphologically, and hence the first to be identified, is mitosis, or M phase, the process of nuclear division that precedes cell division. During this period, the nuclear envelope breaks down. Duplicated chromosomes become condensed and aligned on the mitotic spindle and are segregated to opposite poles of the cell, where nuclei re-form upon chromosome decondensation (top). The end of M phase is marked by cytokinesis, the process by which the cell divides in two. Despite this remarkable reorganization and redistribution of cellular components, M phase occupies only a short period within the cell cycle. During the long interphase from one mitosis to the next, cells grow continuously. Interphase was divided into three parts with the recognition that DNA synthesis takes place only during a specific period, the synthetic or S phase, which begins at about the middle of interphase. The other two periods, which appeared as "gaps" between defined processes, are designated the G_1 and G_2 (for gap) phases. Movie 6.1 (http://bit.ly/Virology_V2_Movie6-1) shows mitosis in HeLa cells that synthesize the microtubule-forming protein tubulin fused to enhanced green fluorescent protein to label the spindles and histone H2B fused to the red fluorescent protein mCherry to label chromosomes. Courtesy of Tim Yen, Fox Chase Cancer Center.

Figure 6.6 The mammalian cyclin-Cdk cell cycle engine. (A) The phases of the cell cycle are denoted on the circle. The progressive accumulation of specific cyclin-Cdks is represented by the broadening arrows, with the arrowheads marking the time of abrupt disappearance. **(B)** The production, accumulation, and activities of both cyclins and cyclin-dependent kinases are regulated by numerous mechanisms. Activating and inhibitory reactions are indicated by green arrows and red bars, respectively. Activation of the kinases can require not only binding to the appropriate cyclin, but also phosphorylation at specific sites and removal of phosphate groups at others. The activities of the kinases are also controlled by association with members of two families of cyclin-dependent kinase-inhibitory proteins, which control the activities of only G_1 (Ink4 proteins) or all (Cip/Kip proteins) cyclin-Cdks. Both types of inhibitor play crucial roles in cell cycle control. For example, the high concentration of p27^{Kip1} characteristic of quiescent cells falls as they enter G_1, and inhibition of synthesis of this protein prevents cells from becoming quiescent.

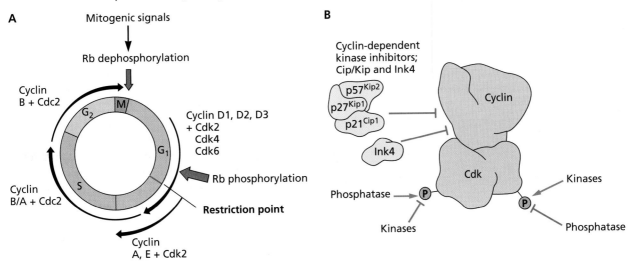

A critical feature is that the individual cyclin-Cdks, the active protein kinases, accumulate in successive waves. The concentration of each increases gradually during a specific period in the cycle, but decreases abruptly as the cyclin subunit is degraded. In mammalian cells, proteolysis is important in resetting the concentrations of individual cyclins at specific points in the cycle, but production of cyclin mRNAs is also regulated. The orderly activation and inactivation of specific kinases govern passage through the cell cycle. For example, cyclin E synthesis is rate limiting for the transition from G_1 to S phase in mammalian cells, and cyclin E-Cdk2 accumulates during late G_1. Soon after cells have entered S phase, cyclin E rapidly disappears from the cell; its task is completed until a new cycle begins.

While the oscillating waves of active Cdk accumulation and destruction are thought of as the ratchet that advances the cell cycle timer, it is important not to interpret this metaphor too literally. The orderly and reproducible sequence of DNA replication, chromosome segregation, and cell division is not determined solely by the oscillating concentrations of individual cyclin-Cdks. Rather, the cyclin-Cdk cycle serves as a device for integrating numerous signals from the exterior and interior of the cell into appropriate responses. The regulatory circuits that feed into and from the cycle are both many and complex (e.g., Fig. 6.6B). These regulatory signals ensure that the cell increases in mass and divides **only** when the environment is propitious or, in multicellular organisms, when the timing is correct. Many signal transduction pathways that convey information about the local environment or the global state of the organism therefore converge on the cyclin-Cdk integrators. In addition, various surveillance mechanisms monitor such internal parameters as DNA damage, problems with DNA replication, and proper assembly and function of the mitotic spindle. Such mechanisms protect cells against potentially disastrous consequences of continuing a cell division cycle that could not be completed correctly. It is primarily these signaling and surveillance (checkpoint) mechanisms that are compromised during transformation by oncogenic viruses.

Oncogenic Viruses

The study of the mechanisms of viral transformation and oncogenesis laid the foundation for our current understanding of cancer, for example, with the identification of oncogenes that are activated or captured by retroviruses (originally known as RNA tumor viruses) and viral proteins that inactivate tumor suppressor gene products (Fig. 6.7). Specific members of a number of different virus families, as well as an unusual, unclassified virus (Box 6.3), have been implicated in naturally occurring or experimentally induced cancers in animals (Table 6.2). It has been estimated that ~20% of all cases of human cancer are associated with infection by one of eight viruses: Epstein-Barr virus, hepatitis B virus, hepatitis C

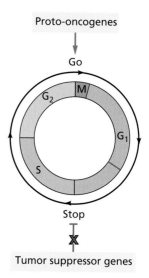

Figure 6.7 A genetic paradigm for cancer. The pace of the cell cycle can be modulated both positively and negatively by different sets of gene products. Cancer arises from a combination of dominant, gain-of-function mutations in proto-oncogenes and recessive, loss-of-function mutations in tumor suppressor genes, which encode proteins that block cell cycle progression at various points. The function of either type of gene product can be affected by oncogenic viruses.

virus, human herpesvirus 8, human immunodeficiency virus type 1, human T-lymphotropic virus type 1, human papillomaviruses, and Merkel cell polyomavirus. In this section, we introduce oncogenic viruses and general features of their transforming interactions with host cells.

Discovery of Oncogenic Viruses

Retroviruses

Oncogenic viruses were discovered more than 100 years ago when Vilhelm Ellerman and Olaf Bang (1908) first showed that avian leukemia could be transmitted by filtered extracts (i.e., viruses) of leukemic cells or serum from infected birds.

Table 6.2 Oncogenic viruses and cancer

Family	Associated cancer(s)
RNA viruses	
Flaviviridae	
Hepatitis C virus	Hepatocellular carcinoma
Retroviridae	Hematopoietic cancers, sarcomas, and carcinomas
DNA viruses	
Adenoviridae	Various solid tumors
Hepadnaviridae	Hepatocellular carcinoma
Herpesviridae	Lymphomas, carcinomas, and sarcomas
Papillomaviridae	Papillomas and carcinomas
Polyomaviridae	Various solid tumors
Poxviridae	Myxomas and fibromas

BOX 6.3

EXPERIMENTS

A cancer virus with genomic features of both papillomaviruses and polyomaviruses

Efforts are under way to prevent the extinction of the western barred bandicoot, an endangered marsupial now found only on two islands in the UNESCO World Heritage Area of Shark Bay, Western Australia. Unfortunately, conservation has been hindered by a debilitating transmissible syndrome, in which wild and captive animals develop papillomas and carcinomas in several areas of the skin. The histological properties of the tumors suggested that a papillomavirus or a polyomavirus might contribute to development of the disease.

In fact, a previously unknown viral genome was discovered in tumor tissues from these animals by multiply primed amplification, cloning, and sequencing, and also by PCR with degenerate primers specific for papillomavirus DNA. This DNA genome exhibits features characteristic of both papillomaviruses and polyomaviruses and includes coding sequences related to those of both families. The papillomavirus-like and polyomavirus-like sequences were shown to be continuous with one another in the viral DNA genome. This property excludes the possibility that the tumor tissues were coinfected with a member of each family, as well as such artifacts as laboratory contamination of samples.

The origin of this unique virus, which was named bandicoot papillomatosis carcinomatosis virus type 1 (and a second, closely related virus isolated from a different bandicoot species), is not known. The virus might have arisen as a result of a recombination event between the genomes of a papillomavirus and a polyomavirus. Alternatively, it might represent the first known member of a new virus family that evolved from a common ancestor of the *Papillomaviridae* and *Polyomaviridae*. Regardless, the viral genome has been detected in 100% of bandicoots with papillomatosis and carcinomatosis syndrome, implicating the virus as a necessary factor in the development of this disease.

Woolford L, Rector A, Van Ranst M, Ducki A, Bennett MD, Nicholls PK, Warren KS, Swan RA, Wilcox GE, O'Hara AJ. 2007. A novel virus detected in papillomas and carcinomas of the endangered western barred bandicoot (*Perameles bougainville*) exhibits genomic features of both the *Papillomaviridae* and *Polyomaviridae*. J Virol **81**:13280–13290.

Perameles bougainville. From J. Gould, *Mammals of Australia*, vol 1 (J. Gould, London, United Kingdom, 1863).

Because leukemia was not recognized as cancer in those days, the significance of this discovery was not generally appreciated. Shortly thereafter (in 1911), Peyton Rous demonstrated that solid tumors could be produced in chickens by using cell extracts from a transplantable sarcoma that had appeared spontaneously. Despite the viral etiology of this disease, the cancer viruses of chickens were thought to be oddities until similar murine malignancies, as well as mouse mammary tumors, were found to be associated with infection by viruses. These oncogenic viruses all proved to be members of the retrovirus family. We now know that retroviruses are endemic in many species, including mice and chickens. When a chicken embryo is infected with avian leukosis virus, immune tolerance is established. In rare cases, tumors arise by a mechanism discussed below.

Early researchers classified the oncogenic retroviruses into two groups depending on the rapidity with which they caused cancer (Table 6.3). The first group comprises rare, rapidly transforming **transducing oncogenic retroviruses**.

These are all highly carcinogenic agents that cause malignancies in nearly 100% of infected animals in a matter of days. They were later discovered to have the ability to transform susceptible cells in culture. The second class, **nontransducing oncogenic retroviruses**, includes less carcinogenic agents. Not all animals infected with these viruses develop tumors, which appear only weeks or months after infection. In the late 1980s, a third type of oncogenic retrovirus, a **long-latency retrovirus**, was identified in humans: tumorigenesis is very rare and occurs months or even years after infection.

Infection by each group of oncogenic retroviruses induces tumors by a distinct mechanism. The long-latency retroviruses encode transforming proteins with no cellular counterparts. As their name implies, the genomes of transducing retroviruses contain cellular genes that become **oncogenes** (genes encoding proteins that cause transformation or tumorigenesis) when expressed in the viral context. The virally transduced versions of these cellular genes are called **v-oncogenes**, and their normal cellular counterparts are called **c-oncogenes**

Table 6.3 The oncogenic retroviruses

Property or characteristic	Transducing viruses	Nontransducing viruses	Nontransducing, long-latency viruses
Example	Rous sarcoma virus	Avian leukosis virus	Human T-lymphotropic virus type 1
Efficiency of tumor formation	High (ca. 100% of infected animals)	High to intermediate	Very low (<5%)
Tumor latency	Short (days)	Intermediate (weeks, months)	Long (months, years)
Infecting viral genome	Viral-cellular recombinant; normally replication defective	Intact; replication competent	Intact; replication competent
Oncogenic element	Cell-derived oncogene carried in viral genome	Cellular oncogene activated in situ by a provirus	Virus-encoded regulatory protein controlling transcription?
Mechanism	Oncogene transduction	*cis*-acting provirus	*trans*-acting protein?
Ability to transform cells in culture	Yes	No	No

or **proto-oncogenes**. The genomes of the nontransducing retroviruses do not encode cell-derived oncogenes. Rather, the transcription of proto-oncogenes is activated inappropriately as a consequence of the nearby integration of a provirus in the host cell genome. In either situation, the oncogene products ordinarily play no role in the reproductive cycle of the retroviruses themselves. With the notable exception of the reproductive cycle of certain epsilonretroviruses (Box 6.4), the oncogenic potential of retroviruses is an accident of their infectious cycles. Nevertheless, the study of v-oncogenes and proto-oncogenes that are affected by retroviruses has been of great importance in advancing our understanding of the origins of cancer.

BOX 6.4

DISCUSSION

Walleye dermal sarcoma virus, a retrovirus with a unique transmission cycle

Episilonretrovirus is the latest genus to be recognized in the family *Retroviridae*. This genus includes retroviruses that infect fish, producing a proliferative disease first identified in walleyes collected in Oneida Lake in New York State in 1969. The genome of the best studied of these viruses, walleye dermal sarcoma virus, includes the conserved *gag*, *pol*, and *env* genes and three open reading frames, designated *orf a*, *orf b*, and *orf c*, which encode accessory proteins.

The most fascinating properties of this virus are its **seasonal reproductive cycle** and its ability to induce both **tumor formation and regression**. Naïve walleyes are infected at the time of spawning, when these fish congregate and the concentration of virus particles in the water is high. The newly infected fish are disease free until the fall, when skin tumors begin to form. The tumors contain ~1 provirus per cell and continue to increase in size through the winter, but only the *orf a* and *orf b* genes are expressed and there is no new virus production. The Orf A protein, called rv-cyclin, probably functions as an ortholog of cellular cyclin C, whereas production of Orf B leads to activation of specific signaling pathways. This protein has the capacity to transform cells *in vitro*. These two viral proteins are thought to work together to promote formation of dermal sarcomas.

With the coming of spring, most likely triggered by change in water temperature, the proviral expression pattern and tumor fate change dramatically. The conserved viral genes are now expressed along with *orf c*. Tumor regression is initiated by the oncolytic Orf C protein in conjunction with the production of 10 to 50 progeny virus particles per cell. The tumors, along with large numbers of new infectious virus particles, are then shed into the water at the next spawning, just in time to initiate a new round of virus infection. Amazingly, the fish develop tumors for only one season and remain tumor free for the rest of their lives, suggesting that immune responses may also participate in tumor regression.

Although oncogenesis is an accidental occurrence with most other retroviruses, walleye dermal sarcoma virus is the exception to this rule, as both tumor production and regression are essential for the successful completion of its reproductive cycle.

Rovnak J, Quackenbush SL. 2010. Walleye dermal sarcoma virus: molecular biology and oncogenesis. *Viruses* **2**:1984–1999.

Walker R. 1969. Virus associated with epidermal hyperplasia in fish. *Natl Cancer Inst Monogr* **31**:195–207.

(A) A walleye salmon carrying a tumor experimentally induced by walleye dermal sarcoma virus. Courtesy of Sandra Quackenbush, Colorado State University. **(B)** Organization of the walleye dermal sarcoma virus genome showing the positions of the genes (*orf a*, *orf b*, and *orf c*) that encode accessory proteins.

A

B

c   gag   pol   env   a   b

Oncogenic DNA Viruses

The first DNA virus to be associated with oncogenesis was the papillomavirus that causes warts (papillomas) in cottontail rabbits; this virus was isolated by Richard Shope in 1933. The lack of cell culture systems for papillomaviruses initially precluded their use as experimental models for oncogenesis. Other viruses, in particular, polyomaviruses, such as simian virus 40, and human adenoviruses, proved much more tractable and soon dominated early studies of transformation and tumorigenesis by DNA viruses. It is important to note that neither simian virus 40 nor adenoviruses are associated with oncogenesis in their natural hosts. However, it was shown soon after their discovery that these viruses can induce tumors in rodents and transform cultured mammalian cells. The possibility that simian virus 40 could have contributed to human cancers continues to be the subject of much debate (Box 6.5). Reproduction of these viruses destroys permissive primate host cells within a few days of infection. In contrast, rodent cells are nonpermissive for viral reproduction or support only limited replication. Consequently, some infected cells survive infection and in rare cases become transformed.

The transforming genes of polyomaviruses and adenoviruses **are** necessary for viral reproduction. However, cellular transformation is a collateral consequence of the activities of the viral transforming proteins. These proteins contribute to transformation by altering the activities of cellular gene products.

BOX 6.5

DISCUSSION

Has simian virus 40 contributed to human cancer?

In 1960, simian virus 40 (SV40) was discovered in the African green monkey kidney cells used to produce poliovirus vaccines; within 2 years it was shown to be tumorigenic in hamsters. These were observations of great concern, because it was realized that many batches of the vaccines contained quite high concentrations of infectious SV40. It has been estimated that 98 million people in the United States, and many more worldwide, were exposed to potentially contaminated poliovirus vaccines before screening to ensure preparation of SV40-free vaccines was introduced in 1963. Ironically, monkey cells had been adopted for poliovirus vaccine production because of the concern that human cells might contain then unknown human cancer viruses!

Epidemiological studies initiated in the 1960s and 1970s monitored thousands of vaccine recipients for up to 20 years, with no evidence for increased cancer risk. The populations studied in this way included more than 1,000 children inoculated with SV40-containing poliovirus vaccine. This group was of particular importance as experimental infection of newborn hamsters with SV40 leads to tumor development in all, while animals older at the time of infection are more resistant to tumor development. The initial alarm raised by the tumorigenicity of SV40 in rodents therefore appeared to be laid to rest. Subsequent reports that the DNA of this virus is present in human tumors, including osteosarcomas and mesotheliomas analogous to those induced by SV40 in hamsters, have led some researchers to reconsider the contribution of this monkey virus to human cancer.

SV40 DNA could also be detected in normal tissue samples from mesothelioma patients and in tumors from individuals who did not receive potentially contaminated vaccine. These observations could indicate that factors in addition to SV40 (notably exposure to asbestos) lead to development of mesothelioma and that SV40 infection of humans is, in fact, quite common. However, careful analysis of the SV40 sequences detected in mesotheliomas (and other tissues) by PCR established that they arose from widespread viral DNA-containing laboratory plasmids that contain an engineered gap that is not present in the viral genome (see the figure). This observation undermined the conclusion of several studies based on PCR. Research and debate continue, for example, with a report in 2012 of an increase in the prevalence of serum antibodies that detect capsid proteins of SV40 (but not of closely related human polyomaviruses) from 15% in healthy individuals to 26% in age-matched mesothelioma patients.

This long-running debate highlights the difficulties of proving the association of a particular virus with human cancer (and the dangers of exquisitely sensitive molecular detection methods like PCR).

Garcea RL, Imperiale MJ. 2003. Simian virus 40 infection of humans. *J Virol* **77:**5039–5045.

Jasani B, Cristaudo A, Emri SA, Gazdar AF, Gibbs A, Krynska B, Miller C, Mutti L, Radu C, Tognon M, Procopio A. 2001. Association of SV40 with human tumours. *Semin Cancer Biol* **11:**49–61.

López-Ríos F, Illei PB, Rusch V, Ladanyi M. 2004. Evidence against a role for SV40 infection in human mesotheliomas and high risk of false-positive PCR results owing to presence of SV40 sequences in common laboratory plasmids. *Lancet* **364:**1157–1166.

Mazzoni E, Corallini A, Cristaudo A, Taronna A, Tassi G, Manfrini M, Comar M, Bovenzi M, Guaschino R, Vaniglia F, Magnani C, Casali F, Rezza G, Barbanti-Brodano G, Martini F, Tognon MG. 2012. High prevalence of serum antibodies reacting with simian virus 40 capsid protein mimotopes in patients affected by malignant pleural mesothelioma. *Proc Natl Acad Sci U S A* **109:**18066–18071.

Poulin DL, DeCaprio JA. 2006. Is there a role for SV40 in human cancer? *J Clin Oncol* **24:**4356–4365.

The results of PCR analysis of DNA samples from human mesotheliomas (designated M plus a number) and an SV40-transformed mouse cell line positive control (1) with a primer pair that distinguishes artificially joined SV40 DNA sequences present in common laboratory plasmids (Plasmid-specific product) from the SV40 genome (Genuine SV40 product). The right lane shows double-stranded DNA markers, with the lengths indicated. Adapted from F. López-Ríos et al., *Lancet* **364:**1157–1160, 2004, with permission.

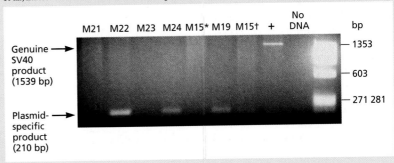

In some cases, such cellular proteins are encoded by the **same** proto-oncogenes transduced or otherwise affected by retroviruses. This important discovery, initially made in studies of the middle T protein (mT) of mouse polyomavirus in the early 1980s, provided the first indication that retroviruses and DNA viruses can transform cells by related mechanisms. Investigation of the biochemical properties of proteins encoded in other transforming genes of these DNA viruses led to equally important insights, notably the characterization of cellular proteins that can block cell cycle progression, the products of **tumor suppressor genes**.

It has been appreciated that herpesviruses can promote the development of tumors in humans and other animals since the discovery in 1966 of Epstein-Barr virus in cells derived from Burkitt's lymphoma. Infection of susceptible cells in culture by members of this family results in immortalization or induction of typical transformed phenotypes. Infection with poxviruses can induce cell proliferation that may be prolonged or rapidly followed by cell death, depending on the virus. Indeed, some members of this family, such as Shope fibroma virus, are associated with tumors of the skin. However, poxviruses do not transform cells in culture, in part because they are highly cytotoxic. The large sizes of their genomes initially presented a major impediment to analysis of the transforming properties of these viruses. It is now clear that herpesviral gene products generally alter cell growth and proliferation by mechanisms related to those responsible for transformation by the smaller DNA viruses or retroviruses. However, the genomes of some of these large DNA viruses also encode micro-RNAs (miRNAs) that contribute to transformation (described in Volume I, Chapter 10).

Contemporary Identification of Oncogenic Viruses

Oncogenic viruses associated with human disease continue to be isolated with some regularity. One discovered in 1994 was a previously unknown member of the family *Herpesviridae*, human herpesvirus 8, which was isolated from tumor cells of patients with Kaposi's sarcoma. Its genome, like those of transducing retroviruses, contains homologs of cellular proto-oncogenes. More recently (in 2008), a polyomavirus associated with a rare form of skin cancer was discovered (Box 6.6). Perhaps an even greater surprise was the realization

---

BOX 6.6

DISCUSSION

A polyomavirus that contributes to development of Merkel cell carcinoma in humans

Mouse polyomavirus and simian virus 40 have been important models for studies of oncogenesis and transformation (see the text). Two human members of this family, BK and JC polyomaviruses, were discovered in 1971. These viruses commonly establish persistent infections, and can be pathogenic in immunosuppressed patients (Appendix, Fig. 20). Eight other polyomaviruses were detected subsequently in human tissues. One, with a genome distantly related to those of other primate polyomaviruses, was detected in tumors from patients with Merkel cell carcinoma, a rare but rapidly metastasizing skin cancer.

Viral DNA sequences initially were identified in tumor tissue by a method based on high-throughput sequencing. Among the unassigned sequences, one from the tumor exhibited significant homology to African green monkey lymphotropic polyomavirus and BK polyomavirus T antigen-coding sequences (see the figure). The 3′ end of this cDNA was shown to include sequences of the human receptor tyrosine phosphatase type G, suggesting that viral DNA sequences were integrated in the genome of tumor cells. Integration of the viral genome was subsequently confirmed by several methods, including Southern blotting. The organization of the viral genome is that typical of polyomaviruses, and includes sequences homologous to the transforming gene products, large and small T antigens, of animal members of the family.

The genome of this virus, which was called Merkel cell polyomavirus, is present in the majority of Merkel cell carcinomas, but the virus has generally not been detected in healthy surrounding tissues or other types of tumors. The pattern of viral DNA integration in the tissues examined indicated that the tumors were monoclonal in origin, implying that viral DNA integration preceded proliferation of the cells. Furthermore, the tumors, but not nearby healthy cells, synthesize T antigen(s), and inhibition of production of these viral proteins by RNA interference in Merkel cell carcinoma-derived cells in culture induces growth arrest or apoptosis. These observations establish that Merkel cell polyomavirus early gene products are required to maintain the oncogenic phenotype of these cells. They therefore provide strong support for causal association between virus infection and the development of Merkel cell carcinoma.

Feng H, Shuda M, Chang Y, Moore PS. 2008. Clonal integration of a polyomavirus in Merkel cell carcinoma. *Science* **319**:1096–1100.

Houben R, Shuda M, Weinkam R, Schrama D, Feng H, Chang Y, Moore PS, Becker JC. 2010. Merkel cell polyomavirus-infected Merkel cell carcinoma cells require expression of viral T antigens. *J Virol* **84**:7064–7072.

The evolutionary relationship of Merkel cell carcinoma polyomavirus to some other mammalian polyomaviruses is shown schematically. Adapted from R. P. Viscidi and K. V. Slak, *Science* **319**:1049–1050, 2008, with permission.

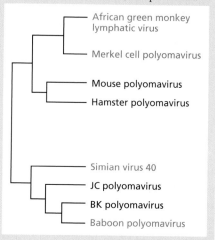

that RNA viruses other than retroviruses can be associated with cancer: hepatitis C virus, a (+) strand RNA virus belonging to the family *Flaviviridae* identified in 1989, is associated with a high risk for hepatocellular carcinoma.

Common Properties of Oncogenic Viruses

Although they are members of different families (Table 6.2), the majority of oncogenic viruses share several general features. In all cases that have been analyzed, transformation is observed to be a single-hit process (defined in Volume I, Chapter 2), in the sense that infection of a susceptible cell with a single virus particle is sufficient to cause transformation. In addition, all or part of the viral genome is usually retained in the transformed cell. With few exceptions, cellular transformation is accompanied by the continuous expression of specific viral genes. On the other hand, transformed cells need not and (except in the case of some retroviruses) **do not** produce infectious virus particles. Most importantly, transforming proteins alter cell proliferation by a limited repertoire of molecular mechanisms.

Viral Genetic Information in Transformed Cells

State of Viral DNA

Cells transformed by oncogenic viruses generally retain viral DNA in their nuclei. These DNA sequences correspond to all or part of the infecting DNA genome, or the proviral DNA made in retrovirus-infected cells. Viral DNA sequences are maintained by one of two mechanisms: they can be integrated into the cellular genome or persist as autonomously replicating episomes.

Integration of retroviral DNA by the viral enzyme integrase is an essential step in the viral reproductive cycle (Volume I, Chapter 7). Although there are some virus-specific biases, integration can occur at many sites in cellular DNA, but the reaction preserves a fixed order of viral genes and control sequences in the provirus (see Volume I, Fig. 7.15). When the provirus carries a v-oncogene, the site at which it is integrated into the cellular genome is of no importance (provided that viral transcription is unimpeded). In contrast, integration of proviral DNA within specific regions of the cellular genome is a hallmark of the induction of tumors by nontransducing retroviruses.

The proviral sequences present in every cell of a tumor induced by nontransducing retroviruses are found in the same chromosomal location, an indication that the tumor arose from a single transformed cell. Such tumors are, therefore, **monoclonal**. The proviruses in the tumor cells have usually lost some or most of the proviral sequences, but at least one long terminal repeat (LTR) containing the transcriptional control region is always present. Viral transcription signals, but not protein-coding sequences, are therefore required for transformation by nontransducing retroviruses. The significance of these properties became apparent when it was discovered that in several tumors proviruses were integrated in the vicinity of some of the same cellular oncogenes that are captured by transducing retroviruses. Because integration of retroviral DNA into the host genome can take place at many sites, there is a limited probability that integration will occur in the vicinity of an oncogene. The long latency for tumor induction by these viruses can be explained in part by the need for multiple cycles of replication and integration.

Integration of viral DNA sequences is not a prerequisite for successful propagation of **any** oncogenic DNA viruses. Nevertheless, integration is the rule in adenovirus- or polyomavirus-transformed cells. Such integration is the result of rare recombination reactions (catalyzed by cellular enzymes) between viral and host DNA sequences with minimal homology. Integration can therefore occur at essentially random sites in the cellular genome. The great majority of cells transformed by these viruses retain only partial copies of the viral genome. The genomic sequences integrated can vary considerably among independent lines of cells transformed by the same virus, but a common, minimal set of genes is always present. The low probability that viral DNA will become integrated into the cellular genome, and the fact that only a fraction of these recombination reactions will maintain the integrity of viral transforming genes, are major factors contributing to the low efficiencies of transformation by these viruses.

A second mechanism by which viral DNA can persist in transformed cells is as a stable, extrachromosomal episome (Volume I, Box 1.7). Such episomal viral genomes are a characteristic feature of B cells immortalized by Epstein-Barr virus, and they can also be found in cells transformed by papillomaviruses. The viral episomes are maintained at concentrations of tens to hundreds of copies per cell, by both replication of the viral genome in concert with cellular DNA synthesis and orderly segregation of viral DNA to daughter cells (described in Volume I, Chapter 9). Consequently, transformation depends on the viral proteins necessary for the survival of viral episomes, as well as those that modulate cell growth and proliferation directly.

Identification and Properties of Viral Transforming Genes

Transforming genes of oncogenic viruses have been identified by classical genetic methods, characterization of the viral genes present and expressed in transformed cell lines, and analysis of the transforming activity of viral DNA fragments directly introduced into cells (Box 6.7). For example, analysis of transformation by temperature-sensitive mutants of mouse polyomavirus established as early as 1965 that the viral early transcription unit is necessary and sufficient to initiate and maintain transformation of cells in culture. Of even greater value were mutants of retroviruses, in particular two mutants of Rous sarcoma virus isolated in the early 1970s. The genome

BACKGROUND

Multiple lines of evidence identified the transforming proteins of the polyomavirus simian virus 40

Early gene products are necessary and sufficient to initiate transformation.

1. Viruses carrying temperature-sensitive mutations in the early transcription unit (*ts*A mutants), but no other region of the genome, fail to transform at a nonpermissive temperature.
2. Simian virus 40 DNA fragments containing only the early transcription unit transform cells in culture; DNA fragments containing other regions of the genome exhibit no activity.

Early gene products are necessary to maintain expression of the transformed phenotype.

1. Many lines of cells transformed by simian virus 40 *ts*A mutants at a permissive temperature revert to a normal phenotype when shifted to a nonpermissive temperature, and vice versa (see figure).
2. Integration of viral DNA sequences disrupts the late region of the viral genome but not the early transcription unit, and early gene products are synthesized in all transformed cell lines.

Both LT and sT contribute to transformation.

1. Simian virus 40 mutants carrying deletions of sequences that are expressed only in sT (Volume I, Appendix, Fig. 23) fail to transform rat cells to anchorage-independent growth.
2. Introduction and expression of LT complementary DNA are sufficient

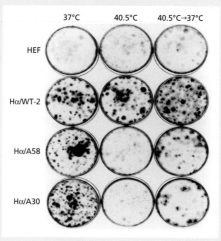

for induction of transformation, but expression of sT can increase efficiency (especially at low LT concentrations), is necessary for expression of specific phenotypes in specific cells, and is required for transformation of resting cells.

Hamster embryonic fibroblasts (HEF) or cells derived by transformation with wild-type simian virus 40 (Hα/WT-2) or with *ts*A mutants (Hα/A58 and Hα/A30) were plated at the temperatures indicated at the top or shifted to the lower temperature after 6 days. Cells were stained 11 days after seeding. In contrast to the wild-type transformed cells, those transformed by *ts*A mutants fail to form colonies at the higher (nonpermissive) temperature. However, this transformed phenotype is exhibited upon shift down to the lower, permissive temperature; that is, it is reversible. Adapted from J. S. Brugge and J. S. Butel, *J Virol* **15:**619–635, 1975. Courtesy of J. Butel, Baylor College of Medicine.

of one mutant carried a spontaneous deletion of ~20% of the viral genome. This mutant could no longer transform the cells it infected, but it could still reproduce. The second mutant was temperature sensitive for transformation, but the virus could reproduce at temperatures both permissive and nonpermissive for transformation. These properties of the mutants showed unequivocally that cellular transformation and viral reproduction are distinct processes. More importantly, the deletion mutant allowed preparation of the first nucleic acid probe specific for a v-oncogene, v-*src* (Box 6.8) (see the interview with Dr. Michael Bishop: http://bit.ly/Virology_Bishop). This *src*-specific probe was found to hybridize to cellular DNA, providing the first conclusive evidence that v-oncogenes are of cellular and **not** viral origin. This finding, for which J. Michael Bishop and Harold Varmus received the 1989 Nobel Prize in physiology or medicine, had far-reaching significance, because it immediately suggested that such cellular genes might become oncogenes by means other than viral transduction.

The presence of cellular oncogenes in their genomes turned out to be the definitive characteristic of transducing retroviruses (Fig. 6.8). As noted earlier, the acquisition of these cellular sequences is a very rare event. In addition, with the

exception of Rous sarcoma virus, the transducing retroviruses are replication defective, having lost all or most of the viral coding sequences during oncogene capture. Such defective transducing viruses can, however, be propagated in mixed infections with replication-competent "helper" viruses, which provide all the proteins necessary for assembly of viral particles.

Viral and cellular protein-coding sequences are fused in many v-oncogenes (Fig. 6.8). The presence of viral sequences can enhance the efficiency of translation of the oncogene mRNA, stabilize the protein, or determine its location in the cell. Unregulated expression or overexpression of the cellular sequence from the viral promoter is sufficient to cause transformation by some v-oncogenes (e.g., *myc* and *mos*). However, in most cases, the captured oncogenes have undergone additional changes that contribute to their transforming potential. These alterations, which include nucleotide changes, truncations at either or both ends, or other rearrangements, affect the normal function of the gene product.

Transformation of primary cells by DNA viruses typically requires the products of two or more viral genes (Table 6.4). The majority of these genes exhibit some ability to alter the properties of the cells in which they are expressed in the absence of

BOX 6.8

TRAILBLAZER
Preparation of the first oncogene probe

In the early 1970s, modern molecular biology was already in full bloom, but some techniques that are currently commonplace, such as PCR amplification of specific genes, had not yet been invented. It was, however, possible at that time to make cDNA copies of RNA with retroviral reverse transcriptase and to separate double-stranded (hybridized) from single-stranded (nonhybridized) nucleic acids. The existence of two genetically related viral genomes, one that contained a transforming gene (Rous sarcoma virus [RSV]) and a deletion mutant (tdRSV) that was replication competent but nontransforming, made it possible to isolate a radioactively labeled probe for the transforming gene, *src*, by exploiting the available techniques, in a strategy known as **subtractive hybridization**.

Complementary (−) strand DNA was prepared by reverse transcription of the (+) strand RSV genome and then hybridized to genomic RNA of the tdRSV mutant. The nonhybridizing DNA (purple) was separated from the double-stranded hybrids by hydoxylapatite chromatography (see the figure). This radioactive DNA was then used as a probe to search for corresponding genetic material in a variety of cells.

Hybridization to chicken genomic DNA and the DNA of other avian species immediately suggested that the *src* sequences and, by inference, other retroviral oncogenes had been captured from the host cells infected by the virus. The observation that *src*-related sequences are conserved among cells from widely different species in the animal kingdom suggested that the proteins they encode play a central role in cell growth and division and that their malfunction could explain the origin of cancers that arise independently of retroviral infection.

Spector DH, **Varmus HE, Bishop JM.** 1978. Nucleotide sequences related to the transforming gene of avian sarcoma viruses are present in DNA of uninfected vertebrates. *Proc Natl Acad Sci U S A* **75:**4102–4106.

Stehelin D, **Varmus HE, Bishop JM.** 1976. DNA related to the transforming gene(s) of avian sarcoma viruses is present in normal avian DNA. *Nature* **260:**170–173.

Genomic RNAs are shown in green and the cDNA products of reverse transcription of common sequences and the unique sequence present on the RSV genome in blue and purple. HAP, hydroxylapatite.

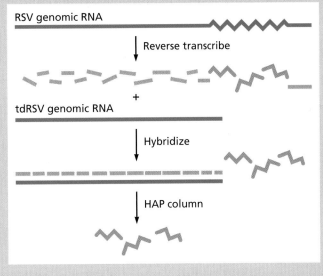

The Origin and Nature of Viral Transforming Genes

other viral proteins. However, some are required only for the induction of specific transformed phenotypes or only under certain conditions (e.g., simian virus 40 small T antigen [sT]), and several exhibit no activity on their own (Table 6.4). A classic example of the latter phenomenon is provided by the adenoviral E1B gene: this gene, together with the E1A gene, was initially shown to be essential for transformation of rodent cells in culture, but it possesses no intrinsic ability to induce **any** transformed phenotype. This apparent paradox has been resolved with elucidation of the molecular functions of the viral gene products: E1A gene products induce apoptosis, but E1B proteins suppress this response and allow cells that synthesize E1A proteins to survive and display transformed phenotypes.

The Origin and Nature of Viral Transforming Genes

Two classes of viral oncogenes can be distinguished on the basis of their similarity to cellular sequences. The oncogenes of transducing retroviruses and certain herpesviruses (e.g., human herpesvirus 8) are so closely related to cellular genes that it is clear that they were captured relatively recently (since the divergence of primates). Such acquisition must be a result of recombination between viral and cellular nucleic acids, a process that has been documented for transducing retroviruses. Retrovirus particles contain some cellular RNAs, and rare recombination reactions during reverse transcription can give rise to transducing retroviruses. The limiting factor appears to be the frequency with which cellular mRNA molecules are encapsidated. Two mechanisms can increase the likelihood of encapsidation, and consequently increase the frequency of gene capture (Fig. 6.9). Both mechanisms depend on integration of a provirus in or near a cellular gene and incorporation of the cellular sequences into a transcript initiated within the LTR. The final step is a recombination reaction(s) between largely nonhomologous sequences in this chimeric transcript and a wild-type viral genome when both are incorporated into a viral particle.

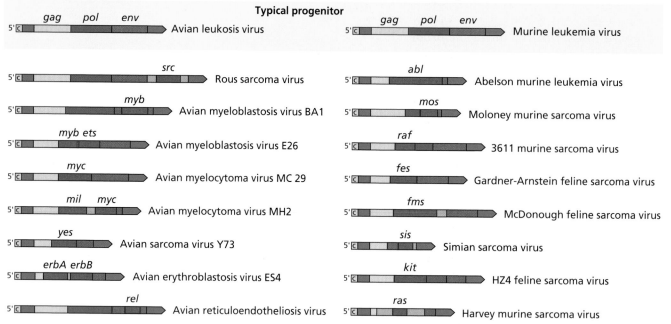

Figure 6.8 Genome maps of avian and mammalian transducing retroviruses. Avian leukosis virus (e.g., Rous-associated virus) and murine leukemia virus are prototypical retroviruses. Their genomes contain the three major coding regions: *gag* (pink), *pol* (blue), and *env* (brown). In Rous sarcoma virus, the oncogene *src* is added to the complete viral genome. In all other avian and mammalian transducing retroviruses, some of the viral coding information is replaced by cell-derived oncogene sequences (red). Consequently, such transducing viruses are defective in replication. The majority of the transducing retroviruses carry a single v-oncogene in their genomes, but some include more than one (e.g., *erbA* and *erbB* in avian erythroblastosis virus ES4). In such cases, one is sufficient for transformation, while the second accelerates this process. In some cases, additional cellular DNA sequences (orange) were also captured in the viral genome. Adapted from T. Benjamin and P. Vogt, p. 317–367, *in* B. N. Fields et al. (ed.), *Fields Virology*, 2nd ed. (Raven Press, New York, NY, 1990), with permission.

Table 6.4 Some transforming gene products of adenoviruses, papillomaviruses, and polyomaviruses

Virus	Gene product	Activities
Adenoviridae		
Human adenovirus type 2	E1A: 243R and 289R	Cooperate with E1B proteins to transform primary cells; not sufficient for establishment of transformed cell lines
	E1B: 55 kDa and 19 kDa	Necessary for E1A-dependent transformation of primary and established cells; counter apoptosis by different mechanisms
Papillomaviridae		
Human papillomavirus types 16 and 18	E6	Required for efficient immortalization of primary human fibroblasts and keratinocytes
	E7	Cooperates with E6 to transform primary rodent cells; required for efficient immortalization of primary human fibroblasts or keratinocytes
Polyomaviridae		
Polyomavirus	LT	Immortalizes primary cells; required to induce but not to maintain transformation of primary cells
	mT	Transforms established cell lines; required to both induce and maintain transformation of primary cells
Simian virus 40	LT	Immortalizes primary cells; required to induce and maintain transformation of primary and established cells
	sT	Required under many conditions, depending on LT concentration, genetic background of recipient cells, and transformation assay

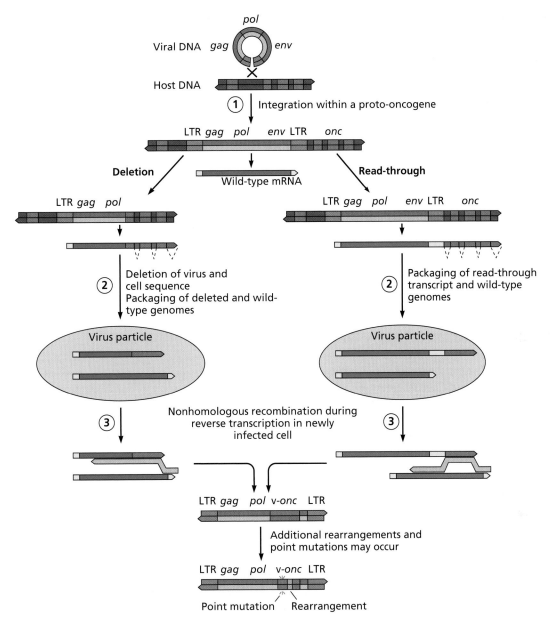

Figure 6.9 Possible mechanisms for oncogene capture by retroviruses. The first step in each of two mechanisms shown is integration of a provirus in or near a cellular gene (*onc*). The deletion mechanism (left) requires removal of the right end of the provirus, thereby linking cellular sequences to the strong viral transcriptional control region in the left LTR. The first recombination step in this mechanism therefore takes place at the DNA level. It leads to synthesis of a chimeric RNA, in which viral sequences from the left end of the provirus are joined to cellular sequences. Chimeric RNA molecules that include the viral packaging signal can be incorporated efficiently into viral particles with a wild-type genome produced from another provirus in the same cell. A second recombination reaction, during reverse transcription (as described in Volume I, Chapter 7), is then required to add right-end viral sequences to the recombinant. At a minimum, these right-end sequences must include signals for subsequent integration of the recombinant viral DNA into the genome of the newly infected host cell, from which the transduced gene is then expressed. The read-through mechanism (right) does not require a chromosomal deletion. Viral transcription does not always terminate at the 3′ end of the proviral DNA, but continues to produce transcripts containing cellular sequences. Such chimeric transcripts can then be incorporated into virus particles together with the normal viral transcript. The cellular sequences can then be captured by recombination during reverse transcription, as indicated. Important additional mutations and rearrangements probably occur during subsequent virus replication. Adapted from J. M. Coffin et al. (ed.), *Retroviruses* (Cold Spring Harbor Laboratory Press, Cold Spring Harbor, NY, 1997), with permission.

Many of the cellular proto-oncogenes from which v-oncogenes are derived have been highly conserved throughout evolution: vertebrate examples often have homologs in yeast. The products of such genes must therefore fulfill functions that are indispensable for a wide variety of eukaryotic cells. Furthermore, as single copies of v-oncogenes are sufficient to transform cells, their functions must override those of the resident, cognate proto-oncogenes. Accordingly, v-oncogenes function as **dominant** transforming genes.

Members of the second class of viral oncogenes, such as adenovirus E1A and polyomavirus LT, are not obviously related to cellular genes. However, the products of these genes may contain short amino acid sequences also present in cellular proteins (for example, see Fig. 6.20). The precise origins of such oncogenes remain shrouded in mystery (see Chapter 10).

Functions of Viral Transforming Proteins

Many approaches have been employed to determine the functions of viral oncogene products. In some cases, the sequence of a viral transforming gene can immediately suggest the function of its protein product. For example, the genomes of certain herpesviruses and poxviruses include coding sequences that are closely related to cellular genes that encode growth factors, cytokines, and their receptors. Or the protein may contain amino acid motifs characteristic of particular biochemical activities, such as tyrosine phosphorylation or sequence-specific DNA binding. In other cases, notably many retroviral v-oncogene products, it has been possible to identify important biochemical properties, such as enzymatic activity, binding to a hormone or growth factor, or sequence-specific binding to nucleic acids (Table 6.5).

The breakthrough to understanding transformation by small oncogenic DNA viruses came with mutational analyses that correlated the transforming activities of viral gene products with binding to specific cellular proteins, notably tumor suppressors. The first such interaction identified was between sequences of adenoviral E1A proteins that are necessary for transformation and the cellular retinoblastoma tumor suppressor protein, Rb. The similar relationship between Rb binding and the transforming activities of simian virus 40 large T antigen (LT) and the E7 protein of oncogenic human papillomaviruses rapidly established the general importance of interaction with Rb in transformation. Interaction of transforming proteins of these small DNA viruses with a second cellular tumor suppressor, the p53 protein, is also required for oncogenesis. These observations established the importance of inactivation of tumor suppressors (Fig. 6.7) in transformation by these DNA viruses. A second common feature is that their transforming proteins affect multiple cellular proteins and pathways (Fig. 6.10).

Viral transforming proteins exhibit great diversity in all their properties, from primary amino acid sequence to biochemical activity. They also differ in the number and nature

Table 6.5 Functional classes of oncogenes transduced by retroviruses[a]

Transduced oncogene[b]	Function of cellular homolog
Growth factors	
sis	Platelet-derived growth factor
Receptor protein tyrosine kinases	
erbB	Epithelial growth factor receptor
Kit	Hematopoietic receptor; product of the mouse W locus
Hormone receptors	
erbA	Thyroid hormone receptor
G proteins	
H-*ras*, K-*ras*	GTPases
Adapter protein	
Crk	Signal transduction
Nonreceptor tyrosine kinases	
src, *abl*	Signal transduction
Serine/threonine kinases	
Mos	Required for germ cell maturation
Akt	Signal transduction
Nuclear proteins	
jun, *fos*	Transcriptional regulator (AP-1 complex)
Myc	Transcriptional regulator

[a]Adapted from T. Benjamin and P. Vogt, *in* B. N. Fields et al. (ed.), *Fields Virology*, 3rd ed. (Lippincott-Raven Publishers, Philadelphia, PA, 1996).

[b]Only some representative examples are listed.

of the cellular pathways they alter. Despite such variation, these viral proteins induce continuous cell proliferation, the definitive characteristic of transformation, by related mechanisms. Indeed, the best characterized fall into one of only two classes, permanent activation of cellular signal transduction cascades or disruption of the circuits that regulate cell cycle progression.

Activation of Cellular Signal Transduction Pathways by Viral Transforming Proteins

The products of transforming genes of both RNA and DNA viruses can alter cellular signal transduction cascades. The consequence is permanent activation of pathways that promote cell growth and proliferation. However, as discussed in subsequent sections, these viral proteins can intervene at various points in these pathways, and they operate in several different ways.

Viral Signaling Molecules Acquired from the Cell

The Transduced Cellular Genes of Acutely Transforming Retroviruses

The v-*src* paradigm. The protein product of v-*src* was the first retroviral transforming protein to be identified,

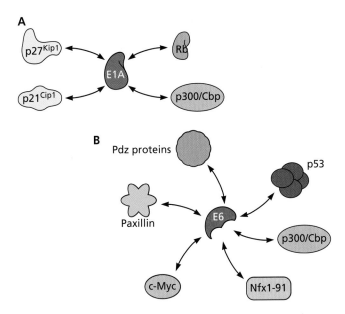

Figure 6.10 Schematic illustration of interactions of DNA virus transforming proteins with multiple cellular proteins. (A) The association of adenoviral E1A proteins with Rb family tumor suppressors, the histone acetyltransferases p300/Cbp, and the cyclin-dependent kinase inhibitors p27^{Kip1} and p21^{Cip1} has been implicated in transformation: E1A protein substitutions that impair these interactions reduce or eliminate transforming activity. **(B)** The human papillomavirus type 16 or 18 E6 protein also interacts with p300/Cbp and tumor suppressors, in this case the p53 protein, as well as several proteins that contain the Pdz domain and are localized at cell junctions (e.g., Dlg1 [discs large 1]) or are phosphatases (e.g., Ptpn3 [tyrosine-protein phosphatase nonreceptor type 3]). In addition, it associates with the transcriptional regulators c-Myc and Nfx1-91 to repress transcription of the gene encoding the protein component of telomerase. These interactions have been implicated in increased production of telomerase in cells synthesizing the E6 protein. In some cases, including p53 and Nfx1-91, the complex includes the cellular ubiquitin ligase E6-Ap, and the other cellular proteins are targeted for proteasomal degradation. Degradation of Pdz domain-containing proteins are also induced by the viral E6 protein.

when serum from rabbits bearing tumors induced by Rous sarcoma virus was shown to immunoprecipitate a 60-kDa phosphoprotein. This v-Src protein was soon found to possess protein tyrosine kinase activity, a property that provided the first clue that phosphorylation of cellular proteins can be critical to oncogenesis. The discovery of this protein tyrosine kinase led to the identification of a large number of other proteins with similar enzymatic activity and important roles in cellular signaling.

The Src protein contains a tyrosine kinase domain (SH1, for Src homology region 1) and two domains that mediate protein-protein interactions (Fig. 6.11A). The first, the SH2 domain, binds to phosphotyrosine-containing sequences, whereas the SH3 domain has affinity for

proline-rich sequences. A fourth Src domain (SH4) includes the N-terminal myristoylation signal that directs Src to the plasma membrane. All four domains are required for Src transforming activity.

The Src kinase phosphorylates itself at specific tyrosine residues. These modifications regulate its enzymatic activity: phosphorylation of Y416 in the kinase domain activates the enzyme, whereas phosphorylation of Y527 in the C-terminal segment inhibits activity. The crystal structures of cellular Src and another member of the Src family revealed the importance of the SH2 and SH3 domains in such regulation. For example, exchange of the intramolecular interaction of SH2 with Y527 for binding of the SH2 domains to phosphotyrosine-containing motifs in **other** proteins initiates conformational changes that activate the kinase (Fig. 6.11B). This autoregulatory mechanism explains earlier findings that transduction and overproduction of the normal Src protein do not lead to cellular transformation, and that the constitutive oncogenic activity of v-*src* requires loss or mutation of the Y527 codon.

Soon after its kinase activity was first discovered, v-Src was shown to localize to focal adhesions, the areas where cells make contact with the extracellular matrix. This observation led to identification of a second protein tyrosine kinase enriched in these areas as a protein that exhibits increased tyrosine phosphorylation in v-Src-transformed cells. This focal adhesion kinase (Fak) and Src family proteins turned out to be crucial components of a signal transduction cascade (normally controlled by cell adhesion) that modulates the properties of the actin cytoskeleton, and hence cell shape and adhesion. It also signals to the Ras/Map kinase pathway that controls cell proliferation (Fig. 6.12). The constitutive activity of v-Src can therefore account for the morphological and growth properties of cells transformed by this oncogene product.

Other transduced oncogenes. The transduced oncogenes of retroviruses are homologs of cellular genes that encode many components of signal transduction cascades, from the external signaling molecules (e.g., v-Sis) and their receptors (v-ErbB and v-Kit) to the nuclear proteins at the end of the relay (v-Fos and v-Myc) (Table 6.5). It therefore seems likely that any positively acting protein in such a cascade has the potential to act as a transforming protein. The oncogenic potential of such transduced oncogenes is realized by two nonexclusive mechanisms: genetic alterations that lead to constitutive protein activity and inappropriate production, or overproduction, of the protein. The former mechanism applies to most of the retroviral oncogenes (Table 6.5). For example, like other small, guanine nucleotide-binding proteins, Ras normally cycles between a conformation that is active (GTP bound) and one that is inactive (GDP bound).

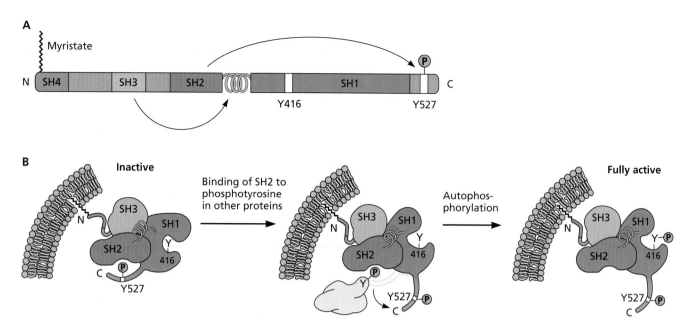

See link: http://www.rcsb.org/pdb/explore/jmol.do?structureId=2SRC&bionumber=1

Figure 6.11 Organization and regulation of the c-Src tyrosine kinase. (A) The functional domains of the protein. The SH4 domain contains the site for addition of the myristate chain that anchors the protein in the cell membrane. The SH2 and SH3 domains mediate protein-protein interactions by binding to phosphotyrosine-containing and proline-rich sequences, respectively. Both domains are found in other proteins that participate in signal transduction pathways. One transforming protein, Crk, is made up of only an SH2 and an SH3 domain: it functions as an adapter, bringing together other proteins in a signal transduction pathway (Table 6.5). Arrows represent intramolecular interactions observed in the repressed-state crystal structures of Src. **(B)** The interactions and their reversal. When Y527 is phosphorylated, the C-terminal region of c-Src in which this residue lies is bound to the SH2 domain. This interaction brings a polyproline helix located between the SH2 and SH1 domains into contact with the SH3 domain, as illustrated at the top (see http://www.rcsb.org/pdb/explore/explore.do?structureId=2src). Binding of SH3 to the helix deforms the kinase domain, accounting for the inactivity of the Y527-phosphorylated form of the protein. Such intramolecular associations maintain the kinase domain (SH1) in an inactive conformation. A conformational change that activates the kinase can be induced as shown, as well as by binding of the SH3 domain to proline-rich sequences in other proteins and probably by dephosphorylation of Y527 (see Fig. 6.16). Once released from the autoinhibited state in this way, Y416 in the kinase domain is autophosphorylated, a modification that stabilizes the active conformation of the SH1 domain. The v-Src protein is not subject to such autoinhibition, because the sequence encoding the C-terminal regulatory region of c-Src was deleted during transduction of the cellular gene.

Such cycling is under the control of GTPase-activating and guanine nucleotide exchange proteins. The latter proteins (e.g., Sos [Fig. 6.3]) stimulate the release of GDP once bound GTP has been hydrolyzed. However, v-Ras proteins fail to hydrolyze GTP efficiently and therefore persist in the active, GTP-bound conformation that relays signals to downstream pathways, such as the Map kinase cascade (Fig. 6.3). Such constitutive activity is the result of mutations that lead to substitution of specific, single amino acids in the protein (at residues 12, 13, or 61) and render the protein refractory to the GTPase-activating protein. Analogous mutations are common in certain human tumors, such as colorectal cancers (Box 6.2), and were the first discrete genetic changes in a proto-oncogene linked to neoplastic disease in humans.

Less commonly, over- or misexpression of the transduced oncogene is sufficient to disrupt normal cell behavior. This type of mechanism is best characterized for *myc*. In normal cells, the expression of this gene is tightly regulated, such that the c-Myc protein is made only during a short period in the G_1 phase of the cell cycle, and is not synthesized when cells withdraw from the cycle or differentiate. The production of even small quantities of Myc or Myc-fusion proteins specified by retroviruses, such as the avian myelocytoma virus MH2 (Fig. 6.8), at an inappropriate time results in cellular transformation.

Viral Homologs of Cellular Genes

The genomes of some larger DNA viruses also contain coding sequences that are clearly related to cellular genes that encode signal transduction molecules. Human herpesvirus 8, a gammaherpesvirus related to Epstein-Barr virus, has been strongly implicated in the etiology of Kaposi's sarcoma, a malignancy common in acquired immunodeficiency syndrome (AIDS) patients, and primary effusion lymphoma (see Chapter 7). Its structural proteins and viral enzymes are closely related to those of other herpesviruses. The genome also contains several homologs of cellular genes that encode signaling proteins, which are clustered in regions interspersed among blocks of genes common to all

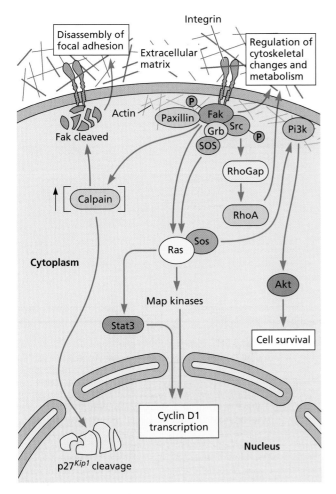

Figure 6.12 Regulation of cell proliferation and adhesion by Src. Both c-Src and v-Src are localized to focal adhesions, where they are associated with focal adhesion kinase (Fak) and adapter proteins, such as Grb2 and paxillin. These protein assemblies normally maintain contacts between the extracellular matrix via integrins and the actin cytoskeleton of the cell. When the Src tyrosine kinase is active, Fak is phosphorylated at specific sites and cleaved into several fragments by the protease calpain. These changes result in disruption of focal adhesions and account for the changes in morphology and motility of v-Src-transformed cells. Calpain-mediated proteolysis of Fak results, at least in part, from increased translation of calpain mRNA induced by v-Src. Another substrate of this protease is the cyclin-dependent kinase inhibitor p27^{Kip1}. As shown, v-Src also induces transcription of genes, including the cyclin D1 gene, via Ras and the Map kinase cascade and the transcriptional activator Stat3. These responses to v-Src result in cell proliferation.

herpesviruses (Fig. 6.13). Among the best characterized is the v-*gpcr* gene, which specifies a guanine nucleotide-binding protein-coupled receptor that is most closely related to a cellular receptor for CXC chemokines. The v-*gpcr* gene induces morphological transformation when introduced into mouse fibroblasts or endothelial cells in culture, and formation of tumors that resemble Kaposi's sarcoma in transgenic mice. Cellular chemokine receptors bind chemokines released at sites of inflammation to activate signal transduction. In con-

trast, v-Gpcr is fully active in the absence of any ligand and can trigger signaling via several cellular pathways to promote cell survival (phosphatidylinositol 3-kinase [Pi3k]/protein kinase B [Akt]) and to activate transcription of cellular genes (via activator protein 1 [Ap-1] and nuclear factor κb [Nf-κb]) (Fig. 6.13B). These genes include several that encode secreted cytokines and growth factors, such as interleukin (IL)-6 and vascular endothelial growth factor (Vegf). These secreted proteins and viral orthologs (for example, vIL-6) are thought to cooperate to induce sustained proliferation of latently infected cells and angiogenesis, the proliferation of new blood vessels (Fig. 6.13B). This characteristic feature of Kaposi's sarcoma is essential for tumor progression.

Alteration of the Production or Activity of Cellular Signal Transduction Proteins

Insertional Activation by Nontransducing Retroviruses

Most tumors induced by nontransducing retroviruses arise as a result of increased transcription of cellular genes located in the vicinity of integrated proviruses. This mechanism of oncogenesis is known as **insertional activation**. It has been implicated in a leukemia-like disease developed by patients participating in a gene therapy trial (Box 6.9). As in the case of the transducing retroviruses, Rous sarcoma-derived avian viruses played a seminal role in delineating the mechanisms of insertional activation. The original stocks of viruses isolated by Peyton Rous included replication-competent leukosis viruses. These viruses do not carry an oncogene, but in young chickens they induce B cell tumors that originate in the bursa of Fabricius, the major lymphoid organ of these birds. A provirus was found integrated in the vicinity of the cellular *myc* gene in each of these tumors. Although the exact integration site varied from tumor to tumor, many integration sites lay in the intron between exon 1 (a noncoding exon) and exon 2 (Fig. 6.14). However, in some tumors proviruses were located upstream or downstream of the cellular *myc* gene. In this avian system, inappropriate synthesis of the normal cellular Myc protein is associated with lymphomagenesis; no changes in the protein are required. Analysis of the sites of proviral DNA integration and the gene products formed in these tumors provided the first evidence for two types of insertional activation: promoter insertion and enhancer insertion (Fig. 6.15).

The first mechanism, **promoter insertion**, results in production of a chimeric RNA in which sequences transcribed from the proviral LTR are linked to cellular proto-oncogene sequences. If transcription originates from the left-end LTR, some viral coding sequences may be included. However, transcription from the right-end LTR seems to be more common, and in these cases the proviral left-end LTR has usually been deleted. Proviral integration often occurs within the cellular

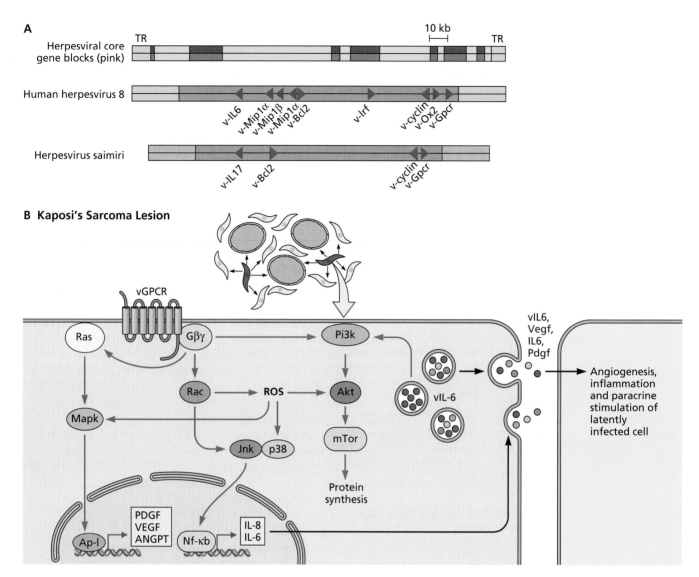

Figure 6.13 Model of paracrine oncogenesis by human herpesvirus 8 gene products. (A) The genomes of human herpesvirus 8 and herpesvirus saimiri contain homologs of various cellular genes. The two viral genomes are shown in orientations that align genes conserved among herpesviruses, the core gene blocks shown at the top. The conserved genes encode proteins needed for virus reproduction and assembly. Homologs of cellular genes (arrowheads) are interspersed among the core gene blocks. Those shown in purple are related to cellular chemokines (v-IL-6, v-IL-17, and macrophage inflammatory factor [v-Mip1α or 1β]), chemokine receptors (v-Gpcr; see the text), or other signaling molecules (interferon-responsive protein [v-Irf] and an N-Cam family transmembrane protein that participates in intercellular signaling [v-Ox2]). The human herpesvirus 8 v-IL-6 protein blocks the action of interferon and can also induce proliferation of B cells. Viral genes shown in red are related to cellular genes that encode proteins that regulate cell proliferation or apoptosis, cyclin D [v-cyclin; see the text] and Bcl-2 [v-Bcl2]. **(B)** As indicated at the top, Kaposi's sarcomas contain human herpesvirus 8-infected cells in which viral lytic genes are expressed (red), as well as latently infected cells (tan) and blood vessels. The former cells produce vGPCR, the product of an early gene, as well as vIL-6, but the latter do not. Rather, latently infected cells synthesize viral proteins that promote survival (vFLIP) or proliferation (v-cyclin). As lytic infection is cytotoxic, a paracrine model for oncogenesis has been proposed. In this model (below), vGPCR made in lytically infected cells triggers signaling via Ras and the β and γ subunits of a trimeric G protein (Gβγ) via Mapk, Pi3k, and Jnk (cJun N-terminal kinase) pathways to stimulate expression of cellular genes that encode cytokines (IL-6, IL-8) and angiogenic growth factors (Vegf and platelet-derived growth factor [Pdgf]). In cooperation with vIL-6 (and other virokines), which are also secreted from lytically infected cells, these cellular proteins act upon neighboring cells (paracrine stimulation) to maintain proliferation of latently infected cells and induce angiogenesis. This model is consistent with the increased incidence of Kaposi's sarcoma in immunosuppressed patients, when lytically infected cells cannot be removed by T cells, and the finding that antiviral drugs that inhibit human herpesvirus 8 reproduction can prevent development of Kaposi's sarcomas in AIDS patients.

BOX 6.9

WARNING
Inadvertent insertional activation of a cellular gene during gene transfer

Retroviruses have long been considered likely to be valuable vectors for gene therapy. One reason is that integration of the retroviral vector into the host genome results in permanent delivery of the potentially therapeutic gene to all infected cells and their descendants. However, an outcome detected in one of the first clinical trials indicates that this property is a double-edged sword.

A French trial was examining the potential of gene therapy using a vector based on mouse Moloney leukemia virus to treat children with severe combined immunodeficiency (SCID). This disease is caused by mutation in a single gene on the X chromosome, and the only therapies available are associated with severe, often fatal, side effects. A trial with 10 children with the disease, who were given gene transfer as early as possible, initially appeared to be very successful: in most cases, the immune system was restored without side effects. But, early in 2002, one patient was found to have developed a T cell leukemia-like disease. The overproliferating T cells were monoclonal: all carried a provirus integrated into the same site on chromosome 11, near a gene (*Lmo2*) that is expressed abnormally in a form of childhood acute lymphoblastic leukemia.

The monoclonal origin of the T cells that proliferated in this child indicates that proviral insertion contributed to the development of the disease. It initially seemed likely that other factors also did so: a predisposition to childhood cancers was evident in other members of the child's family. However, other children participating in the same trial or a similar trial in the United Kingdom were later diagnosed with leukemia associated with insertion of the provirus in the same chromosome 11 site (see the figure). This unfortunate outcome temporarily halted these and numerous other clinical trials of gene transfer using retroviral vectors in the United States and Europe.

Subsequent follow-up studies showed that while acute leukemia developed in four patients, three were treated successfully by chemotherapy. Furthermore, seven patients, including three survivors of leukemia, had sustained immune reconstitution: all were able to live in nonprotected environments, controlling microorganisms successfully, and are developing normally. These results demonstrate the therapeutic potential of gene therapy.

Since these first trials, Moloney leukemia virus vectors that lack the enhancers and/or are not targeted to promoters in the cellular genome have been developed. Vectors for gene transfer have also been derived from the lentivirus simian immunodeficiency virus, which shows no preference for integration into promoter regions. When pseudotyped with the envelope G protein from vesicular stomatitis virus, this lentivirus vector infects almost all types of cells.

Numerous trials with these vectors are currently ongoing in the United States and elsewhere.

See http://www.genetherapynet.com/clinicaltrialsgov.html for information about gene therapy and viral vectors.

Deichmann A, Hacein-Bey-Abina S, Schmidt M, Garrigue A, Brugman MH, Hu J, Glimm H, Gyapay G, Prum B, Fraser CC, Fischer N, Schwarzwaelder K, Siegler ML, de Ridder D, Pike-Overzet K, Howe SJ, Thrasher AJ, Wagemaker G, Abel U, Staal FJ, Delabesse E, Villeval JL, Aronow B, Hue C, Prinz C, Wissler M, Klanke C, Weissenbach J, Alexander I, Fischer A, von Kalle C, Cavazzana-Calvo M. 2007. Vector integration is nonrandom and clustered and influences the fate of lymphopoiesis in SCID-XI gene therapy. *J Clin Invest* **117**:2225–2232.

Hacein-Bey-Abina S, Hauer J, Lim A, Picard C, Wang GP, Berry CC, Martinache C, Rieux-Laucat F, Latour S, Belohradsky BH, Leiva L, Sorensen R, Debré M, Casanova JL, Blanche S, Durandy A, Bushman FD, Fischer A, Cavazzana-Calvo M. 2010. Efficacy of gene therapy for X-linked severe combined immunodeficiency. *N Engl J Med* **363**:355–364.

Touzot F, Hacein-Bey-Abina S, Fischer A, and Cavazzana M. 2014. Gene therapy for inherited immunodeficiency. *Expert Opin Biol Ther* **14**:789–798.

Sites of the retroviral vectors for the γ subunit of the interleukin-2 receptor carrying the gene (γC) in the first two children in the French trial who developed T cell leukemia. Both proviruses integrated close to the promoter of the lmo2 gene. Adapted from S. Hacein-Bey-Abina et al., *Science* **302**:415–419, 2003, with permission.

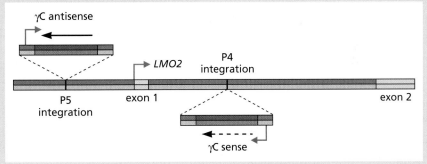

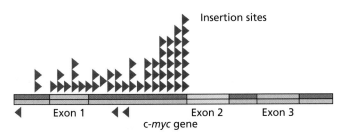

Figure 6.14 Insertional activation of c-*myc* by avian leukosis viruses. In avian cells derived from avian leukosis virus-induced B cell lymphomas, individual proviral integration sites are clustered as shown (arrowheads) within noncoding exon 1 and intron 1 of the *myc* gene. Most integrated proviruses are oriented in the direction of *myc* transcription (arrowheads pointing to the right). Adapted from J. Nevins and P. Vogt, p. 301–343, *in* B. N. Fields et al. (ed.), *Fields Virology*, 3rd ed. (Lippincott-Raven Publishers, Philadelphia, PA, 1996), with permission.

Promoter insertion

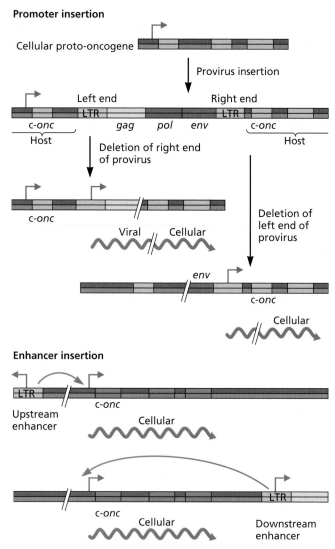

Enhancer insertion

Figure 6.15 Mechanisms for insertional activation by nontransducing oncogenic retroviruses. During promoter insertion (top), the typical deletion of the proviral left-end LTR is probably important, because transcriptional read-through from the left-end promoter reduces transcription from the right-end LTR.

proto-oncogene, truncating cellular coding sequences and eliminating noncoding domains that may include negative regulatory sequences. Some chimeric transcripts formed in this way are analogous to the intermediates that give rise to oncogene capture by the transducing retroviruses (compare Fig. 6.15 and 6.9). Indeed, it has been possible to isolate newly generated, oncogene-transducing retroviruses from tumors arising as a result of promoter insertion.

In the second type of insertional activation, **enhancer insertion**, viral and cellular transcripts are not fused. Instead, activation of the cellular gene is mediated by the strong viral enhancers, which increase transcription from the cellular promoter (Fig. 6.15). Because enhancer activity is indepen-

dent of orientation and can be exerted over long distances, the provirus need not be oriented in the same direction as the proto-oncogene, and may lie downstream of it.

Viral Proteins That Alter Cellular Signaling Pathways

Some viruses alter the growth and proliferation of infected cells by the action of viral signal transduction proteins that are not obviously related in sequence to cellular proteins. Some of these viral proteins operate by mechanisms well established in studies of cellular signaling cascades, but others function in different ways.

Constitutively active viral "receptors." The genomes of several gammaherpesviruses encode membrane proteins that initiate signal transduction. The best-understood example of this mechanism is provided by Epstein-Barr virus latent membrane protein 1 (LMP-1), one of several viral gene products implicated in immortalization of human B lymphocytes (Table 5.3) and synthesized in the majority of tumors associated with the virus. LMP-1 induces typical transformed phenotypes when synthesized in fibroblasts or epithelial cells in culture and induces lymphomas in transgenic mice. It is an integral plasma membrane protein that functions as a constitutively active receptor. In the absence of any ligand, LMP-1 oligomerizes to form patches in the cellular membrane and binds to the same intracellular adapter proteins as the active, ligand-bound form of members of the tumor necrosis factor receptor family (Fig. 6.16). This viral protein induces release of Nf-κb from association with cytoplasmic inhibitors by multiple mechanisms, and activates a second transcriptional regulator, Ap-1, as well as signaling via the lipid kinase Pi3k and protein kinase Akt (Fig. 6.16). Activation of these pathways accounts for the increased expression of most of the cellular genes that is observed in LMP-1-producing cells, and the alterations in the properties of these cells. These changes include increased production of certain cell adhesion molecules and cell proliferation.

It has long been known that LMP-1 is not synthesized in all cells within tumors associated with Epstein-Barr virus. Remarkably, recent studies suggest that such cells would nevertheless be influenced by LMP-1 secreted in exosomes from cells in which the viral protein is made efficiently, a previously unrecognized form of "bystander effect" (Box 6.10).

Viral adapter proteins. Members of both the *Polyomaviridae* and the *Herpesviridae* encode proteins that permanently activate signal transduction pathways as a result of binding to Src family tyrosine kinases. This mechanism was first encountered in studies of the mouse polyomavirus mT protein, a viral early gene product with no counterpart in the genome of the related polyomavirus simian virus 40 (Fig. 6.17A). This pro-

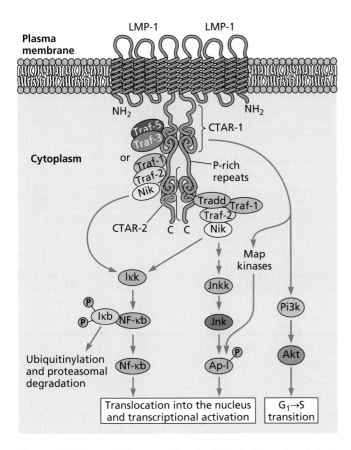

Figure 6.16 Constitutive signaling by Epstein-Barr virus latent membrane protein 1. LMP-1, which possesses six membrane-spanning segments but no large extracellular domain, oligomerizes in the absence of ligand, a property represented by the LMP-1 dimer depicted. When localized to the plasma membrane, the C-terminal segment of LMP-1 to which these proteins bind is sufficient for both immortalization of B cells and activation of cellular transcriptional regulators. The long cytoplasmic C-terminal domain of the viral protein contains three segments implicated in the activation of signaling, designated C-terminal activation regions (CTARs) 1 and 2, and proline (P)-rich repeats. As shown, multiple members of the tumor necrosis factor receptor-associated protein family (Trafs) bind to CTAR-1. Binding of Trafs leads to activation of the protein kinase Nik and Iκ-kinase (Iκk), and ultimately of Nf-κb, via induction of release of Nf-κb from association with its cytoplasmic inhibitors. The same pathway is activated in uninfected cells by binding of tumor necrosis factor to its receptor. The CTAR-2 domain of LMP-1 is responsible for activation of Ap-1 via the Jun N-terminal kinase (Jnk) pathway. The first reaction appears to be indirect association of this region of LMP-1 with Trafs via a second cellular protein (Tradd), a reaction that may lead to activation of both Nf-κb and Ap-1, as shown. The Traf-binding domain of CTAR-1 also induces activation of signaling via Pi3k and the protein kinase Akt, and of the Map kinase cascade. These responses to LMP-1 are required for transformation of rat fibroblasts.

tein can transform established rodent cell lines (Table 6.4) and induce endotheliomas (Box 6.1) when overproduced in transgenic animals.

mT becomes inserted into cellular membranes by means of a C-terminal transmembrane domain, and associates with

cellular signaling proteins en route to the plasma membrane via the secretory pathway. An N-terminal mT sequence that is also present in sT (Fig 6.17A) becomes bound to cellular protein phosphatase 2A (Pp2A) in the cytoplasm, an interaction that is necessary for subsequent interaction with c-Src in the endoplasmic reticulum. This requirement ensures that the phosphatase is brought into close association with c-Src. When c-Src is bound to mT, its catalytic activity is increased by an order of magnitude, because autoinhibition of c-Src kinase activity (Fig 6.9) is reversed (Fig 6.17B).

It was initially surprising that mT-transformed cells do not contain elevated levels of phosphotyrosine, despite activation of c-Src family kinases. It is now clear that mT itself is a critical substrate of the cellular enzyme: phosphorylation of specific mT tyrosine residues by activated c-Src allows a number of cellular proteins that contain phosphotyrosine-binding domains to bind to mT (Fig. 6.17B). When bound to mT, these signaling proteins are phosphorylated by the activated c-Src kinase to trigger signal transduction, for example, by activation of Ras and the Map kinase pathway. Consequently, mT both bypasses the normal mechanism by which the kinase activity of c-Src is regulated, and also serves as a virus-specific adapter, bringing together cellular signal transduction proteins when they would not normally be associated.

Alteration of the Activities of Cellular Signal Transduction Molecules

Activation of plasma membrane receptors. Many signal transduction cascades are initiated by binding of external growth factors to the extracellular portions of cell surface receptor tyrosine kinases. Ligand-bound receptors are internalized rapidly (within 10 to 15 min) by endocytosis. Following acidification of the endosomes, the ligand is released and all but a small fraction of the receptor molecules are usually degraded. As a result, the initial signal is short-lived. The E5 protein of papillomaviruses that cause fibropapillomas, such as bovine papillomavirus type 1, interferes with the mechanisms that control the function of this class of receptor.

This E5 protein, a hydrophobic molecule of only 44 amino acids, efficiently transforms mammalian fibroblasts in culture in the absence of any other viral proteins (Table 6.4). This activity depends on binding to platelet-derived growth factor receptor β (Pdgfr-β). The E5 protein is a dimer that accumulates in host cell membranes, where it induces ligand-independent dimerization of the receptor, and hence activation of its tyrosine kinase and downstream signaling relays (Fig. 6.3). The E5 protein binds stably and with high specificity to the transmembrane domain and an adjacent internal segment of the receptor, in contrast to the natural ligand, which binds to the extracellular domain. This mechanism is likely to be important in the oncogenicity of the virus in its natural hosts: in bovine tumors, the E5 protein

DISCUSSION
Transformation by remote control?

Epstein-Barr virus contributes to the development of several cancers of B lymphocytes and epithelial cells, including nasopharyngeal carcinoma. The viral genome is present in all such tumors, which are monoclonal in origin, but expression of the gene that encodes the transforming LMP-1 protein is variable and often difficult to detect in tumor samples. Nevertheless, this viral protein may stimulate the proliferation of infected cells in which it is not made by an unusual mechanism, transfer from cells that **do** produce LMP-1 via exosomes.

Exosomes are small (40 to 100 nm in diameter) vesicles that are secreted by many types of cell and permit intercellular communication. They form initially as intraluminal vesicles by inward budding of the membranes of multivesicular bodies, in which they accumulate prior to release by fusion of these bodies with the plasma membrane (see the figure). Exosomes have been implicated in several normal processes, including antigen presentation, maturation of sperm, and communication among neurons, as well as in transformation and tumorigenesis. They are thought to exert their effects by both interaction with target cells and direct transfer of cargo into cells following fusion with the plasma membrane. Exosomes can transfer not only numerous soluble and membrane proteins but also RNAs (mRNAs and miRNAs) from one cell to another.

Exosomes carrying the viral LMP-1 protein in their membranes are secreted from Epstein-Barr virus-transformed epithelial and B cells in culture, and have been observed in sera from nasopharyngeal carcinoma patients. The mechanism by which LMP-1, which normally resides in the plasma membrane, is recruited to exosomes is not well understood. However, the fusion of such exosomes with uninfected cells has been reported to stimulate signal transduction pathways that promote cell proliferation and survival, for example, signaling via Map kinases and Akt. Furthermore, these LMP-1-containing vesicles appear to be enriched in other signaling molecules and to contain viral miRNAs. These properties suggest that intercellular transfer of LMP-1 (and perhaps other molecules) via exosomes could contribute to viral transformation and tumorigenesis.

Meckes DG, Jr, Shair KH, Marquitz AR, Kung CP, Edwards RH, Raab-Traub N. 2010. Human tumor virus utilizes exosomes for intercellular communication. *Proc Natl Acad Sci U S A* **107:**20370–20375.

Exosomes formed in Epstein-Barr virus-infected cells carry the viral LMP-1 protein and have the potential to transfer this protein, and internal cargo such as viral miRNAs, to recipient cells.

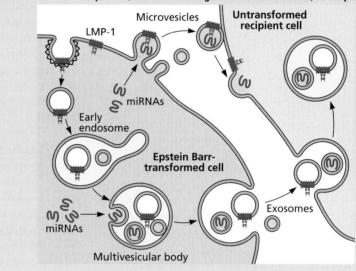

and Pdgfr-β are colocalized, and both the receptor and downstream signaling pathways are activated.

Nontransducing retroviruses can also activate cell surface receptors, because these cellular gene products may be altered by provirus integration. In certain chicken lines, Rous-associated virus 1 induces erythroblastosis instead of lymphomas (Box 6.1). These tumors contain intact, nondefective proviruses integrated in the cellular *erbB* gene, which encodes the cell surface receptor for epidermal growth factor. The proviral integrations are clustered in a region that encodes the extracellular portion of this receptor, and readthrough transcription produces chimeric RNAs (Fig. 6.15). The proteins synthesized from these RNAs are truncated growth factor receptors that lack the ligand-binding domain and produce a constitutive mitogenic signal. The v-*erbB* gene captured by transducing retroviruses encodes a protein with a similar truncation.

Inhibition of protein phosphatase 2A. In preceding sections, we have discussed transformation by viral gene products as a result of permanent or prolonged activation of signal transduction pathways that control cell proliferation. In normal cells, such signaling is a transient process, because the molecular components are reset once they have transmitted the signal. Inhibition of the reactions that terminate signaling therefore can also contribute to transformation, a mechanism

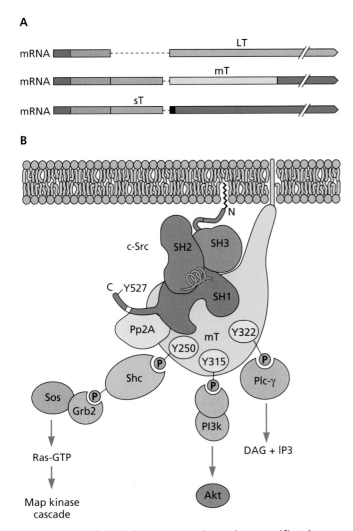

A

mRNA

LT

mRNA

mT

sT

mRNA

B

N

c-Src

SH2

SH3

C Y527

SH1

Pp2A

mT Y322

Y250

Shc

P

Y315

Sos

P

Plc-γ

Grb2

P

PI3k

Ras-GTP

DAG + IP3

Map kinase cascade

Akt

Figure 6.17 Polyomavirus mT protein, a virus-specific adapter. **(A)** The mouse polyomavirus early protein-coding sequences are shown as boxes within the mRNAs from which the proteins are synthesized. The mRNAs are drawn as arrows, in which the arrowheads indicate the site of polyadenylation and the dashed pink lines indicate the introns removed during RNA splicing. The three proteins produced from these mRNAs, LT, mT, and sT, share an N-terminal sequence but carry unique C-terminal sequences as a result of alternative splicing of early transcripts. **(B)** mT binds to c-Src (or the related tyrosine kinases c-Yes or c-Fyn) at cellular membranes and to Pp2A. As a result of formation of the ternary complex, c-Src is trapped in the active conformation and Y527 is unphosphorylated: mT sequesters the Y527-containing segment of c-Src for dephosphorylation of the tyrosine residue by Pp2A, thereby stabilizing the active conformation of the enzyme. Consequently, mT-bound Src is catalytically active and phosphorylates specific tyrosines in mT. These phosphorylated residues are then bound by cellular proteins that contain phosphotyrosine-binding motifs, such as Shc, phospholipase C-γ (Plc-γ) (an enzyme that catalyzes synthesis of lipid second messengers), and Pi3k. These proteins can then be phosphorylated by Src and activated. The lipids produced upon activation of Plc-γ act as second messengers, relaying signals to various pathways, while Pi3k activates signaling via the protein kinase Akt. In all cases, substitutions that disrupt binding of the cellular protein to mT impair the transforming activity of the viral protein.

exemplified by the sT protein of polyomaviruses such as simian virus 40.

This protein is not necessary for transformation of many cell types, but can stimulate transformation by simian virus 40 LT and is required for the transformation of resting cells. In both infected and transformed cells, the sT protein binds to protein phosphatase 2A, a widespread, abundant serine/threonine protein phosphatase. This protein is a heterotrimer, composed of a core enzyme comprising a scaffolding and a catalytic subunit bound to one of a substantial number of regulatory subunits. sT binds via two domains to the scaffolding subunit of the core enzyme to block access of substrates to the active site in the catalytic subunit and binding of regulatory subunits (Fig. 6.18A). One important consequence of sT binding is failure of the phosphatase to inactivate Map kinases, a process normally accomplished by the dephosphorylation of serine/threonine or tyrosine residues (Fig. 6.18B). Consequently, sT increases the activity of sequence-specific transcriptional activators that are substrates of Map kinases. The increased activities of these transcriptional stimulators lead to synthesis of G_1-phase and S-phase cyclins, thereby circumventing the need for growth factors or other mitogens during transformation by simian virus 40.

Disruption of Cell Cycle Control Pathways by Viral Transforming Proteins

One end point of many signal transduction pathways is the transcription of genes coding for proteins that regulate cell cycle progression and the metabolic activity of the cell. Consequently, **permanent activation** of such pathways by viral proteins, by any of the mechanisms described in the previous section, can result in an increased rate of cell growth and division or in proliferation of cells that would normally be in the resting state. Other viral proteins intervene directly in the intricate circuits by which cell cycle progression is mediated and regulated.

Abrogation of Restriction Point Control Exerted by the Rb Protein

The Restriction Point in Mammalian Cells

In mammalian cells, passage through G_1 into S and reentry into the cell cycle from G_0 depend on extracellular signals that regulate proliferation, termed **mitogens**. Late in G_1, cells that respond to such external cues become committed to enter S and to divide and complete the cell cycle; during this period, they are refractory to mitogens. Cells that have entered this state are said to have passed the G_1 **restriction point** (Fig. 6.6). Normal cells respond to mitogenic signals by mobilization of the G_1 Cdks that contain D-type cyclins. Expression of genes that encode one or more of these cyclins is induced by such signals via the Ras/Map kinase pathways

A

Active site

Hsp70

B

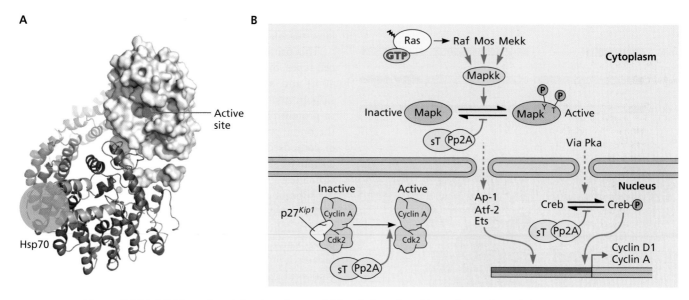

Figure 6.18 Inhibition of protein phosphatase 2A by simian virus 40 small T antigen. (A) Model of small T antigen bound to the core enzyme, which comprises a catalytic (gray, shown as a surface model) and a scaffolding (blue) subunit, with the small T J and unique C-terminal domains shown in green and magenta, respectively. This model was derived by superimposing X-ray crystal structures of a complete Pp2A (scaffolding, regulatory, and catalytic subunits) and of the sT-scaffolding subunit complex. The viral protein binds to the scaffolding subunit in place of the regulatory subunit and likely makes contact with the catalytic subunit via its J domain, which is necessary for efficient inhibition of the catalytic activity of Pp2A. Adapted from U. S. Cho et al., *PLoS Biol* **5:**e202, 2007, with permission. Courtesy of W. Xu, University of Washington, Seattle. **(B)** Inhibition of the activity of Pp2A by sT results in activation of cellular transcriptional regulators via the Map kinase pathway (e.g., activator protein 1 [Ap-1] and activating transcription factor 2 [Atf-2]). In addition, dephosphorylation of activated cAMP response element binding protein (Creb) within the nucleus is inhibited. Production of sT within cells stimulates cyclin D1 and cyclin A transcription. Binding of sT to Pp2A also induces a large increase in the activity of cyclin A-dependent Cdk2, concomitant with inhibition of dephosphorylation of the cyclin-Cdk inhibitor p27^{Kip1} (Fig. 6.6) and degradation of this protein.

(Fig. 6.19A). When such stimulation is continuous, Cdk activity appears at mid-G$_1$ and increases to a maximum near the G$_1$-to-S-phase transition. Such activity must be maintained until the restriction point has been passed, but then becomes dispensable. This property implies that the kinase activity of the cyclin D-dependent Cdks is necessary for exit from G$_1$. The best-characterized substrates of these kinases are the Rb protein and the related p107 and p130 proteins. The Rb protein controls the activity of members of the E2f family of sequence-specific transcriptional regulators (described in Volume I, Chapter 8).

Hypophosphorylated Rb present at the beginning of G$_1$ binds to specific members of the E2f family. These complexes inhibit transcription of E2f-responsive genes (Fig. 6.19B). The Rb protein is phosphorylated at numerous sites by G$_1$ cyclin-Cdks. Phosphorylated Rb can no longer bind to E2f, which therefore becomes available to activate transcription from E2f-responsive promoters. These promoters include those of the genes encoding the kinase Cdk2, the cyclins that associate with this kinase, and E2f proteins themselves. The initial release of E2fs from association with Rb therefore triggers a positive feedback loop that augments both

Rb phosphorylation and release of E2fs. The result is a rapid increase in the concentrations of E2fs and cyclin E-Cdk2. In this way, cell cycle progression becomes independent of the mitogens necessary for the production of cyclin D-Cdks. These regulatory circuits account well for passage through the restriction point and commitment of a cell to divide. Nevertheless, there is accumulating evidence for functional redundancy among the Rb family proteins and mechanisms for detecting mitogenic stimuli that do not operate via D-type cyclins.

The E2f proteins that accumulate upon Rb phosphorylation also stimulate transcription of genes that encode proteins needed for DNA synthesis (Volume I, Chapter 9), allowing genome replication to take place in S phase. The cyclin A-Cdk2 produced in response to E2f phosphorylates and thereby inhibits the ubiquitin ligase that marks cyclin B for proteasomal degradation throughout much of the cell cycle (Fig. 6.19B). Consequently, cyclin B, which is required for entry into mitosis, accumulates as S phase progresses. Phosphorylation of the Rb protein therefore ensures not only passage through the restriction point and entry into S phase, but also the coordination of these processes with later

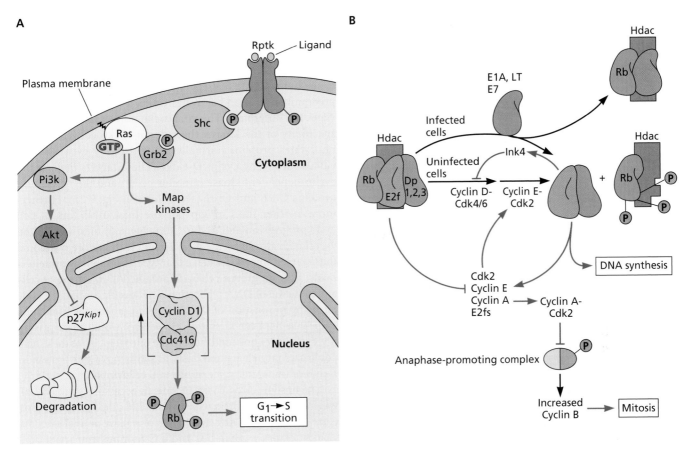

Figure 6.19 Passage through the restriction point in mammalian cells. (A) Mitogenic activation of cell cycle progression is initiated by binding of a growth factor to its cognate receptor protein (Rptk). Signaling via Ras and Map kinase cascades leads to increased transcription of the cyclin D1 gene and accumulation of cyclin D1-Cdc4/6 in the nucleus. Activation of this G_1 cyclin is facilitated by the degradation of its inhibitor p27^{Kip1}, which is induced by signaling via Pi3k and the protein kinase Akt. The active G1 cyclin phosphorylates the negative regulator of cell cycle progression Rb (and the related p107 and p130 proteins). Many lines of evidence indicate that cyclin D-dependent Cdks initiate the transition through the restriction point by phosphorylation of Rb. For example, inhibition of cyclin D synthesis or function prevents entry into S phase in Rb-containing cells, but this cyclin is not required in Rb-negative cells. **(B)** When cells enter G_1, hypophosphorylated Rb is bound to transcriptional regulators of the E2f family, which are heterodimers of an E2f and a Dp protein. The E2f-Rb complex represses transcription when bound to E2f recognition sites in promoters, because Rb is a transcriptional repressor. This function requires the binding of histone deacetylases (Hdacs). Phosphorylation of Rb by cyclin D- and cyclin E-dependent Cdks disrupts the binding of Rb to E2fs. Rb is phosphorylated at many sites by both cyclin D-Cdk4/6 and cyclin E-Cdk2. The latter cyclin, which appears in mid to late G_1 (Fig. 6.6A), is required for entry into S phase. Its modification of Rb depends on the prior action of cyclin D-Cdk4/6. Free E2f-Dp heterodimers activate transcription from E2f-responsive promoters, including those of the genes encoding cyclins E and A, Cdk2, and E2f proteins themselves, to establish a positive autoregulatory loop. The positive feedback loop for activation of cyclin E-dependent kinases and E2fs late in G_1 is subject to several checks and balances imposed by inhibitory proteins (Fig. 6.6B). These inhibitory proteins therefore must be inactivated or destroyed to allow progression into S phase (see panel A). The synthesis of at least one member of the Ink4 family of cyclin-Cdk inhibitors is also induced in response to free E2f. It is therefore thought that the accumulation of this inhibitor establishes a feedback loop that blocks the activity of the cyclin D-Cdks and hence the ability of cells to respond to mitogens, a characteristic property of cells that have passed the G_1 restriction point. Although E2f proteins are the best-characterized targets of Rb, the latter protein can also bind to numerous other proteins that mediate or stimulate transcription, as well as to regulatory proteins such as the Abl tyrosine kinase (Table 6.5). These interactions can lead to activation or repression of transcription and, at least in some cases, have been implicated in inhibition of cell cycle progression by Rb.

events in the cell cycle. Indeed, the results of genome-wide approaches to identify genes regulated by E2f family members suggest that E2fs contribute broadly to orderly progression through the cell cycle.

Viral Proteins Inhibit Negative Regulation by Rb and Related Proteins

The products of transforming genes of several DNA viruses bypass the sophisticated circuits that impose restriction point control, and hence the dependence on environmental cues for passage into S phase. The adenoviral E1A proteins, simian virus 40 LT, and the E7 proteins of oncogenic human papillomavirus (types 16 and 18) can induce DNA synthesis and cell proliferation. All three viral proteins make contacts with the two noncontiguous regions by which hypophosphorylated Rb associates with E2f family members (regions A and B in Fig. 6.20A) to disrupt Rb-E2f complexes. As a result, they induce transcription of E2f-dependent genes and inappropriate entry of cells into S phase (Fig. 6.19B), when cellular proteins needed for replication and transcription of viral genomes are synthesized.

The adenoviral E1A, papillomaviral E7, and simian virus 40 LT proteins share a sequence motif necessary for binding Rb (Fig. 6.20A). Nevertheless, they induce removal of Rb from its association with E2f by different mechanisms. The E1A CR1 segment (Fig. 6.21) is structurally similar to the Rb-binding site of E2Fs, and the viral protein competes efficiently for Rb. In contrast, dismantling of Rb-E2f complexes by LT appears to be an active process: the N-terminal J domain of LT, which recruits the cellular, ATP-dependent chaperone Hsc70, and ATP are also required (Fig. 6.20B). The E7 proteins interact with not only Rb but also a cellular cullin 2-containing E3 ubiquitin ligase to target Rb for degradation by the proteasome.

The Rb protein is the founding member of a small family of related gene products, which includes the proteins p107 and p130. The latter two proteins were discovered by virtue of their interaction with adenoviral E1A proteins (Fig. 6.21), but they also bind to both LT and E7 proteins. The Rb, p107, and p130 proteins bind preferentially to different members of the E2f family during different phases of the cell cycle. For example, hypophosphorylated Rb binds primarily to E2f-1, E2f-2, or E2f-3 during the G_0 and G_1 phases, and p130 binds E2f-4 and E2f-5 in G_0. Binding of p130 to these E2f family members appears to be critical for maintaining cells in the quiescent state, and such complexes predominate in mammalian cells in G_0. Their disruption by adenoviral, papillomaviral, or polyomaviral transforming proteins is thought to allow such cells to reenter the cycle, in part via stimulation of the transcription of genes encoding both the E2f proteins and the cyclin-dependent kinase (Cdk2) needed for entry into S phase.

Production of Virus-Specific Cyclins

Human herpesvirus 8 and its close relative herpesvirus saimiri encode functional cyclins. The cyclin gene of human herpesvirus 8, designated v-*cyclin*, has 31% identity and 53% similarity to the human gene that encodes cyclin D2. Its product binds predominantly to and activates Cdk6, which then phosphorylates the Rb protein. The viral cyclin also alters the substrate specificity of the kinase: the v-*cyclin*–Cdk6 complex phosphorylates proteins normally recognized by cyclin-bound Cdk2, but not by cyclin D-Cdk6. These targets include the cyclin-dependent kinase inhibitor p27^{Kip1} and the replication proteins Cdc6 and Orc1 (see Volume I, Fig. 9.25). Furthermore, neither the Cip/Kip nor the Ink4 inhibitors of cell cycle progression and cyclin-Cdks (Fig. 6.6) bind well to the v-cyclin. Synthesis of the viral cyclin can therefore overcome the G_1 arrest imposed when either type of inhibitory protein is made, and can induce cell cycle progression in quiescent cells and initiation of DNA replication. The specific advantages conferred by production of the viral cyclin during the infectious cycle have not been identified. However, it would be surprising if v-cyclin does not contribute to the oncogenicity of these herpesviruses in their natural hosts, as synthesis of this protein in B cells of transgenic mice results in B cell lymphoma.

The epsilonretroviruses encode an accessory gene (*orf a*), which specifies a protein with a cyclin fold called rv-cyclin (Box 6.4). The best studied rv-cyclin, that of walleye dermal sarcoma virus, includes a cyclin box and a C-terminal transcriptional activation domain. The viral protein accumulates in the nucleus of infected cells, and its location and physical association with transcriptional regulators are consistent with a function in transcriptional control. The viral rv-cyclin can promote cell cycle progression when produced in G_1 cyclin-deficient yeast cells. This property, and interaction of rv-cyclin with Cdk3, suggests that this viral protein functions as an ortholog of cellular cyclin C to promote proliferation and oncogenesis in the piscine host, but the mechanistic details have yet to be elucidated.

Inactivation of Cyclin-Dependent Kinase Inhibitors

The production of viral cyclins in infected cells appears to be a property of only certain herpesviruses and the epsilonretroviruses, but other DNA viruses encode proteins that facilitate cell cycle progression by inactivating specific inhibitors of Cdks (Fig. 6.22). One example is the E7 protein of human papillomavirus type 16, which binds to the p21^{Cip1} protein and inactivates it. This member of the cellular Cip/Kip family inhibits G_1 cyclin-Cdk complexes (Fig. 6.6). The increase in intranuclear concentrations of p53 triggered by unscheduled inactivation of the Rb protein (see next section) results in accumulation of p21^{Cip1}. The ability of the E7 protein to inactivate both Rb and p21^{Cip1} is necessary to induce differentiated human epithelial cells to enter S phase.

A

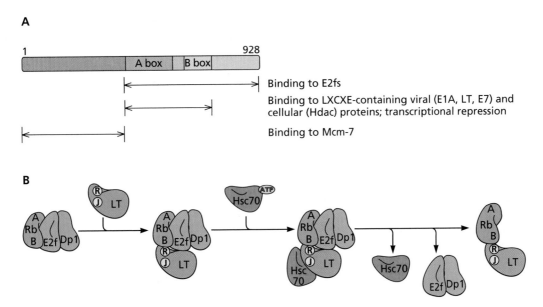

Binding to E2fs

Binding to LXCXE-containing viral (E1A, LT, E7) and cellular (Hdac) proteins; transcriptional repression

Binding to Mcm-7

B

Figure 6.20 Model for active dismantling of the Rb-E2f complex by simian virus 40 LT. (A) Functional domains of the human Rb protein are shown to scale. The A- and B-box regions form the so-called pocket domain, which is necessary for binding of Rb to both E2fs and the viral proteins described in the text. The similarity of p107 and p130 to Rb is most pronounced in the A and B sequences, and the residues in Rb that contact the common binding motif of the viral proteins are invariant among the other family members. This segment is also sufficient to repress transcription when fused to a heterologous DNA-binding domain, and it is required for binding to histone deacetylases (Hdacs). Like the viral Rb-binding proteins, Hdacs contain the motif LXCXE within the region that binds to Rb. The N-terminal segment of the protein, which is also important for suppression of cell proliferation, binds to human Mcm-7, a component of a chromatin-bound complex required for DNA replication and control of initiation of DNA synthesis. **(B)** The LT protein binds to the Rb A- and B-box domains via the sequence that contains the LXCXE motif, designated R. The adjacent, N-terminal J domain of LT is not necessary for binding to Rb, but is required for induction of cell cycle progression. It has been proposed that the J domain recruits the cellular chaperone Hsc70. The chaperone then acts to release E2f-Dp1 heterodimers from their association with Rb, by a mechanism that is thought to depend on ATP-dependent conformational change.

Figure 6.21 Organization of the larger adenoviral E1A protein. Regions of the protein are shown to scale. Those designated CR1 to CR3 are conserved in the E1A proteins of human adenoviruses. The CR3 region, most of which is absent from the smaller E1A protein because of alternative splicing, is not necessary for transformation. The locations of the Rb-binding motif and of the regions required for binding to the other cellular proteins discussed in the text are indicated.

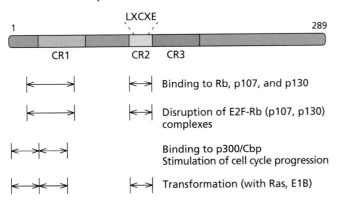

Binding to Rb, p107, and p130

Disruption of E2F-Rb (p107, p130) complexes

Binding to p300/Cbp Stimulation of cell cycle progression

Transformation (with Ras, E1B)

Figure 6.22 Inactivation of cyclin-dependent kinase inhibitors by viral proteins. The human papillomavirus (HPV) 16 E7 and adenovirus E1A proteins bind to and inactivate p21^{Cip1} and, in the case of E1A proteins, also p27^{Kip1}. The viral proteins interact with the regions of the inhibitors that mediate association with G1 cyclins. They also block activation of transcription of the p21^{Cip1} gene in response to DNA damage as a result of sequestration of the transcriptional coactivators p300 and Cpb. Rather than simply blocking the inhibitor-cyclin interaction, the Epstein-Barr virus (EBV) ENBA3C protein, which is necessary for transformation of B cells in culture, recruits a cellular E3 ubiquitin ligase that marks p27^{Kip1} for degradation by the proteasome. This viral protein inactivates Rb in the same manner.

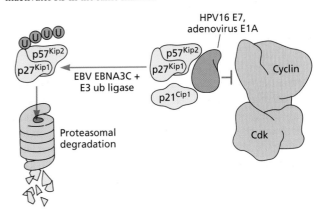

Transformed Cells Must Grow and Survive

The rapid proliferation of cells transformed by viral proteins depends on high rates of metabolism and growth during each cell cycle. It seems likely that any viral oncogene product that results in activation of Ras (or Akt) promotes cell growth (Fig. 6.4), as well as proliferation. How viral transforming proteins that impinge directly on the nuclear circuits that govern cell cycle progression accelerate cell growth is less clear. However, the actions of many of these proteins lead to changes in the transcription of numerous cellular genes, responses that might increase the concentrations of biosynthetic and other metabolic enzymes.

Mechanisms That Permit Survival of Transformed Cells

As discussed in Chapter 3, metazoan cells can undergo programmed cell death (apoptosis). This program is essential during development and serves as a powerful antiviral defense of last resort. Apoptosis can be activated not only by external cues, but also by intracellular events, notably damage to the genome or unscheduled DNA synthesis. Consequently, viral transforming proteins that induce cells to enter S phase and proliferate when they would not normally do so will also promote the apoptotic response. This potentially fatal side effect is foiled by a variety of mechanisms that allow survival of infected and transformed cells.

Viral Inhibitors of the Apoptotic Cascade

Many viral genomes encode mimics of cellular proteins that hold apoptosis in check (see Table 3.3). Such viral inhibitors of apoptosis can contribute to transformation. For example, the human adenovirus E1B 19-kDa protein, one of the first viral homologs of cellular antiapoptotic proteins to be identified, allows survival, and hence transformation, of rodent cells that also contain the proliferation-promoting viral E1A protein (Table 6.4).

Integration of Inhibition of Apoptosis with Stimulation of Proliferation

Cells must continually interpret the numerous internal and external signals that impinge upon them to execute an appropriate response. Not all the mechanisms that integrate the many types of information that cells receive have been elucidated. However, it is well established that signal transduction cascades that induce cell proliferation can simultaneously promote cell survival by blocking the apoptotic response. For example, signaling via the small G protein Ras results in activation of not only the cyclin-dependent kinases that drive the G_1-to-S-phase transition (Fig. 6.19A), but also Pi3k and the protein kinase Akt. Akt induces transcriptional and post-transcriptional mechanisms that inhibit the production and

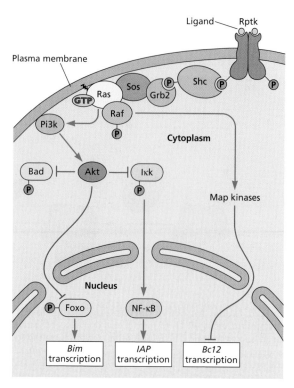

Figure 6.23 Signaling pathways that facilitate cell survival. Activation of Ras promotes cell survival by inhibition of synthesis or activity of proapoptotic proteins and by stimulation of production of inhibitors of programmed cell death. Substrates of activated Akt include the proapoptotic protein Bad and the transcriptional regulator Foxo, which are inactivated by phosphorylation. Akt also phosphorylates the inhibitor of Nf-κb (Iκk) to promote transcription of genes that encode other inhibitors of apoptosis. As shown in Fig. 6.19A, signaling via Ras and the Map kinase cascade induces cell proliferation. The Pi3k/Akt pathway also promotes proliferation, for example, by phosphorylation and inactivation of cyclin-dependent kinase inhibitors. Consequently, these (and other) signaling networks integrate cell proliferation and survival.

activity of proapoptotic proteins, such as Bad and Bim, and stimulate synthesis of inhibitors of apoptosis, including Bcl-2 (Fig. 6.23). Consequently, any viral transforming protein that elicits activation of Akt will also induce protection against apoptosis. Such proteins include the many retroviral gene products that function in the receptor protein tyrosine kinase pathway (Fig. 6.3; Table 6.5), v-Src (Fig. 6.12), and Epstein-Barr virus LMP-1 (Fig. 6.16).

Inactivation of the Cellular Tumor Suppressor p53

Transformation by several DNA viruses requires inactivation of a second tumor suppressor, the p53 protein, first identified by virtue of its binding to simian virus 40 LT. This protein is a critical component of regulatory circuits that determine the response of cells to damage to their genomes, as well as to low concentrations of nucleic acid precursors, hypoxia, and other forms of stress. Its importance in the appropriate response to

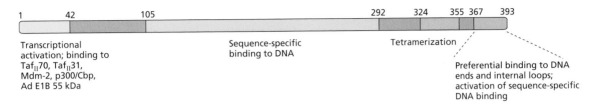

Transcriptional
activation; binding to
Taf$_{II}$70, Taf$_{II}$31,
Mdm-2, p300/Cbp,
Ad E1B 55 kDa

Sequence-specific
binding to DNA

Tetramerization

Preferential binding to DNA
ends and internal loops;
activation of sequence-specific
DNA binding

Figure 6.24 The human p53 protein. The functional domains of the protein are shown to scale. TafII70 and TafII31 are TATA-binding protein-associated proteins present in transcription initiation protein IID (TfIId) (Volume I, Chapter 8); other proteins are defined in the text.

such damage or stress is emphasized by the fact that *p53* is the most frequently mutated gene in human tumors.

The accumulation and activity of p53 are tightly regulated.

The intracellular concentration of p53 is normally very low, because the protein is targeted for proteasomal degradation, for

example, by binding of the Mdm-2 protein (Fig. 6.24 and 6.25). However, DNA damage, such as double-strand breaks produced by γ-irradiation, the collapse of replication forks, or the accumulation of DNA repair intermediates following UV irradiation, leads to the stabilization of p53 and a substantial increase in its concentration. The rate of translation of the protein may

Figure 6.25 Regulation of the stability and activity of the p53 protein. Under normal conditions (left), cells contain only low concentrations of p53. This protein is unstable, turning over with a half-life of minutes, because it is targeted for proteasomal degradation by the Mdm-2 protein. Mdm-2 is a p53-specific E3 ubiquitin ligase that catalyzes polyubiquitinylation of p53, the signal that allows recognition by the proteasome, to maintain inactive p53 at low concentrations. The availability and activity of Mdm-2 are also regulated, for example, by Arf proteins encoded by the *ink4a/arf* tumor suppressor gene and by stimulation of Mdm-2 transcription by the p53 protein itself. Signaling pathways initiated in response to damage to the genome or other forms of stress lead to posttranslational modification and stabilization of p53. Such posttranscriptional regulation is thought

to allow a very rapid response to conditions that could be lethal to the cell. As illustrated with pathways operating in response to DNA damage (double-strand [ds] breaks caused by ionizing radiation), p53 is stabilized in multiple ways. These mechanisms include phosphorylation of p53 at specific serines by Atm (see the text) and checkpoint kinase 2 (Chk2), binding to the c-Abl tyrosine kinase, sequestration of the Mdm-2 protein by Arf, and deubiquitinylation of p53 (in the presence of Mdm-2) by the herpesvirus-associated ubiquitin-specific protease (Hausp). Multiple mechanisms, including various modifications within the C-terminal domain (e.g., acetylation), also stimulate the sequence-specific DNA-binding activity of p53 or its association with the transcriptional coactivators p300/Cbp, and hence transcription from p53-responsive promoters.

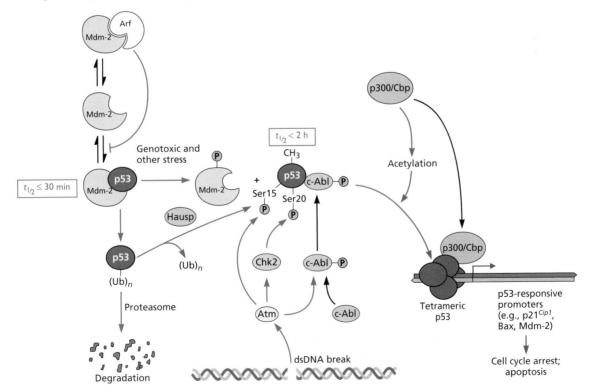

also increase. Various proteins that appear to be important for stabilization of p53 have been identified, including the product of a human gene called *Atm* (ataxia telangiectasia mutated), which recognizes potentially genotoxic DNA damage. Cells lacking the Atm protein do not accumulate the p53 protein and fail to arrest at the G$_1$/S boundary in response to DNA damage.

The p53 protein is a sequence-specific transcriptional regulator containing an N-terminal activation domain and a central DNA-binding domain (Fig. 6.24). This protein also operates in the cytoplasm, where it binds to proteins associated with mitochondria that inhibit apoptosis to induce release of proapoptotic effectors, such as Bax and Bak. The intracellular location, stability, and activities of p53, such as binding to DNA and stimulation or repression of transcription of particular sets of p53-responsive genes, are regulated by both multiple types of posttranslational modification of numerous residues and the constellation of associated proteins. The very large repertoire of such mechanisms provides the means to integrate the multiple signals that are monitored to ensure that this potent protein alters cell physiology only under extreme conditions.

In response to damage to the genome, or other inducing conditions, p53 prevents further cell proliferation by eliciting G$_1$/S arrest, apoptosis, or senescence (irreversible G$_1$ arrest). One important component of the pathway that leads to cell cycle arrest is the p53-dependent stimulation of transcription of the gene that encodes p21^{Cip1}, the G$_1$ cyclin-dependent kinase inhibitor (Fig. 6.6). The p53 protein promotes apoptosis both directly, by interaction with mitochondrial proteins that block this response, and indirectly, by stimulation of transcription of genes that encode proapoptotic proteins, such as Apaf-1 and Bax. It also impairs mechanisms that promote cell survival by increasing transcription of genes that encode inhibitors of certain signaling pathways. For example, increased production of the protein Pten leads to impaired signaling via Pi3k to Akt (Fig. 6.23), as Pten is a phosphatase that dephosphorylates phosphoinositides. The ability of p53 to repress transcription of genes for antiapoptotic proteins, such as *survivin*, may also be important. Whether p53 promotes cell cycle arrest, apoptosis, or senescence is determined by numerous parameters, including the cell type, the nature of extracellular stimuli, and the concentration of the p53 protein itself. However, the apoptotic response prevails in many cell types under many circumstances, in particular following expression of viral oncogenes that induce entry into S phase.

Viral proteins inactivate p53. The genomes of many viruses encode proteins that interact with p53. However, the mechanisms by which the functions of this critical cellular regulator can be circumvented are best understood for the small DNA tumor viruses. As we have seen, transforming proteins of these viruses induce release of E2f family members from association with Rb to promote cell cycle progression. Stabilization of p53 appears to be an inevitable consequence: E2f activates transcription from the promoter of the *Ink4/Arf* gene, which encodes a negative regulator of Mdm-2 (Fig. 6.26).

Figure 6.26 Stabilization of p53 by viral transforming proteins that bind to Rb. As described previously, binding of the adenoviral E1A (or polyomavirus LT or human papillomavirus E7) proteins to Rb allows transcription of E2f-responsive genes. This large set includes the *ink4/arf* gene, and Arf therefore accumulates. Binding of Arf to Mdm-2 sequesters this ubiquitin ligase, and hence leads to accumulation of the p53 protein. The E1A proteins also stabilize p53 via p300/Cbp-mediated acetylation of Rb. Acetylated Rb forms a ternary complex with p53 and Mdm-2 and blocks p53 degradation. The N-terminal transcriptional activation domain of p53 remains blocked by Mdm-2, but in this form p53 can repress transcription and promote apoptosis.

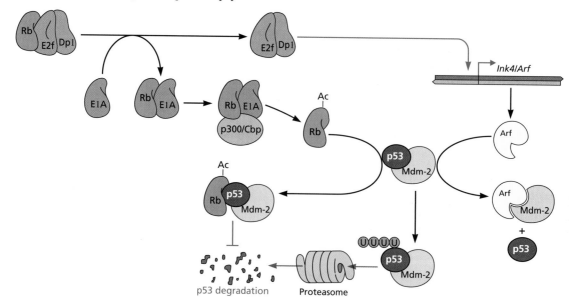

Viral proteins block p53 function in different ways (Fig. 6.27). The human papillomavirus type 16 or 18 E6 proteins bind to both p53 and a cellular E3 ubiquitin protein ligase (the E6-associated protein), and thereby target p53 for proteasome-mediated destruction. In conjunction with the viral E4 Orf6 protein, the adenoviral E1B 55-kDa protein also induces increased turnover of p53, but by directing it to a different E3 ubiquitin ligase. In contrast, simian virus 40 LT actually stabilizes the p53 protein, but sequesters this cellular regulator in inactive complexes.

Among the enzymes that acetylate p53 to increase its stability and activity are the histone acetyltransferases p300 and the closely related transcriptional activator cellular cAMP response element binding protein (Creb)-binding protein (Cbp). The former protein was first identified by virtue of its binding to adenoviral E1A proteins. This interaction with the cellular histone acetyltransferases blocks acetylation of p53, as does that of the human papillomavirus type 16 E6 protein, to limit activation of the tumor suppressor.

Tumorigenesis Requires Additional Changes in the Properties of Transformed Cells

The mechanisms described in the preceding sections account for the sustained proliferation and survival of cells transformed by viral oncogenes. However, they are not necessar-

Figure 6.27 Inactivation of the p53 protein by adenoviral, papillomaviral, and polyomaviral proteins. The synthesis of transforming proteins in infected or transformed cells leads to accumulation of p53 (Fig. 6.26). Each of these viral genomes encodes proteins that interfere with the normal function of this critical cellular regulator. Binding of simian virus 40 LT to p53, an interaction that is facilitated by sT, sequesters the cellular protein in inactive complexes. The E1B 55-kDa and E4 Orf6 proteins assemble with the cellular cullin 5, elongins B and C, and Rbx proteins to form a virus-specific E3 ubiquitin ligase. This enzyme ubiquitinylates p53 and marks it for proteasomal degradation. The E6 proteins of human papillomavirus types 16 and 18 bind to p53 via the cellular E6-associated protein (E6-Ap). The latter protein is a ubiquitin protein ligase that polyubiquitinylates p53 in the presence of the viral E6 protein, targeting p53 for degradation by the proteasome. The adenoviral E1B 55-kDa protein can also bind to the N-terminal activation domain of p53 to convert p53 from an activator to a repressor of transcription. This function of the E1B 55-kDa protein correlates with its ability to transform rodent cells in cooperation with E1A proteins. In transformed rodent cells, the E1B 55-kDa protein also induces relocalization of p53 from the nucleus to a perinuclear, cytoplasmic body. The results of mutational analyses have correlated the changes in concentration or activity in p53 induced by these viral proteins with their transforming activities.

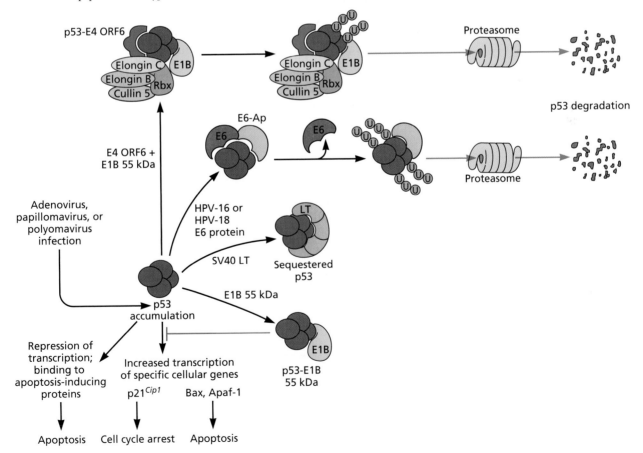

ily sufficient for the induction of tumors or other types of cancer: tumorigenesis generally also requires the ability of transformed cells to survive in the face of immune defenses. In some cases, induction of the growth of new blood vessels (angiogenesis) is also necessary (see "Viral Homologs of Cellular Genes" above).

Inhibition of Immune Defenses

The crucial contribution of mechanisms that protect transformed cells from immune defenses is illustrated by the properties of rodent cells transformed by oncogenic or nononcogenic human adenoviruses (Box 6.11). As discussed in Chapters 3 and 4, mechanisms that render infected cells refractory to immune defenses are important for the ability of many viruses to reproduce in immunocompetent animals. How such mechanisms facilitate the survival of transformed cells and oncogenesis is best understood for herpesviruses associated with human cancers: Epstein-Barr virus and human herpesvirus 8.

Epstein-Barr virus is associated with Burkitt's lymphoma (a B cell lymphoma) and nasopharyngeal carcinoma. Although LMP-1 is the only viral gene product that can transform cells in culture, other viral proteins are made in such tumor cells. These products include Epstein-Barr virus nuclear antigen 1 (EBNA-1), which is necessary for replication and maintenance of the episomal viral genome (Volume I, Chapter 9). This protein also contains a sequence that inhibits presentation of EBNA-1 epitopes by major histocompatibility complex (MHC) class I proteins. Consequently, tumor cells cannot be detected so readily by T cells of the adaptive immune system. Similarly, several of the human herpesvirus 8 genes that have been implicated in transformation or tumorigenicity encode proteins that inhibit innate or adaptive immune responses. For example, the viral cytokine v-IL-6, which is a B cell mitogen, also blocks the action of interferon by inhibiting phosphorylation of substrates of the interferon receptor, such as Stat2. In addition, the vFLIP protein, which can enhance the tumorigenicity of murine B cells, inhibits killing by natural killer cells.

Other Mechanisms of Transformation and Oncogenesis by Human Tumor Viruses

The mechanisms by which some viruses associated with human cancers transform cells and contribute to tumor development cannot be subsumed within the general paradigms discussed in the preceding sections. Our current understanding of the development of these neoplastic diseases is described in this section.

Nontransducing Oncogenic Retroviruses: Tumorigenesis with Very Long Latency

The prototype for the nontransducing oncogenic retroviruses with complex genomes is human T-lymphotropic virus type 1, which is associated with adult T cell leukemia (ATL). This

---

BOX 6.11

BACKGROUND
Escape from immune surveillance and the oncogenicity of adenovirus-transformed cells

One of the earliest classifications of human adenovirus serotypes was based on the ability of the viruses to induce tumors in laboratory animals (see the table). Rodent cells transformed with the viral E1A and E1B genes in culture exhibit the tumorigenicity characteristic of the virus: cells transformed by the adenovirus type 12 genes form tumors efficiently when inoculated into syngeneic, immunocompetent animals, whereas cells transformed by adenovirus type 5 DNA induce tumors only in immunocompromised animals, such as nude mice. This difference was exploited to map the ability of transformed cells to form tumors efficiently in normal animals to a small region of the E1A gene, unique to adenovirus type 12 (and other highly oncogenic adenoviruses). The tumorigenicity of transformed cells was also correlated with repression of transcription of MHC class I genes: the adenovirus type

12 E1A proteins, but not those of adenovirus type 5, inhibit transcription of MHC class I genes by stimulating the binding of a translational repressor and histone deacetylases to the MHC class I enhancers.

The inhibition of MHC class I transcription induced by adenovirus type 12 E1A proteins results in reduced protein concentration on the cell surface, and hence in impaired presentation of antigens to cells of the immune system. Adenovirus type 12-transformed cells therefore escape immune surveillance and

destruction, whereas those transformed by adenovirus type 5 do not.

Yewdell JW, Bennink JR, Eager KB, Ricciardi RP. 1998. CTL recognition of adenovirus-transformed cells infected with influenza virus: lysis by anti-influenza CTL parallels adenovirus-12-induced suppression of class I MHC molecules. *Virology* **162:** 236–238.

Zhao B, Huo S, Ricciardi RP. 2003. Chromatin repression by COUP-TFII and HDAC dominates activation by NF-κB in regulating major histocompatibility complex class I transcription in adenovirus tumorigenic cells. *Virology* **306:**68–76.

Classification of human adenoviruses on the basis of oncogenicity

Subgroup	Representative serotypes	Oncogenicity in animals
A	12, 18, 31	High: induce tumors rapidly and efficiently
B	3, 7, 21	Low: induce tumors in only a fraction of infected animals, with a long latent period
C	1, 2, 5	None

disease was first described in Japan in 1977, and has since been found in other parts of the world, including the Caribbean and areas of South America and Africa. The virus entered the human population as a zoonosis from infected primates some 30,000 to 40,000 years ago. The human virus was isolated in 1980 and is now classified as a deltaretrovirus (Volume I, Appendix, Fig. 29).

Human T-lymphotropic virus is transmitted via the same routes as human immunodeficiency virus: during sexual intercourse, by intravenous drug use and blood transfusions, and from mother to child. Infection is usually asymptomatic, but can progress to ATL in about 5% of infected individuals over a period of 30 to 50 years (see Chapter 5). Although some progress has been made using stem cell transplantation and antiviral drugs, there is no effective treatment for the disease, which is usually fatal within a year of diagnosis. The mechanism(s) by which the virus induces malig-

nancies is still uncertain, but some of the features of ATL are consistent with a role for a viral regulatory protein. A provirus is found at the same site in all leukemic cells from a given case of ATL, indicating clonal origin, but there are **no** preferred chromosomal locations for these integrations. Activation or inactivation of a specific cellular gene is not, therefore, a likely mechanism of transformation. As the genome of human T-lymphotropic virus type 1 does not contain **any** cell-derived nucleic acid, some viral sequences must be responsible for this activity. The search for such sequences rapidly focused on the region denoted X, which encodes a number of regulatory and accessory proteins (Fig. 6.28). One of the best studied among these is the multifunctional transcriptional activator Tax. This protein is required for efficient proviral gene expression, and it also regulates the expression and function of a number of cellular genes and proteins that regulate T cell physiology, a feature consistent with its desig-

Figure 6.28 Transcription map of human T-lymphotropic virus type 1 (HTLV-1) proviral DNA showing gene-coding regions and their functions. Transcription of all but the *Hbz* gene is initiated between the unique 3′ (U3) and R regions in the 5′ LTR. Dotted lines indicate spliced introns. The *Hbz* gene is transcribed either from multiple Sp1-promoted initiation sites in the unique 5′ (U5) and R regions of the 3″ LTR, which produce spliced mRNA transcripts (*sHbz*), or from an initiation site within the *Tax* gene to form the unspliced transcript (*usHbz*). Translation of the spliced and unspliced mRNAs produces two proteins of 206 and 209 amino acids, respectively, both containing three functional domains. However, the latter protein has a very short half-life, and only products of the spliced transcript can be detected readily in ATL cells. Adapted from P. Kannian and P. L. Green, *Viruses* **2:**2037–2077, 2010, with permission.

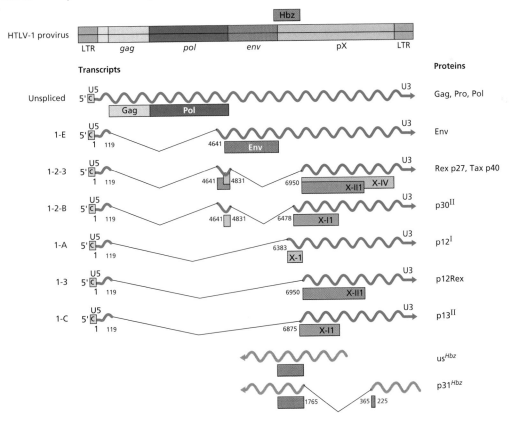

nation as a viral oncoprotein. It stimulates transcription indirectly, by interaction with Creb and by activation of Nf-κb (Fig. 6.29).

The latter effect leads to inhibition of apoptosis in infected T cells and enhanced transcription of a number of genes that encode cytokines, their receptors, and other regulatory proteins. Tax interactions with particular cellular proteins have also been shown to promote cell cycle progression and block cellular DNA repair, resulting in genome instability and evolution of malignant clones. Given the numerous and important functions attributed to this viral protein, it was surprising that *Tax* gene expression could be detected in the leukemic cells of only some 60% of ATL patients. This finding suggested that Tax might be required for the initiation of oncogenesis, but not for its maintenance. This notion was confirmed following the discovery and analysis of the Hbz open reading frame in the X region of the proviral DNA genome (Fig. 6.28).

Hbz was first identified as a Creb-binding protein that inhibits Tax-mediated transcription from the human

T-lymphotropic virus type 1 LTR. Large quantities of the potent immunogen Tax are synthesized and even secreted by these cells. As Hbz is not very immunogenic, its inhibition of Tax reduces the host's immune response to ATL cells. Most importantly, Hbz is detected in all ATL cells, and although antagonistic to some Tax functions or activities, it also activates some of the same signaling and proliferation pathways that are required for tumor maintenance (Fig. 6.29). Hbz inhibition of the Nf-κB pathway also enhances cell proliferation and progression of disease by blocking Tax-induced senescence and by reducing host innate and inflammatory immune responses. In addition, *Hbz* RNA itself has been found to promote T cell proliferation.

Although Tax and Hbz are clearly major players in the "yin and yang" of initiation and maintenance of oncogenesis by human T-lymphotropic virus type 1, other accessory genes encoded in the X region are likely to contribute to viral pathogenesis. The products of these genes, p12/p8, p13, and p30, are dispensable for viral replication and transformation of cells in culture, but they are required for efficient

Figure 6.29 Domains and interactions of the human T-lymphotropic virus type 1 Tax and Hbz proteins. (A) Tax-1 protein. The protein includes two leucine zipper domains (LZR) and nuclear localization (NLS) and nuclear export (NES) sequences. As indicated, Tax interacts with numerous cellular proteins to regulate transcription. The Tax response element in the U3 region of the viral LTR is a 21-bp triple repeat, containing an octamer motif that is homologous to the binding site for Creb. Tax does not bind directly to the TRE, or the numerous cellular promoters that contain cAMP response element, but interacts with the basic leucine zipper domain of Creb, thereby enhancing its dimerization and its affinity to these response elements. Coactivators such as Cbp/p300 are also recruited to these complexes, inducing the chromatin remodeling that facilitates transcription. **(B)** The Hbz protein has three major domains: an N-terminal activating domain (AD), a C-terminal basic zipper domain (bZIP), and a central domain with three basic regions that include nuclear localization signals. The (+) and (−) symbols indicate activation or inhibition, respectively, upon binding. For example, binding of the bZIP domain to Creb blocks Creb dimerization and hence stimulation of Tax-dependent transcription.

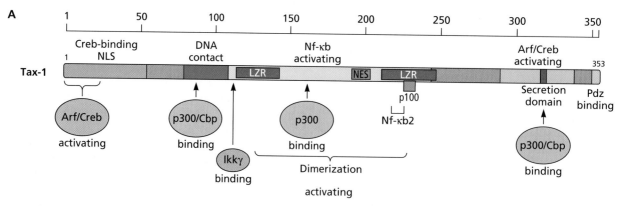

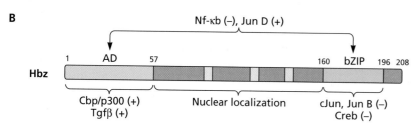

replication and persistence of the virus in rabbits and non-human primates. Because virus-induced oncogenic events occur a long time before ATL appears, it is difficult to sort out and evaluate the multiple effects of all of these viral proteins. Furthermore, the lack of a suitable animal model for the disease and the inefficiency of infection of T cells have represented significant challenges to research with the human lymphotropic viruses.

Oncogenesis by Hepatitis Viruses

Hepatitis B Virus

Hepatitis B virus is a member of the family *Hepadnaviridae*. The major site of reproduction for all hepadnaviruses is the hepatocyte, the major cell of the liver. Infections by these viruses can be acute (3 to 12 months) or lifelong. In humans, the frequency of persistent infection ranges from 0.1 to 25% of the population in different parts of the world. Most persistent infections are acquired neonatally or during the first year of life. Chronic infection of hepatocytes leads to their persistent destruction by the immune system and formation of fibrotic scars that obstruct the passage of blood (a life-threatening condition known as cirrhosis). Long-term carriers are also at high risk for developing hepatocellular carcinoma (Box 6.1), which generally leads to death within 5 years of diagnosis. As many as 1 million people die of hepatocellular carcinoma each year, despite the fact that a vaccine that prevents infection with this virus, actually the very first "anticancer" vaccine, has been available for several decades.

Sustained low-level liver damage is characteristic of persistent infection by hepatitis B viruses. Almost all such damage can be attributed to attack on infected hepatocytes by the host's immune system. The rate of hepatocyte proliferation must increase in such cases to compensate for cell loss. It is generally accepted that such an increased rate of proliferation over long periods is a major contributor to the development of both cirrhosis and liver cancer. In addition, the inflammation and phagocytosis that are integral to the immune response can result in high local concentrations of superoxides and free radicals. It is therefore possible that DNA damage and the resulting mutagenesis also contribute to hepadnavirus-induced hepatocellular carcinoma. Consequently, there is considerable incentive for developing new antiviral therapies to treat persistent hepatitis B virus infection. While several reverse transcriptase inhibitors are available for treatment, current antiviral therapies cannot cure infections.

The almost universal presence of integrated fragments of hepadnaviral DNA in tumor genomes suggests that this feature plays a role in oncogenesis. However, there is no known association of viral sequences with proto-oncogenes. Other studies suggest that viral proteins, such as the X protein or a truncated form of the envelope protein encoded by integrated viral DNA sequences, may contribute to carcinogenesis in humans. In cell culture systems, and by inference in the infected liver, the hepatitis B virus X protein stimulates transcription from many cellular genes (including proto-oncogenes), both by altering the DNA binding of cellular transcriptional regulators and by activation of signaling via Nf-κb and other pathways. However, the long time required for development of human liver cancer implies that several low-probability reactions must take place over an extended period. The relative importance of X or other viral proteins and the indirect effects of immune damage to the process remain to be determined.

Hepatitis C Virus

Hepatitis C virus is a (+) strand RNA virus in the family *Flaviviridae*. Its discovery in 1989 established the etiology of what had been known previously as non-A, non-B hepatitis, a disease contracted by a small fraction of transfusion recipients who developed acute and chronic hepatitis and, years later in some cases, liver cancer. Routine screening of the blood supply has since reduced this mode of infection. Approximately 75 to 85% of infected individuals develop a persistent infection. An estimated 170 million people are still chronically infected worldwide; among these, 1 to 5% will develop hepatocellular carcinoma. Although a small percentage, this still amounts to 1.3 million to 7.2 million cases and approximately a third of all liver cancer cases worldwide. Fortunately, owing to recent success in the development of potent antiviral therapies that can cure chronic infections (see below), it can be expected that the number of liver cancers diagnosed each year will soon begin to decline.

Like hepatitis B virus, hepatitis C virus is hepatotropic. Chronic infection of hepatocytes leads to their destruction by the immune system, resulting in formation of fibrotic scars that obstruct the passage of blood. Not all patients with such cirrhosis develop cancer, and genome-wide association studies have suggested that the genetic background of the host influences the course of infection. Some evidence indicates that certain viral proteins (capsid, envelope, and several non-structural proteins) can block the normal response of hepatocytes to apoptotic signals, affect signal transduction, and increase the concentration of damaging reactive oxygen species. Deregulation of cellular miRNA production has also been associated with hepatitis C infection. The importance of these activities to oncogenesis has been difficult to test. Although chimpanzees are susceptible to hepatitis C (and hepatitis B) virus, infection has not been shown to cause hepatocellular carcinoma. In addition, current guidelines restrict the use of chimpanzees for research in the United States, and there is no good small-animal model for hepatitis C virus-mediated hepatocellular carcinogenesis. However, as with hepatitis B virus, the indirect effects of immune-mediated inflammation and oxidative damage induced by infection are thought to be the major contributors to cirrhosis and cancer. Whether viral proteins have a modulating role is yet to be determined.

In contrast to hepatitis B, infection with hepatitis C is curable because the virus reproduces entirely within the cytoplasm of infected cells and its persistence depends on continuous genome amplification. Early on, interferon-based therapies led to the cure of chronic infections in about 50 to 70% of patients. The establishment of cell culture systems for hepatitis C infection has enabled detailed study of the viral reproduction cycle and also served as an important tool for the development of direct-acting antiviral therapies. Selective inhibitors of the viral serine protease (product of the *NS3-NS4A* gene), the viral polymerase (product of the *NS5B* gene), and the NS5A protein are very potent antivirals that, in combination with interferon, can cure ~90% of infected patients. Some of these inhibitors are already approved by the Food and Drug Administration, and others are expected to gain approval in the near future.

Perspectives

The discovery that viruses can cause cancer, initially made over a century ago, was the harbinger of the spectacular progress in understanding the molecular basis of transformation and oncogenesis that has occurred within the past 4 decades. Because tumor cells grow and divide when normal cells do not, elucidation of the mechanisms of transformation has inevitably been accompanied by the tracing of the intricate circuits that regulate cell proliferation in response to both external and internal signals. The remarkable discovery that the transforming gene of the retrovirus Rous sarcoma virus was a transduced cellular gene paved the way for identification of many cellular proto-oncogenes, and the elucidation of the signal transduction pathways in which the proteins encoded by them function. Indeed, in several cases, we can now describe in atomic detail the mechanisms by which mutations introduced into these genes during or following their capture into retroviral genomes lead to constitutive activation of signaling. These viral genes and their cellular counterparts that have acquired specific mutations in tumors are dominant oncogenes. In contrast, studies of a hereditary juvenile cancer in humans, retinoblastoma, had indicated that neoplastic disease can also develop following the loss of function of specific genes, which were therefore named tumor suppressor genes. Our current appreciation of the critical roles played by the products of such tumor suppressor genes in the control of cell cycle progression stems directly from studies of transforming proteins of adenoviruses, papillomaviruses, and polyomaviruses.

The initial cataloging of viral transforming genes and the properties of the proteins they encode suggested a bewildering variety of mechanisms of viral transformation. With the perspective provided by our present understanding of the circuits that control cell proliferation, we can now see that the great majority of these mechanisms fall into one of two general classes: viral transformation can be the result of either constitutive activation of signal transduction cascades or disruption of pathways that negatively regulate cell cycle progression. In both cases, viral proteins or transcriptional control signals override the finely tuned mechanisms that normally ensure that cells grow, duplicate their DNA, and divide only when external and internal conditions are propitious and, in many cases, also promote cell growth and survival.

Such an integrated view of the mechanisms by which viruses belonging to very different families can transform cells is intellectually satisfying. However, transformation of cells in culture is not necessarily accompanied by acquisition of the ability to form tumors in animals. This dissociation is evident in the etiology of some human cancers associated with viral infections. For example, infection by Epstein-Barr virus, which immortalizes human B cells in culture by mechanisms that we can describe in detail, is but one of several factors implicated in the development of Burkitt's lymphoma. A deeper appreciation of the parameters that determine a host's response to transformed cells will clearly be necessary if we are to understand the complex process of tumorigenesis.

References

Book Chapters

Lairmore MD, Franchini G. 2007. Human T-cell leukemia virus types 1 and 2, p 2071–2105. *In* Knipe DM, Howley PM Griffin DE, Lamb RA, Martin MA, Roizman B, Straus SE (ed), *Fields Virology*, 5th ed. Lippincott Williams & Wilkins, Philadelphia, PA.

Seeger C, Zoulim F, Mason WS. 2007. Hepadnaviruses, p 2977–3029. *In* Knipe DM, Howley PM, Griffin DE, Lamb RA, Martin MA, Roizman B, Straus SE (ed), *Fields Virology*, 5th ed. Lippincott Williams & Wilkins, Philadelphia, PA.

Review Articles

Berk AJ. 2005. Recent lessons in gene expression, cell cycle control, and cell biology from adenovirus. *Oncogene* **24:**7673–7685.

Blume-Jensen P, Hunter T. 2001. Oncogenic kinase signaling. *Nature* **411:**355–365.

Bracken AP, Ciro M, Cocito A, Helin K. 2004. E2F target genes: unraveling the biology. *Trends Biochem Sci* **29:**409–417.

DeCaprio JA. 2009. How the Rb tumor suppressor structure and function was revealed by the study of Adenovirus and SV40. *Virology* **384:**274–284.

Fallot G, Neuveut C, Buendia MA. 2012. Diverse roles of hepatitis B virus in liver cancer. *Curr Opin Virol* **2:**467–473.

Frame MC, Fincham VJ, Carragher NO, Wyke JA. 2002. v-Src's hold over actin and cell adhesions. *Nat Rev Mol Cell Biol* **3:**233–245.

Goodman RH, Smolik S. 2000. CBP/p300 in cell growth, transformation, and development. *Genes Dev* **14:**1553–1577.

Harbour JW, Dean DC. 2000. The Rb/E2F pathway: expanding roles and emerging paradigms. *Genes Dev* **14**:2393–2409.

Kannian P, Green PL. 2010. Human T lymphotropic virus type 1 (HTLV-1): molecular biology and oncogenesis. *Viruses* **2**:2037–2077.

Levine AJ. 2009. The common mechanisms of transformation by the small DNA tumor viruses: the inactivation of tumor suppressor gene products: *p53*. *Virology* **384**:285–293.

Massagué J. 2004. G1 cell-cycle control and cancer. *Nature* **432**:298–306.

Mesri EA, Cesarman E, Boshoff C. 2010. Kaposi's sarcoma and its associated herpesvirus. *Nat Rev Cancer* **10**:707–719.

Mesri EA, Feitelson MA, Munger K. 2014. Human viral oncogenesis: a cancer hallmarks analysis. *Cell Host Microbe* **15**:266–282.

Moore PS, Chang Y. 2010. Why do viruses cause cancer? Highlights of the first century of tumour virology. *Nat Rev Cancer* **10**:878–889.

Morales-Sánchez A, Fuentes-Pananá EM. 2014. Human viruses and cancer. *Viruses* **6**:4047–4079.

Münger K, Baldwin A, Edwards KM, Hayakawa H, Nguyen CL, Owens M, Grace M, Huh K. 2004. Mechanisms of human papillomavirus-induced oncogenesis. *J Virol* **78**:11451–11460.

Raab-Traub N. 2012. Novel mechanisms of EBV-induced oncogenesis. *Curr Opin Virol* **2**:453–458.

Scheel TK, Rice CM. 2013. Understanding the hepatitis C virus life cycle paves the way for highly effective therapies. *Nat Med* **19**:837–849.

Sullivan CS, Pipas JM. 2002. T antigens of simian virus 40: molecular chaperones for viral replication and tumorigenesis. *Microbiol Mol Biol Rev* **66**:179–202.

Thomas M, Pim D, Banks L. 1999. The role of the E6-p53 interaction in the molecular pathogenesis of HPV. *Oncogene* **18**:7690–7700.

Vousden KP, Prives C. 2009. Blinded by the light: the growing complexity of p53. *Cell* **137**:413–431.

White MK, Khalili K. 2004. Polyomaviruses and human cancer: molecular mechanisms underlying patterns of tumorigenesis. *Virology* **324**:1–16.

Papers of Special Interest

Transformation and Tumorigenesis by Retroviruses

Collett MS, Erikson RL. 1978. Protein kinase activity associated with the avian sarcoma virus *src* gene product. *Proc Natl Acad Sci U S A* **75**:2021–2024.

Dhar R, Ellis RW, Shih TY, Oroszlan S, Shapiro B, Maizel J, Lowy D, Scolnick E. 1982. Nucleotide sequence of the p21 transforming protein of Harvey murine sarcoma virus. *Science* **217**:934–936.

Ellermann V, Bang O. 1908. Experimentelle Leukämie bei Huhnern. *Zentbl Bakteriol* **46**:595–609.

Fasano O, Taparowsky E, Fiddes J, Wigler M, Goldfarb M. 1983. Sequence and structure of the coding region of the human H-ras-1 gene from T24 bladder carcinoma cells. *J Mol Appl Genet* **2**:173–180.

Hayward WS, Neel BG, Astrin SM. 1981. Activation of a cellular *onc* gene by promoter insertion in ALV-induced lymphoid leukosis. *Nature* **290**:475–480.

Hunter T, Sefton B. 1980. Transforming gene product of Rous sarcoma virus phosphorylates tyrosine. *Proc Natl Acad Sci U S A* **77**:1311–1315.

Levinson AD, Oppermann H, Levintow L, Varmus HE, Bishop JM. 1978. Evidence that the transforming gene of avian sarcoma virus encodes a protein kinase associated with a phosphoprotein. *Cell* **15**:561–572.

Nusse R, Varmus HE. 1982. Many tumors induced by the mouse mammary tumor virus contain a provirus integrated in the same region of the host genome. *Cell* **31**:99–109.

Rous P. 1910. A transmissible avian neoplasm: sarcoma of the common fowl. *J Exp Med* **12**:696–705.

Sicheri F, Moarefi I, Kuriyan J. 1997. Crystal structure of the Src family tyrosine kinase Hck. *Nature* **385**:602–609.

Stehelin D, Varmus HE, Bishop JM, Vogt PK. 1976. DNA related to the transforming gene(s) of avian sarcoma viruses is present in normal avian DNA. *Nature* **260**:170–173.

Xu W, Harrison SC, Eck MJ. 1997. Three-dimensional structure of the tyrosine kinase c-Src. *Nature* **385**:595–602.

Transformation by Polyomaviruses

Bolen JR, Thiele CJ, Israel MA, Yonemoto W, Lipsich LA, Brugge JS. 1984. Enhancement of cellular *src* gene product associated tyrosyl kinase activity following polyoma virus infection and transformation. *Cell* **38**:367–377.

DeCaprio JA, Ludlow JW, Figge J, Shew JY, Huang CM, Lee WH, Marsilio E, Paucha E, Livingston DM. 1988. SV40 large tumor antigen forms a specific complex with the product of the retinoblastoma susceptibility gene. *Cell* **54**:275–283.

Lane DP, Crawford LV. 1979. T antigen is bound to a host protein in SV40-transformed cells. *Nature* **278**:261–263.

Linzer DI, Levine AJ. 1979. Characterization of a 54K dalton cellular SV40 tumor antigen present in SV40-transformed cells and uninfected embryonal carcinoma cells. *Cell* **17**:43–52.

Sontag E, Fedorov S, Kamibayashi C, Robbins D, Cobb M, Mumby M. 1993. The interaction of SV40 small tumor antigen with protein phosphatase 2A stimulates the Map kinase pathway and induces cell proliferation. *Cell* **75**:887–897.

Talmage DA, Freund R, Young AT, Dahl J, Dawe CJ, Benjamin TL. 1989. Phosphorylation of middle T by pp60$^{c\text{-}src}$: a switch for binding of phosphatidylinositol 3-kinase and optimal tumorigenesis. *Cell* **54**:55–65.

Zalvide J, Stubdal H, DeCaprio J. 1998. The J domain of simian virus 40 large T antigen is required to functionally inactivate Rb family proteins. *Mol Cell Biol* **18**:1408–1415.

Transformation by Papillomaviruses

Dyson N, Howley PM, Harlow E. 1988. The human papillomavirus E7 oncoprotein is able to bind to the retinoblastoma gene product. *Science* **243**:934–937.

Huh H, Zhou X, Hayakawa H, Cho JY, Libermann TA, Jin J, Harper JW, Munger K. 2007. Human papillomavirus type 16 E7 oncoprotein associates with cullin 2 ubiquitin ligase complex, which contributes to degradation of the retinoblastoma tumor suppressor. *J Virol* **81**:9737–9747.

Lai CC, Henningson C, DiMaio D. 1998. Bovine papillomavirus E5 protein induces oligomerization and trans-phosphorylation of the platelet-derived growth factor β receptor. *Proc Natl Acad Sci U S A* **95**:15241–15246.

Nguyen ML, Nguyen MM, Lee D, Griep AE, Lambert PF. 2003. The PDZ ligand domain of the human papillomavirus type 16 E6 protein is required for E6's induction of epithelial hyperplasia in vivo. *J Virol* **77**:6957–6964.

Scheffner M, Werness BA, Huibregtse JM, Levine AJ, Howley PM. 1990. The E6 oncoprotein encoded by human papillomavirus types 16 and 18 promotes the degradation of p53. *Cell* **63**:1129–1136.

Transformation by Adenoviruses

Egan C, Bayley ST, Branton PE. 1989. Binding of the Rb1 protein to E1A products is required for adenovirus transformation. *Oncogene* **4**:383–388.

Liu X, Marmorstein R. 2007. Structure of the retinoblastoma protein bound to adenovirus E1A reveals the molecular basis for viral oncoprotein in activation of a tumor suppressor. *Genes Dev* **21**:2711–2716.

Querido E, Blanchette P, Yan Q, Kamura T, Morrison M, Boivin D, Kaelin WG, Conaway RC, Conaway JW, Branton PE. 2001. Degradation of p53 by adenovirus E4orf6 and E1B55K proteins occurs via a novel mechanism involving a Cullin-containing complex. *Genes Dev* **15**:3104–3117.

Rao L, Debbas M, Sabbatini D, Hockenberry D, Korsmeyer S, White E. 1992. The adenovirus E1A proteins induce apoptosis, which is inhibited by the E1B 19-kDa and Bcl-2 proteins. *Proc Natl Acad Sci U S A* **89**:7742–7746.

Van der Eb AJ, Mulder C, Graham FL, Houweling A. 1977. Transformation with specific fragments of adenovirus DNAs. I. isolation of specific fragments with transforming activity of adenovirus 2 and 5 DNA. *Gene* **2**:115–132.

Whyte P, Williamson NM, Harlow E. 1989. Cellular targets for transformation by the adenovirus E1A proteins. *Cell* **56**:67–75.

Transformation by Herpesviruses

Bais C, Santomasso B, Coso O, Arvanitakis L, Raaka EG, Gutkind JS, Asch AS, Cesarman E, Gershengorn MC, Mesri EA. 1998. G-protein-coupled receptor of Kaposi's sarcoma-associated herpesvirus is a viral oncogene and angiogenesis activator. *Nature* **391**:86–89.

Ciufo DM, Cannon JS, Poole LJ, Wu FY, Murray P, Ambinder RF, Hayward GS. 2001. Spindle cell conversion by Kaposi's sarcoma-associated herpesvirus: formation of colonies and plaques with mixed lytic and latent gene expression in infected primary dermal microvascular endothelial cell cultures. *J Virol* **75**:5614–5626.

Dirmeier U, Hoffmann R, Kilger E, Schultheiss U, Briseño C, Gires O, Kieser A, Eick D, Sugden B, Hammerschmidt W. 2005. Latent membrane protein 1 of Epstein-Barr virus coordinately regulates proliferation with control of apoptosis. *Oncogene* **24**:1711–1717.

Gires O, Zimber-Strobl U, Gonnella R, Ueffing M, Marschall G, Zeidler R, Pich D, Hammerschmidt W. 1997. Latent membrane protein 1 of Epstein-Barr virus mimics a constitutively active receptor molecule. *EMBO J* **16**:6131–6140.

Kulwichit W, Edwards RH, Davenport EM, Baskar JF, Godfrey V, Raab-Traub N. 1998. Expression of the Epstein-Barr virus latent membrane protein 1 induces B cell lymphoma in transgenic mice. *Proc Natl Acad Sci U S A* **95**:11963–11968.

Mainou BA, Everly DN, Jr, Raab-Traub N. 2007. Unique signaling properties of CTAR1 in LMP1-mediated transformation. *J Virol* **81**:9680–9692.

Swanton C, Mann DJ, Fleckenstein B, Neipel F, Peters G, Jones N. 1997. Herpes viral cyclin/Cdk6 complexes evade inhibition by CDK inhibitor proteins. *Nature* **390**:184–187.

Sylla BS, Hung SC, Davidson DM, Hatzivassiliou E, Malinin NL, Wallach D, Gilmore TD, Kieff E, Mosialos G. 1998. Epstein-Barr virus-transforming protein latent infection membrane protein 1 activates transcription factor NF-κB through a pathway that includes the NF-κB-inducing kinase and the IκB kinases IKKα and IKKβ. *Proc Natl Acad Sci U S A* **95**:10106–10111.

7 Human Immunodeficiency Virus Pathogenesis

Introduction
Worldwide Impact of AIDS

HIV Is a Lentivirus
Discovery and Characterization
Distinctive Features of the HIV Reproduction Cycle and the Functions of Auxiliary Proteins
The Viral Capsid Counters Intrinsic Defense Mechanisms

Cellular Targets

Routes of Transmission
Modes of Transmission
Mechanics of Spread

The Course of Infection
The Acute Phase
The Asymptomatic Phase
The Symptomatic Phase and AIDS
Variability of Response to Infection

Origins of Cellular Immune Dysfunction
CD4+ T Lymphocytes
Cytotoxic T Lymphocytes
Monocytes and Macrophages
B Cells
Natural Killer Cells
Autoimmunity

Immune Responses to HIV
Innate Response
The Cell-Mediated Response
Humoral Responses
Summary: the Critical Balance

Dynamics of HIV-1 Reproduction in AIDS Patients

Effects of HIV on Different Tissues and Organ Systems
Lymphoid Organs
The Nervous System
The Gastrointestinal System
Other Organs and Tissues

HIV and Cancer
Kaposi's Sarcoma
B Cell Lymphomas
Anogenital Carcinomas

Prospects for Treatment and Prevention
Antiviral Drugs
Confronting the Problems of Persistence and Latency
Gene Therapy Approaches
Immune System-Based Therapies
Antiviral Drug Prophylaxis

Perspectives

References

LINKS FOR CHAPTER 7

▶▶ *Video: Interview with Dr. Beatrice Hahn*
http://bit.ly/Virology_Hahn

▶▶ *Movie 7.1: Molecular Model for Apobec3F Degradation by the Vif/Cbf-b Ubiquitin Ligase*
http://bit.ly/Virology_V2_Movie7-1

▶▶ *Movie 7.2: HIV particles in a virological synapse between mature dendritic cells and susceptible T cells.*
http://bit.ly/Virology_V2_Movie7-2

▶▶ *Does a gorilla shift in the woods?*
http://bit.ly/Virology_Twiv327

▶▶ *Joint United Nations Programme on HIV/AIDS*
http://www.unaids.org

▶▶ *United States Centers for Disease Control and Prevention*
http://www.cdc.gov/hiv/

Introduction

Worldwide Impact of AIDS

Acquired immunodeficiency syndrome (AIDS) is the name given to the end-stage disease caused by infection with human immunodeficiency virus (HIV). By almost any criteria, HIV qualifies as one of the world's deadliest scourges. First recognized as a clinical entity in 1981, by 1992 AIDS had become the major cause of death in individuals 25 to 44 years of age in the United States. Although the rates of both infection and deaths have been reduced substantially since 2000, the current worldwide statistics are still staggering, with the low-income countries of Africa and parts of Asia being especially hard hit (Fig. 7.1). An end-of-year report from the United Nations' AIDS program estimated the number of new HIV infections in 2012 to be 2.3 million, bringing the total number of infected people worldwide to approximately 35.3 million. This number corresponds to almost 1 in every 100 adults aged 15 to 49 in the world's population. The availability of effective drugs to treat HIV infection has decreased the annual death toll in high-income countries. In some lower-income countries, up to 90% of infected individuals have access to such drugs, but in others only about 30% are being treated, and HIV/AIDS is still the leading cause of death in many of these regions. AIDS still kills more people than any other infectious disease (Table 11.4).

The emergence and spread of HIV was probably the consequence of a number of political, economic, and societal changes, including the breakdown of national borders, migration of large populations because of military conflicts and economic distress, and the ease and frequency of travel throughout the world. International efforts, including large investments from the U.S. President's Emergency Plan for AIDS Relief (known as PEPFAR) and the Global AIDS Program (GAP) that started in 2004, have focused on bringing funds and expertise to bear on the pandemic in Africa and elsewhere. While the task is enormous, much progress has been made in the past decade (see Fig. 9.22), fueling hopes that it will be possible to end the AIDS pandemic in the next decade.

Because of its medical importance, HIV has become the most intensely studied infectious agent. Research with the virus has not only contributed to our understanding of AIDS and related veterinary diseases, but has also provided new insights into principles of virology, cell biology, and immunology. This chapter describes the many facets of HIV-induced pathogenesis and what has been learned from their analysis. The complexities illustrate the enormous scope of the challenges faced by biomedical researchers and physicians in their efforts to control this agent, which strikes at the very heart of the body's defense systems.

HIV Is a Lentivirus

Discovery and Characterization

The first clue to the etiology of AIDS came in 1983 with the isolation of a retrovirus from the lymph node of a patient with lymphadenopathy at the Pasteur Institute in Paris. Although initially not fully appreciated, the significance of this finding became apparent in the following year with the isolation of a cytopathic, T cell-tropic retrovirus from combined blood cells of AIDS patients by researchers at the U.S. National Institutes of Health, and of a similar retrovirus from blood cells of an AIDS patient at the University of California, San Francisco. Although the National Institutes of Health isolate was later shown to originate from a sample received from the Pasteur Institute (Box 7.1), the virus isolated at the University of California, San Francisco, and subsequent isolates at the

---

PRINCIPLES *HIV pathogenesis*

- The disease associated with human immunodeficiency virus (HIV) infection, AIDS, still kills more people than any other disease of viral origin.

- HIV is transmitted from person to person by sexual contact, blood exchange, or from mother to child.

- The course of HIV infection is characterized by three phases: an initial acute phase, a variable asymptomatic phase, and eventual end-stage disease.

- The major target of HIV infection is the CD4+ T cell. Both the CD4 receptor and chemokine coreceptors, which are required for viral entry, are displayed on the surface of these cells.

- Individuals with mutations in the genes encoding chemokine coreceptors are resistant to HIV infection.

- Accessory proteins of HIV contribute to pathogenesis via a common mechanism of action; all function as adapter proteins that disrupt the normal trafficking of particular cellular proteins and, in most cases, facilitate their degradation.

- The gut-associated lymphoid tissue (GALT), which contains ~40% of the body's lymphocytes, is a major site of HIV reproduction.

- The defining feature of HIV disease is impaired function of immune cells, and most AIDS patients succumb to opportunistic infections with microorganisms that are little threat to individuals with healthy immune systems.

- The inability to identify with certainty the cell types, number, and body compartments that comprise the total HIV latent reservoir is a serious barrier to devising strategies for complete clearance of the virus, and effective cure of the infection.

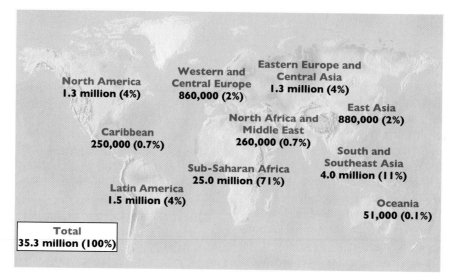

Figure 7.1 Estimated number of people living with HIV infection worldwide, 2012. Data from the UNAIDS Report on Global AIDS Epidemic, 2013.

National Institutes of Health laboratory were unique. Electron microscopic examination revealed that these viruses were morphologically similar to a known group of retroviruses, the lentiviruses, and further characterization confirmed this relationship

Lentiviruses comprise a separate genus of the family *Retroviridae* (Table 7.1). The equine infectious anemia lentivirus was one of the first viruses to be identified. Discovered in 1904, this virus causes episodic autoimmune hemolytic anemia in horses. Lentiviruses of sheep and goats have also been known for many years. All these viruses are associated with long incubation periods and are therefore called **slow viruses** (Chapter 5). A distinct type of HIV that is prevalent in certain regions of West Africa was discovered in 1986, and given the name HIV-2. Individuals infected with HIV-2 also develop AIDS, but with a longer incubation period and lower morbidity.

Many independent isolates of both HIV-1 and HIV-2 have been characterized over the past two decades. Nucleotide sequence comparisons allow us to distinguish four distinct groups among HIV-1 isolates (Fig. 7.2). Group M includes most HIV-1 isolates, and is the cause of greater than 95% of HIV infections worldwide. Eleven distinct M subtypes are currently recognized (called **clade** A to K), each of which is prevalent in a different geographic area. For example, clade B is the most common subtype in North America and Europe. HIV-1 group O (for "outliers") also includes diverse subtypes but is relatively rare, and groups N and P have been identified only in a few individuals in Cameroon. Eight distinct groups of HIV-2 have also been identified. Of these, groups A

and B (found in different parts of West Africa) account for most infections. Sequence analyses and identification of related nonhuman primate strains have indicated that these viruses entered humans by cross-species transmission on several occasions.

The African monkey and ape isolates are endemic to each of the species from which they were obtained and do not cause disease in their native hosts, most likely because of intrinsic defense mechanisms (Chapter 3). However, a fatal AIDS-like disease is caused by infection of Asian macaques with virus originating from the African sooty mangabey (SIV$_{sm}$). Close contact between sooty mangabeys and humans is common, as these animals are hunted for food and kept as pets. Such interactions, and the observation that several isolates of HIV-2 are nearly indistinguishable in nucleotide sequence from SIV$_{sm}$, support the hypothesis that HIVs emerged via interspecies transmission from nonhuman primates to humans. Other studies indicate that the M, N, and O HIV-1 groups arose via at least three independent transmissions from chimpanzees (Box 1.2). The strains of SIV$_{cpz}$ from the chimpanzee *Pan troglodytes troglodytes* are closest in sequence to HIV-1, implicating this subspecies as the origin of the human virus (Box 1.2). Analyses of stored blood and tissue samples indicate that the common ancestor of group M viruses may have been transmitted to humans as early as 1900 (Box 7.2). In this chapter we use the abbreviation HIV to describe properties shared by HIV-1 and HIV-2, and specify the type when referring to one or the other.

As summarized in Table 7.1, lentiviruses cause immune deficiencies and disorders of the hematopoietic and central

BOX **7.1**

DISCUSSION
Lessons from discovery of the AIDS virus(es)

The first AIDS virus was obtained from a patient with lymphadenopathy by Françoise Barré-Sinoussi in collaboration with Jean-Claude Chermann and Luc Montagnier at the Pasteur Institute (1983) and named LAV, for lymphadenopathy virus. The isolate, named Bru, grew only in primary cell cultures. We now know that Bru belonged to a class of slow-growing, low-titer viruses that are common in early-stage infection.

Between 20 July and 3 August 1983, Bru-infected cultures at the Pasteur Institute became contaminated with a second AIDS virus, called Lai, which had been isolated from a patient with full-blown AIDS, and which belonged to a class of viruses that grow well in cell culture. HIV-1 Lai rapidly overtook the cultures.

Unaware of this contamination, Pasteur scientists subsequently sent out virus samples from these cultures as "Bru" to several laboratories, including those of Robin Weiss in Britain and Malcolm Martin and Robert Gallo in the United States. Unlike earlier samples of Bru, this virus grew robustly in the laboratories to which it was distributed. Indeed, Lai was later discovered to have contaminated some AIDS patient "isolates" obtained by Weiss. In retrospect, such contamination is not surprising, as biological containment facilities were limited at the time, with the same incubators and hoods being used for maintaining HIV stocks and making new isolates.

Lai also contaminated cultures of blood cells combined from several AIDS patients in the Gallo laboratory at the National Institutes of Health. Because the properties of this virus were found to be different from those described for Bru, Gallo and coworkers reported the discovery of a second type of AIDS virus, which they believed to have originated from one of their AIDS patients. They called the virus HTLV-III, for human T cell lymphotrophic virus, believing it was unique, but probably related to the human

T cell lymphotropic viruses I and II, which they had also been studying.

This second claim, a race to develop blood-screening tests, and the later revelation from DNA sequence analyses that the French and the Gallo viruses were one and the same (Lai) led to a much publicized scientific controversy in which patenting agencies, lawyers, businesses, and even governments were embroiled.

Simon Wain-Hobson and colleagues at the Pasteur Institute eventually sorted out the chain of events (in 1991) by comparing nucleotide sequences of stored samples of the original stocks of Bru and Lai. The controversy has since subsided, and the nomenclature was simplified in 1986, when the International Committee on Taxonomy of Viruses recommended that the current name, human

immunodeficiency virus (HIV) replace LAV, HTLV-III, and ARV, a third name used by investigators in San Francisco who had obtained an independent isolate.

What remains from this story are three important lessons in virology: that contamination can be a real problem, that passage in the laboratory tends to select for viruses that reproduce rapidly, and that rigorous characterization (nowadays by genome sequencing) is a prudent safeguard against costly mistakes.

Goudsmit J. 2002. Lots of peanut shells but no elephant. A controversial account of the discovery of HIV. *Nature* **416:**125–126.

Wain-Hobson SJ, Vartanian P, Henry M, Chenciner N, Cheynier R, Delassus S, Martins LP, Sala M, Nugeyre MT, Guetard D, et al. 1991. LAV revisited: origins of the early HIV-1 isolates from Institut Pasteur. *Science* **252:**961–965.

Weiss R, Martin M. Personal communication.

Reagan HHS Secretary Margaret Heckler, shown with Robert Gallo. April 23, 1984. National Cancer Institute (NCI) researcher Robert Gallo reports isolation of an AIDS virus he calls HTLV-III. Later, it turns out to be LAV from a sample sent by the Montagnier lab, but not before HHS Secretary Margaret Heckler gives Gallo full credit. Heckler predicts a vaccine in 2 years. From WebMD http://www.webmd.com/hiv-aids/ss /slideshow-aids-retrospective, with permission.

nervous systems and, sometimes, arthritis and autoimmunity. Lentiviral genomes are relatively large. In addition to the three structural polyproteins Gag, Pol, and Env, common to all retroviruses, these genomes encode a number of additional **auxiliary proteins** (Fig. 7.3). Two HIV auxiliary proteins, Tat and Rev, perform regulatory functions that are essential for

viral reproduction. The remaining four additional proteins of HIV-1, Nef, Vif, Vpr, and Vpu, are not essential for viral reproduction in most immortalized T cell lines and hence are known as **accessory proteins**. However, these proteins do modulate virus reproduction, and they are essential for efficient virus production *in vivo* and the ensuing pathogenesis.

Table 7.1 Lentiviruses[a]

Virus	Host infected	Primary cell type infected	Clinical disorder(s)
Equine infectious anemia virus	Horse	Macrophages	Cyclical infection in the first year, autoimmune hemolytic anemia, sometimes encephalopathy
Visna/maedi virus	Sheep	Macrophages	Encephalopathy/pneumonitis
Caprine arthritis-encephalitis virus	Goat	Macrophages	Immune deficiency, arthritis, encephalopathy
Bovine immunodeficiency virus	Cow	Macrophages	Lymphadenopathy, lymphocytosis
Feline immunodeficiency virus	Cat	T lymphocytes	Immune deficiency
Simian immunodeficiency virus	Primate	T lymphocytes	Immune deficiency and encephalopathy
Human immunodeficiency virus	Human	T lymphocytes	Immune deficiency and encephalopathy

[a]Adapted from Table 1.1 (p. 2) of J. A. Levy, *HIV and the Pathogenesis of AIDS,* 3rd ed. (ASM Press, Washington, DC, 2007).

Figure 7.2 Phylogenetic relationships among primate lentiviruses. The tree was reconstructed by computational methods and alignment of 34 published nucleotide sequences of the retroviral *pol* genes, taken from GenBank. SIVcol signifies black and white colobus; SIVdrl, drill; SIVgsn, greater spot-nosed monkey; SIVlhoest, L'Hoest monkey; SIVmac, macaque; SIVmnd, mandrill; SIVmon, Campbell's mona monkey; SIVrcm, red-capped monkey; SIVsab, Sabaeus monkey; SIVsun, sun-tailed monkey; SIVsyk, Sykes' monkey; SIVtan, tantalus monkey; SIVver, vervet monkey; SIVcpz, chimpanzee. The species of transmission from chimps to humans, giving rise to HIV-1 groups M, N, and O, are identified. For clarity, only some subtypes of HIV-1 and HIV-2 are shown. From Figure 2 of A. Rambaut, D. Posada, K. A. Crandall, and C. Holmes, The causes and consequences of HIV evolution. *Nat Rev Genet* **5:**52–56, 2004, with permission.

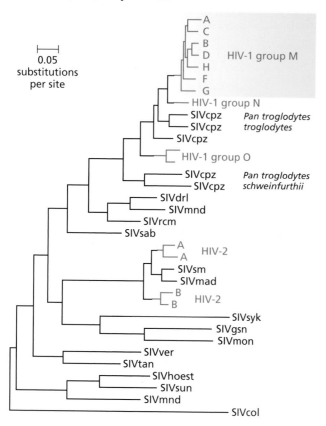

Distinctive Features of the HIV Reproduction Cycle and the Functions of Auxiliary Proteins

Much of what we know about the function of the auxiliary proteins of HIV comes from studies of their effects on cells in culture, often produced transiently from plasmid expression vectors in the absence of other viral components (Volume I, Box 8.8). Although these methods are simple and sensitive, they do not necessarily reproduce the conditions of viral infection. Preparation and analysis of viral mutants have also been used to investigate the functions of these proteins in cell culture. However, as the hosts for these viruses are humans, it is difficult to evaluate the significance of many of the functions deduced from cell culture to pathology in the whole organism.

The Regulatory Proteins Tat and Rev

Tat (for transactivator of transcription) stimulates processive transcription. As in all retroviruses, expression of integrated HIV DNA is regulated by sequences in the transcriptional control region of the viral long terminal repeat (LTR), which are recognized by the cellular transcriptional machinery. The HIV LTR functions as a promoter in a variety of cell types, but its basal level is very low. As described in Volume I, Chapter 8 (Fig. 8.10), the LTR of HIV includes an enhancer sequence that binds a number of cell type-specific transcriptional activators, among them Nf-κb (Volume I, Fig. 8.11). The fact that Nf-κb enters the nucleus to promote T cell activation may explain why HIV reproduction requires T cell stimulation.

Just downstream of the site of initiation of transcription in the HIV LTR is a unique viral regulatory sequence, the *trans*-activating response element TAR (Fig. 7.4). TAR RNA forms a stable, bulged stem-loop structure that binds Tat together with a number of host proteins (Volume 1, Chapter 8, Fig. 8.13). The principal role of Tat is to stimulate processivity of transcription and thereby facilitate the elongation of viral RNA.

BOX 7.2

BACKGROUND
The earliest record of HIV-1 infection

For some time, the earliest record of HIV-1 infection came from a serum sample obtained in 1959 from a Bantu male in the city of Leopoldville, now known as Kinshasa, in the Democratic Republic of Congo (DRC). Phylogenetic analyses placed this viral sequence (ZR59) near the ancestral node of clades B and D. As this is not at the base of the M group, this group must have originated earlier, and back calculations suggested that the M group of viruses arose via cross-species transmission from a chimpanzee into the African population around 1930. Subsequent characterization of viral sequences in a paraffin-embedded lymph node biopsy specimen prepared in 1960 from another individual in Kinshasa (DRC60) led to a revision of that estimate. The ZR59 and DRC60 sequences differ by a degree (12%) seen in the most divergent strains within subtypes. Results from a variety of statistical analyses with these and additional archived samples indicate that the epidemic was well established by 1959/1960 and that the common ancestor was probably circulating as early as 1910. The initial transmission event could have occurred even earlier. Because the human strains shared a common ancestor with the chimp strains in about 1850, the period between ~1850 and ~1910 is the most likely window for the fateful first jump of what became the pandemic HIV/AIDS lineage. As the greatest diversity of group M subtypes has been found in Kinshasa, it seems likely that all of the early diversification of HIV-1 group M viruses occurred in the Leopoldville area, which was one of the largest urban centers at the time.

Sharp PM, Hahn BH. 2008. Prehistory of HIV. *Nature* **455:**605–606. (A personal account of the efforts to determine the routes of transmission from primates to humans can be found in an interview with Dr. Beatrice Hahn: http://bit.ly/Virology_Hahn)

Worobey M, Gemmel M, Teuwen DE, Haselkorn T, Kunstman K, Bunce M, Muyembe J-J, Kabongo J-MM, Kalengayi RM, Van Marck E, Gilbert MTP, Wolinsky SM. 2008. Direct evidence of extensive diversity of HIV-1 in Kinshasa by 1960. *Nature* **455:**661–665.

Branch lengths are depicted in unit time (years) and represent the median of those nodes that were present in at least 50% of the sampled trees. DRC60 (red), ZR59 (black), and the three control sequences from paraffin-embedded specimens from known AIDS patients (gray) are depicted in bold. Sequences sampled in the DRC are highlighted with a bullet at the tip. Nodes (sub-subtype and deeper) are marked with gray circles. DRC60 and the two control sequences from the DRC each form monophyletic clades with previously published sequences from the DRC, whereas the Canadian control sequence clusters, as expected, with subtype B sequences. Unclassified strains are labeled U. The dashed circle and shaded area show the extensive HIV-1 diversity in Kinshasa in the 1950s. From Fig. 2 of M. Worobey et al. *Nature* **455:**661–664, 2008, with permission.

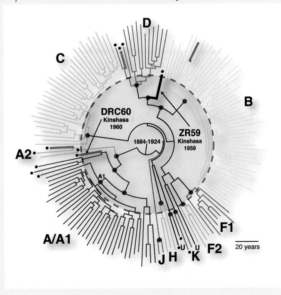

Tat is released by infected cells; it can then be taken up by other cells and influence their function. Tat can act as a chemo-attractant for monocytes, basophils, and mast cells. It also induces synthesis of a variety of important proteins in the cells that it enters, and some of these can have a profound effect on virus spread and immune cell function. For example, in transient-expression assays Tat can stimulate the expression of genes encoding the CXCr4 and CCr5 coreceptors in target cells and also enhance the synthesis of a number of chemokines. The Tat protein is cytotoxic to some cultured cells and is neurotoxic when inoculated intracerebrally into mice.

Multiple splice sites and the function of Rev. In contrast to the oncogenic retroviruses with simpler genomes, the full-length HIV transcript contains numerous 5′ and 3′ splice sites. The regulatory proteins Tat and Rev (for regulator of expression of virion proteins) and the accessory protein Nef are synthesized early in infection from multiply spliced mRNAs (Volume 1, Appendix Fig. 29). As Tat then stimulates transcription, these mRNAs are found in abundance at this early time. However, accumulation of the Rev protein brings about a change in the pattern of mRNAs, leading to a temporal shift in viral gene expression.

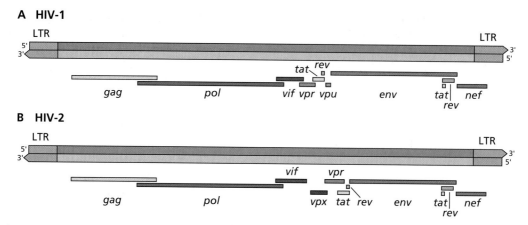

Figure 7.3 Organization of HIV-1 (A) and HIV-2 proviral DNA (B). Vertical positions of the colored bars denote each of the three different reading frames that encode viral proteins. The LTRs contain sequences necessary for transcriptional initiation and termination, reverse transcription, and integration.

Rev is an RNA-binding protein that recognizes a specific sequence within a structural element in the *env* region of the elongated transcript, called the **Rev-responsive element (RRE)** (Fig. 7.4). Rev mediates the nuclear export of any RRE-containing RNA by a mechanism discussed more fully in Volume I, Chapter 10 (Figs. 10.14 to 10.16). As the concentration of Rev increases, unspliced or singly spliced transcripts containing the RRE are exported from the nucleus. In this way, Rev promotes synthesis of the viral structural proteins and enzymes and ensures the availability of full-length

Figure 7.4 Mechanisms of Tat and Rev function. Some regulatory sequences in the HIV LTR are depicted in the expanded section at the top. The numbers refer to positions relative to the site of initiation of transcription. The opposing arrows in R represent a palindromic sequence that folds into a stem-loop structure (TAR) in the transcribed mRNA to which Tat binds (center). Tat recruits cellular proteins that are required for efficient elongation during HIV-1 RNA synthesis. The position of the RRE in the *env* transcript (with bound Rev dimers) and the *cis*-acting repressive sequences (instability sequences, INS) in the unspliced or singly spliced transcripts are also illustrated. Mutations in the A+U-rich INS increase the stability, nuclear export, and translation efficiency of the transcripts in the absence of Rev. Response to INS appears to be cell type-dependent, but the mechanisms by which they act, and exactly how Rev counteracts their effects, are not understood.

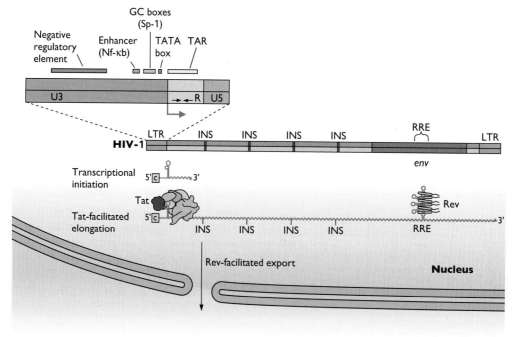

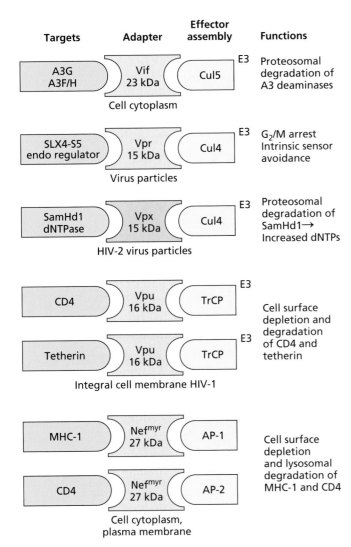

Targets	Adapter	Effector assembly	Functions

A3G / A3F/H — Vif 23 kDa — Cul5 — E3 — Proteosomal degradation of A3 deaminases

Cell cytoplasm

SLX4-S5 endo regulator — Vpr 15 kDa — Cul4 — E3 — G₂/M arrest Intrinsic sensor avoidance

Virus particles

SamHd1 dNTPase — Vpx 15 kDa — Cul4 — E3 — Proteosomal degradation of SamHd1→ Increased dNTPs

HIV-2 virus particles

CD4 — Vpu 16 kDa — TrCP — E3 — Cell surface depletion and degradation of CD4 and tetherin

Tetherin — Vpu 16 kDa — TrCP — E3

Integral cell membrane HIV-1

MHC-1 — Nef^myr 27 kDa — AP-1 — Cell surface depletion and lysosomal degradation of MHC-1 and CD4

CD4 — Nef^myr 27 kDa — AP-2

Cell cytoplasm, plasma membrane

Figure 7.5 Adapter functions of HIV-1 accessory proteins. The major targets for the HIV accessory proteins, their locations, and the effector assemblies with which they interact are noted. Most effector assemblies function by mediating destruction of the targets via reactions in which the viral proteins are recycled: In a reaction that requires binding of the transcription regulator, cellular core binding factor beta (Cbfβ), **Vif** assembles with additional cellular proteins (Cul5, Elongins B and C, and Rbx1) in an E3 ubiquitin ligase that then targets Apobec proteins (A3G, F, and H) for ubiquitination and proteosomal degradation. While less potent than A3G, A3F and A3H have antiviral activities and are produced in abundance in lymphoid cells (Chapter 3). **Vpr** functions in a similar way, but interacts with a different member of the cullin family (Cul 4), in a damaged DNA-binding protein (Ddb1) and Cul-associated factor 1 (Dcaf1)-E3-ligase assembly. This interaction leads to dysregulation of cellular endonuclease activities, triggering a DNA damage response and cell cycle arrest. The **Vpx** protein of HIV-2 engages the same Cul 4-E3 complex as Vpr, but their targets are quite distinct: Vpx targets the dGTP-dependent deoxynucleoside triphosphohydrolase, SamHd1, for ubiquitination and proteosomal degradation. Phosphorylation of **Vpu** leads to recruitment of another multiprotein E3 ubiquitin ligase, Scf (which includes Cul1, Skp1, and Roc1) via interaction with the adapter, β-transducin repeat-containing protein (β-TrCP), and targets both CD4 and tetherin degradation. **Nef** binds to the cytoplasmic tail of CD4 and links this receptor to components of a clathrin-dependent

genomic RNA for incorporation into progeny virus particles. The accessory proteins Vif, Vpr, and Vpu (for HIV-1) or Vif, Vpr, and Vpx (for HIV-2) are also produced later in infection from singly spliced mRNAs that are dependent on Rev for export to the cytoplasm (Volume I, Appendix Fig. 29).

The Accessory Proteins

While a very large number of seemingly disparate activities have been attributed to the HIV accessory proteins, recent studies have revealed a common mechanism for their action: all are **adapter proteins** that disrupt the normal trafficking of particular cellular proteins and, in most cases, lead to their destruction (Fig. 7.5). In this way, HIV accessory proteins function as antagonists of cellular intrinsic defense mechanisms that detect infection and counteract virus reproduction.

Vif protein. Vif (viral infectivity factor) is a 23-kDa protein that accumulates in the cytoplasm and at the plasma membrane of infected cells. Early studies showed that mutant virus particles lacking the *vif* gene were approximately 1,000 times less infectious than the wild type in certain CD4⁺ T cell lines and peripheral blood lymphocytes and macrophages. It was discovered that production of Vif from a plasmid vector in susceptible host cells did not compensate for its absence in the cell that produced virus particles. Rather, Vif was needed at the time of virus assembly.

Vif is an RNA-binding protein and small quantities can be detected in HIV particles. Virus particles produced from *vif*-defective HIV genomes contain the normal complement of progeny RNA, and they are able to enter susceptible cells and to initiate reverse transcription, but full-length double-stranded viral DNA is not detected. These observations demonstrated that Vif is required in a step that is essential for completion of reverse transcription. The requirement for Vif is strikingly cell type-dependent, and experiments in which cells that are permissive for *vif* mutants were fused with cells that are nonpermissive established that the nonpermissive phenotype is dominant. The infectivity of virus particles produced in such heterokaryons was enhanced by Vif production. This observation suggested that Vif may suppress a host cell function that otherwise inhibits progeny virus infectivity.

All of these seemingly unusual properties were demystified with the discovery that Vif blocks the antiviral action of members of an RNA-binding family of cellular cytidine deaminases, called apolipoprotein B mRNA editing enzyme catalytic peptides 3 (Apobec3). In nonpermissive cells, these enzymes are incorporated into progeny virus particles via interactions with

trafficking pathway at the plasma membrane, leading to its internalization and delivery to lysosomes for degradation. Nef decreases cell surface expression of MHC class I molecules by a different pathway, mediating an interaction between the cytoplasmic domain of the MHC class I molecules and the clathrin adapter protein complex (AP-1).

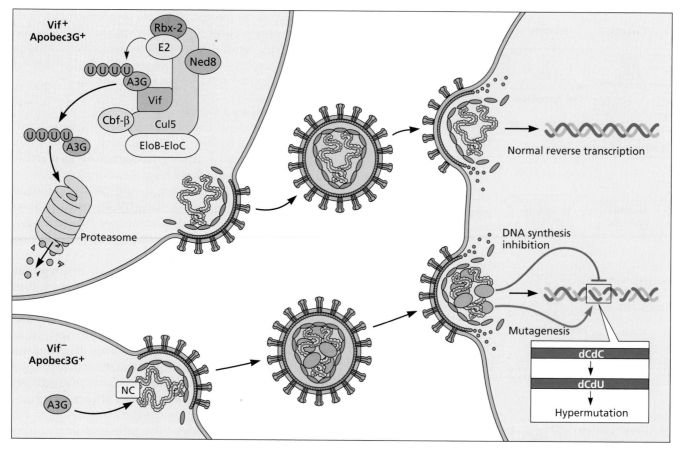

Figure 7.6 Mechanisms of action of Vif and Apobec3G. (Top) Vif counteracts the antiviral effects of Apobec3G (A3G) by mediating its polyubiquitination, which leads to proteosomal degradation. **(Bottom)** In the absence of Vif, A3G is incorporated into newly formed virus particles through interaction between viral RNA and NC protein. In the newly infected cell, reverse transcription is inhibited by A3G, and cytosines in the newly synthesized DNA are converted to uracil, causing hypermutation through eventual C to A transversions. Adapted from B. Cullen, *J Virol* **80:**1067–1076, 2006, with permission.

the viral RNA and possibly NC protein (Fig. 7.6). Apobec3G was the first family member to be identified as a Vif target. It was subsequently shown that Vif prevents its incorporation into virus particles by binding to the protein and inducing its depletion. In a reaction that requires binding of a transcriptional regulator, Vif assembles with additional cellular proteins to form an E3 ubiquitin ligase that recognizes Apobec3G as a substrate for polyubiquitination, a signal for its subsequent degradation in proteasomes (Fig. 7.7).

Apobec3G appears to inhibit virus reproduction in a number of ways. It has been proposed that binding to viral RNA may account, in part, for the deaminase-independent inhibition of reverse transcription in newly infected cells. The deaminase activity of Apobec3G leads to formation of deoxyuridine (dU), most frequently at preferred deoxycytidine (dC) sites in the first (−) strand of viral DNA to be synthesized by reverse transcriptase. Consequently, the (+) strand complement of the deaminated (−) strand will contain deoxyadenosine in

place of the normal deoxyguanosine at such sites (Fig.7.6). Indeed, the frequency of G→A transitions is abnormally high in the genomes of *vif*-defective particles produced in nonpermissive cells, and incomplete protection from Apobec3 proteins by Vif may explain why such transitions are the most frequent point mutations in HIV genomes. It has been suggested that the Apobec3 proteins represent an ancient intrinsic cellular defense against retroviruses (see Chapter 3).

Vpr protein. The 15-kDa viral protein R, Vpr, derives its name from the early observation that it affects the **rapidity** with which the virus reproduces in, and destroys, T cells. Most T cell-adapted strains of HIV-1 carry mutations in *vpr*. The SIV and HIV-2 genomes include a second, related gene *vpx*, which is discussed below and appears to have arisen as a duplication of *vpr*. The other lentiviruses do not contain sequences related to *vpr* but do include small open reading frames that might encode proteins with similar functions.

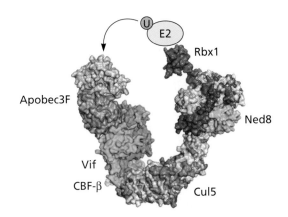

Figure 7.7. Molecular model for Apobec3F degradation by the Vif/Cbf-β ubiquitin ligase. Existing crystal structures of analogous proteins and complexes were used to model the Apobec3F degradation complex. In the model, Vif (green and blue) functions as a scaffold to assemble the E3 ligase by binding to Cbf-β (yellow and pink) and to Elo C and Cul5 (N-terminal domain). Activation of the E3 ligase is thought to occur by neddylation by Ned8 (lime green) of a lysine on the Cul5 (C-terminal domain). Subsequently, Rbx1 (magenta) adopts a conformation that facilitates polyubiquitination of Vif-bound Apobec3F (N-terminal domain, light orange and C-terminal domain, dark orange) via a cellular E2 ubiquitination enzyme. The model and Movie 7.1 was created by Drs. Nadine Shaban and Reuben Harris, University of Minnesota. For more details see J. S. Albin et al., A structural model of the APOBEC3F-Vif interaction informed by biological and computational studies. In preparation.

Vpr is incorporated into HIV-1 particles via specific interactions with a proline-rich domain at the C terminus of the Gag polyprotein. The host's uracil DNA glycosylase, Ung2, is then incorporated into particles by binding to Vpr. A substantial quantity, about 100 to 200 molecules of Vpr, is present in capsids. Its presence in virus particles is consistent with the observation that Vpr function is required at an early stage in the virus reproduction cycle.

Like Vif, Vpr functions as an adapter protein in an E3-ligase, but via interaction with a different member of the cullin family (Cul4, Fig. 7.5). Proteomic and biochemical studies have demonstrated that Vpr in this protein assembly facilitates recruitment of the cellular Slx4 structure-specific endonuclease regulator and untimely activation of the endonuclease activities that it controls. Such unleashed endonuclease activity could lead to the degradation of excess viral DNA, and it has been proposed that this feature may prevent detection by the intrinsic immune system (Chapter 3) and limit a defensive interferon response at early times after infection. Loss of endonuclease regulation at replication forks in the host DNA induces a damage response that may explain the G2/M arrest and apoptosis responses that are known to be triggered by Vpr. The possible advantage of preventing infected cells from entering mitosis is not apparent, especially as the requirement for Vpr function is most evident in HIV infection of macrophages, cells that do not divide. One idea is that the increased activity of the LTR promoters in the G_2 phase of the cell cycle may lead to enhanced virus production in the presence of Vpr.

In addition to cell cycle arrest and apoptosis, numerous other functions and interactions have been ascribed to this tiny protein, including modulation of the transcription of host and viral genes, maintenance of reverse transcriptase fidelity, recruitment of Ung2, and nuclear import of preintegration complexes in nondividing cells. Vpr has been shown to bind to nuclear pore proteins. Although these interactions are not essential for nuclear import, they may facilitate docking of the HIV-1 preintegration complex at the nuclear pore in preparation for import (Volume I, Chapter 5).

Vpx protein. Vpx is also packaged specifically via interaction with the Gag polyprotein. Vpx functions as an adapter that engages the same Cul4-E3 ubiquitin ligase as Vpr, but targets quite distinct proteins: Vpx recruits the dGTP-dependent deoxynucleoside triphosphohydrolase, SamHd1, for ubiquitination and proteosomal degradation (Fig. 7.5). SamHd1 blocks lentiviral DNA synthesis in myeloid cells by hydrolyzing cellular deoxynucleotide triphosphates to reduce concentrations to below those required for reverse transcription. The finding that Vpx can mediate degradation of this enzyme helped to explain why HIV-2 but not HIV-1 can propagate efficiently in macrophages or dendritic cells. Indeed, ectopic production of Vpx enhances HIV-1 infection in myeloid and CD4+ T cells, as does RNA interference-mediated knockdown of SamHd1.

Vpu protein. This small, 16-kDa viral protein is unique to HIV-1 and the related SIV_cpz (Fig. 7.3), hence the name viral protein U. The predicted sequence of Vpu includes an N-terminal stretch of 27 hydrophobic amino acids comprising a membrane-spanning domain and a cytoplasmic domain containing two α-helices. Biochemical studies show that Vpu is an integral membrane protein that self-associates to form oligomers. In infected cells, the protein is located on all major membranes, but is concentrated mainly in the endoplasmic reticulum, the *trans*-Golgi network, and endosomes.

Synthesis of Vpu is required for the proper maturation and targeting of progeny virus particles and for their efficient release. In its absence, particles containing multiple cores are produced, and budding is targeted to multivesicular bodies rather than to the plasma membrane (Fig 7.8). Vpu also reduces the syncytium-mediated cytopathogenicity of HIV-1, perhaps because the inefficient release of virus particles prevents the accumulation of sufficient Env protein at the cell surface to promote cell fusion.

A major function of Vpu in the pandemic group M strains of HIV-1 is to block the activity of a cellular membrane protein initially called bone stromal antigen 2 (Bst-2), but now more commonly known by the more descriptive name, tetherin. Tetherin is a dimeric type II membrane protein with an N-terminal cytoplasmic tail, a transmembrane region, and a C-terminal glycophosphatidyl inositol membrane anchor.

A   B

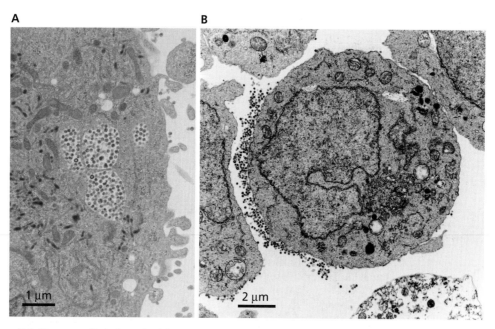

Figure 7.8 Human cells infected with an HIV-1 virus lacking Vpu. (A) Electron microscope image of a human macrophage showing intracellular accumulation of virus particles. **(B)** Electron microscope image of an infected H9 T cell showing accumulation of virus particles at the cell surface. Images courtesy of Drs. Jaang-Jiun Wang and Paul Spearman.

An interferon-inducible protein, tetherin restricts the propagation of enveloped viruses by inhibiting their release from infected cells (Fig. 7.9) and has been shown to act as a pattern-recognition receptor that induces Nf-κb-dependent proinflammatory gene expression in infected cells (Chapter 3).

Several residues in the transmembrane domain of Vpu interact directly with the transmembrane domain of tetherin. Vpu binds tetherin in the *trans*-Golgi network, inhibiting the transport of tetherin to the plasma membrane. Phosphorylation of the Vpr then leads to recruitment of another multiprotein E3 ubiquitin ligase, Scf, and the subsequent ubiquitinylation and degradation of tetherin in an endolysosomal pathway (Fig. 7.5). Vpu also facilitates the degradation of CD4: the viral protein traps newly formed CD4 receptor molecules in the endoplasmic reticulum, mediates ubiquitinylation by the Scf ubiquitin ligase, and entry of CD4 into the endoplasmic

reticulum-associated proteasome degradation pathway. Reducing the quantity of CD4 at the cell surface limits superinfection by HIV-1. It also reduces loss of Env protein via CD4 binding, thereby enhancing production of infectious particles.

Oligomerization of the membrane-spanning domains of Vpu is the basis of yet another property of Vpu, namely formation of an ion-conducting channel known as a **viroporin**, similar to that of the influenza A virus protein M2. While cell membrane integrity is disrupted and permeability to small molecules is increased when Vpr is produced in *Escherichia coli* or cultured mammalian cells, the relevance of Vpr viroporin activity to HIV pathogenesis is still unclear. Such activity could certainly affect the function of internal membranes, which are the major sites of Vpr accumulation. It has also been proposed that virus particle release may be enhanced by changes in the membrane potential across the budding plasma membrane.

Figure 7.9 Antagonism by viral proteins. Illustration of ways in which simian immunodeficiency virus (SIV) Nef protein **(A)**, HIV-1 Vpu **(B)**, and HIV-2 Env **(C)** target the cellular protein tetherin. Protein interactions are indicated by the double-headed arrows; Ap2, clathrin adapter protein complex-2.

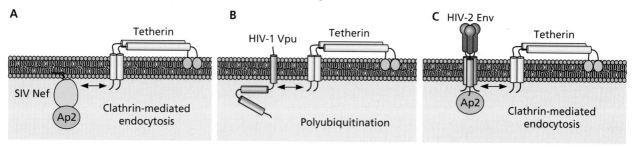

Genetic studies and experiments with a viroporin-specific inhibitor (BIT225) indicate that the ion channel and tetherin-antagonizing activities of the membrane-spanning domain of Vpu are independent of one another.

One might wonder: if tetherin antagonism is so important to HIV reproduction, why is Vpu only found in HIV-1? The answer is that different retroviral proteins have assumed this function during evolution (Fig. 7.9). The envelope protein of HIV-2 has evolved to include this function and the Nef proteins of several primate viruses are antagonists of the tetherin orthologs in their host species

Nef protein. Most laboratory strains of HIV-1, which have been adapted to grow well in T cell lines, contain deletions or other mutations in the *nef* gene. Restoration of *nef* reduces the efficiency of virus reproduction in these cells, hence the name "negative factor." Multiple functions have been attributed to Nef, and it is now clear that its importance may vary in different cell types.

Nef is translated from multiply spliced early transcripts. The 5′ end of Nef mRNA includes two initiation codons and, as both are utilized, two forms of Nef are produced in infected cells. The apparent size of these proteins can vary because of differences in posttranslational modification. Like Vpr, Nef is incorporated into virus particles via interaction with the Gag polyprotein. Nef molecules in virus particles appear to contribute to capsid disassembly following infection and may also enhance reverse transcription. The protein is myristoylated posttranslationally at its N terminus and thereby anchored to the inner surface of the plasma membrane.

Nef includes a protein-protein interaction domain (SH3) that mediates binding to components of intracellular signaling pathways, eliciting a program of gene expression similar to that observed after T cell activation. Such gene expression is thought to provide an optimal environment for viral reproduction. Among the best-studied, and clearly physiologically relevant activities of Nef is its downregulation of surface concentrations of CD4 and major histocompatibility complex (MHC) class I molecules (Fig. 7.5). The former activity is shared with Vpu. At the plasma membrane Nef binds to the cytoplasmic tail of CD4 and components of a clathrin-dependent trafficking pathway, leading to its internalization and delivery to lysosomes for degradation (Fig. 7.10, left). As with Vpu, the ensuing reduction in the number of CD4 molecules at the cell surface facilitates virus particle release and limits reinfection.

Nef decreases cell surface MHC class I molecules by a different pathway. It mediates interaction between the cytoplasmic domain of the MHC class I molecules and the clathrin adapter protein complex (AP-1) in the *trans*-Golgi network, prior to their transport to the cell surface (Fig. 7.10, right). The Nef-induced complex is retained in this Golgi compartment and MHC class I molecules are subsequently diverted to lysosomes for degradation. As a strong cytotoxic T lymphocyte (CTL) response against viral infection requires

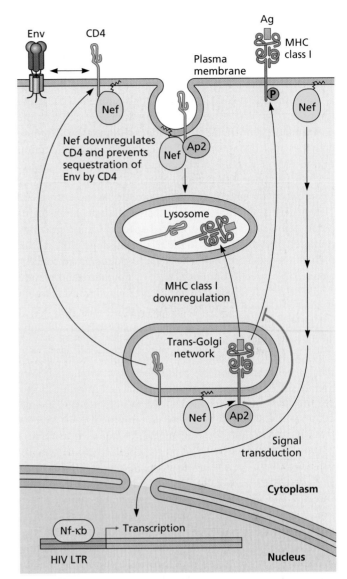

Figure 7.10 Intracellular functions attributed to Nef. Nef is myristoylated posttranslationally; the jagged protrusion represents myristic acid covalently linked to the glycine residue at position 2. Myristoylation enables Nef to attach to cell membranes, where it can interact with membrane-bound cellular proteins. Nef reduces the cell surface expression of CD4 by binding to sequences in the cytoplasmic domain of this receptor and enhancing clathrin-dependent endocytosis and the subsequent degradation of CD4 within lysosomes; Ap2, clathrin adapter protein complex-2. This activity reduces interaction of CD4 with surface Env proteins, thereby enhancing Env incorporation into budding virus particles. In contrast, MHC class I expression on the cell surface is reduced by Nef binding in the membrane of the *trans*-Golgi network. This interaction interferes with the normal vesicular sorting required for passage of the receptor to the cell surface, and MHC class I is directed to the lysosome for degradation. Nef also affects signal transduction by increasing the activity of the cellular transcriptional activator Nf-κb and perhaps other cellular transcription proteins.

recognition of viral epitopes presented by MHC class I molecules, this inhibitory activity of Nef allows infected cells to escape lysis by CTLs and is probably a major factor contributing to HIV-1 pathogenesis. Nef induces decreased concentrations of a number of other cell surface molecules including a component of the T cell receptor complex (CD3), the lymphocyte-specific protein tyrosine kinase (Lck), and the costimulatory molecule for T cell activation (CD28). These activities of Nef contribute to T cell activation and recognition of infected cells by cells of the immune system (Chapter 4).

Other activities ascribed to Nef also seem likely to contribute to pathogenesis in important ways. One example is Nef-mediated inhibition of endocytosis of the type II transmembrane lectin, DC-Sign (dendritic-cell-specific, Icam-3-grabbing nonintegrin), which binds to the HIV-1 envelope protein with high affinity. Such inhibition by Nef leads to **increased** concentration of this lectin on the surfaces of immature and mature dendritic cells. DC-Sign facilitates dendritic cell transmigration through the vascular and lymphoid endothelium, as well as the adhesion of these cells to T cells during antigen presentation. Studies with dendritic cells infected with HIV-1 have shown that such cells form many more clusters with activated primary T cells than dendritic cells infected with a *nef* − HIV-1 mutant. As dendritic cells can retain attached infectious virus particles for several days, the increase in surface accumulation of DC-Sign cell surface concentration induced by Nef may be important for both viral spread and transmission to T cells.

Although the initial cell culture experiments suggested a negative effect on virus production, subsequent experiments with animals showed that Nef augments HIV pathogenesis quite significantly. Rhesus macaques inoculated with a Nef-defective mutant of SIV had low virus titers in their blood during early stages of infection, and the later appearance of high titers was associated with reversion of the mutation. More importantly, adult macaques inoculated with a virus strain containing a deletion of *nef* did not progress to clinical disease and were, in fact, immune to subsequent challenge with wild-type virus. The observation that *nef* had been deleted in HIV-1 isolates from some individuals who remained asymptomatic for long periods also suggests that this viral protein can contribute to pathogenesis. Initial hopes that intentional deletion of *nef* might facilitate the development of a vaccine strain for humans were dashed when it was discovered that the humans infected with *nef* deletion mutants eventually developed AIDS.

The Viral Capsid Counters Intrinsic Defense Mechanisms

Following entry of HIV-1, capsid proteins remain associated with the reverse transcription machinery (Volume I, Chapter 7). This subviral structure moves through the cytoplasm to the nuclear pore via interaction with the host cell cytoskeletal fibers, as viral DNA is synthesized and an integration-competent nucleoprotein assembly is formed. Genetic and biochemical studies have identified two host proteins that bind to the HIV-1 capsid protein, Cpsf 6 (a cleavage and polyadenylation factor) and CypA (the peptidyl-prolyl isomerase cyclophilin A). Such binding imparts stability to the capsid structure and helps to suppress its premature disassembly by host proteins such as Trim5α and TrimCypA. The stabilized capsid structure also shields viral nucleic acids from detection by intrinsic immune sensors in the cytoplasm, such as double-stranded RNA helicase, Rig-I (viral RNA), and the cyclic GMP-AMP synthase, cGAs (viral DNA) (Chapter 3). The biological importance of these capsid protein interactions early in infection is emphasized by the finding that HIV-1 mutants with capsids that are either fragile, or abnormally stable, are replication defective.

Capsid protection of viral nucleic acids may account for the observation that there is little or no interferon response at early times after HIV infection. At late times, however, when large quantities of viral RNAs are produced, there is a robust interferon response. One interferon-induced human protein, Mx2, was found to be a potent inhibitor of HIV-1 reproduction in certain cell types, particularly macrophages. Mx2 also targets the HIV-1 capsid, inhibiting nuclear import of viral DNA by a mechanism yet to be discovered. Mx2 derives its name from its close sequence relationship to the myxovirus resistance 1 protein (Mx1), which is a broadly acting inhibitor of RNA and DNA viruses, including the orthomyxovirus influenza A virus (Chapter 3). However, these proteins have distinct activities as Mx1 is not an inhibitor of HIV-1, and Mx2 is ineffective against influenza A virus.

Cellular Targets

Attachment and entry into host cells depend on the interaction between viral proteins and cellular receptors (Volume I, Chapter 5). While the major receptor for the HIV envelope protein is the cell surface CD4 molecule, the envelope protein must also interact with a coreceptor to trigger fusion of the viral and cellular membranes and gain entry into the cytoplasm. The two major coreceptors for HIV-1 are the α- and β-chemokine receptors, CXCr4 and CCr5. Strains of HIV that bind to CXCr4 or CCr5 are commonly referred to as X4 and R5 strains, respectively (Fig. 7.11). For reasons that are still not completely understood, R5 viruses are transmitted preferentially during infection. X4 viruses predominate in the late stages, following extensive evolution of the virus population within an infected individual and concomitant with immune system breakdown (Box 10.5). The importance of these two chemokine receptors to HIV pathogenesis is demonstrated by two findings. People who carry a particular mutation in the gene encoding CCr5 produce a defective receptor protein and are resistant to HIV-1 infection. So too are individuals who carry a mutation in the gene for the ligand of CXCr4

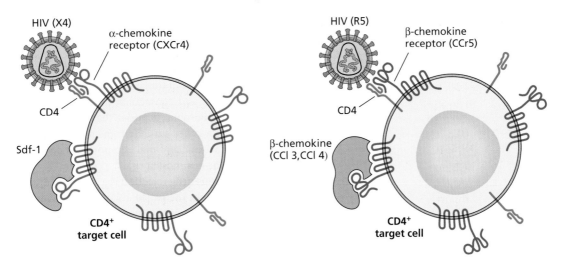

Figure 7.11 Coreceptors for HIV-1. CxCr4 is the coreceptor for HIV-1 variants that predominate during the late phase of infection; entry of such strains (denoted X4) is inhibited by the coreceptor's natural ligand, Sdf-1. CCr5 is the coreceptor for HIV-1 variants that predominate early in infection (denoted R5); their entry is inhibited by the coreceptor's natural ligands, CCl5 (Rantes), and the macrophage inflammatory proteins CCl3 and CCl4 (Mip-1α and Mip-1β). Primary T cells and monocytes can produce both coreceptors. Adapted from Fig. 3 of A. S. Fauci, *Nature* **384:**529–533, 1996, with permission.

(Table 3.7). The latter mutation may lead to increased availability of the ligand, which then blocks virus entry by competing for coreceptor binding. This idea is consistent with earlier observations that chemokine binding to the receptors inhibits the infectivity of specific strains of HIV in cell culture (Fig. 7.11). Cells of the hematopoietic lineage that bear CD4 and one or more of these chemokine receptors are the main targets of infection, and they produce the highest titers of progeny virus particles.

Several additional chemokine receptors have been identified as coreceptors for HIV and SIV in cell culture experiments, but their roles in natural infection remain to be determined. These additional coreceptors may allow the virus to enter a broader range of cells than first appreciated. Some are found on cells of the thymus gland and the brain, and some could play a role in infection in infancy or of cells in the central nervous system. It has also been proposed that binding to these additional coreceptors may trigger signals that affect virus reproduction in target cells, or that harm nonpermissive cells, producing a "bystander" effect.

Experiments in cell cultures have identified additional mechanisms by which HIV may enter cells. For example, the virus can be transmitted very efficiently through direct cell contact. In addition, cells may be infected by virus particles that are endocytosed after binding to cell surface galactosyl ceramide or to Fc receptors (as antibody-virus complexes). HIV can infect many different types of human cells in culture and has been found in small quantities in several tissues of the body. As discussed below, infection of these cells and tissues is likely to be relevant to HIV-1 pathogenesis.

Routes of Transmission

Even before HIV-1 was identified, epidemiologists had established the most likely routes of the agent's transmission to be sexual contact, blood exchange, and from mother to child. As might be anticipated, the efficiency of transmission is influenced greatly by the concentration of the virus particles in the body fluid to which an individual is exposed. Estimates of the percentage of infected cells and the concentration of HIV-1 in different body fluids indicate that highest quantities are observed in peripheral blood monocytes, in blood plasma, and in cerebrospinal fluid (Table 7.2), but semen and female genital secretions also appear to be important sources of the virus.

Other routes of transmission are relatively unimportant or nonexistent. Among these are nonsexual physical contact, exposure to saliva or urine from infected individuals, and exposure to blood-sucking insects. Fortunately, HIV-1 infectivity is reduced upon air-drying (by 90 to 99% within 24 h), by heating (56 to 60°C for 30 min), by exposure to standard germicides (such as 10% bleach or 70% alcohol), or by exposure to pH extremes (e.g., <6 or >10 for 10 min). This information and results from epidemiology studies have been used to establish safety regulations to prevent transmission in the public sector and the health care setting.

Modes of Transmission

Modes of HIV transmission vary in different geographic locations and among different populations within the same locations (Fig. 7.12). In the United States, the major overall route is via homosexual contact, specifically among men who have sex with men. Heterosexual contact is the predominant manner in

Table 7.2 Isolation of infectious HIV-1 from body fluids[a]

Fluid	Virus isolation[b]	Estimated quantity of virus[c]
Cell-free fluid		
Cerebrospinal fluid	21/40	10–10,000
Ear secretions	1/8	5–10
Feces	0/2	None detected
Milk	1/5	<1
Plasma	33/33	1–5,000[d]
Saliva	3/55	<1
Semen	5/15	10–50
Sweat	0/2	None detected
Tears	2/5	<1
Urine	1/5	<1
Vaginal-cervical	5/16	<1
Infected cells		
Bronchial fluid	3/24	Not determined
PBMC	89/92	0.001–1%[d]
Saliva	4/11	<0.01%
Semen	11/28	0.01–5%
Vaginal-cervical fluid	7/16	Not determined

[a]Adapted from Table 2.1 (p. 28) of Levy JA, *HIV and the Pathogenesis of AIDS,* 3rd ed. ASM Press, Washington, DC, 2007.

[b]Number of samples positive/number analyzed.

[c]For cell-free fluid, units are infectious particles per milliliter; for infected cells, units are percentages of total cells capable of releasing virus. Results from studies in the laboratory of J. A. Levy are presented.

[d]High levels associated with acute infection and advanced disease ($\sim 5 \times 10^6$ PBMCs/ml of blood).

which the virus is transmitted in other parts of the world and also among females in the United States. Transmission of HIV from an infected to an uninfected person is generally characterized as being relatively inefficient. For example, the probability is reported to range from 0.005 to 0.0001 per heterosexual contact. However, a single interaction can be sufficient for transmission if the infected partner is highly viremic. The presence of other sexually transmitted diseases also increases the probability of transmission, presumably because infected inflammatory cells and mediators may be present in both seminal and vaginal fluids. The likelihood of transmission can also be increased by genital ulceration and consequent direct exposure to blood cells, and by antimicrobial peptides that may be produced by cells infected with other sexually transmitted viruses (Box 7.3). In both heterosexual and male homosexual contact, the recipient partner is the one most at risk.

Intravenous drug injection is a common mode of transmission throughout the world, owing to the widespread practice of sharing contaminated needles. Here again, the probability of transmission is a function of the frequency of exposure and the degree of viremia among a drug user's contacts. Of course, sexual partners of drug users are also at increased risk.

Until 1985, when routine HIV antibody testing of donated blood was established in the United States and other industrialized countries, individuals who received blood transfusions or certain blood products, such as clotting factors VIII and IX, were at high risk of becoming infected. Transfusion of a single unit (500 ml) of blood from an HIV-1-infected individual nearly always led to infection of the recipient. Appropriate heat treatment of clotting factor preparations and, more recently, *ex vivo* production of these proteins by modern biotechnology methods, have eliminated transmission from this source. Fortunately, other blood products, such as pooled immunoglobulin, albumin, and hepatitis B vaccine, were not implicated in HIV-1 transmission, presumably because their production methods include steps that destroy the virus.

Transmission of HIV from mother to child can occur across the placenta (5 to 10%) or, more frequently, at the time of delivery as a consequence of exposure to a contaminated genital tract (ca. 20%). The virus can also be transmitted from infected cells in the mother's milk during breast-feeding (ca.15%).

Figure 7.12 Modes of transmission of HIV in the United States and worldwide. Data for the United States includes adults and adolescents in the mainland and six dependent areas, and was compiled in 2011 by the U.S. Centers for Disease Control and Prevention. Data for the other areas of the world are from the World Heath Organization (WHO) AIDS Epidemic Update of December 2006.

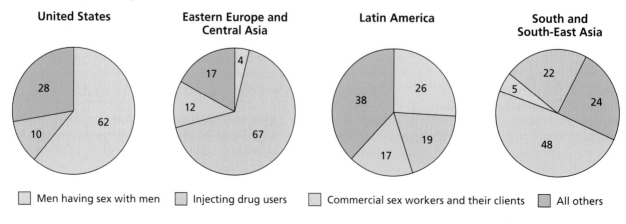

BOX 7.3

BACKGROUND
Antimicrobial peptides induced by herpesvirus can enhance HIV-1 infection

The risk of being infected with human immunodeficiency virus type 1 (HIV-1) is enhanced substantially in individuals with other sexually transmitted diseases. For example, infection with the genital herpes simplex virus type 2 doubles the risk of acquiring HIV-1. Explanations for this increased risk include direct inoculation of HIV-1 into the blood through herpesvirus-induced genital ulcers and the recruitment of inflammatory cells that are targets for infection by HIV-1. The results of infections carried out in skin explants and cultured cells suggest another mechanism for the enhancement of HIV-1 infection by herpes simplex virus type 2.

Langerhans (LC) and other dendritic cells patrol the mucosal epithelium, taking up and processing antigens and presenting them to T cells. Such dendritic cells are believed to be one of the first to encounter HIV-1 after sexual exposure (Fig. 7.13). Whether they can support HIV-1 reproduction in a natural infection is not yet clear. However, it was shown that LCs can be infected with HIV-1 in human skin explants, and coinfection with herpes simplex virus type 2 increased the number of HIV-1 infected LCs substantially. This observation could not be explained by infection of the same cell by both viruses because very few such doubly infected cells were observed. Furthermore, application of virus-free supernatant from herpesvirus-infected cultured cells led to an increase in the number of monocyte-derived, cultured LCs (mLCs) that were infected with HIV-1. These results suggested that the herpesvirus-infected cells produce one or more substances that can increase the efficiency of HIV-1 infection of LCs.

Human epithelial and epidermal cells are known to produce antimicrobial peptides such as defensins and cathelicidin. These are short, evolutionarily conserved peptides that inhibit the growth of bacteria, viruses, and fungi. Herpes simplex virus type 2-infected keratinocytes produce a number of such peptides, but the most important in the current context is LL-37. This peptide was shown to increase the concentration of the HIV-1 receptors, CD4 and CCr5, on the surface of LCs. Furthermore, removing LL-37 from the supernatant of herpesvirus-infected cells reduced its ability to stimulate HIV-1 infection of LCs.

It is unclear what, if any, advantage upregulation of these surface molecules on LCs might confer on herpes simplex virus type 2, and caution is needed when extrapolating results from cultured cells or even tissue explants, to a natural infection. Nevertheless, the proposed mechanism provides one plausible way in which HIV-1 infection may be enhanced by herpes simplex virus type 2. The proposal is supported by the observation that elevated levels of LL-37 correlate with HIV-1 infection in sex workers.

Ogawa Y, Kawamura T, Matsuzawa T, Aoki R, Gee P, Yamashita A, Moriishi K, Yamasaki K, Koyanagi Y, Blauvelt A, Shimada S. 2013. Antimicrobial peptide LL-37 produced by HSV-2-infected keratinocytes enhances HIV infection of Langerhans cells. *Cell Host Microbe* 13:77–86.

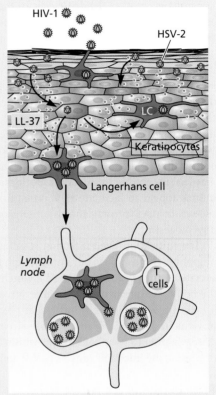

Model for human genital herpes simplex virus 2 (HSV-2) stimulation of HIV-1 infection. Upon exposure to HSV-2, infected keratinocytes (green cells) release antimicrobial peptides, including LL-37 (blue dots). LL-37 stimulates the upregulation of HIV-1 receptors CD4 and CXCr5 in Langerhans cells (LCs). It is proposed that such upregulation augments the efficiency with which LCs disseminate HIV-1 to CD4⁺ T cells in the local lymph nodes.

Without antiviral drug intervention, rates of transmission from an infected mother to a child range from as low as 11% to as high as 60%, depending on the severity of infection (concentration of virus particles present) in the mother and the prevalence of breast-feeding (frequency of the infant's exposure). Administration of antiviral drug therapy during pregnancy is a very effective measure for decreasing the frequency of transmission to newborns. Pediatric HIV infection is no longer a major public health problem in high-income countries where infected pregnant women are treated with combinations of anti-HIV drugs. However, even a single treatment with one antiviral drug early in labor can reduce the incidence significantly. In lower-income countries such treatment (with AZT or nevirapine) is becoming increasingly more common, and the number of HIV-infected infants is decreasing (Chapter 9 and Fig 9.22). Unfortunately, in some areas, testing facilities and drug treatment remained unavailable as recently as 2012, when an estimated 300,000 newborns were infected with HIV in lower-income countries where the AIDS burden is still high.

Mechanics of Spread

Except in cases of direct needle sticks or blood transfusion, HIV enters the body through mucosal surfaces, as do most viruses (Chapter 2). In the case of sexual transmission, a likely source is virus-infected cells, as they can be present in much larger numbers than free virus particles in vaginal or seminal fluids. Results from various studies, including observations of SIV infection of macaques, indicate that partially activated CD4$^+$ cells of the genital mucosa in the recipient partner are the initial targets of infection (Fig. 7.13). Infection may be facilitated by interaction with antigen-presenting CD4$^+$ Langerhans or other dendritic cells in the vaginal and cervical epithelia, which can capture infectious virus particles and transport them to target CD4$^+$ T cells (Fig. 7.14).

Activated T cells, prevalent in genital lesions caused by infections, are also likely targets of the virus. Although the insertive partner is at relatively low risk for infection, transmission to the male can occur through cells in the lining of the urethral canal of the penis, presumably from infected cells in the cervix or the gastrointestinal mucosa of the infected partner. Uncircumcised males have a twofold-increased risk of infection, suggesting that the mucosal lining of the foreskin may be susceptible to infection. Both male and female hormones appear to facilitate HIV transmission by stimulating cell-cell contact (prostaglandins) or erosion of the vaginal lining (progestin).

The initial, localized infection is followed within days by migration of the virus via draining lymph nodes to the gut-associated lymphoid tissue (GALT), which contains ~40% of the body's lymphocytes, and, where susceptible, CCr5 CD4$^+$ T cells are abundant. The subsequent explosive production of virus in the GALT leads to acute viremia and widespread dissemination to lymphoid organs within 10 to 20 days. From this point, the infection runs its protracted course.

The Course of Infection
The Acute Phase
In the first few days after infection, virus particles are produced in large quantities by the activated lymphocytes in lymph nodes, sometimes causing the nodes to swell (lymphadenopathy) and/or producing flulike symptoms. Particles released into the blood can be detected by infectivity of appropriate cell cultures or by screening directly for viral RNA or proteins (Fig. 7.15). As many as 5×10^3 infectious particles or 1×10^7 viral RNA molecules (i.e., ~5×10^6 particles) per ml of plasma can be found at this stage. During this time, some 30 to 60% of CD4$^+$ T cells in the gut are destroyed, either directly or as a result of bystander effects. The associated loss of mucosal integrity leads to the translocation of microbial products into the circulation and disruption of metabolic and digestive functions (Figure 7.16).

A percentage of quiescent memory T cells in the GALT can survive, because replication-competent proviruses cannot be transcribed in these cells. These cells then form a long-lived, latent viral reservoir. Interaction of such memory cells with their cognate antigens, sometimes many years after the initial HIV infection, will lead to their activation and subsequent transcription of the latent provirus. If antiviral treatments

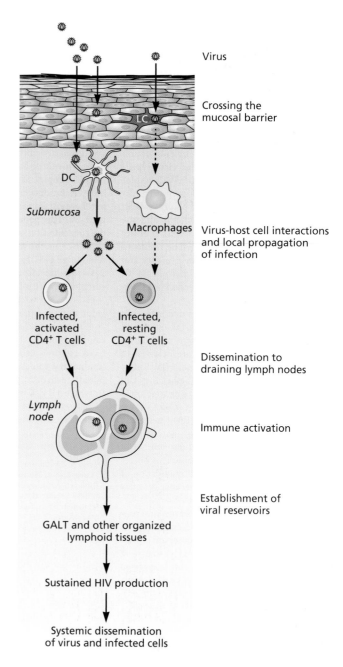

Figure 7.13 Mechanics of viral spread. Infection begins with transmission across a mucosal barrier either by a free virus particle (via transcytosis or following physical abrasion), an infected cell, or particles attached to dendritic (DC) or Langerhans (LC) cells. Virus propagation in partially activated CD4$^+$ T cells is followed by the transfer of virus particles to draining lymph nodes and to the gastrointestinal-associated lymphoid tissue (GALT) where massive propagation of the virus occurs (see Fig. 7.22). The virus is then disseminated to other lymphoid tissues, with the establishment of stable viral reservoirs and latently infected cells. Virus production and concomitant destruction of epithelial barriers in the GALT lead to sustained immune activation and continued dissemination of virus and infected cells.

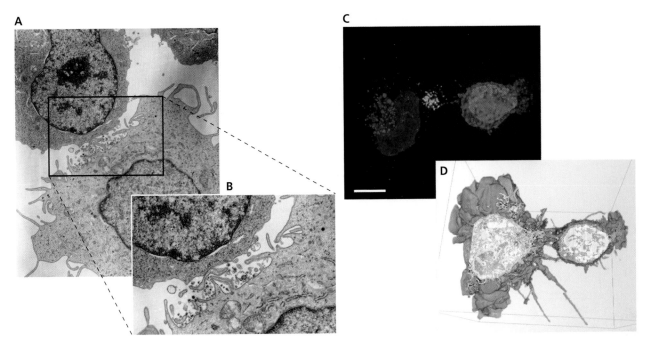

Figure 7.14 HIV particles in a virological synapse between mature dendritic cells and susceptible T cells. (A) Large numbers of HIV particles are concentrated at the mature dendritic–T cell junction (dendritic cell is at bottom right). **(B)** The inset shows a higher-magnification image of the boxed area. Electron micrographic images in A and B were produced by Clive Wells and provided through the courtesy of Li Wu, Medical College of Wisconsin. Reprinted from J.-H. Wang et al., *J Virol* **81:**8933–8943, 2007, with permission. **(C)** Fluorescence micrograph of a CD4 T cell (red) extending membranes into a pocket in a dendritic cell in which HIV-1 particles (green) are accumulated. Nuclei are depicted in blue, and the dendritic cell membrane stain is not shown for clarity. **(D)** Electron microscopic image of the cells in C. A midcell section is shown with half of the data rendered in volume (Voltex) view. The cells are displayed semitransparently to allow a view of the interior structures. These images indicate that virus particles can be stored within invaginated membrane "pockets" that remain contiguous with the surface of the dendritic cell. T cells appear to extend membranous projections into the pocket, where stored virus particle may be captured by binding to the CD4 receptor on the T cell. Images in C and D were prepared by K. Olszens and D. McDonald, and kindly provided by Dr. McDonald (Ohio State University). See also Movie 7.2.

have not been continued, the resulting progeny particles can initiate a new round of infection.

The initial peak of viremia is greatly curtailed within a few weeks after initial infection, as the susceptible T cell population is depleted and a cell-mediated (CTL) immune response is mounted. The number of CTLs increases before neutralizing antibodies can be detected. The inflammatory response that occurs upon primary infection stimulates the production of additional CD4$^+$ T cells, stemming the depletion of this population. Consequently, the CD4$^+$ T cell count returns to near normal levels, but these cells represent a new source of susceptible targets and their infection produces chronic immune stimulation.

During the period of acute infection, the virus population is relatively homogeneous. In most cases, the predominant virus in the infected individual is a minor variant in the population present in the source of the infection (Box 10.5). For example, in approximately 80% of infections resulting from heterosexual contact, the infection was found to originate from a single viral genome. The reason for this apparent genetic bottleneck, and selective transmission, is a topic of intense investigation, as it has great bearing on vaccine development.

The Asymptomatic Phase

By 3 to 4 months after infection, viremia is usually reduced to low levels (Fig. 7.15), with small bursts of virus particles appearing from time to time. It is known that the degree of viremia at this stage of infection, the so-called **virologic set point**, is a direct predictor of how fast the disease will progress in a particular individual: the higher the set point, the faster the progression. During this period, CD4$^+$ T cell numbers decrease at a steady rate, estimated to be approximately 60,000 cells/ml/year. Cytopathogenicity induced by the virus and apoptosis due to continued immune stimulation and inappropriate cytokine production seem likely explanations. In this protracted asymptomatic period, which can last for years, the CTL count remains slightly elevated, but virus reproduction continues at a low rate, mainly in the lymph

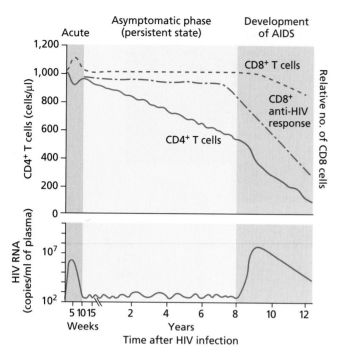

Figure 7.15 Schematic diagram of events occurring after HIV-1 infection. Adapted from Fig. 13.1 of J. A. Levy, *HIV and the Pathogenesis of AIDS*, 3rd ed. (ASM Press, Washington, DC, 2007).

nodes. In lymphoid tissues, a relatively large, stable pool of virus particles bound to the surface of follicular dendritic cells can be detected. Small numbers of infected T cells are also observed. During this phase of persistent infection, also known as **clinical latency**, only 1 in 300 to 400 infected cells in the lymph nodes may actually release virus particles. It is thought that, as in acute infection, virus propagation is suppressed at this stage by the action of antiviral CTLs. The number of these specific lymphocytes decreases toward the end of this stage. During the asymptomatic phase, viral genetic diversity is increased as a consequence of continuous, positive

selection for mutants that can evade the host's immune responses (Chapter 4).

The Symptomatic Phase and AIDS

The end stage of disease, when the infected individual develops symptoms of AIDS, is characterized by vastly increased quantities of virus particles and a CD4+ T cell count below 200 cells per ml (Fig. 7.15, Fig. 7.16). The total CTL count also decreases, probably because of the precipitous drop in the number specific for HIV. In the lymph nodes, virus reproduction increases, with concomitant destruction of lymphoid cells and of the normal architecture of lymphoid tissue. The cause of such lymph node degeneration is not clear; it may be the result of virus reproduction, indirect effects of chronic immune stimulation, or both. It has been proposed that chronic antigenic stimulation, which induces rapid turnover and differentiation in the various lymphocyte populations of an infected individual, ultimately culminates in progressive loss of their regenerative potential.

In this last stage, the virus population again becomes relatively homogeneous and specific for the CXCr4 receptor. Properties associated with increased virulence predominate, including an expanded cellular host range, ability of the virus to cause formation of syncytia, rapid reproduction kinetics, and CD4+ T cell cytopathogenicity. Late-emerging virus also appears to be less sensitive to neutralizing antibodies and more readily recognized by antibodies that enhance infectivity. In some cases, strains with enhanced neurotropism or increased pathogenicity for other organ systems emerge. In some cases, these changes have been traced to specific mutations, for example, in the viral envelope gene or in a regulatory gene (e.g., *tat*).

Variability of Response to Infection

Studies of large cohorts of HIV-1-infected adults show that in the absence of antiviral therapy, approximately 10% progress to AIDS within the first 2 to 3 years of infection.

Figure 7.16 Pathological conditions associated with HIV-1 infection.

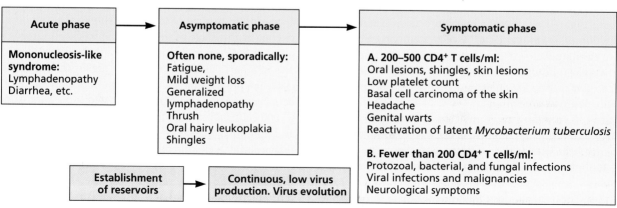

Over a period of 10 years, approximately 80% of untreated, infected adults will show evidence of disease progression and, of these, 50% will have developed AIDS. Of the remainder, 10 to 17% are AIDS free for over 20 years; a very small percentage of these individuals are completely free of symptoms, with no evidence of progression to disease. What are the parameters that contribute to such variability?

One is the degree to which an individual's immune system may be stimulated by infection with other pathogens. HIV-1 reproduces most efficiently in activated T cells, and it is known that virus concentrations increase when the immune system is activated by opportunistic infections with other microorganisms. Such activation can explain the fact that HIV-1 disease is generally more aggressive in sub-Saharan Africa, where chronic infection by parasites and other pathogens is frequent. As might be expected for an outbred population, variations in an individual's genetic makeup can modulate the immune response to infection and affect survival (Chapter 1). Genetic differences in chemokines or chemokine receptors and, probably, in any one of several components of the immune system can have an impact on the course of the disease. Genomic analyses have identified specific polymorphisms that are associated with variations in viral load in infected individuals, and, as might be expected, major histocompatibility alleles are among the genes affected.

Clearly, accumulation of mutations in the genomes of the virus also influences the course of the disease. As noted above, some long-term survivors of HIV-1 infection harbor viruses with deletions in the *nef* gene. Others appear to be infected with differently attenuated strains that produce low titers in cells in culture and have restricted cell tropism. The presence of virus-neutralizing antibodies, and the absence of virus-enhancing antibodies, are other characteristics of these infections. The fortunate individuals who are nonprogressors or long-term survivors of HIV-1 infection have been under intensive study: such investigations can provide a better understanding of the critical correlates of viral pathogenesis, and may suggest new strategies for effective prevention or therapy.

Origins of Cellular Immune Dysfunction

The defining feature of HIV disease is impaired immune cell function. This defect is eventually devastating because immune defenses are vital in the body's battle against the virus as well as other pathogens. At first, the immune system keeps the HIV infection in check. However, the virus is not eliminated, and the infection that persists in the asymptomatic stage leads to increasing dysfunction among immune cells. In the end, most AIDS patients actually succumb to opportunistic infections with microorganisms that are little threat to individuals with healthy immune systems. Impairment in immune cell function results from both direct attack of the virus on particular cell types and the responses of uninfected cells to viral gene products or specific proteins made in infected or stimulated cells.

CD4+ T Lymphocytes

The major target of HIV, the CD4+ T cell, is a critical regulator of the adaptive immune response (Chapter 4). Even before the profound depletion of these cells, which is a signal of end-stage disease, abnormalities in CD4+ T cell function can be detected in the peripheral blood of HIV-infected individuals. These abnormalities include decreased ability to form colonies when grown in tissue culture, decreased expression of the cytokine interleukin-2 (IL-2) and its receptor, and reduced proliferative responses to various antigens (Fig. 7.17). The precise causes of these impairments are unknown. The MHC class II interactions required for antigen-specific responses in infected individuals may become less frequent, because the surface concentration of CD4 is reduced by some viral proteins in infected cells (SU, Vpu, and Nef). Furthermore, noninfectious virus particles and viral proteins (e.g., Tat or SU) shed from infected cells can bind

Figure 7.17 Immune cell dysfunction associated with AIDS.

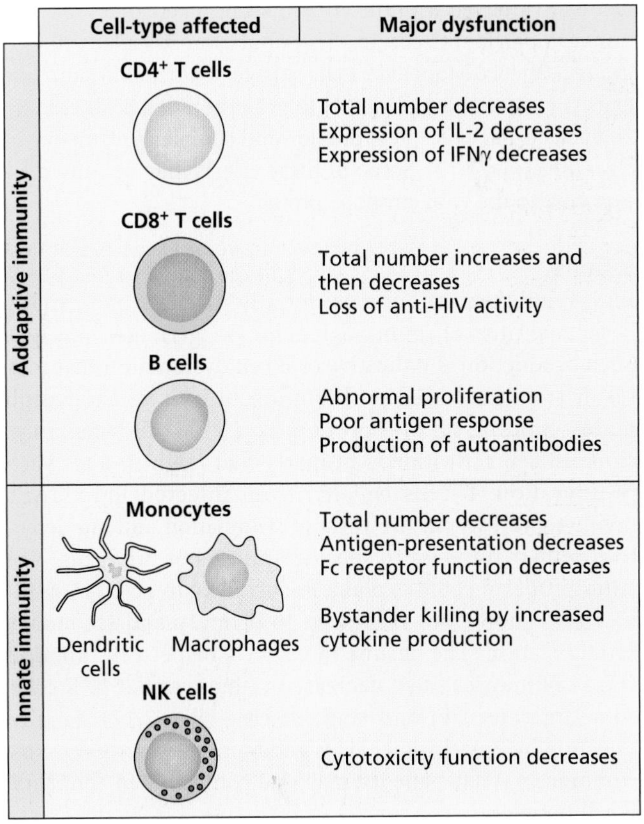

Cell-type affected	Major dysfunction
Adaptive immunity — CD4+ T cells	Total number decreases Expression of IL-2 decreases Expression of IFNγ decreases
CD8+ T cells	Total number increases and then decreases Loss of anti-HIV activity
B cells	Abnormal proliferation Poor antigen response Production of autoantibodies
Innate immunity — Monocytes / Dendritic cells / Macrophages	Total number decreases Antigen-presentation decreases Fc receptor function decreases Bystander killing by increased cytokine production
NK cells	Cytotoxicity function decreases

to or enter uninfected CD4$^+$ T cells, triggering inappropriate (bystander) responses, such as the inhibition of synthesis of IL-2 and its receptor by SU. It is also possible that SU bound to CD4 on uninfected T cells interferes sterically with their interaction with MHC class II molecules. Finally, changes in cytokine production by HIV-1-infected macrophages can trigger programmed cell death in uninfected CD4$^+$ T cells.

Cytotoxic T Lymphocytes

The number of CTLs in the blood is abnormally high following the acute phase and decreases precipitously during the end stage of the disease (Fig. 7.17). The early increase may be the result of an imbalance brought about as the immune system attempts to achieve homeostasis of CD8$^+$ and CD4$^+$ cells while CD4$^+$ cells are being destroyed. The reduced numbers of anti-HIV-1 CTLs at late times can be explained in part by the direct infection and killing of their progenitors ("double positive" CD4$^+$ CD8$^+$ T cells). Additionally, and most importantly, because CTL proliferation and function depend on inductive signals from CD4$^+$ T cells (Fig. 4.7), the decline of the CD4$^+$ population also contributes to CTL dysfunction.

Monocytes and Macrophages

HIV-1-infected macrophages can be detected readily in tissues throughout the body of an infected individual. However, as only a small proportion of monocytes/macrophages in the blood are infected with the virus, it seems likely that the functional impairment seen in this population of cells is due to indirect effects. Monocyte/macrophage abnormalities include defects in chemotaxis, inability to promote T cell proliferation, and defects in Fc receptor function and complement-mediated clearance (Fig. 7.17). Some of these effects may be caused by exposure to the viral envelope protein.

B Cells

HIV-1-infected individuals initially produce abnormally large quantities of immunoglobulin G (IgG), IgA, and IgD. Such production is indicative of B cell dysfunction that may result from increased proliferation of cells of the lymph nodes. Binding of viral proteins (e.g., TM) induces polyclonal B cell activation, a property that might induce such proliferation. B cells isolated from infected individuals divide rapidly in culture without stimulation and are defective in their response to specific antigens or mitogens. The latter property could explain, in part, why infected individuals also show poor responses to primary and secondary immunization. The decline in CD4$^+$ T helper cell function (Fig. 4.7) may lead to a decrease in total number of B cells, sometimes seen in end-stage disease (Fig. 7.17). Finally, infection by Epstein-Barr virus and human cytomegalovirus, common in AIDS patients, may also contribute to abnormal B cell function.

Natural Killer Cells

Impairment of natural killer (NK) cell function is observed throughout the course of infection, becoming more severe during end-stage disease (Fig. 7.17). The reduction in NK cell function cripples the innate immune response to infection by HIV and other microorganisms. As NK cell cytotoxicity depends on IL-2, these abnormalities may be a consequence of impaired CD4$^+$ T cell function and the reduced production of this cytokine.

Autoimmunity

Because of the imbalance in the immune system, it is not surprising that immune disorders, such as a breakdown in the system's ability to distinguish self from nonself, accompany infection. In early studies, antibodies against platelets, T cells, and peripheral nerves were detected in AIDS patients. Subsequently, autoantibodies to a large number of normal cellular proteins have been found in infected individuals. The specific reason for the appearance of such antibodies is not clear, but their production might be stimulated in part by cellular proteins on the surface of viral particles or by viral proteins, regions of which may resemble cellular proteins (molecular mimicry) (Chapter 5).

Immune Responses to HIV

Innate Response

The rate of reproduction of HIV in the acute phase of infection is often reduced before induction of the adaptive immune response, suggesting that the innate immune system plays an early role in antiviral defense (Chapter 3). Recognition of viral components by dendritic cells is associated with the release of type I interferons and Tnf-α, robust induction of additional cytokines, and the inhibition of virus reproduction in infected cells. An array of additional cells, including phagocytes (e.g., macrophages) and cytolytic (e.g., NK) cells respond to this cytokine cascade and participate in the destruction of infected cells and the capture of viral antigens that can be presented to the adaptive immune system. The finding that dendritic cells from females produce larger quantities of IFN-α than do those from males may explain, in part, the fact that women generally show a lower viral set point than men.

The Cell-Mediated Response

Antigen-specific cellular immune responses include activities of CTLs and T-helper cells. The great majority of CTLs are CD8$^+$ and their general role in limiting or suppressing viral reproduction is discussed in Chapter 4. CTLs can also exert a suppressive effect by producing antiviral chemokines and small peptides called defensins that inhibit transcription of viral genes.

CTLs programmed to recognize virtually all HIV-1 proteins have been detected in infected individuals. There is a direct correlation among a good CTL response, low virus load, and slower disease progression. Furthermore, a broadly reactive response appears to correlate with a less fulminant course of disease, and end-stage disease is characterized by a rapid drop in the number of anti-HIV-1 CTLs. Results of studies with severe combined immunodeficient (SCID) mice that have been reconstituted with human lymphoid cells show that adoptive transfer of human anti-HIV-1 CTLs provides some protection against subsequent challenge with the virus. These findings all demonstrate a significant role for CTLs in fighting HIV-1 disease.

Humoral Responses

Antibodies to the infecting strain of HIV-1 can be detected generally within 1 to 3 months after acute infection. These antibodies, which are secreted into the blood and are present on mucosal surfaces of the body, can be detected in various body fluids. This phenomenon has been exploited in the design of home kits for detecting anti-HIV antibodies in the blood or urine. Among the various isotypes, IgG1 antibodies are known to play a dominant role at all stages of infection, giving rise to an antibody-dependent cellular cytotoxicity response (called ADCC), complement-dependent cytotoxicity, and neutralizing and blocking responses (Fig. 7.18; Chapter 4).

Neutralizing antibodies that can block viral infection of susceptible cells may contribute to limiting viral reproduction during the early, asymptomatic stage of infection. However, the titers of these antibodies are generally very low and, as such, may favor selection of resistant mutants. Indeed, many individuals produce antibodies that neutralize earlier virus isolates but not isolates present at the time of serum collection. Some studies show loss of neutralizing antibodies with progression to AIDS, but the clinical relevance of this change during the later stages of infection remains obscure.

Neutralizing antibodies generally bind to specific sites on the viral envelope protein (Fig. 7.18). Antibodies that bind to epitopes on the variable region 3 (V3) in HIV-1 SU can be detected early in infection. Due to the high sequence variation within the V3 loop (hence the name "variable"), such antibodies are usually strain specific. Consequently, despite its relatively strong antigenicity, V3 is not a good target for the development of vaccines or broadly specific antiviral drugs. Some neutralizing antibodies bind to conserved sites on SU or TM and hence can react with many strains of HIV-1; broad neutralizing activity against carbohydrate-containing regions of the viral envelope protein has also been detected.

Some antibodies (**interfering antibodies**) can bind to virus particles or infected cells and block interaction with neutralizing antibodies (Fig. 7.18, left). Others (**enhancing antibodies**) can actually facilitate infection by allowing particles coated with them to enter susceptible cells (Fig. 7.18, right). In complement-mediated antibody enhancement, the

complement receptors Cr1, Cr2, and Cr3 provide a critical function in attaching such complexes to susceptible cells. In Fc-mediated enhancement, attachment is via Fc receptors that are abundant not only on monocytes/macrophages and NK cells but also on other human cell types. As HIV-1 has been shown to replicate in cells that lack CD4 but produce an Fc receptor, binding to the CD4 receptor is probably not required for Fc-mediated enhancement. It is noteworthy that the same cellular receptors (for complement and Fc) are implicated in infection enhancement and ADCC responses (Fig. 7.18, right). In the case of enhancement, the receptors allow antibody-coated virus particles to enter susceptible cells bearing such receptors. In the case of ADCC, such receptors on CTLs, NK cells, or monocytes/macrophages mediate the recognition for subsequent killing of antibody-coated infected cells.

As both neutralizing and enhancing antibodies recognize epitopes on SU and TM, it has been difficult to identify the features that specify the response. Indeed, polyclonal antibodies against SU possess both neutralizing and enhancing activities. The clinical importance of antibody-dependent

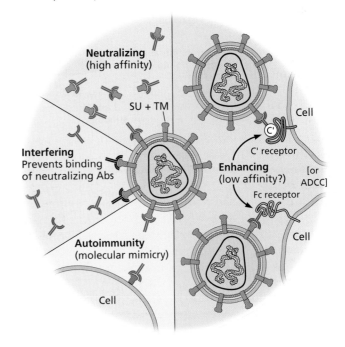

Figure 7.18 Antibody (Ab) responses to HIV infection. A summary of the various responses described in the text is presented. One idea is that the relative affinities of the antibodies may be critical determinants of the response. According to this hypothesis, high-affinity antibodies neutralize infectivity by binding tightly to SU, causing it to detach from the virus particles (**Top Left**); low-affinity antibodies bind to SU but not tightly enough to cause its detachment (**Right**). Conformational changes in SU that might occur as a consequence of such low-affinity binding would then facilitate viral entry. C′, complement; ADCC, antibody-dependent cellular cytotoxicity.

enhancement is uncertain, but the fact that circulating infectious virus-antibody complexes have been described suggests that this phenomenon may contribute to HIV-1 pathogenesis. In addition, infectivity-enhancing antibodies have been demonstrated in individuals who progress to disease. Results from studies of other lentiviral infections, as well as infections with other viruses (dengue viruses, coronaviruses, and others), have established a correlation between increased symptoms of disease and increased quantities of enhancing antibodies. These observations, together with the finding that enhancing antibodies have been found in individuals who generated immune memory to Env protein from a different HIV-1 strain, certainly complicate strategies for the development of an effective vaccine.

The discovery that strong **broadly neutralizing antibody** responses do develop in a subset of individuals after primary infection has stimulated efforts to develop an effective vaccine against HIV. The establishment of high-throughput assays has allowed the screening of sera from large numbers of HIV-infected individuals and identification of several potent and broadly neutralizing antibodies. Cocrystal structures of such antibodies and their viral envelope epitopes have been determined (Fig. 7.19). One type binds to a conserved site within the V2-V3 region of SU, and another to a conserved region in SU that interacts with the CD4 receptor. The TM portion of the HIV-1 envelope is also a target for some of these antibodies. It is hoped that these and other newly identified, broadly neutralizing human antibodies will be useful in the development of effective vaccination strategies. However, these high-affinity molecules appear only years after infection and an extended period of antigen exposure. Furthermore, extensive somatic mutation is apparently required to achieve their effective potency. Consequently, vaccination strategies will need to consider not only the nature/structure of the immunogen, but also the number of exposures required to reach the appropriate antibody affinity (see Chapter 8).

Summary: the Critical Balance

HIV-1 reproduction is controlled by what may be thought of as a finely balanced scale that can be tipped in either direction by a number of stimulatory or inhibitory host proteins. Among these, various cytokines have important but opposing

Figure 7.19 Structure of the HIV-1 envelope trimer and binding of broadly neutralizing human antibodies. An image of the viral spike obtained by cryo-electron microscopy (light gray) is shown with superimposed atomic-level ribbon models (red) for three portions of the HIV-1 envelope glycoprotein (Env): The membrane-proximal external region (MPER) of the transmembrane portion, TM (gp41), is toward the top of the image, the core of SU (gp120) with intact amino- and carboxyl termini is in the middle of the image, and the V1/V2 domain is toward the bottom. Antibodies that effectively neutralize HIV-1 target primarily four specific regions in Env: the MPER, which is bound by antibody 10E8 (cyan); the CD4-binding site on SU, which is targeted by antibody VRC01 (fuchsia); and two sites of *N*-linked glycosylation, one of which is in the V1/V2 region at residue Asn160 and is bound by antibody PG9 (green), and the other which is a glycan V3 epitope that generally includes residue Asn332 and is bound by antibody PGT128 (blue). From Figure 1 in P. D. Kwong et al., Broadly neutralizing antibodies and the search for an HIV-1 vaccine: the end of the beginning. *Nat Rev Immunol* **13:**693–701, 2013, with permission.

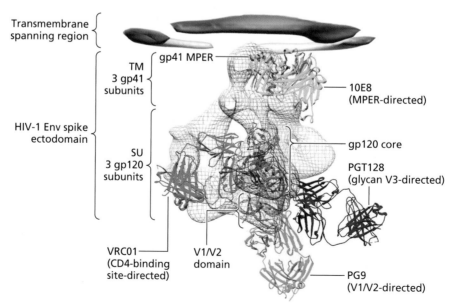

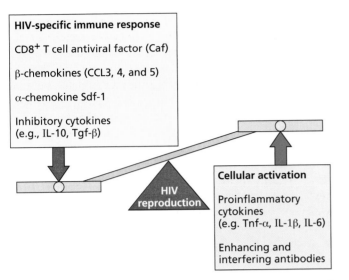

Figure 7.20 Control of HIV-1 reproduction and the progress of HIV-induced disease by the balance of host proteins. Cellular activation and proinflammatory cytokines stimulate viral reproduction. These stimulatory effects are counterbalanced by inhibitory proteins. Adapted from Fig. 4 of A. S. Fauci, *Nature* **384:**529–533, 1996, with permission.

effects. Immune responses and the production of specific chemokines can inhibit virus reproduction, whereas immune cell activation and certain antibodies can be stimulatory (Fig. 7.20). The challenge has been to identify practical interventions that tip the balance in the patient's favor.

Dynamics of HIV-1 Reproduction in AIDS Patients

The availability of potent drugs that block HIV-1 reproduction by inhibiting the activity of the viral enzymes reverse transcriptase and protease made it possible to measure the dynamics of virus production in humans. Clinical studies performed with patients near end stage, whose CD4$^+$ T cell counts are in decline, have revealed the magnitude of the battle between the virus and the immune system in HIV-1 disease. Within the first 2 weeks of treatment with a combination of these drugs, an exponential decline in viral RNA in the plasma was observed, followed by a second, slower decline. The initial drop represented clearance of free virus and loss of virus-producing CD4$^+$ lymphocytes from the blood. The most important contributor to the second drop is presumed to be the loss of longer-lived infected cells, such as tissue macrophages and dendritic cells, with a minor but lingering contribution from the clearance of latently infected, nonactivated T lymphocytes. The existence of the latter population, comprising infected CD4$^+$ T cells that have returned to a quiescent state (i.e., memory T cells), is the major barrier to eradication of the virus by treatment with antiviral drugs.

At steady state in the absence of drugs, the rate of virus production must equal the rate of virus clearance. Mathematical analyses of the data from clinical studies can therefore provide estimates of the rates of HIV-1 appearance in the blood and other compartments of the body, as well as the rate of loss of virus and virus-infected cells. The results are nothing less than astonishing. The minimal rate estimated for release into the blood is on the order of 10^{10} virus particles per day. This minimal number computes to approximately 1 cycle per infected cell per day. Continuous high-reproduction capacity is undoubtedly the principal engine that drives viral pathogenesis at this stage. Because of the high mutation rate of HIV-1, on average every possible change at every position in the genome is predicted to occur numerous times each day. It has therefore been estimated that the genetic diversity of HIV produced in a single infected individual can be greater than the worldwide diversity of influenza virus during a pandemic. This enormous variation and the continuous onslaught of infection must present a colossal challenge to the immune system.

More than 90% of the virus particles in the blood come from infected activated CD4$^+$ lymphocytes that have average half-lives of only ca. 1.1 days (Fig. 7.21). A smaller percentage, approximately 1 to 7%, comes from longer-lived cells in other compartments, with half-lives from 8.5 to 145 days. Consequently, even if *de novo* synthesis of virus could be blocked **completely** by drug treatment, it would take approximately 3 to 5 years before these longer-lived compartments were free of cells with the potential to produce virus. Sadly, this can be considered only a minimal estimate. Complete eradication will not be possible until proviruses are eliminated from long-lived quiescent memory T cells and cells residing in "sanctuary" compartments that are not readily accessible to drugs, such as the brain.

The other dramatic change that takes place in patients treated with antiviral drugs is a resurgence of the CD4$^+$ lymphocyte count in the blood. From the initial rates of recovery, it has been calculated that, during an ongoing infection, as many as 4×10^7 of these cells are replaced in the blood each day. Lymphocytes in the peripheral blood comprise a relatively small fraction (ca. 1/50) of the total in the body, and lymphocyte trafficking, homing, and recirculation are complicated processes. It is still uncertain whether such CD4$^+$ lymphocyte replacement following drug treatment represents new cells, or simply redistribution from other compartments. If one assumes that the increase in the circulation is proportional to the total, then as many as 2×10^9 new CD4$^+$ T cells are produced each day. This estimate is controversial because it seems to exceed significantly the normal proliferative capacity of these cells. However, some studies suggest that HIV-1-infected individuals who are treated with potent drug combinations do produce new CD4$^+$ T cells at rates higher

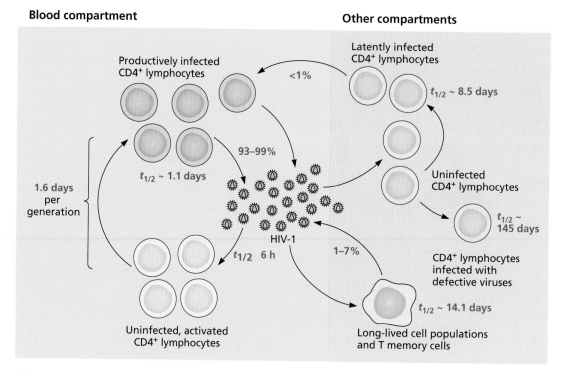

Blood compartment

Other compartments

Figure 7.21 Summary of kinetics of HIV-1 production in the blood and other compartments. The percentages indicate the relative quantities of virus particles calculated to be produced in blood plasma by the various cell populations illustrated. The average time in days for 50% of the cells in each population to be destroyed or eliminated is indicated as the half-life ($t_{1/2}$). The average time in hours (h) for 50% of the particles to be eliminated from the plasma ($t_{1/2}$) is also shown. Adapted from Fig. 1 of D. D. Ho, *J Clin Investig* **99**:2565–2567, 1997, with permission.

than normal. Perhaps increased rates of both synthesis and redistribution contribute to the observed resurgence in CD4$^+$ T cells in the periphery.

Large-scale sequencing studies have established that subpopulations of cells in the blood of patients receiving antiviral therapy bear identical sites of HIV-1 provirus insertion. Furthermore, integrations into genes associated with cancer and cell cycle control are overrepresented in these subpopulations. These observations suggest that some latently infected cells may be driven to proliferate via insertional activation of growth-promoting genes by HIV proviruses.

Effects of HIV on Different Tissues and Organ Systems

Lymphoid Organs

Most of the critical steps in HIV pathogenesis occur not in the blood but in lymphoid tissues, which contain the majority of the body's lymphocytes (Chapter 4). Many take place within the first few weeks of infection and occur in tissues such as the GALT of the intestinal mucosa and lymph nodes. The function of these tissues is to retain invading microbes, such as HIV, and to present them to immunocompetent cells. Studies in primate model systems indicate that, within 2 weeks after infection,

about 90% of CD4$^+$ T cells are depleted from the GALT, which contains the majority of all lymphocytes and macrophages in the body. Immune dysfunction in the intestine results in structural damage to the mucosa and breakdown of the epithelial barrier (Fig. 7.22). As a result, microbes and microbial antigens enter the bloodstream. Their passage through the now "leaky gut" stimulates release of inflammatory cytokines promoting sustained and systemic activation of the host response, which is a major source of the acquired immune deficiency. Intestinal mucosal tissues appear to be a primary site of HIV-1 persistence, even with antiviral drug treatment.

By 2 weeks after infection in primate models, virus is widely distributed in all lymphoid organs including lymph nodes, spleen, and thymus. Most HIV-1 particles in the lymph nodes of infected humans are trapped within the germinal centers that comprise networks of follicular dendritic cells with long interdigitating processes that surround lymphocytes. Follicular dendritic cells also trap antibodies and complement and present antigens to B cells. The subsequent activation of resident and recruited CD4$^+$ T cells makes them permissive for HIV-1 reproduction. Eventually, infection of macrophages is also observed. Lymph nodes appear to contain a much larger percentage of virus-infected cells than does the peripheral blood.

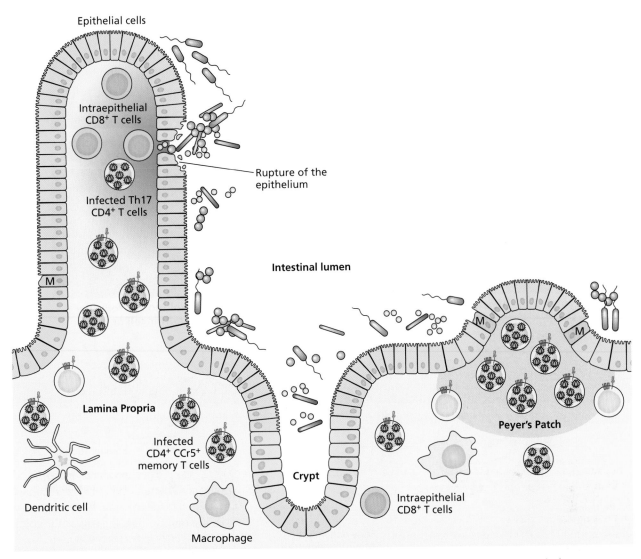

Figure 7.22. Effects of HIV-1 infection of the intestinal mucosa. The intestinal immune system is the largest immunological organ in the body. Terminally differentiated CD4$^+$ T memory cells in both the diffuse and organized (Peyer's patches) gastrointestinal-associated lymphoid tissue are massively depleted within just days after initial infection with HIV. T cells (Th17), which are essential for maintaining the integrity of the intestinal mucosa and defending against pathogens in the intestinal lumen, are also depleted. Epithelial cells are damaged in the wake of this T cell destruction, allowing entry of diverse luminal pathogens, sustained stimulation of the immune response, and persistent inflammation. Adapted from A. A. Lackner et al., HIV Pathogenesis: the host. Figure 2, p. 13. *Cold Spring Harb Perspect Med* **2:**a007005, 2012, doi: 10.1101/cshperspect.a007005, with permission. For an instructive, animated view of the GALT, see the following link: http://link.brightcove.com/services/player/bcpid1966016696001?bckey=AQ~~,AAAByWTdmvk~, YEX2I6TuT0mdQPquhJg1bWcq9Ufv7FQ_&bclid=0&bctid=2144234478001

Early in infection the germinal centers in lymph nodes appear to remain intact, although there is some proliferation of activated immune cells (Fig. 7.23). At the beginning of the asymptomatic stage, in addition to intestinal problems and diarrhea, infected individuals often have palpable lymphadenopathy at two or more sites as a result of follicular dendritic cell hyperplasia and capillary endothelial cell proliferation. Later, during an intermediate stage of disease (CD4$^+$ T cell counts of 200 to 500 per ml), the nodes begin to deteriorate, there is evidence of cell death, and the trapping efficiency of the follicular dendritic cells declines. At a more advanced stage of disease (<200 CD4$^+$ cells per ml), the architecture of the lymphoid tissue is almost completely destroyed and the follicular dendritic cells disappear (Fig.7.23).

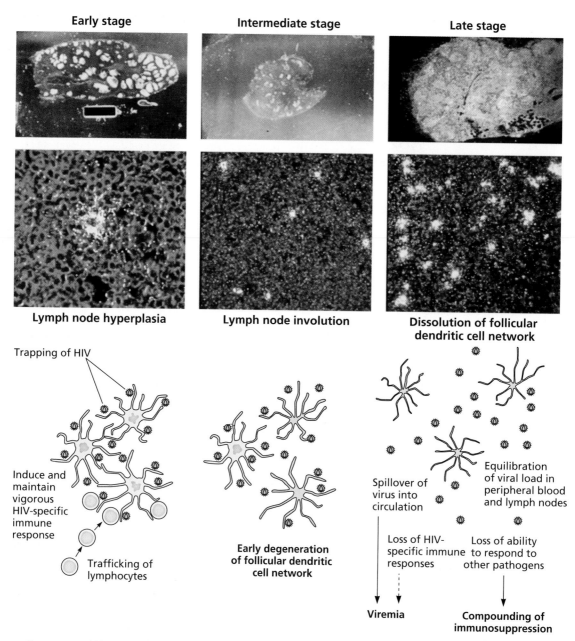

Figure 7.23 Effects of HIV-1 infection on lymphoid tissue. (Top) Shows changes in lymph node germinal centers, as determined by selective staining (above) and by the location of viral replication, as blue-white dots, in lymph node tissue (below). HIV-1 nucleic acid was detected by polymerase chain reaction (PCR). The examples illustrate conditions in the early and late stages of HIV infection when connective tissue replaces much of the normal cell population. Reprinted from J. A. Levy, *HIV and the Pathogenesis of AIDS*, 3rd ed. (ASM Press, Washington, DC, 1998). **(Bottom)** Illustrates events that take place in lymph node germinal centers during various stages of HIV-1 disease (see text). Adapted from Fig. 10 of A. S. Fauci and R. C. Desrosiers, p. 587–635. *In* J. M. Coffin et al. (ed), *The Retroviruses* (Cold Spring Harbor Laboratory Press, Cold Spring Harbor, NY, 1998), with permission.

The Nervous System

HIV-1 can be detected in the spinal fluid of affected individuals early after infection. In the absence of antiviral therapy, nearly two-thirds of all HIV-1-infected individuals ultimately develop AIDS dementia. The disease progresses slowly over a period of up to 1 year, but mean survival time from the onset of severe symptoms is less than 6 months. Several AIDS-associated opportunistic infections can also produce neurological damage. A substantial proportion of patients on optimal antiviral therapy also exhibit neurological

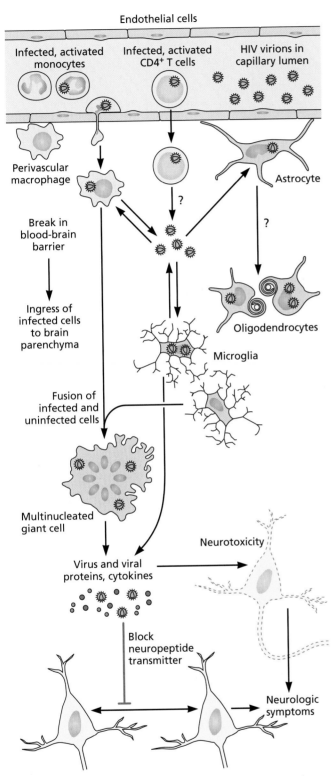

Figure 7.24 HIV-1 neuropathogenesis. It is proposed that infected activated macrophages or T cells may transfer HIV-1 across the blood-brain barrier, although infected CD4$^+$ T cells may also enter via transcytosis. The infected cells can produce virus particles that then infect microglial cells. Microglial cell fusion results in the formation of

multinucleated giant cells, which is a hallmark of HIV neuropathology. Astrocytes are affected by cytokines from infected cells, but are not thought to support viral propagation. Production of virus particles and proteins, and the release of various cytokines and other cellular products from infected cells, may lead to an interruption of neuronal cell-to-cell transmission by blocking the production of neurotropic factors. High concentrations of virus proteins (e.g., SU, TM, Tat, and Nef) and cytokines could lead to direct neurotoxicity through detrimental effects on the cell membrane.

abnormalities (called HIV-associated neurocognitive disorder, or simply HAND).

In general, brain-derived isolates of HIV-1 bind to the CCr5 receptor and reproduce effectively in cultured macrophages. This observation, and the fact that cells of the macrophage lineage in the brain (microglia) are routinely found to contain viral proteins and RNA, suggest that HIV-1-infected monocytes may be the main vehicle for virus transmission to the central nervous system. However, other potential sources including infected CD4$^+$ T cells or even free virus particles have not been excluded (Fig 7.24). The entry of virus particles and infection of macrophages and microglial cells triggers an inflammatory response and, eventually, neuronal cell destruction. As HIV-1 does not appear to infect neurons, viral reproduction in these cells is unlikely to explain their loss. It is more probable that the release of toxic cellular products and viral proteins (e.g., SU, TM, or Tat) from infected macrophages or microglial cells is responsible for the damage to neurons as well as astrocytes (Fig. 7.25). There is ample evidence that these viral proteins could contribute to neuropathogenesis. Transgenic mice expressing HIV-1 *env* under the control of neuronal promoters show abnormalities in astrocytes and neuronal processes similar to those seen in HIV-1-infected individuals. It has also been proposed that the systemic activation of macrophages caused by microbial translocation through the leaky gut predisposes these cells to invade the perivascular spaces in the central nervous system.

The Gastrointestinal System

Infection of the GALT leads to disruption of gastrointestinal epithelia early after infection. The more advanced stages of HIV disease are often associated with severe damage to the gastrointestinal system. Diarrhea and chronic malabsorption, with consequent malnourishment and weight loss, are frequently observed. In Africa, this condition has been called "slim disease." In some cases, the disorders are associated with opportunistic infections with other microbial agents. Nevertheless, in cases where no opportunistic agent can be identified, HIV-1 itself is a likely primary cause of gastrointestinal pathogenesis.

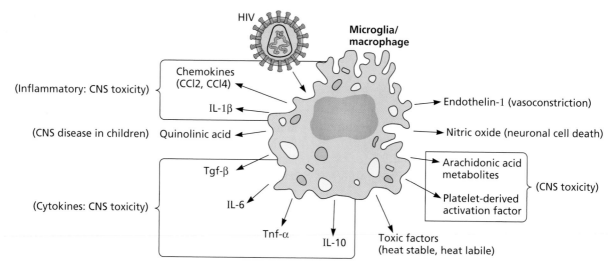

Figure 7.25 The role of infected microglia/macrophages in neural pathogenesis. After exposure to HIV or to viral envelope proteins, microglia/macrophages may be induced to produce cytokines and the macrophage inflammatory chemokines that can be toxic to cells of the central nervous system (CNS).

Other Organs and Tissues

HIV has been found in the lungs of patients with pneumonia, in the hearts of some with heart muscle dysfunction (cardiomyopathy), in the kidneys of some with renal injury, and in the joint fluid of patients with arthritis. It also has been identified in the adrenal glands of infected individuals. The contribution of the virus to these pathologies is not clearly understood. Opportunistic infection of the lungs is common; *Pneumocystis jiroveci,* a yeast-like fungus that is usually dormant in the host lung, causes pneumonia in approximately 50% of AIDS patients. Other microorganisms, most notably *Mycobacterium tuberculosis, Mycobacterium avium,* and human cytomegalovirus, may also cause pulmonary infections. The known effects of SU on membrane permeability or other toxic effects of viral proteins might explain some of the lung damage and the electrophysiological abnormalities associated with heart disease. Direct infection of endothelial cells or other cells in the kidneys has been proposed as a potential cause of tubular destruction. Deposition of antigen-antibody complexes could also account for some kidney damage.

HIV-1 is also found in the male and female genital tracts and in the breasts. There is evidence of genetic compartmentalization of the virus isolated from these locations, suggesting that virus may reproduce at these sites.

HIV and Cancer

HIV-1 infection leads to an increased incidence of neoplastic malignancies: some form of cancer eventually occurs in approximately 40% of untreated infected individuals. The mechanism of oncogenesis in this case is quite different from that of other retroviruses (Chapter 6). HIV-1 kills its major target cell, rather than promoting the immortalization and unrestrained proliferation typical of oncogenesis. HIV-associated malignancies arise from the indirect effects of deregulation of the host's immune system. Contributory factors probably include the absence of proper immune surveillance directed against other (oncogenic) viruses or transformed cells. High levels of cytokine production associated with HIV infection might induce inappropriate proliferation of uninfected cells and promote the generation of blood vessels (angiogenesis) in developing tumors. Indeed, cancers that develop in HIV-infected individuals generally are more aggressive than those in uninfected people. These malignancies can develop in a number of tissues and organs, but certain types, such as Kaposi's sarcoma and B cell lymphoma, are especially prevalent. It may be important that in these cases the neoplastic cells are derived from the immune system and the endothelial cells thought to give rise to Kaposi's sarcoma can act as accessory cells in lymphocyte activation. One reasonable hypothesis is that proliferation of endothelial cells, B cells, and the epithelial cells that give rise to carcinomas may be promoted by cytokines produced by immune cells. As discussed in this section, many malignancies that develop in HIV-1-infected individuals are associated with infection by oncogenic viruses.

Kaposi's Sarcoma

Kaposi's sarcoma was first described by the Hungarian physician Moritz Kaposi in 1872. It is a multifocal cancer in that the lesions contain many cell types; the dominant type is a called a spindle cell, thought to be of endothelial origin. The tumors contain infiltrating inflammatory cells and many newly formed blood vessels. Kaposi's sarcoma was typically found in older men from the Mediterranean region and Eastern

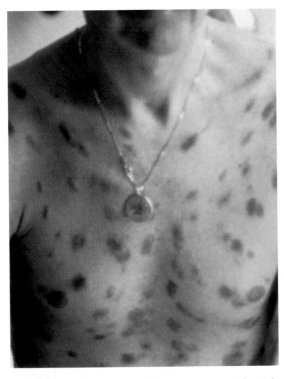

Figure 7.26 Kaposi's sarcoma in a young man infected with HIV-1. Note the distribution of the lesions, suggesting lymphatic involvement. Reprinted from Color Plate 19 of J. A. Levy, *HIV and the Pathogenesis of AIDS*, 3rd ed. (ASM Press, Washington, DC, 2007). Photo courtesy of P. Volberding.

Europe. In these areas, Kaposi's sarcoma normally appears in a nonaggressive (classical) form confined to the skin and extremities, and is rarely lethal. The classical form, as well as a more aggressive and sometimes lethal form, is found in sub-Saharan Africa, where there are more immunocompromised individuals. In HIV-1-infected men, Kaposi's sarcoma appears in the aggressive form, affecting both mucocutaneous and visceral areas. This disease occurs in about 20% of HIV-1-infected homosexual men (Fig. 7.26) but in only about 2% of HIV-1-infected women, transfusion-infected recipients, and blood product-infected hemophiliacs in the United States.

Spindle cell cultures established from Kaposi's sarcoma tumors are not fully transformed according to the criteria defined in Chapter 6, but they do produce a variety of proinflammatory and angiogenic proteins. It is thought that these products are responsible for recruiting other cell types in these tumors. Spindle cells from AIDS patients are also not infected with HIV-1, and the disease occurs in individuals who are not infected with the virus. Consequently, HIV cannot be the sole factor in the development of Kaposi's sarcoma. Epidemiologic studies suggested that another sexually transmitted virus was the inducing agent. Subsequently, a new member of the gammaherpesviruses, called human herpesvirus 8, that can infect spindle cells was found to be associated with Kaposi's sarcoma. The results of *in situ* hybridization studies show RNA transcripts from this virus in the vast majority of Kaposi's sarcoma lesions, irrespective of the presence or absence of HIV-1. Human herpesvirus 8 can also infect B cells and has been linked to certain AIDS-associated B cell lymphomas (Fig. 7.27).

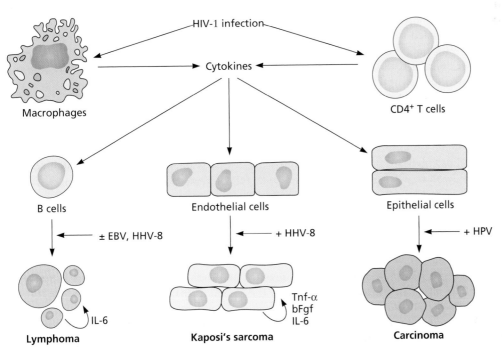

Figure 7.27 Induction of cancers in HIV-1 infection. Infection of macrophages, CD4$^+$ T cells, or other cells leads to the production of cytokines that can enhance the proliferation of certain other cells, such as B cells, endothelial cells, and epithelial cells. The enhanced proliferation of these cells, as a result of either cytokine production or subsequent viral infection, could lead to development of the malignancies noted. In some cases, such as B cell lymphomas and Kaposi's sarcoma, ongoing cytokine production by the tumor cells maintains the malignant state. Adapted from Fig. 12.12 of J. A. Levy, *HIV and the Pathogenesis of AIDS*, 3rd ed. (ASM Press, Washington, DC, 2007).

Antibodies to human herpesvirus 8 can be found in up to 6% of the general population, but in 25% of the population in areas of endemic classical infection, indicating widespread exposure. As very few people develop this disease, other parameters must be important in its etiology. The immune deficiency associated with HIV-1 is certainly one explanation for its prevalence in AIDS patients and synthesis of viral proteins, such as Tat, may be another. It has been reported that transgenic expression of Tat in mice causes a disease that resembles Kaposi's sarcoma. Although the human disease is certainly caused by herpesvirus 8, Tat contributes to the aggressive nature of this malignancy in AIDS patients by promoting the growth of spindle cells in the Kaposi's sarcoma lesions. In addition, the human herpesvirus 8 genome includes a number of open reading frames with sequence homology to genes encoding cellular proteins known to be important in growth control, cell signaling, and immunoregulation (Fig. 6.13).

It is notable that one of the first reports of AIDS in 1981 described seven patients with Kaposi's sarcoma. While HIV-1 was identified soon afterward, the discovery of human herpesvirus 8 did not occur until many years later, even though both viruses were present and contributed to the pathology in these first patients.

B Cell Lymphomas

B cell lymphomas are 60 to 100 times more common in AIDS patients than in the general population. The incidence is especially high among patients whose survival has been prolonged by anti-HIV-1 drugs. Tumors can be found in many locations, including lymph nodes, the intestine, the central nervous system, and the liver. B cell lymphomas in the peritoneal or other body cavities are almost always associated with human herpesvirus 8. Both polyclonal and monoclonal B cell lymphomas are found in the central nervous system, with monoclonal types being more common. Epstein-Barr virus is found in all AIDS-associated primary lymphomas in the brain. Why these two B cell lymphoma-inducing herpesviruses show such site specificity remains unknown. On the other hand, approximately 60% of the tumors outside the brain show no evidence of infection with either Epstein-Barr virus or human herpesvirus 8, indicating that B cell transformation in HIV-1-infected individuals does not require infection with such herpesviruses.

Lymphomas may arise because of the destruction of germinal centers in the lymphatic system. Lysis of antigen-presenting follicular dendritic cells would render B cells less sensitive to normal apoptotic processes, allowing them to live longer and to replicate. Epstein-Barr virus latent membrane protein 1 also inhibits apoptosis. Proliferation of B cells as a result of production of cytokines by macrophages or $CD4^+$ T cells, or even some viral proteins (e.g., SU or TM),

may also play a role in this process. Uncontrolled proliferation, by whatever mechanism, could lead to the chromosomal changes required for cell transformation and malignancy.

Anogenital Carcinomas

Anogenital carcinomas are two to three times more frequent in HIV-1-infected individuals than in the general population. They are associated with human papillomavirus infections that are typically spread through sexual contact. The high-risk human serotypes 16 and 18 are associated with both cervical and anal carcinomas. Such cancers often arise in areas of squamous metaplasia near the glandular epithelium and reach more advanced stages in immunosuppressed individuals.

Prospects for Treatment and Prevention

Antiviral Drugs

The decrease in rate of new HIV infections worldwide is a consequence of more widespread availability of antiviral drugs, as well as improved methods for blood testing and screening and successful campaigns focused on behavioral changes to aid prevention. Treatment with highly effective antiviral drugs and drug combinations (Chapter 9) has reduced the problem of emerging resistant viruses, lowered the rates of transmission, and afforded HIV-infected individuals longer and more normal lives. While clearly a triumph, several areas of concern remain. Even though the overall life expectancy of treated individuals is approximately 10 years less than that of uninfected individuals, HIV-infected patients are now living longer than ever before. Complications in these patients include accelerated appearance of typical age-related ailments, such as cardiovascular disease, liver and renal failure, and neurocognitive dysfunction. Some pathology may be explained by incomplete recovery of the immune system following drug therapy: most individuals never achieve complete reconstitution. Secondary complications from chronic infections and reduced ability to suppress oncogenic viruses may account for the increased incidence of some types of cancer in these individuals. Other problems are connected to the antiviral drugs that must be taken constantly. For example, the risk of cardiovascular disease is compounded by effects of prolonged therapy with HIV reverse-transcriptase inhibitors on lipid metabolism. These considerations underscore the need for better drugs and treatment strategies.

Although at most times the virus is undetectable in treated patients, it is not gone: small, intermittent bursts of viremia occur. The hope for total "clearance," or at least a "functional cure" (in which no drug treatment would be necessary), was buoyed by the case of a particular AIDS patient (the "Berlin patient") who received a hematopoietic stem cell transplant from a donor with the $CCr5\Delta32$ mutation (Box 7.4). The patient stopped taking antiretroviral drugs in 2007 and remains

BOX 7.4

BACKGROUND
The Berlin patient

Timothy Ray Brown, a native of Seattle, Washington, is the first, and as yet **sole**, individual to be cured of HIV.

While living in Berlin, Germany, in 1995, Brown was diagnosed with HIV and treated successfully with antiviral drugs for more than a decade. In 2007, he was diagnosed with acute myeloid leukemia and, after unsuccessful chemotherapy, discontinued antiviral drug therapy and underwent two stem cell transplantation procedures within a period of about a year. The second transplant followed a relapse of his leukemia and was preceded by a cytotoxic drug and whole-body irradiation regimen to ablate all or most of his leukemic and immune cells. Peripheral blood stem cells from the same donor were used for both transplant procedures. Most importantly, hoping to "kill two birds with one stone," Brown's Berlin physician screened 62 possible donors to identify an individual who carried a homozygous CCR5Δ32 mutation, which confers resistance to infection with CCr5-tropic HIV strains.

Despite enduring complications and undergoing two transplants, Brown's treatment was a success: he was cured both of his leukemia and HIV infection. Even though he had stopped taking antiviral drugs, there was no evidence of the virus in his blood following his treatment, and his immune system gradually rallied. Follow-up studies in 2011, including biopsies from his brain, gut, and other organs, showed no signs of viral RNA or DNA, and also provided evidence for the replacement of long-lived host cells with donor-derived cells. Recent studies also showed that Brown had no detectable CXCr4-tropic virus prior to transplantation. Brown remains HIV-1 free as of this writing (2015).

Although clearly somewhat of a "medical miracle," and by no means a practical road map for HIV treatment, the example of the "Berlin patient" has galvanized research efforts and continues to inspire hope that a simpler and more general cure for infection may someday be achieved.

Allers K, Hütter G, Hofmann J, Loddenkemper C, Rieger K, Thiel E, Schneider T. 2011. Evidence for the cure of HIV infection by CCR5 Δ32/Δ32 stem cell transplantation. *Blood* **117**:2791–2799.

Hütter G, Nowak D, Mossner M, Ganepola S, Müssig A, Allers K, Schneider T, Hofmann J, Kücherer C, Blau O, Blau IW, Hofmann WK, Thiel E. 2009. Long-term control of HIV by CCR5 Delta32/Delta32 stem-cell transplantation. *New Engl J Med* **360:** 692–698.

Berlin gate.

---

virus-free. Unfortunately, despite further efforts, this result has not been repeated: the virus reemerged consistently following transplants that were subsequently performed in other patients using a variety of strategies. Transplantation would certainly not be a practical treatment for the majority of HIV-1-infected individuals. Nevertheless, the successful Berlin case has encouraged efforts aimed at addressing the challenge of virus persistence and latency and at devising new gene therapy approaches.

Confronting the Problems of Persistence and Latency

Eradicating all traces of HIV from an infected individual is particularly challenging. Although small, the reservoir of long-lived, latently infected cells is established early in infection, can be replenished during short bursts of viremia and, in some cases, may be expanded by provirus-induced cell proliferation. The failure of early attempts to eliminate the reservoir by intensifying drug therapy is consistent with the notion that some of these cells may reside in drug-inaccessible sanctuaries, such as the brain. Most long-lived latently infected cells appear to have been derived from infected quiescent CD4[+] memory or activated T cells that survive infection long enough to revert to the resting memory state. A variety of mechanisms that inhibit proviral gene transcription in such cells have been described, including epigenetic suppression and deficiency of essential host transcriptional regulators,

such as nuclear Nf-κb. Such knowledge underlies the treatment strategy known as "shock and kill," in which provirus expression is induced in latently infected cells, while virus infection of new cells is prevented by treatment with antiviral drugs and/or neutralizing antibodies. The first implementations of this strategy, in which patients were treated with IL-2 or FDA-approved epigenetic drugs, were not successful: although some increase in virus production could be detected after treatment, there was no apparent decrease in the size of the latent reservoir. However, research in this area continues with the expectation that a fuller appreciation of the cell types that comprise the latent reservoir and more detailed understanding of their biology may lead to more effective methods for their activation and elimination.

Gene Therapy Approaches

Modern biotechnology has provided a number of methods, including direct gene editing, by which CD4[+] T cells and hematopoietic stem cells may be modified to make them resistant to HIV-1 infection. The use of cells from a donor with the CCr5Δ32 mutation in the case of the Berlin patient inspired a variety of trial strategies in which CD4[+] T or hematopoietic stem cells are obtained from a patient, the CCR5 gene is either mutated or its expression blocked by RNA interference, and then the resulting virus-resistant cells are returned to the patient. In some cases genes that encode proteins that inhibit

HIV reproduction have also been added to such cells. Several clinical trials using these approaches have been initiated, and the initial results appear promising. Other gene therapy approaches include deletion of both CCr5 and CXCr4 coreceptors or infection of CD4[+] T cells or hematopoietic stem cells with viral vectors that encode genes for small peptides that block viral entry or that express a site-specific recombinase that has been tailored to excise the proviral LTR.

Immune System-Based Therapies

Because immunization seems to be protective against subsequent infection of rhesus macaques with simian immunodeficiency virus, treatment strategies that combine antiviral drug treatment with augmentation of HIV-specific immunity have been proposed. Thus far, interventions based on administration of various cytokines (IL-2, IL-7, IL-15) aimed at improvement in T cell function have not shown significant benefit, when administered singly. Another strategy is targeted at immune checkpoint molecules. Ligands for one such molecule, programmed cell death protein 1 (Pd-1), are produced widely in tissues and, when engaged, result in suppression of T cell function and return to an inactivated state. Inhibitors of the Pd-1 pathway restore T cell function. Clinical trials for the use of such inhibitors to augment host immune control in HIV-1-infected individuals have been initiated. It is possible that inhibitors of additional immune checkpoint molecules (Ctla-4 and others) may also be effective in this context.

The most potent immunological defense against viral infection is a vaccine, a topic discussed in more detail in Chapter 8. While HIV-1 vaccine development presents unique challenges, it is important to appreciate that an AIDS vaccine need not be 100% effective to be useful. With such a deadly disease, even partial protection that might spare 20 to 40% of potential victims would save millions of lives and reduce transmission significantly. In the absence of a vaccine, recent success in identifying potent broadly neutralizing antibody molecules suggests that such reagents could be generated *ex vivo* and used for passive immunization in certain situations. Broadly neutralizing antibodies may also be administered via viral vectors. Studies with primates and humanized mice have demonstrated that inoculation with viral vectors that produce such antibodies can provide long-lasting resistance to infection with simian immunodeficiency virus or HIV, respectively (Box 7.5). Other studies have shown that the affinity of these antibodies can be increased substantially by using recombinant DNA methods to produce bispecific molecules (called immunoadhesins) that can bind simultaneously to two separate epitopes on the HIV-1 envelope (e.g., in TM and SU). This strategy could improve both antibody recognition and viral neutralization activities, as the relatively low density of envelope protein on HIV-1 particles (ca. 7 to 17 spikes/particle) is thought to reduce the efficiency of bivalent binding by monospecific antibodies.

Antiviral Drug Prophylaxis

Postexposure prophylaxis (PEP)

It is well established that treating individuals with antiviral drugs within hours after exposure to HIV-1 (e.g., from needle sticks) reduces the risk of infection substantially. The efficacy of postexposure prophylaxis suggests that the first cells that are infected with HIV can be eliminated. Furthermore, treating infected individuals early, during acute infection, reduces the viral set point and the size of the latent reservoir and preserves immune function. The potential efficacy of early treatment was suggested by the case of the Mississippi baby who was infected *in utero* and started on aggressive antiviral treatment before she was 2 days old. The latent reservoir in this infant was so small that her body appeared to clear the virus while on drug treatment. Therapy was discontinued after 18 months. Although the virus reemerged 27 months later, the long delay suggested that the reservoir was indeed small and/or controlled for a relatively long time following such early treatment.

Preexposure prophylaxis (PrEP)

Infection with HIV-1 can be prevented in uninfected people who are at substantial risk of acquiring it by adherence to a regular regimen of antiviral drug treatment. The currently accepted regimen comprises daily ingestion of a single pill that contains a combination of antiviral drugs (e.g., tenofovir and emtricitabine). If an individual is then exposed to HIV, through sex activity or drug use, PrEP can keep the virus from establishing a permanent infection. According to the U.S. Centers for Disease Control and Prevention, "PrEP has been shown to reduce the frequency of HIV infection in people who are at high risk by up to 92%." Unfortunately, clinical trials of PrEP have had limited success, mainly because of lack of adherence to the treatment.

Perspectives

Pneumocystis Pneumonia—Los Angeles. In the period October 1980–May 1981, 5 young men, all active homosexuals, were treated for biopsy-confirmed *Pneumocystis carinii* pneumonia at three different hospitals in Los Angeles, California. Two of the patients died....

M. S. Gottlieb et al. (Centers for Disease Control)
Morb Mortal Wkly Rep **30**:250–252, 1981

So began the first warning, soon echoed in large urban centers throughout the United States and Europe, where physicians were being confronted by a puzzling and ominous new disease that was killing young homosexual men. In a deceptively low-key editorial note with this report, it was observed that "*Pneumocystis* pneumonia is almost exclusively limited to severely immunosuppressed patients," that "the occurrence of the disease in these five previously healthy individuals is unusual," and that "the fact that these patients were all

BOX **7.5**

EXPERIMENTS
Vector-derived immunoprophylaxis

The identification of broadly neutralizing human antibodies against HIV-1 has renewed vaccine development efforts. While many obstacles must be overcome before a vaccine is available, strategies for use of these reagents seem more immediately practical. One such strategy employs gene transfer technology with adenovirus-associated viral (AAV) vectors that encode neutralizing antibodies or engineered derivatives known as immunoadhesins.

Such studies, first with mice and then with Rhesus macaques, showed this approach to be quite promising. Long-lasting neutralizing activity was detected in the serum of animals injected intramuscularly with the AAV vectors. Furthermore, macaques injected with AAV vectors encoding antibodies that targeted simian immunodeficiency virus were protected against intravenous challenge with this virus. Six of nine animals that received the gene transfer were not infected, and none of the nine developed AIDS.

To test the efficacy of this approach for HIV-1, "humanized" mice that had been engrafted with human bone marrow, liver, and thymus tissue, were injected with AAV vectors encoding broadly neutralizing antibodies against HIV-1. After 4 weeks, high concentrations of the human IgG antibodies could be detected in the serum and even in vaginal washes of these animals. The mice were then challenged weekly with low concentrations of CXCr5 virus particles, administered vaginally to approximate the predominant mode of sexual infection in humans. The results showed that these mice were highly resistant to such challenge. Animals injected with a vector that encoded an optimized derivative of the SU binding antibody VRC01 (Fig. 7.19) were completely resistant to virus infection even after 21 weekly challenges, conditions in which all control animals became infected.

The investigators acknowledge that there are substantial anatomical differences between mice and people, and that the humanized mouse does not possess a fully functional human immune system. Nevertheless, this kind of immunoprophylaxis would seem to hold considerable promise as a strategy to reduce the probability of sexual transmission of the HIV between humans, especially in situations in which PrEP is impractical due to antiviral drug costs or lack of adherence to the prescribed regimen.

Balasz AB, Chen J, Hong CM, Rao DS, Yang L, Baltimore D. 2011. Antibody-based protection against HIV infection by vectored immunoprophylaxis. *Nature* **481:**81–84.

Balazs AB, Ouyang Y, Hong CM, Chen J, Nguyen SM, Rao DS, An DS, Baltimore D. 2014. Vectored immunoprophylaxis protects humanized mice from mucosal HIV transmission. *Nat Med* **20:**296–300.

Johnson PR, Schnepp BC, Zhang J, Connell MJ, Greene SM, Yuste E, Desrosiers RC, Clark KR. 2009. Vector-mediated gene transfer engenders long-lived neutralizing activity and protection against SIV infection in monkeys. *Nat Med* **15:**901–906.

Vector-derived immunoprophylaxis protects humanized mice from HIV-1 infection. (A) Experimental design. The optimized adenovirus-associated viral vector genome includes a promoter (CASI) that combines a cytomegalovirus enhancer and chicken β-actin promoter followed by a splice donor and splice acceptor flanking the ubiquitin enhancer region. Sequences encoding heavy (HC) and light chain (LC) V-regions from broadly neutralizing antibodies were inserted into the vector DNA, separated by a self-processing sequence (2A). A sequence for improved nuclear export of transcripts (WPRE), and the simian virus 40 late-polyadenylation signal (SV40pA), are also included to increase the efficiency of gene expression (Volume I, Chapter 10). ITR indicates the positions of the AAV inverted terminal repeats. Encapsidation by a protein from a rhesus macaque-derived subtype of the virus enhances the effectiveness of gene transfer by this vector. Humanized mice were injected in the gastrocnemius calf muscle with 1×10^{11} genome copies of the vector and challenged 4 weeks later with weekly vaginally administered HIV. **(B)** Viral load detected in plasma of vector-treated humanized mice following weekly intravaginal challenge with HIV-1. Limit of detection = 1000 copies ml⁻¹. **(C)** Fraction of uninfected vector-treated humanized mice over the course of repetitive intravaginal challenge. Black lines and open symbols show results with mice injected with a control vector that encodes a luciferase gene (Luc); Red lines and filled circles show results with mice injected with a vector encoding a potent, broadly neutralizing antibody directed against the CD4 binding site in SU (VRCO7W). Positive infection is defined by two consecutive viral load measurements above the limit of detection. From Figure 3 of A. B. Balazs et al., *Nat Med* **20:**296–300, 2014, with permission.

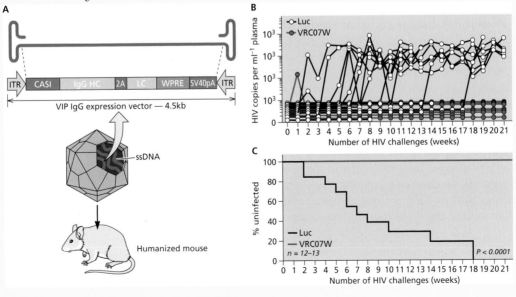

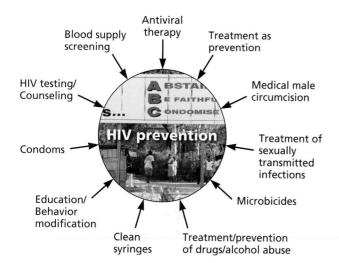

Figure 7.28. The multifaceted approach to prevention of infection with HIV.

homosexuals suggests an association between some aspect of a homosexual lifestyle or disease acquired through sexual contacts . . ." Soon the disease was to ravage this group and also the hemophiliac community, whose lives depended on blood products. This state of distress led to unprecedented social activism that demanded significant public investment for AIDS research and put patient advocates on scientific review panels for the first time.

Initial progress in AIDS research was impressive. The etiological agent of the disease was identified within 2 years, and screening assays to safeguard the blood supply were developed shortly thereafter. As a broad base of knowledge about retroviruses already existed, it seemed that treatment and a vaccine should soon be available. Unfortunately, more than 3 decades later, AIDS is still with us. Nevertheless, the prospect is much more upbeat today than it was in the last edition of this book. The use of potent drugs and drug combinations has reduced the rates of transmission and prolonged the lives of those infected with HIV throughout the globe. At the same time, a multifaceted approach has been applied to prevent new infections, including improved testing and counseling, and public education campaigns that discourage risky behavior (Fig. 7.28). While the two major challenges of finding a "cure" and developing a vaccine to prevent infection remain, there has been substantial progress in our understanding of the many aspects of HIV disease, from molecular to the epidemiological aspects. The continuously expanding knowledge base and technology toolbox have motivated renewed efforts to meet both of these challenges. The more optimistic among us can even envision the possibility of a world without AIDS in the future.

References

Books

Bushman FD, Nabel GJ, Swanstrom R (ed). 2012. *HIV: From Biology to Prevention and Treatment.* Cold Spring Harbor Laboratory Press, Cold Spring Harbor, NY.

Levy JA. 2007. *HIV and the Pathogenesis of AIDS,* 3rd ed. ASM Press, Washington, DC.

Reviews

Freed EO, Martin MA. 2013. Human immunodeficiency viruses: replication, p 1502–1560. *In* Knipe DM, Howley PM (ed), *Fields Virology,* 6th ed, Vol 2. Lippincott Williams & Wilkins, Philadelphia, PA.

Harris RH, Hultquist JF, Evans DT. 2012. MiniReview: the restriction factors of human immunodeficiency virus. *J Biol Chem* **287:**40875–40883.

Killian MS, Levy JA. 2011. HIV/AIDS: 30 years of progress and future challenges. *Eur J Immunol* **41:**3401–3411.

Kuritzkes DR, Koup RA. 2013. HIV-1: pathogenesis, clinical manifestations, and treatment, p 1561–1583. *In* Knipe DM, Howley PM (ed), *Fields Virology,* 6th ed, Vol 2. Lippincott Williams & Wilkins, Philadelphia, PA.

Moir S, Chun T-W, Fauci AS. 2011. Pathogenic mechanisms of HIV disease. *Annu Rev Pathol* **6:**223–248.

Ruelas DS, Greene WC. 2013. An integrated overview of HIV-1 latency. *Cell* **155:**519–529.

Expert's Current Opinions

Cell Press. 2013. Leading edge voices: the possibility of an AIDS free world. *Cell* **155:**491–492.

Research Articles of Historical Interest

Barré-Sinoussi F, Chermann JC, Rey F, Nugeyre MT, Chamaret S, Gruest J, Dauguet C, Axler-Blin C, Vézinet-Brun F, Rouzioux C, Rozenbaum W, Montagnier L. 1983. Isolation of a T-lymphotropic retrovirus from a patient at risk for acquired immune deficiency syndrome (AIDS). *Science* **220:** 868–871.

Keele BF, Van Heuverswyn F, Li Y, Bailes E, Takehisa J, Santiago ML, Bibollet-Ruche F, Chen Y, Wain LV, Liegeois F, Loul S, Mpoudi Ngole E, Bienvenue Y, Delaporte E, Brookfield JFY, Sharp PM, Shaw GM, Peters M, Hahn BH. 2006. Chimpanzee reservoirs of pandemic and nonpandemic HIV-1. *Science* **313:**523–526.

Simon F, Mauclère P, Roques P, Loussert-Ajaka I, Müller-Trutwin MC, Saragosti S, Georges-Courbot MC, Barré-Sinoussi F, Brun-Vézinet F. 1998. Identification of a new human immunodeficiency virus type 1 distinct from group M and group O. *Nat Med* **4:**1032–1037.

Zhu T, Korber BTM, Nahmias AJ, Hooper E, Sharp PM, Ho DD. 1998. An African HIV-1 sequence from 1959 and implications for the origin of the epidemic. *Nature* **391:**594–597.

Websites

http://www.unaids.org *Joint United Nations Programme on HIV/AIDS*

http://www.cdc.gov/hiv/ *United States Centers for Disease Control and Prevention*

8 Vaccines

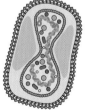

Introduction

The Origins of Vaccination

Smallpox: a Historical Perspective

Large-Scale Vaccination Programs Can Be Dramatically Effective

Vaccine Basics

Immunization Can Be Active or Passive

Active Vaccination Strategies Stimulate Immune Memory

The Fundamental Challenge

The Science and Art of Making Vaccines

Inactivated or "Killed" Virus Vaccines

Attenuated Virus Vaccines

Subunit Vaccines

Recombinant DNA Approaches to Subunit Vaccines

Virus-Like Particles

DNA Vaccines

Attenuated Viral Vectors and Foreign Gene Expression

Vaccine Technology: Delivery and Improving Antigenicity

Adjuvants Stimulate an Immune Response

Delivery and Formulation

Immunotherapy

The Quest for an AIDS Vaccine

Formidable Challenges and Promising Leads

Perspectives

References

LINKS FOR CHAPTER 8

▶▶ *Video: Interview with Dr. Gary Nabel*
http://bit.ly/Virology_Nabel

▶▶ *Ebola lite*
http://bit.ly/Virology_Twiv335

▶▶ *An unexpected benefit of inactivated poliovirus vaccine*
http://bit.ly/Virology_1-6-15

> *An ounce of prevention is worth a pound of cure.*
> BENJAMIN FRANKLIN

Introduction

Imagine that, while walking down your street, you encounter a new dog. You offer your hand in hope of a pleasant exchange, but you are greeted with snarls, bared teeth, and a menacing glare. When you meet this same dog the next time, it is most likely that you will walk past quietly and quickly, recalling your previous negative interaction. Recollection of a former encounter changes one's future response to that same stimulus. This simple example is the basis of immunological memory, described in Chapter 4: your immune system does more than "remember" a former pathogen; it responds to a second challenge differently from the first. Following an initial encounter with a pathogen, memory immune cells are established. Reexposure to that same pathogen reawakens these memory cells to control the secondary infection quickly and prevent subsequent disease. For many centuries, the illness that accompanied a primary infection was unavoidable, and for diseases such as smallpox, infected people often did not survive to face a second exposure. The goal of vaccination is to trigger an immune response more rapidly and with less harm than a natural infection: in essence, to avoid the disease that often accompanies the first exposure while enabling establishment of long-lasting immunological memory.

Vaccines against viral and bacterial pathogens prevent catastrophic losses of life in humans, other animals, and plants, and are considered among the greatest public health achievements (Fig. 8.1) (http://www.historyofvaccines.org/content/timelines/diseases-and-vaccines). But vaccines are not without their limitations and potential side effects. The history of vaccinology is therefore also the story of how the formulation and delivery of vaccines developed and improved to preserve efficacy while increasing safety and durability. In this chapter, we begin with a recounting of the fortuitous observations that catalyzed this field; highlight specific examples of how the use of vaccines has led to the eradication of devastating viruses; and discuss the differences among various vaccination strategies, comparing the benefits and challenges of each. (For more, see the interview with Dr. Gary Nabel: http://bit.ly/Virology_Nabel.)

The Origins of Vaccination
Smallpox: a Historical Perspective

Smallpox is the most destructive disease in history, and has probably been part of human existence since 10,000 BC or before. It has been estimated that infection by smallpox virus killed, crippled, or disfigured more than 1 in 20 of all humans who ever lived. In the 20th century alone, between 300 million and 500 million people died as a consequence of infection. As recently as 1967, there were >15 million cases worldwide, with 2 million deaths (note the high case-fatality ratio). Yet thanks to a worldwide vaccination campaign, a little more than a decade later the World Health Organization declared that smallpox had been eradicated. Smallpox remains the only infectious disease of humans for which this is true, and its eradication is surely among the greatest achievements of modern medicine.

While we correctly credit Edward Jenner with the development of the first vaccine, efforts to prevent infection and disease were made for many centuries before Jenner's contribution. Chinese and Indian physicians of the 11th century injected pus from smallpox lesions into healthy individuals, or blew a powder made from dried smallpox scabs into the nostrils of such individuals, with the hope of inducing mild disease that would provide lifelong protection (a process later called **variolation**). The word "variola," the name by which the disease was known for centuries, derives from the Latin

---

PRINCIPLES *Vaccines*

- Following an initial encounter with a pathogen, memory immune cells are established; reexposure to the same pathogen reawakens these memory cells to control the infection and prevent disease.

- The goal of vaccination is to trigger an immune response more rapidly and with less harm than a natural infection.

- Smallpox virus, which caused infections that killed, crippled, or disfigured more than 1 in 20 of all humans who ever lived, is the only human virus to be eradicated.

- Viral candidates for eradication must possess two essential features: the infectious cycle must take place in a single host, and infection (or vaccination) must induce lifelong immunity.

- Vaccination can be active (the host makes its own response to a viral preparation) or passive (components of the immune response are obtained from an appropriate donor or donors and injected directly into the patient).

- To be effective, a vaccine must induce protective immunity in a fraction of the population that is sufficient to impede person-to-person transmission, a concept called herd immunity.

- Active vaccination can occur by administration of virus preparations that have been inactivated or attenuated or by delivery of individual immunogenic proteins or recombinant DNA vectors that encode them.

- Inactivated virus particles or purified proteins often do not induce the same immune response as attenuated preparations, unless mixed with adjuvants that stimulate the early inflammatory response.

- The failure to develop a human immunodeficiency virus vaccine can be explained by both the biology of this virus and its interaction with the host immune system.

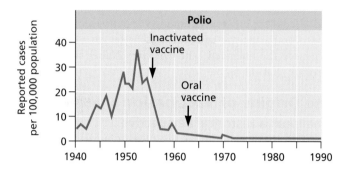

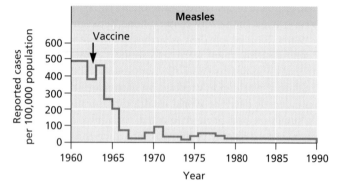

Figure 8.1 Profiles of successful vaccination campaigns. The number of reported cases of poliovirus (top) and measles virus (bottom) infection in the United States has been greatly reduced after massive vaccination programs. Adapted from C. A. Janeway, Jr., et al., *Immunobiology: the Immune System in Health and Disease* (Current Biology Limited, Garland Publishing Inc., New York, NY, 2001), with permission.

Figure 8.2 Irrational fears of the effects of vaccines. Some believed that vaccination using a virus that infected cows would cause cow-like features to appear in the recipient. The preponderance of data suggests this is not the case.

varius, meaning "mark on the skin." Variolation did confer a milder disease in some: the case-fatality ratio was ∼10 times lower than in people infected directly. Nevertheless, this approach caused infections, not prevented them. Despite the dangers of variolation, little additional progress in controlling the disease was made for centuries, and a virtually unchanged protocol was widely in use in Europe in the 1800s. The process was always controversial and was banned in many countries. While part of the rationale for banning variolation was that people became infected, there was also a strong sense that this procedure was "sinful," interfering with God's intended plan for individuals. As we will see later in this chapter, cultural and religious beliefs continue to influence some individuals' willingness to be vaccinated.

The vaccination story begins with Edward Jenner (1749–1823), a country doctor and naturalist, who was well known at the time for a seminal paper titled "Observations on the Natural History of the Cuckoo." At first glance, Jenner seems an unlikely candidate to conceive of, and establish, the means by which natural infection by smallpox was eventually eradicated. However, he was a careful and thoughtful observer of both cuckoos and his patients, and one day overheard

a dairymaid commenting that she would never get smallpox because she had been previously infected with cowpox. Jenner put this assertion to the test on May 14, 1796, when he injected fluid from a cowpox lesion on the finger of milkmaid Sarah Nelmes under the skin of James Phipps, a healthy 8-year-old boy. As expected, the boy developed a fever and a lesion typical of cowpox at the site of the injection. Two weeks later, Jenner then deliberately infected Phipps with smallpox. The young boy survived this potentially lethal challenge; needless to say, such an experiment would not be possible today.

Despite this promising result, the Royal Society in England rejected Jenner's paper, concerned that it was too anecdotal, leading Jenner to publish his work privately. While it is Jenner's name that is remembered, his colleague William Woodville, a prominent physician, was responsible for the first large-scale test that confirmed Jenner's observations. Introduction of the vaccine met with public skepticism and irrational fears (Fig. 8.2). Nevertheless, the smallpox vaccine was put into widespread use in 1800, and the disease was declared eradicated by the World Health Organization in 1979. Despite this monumental achievement, the specter of bioterrorism in the late 20th century has renewed interest in the virus and its vaccine (Box 8.1).

As is the case for many early discoveries, the scientific world was not prepared initially to exploit Jenner's approach and apply it to other pathogens. It took more than a century before the next practical vaccine for a viral disease appeared. Louis Pasteur, known for the germ theory and developing a technique to limit food spoilage due to microbes (pasteurization), prepared a rabies vaccine from the dehydrated spinal cord of an infected rabbit, and introduced the term **vaccination** (from *vacca*, Latin for "cow") in honor of Jenner's

BOX 8.1

BACKGROUND
The current U.S. smallpox vaccine

Prior to 2010, the vaccine stockpiled in the United States to protect civilian and military personnel against deliberate dissemination of smallpox virus was Dryvax, a freeze-dried, replication-competent vaccinia virus that was grown in calf lymph. Widespread distribution of this vaccine was discontinued by Wyeth in 1983, soon after the virus was declared eradicated. In 2008, the remaining stocks of Dryvax were destroyed by the Centers for Disease Control and Prevention and were replaced by a similar preparation, Sanofi-Pasteur's ACAM2000, which is prepared by infecting kidney epithelial cells in culture, rather than growing it in the skin of calves. A safer alternative was introduced in 2010 by Bavarian Nordic. This vaccine, Imvamune, is a nonreplicating strain that would eliminate the inherent risks of the replication-competent vaccine, which caused severe illness in 1 to 2% of recipients.

Development of a safer vaccine against smallpox is an important advance because the original vaccine caused rare but serious side effects, including severe skin reactions and central nervous system disorders. During the period in which every child in the United States was vaccinated, ~7 to 9 deaths per year were attributed to vaccination, with the highest risk occurring in infants. Inadvertent administration of the vaccine to immunodeficient individuals or to people with preexisting skin diseases resulted in a significantly larger number of adverse reactions. In 2002, following fears of bioterrorism in the wake of 9/11, the U.S. government announced that it would immunize military personnel and frontline civilian health care workers.

Harrison SC, Alberts B, Ehrenfeld E, Enquist L, Fineberg H, McKnight SL, Moss B, O'Donnell M, Ploegh H, Schmid SL, Walter KP, Theriot J. 2004. Discovery of antivirals against smallpox. *Proc Natl Acad Sci U S A* **101:**11178–11192.

Henderson DA. 1999. Smallpox: clinical and epidemiologic features. *Emerg Infect Dis* **5:**537–539.

Kaiser J. 2007. Smallpox virus. A tame virus runs amok. *Science* **316:**1418–1419.

Rosenthal SR, Merchlinsky M, Kleppinger C, Goldenthal KL. 2001. Developing new smallpox vaccines. *Emerg Infect Dis* **7:**920–926.

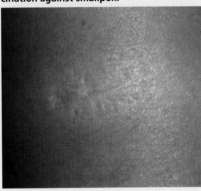

A scar remaining following a successful vaccination against smallpox.

pioneering work. Even with Pasteur's success against rabies, other antiviral vaccines were slow to follow, largely because viruses were difficult to identify, propagate, and study. Consequently, the next vaccines (against yellow fever and influenza viruses) did not appear until the mid-1930s.

Large-Scale Vaccination Programs Can Be Dramatically Effective

Ideally, vaccination mobilizes the host immune system to prevent a pathogenic outcome upon reinfection. As progressively more individuals in a population become immunized, the transmission cycle of host-to-host spread in a population is disrupted. As illustrated by the eradication of smallpox and the steadily decreasing rates of some common viral diseases as a consequence of sustained vaccination efforts, vaccination provides a remarkably effective antiviral defense. The World Health Organization reported in 2012 that ~84% of all children received at least one dose of measles vaccine before their first birthday, a 12% increase since 2000. Increased vaccine coverage resulted in a 78% drop in measles deaths worldwide between 2000 and 2012. Such achievements require massive, coordinated efforts of public health workers, governments, local clinics, vaccine providers, and funding agencies. As an example of the magnitude of the effort, in a **single** day, the World Health Organization once administered 127 million poliovirus vaccines to children in more than 650,000 villages in India.

Eradicating a Viral Disease: Is It Possible?

Viruses have survived countless bouts of selection during evolution, so the objective to eliminate a virus from the planet may seem naïve and unattainable. However, since the proclamation from the Director General of the World Health Organization that smallpox was eradicated, no natural cases of smallpox have been reported. The debate continues: should existing laboratory stocks of smallpox be destroyed (Box 8.2)?

The second virus to be vanquished was rinderpest virus, which infects cattle, buffalo, and other hoofed animals. This morbillivirus is a relative of measles virus and can be transmitted by aerosol or through drinking contaminated water. Outbreaks would often devastate entire herds, with deaths approaching 100% in immunologically naïve populations. The vaccine, developed in 1962, resulted in global eradication in June 2011.

These successes have prompted anticipation that other devastating infections, including poliovirus and measles virus, may be next on the horizon for elimination. To have any rational hope of eradication, a viral candidate must possess two essential features: the infectious cycle must take place in a single host, and infection (or vaccination) must induce lifelong immunity. By definition, a vaccine that renders the host population immune to subsequent infection by a virus that can grow **only** in that host effectively eliminates the virus. In contrast, a virus with alternative host species in which to propagate cannot be eliminated by vaccination of a single host population; other means of blocking viral spread are required.

BOX 8.2

DISCUSSION
Should laboratory stocks of smallpox virus be destroyed?

Samples of smallpox still exist in carefully regulated and locked freezers in the United States and Russia, and there is much debate about whether these stocks should be destroyed or preserved for future, potential scientific research. Since the disease was eradicated in the late 1970s, the World Health Organization has, on several occasions, delayed destroying the virus to permit research on smallpox vaccines and treatments, particularly in light of concerns about bioterrorist attacks.

The important issues are the following:

- Should we destroy biodiversity and gene pools that are not well understood?
- Are stocks of smallpox virus necessary for development of new vaccines and antivirals?
- If we do move forward with "lab eradication," how do we ensure that all reserves have been destroyed?

While some argue that the presence of frozen smallpox samples leaves open the possibility for nefarious groups to amplify and use them to harm a nonvaccinated population, others, including most scientists, believe that these stocks would be of extraordinary value should the virus reemerge in the human population. These reserves would also be useful in the licensure of new antivirals and vaccines and for the development of accurate diagnostics that can distinguish smallpox from other poxvirus relatives.

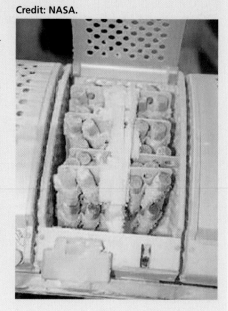

Fenner F. 1996. History of smallpox, p 25–37. *In* Koprowski H, Oldstone MBA (ed), *Microbe Hunters, Then and Now.* Medi-Ed Press, Bloomington, IL.

Henderson DA, Fenner F. 2001. Recent events and observations pertaining to smallpox virus destruction in 2002. *Clin Infect Dis* **33:**1057–1059.

Even when these requirements are fulfilled, global eradication can remain a formidable challenge. For example, widespread use of inexpensive, effective poliovirus vaccines has severely minimized this virus's impact on human health, and the vaccine is effective and inexpensive. Fewer than 100 cases of acute poliomyelitis were diagnosed in 2014. As the virus has no host other than humans, it should be possible to eliminate it by vaccinating a sufficient number of people and thereby ending the spread of the virus. Accordingly, the World Health Organization targeted the eradication of poliovirus by 2005 with a massive worldwide vaccination program. Sadly, the goal was not achieved, not because of lack of vaccine efficacy, but rather as a result of human variables that are difficult if not impossible to control, and which collectively limited complete coverage. Poverty, societal views of health care, lack of trust in physicians or the government, poor local health system infrastructure, and economic challenges often conspire to prevent vaccination of children in inner cities and in resource-poor countries. As a result, the virus remains endemic in Afghanistan, Nigeria, and Pakistan. Given that it may be impossible to eliminate all sources of the virus, poliovirus vaccination may be part of public health programs indefinitely (Box 8.3).

After poliovirus, measles virus is next on the World Health Organization's list for eradication. Measles virus is historically one of the world's leading infectious causes of childhood mortality, resulting in >100,000 deaths worldwide each year. Vaccination campaigns have reduced measles virus deaths by >75% since 2000, and the incidence of acute infections in the United States is steadily declining as a result of the efficacy of the vaccine (Fig. 8.3). However, unlike the polio vaccine, which can be given orally, the current measles vaccine requires two injections for maximal efficacy. This requirement alone imposes logistical and practical problems that will complicate global elimination of this virus. In addition, other virus properties of measles complicate its eradication. The infection is highly contagious, and the long period during which an individual is asymptomatic but shedding virus particles makes it difficult to identify and quarantine infected individuals. The reproduction number (R_0; see Chapter 10) of measles virus provides a measure of its contagious nature. R_0, the number of secondary infections produced by an infected person in a fully susceptible population, is estimated to be between 5 and 7 for smallpox virus, but for measles virus it is between 12 and 18. Amazingly, one person infected with measles infects 2 to 3 times as many susceptible people as does someone infected with smallpox. Whether a vaccination program can stay ahead of this pervasive foe remains to be determined.

National Programs for Eradication of Agriculturally Important Viral Diseases Differ Substantially from Global Programs

National vaccination and disease control programs are typically established for economically important livestock diseases. The goal is to keep a country free of a particular viral disease even though that disease may still be present in other countries. For example, the United States and Canada have been declared free of foot-and-mouth disease, but outbreaks still occur in parts of Europe, South America, and Asia. National programs can be successful only when augmented with broad governmental enforcement and border security,

DISCUSSION
The poliomyelitis eradication effort: should vaccine eradication be next?

The worldwide effort to eradicate poliomyelitis, launched in 1988 by the World Health Assembly (the decision-making body of the World Health Organization), remains stalled. The goal for eradication was set to occur in 2000, but setbacks necessitated shifting the target date forward to 2010. The current objective is for global eradication by 2018.

Enthusiasm was high during the initial years of the campaign; with the then-recent achievement of smallpox elimination, there was much optimism that the success could be duplicated for poliovirus. Indeed, the number of cases of the disease had been steadily falling, from a pre-vaccination estimate of 350,000 to <100 cases in 2014. The initial optimism has been replaced by doubt over whether eradication is realistic in light of the biological and political realities that have emerged in the course of the campaign.

The strategy to eradicate polio makes use of large-scale immunization campaigns with live attenuated poliovirus vaccine. These vaccine strains were known to revert to neurovirulence and cause vaccine-associated poliomyelitis. However, it was thought initially that vaccine-derived poliovirus strains do not circulate efficiently in the population, and that once wild-type poliovirus was eradicated, cessation of vaccination would eliminate vaccine-associated disease. Unfortunately, the 2000 outbreak of poliomyelitis in Hispaniola revealed this assumption to be incorrect. In this outbreak, 21 confirmed cases, all but 1 of which occurred in unvaccinated or incompletely vaccinated children, were reported. Subsequent analyses showed that the viruses responsible for the outbreak were derived from the Sabin poliovirus type 1 vaccine administered in 1998 and 1999. The neurovirulence and transmissibility of these viruses were indistinguishable from those of wild-type poliovirus type 1. Evidence of circulating vaccine-derived poliovirus was subsequently identified in Egypt and Nigeria. The previously underestimated threat of vaccine-derived polioviruses now makes the plan to cease vaccination unacceptable.

During the eradication campaign, regions are certified as free from wild-type polioviruses when the virus cannot be isolated for a 3-year period. As of 2015, poliovirus is endemic in only three countries: Afghanistan, Nigeria, and Pakistan. Since the last edition of this textbook, India has successfully transitioned to a "polio-free" country. Failure to eliminate transmission of wild polioviruses in these remaining countries is probably a consequence of insufficient vaccine coverage due to politics and war. For example, it is difficult to deliver polio vaccine to the border of Pakistan and Afghanistan, where skirmishes occur regularly and where health care workers are at great peril for kidnapping or murder. Even when the vaccine is administered, children continue to contract polio, probably as a result of poor sanitation, crowding, poverty, and infection with other microbes.

These considerations lend strength to the conclusion that polio eradication, followed by cessation of vaccination, is not a realistic goal, and that the program should be modified to ensure the protection of as many individuals as possible from poliomyelitis. The vaccine of choice would be one that does not revert to neurovirulence, such as the inactivated vaccine. However, this vaccine provides poor gut immunity compared to the attenuated vaccine, indicating that further improvements should be sought.

Chumakov K, Ehrenfeld E, Wimmer E, Agol VI. 2007. Vaccination against polio should not be stopped. *Nat Rev Microbiol* **5:**952–958.

Dove AW, Racaniello VR. 1997. The polio eradication effort: should vaccine eradication be next? *Science* **277:**779–780.

Minor PD. 2004. Polio eradication, cessation of vaccination and re-emergence of disease. *Nat Rev Microbiol* **2:**473–482.

Globally reported incidence of poliomyelitis in 1988 and 2012. The Americas, Western Pacific, European regions, and recently India have been declared poliomyelitis free by the World Health Organization. The number of cases has declined from an estimated 350,000 in 1988 to fewer than 100 cases in 2014. At the same time, the number of countries in which poliovirus is endemic has decreased from >125 to 3: Afghanistan, Nigeria, and Pakistan. Credit: Centers for Disease Control and Prevention.

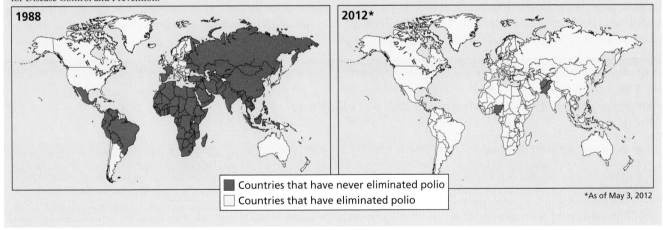

as animals in the virus-free country are constantly at risk for exposure from sources elsewhere. The perpetual concern of accidental import of these agricultural viruses is why customs officials inquire if residents of disease-free countries were exposed to livestock when traveling abroad.

Countries in which the disease is still present must have other means of control to limit outbreaks. Surveillance and containment strategies must be mobilized quickly and aggressively to identify and stop the spread from localized outbreaks. A common practice is to slaughter every host

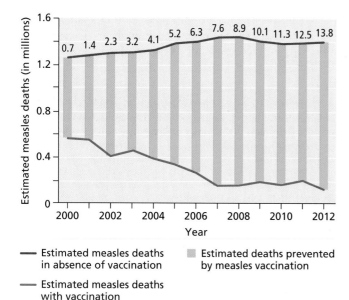

Figure 8.3 Decline in worldwide measles deaths due to vaccination. Estimated worldwide measles mortality and measles-induced deaths that were prevented by vaccination during 2000 to 2012 are shown. Compared with a scenario of no vaccination against measles, an estimated 13.8 million deaths were prevented by measles vaccination during this period.

— Estimated measles deaths in absence of vaccination
— Estimated measles deaths with vaccination
▨ Estimated deaths prevented by measles vaccination

animal in farms at increasing distances surrounding an outbreak site (the so-called "ring-slaughter" approach [Box 8.4]). Because acute infections spread rapidly from the outbreak site by many routes, and do so before identifiable symptoms are visible, the ring-slaughter containment often is breached unknowingly. As a result, preemptive slaughter of **all** animals on "at-risk" farms may be required. For example, South Korea faced a major foot-and-mouth disease outbreak on pig farms in 2010, in which >100 confirmed cases were identified. To halt the outbreak, >3 million animals (12% of the domestic pig population) were destroyed. Consequently, the devastating economic ramifications of such outbreaks have an impact even on those farms where no foot-and-mouth disease is present. Obviously, the faster an outbreak is identified, the more likely the success of containment efforts. Unfortunately, on-farm diagnostic tools that provide reliable identification of pathogens before symptoms are visible are not yet available. When developed, such tools need to be quick, easy, and accurate: false-positive identification of an outbreak could result in unnecessary sacrifice of many farm animals, with attendant economic loss.

Although prevention through vaccination is a powerful tool, and foot-and-mouth disease vaccines are available, other considerations limit the efficacy of this vaccine. Many serotypes for foot-and-mouth disease are currently in circulation, and vaccination against one serotype does not necessarily provide protection against others. Even strains within a given serotype may possess small sequence changes that invalidate the vaccine's efficacy. Moreover, standard blood tests to identify antibodies cannot distinguish between an infected animal and a vaccinated animal. Consequently, many farmers are reluctant to vaccinate for fear that their meat products will not be exportable to other markets.

Vaccine Basics
Immunization Can Be Active or Passive
Active immunization with attenuated or killed virus preparations or with purified viral proteins induces immunologically mediated resistance to infection or disease. In contrast, **passive immunization** introduces components of the immune response (e.g., antibodies or stimulated immune cells) obtained from an appropriate donor(s) directly into the patient. All neonates benefit from passive immunization fol-

BOX 8.4
BACKGROUND
Stopping epidemics in agricultural animals by culling and slaughter

Vaccination of agriculturally important animals such as cattle and pigs may not be cost-effective or may run afoul of government rules that block the shipping and sale of animals with antibodies to certain viruses. The 2001 foot-and-mouth disease epidemic in the United Kingdom provides a dramatic example of how viral disease is controlled when vaccination is not possible. The solution that stopped the epidemic was mass slaughter of **all animals** (infected or not) surrounding the affected areas and chemical decontamination of farms. It is estimated that >6 million animals were destroyed in less than a year before the spread of foot-and-mouth disease virus was contained, and similar mass killing of chickens and pigs was employed in South Korea in 2013 to prevent spread of influenza A virus and Nipah virus, respectively.

Animal slaughter is often the only action available to officials dealing with potential epidemic spread. For example, in recent years, millions of chickens in Hong Kong were killed to stop an influenza virus epidemic with potential to spread to humans. Influenza is not the only agricultural threat to poultry: chickens in California poultry farms were slaughtered in the 1970s to stem the spread of Newcastle disease virus.

Keeling MJ, Woolhouse ME, May RM, Davies G, Grenfell BT. 2003. Modelling vaccination strategies against foot-and-mouth disease. *Nature* **421:**136–142.

Kitching RP, Hutber AM, Thrusfield MV. 2005. A review of foot-and-mouth disease with special consideration for the clinical and epidemiological factors relevant to predictive modelling of the disease. *Vet J* **69:**197–209.

Woolhouse M, Donaldson A. 2001. Managing foot-and-mouth. *Nature* **410:**515–516.

lowing birth, as some of the mother's antibodies pass into the fetal bloodstream via the placenta to provide transient protection to the immunologically naïve newborn. Newborns who are breastfed are further protected by transfer of a particular type of antibody (IgA) in the antibody-rich colostrum. This protective effect can be detrimental if active immunization of infants is attempted too early, as maternal antibody may block a vaccine from stimulating immunity in the infant (Fig. 8.4). For this reason, most vaccines are not administered until 6 to 12 months postbirth.

Passive immunization is a preemptive response, usually adopted when a virus epidemic is suspected, because it provides immediate protection and does not require the host to mount an effective memory response. In 1997, consumption of contaminated fruit led to a widespread outbreak of hepatitis A virus infections in the United States. Pooled human antibodies (also called immunoglobulin) were administered in an attempt to block the spread of infection and reduce disease. This antibody cocktail contains the collective immunological experience of many individual infections and provides instant protection against some viruses. Passive immunization and immunotherapy are used for multiple virus infections, but the best-known instance is for rabies, in which a preparation of human immunoglobulin is delivered as soon as possible after a bite from a rabid animal to contain the virus before it can be disseminated. In addition, the standard procedure

Figure 8.4 Passive transfer of antibody from mother to infant. The fraction of the adult concentration of various antibody classes is plotted as a function of time, from conception to adulthood. Newborn babies have high levels of circulating IgG antibodies derived from the mother during gestation (passively transferred maternal IgG), enabling the baby to benefit from the broad immune experience of the mother. This passive protection falls to low levels at about 6 months of age as the baby's own immune response takes over. Total antibody concentrations are low from about 6 months to 1 year after birth, a property that may increase susceptibility to disease. Premature infants are particularly at risk for infections because the level of maternal IgG is lower and their immune system remains underdeveloped. The time course of production of various isoforms of antibody (IgG, IgM, and IgA) synthesized by the baby is indicated. Adapted from C. A. Janeway, Jr., et al., *Immunobiology: the Immune System in Health and Disease* (Current Biology Limited, Garland Publishing Inc., New York, NY, 2001), with permission.

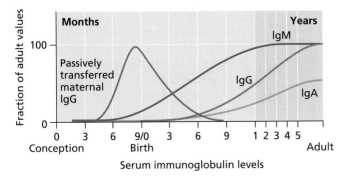

for smallpox vaccination with live vaccinia virus requires that so-called "vaccine immunoglobulin" be available should disseminated vaccinia occur in the vaccine recipient. When stimulated immune cells (e.g., T cells) are used, the process is called **adoptive transfer**; transfer of memory T cells may provide longer-lasting protection than antibody transfer, as the relatively short half-life of the antibody proteins limits sustained efficacy. Either way, passive immunization produces short-term effects, whereas active immunization can be lifelong.

Active Vaccination Strategies Stimulate Immune Memory

Vaccines work because they educate the host's immune system to recall the identity of a specific virus years after the initial encounter, a phenomenon called **immune memory** (Box 8.5). The resounding practical success of immunization in stimulating long-lived immune memory is among the greatest medical achievements.

Immune memory is maintained by dedicated T and B lymphocytes that remain after an infection has been resolved and most activated immune cells have died. These memory cells are able to respond quickly to a subsequent infection (Fig. 8.5). Antiviral vaccines establish immunity and memory without the pathogenic effects typical of the initial encounter with a virulent virus. Ideally, an effective vaccine is one that induces and maintains significant concentrations of memory cells in serum or at points of viral entry, such as mucosal surfaces and skin. When the pathogen to which they are specific returns, the memory B and T cells spring to action, unleashing an aggressive response that rapidly controls the pathogen before pathogenesis can ensue.

Protection from Infection or Protection from Disease?

It is important to consider that there are different possible outcomes following vaccine administration. In some cases, the antibody and memory T cells established by vaccination are maintained for long periods, and their mobilization will be sufficient to stop a subsequent infection before the virus can spread beyond the site of entry. Disease is prevented because the virus cannot reproduce or spread. In other cases, virus reproduction and spread may not be blocked immediately. Such infections can **only** be cleared by the coordinated action of vaccine-induced immune effectors **and** infection-induced immune responses (e.g., interferon production). In this case, disease may not be prevented, but its onset can be delayed or its severity lessened. In a third, less optimal outcome, the virus will not be eliminated because the host's response to the vaccine or to subsequent infection (or both) is inadequate. Consequently, disease is not prevented and vaccination may confer only a modest delay in the appearance of disease.

BOX 8.5

EXPERIMENTS
A natural "experiment" demonstrating immune memory

A striking example of immune memory is provided by a natural "experiment" in the 18th and 19th centuries on the Faroe Islands in the northern Atlantic Ocean. These islands were an ideal stopping point for cargo ships that transported goods between America and Europe. In 1781, measles, probably introduced by an infected sailor, infected many islanders and drastically reduced the islands' population. Subsequently, changes in shipping routes made the Faroes a less ideal stopping point, and for the next 65 years, the islands remained measles virus free and the surviving population flourished. In 1846, as the islands were revisited by these vessels, measles struck again, infecting >75% of the population with similar devastating results. In a personal and entertaining diary-like article titled "Observations Made during the Epidemic of Measles on the Faroe Islands in the Year 1846" (http://www.deltaomega. org/documents/PanumFaroeIslands.pdf), the Danish physician Peter Panum noted that none of those individuals who survived the 1781 epidemic became infected in 1846. As a perfect age-matched control, their peers who had **not** been infected earlier were ravaged by measles in this second outbreak. This natural experiment illustrates two important points: immune memory can last for decades, and it can be maintained without ongoing exposure to the virus.

Ahmed R, Gray D. 1996. Immunological memory and protective immunity: understanding their relation. *Science* **272:**54–60.

Panum PL. 1847. Observations made during the epidemic of measles on the Faroe Islands in the year 1846. *In Bibliothek for Laeger, Copenhagen, 3R.,* **1:**270–344.

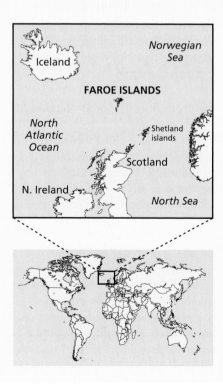

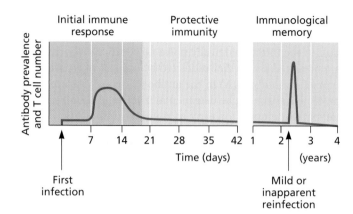

Figure 8.5 Antibody and effector T cells are the basis of protective immunity. The relative concentrations of antibody and T cells are shown as a function of time after first (primary) infection. Antibody levels and numbers of activated T cells decline after the primary viral infection is cleared. Reinfections at later times (years later), even if mild or inapparent, are marked by rapid and robust immune response because of the reanimation of a memory response. Adapted from C. A. Janeway, Jr., et al., *Immunobiology: the Immune System in Health and Disease* (Current Biology Limited, Garland Publishing Inc., New York, NY, 2001), with permission.

Vaccines Must Be Safe, Efficacious, and Practical

The two prerequisites for an effective vaccine are that it is safe and effective. Vaccines that are based on inactivated virus particles or immunogenic viral proteins must not contain infectious particles or viral nucleic acids, respectively. If a live vaccine is used, virulent revertants must be exceedingly rare. In addition, there can be no contamination of vaccines with other microbes introduced during production. These ideals may seem obvious, but given the potential for human error, absolute safety is impossible to guarantee. Furthermore, when rare side effects do appear, they are often identified only after millions of people have been vaccinated (Box 8.6). In addition, attenuated (replication-competent) vaccines have the potential to spread to individuals in a population who have not been vaccinated. The smallpox vaccine is not given to immunosuppressed individuals, because they cannot contain vaccine-associated infection caused by the live vaccine.

To be effective, a vaccine must induce protective immunity **in a significant fraction of the population**. Not every individual in the population need be immunized to stop viral spread, but the number must be sufficiently high to impede virus transmission. Person-to-person transmission stops when the probability of infection drops below a critical threshold. This effect has been called **herd immunity**. To appreciate the importance of this effect, one might consider the likely outcome in two hypothetical elementary schools: in elementary school A of some 500 students, all but 10 have been vaccinated against measles virus, whereas in school B, only half of the same number have been vaccinated. Most

BOX 8.6

DISCUSSION
The public's view of risk-taking is a changing landscape

Whooping cough was a major lethal disease of children until the introduction of the DPT (diphtheria, pertussis, and tetanus) vaccine, which virtually eliminated the disease. Immunization resulted in frequent but mild side effects: $20% of children experienced local pain and some tiredness. However, about 1 immunized child in 1,000 had more-severe side effects, including seizures and sustained high fever. Given that whooping cough was well known to be a child killer, these side effects were generally deemed acceptable. Because whooping cough is perceived as a "disease of the past," some parents now feel that the risk of immunization side effects is unacceptable and are electing not to vaccinate their children. This reluctance to vaccinate has resulted in the predictable presence of whooping cough victims in clinics and alarming increases in the frequency of this disorder in California, Michigan, and the United Kingdom, among other places. The risk posed by the vaccine has not changed, but in the face of reduced threat of natural disease, the perceived risk of vaccination is elevated.

A quote from a 2004 article on game theory defines the problem in succinct and powerful terms: "Voluntary vaccination policies for childhood diseases present parents with a subtle challenge: if a sufficient proportion of the population is already immune, either naturally or by vaccination, then even the slightest risk associated with vaccination will outweigh the risk from infection. As a result, individual self-interest might preclude complete eradication of a vaccine-preventable disease."

Bauch CT, Earn DJ. 2004. Vaccination and the theory of games. *Proc Natl Acad Sci U S A* **101:**13391–13394.

Johnson B. 2001. Understanding, assessing, and communicating topics related to risk in biomedical research facilities. *ABSA Anthology of Biosafety IV—Issues in Public Health*, chapter 10. http://www.absa.org/0100johnson.html.

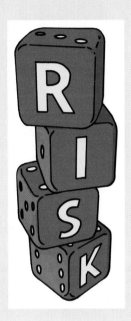

would intuitively and correctly judge that an unvaccinated child is much less likely to become infected with measles virus in school A than in school B. This is true because the 490 vaccinated students in school A provide an immunological wall around the unvaccinated children. In school B, the wall is leakier, and consequently there are more opportunities for the virus to gain a foothold in this population and to be transmitted to those who have not been vaccinated.

The actual calculation for herd immunity is pathogen and population specific, but generally corresponds to 80 to 95% of the population acquiring vaccine-induced immunity to provide protection to all members of that community. The herd immunity threshold is calculated as $1 - 1/R_0$. Recall that R_0 is the number of nonimmune individuals that would get infected upon encounter with an actively infected individual. As this number increases (that is, as the virus is transmitted to more individuals), the value of $1/R_0$ decreases, and thus $1 - 1/R_0$ gets closer to 1, or 100%. For smallpox virus, the herd immunity threshold is 80 to 85%, while for measles virus (which has a high R_0), it is 93 to 95%. Subtle changes in herd immunity have direct implications for the risks of outbreaks of infection (Fig. 8.6). No vaccine is 100% effective in a population; consequently, the level of immunity is not equal to the number of people immunized. In fact, we know that when 80% of a population is immunized with measles vaccine, about 76% of the population is actually immune, clearly well below the 93 to 95% required to prevent measles virus from infecting this community. Obviously, achieving such high levels of immunity by vaccination is a daunting task. Moreover, if the virus remains in other populations or in alternative hosts, reinfection is always possible. In closed populations (e.g., military training camps or animal herds), high levels of immunity can be achieved by vaccination of all individuals, but larger or less-controlled populations in widespread areas present serious logistical challenges. In addition, public complacency or reluctance to be immunized is dangerous to any vaccine program (Box 8.6).

The protection provided by a vaccine must also be long-term, lasting many years. While some vaccines cannot provide lifelong immunity after a single administration, subsequent inoculations (booster shots) given after the initial dose can stimulate waning immunity. However, this practice may be impractical for administration to large populations and can pose serious record-keeping challenges in resource-poor areas. Mounting the "proper" immune response is also required. For example, primary infection by some viruses, such as poliovirus, can be blocked only when a robust antibody response is evoked by vaccination. On the other hand, a potent cellular immune response is required for protection against herpesviral disease. To maximize effectiveness, a vaccine must be tailored to elicit the same type of immune responses as the natural virus infection it is designed to prevent.

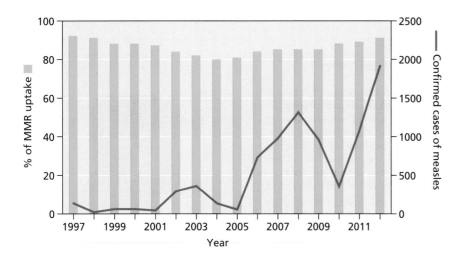

Figure 8.6 The correlation between herd immunity and the potential for outbreaks. Shown is the period between 1997 and 2012, with vaccination rates in Great Britain indicated on the left (gold bars expressed as a percentage of the population) and the number of reported cases of measles virus shown on the right (the solid blue line showing total number of cases). As the percentage of vaccinated individuals dips below ~90%, a corresponding rise in the number of acute cases is observed.

Outbred populations always have varied responses to vaccination. Some individuals exhibit a robust response, while others may not respond as well (a "poor take"). While many parameters influence such variability, the age and health of the recipient are major contributors. For example, the influenza virus vaccine available each year is far more effective in young adults than in the elderly. Weak immune responses to vaccination pose several problems. Obviously, protection against subsequent infection may be inadequate, but another concern is that upon such subsequent infection, viral reproduction will occur in the presence of weak immune effectors. Mutants that can escape the host's immune response can then be selected, and may spread in the immunized population. Indeed, vaccine "escape" mutants are well documented.

Once safety and efficacy are assured, other practical requirements including stability, ease of administration, and cost must be considered. If a vaccine can be stored at room temperature rather than refrigerated or frozen, it can be used where cold storage facilities are limited. One of the abiding challenges of measles virus eradication is that the vaccine, which is a live attenuated virus, must be kept cold from its synthesis to inoculation in the recipient host (the "**cold chain**"). Failure to keep the vaccine cold inactivates the attenuated virus and weakens its ability to induce immunity. In countries where electricity (and therefore, refrigeration) is not ubiquitous, this poses a substantial problem. But the science and engineering community loves a challenge: recently, the Gates Foundation supported the development of new thermoses that can keep vaccines frozen for more than a month without the need of any electrical source (Fig. 8.7). Continued collaborations with colleagues in the engineering field will be critical for the development of creative solutions to some of the practical challenges of worldwide vaccination (Box 8.7).

The route of administration and cost per dose are important considerations as well: when a vaccine can be administered orally rather than by injection, it will be more widely accepted.

Similarly, the World Health Organization estimates that a vaccine must cost less than $1 per dose if its global use is to be meaningful. However, the research and development costs for a modern vaccine are in the range of hundreds of millions of dollars. Another, often prohibitive expense is covering the liability of the vaccine producer. Liability expenses can be astronomical in a litigious society and have forced many companies to abandon vaccine development completely. Unfortunately, there is an inherent conflict between providing a good return on investment to vaccine developers and supplying vaccines to people and government agencies with a limited ability to bear the cost. Nongovernmental organizations such as the Red Cross, the Global Vaccine Fund, and others have been instrumental in ensuring effective vaccine disbursal.

The Fundamental Challenge

Given the remarkable success of vaccines against smallpox, measles, and polioviruses, it might seem feasible to prepare vaccines that prevent all common viral diseases. Unfortunately, as described above, designing and producing an effective

Figure 8.7 Vaccine thermoses. Development of novel chambers that keep vaccines cold for extended periods without electricity may revolutionize the efficacy of delivery of some attenuated vaccines for which the cold chain must be maintained. Credit: Intellectual Ventures.

A

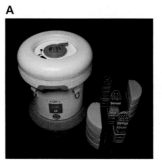

B

BOX 8.7

DISCUSSION
Development of new delivery vehicles for vaccines

The need for trained professionals to administer traditional vaccines limits speedy distribution in some developing countries. A new patch that does not require any medical training to administer has been developed: this microneedle patch contains hundreds of microscopic needles that penetrate the outer epidermis and dissolve into the skin once the vaccine has been delivered. These patches can be self-administered, do not require costly and dangerous disposal of hypodermic needles, have enhanced stability, and may actually confer stronger protection than classical needle-based vaccination.

An additional recent development uses the sugars sucrose and trehalose to put vaccines into a kind of suspended animation, where stability can be maintained for >6 months, even if unrefrigerated. As noted in the text, freedom from the cold chain would almost certainly revolutionize vaccine administration in developing regions with poor or unreliable electricity. The sugar method, developed in the United Kingdom, allows the vaccine to gradually dry into a syrup, and ultimately a thin film, which can be rehydrated immediately before injection.

Alcock R, Cottingham MG, Rollier CS, Furze J, De Costa SD, Hanlon M, Spencer AJ, Honeycutt JD, Wyllie DH, Gilbert SC, Bregu M, Hill AV. 2010. Long-term thermostabilization of live poxviral and adenoviral vaccine vectors at supraphysiological temperatures in carbohydrate glass. *Sci Transl Med* 2:19ra12. doi:10.1126/scitranslmed.3000490.

Norman JJ, Arya JM, McClain MA, Frew PM, Meltzer MI, Prausnitz MR. 2014. Microneedle patches: usability and acceptability for self-vaccination against influenza. *Vaccine* 32:1856–1862.

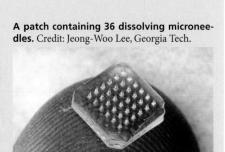

A patch containing 36 dissolving microneedles. Credit: Jeong-Woo Lee, Georgia Tech.

vaccine are exceedingly difficult, and the enthusiasm for creating a new vaccine is dependent on medical need and the economic market. For example, during the 2014 Ebola outbreak, there was a massively accelerated effort to create a new vaccine, but it took a worldwide crisis to catalyze this initiative. Furthermore, despite considerable research progress, we cannot predict with confidence the efficacy or undesirable side effects of different vaccine preparations. Because we lack sufficiently detailed knowledge of the important mechanisms of immune protection against most viral infections, the optimal design of a vaccine is not always obvious. Questions such as "Is a neutralizing antibody response important?" or "Is a cytotoxic-T-lymphocyte response essential?" cannot be answered with certainty, even for the most common viral infections. In fact, only when a vaccine is effective (or more often, when it fails) can we learn what immune features constitute a protective response. To complicate the situation, even when an experienced vaccine manufacturer sets out to develop, test, and register a new vaccine, the process can take years and millions of dollars. For example, it took 22 years to develop and license a relatively straightforward hepatitis A virus vaccine. The fundamental challenge is to find ways to capitalize on the discoveries in molecular virology and medicine to expedite vaccine development.

The Science and Art of Making Vaccines

There are four basic approaches to produce vaccines (Fig. 8.8). Each uses components of the pathogenic virus that the vaccine is intended to target. A vaccine developer may produce large quantities of the virus of interest and chemically inactivate it (**inactivated vaccine**), attenuate the pathogenicity through laboratory manipulation (**replication-competent, attenuated vaccine**), produce individual proteins free of the viral nucleic acid (**subunit vaccine**), or molecularly clone all or portions of the viral genome for preparation of recombinant DNA vaccines (**recombinant vaccine**). The methods are designed to adhere to principles that would be understood by Pasteur (Box 8.8). The most common, commercially successful vaccines simply comprise attenuated or inactivated virus particles, though as we will see at the end of this chapter, cheaper, cleverer, and more effective strategies are in the pipeline (Box 8.9).

Inactivated or "Killed" Virus Vaccines

The inactivated poliovirus, influenza virus, hepatitis A virus, and rabies virus vaccines are examples of effective inactivated vaccines administered to humans (Table 8.1). Moreover, inactivated vaccines, such as those which prevent equine influenza virus and porcine circovirus infections, are widely used in veterinary medicine. To prepare such a vaccine, virulent virus particles are isolated and inactivated by chemical or physical procedures. These treatments eliminate the infectivity of the virus, but not its antigenicity (i.e., the ability to induce the desired immune response). Common methods to inactivate virions include treatment with formaldehyde or β-propriolactone, or extraction of enveloped virus particles with nonionic detergents. These vaccines are safe for immunodeficient individuals, as the treated viruses cannot reproduce. Immunization by inactivated vaccines, however, often requires the administration of multiple doses, as the first dose is generally insufficient to produce a protective response.

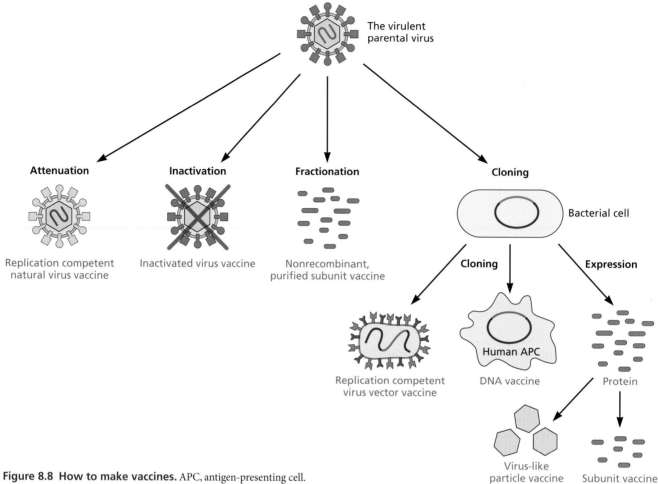

Figure 8.8 How to make vaccines. APC, antigen-presenting cell.

BACKGROUND
Most contemporary vaccines are based on old technology

In this age of modern biology, it seems ironic that the mutations in many of our common vaccine strains were not introduced by site-directed mutagenesis of genes known to be required for viral virulence, but rather were isolated by selection of mutants that could replicate in various cell types. The vaccines were produced with little bias for reduction of virulence *per se*.

Despite the old technology, the vaccines are relatively safe and remarkably effective. Consequently, their analysis has led to the identification of important attenuating mutations, as well as parameters that affect the protective immune response. The current vaccines not only provide protection from the intended virus, but are also the foundation for vaccines targeting other viruses. One example is the use of the yellow fever virus vaccine strain as a vehicle for West Nile and dengue viruses.

Painting of Louis Pasteur examining the dried spinal cord of an infected rabbit used to prepare an attenuated strain of rabies virus. Image courtesy of the Pasteur Institute (Photothèque/ Relations Presse et Communication externe, Institut Pasteur, Paris, France).

BOX 8.9

TERMINOLOGY
Live and let die

The authors of this textbook can be quite particular about word choice: for each edition, we sit around a table and read every word out loud. Even with the fourth edition, we still discovered long-standing inconsistencies and errors. While some of us are more aggrieved by the use of some words or phrases than others of us, we all concur that "live" and "dead" are misleading shorthands when appended to viruses. Alas, the vaccine community is less attentive to this issue, and so the terms "killed" and "live attenuated" vaccines are used generously in the discussion of vaccine types. Simply put: we will refer to such "killed" vaccines as "inactivated," and "live" vaccines will be termed "replication competent." We recognize that, while more accurate, it is also modestly more cumbersome: few would go to a James Bond movie called "Replication Competent and Let Inactivated." (See: http://www.imdb.com/title/tt0070328/ for the 007 classic that this Box is referring to.)

In principle, inactivated vaccines are very safe, but accidents can and do happen. In the 1950s, a manufacturer of Salk poliovirus vaccine, Cutter Laboratories, did not inactivate the virus completely, and >200 children developed disease as a result of vaccination (either by direct inoculation or contact with an infected child). Incomplete inactivation and contamination of vaccine stocks with potentially infectious viral nucleic acids have been singled out as major problems with this type of vaccine, though improved methods to detect residual infectious virus have reduced this risk substantially.

Administration of an inactivated vaccine annually is currently the most important measure for reducing influenza

Table 8.1 Viral vaccines licensed in the United States

Disease or virus	Type of vaccine	Indications for use	Schedule
Adenovirus	Attenuated, oral	Military recruits	One dose
Hepatitis A	Inactivated whole virus	Travelers, other high-risk groups	0, 1, and 6 mo
Hepatitis B	Yeast-produced recombinant surface protein	Universal in children, exposure to blood, sexual promiscuity	0, 1, 6, and 12 mo
Influenza	Inactivated viral subunits	Elderly and other high-risk groups	One dose seasonally
	Recombinant proteins	Elderly; those with egg allergies	One dose seasonally
Influenza	Attenuated	Children 2–8 yr old, not previously vaccinated with influenza vaccine	Two doses at least 1 mo apart
		Children 2–8 yr old, previously vaccinated with influenza vaccine	One dose
		Children, adolescents, and adults 9–49 yr old (e.g., FluMist, FluBlo)	One dose
Japanese encephalitis	Inactivated whole virus	Travelers to or inhabitants of high-risk areas in Asia	0, 7, and 30 days
Measles	Attenuated	Universal vaccination of infants	12 mo of age; 2nd dose, 6 to 12 yr of age
Mumps	Attenuated	Universal vaccination of infants	Same as measles, given as MMR
Papilloma (human)	Yeast- or SF9-produced virus-like particles	Females 9–26 yr old	Three doses
Rotavirus	Reassortant	Healthy infants	2, 3, and 6 mo or 2 and 4 mo of age depending on vaccine
Rubella	Attenuated	Universal vaccination of infants	Same as measles, given as MMR
Polio (inactivated)	Inactivated whole viruses of types 1, 2, and 3	Changing: commonly used for immunosuppressed where live vaccine cannot be used	2, 4, and 12–18 mo of age, then 4 to 6 yr of age
Polio (attenuated)	Attenuated, oral mixture of types 1, 2, and 3	Universal vaccination; no longer used in United States	2, 4, and 6–18 mo of age
Rabies	Inactivated whole virus	Exposure to rabies, actual or prospective	0, 3, 7, 14, and 28 days postexposure
Smallpox	Vaccinia virus	Certain laboratory workers	One dose
Varicella	Attenuated	Universal vaccination of infants	12 to 18 mo of age
Varicella-zoster	Attenuated	Adults 60 yr old and older	One dose
Yellow fever	Attenuated	Travel to areas where infection is common	One dose every 10 yr

virus-induced morbidity and mortality. In the United States alone, influenza virus infections cause as many as 50,000 deaths every year and consume at least $12 billion in health care, although epidemics can cost as much as $150 billion. Each year, millions of citizens seeking to avoid infection receive their flu shot, which contains several strains of influenza virus that have been predicted to reach the United States in the next flu season. The magnitude of this undertaking is noteworthy: >150 million doses of inactivated vaccine must be manufactured every year. Typically, these vaccines are formalin-inactivated or detergent- or chemically disrupted virus particles. The viruses, which are mass-produced in embryonated chicken eggs, can be natural isolates or reassortant viruses constructed to contain the appropriate hemagglutinin (HA) or neuraminidase (NA) genes from the expected virulent strain.

Currently, a typical influenza vaccine dose is standardized to comprise 15 µg of each viral HA protein, but it contains other viral structural proteins as well. The efficacy of these vaccines varies considerably. They are reportedly 60 to 90% effective in protecting healthy children and adults younger than 65 years who are exposed to virus strains in the vaccine; they are less effective in the elderly, immunosuppressed individuals, and people with chronic illnesses. Protection against illness correlates with the concentration of antibodies that react with viral HA and NA proteins produced after vaccination. Immunization may also stimulate limited mucosal antibody synthesis and cytotoxic-T lymphocyte activities, but these responses vary widely.

The envelope proteins of influenza viruses change by antigenic drift and shift as the virus reproduces in various animal hosts around the world. Consequently, protection one year does not guarantee protection the next, which is why annual flu shots are recommended. More than 140 national influenza centers conduct year-round surveillance of influenza virus trends and types and relay this information to a number of World Health Organization agencies, including the Centers for Disease Control and Prevention in the United States, the National Institute for Medical Research in the United Kingdom, and the National Institute for Viral Disease Control and Prevention in China. The World Health Organization makes general recommendations about which influenza strains to include in the vaccine, but it is up to each country to make its own decision. Within the United States, this job falls to the Food and Drug Administration. Timing is critical, as the final decision for the virus composition in the vaccine must be made within the first few months of each year to allow sufficient time for production of the vaccine. Any delay or error in the process, from prediction to manufacture, has far-reaching consequences, given the millions of people who are vaccinated and expect safe protection (Fig. 8.9). Even if the vaccine contains the appropriate viral antigens and is made promptly and safely, inactivated influenza virus vaccines have the potential to cause side effects in some individuals who are allergic to the eggs in which the vaccine strains are grown. As an example of other problems, the H5N1 avian virus that first infected humans in Hong Kong in the 1990s was extraordinarily cytopathic to chicken embryos, making it difficult to propagate. Reassortants had to be constructed by placing the new H5N1 segments in less cytopathic viruses.

In addition to the antigens intended to stimulate protective immunity, inactivated vaccines also contain trace amounts of other ingredients that are introduced during creation of the vaccine or to improve safety and/or efficacy. These include residual egg proteins, residual antibiotics (present to prevent contamination by bacteria during the manufacturing process), preservatives, and stabilizers such as gelatin and sugars that maintain potency during transportation and storage. A vaccine that is propagated in cell culture avoids some of these concerns.

Figure 8.9 Annual timeline for creating an influenza virus vaccine. Data are collected from many surveillance centers by the World Health Organization, and plans are in place early in the calendar year to determine against which strains the annual influenza vaccine will be created. In spring and summer, the vaccine is mass-produced in time for vaccination in late fall, when the process begins again.

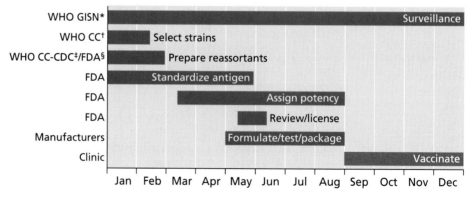

*World Health Organization Global Influenza Surveillance Network
†WHO Collaborating Centres
‡US Centers for Disease Control and Prevention
§US Food and Drug Administration

Attenuated Virus Vaccines

Replication-competent, attenuated vaccines are effective for at least two reasons. Progeny virus particles are generally restricted to tissues around the site of inoculation, and this focal restriction generally results in mild or inapparent disease (Fig. 8.10). However, the limited virus reproduction stimulates a potent and lasting immune response. Attenuated (less virulent) viruses are selected by growth in cells other than those of the normal host or by propagation at nonphysiological temperatures (Fig. 8.11). Mutants able to propagate under these conditions are isolated, purified, and subsequently tested for pathogenicity in appropriate models. Temperature-sensitive and cold-adapted mutants are often less pathogenic than the parental viruses because of reduced capacity for reproduction and spread in the warm-blooded host. In the case of viruses with segmented genomes (e.g., arenaviruses,

orthomyxoviruses, bunyaviruses, and reoviruses), attenuated, reassortant viruses may be obtained after mixed infections with pathogenic and nonpathogenic viruses.

Replication-competent oral poliovirus vaccines in use today comprise three attenuated strains selected for reduced neurovirulence. Type 1 and 3 vaccine strains were isolated by passage of virulent viruses in different cells and tissues until mutants with reduced neurovirulence in laboratory animals were obtained (Fig. 8.12A). The type 2 component was derived from a naturally occurring attenuated isolate. The mutations responsible for the attenuation phenotypes of all three serotypes are shown in Fig. 8.12B.

The attenuated measles virus vaccine currently in use was derived from a virulent virus called the Edmonston strain, isolated in 1954 by John Enders. Attenuated virus particles were isolated following serial passage of this virus through various cell types. Even though this approach was undirected, the viruses that were isolated could propagate only poorly at body temperature and caused milder signs of infection in primates. As one would expect, the vaccine strain so derived harbors a number of mutations, including several that affect the viral attachment protein, hemagglutinin.

The attenuated varicella-zoster virus vaccine is currently the only licensed human herpesvirus vaccine. It has proven to be safe and effective in children and adults, providing significant protection against infection by varicella-zoster virus, which causes chickenpox. Because this virus establishes a latent infection in all unvaccinated infected hosts, even if the initial infection is resolved, the virus can be reactivated at later times in life, resulting in painful and often serious conditions (shingles and postherpetic neuralgia). Subsequently, a much more concentrated (by at least 14-fold) formulation of the vaccine was licensed for use in previously infected adults (>60 years of age) to protect against recurrent disease.

Live attenuated viruses are administered by injection (e.g., measles-mumps-rubella [MMR] and varicella-zoster vaccines), by mouth (e.g., poliovirus, rotavirus, and adenovirus vaccines), or by nasal spray (influenza virus vaccine). The highly effective Sabin poliovirus vaccine is given as drops to be swallowed, and enteric adenovirus vaccines are administered as virus-impregnated tablets. One virtue of the oral delivery method for enteric viruses is that it mimics the natural route of infection and, as such, has greater potential to induce an immune response similar to that of the natural infection. A second advantage is that it bypasses the traditional need for hypodermic needles, which creates undue anxiety in many young, and some adult, vaccine recipients.

Attenuated virus vaccines have some inherent risks. Despite reduced spread in the vaccinee, we know that in the case of poliovirus vaccination, some shedding of the vaccine strain occurs, and these virus particles then have the potential to infect unvaccinated individuals. In most cases,

Figure 8.10 Comparison of the predicted immune responses to attenuated and inactivated viruses used in vaccine protocols. (**Top**) Immune responses plotted against time after injection of an inactivated virus vaccine (red curve). Three doses of inactivated virus particles were administered as indicated. (**Bottom**) Results after injection of a replication-competent, attenuated virus vaccine. A single dose was administered at the start of the experiment. The filled histogram (lavender-colored area) under the curve displays the titer of infectious attenuated virus. Redrawn from C. A. Mims et al., *Mims' Pathogenesis of Infectious Disease*, 4th ed. (Academic Press, Inc., Orlando, FL, 1995), with permission.

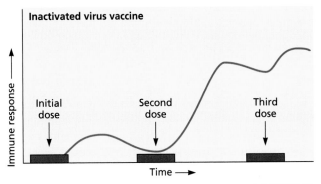

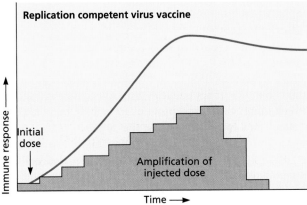

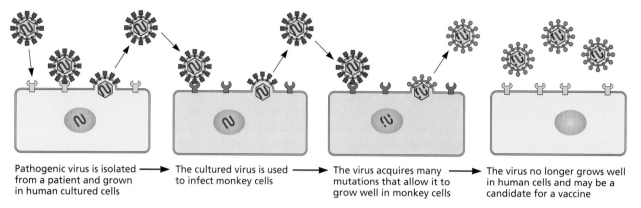

Pathogenic virus is isolated → The cultured virus is used → The virus acquires many → The virus no longer grows well
from a patient and grown   to infect monkey cells   mutations that allow it to   in human cells and may be a
in human cultured cells   grow well in monkey cells   candidate for a vaccine

Figure 8.11 Viruses specific for humans may become attenuated by passage in nonhuman cell lines. The four panels show the process of producing an attenuated human virus by repeated transfers in cultured cells. The first panel depicts isolation of the virus from human cells (yellow). The second panel shows passage of the new virus in monkey cells (lavender). During the first few passages in nonhuman cells, virus yields may be low. Viruses that grow better can be selected by repeated passage, as shown in the third panel. These viruses usually have several mutations, facilitating growth in nonhuman cells. The last panel shows one outcome in which the monkey cell-adapted virus now no longer grows well in human cells. This virus may also be attenuated (have reduced ability to cause disease) after human infection. Such a virus may be a candidate for an attenuated vaccine if it induces immunity but not disease. Adapted from C. A. Janeway, Jr., et al., *Immunobiology: the Immune System in Health and Disease* (Current Biology Limited, Garland Publishing Inc., New York, NY, 2001), with permission.

this would be akin to "bystander vaccination," but given the high rate of mutation associated with RNA virus replication, reversion to virulence is expected. Shedding of a virulent revertant virus is one of the main obstacles to developing effective attenuated vaccines, and is formally equivalent to the emergence of drug-resistant mutants (see "Drug Resistance" in Chapter 9). While such revertants are a serious problem, considerable insight into virus biology and pathogenesis can be obtained by identifying the changes responsible for increased virulence (Fig. 8.12C). Moreover, when one considers that these vaccines are safe and afford lifelong protection for the majority of the recipients, some degree of public health risk may seem acceptable. How such risk is determined and tolerated within a community becomes more of a sociological and ethical discussion rather than a virological question (Box 8.10).

Ensuring purity and sterility of the product is a problem inherent in the production of biological reagents on a large scale. If the cultured cells used to propagate attenuated viruses are contaminated with unknown viruses, the vaccine may well contain these adventitious agents. Sensitive detection methods, such as polymerase chain reaction, have minimized this threat in most of today's vaccines, but in the 1950s, early batches of poliovirus vaccine were grown in monkey cells that were unknowingly infected with the polyomavirus simian virus 40. It is estimated that 10 million to 30 million individuals received one or more doses of simian virus 40 with their poliovirus vaccine, and many developed antibodies to simian virus 40 proteins. Some concern existed that rare tumors may

be linked to this inadvertent infection, but this connection has since been discounted.

Alternatives to the classical empirical approach to attenuation can now be applied based on modern virological and recombinant DNA technology. For example, deletion mutations with exceedingly low probabilities of reversion can be created, though none yet exist (Fig. 8.13). In another approach that relies on genome segment reassortment of influenza viruses and reoviruses, genes encoding proteins that contribute to virulence are replaced with those from related but nonpathogenic viruses. No matter which technology is applied to achieve attenuation, the genetic engineer and the classical virologist must satisfy the same fundamental requirements: isolation or construction of an infectious agent with low pathogenic potential that is, nevertheless, capable of inducing a long-lived, protective immune response.

Subunit Vaccines

A vaccine may consist of only a subset of viral proteins, as demonstrated by the highly successful hepatitis B vaccine. Vaccines formulated with purified components of viruses, rather than the intact particles, are called subunit vaccines. Determining which viral proteins to include in a vaccine is accomplished by selecting those that are recognized by antibodies and cytotoxic T lymphocytes; this selection can be determined by assessing the immune responses of individuals who have recovered from the disease. Although the most obvious proteins would be those present on the virus surface, in fact, any viral protein could be a good immunological

A Derivation of Sabin type 3 attenuated poliovirus

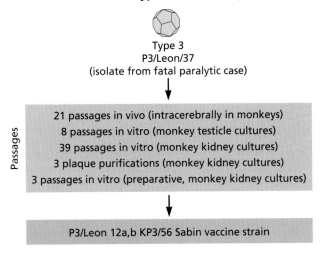

B Determinants of attenuation in the Sabin vaccine strains

Virus	Mutation (location/nucleotide position)
P1/Sabin	5'-UTR nt 480 VP1 aa 1106 VP1 aa 1134 VP3 aa 3225 VP4 aa 4065
P2/Sabin	5'-UTR nt 481 VP1 aa 1143
P3/Sabin	5'-UTR nt 472 VP3 aa 3091

C Reversion of P3/Sabin

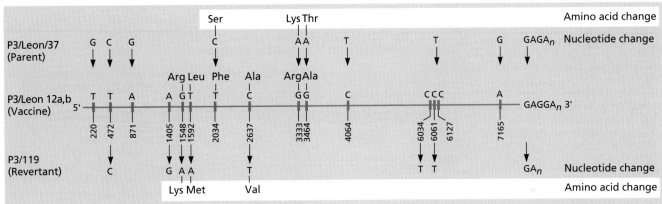

Figure 8.12 Replication-competent, attenuated Sabin oral poliovirus vaccine. (A) All three viral serotypes may cause poliomyelitis. Therefore, the Sabin vaccine is administered as a mixture of three different strains that are representatives of poliovirus serotypes 1, 2, and 3. Shown is the derivation of the type 3 vaccine strain, called P3/Sabin (the letter P means poliovirus). The parent of P3/Sabin is P3/Leon, a virus isolated from the spinal cord of an 11-year-old boy named Leon, who died of paralytic poliomyelitis in 1937 in Los Angeles. P3/Leon virus was passaged serially as indicated. At various intervals, viruses were cloned by limiting dilution, and the virulence of the virus was determined in monkeys. An attenuated strain was selected to be the final P3/Sabin strain included in the vaccine. **(B)** Determinants of attenuation in all three strains of the Sabin vaccine. The mutations responsible for the reduced neurovirulence of each serotype of the poliovirus vaccine are indicated (5'-UTR is the 5' untranslated region; VP1 to VP4 are the viral structural proteins; aa, amino acid; nt, nucleotide). **(C)** Reversion of the P3/Sabin vaccine strain. Differences in the nucleotide sequences of the virulent P3/Leon strain and the attenuated P3/Sabin vaccine strain are shown above the depiction of the parental viral RNA (green) and of a virulent virus (P3/119) isolated from a case of vaccine-associated poliomyelitis (below). Note that "revertant" does not mean that the sequence of the virus "reverts" to the sequence of the pathogenic precursor; rather, changes in other residues enable the virus to regain its potential to reproduce at wild-type virus levels.

target. A further critical parameter is that the viral proteins must be recognized by most individuals: as humans are an outbred population, the specificities of immune recognition differ from person to person, though some antigens can be recognized by a large proportion of the human population.

Even fragments of viral proteins may be sufficient to induce protective immunity. Synthetic peptides of about 20 amino acids or more in length can induce specific antibody responses when chemically coupled to protein carriers that can be taken up, degraded, and presented by major histocompatibility complex class II proteins (discussed in Chapter 4). In principle, synthetic peptides should be the basis for an extremely safe, well-defined vaccine, in which reversion or contamination with infectious virus is impossible. To date, however, peptide vaccines have had little success, mainly because synthetic peptides are expensive to make in sufficient quantity, and the antibody response they elicit is often weak and short-lived. Furthermore, given the likely simplicity of

BOX 8.10

DISCUSSION
Vaccine risks

Because our knowledge about virulence is limited, it is difficult to predict how a replication-competent, attenuated virus will behave in individuals and in the population. The attenuating mutations may lead to unexpected diversions from the natural infection and expected host response: the attenuated virus may be eliminated from the vaccinated individual before it can induce a protective response; it may infect new tissues or cells in the host with unpredictable effects; or it may initiate atypical infections (e.g., slow or chronic infections) that can trigger immunopathological responses of unknown etiology, such as **Guillain-Barré syndrome**. While this syndrome is most typically associated with a bacterial infection (*Campylobacter jejuni*), human cytomegalovirus and influenza virus have also been implicated as potential causative agents. Bacterial or viral infections (or vaccinations) can trigger a host immune response that then may cross-react with proteins present in human peripheral nerves. This process, termed molecular mimicry (Chapter 5), results in autoimmunity that, in the case of Guillain-Barré syndrome, is characterized by rapidly progressing, symmetric weakness of the extremities. Vaccine side effects, whether real or not, often have a detrimental effect on public acceptance of national vaccine programs.

Given that all procedures have risks, a fascinating ethical question that we must address as a society is: "How much harm can we tolerate for the global good?" That is, if millions of people each year benefit from the protective effects of a vaccine (or some other medical intervention), how many "side effects" are we willing to tolerate? Questions like these—which, of course, have no single answer—lie at the heart of bioethics.

Destruction of myelin layer in the peripheral nerve processes of individuals with Guillain-Barré syndrome.

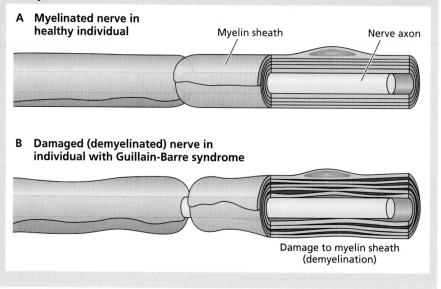

A **Myelinated nerve in healthy individual** — Myelin sheath — Nerve axon

B **Damaged (demyelinated) nerve in individual with Guillain-Barre syndrome** — Damage to myelin sheath (demyelination)

the host's "antipeptide" response (usually a single epitope is contained within the peptide), selection of escape mutants is highly probable.

Recombinant DNA Approaches to Subunit Vaccines

Recombinant DNA methods allow cloning of selected viral genes into nonpathogenic viruses, bacteria, yeasts, insect cells, or plant cells to produce the immunogenic protein(s). As only a portion of the viral genome is required for such production, there can be no contamination of the resulting vaccine with the original virus, solving a major safety problem inherent in inactivated virus vaccines. Viral proteins can be made inexpensively in large quantities by engineered organisms under conditions that simplify purification and quality control. For example, complications due to egg allergies after vaccination can be eliminated completely when influenza virus proteins are synthesized in *Escherichia coli*, insect cells, or yeasts. Baculoviruses, which infect insect cells in nature, can infect a large number of mammalian cell types in culture. Because this virus is nonpathogenic to humans and can be modified to express heterologous (and immunogenic) proteins from human viral pathogens, its use as a vaccine vector holds great promise.

Unfortunately, most candidate subunit vaccines fail because they do not induce an immune response sufficient to protect against infection, the gold standard of any vaccine. The immune repertoire evoked by an infectious virus infection may be only partially represented in a response to a subunit vaccine. In particular, purified protein antigens rarely stimulate the appearance of mucosal antibodies, particularly IgA. To date, the single exception of a successfully engineered subunit vaccine is Flublok, in which large quantities of the hemagglutinin protein of the three most prominent influenza viruses are synthesized from baculovirus vectors in a preservative- and egg-free cell culture system.

Virus-Like Particles

The capsid proteins of nonenveloped and of some enveloped virus particles may self-assemble into virus-like particles. These particles have capsid-like structures that are virtually identical to those in virus particles, but unlike authentic

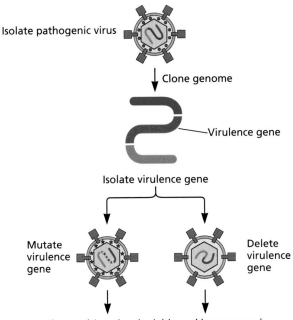

Isolate pathogenic virus

Clone genome

Virulence gene

Isolate virulence gene

Mutate virulence gene

Delete virulence gene

The resulting virus is viable and immunogenic but not virulent. It may be used as a vaccine.

Figure 8.13 Construction of attenuated viruses by using recombinant DNA technology. Once the genome of a pathogenic virus is cloned in a suitable system, deletions, insertions, and point mutations can be introduced by standard recombinant DNA techniques. If the cloned genome is infectious or if mutations in plasmids can be transferred to infectious virus, it is possible to mutate viral genes systematically to find those required for producing disease. The virulence gene can then be isolated and mutated, and attenuated viruses can be constructed. Such viruses can be tested for their properties as effective vaccines. The mutations in such attenuated viruses may be point mutations (e.g., temperature-sensitive mutations) or deletions. Multiple point mutations or deletions are preferred to reduce or eliminate the probability of reversion to virulence. Adapted from C. A. Janeway, Jr., et al., *Immunobiology: the Immune System in Health and Disease* (Current Biology Limited, Garland Publishing Inc., New York, NY, 2001), with permission.

particles, these capsids are empty: they contain no genetic material and cannot propagate. Because the empty capsids retain most of the conformational epitopes not found on purified or unstructured proteins, virus-like particle vaccines often induce durable neutralizing antibodies and other protective responses after injection. Furthermore, as the particles are completely noninfectious, inactivation with formalin or other agents is not required. This feature affords at least two additional advantages: immunogenicity is not compromised (formalin and other alkylating chemicals often alter the conformation of epitopes in inactivated vaccines), and concerns about efficiency of inactivation are avoided. Virus-like particle vaccines have proven to be particularly attractive for viruses that are propagated poorly in cell culture.

The highly successful hepatitis B virus subunit vaccine comprises virus-like particles produced in yeast. This vaccine contains a single viral structural protein (the surface antigen) that assembles spontaneously into virus-like particles, whether made in yeast, *E. coli*, or cultured mammalian cells. Formation of particles is critical, as purified monomeric capsid protein does not induce a protective immune response. Typically, 10 to 20 μg of virus-like particles is administered in each of three doses over a 6-month period, and >95% of recipients develop antibody against the surface antigen. The hepatitis B vaccine was the first anticancer vaccine, as a portion of chronically infected individuals develop fatal liver cirrhosis and hepatocellular carcinoma.

The virus-like particle vaccine effective against papillomavirus infections is among the newest vaccines developed for humans. More than 80% of sexually active women will be infected with several serotypes of human papillomavirus during their lifetime. As a result, many will develop genital warts and/or cervical cancer. There are numerous serotypes of this virus, but serotypes 6, 11, 16, and 18 cause 70% of cervical cancers and 90% of genital warts. Men can develop anogenital warts as well, and both sexes may develop head and neck cancers as a consequence of oral sex with an infected individual (Chapter 2).

It had been known for some time that the human papillomavirus L1 capsid protein forms virus-like particles when synthesized in a variety of heterologous systems. These empty capsids proved to be exceptional inducers of a protective immune response. As a result, a quadrivalent, virus-like particle vaccine effective against the four major cancer-causing serotypes of the virus was formulated, and in 2006, the Food and Drug Administration approved this formulation as the first vaccine to be developed to prevent cervical cancer induced by a virus. As with any new vaccine, there were, and still are, attendant societal discussions about its use (Box 8.11).

DNA Vaccines

DNA vaccines, a variant of the subunit protein approach to immunization, consist simply of plasmids encoding viral genes that can be expressed in cells of the animal to be immunized. In the simplest case, the plasmid encodes only the immunogenic viral protein under the control of a strong eukaryotic promoter. The plasmid DNA, usually produced in bacteria, can be prepared free of contaminating protein and has no capacity to replicate in the vaccinated host, but can be the template for expression of the immunogenic protein. Remarkably, no adjuvants or special formulations are necessary to stimulate an immune response.

The main challenge when this technology was first described in 1992 was how to introduce the plasmids into host cells such that the cell's transcriptional and translational machinery would take over. Direct intramuscular delivery of the vaccine in an aqueous solution containing a few

DISCUSSION
Should men be encouraged to get the human papillomavirus vaccine?

The answer, simply, is yes. The vaccine is approved for use in males in several countries, including the United States. It has been shown to be effective for prevention of infection by those papillomavirus strains that can cause both genital warts and anal cancer. Beyond these direct benefits, immunization of males with the human papillomavirus vaccine (e.g., Gardasil) limits the male-to-female transmission of strains that are most typically associated with cervical cancer.

Unfortunately, not everyone supports this recommendation. Some believe that vaccination will give a false sense of security and promote promiscuity among young people, while others view the recommendations as an intrusion on parental rights. Unlike other vaccines that are mandated (as a requirement for entry into public school, for example), the papillomavirus vaccine remains elective. Consequently, individuals must make their own decisions about whether to vaccinate or not. What is not debatable is the value of vaccination, which provides protection both for the individual and for the community in which that individual resides.

microgram of plasmid DNA was only moderately successful. A more effective delivery method uses a "gene gun" that literally shoots DNA-coated microspheres, inert particles coated with the DNA of interest, through the skin into dermal tissue. The goal of either approach is to ensure that plasmid DNA is engulfed by a macrophage or dendritic cell, such that the epitopes of the newly made viral protein are appropriately presented in the context of class I major histocompatibility complex molecules needed for T cell recognition and amplification. Both antibodies and cytotoxic T lymphocytes can be stimulated by DNA vaccination. The striking property of DNA vaccination is that a relatively low dose of DNA appears sufficient to induce long-lasting immune responses, and the cost of this approach is a fraction of that required to generate a protein-based vaccine. Moreover, the stability of DNA and its ability to withstand drying make this strategy particularly attractive for vaccine delivery in resource-poor areas where refrigeration is limited or unreliable.

It was subsequently shown that the method of inoculation could dictate the type of immune response that is generated by a DNA vaccine. A T_h1 response predominates after injection of an aqueous DNA solution into muscle. In contrast, after DNA immunization by gene gun, a T_h2 response predominates. Ensuring that the "correct" response is made following vaccination will be a key challenge as this vaccination technology moves forward.

Despite 30 years of study and enthusiasm that this vaccination strategy could be transformative for vaccine design, much remains to be investigated. Variations on this approach are showing promise. A technique called **gene shuffling**, which can be applied to produce diverse coding sequences, may have utility in DNA vaccine technology. Another variation on the single-gene DNA vaccine is a **genomic vaccine**, in which a library of all the genes of a particular pathogen is prepared in multiple DNA vaccine vectors. The entire plasmid mixture is injected into an animal. Such a vaccine has the potential to present every gene product of the pathogen to the immune system.

While DNA vaccines do not carry many of the risks of more traditional vaccines (such as reversions or the consequences of adjuvant use), other concerns about their safety must be addressed. Some possible dangers include unintentional triggering of autoimmune responses to the plasmid DNA (including the synthesis of anti-DNA antibodies) and induction of immune tolerance to the protein produced. However, an intramuscular DNA vaccine to prevent West Nile virus infection of horses has been approved, and there are promising leads for human immunodeficiency virus and hepatitis C virus DNA vaccines.

Attenuated Viral Vectors and Foreign Gene Expression

Genes from a pathogenic virus can be inserted into a nonpathogenic viral vector to produce viral proteins that can immunize a host against the pathogenic virus. In principle, the vector provides the benefits of a viral infection with respect to stimulating an immune response to the exogenous proteins, but without the attendant pathogenesis associated with a virulent virus. However, any replicating viral vector has the potential to produce pathogenic side effects, particularly if injected directly into organs or the bloodstream. The immune response to such hybrid viruses is not always predictable, particularly in more vulnerable populations (children, the elderly, and immunocompromised individuals).

Poxviruses, such as vaccinia virus, often are used as vaccine vectors. A wide variety of systems are available for the construction of vaccinia virus recombinants that cannot replicate in mammalian cells, but that allow the efficient synthesis of cloned gene products that retain their immunogenicity. Such vectors can accommodate >25 kb of new genetic information. Vaccinia virus recombinants can also be used to dissect the immune response to a given protein from a pathogenic virus. This application is illustrated in Fig. 8.14. Other poxviruses, including raccoonpox, canarypox, and fowlpox viruses, are possible alternatives because they are able to infect, but not propagate in, humans. Moreover, these viruses can be used serially: the individual can be inoculated with a

different vector expressing the same antigens, overcoming the limitations of making an immune response to the vector itself.

The successful use of an oral rabies vaccine for wild animals in Europe and the United States demonstrates that recombinant vaccinia virus vaccines have considerable potential. Recombinant vaccinia virus genomes encoding the major envelope protein of rabies virus yield virus particles that are formulated in edible pellets to be spread in the wild. The pellets are designed to attract the particular animal to be immunized (for example, foxes or raccoons). The animal eats the pellet, is infected by the recombinant virus, and becomes vaccinated. While effective, this clever approach must be applied with care. As vaccinia virus infection of humans is associated with rare but serious side effects, inadvertent human infection by these wildlife vaccines poses a risk (Box 8.12).

Vaccine Technology: Delivery and Improving Antigenicity
Adjuvants Stimulate an Immune Response
Charles Janeway once revealed what he called "the immunologist's dirty little secret": inactivated virus particles or purified proteins often do not induce the same immune response

Figure 8.14 Use of recombinant vaccinia viruses to identify and analyze T and B cell epitopes from other viral pathogens. As illustrated, it is possible to determine if a particular viral protein contains a B cell epitope (binds antibodies), a T cell epitope (recognized by cytotoxic T lymphocytes), or both. Subsequent site-directed mutational analysis of the viral genes enables precise localization of these epitopes on the viral protein.

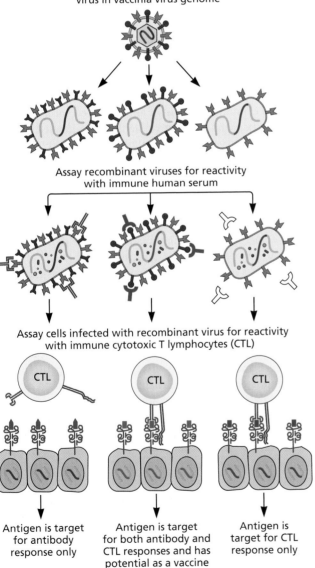

Clone individual genes from pathogenic virus in vaccinia virus genome

Assay recombinant viruses for reactivity with immune human serum

Assay cells infected with recombinant virus for reactivity with immune cytotoxic T lymphocytes (CTL)

CTL CTL CTL

Antigen is target for antibody response only

Antigen is target for both antibody and CTL responses and has potential as a vaccine

Antigen is target for CTL response only

as replication-competent, attenuated preparations, unless mixed with a substance that stimulates the early inflammatory response. Such immunostimulants are called **adjuvants**. Their development has been largely empirical, although as our understanding of the various regulators of immune responses increases, more-specific and -powerful adjuvants are being discovered and employed. Vaccine researchers can optimize a vaccine by using different combinations of adjuvant and immunogen to induce a protective immune response.

Adjuvants act by stimulating early intrinsic and innate defense signals, which then shape subsequent adaptive responses. These immunostimulators function in at least three distinct ways: by presenting antigens as particles, by sequestering antigen at the site of inoculation, and by directly stimulating the intrinsic and innate immune responses. The latter occurs when adjuvants mimic or induce cellular damage or alter homeostasis (sometimes called "danger" signals), or when they engage intrinsic cellular defense receptors.

Adjuvants vary in composition, from complex mixtures of killed mycobacteria and mineral oil (complete Freund's adjuvant) to lipid vesicles or mixtures of aluminum salts. Some adjuvants, like alum (microparticulate aluminum hydroxide gel), are widely used for human vaccines such as the papillomavirus, hepatitis A, and hepatitis B vaccines. Others, such as complete Freund's adjuvant, are used only in research. This adjuvant is extremely potent, but causes extensive tissue damage and toxicity. Two of the active components in Freund's adjuvant have been identified as muramyl dipeptide and lipid A, both potent activators of the inflammatory response. We now understand that the strong adjuvant effects of this complex cocktail are, at least in part, a result of the presence in an emulsion of mycobacterial DNA that activates the Toll-like receptor 9 (Tlr9) pathogen recognition protein. Less toxic derivatives, along with saponins and linear polymers of clustered hydrophobic and hydrophilic monomers, are promising and far safer alternatives.

Delivery and Formulation

Delivery by injection has many disadvantages, and therefore improvement of the administration of vaccines is an important goal of manufacturers. In some cases, alternative delivery methods require unique formulations. At present, vaccines are delivered by a limited number of methods, including the traditional hypodermic needle injection, oral administration, and the "air gun" injection of liquid vaccines under high pressure through the dead layers of skin to reach dendritic cells. Other methods under consideration include new emulsions, artificial particles, and direct injection of fine powders through the skin. Oral delivery of vaccines can be effective in stimulating IgA antibodies at mucosal surfaces of the intestine and in inducing a more systemic response. Genetically engineered edible plants that synthesize immunogenic viral proteins represent an attractive approach to designing potent and cost-effective oral vaccines. Transgenic plants expressing viral antigens can be developed, or plant viruses with genomes encoding immunogenic proteins can be used to infect food plants. Early experiments are promising: when such a plant is eaten, antibodies to the viral structural protein can be demonstrated in the animal's serum. Oral vaccination, by whatever methodology, is not always possible, because the enzymes of the oral cavity, coupled with the high acidity of the alimentary tract, destroy many vaccines.

Immunotherapy

Vaccination of patients who are already infected with viruses that cause persistent infections, or that are reactivated from latency by an immune response, presents special problems. One approach to resolve an established infection is via immunotherapy. **Immunotherapy** is a strategy to provide the already-infected host with antiviral cytokines, antibodies, or lymphocytes over and above those provided by the normal immune response. Immunotherapy can be administered by introduction of purified compounds or of a gene encoding the immunotherapeutic molecule. An attenuated virus or a DNA vaccine can be modified to synthesize cytokines that stimulate a desired immune response. If an attenuated vaccine is used, care must be taken, as it is possible that the intended immune response will have unexpected effects, such as increased virulence, persistence, and/or pathogenesis of the vaccine strain.

It is also possible to isolate lymphocytes from patients, infect these cells with a defective virus vector (e.g., a retrovirus) encoding an immunoregulatory molecule, and then infuse the transduced cells back into the patient. If these cells survive and synthesize the transduced protein, the patient's immune response may be boosted. When stem cells are transduced, a long-term effect may be achieved, as these cells will continue to divide in the transfused patient, producing daughter cells that propagate the transgene.

Immunotherapy with cytokines can also be effective. For example, the cytokine interferon α is approved in the United States for treatment of chronic hepatitis caused by hepatitis B and C viruses. Its effect on chronic hepatitis B virus infection is remarkable: as many as 50% of treated patients have no detectable infection after treatment. However, similar treatment of hepatitis C virus-infected patients has been less successful, for reasons that are not clear. Limitations of interferon therapy (and probably cytokine therapy in general) are that the biological activity of the therapeutic interferon is not sustained for a prolonged time, side effects are significant and often so extreme that patients elect to stop therapy, and treatment is expensive.

Immunomodulating agents, including interferon, cytokines that stimulate the T_h1 response (e.g., interleukin-2), and certain immune cell-attracting chemokines, are being studied individually and in combinations for their ability to reduce virus load and to moderate complications of persistent infections caused by papillomavirus and human immunodeficiency virus. Cytokines that stimulate natural killer cells (e.g., interleukin-12 and interferon γ) may hold promise as well.

The Quest for an AIDS Vaccine

In 1984, several years after human immunodeficiency virus was identified, officials in the U.S. government predicted that an AIDS vaccine would be available within 3 years. Despite 30 years of intensive work by laboratories across the world, substantial recent progress in identifying new epitopes, and developing strategies to overcome epitope diversity, a vaccine that protects against AIDS is still not on the horizon. Inaccurate predictions are nothing new in vaccine development: 3 years after poliovirus was isolated in 1908, Simon Flexner of the Rockefeller Institute confidently announced that a vaccine would be prepared in 6 months. Almost 50 years of research on basic poliovirus biology was necessary to provide the knowledge of pathogenesis and immunity that allowed the development of effective poliomyelitis vaccines (Table 8.2). The prevalence of human immunodeficiency virus throughout the world and the destruction that it causes to its victims and families make the development of an AIDS vaccine the new holy grail for vaccinologists.

The past 30 years has been a roller coaster of promise and disappointment. It was appreciated even in the 1990s that human immunodeficiency virus would be a formidable challenge for vaccine development. The lack of success can be explained by both the biology of this virus and its interaction with the host immune system. Although a vigorous immune response is induced after infection, the virus is not cleared. Infection and death of the very cells that coordinate an effective adaptive immune response, the CD4+ T lymphocytes, frustrate a coordinated host response. Equally frustrating is the relationship of the virus with its host cell: lentivirus reproduction requires integration of a DNA copy of the viral genome into the cellular genome. In some cells, a latent infection can be established in which the integrated proviral DNA produces no proteins, and the infection thus remains invisible to immune recognition. A third obstacle to vaccine development is the high mutation frequency of the virus: the host may make a suitable response to a particular epitope, but when the epitope is altered by mutations in its coding sequence, this response will be rendered useless. Consequently, while most vaccines are intended to mimic the natural host response to infection, to be effective, a vaccine against human immunodeficiency virus must improve on that response.

In addition to these technical challenges, complicated social, ethical, and political issues arise when vaccines for viruses in which humans are the only hosts are to be tested. For example, should at-risk groups, including gay men or sex workers, be preferentially selected as candidates for vaccination tests? On a larger scale, significant political issues related to mutual trust arise when vaccines are to be tested in resource-poor countries. In northern Nigeria in 2003, the political and religious leaders of three states called on parents not to vaccinate their children with the "Western" poliovirus vaccine, stating that the vaccine could be contaminated with antifertility agents or cancer-causing compounds. Such scare tactics are effective: Nigeria remains one of only three countries in which poliovirus is still endemic.

At a minimum, we can make a short list of our expectations for a human immunodeficiency virus vaccine. Of course it must be effective and safe. The protection should last for many years and protect against as many of the diverse human immunodeficiency virus strains as possible. The vaccine should not be so complicated that it cannot be produced on a large scale at a reasonable price. It should be stable, with a significantly long shelf life, so that it can be distributed, stored, and delivered when needed, especially in underdeveloped regions of the world where human immunodeficiency virus infections are prevalent. As the window of opportunity to block a primary infection, integration, and dissemination within a host is very short, the vaccine must act quickly and result in complete protection before the virus can go into hiding.

Formidable Challenges and Promising Leads

In the early days of human immunodeficiency virus vaccine research, a number of approaches, including the use of inactivated virus particles, or subunit vaccines based on single viral

Table 8.2 When can we expect an HIV vaccine?

Viral vaccine	Yr when etiologic agent was discovered	Yr when vaccine was developed in the United States	No. of yr elapsed
Polio	1908	1955	47
Measles	1953	1983	30
Hepatitis B	1965	1981	16
Rotavirus	1970	1998	28
Hepatitis A	1973	1995	22
HIV	1983	None yet	>30

BOX 8.13

DISCUSSION

National vaccine programs depend on public acceptance of their value

The measles-mumps-rubella (MMR) vaccine, a cocktail of three attenuated virus strains, has proven to be remarkably effective in reducing the incidence of these highly contagious and serious diseases. The economic benefit in the United States alone from use of the MMR vaccine has been estimated to exceed $5 billion per year.

In 1998, a publication in the prestigious medical journal *The Lancet* raised the specter that vaccines may contribute to the development of childhood autistic spectrum disorders and the associated colitis that affects many children with these disorders. Because the rates of diagnosis of childhood autism were increasing without a defined etiological basis, this report was quickly embraced by news media, raising parental concerns about vaccine safety.

The controversial report triggered more than a dozen retrospective and prospective epidemiological studies across the world, all of which concluded that vaccination was **not** linked to the escalation in autism diagnoses. The Centers for Disease Control and Prevention, the American Academy of Pediatrics, the Institute of Medicine, the National Institutes of Health, and other global health organizations have unambiguously and repeatedly affirmed the vaccine's safety.

Moreover, further investigation into the *Lancet* report revealed conflicts of interest, scientific inconsistencies, and ethically questionable practices in the research methodology used in this paper. As a result, all but the lead author, Andrew Wakefield, removed their names from the manuscript, which was partially retracted in 2004 and fully retracted in 2010. Wakefield, a practicing physician in the United Kingdom, was found guilty of professional misconduct by the General Medical Council and was struck from the Medical Register.

Nevertheless, as a consequence of this misinformation, the measles immunization rate has fallen significantly in resource-rich countries that had previously had high levels of vaccine coverage and negligible cases of infection. In many communities, rates of

protection have dipped below that required to maintain herd immunity, and predictably, measles virus infections are now appearing for the first time since the 1970s (see the figure). A measles outbreak in 2015 that began in Disneyland in California brought the risks of not vaccinating into clear focus, and began to change the conversation from parental choice to public obligation.

What can be learned from this sad chapter in vaccine history? First, it is heartening that theories proposed in the literature are subjected to repeated testing and revision by other groups: the rapid mobilization of many laboratories (at considerable effort and expense) provided an analytical and controlled counterpoint to the *Lancet* study. More sobering, however, is that, even when the data are unambiguous, the public response cannot be predicted. There remains a critical need to improve the lines of communication between scientists and the lay community.

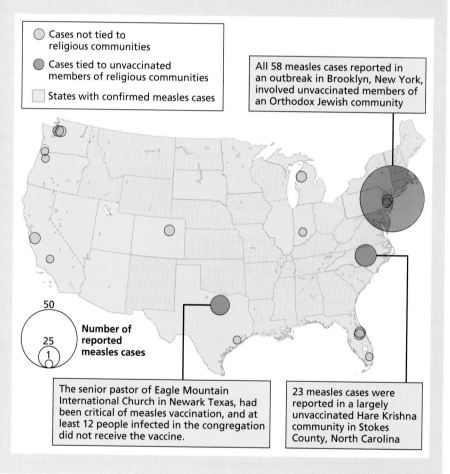

Cases not tied to religious communities

Cases tied to unvaccinated members of religious communities

States with confirmed measles cases

All 58 measles cases reported in an outbreak in Brooklyn, New York, involved unvaccinated members of an Orthodox Jewish community

50
25
1

Number of reported measles cases

The senior pastor of Eagle Mountain International Church in Newark Texas, had been critical of measles vaccination, and at least 12 people infected in the congregation did not receive the vaccine.

23 measles cases were reported in a largely unvaccinated Hare Krishna community in Stokes County, North Carolina

proteins, as well as passive immunization, were tested with no success. In particular, subunit vaccines, although capable of inducing strong antibody responses, were markedly inefficient in eliciting a cytotoxic-T-lymphocyte response. Furthermore, attenuated human immunodeficiency virus type

1 vaccines, modeled after the successful poliovirus vaccine, presented not only difficult scientific challenges but also ethical problems: the risks associated with injecting thousands of healthy, uninfected volunteers with an infectious (attenuated) virus were simply too great.

In 2013, the field of human immunodeficiency virus vaccine research suffered a major and highly publicized setback with the failure of a trial designed to test the hypothesis that high levels of $CD8^+$ T cells could protect against transmission. Despite strong T cell responses and appreciable antibody levels in the vaccinees, there was no added protection against infection. But hopes were rekindled when it was shown that a cytomegalovirus vector that expressed simian immunodeficiency virus antigens could eliminate infection in half of the infected rhesus macaques. The search is now on to develop a similar strategy for use in humans. To date, the only trial in humans that showed any efficacy was one done in Thailand using a canarypox vector encoding the viral glycoprotein gp120, which provided protection for a third of the vaccinees.

Given the high mutation frequency, many believe that a vaccine will need to induce broadly neutralizing antibodies that can inhibit multiple viral strains to be efficacious. Technological advances have enabled the isolation of many such antibodies from infected patients (see Chapter 7), and this success has generated newfound enthusiasm that a vaccine to prevent infection and AIDS may, at last, be possible.

Perspectives

Historic successes with vaccines for smallpox, measles, polio, and other viral infections, combined with promising new formulations such as the virus-like particle preparations that protect against hepatitis B virus and papillomavirus infections, have transformed modern medicine. However, our lack of progress with other viruses, including the failure as yet to develop a human immunodeficiency virus vaccine, reminds us that viral infection and immunity are intricate and poorly understood processes. Differences in a host's immune response to individual viruses, and the difficulty in developing safe and efficacious vaccines without a complete grasp of how immunity "works" in each case, pose significant roadblocks to progress. We have learned that intervening in any complex host-parasite interaction often produces unanticipated effects. Indeed, we frequently find out how little we know when vaccines are tested in the real world: formulations that worked in the lab fail in the field.

Sadly, even when vaccines are available, people may refuse to accept vaccines, and societies may not use or be able to pay for them. These issues, which are more about public perception and education, determine how effective a vaccine campaign will be (Box 8.13). In some cases, individuals or certain communities may not be able to pay for vaccines. As viruses do not respect country boundaries, this is a problem that international organizations must address.

It is noteworthy, and somewhat humbling, that essentially all successful vaccines on the market today were developed empirically: think back to the high-risk experiment that Jenner attempted, and how this was based more on a hunch than on actual scientific evidence. As we learn more about the molecular mechanisms of antiviral immune defenses and the epidemiology of infections, we anticipate that the development of future generations of vaccines will be based on scientifically sound principles, rather than conversations overheard between milkmaids.

References

Books

Koff WC, Kahn P, Gust ID (ed). 2007. *AIDS Vaccine Development: Challenges and Opportunities.* Caister Academic Press, Wymondham, United Kingdom.

Mims C, Nash A, Stephen J. 2001. *Mims' Pathogenesis of Infectious Disease,* 5th ed. Academic Press, Inc, Orlando, FL.

Murphy K, Travers P, Walport M. 2007. *Janeway's Immunobiology,* 7th ed. Garland Science, Garland Publishing Inc., New York, NY.

Plotkin SA, Orenstein WA, Offit PA (ed). 2012. *Vaccines,* 6th ed. WB Saunders Co, Philadelphia, PA.

Richman DD, Whitley RJ, Hayden FG (ed). 2009. *Clinical Virology,* 3rd ed. ASM Press, Washington, DC.

Historical Papers and Books

Bodian D. 1955. Emerging concept of poliomyelitis infection. *Science* **122:**105–108.

Enders JF, Weller TH, Robbins FC. 1949. Cultivation of the Lansing strain of poliomyelitis virus in cultures of various human embryonic tissues. *Science* **109:**85–87.

Jenner E. 1788. Observations on the natural history of the cuckoo. *Philos Trans R Soc Lond B* **78:**219–237.

Jenner E. 1798. *An Inquiry into the Causes and Effects of the Variolae Vaccinae, a Disease Discovered in Some of the Western Counties of England, Especially Gloucestershire, and Known by the Name of Cow Pox.* Sampson Low, London, England.

Metchnikoff E. 1905. *Immunity in the Infectious Diseases.* Macmillan Press, New York, NY.

Salk J, Salk D. 1977. Control of influenza and poliomyelitis with killed virus vaccines. *Science* **195:**834–847.

Woodville W. 1796. *The History of the Inoculation of the Smallpox in Great Britain; Comprehending a Review of All the Publications on the Subject: with an Experimental Inquiry into the Relative Advantages of Every Measure Which Has Been Deemed Necessary in the Process of Inoculation.* James Phillips, London, England.

Reviews

Angel J, Franco MA, Greenberg HB. 2007. Rotavirus vaccines: recent developments and future considerations. *Nat Rev Microbiol* **5:**529–539.

Corti D, Lanzavecchia A. 2013. Broadly neutralizing antiviral antibodies. *Annu Rev Immunol* **31:**705–742.

Donnelly JJ, Wahren B, Liu MA. 2005. DNA vaccines: progress and challenges. *J Immunol* **175:**633–639.

Flego M, Ascione A, Cianfriglia M, Vella S. 2013. Clinical development of monoclonal antibody-based drugs in HIV and HCV. *BMC Med* **11:**4. doi:10.1186/1741-7015-11-4.

Guy B. 2007. The perfect mix: recent progress in adjuvant research. *Nat Rev Microbiol* **5:**505–517.

Haynes BF, Moody MA, Alam M, Bonsignori M, Verkoczy L, Ferrari G, Gao F, Tomaras GD, Liao HX, Kelsoe G. 2014. Progress in HIV-1 vaccine development. *J Allergy Clin Immunol* **134:**3–10.

Ishii KJ, Koyama S, Nakagawa A, Coban C, Akira S. 2008. Host innate immune receptors and beyond: making sense of microbial infections. *Cell Host Microbe* **3:**352–363.

Kwong PD, Mascola JR. 2012. Human antibodies that neutralize HIV-1: identification, structures, and B cell ontogenies. *Immunity* **37:**412–425.

Lefebvre JS, Haynes L. 2013. Vaccine strategies to enhance immune responses in the aged. *Curr Opin Immunol* **25:**523–528.

Madhan S, Prabakaran M, Kwang J. 2010. Baculovirus as vaccine vectors. *Curr Gene Ther* **10:**201–213.

Moss B. 1996. Genetically engineered poxviruses for recombinant gene expression, vaccination, and safety. *Proc Natl Acad Sci U S A* **93:**11341–11348.

Moss WJ, Griffin DE. 2006. Global measles elimination. *Nat Rev Microbiol* **4:**900–908.

Neutra MR, Kozlowski PA. 2006. Mucosal vaccines: the promise and the challenge. *Nat Rev Immunol* **6:**148–158.

Palesch D, Kirchhoff F. 2013. First steps toward a globally effective HIV/AIDS vaccine. *Cell* **155:**495–497.

Roehrig JT. 2013. West Nile virus in the United States: a historical perspective. *Viruses* **5:**3088–3108.

Stephenson KE, Barouch DH. 2013. A global approach to HIV-1 vaccine development. *Immunol Rev* **254:**295–304.

Wherry EJ, Ahmed R. 2004. Memory CD8 T-cell differentiation during viral infection. *J Virol* **78:**5535–5545.

Selected Papers

Baden LR, Curfman GD, Morrissey S, Drazen JM. 2007. Human papillomavirus vaccine—opportunity and challenge. *N Engl J Med* **356:**1990–1991.

Brochier B, Kieny MP, Costy F, Coppens P, Bauduin B, Lecocq JP, Languet B, Chappuis G, Desmettre P, Afiademanyo K, Libois R, Pastoret PP. 1991. Large-scale eradication of rabies using recombinant vaccinia-rabies vaccine. *Nature* **354:**520–522.

Chen D, Endres RL, Erickson CA, Weis KF, McGregor MW, Kawaoka Y, Payne LG. 2000. Epidermal immunization by a needle-free powder delivery technology: immunogenicity of influenza vaccine and protection in mice. *Nat Med* **6:**1187–1190.

Clark HF, Offit PA, Ellis RW, Eiden JJ, Krah D, Shaw AR, Pichichero M, Treanor JJ, Borian FE, Bell LM, Plotkin SA. 1996. The development of multivalent bovine rotavirus (strain WC3) reassortant vaccine for infants. *J Infect Dis* **174**(Suppl 1):S73–S80.

Eisenbarth SC, Colegio OR, O'Connor W, Sutterwala FS, Flavell RA. 2008. Crucial role for the Nalp3 inflammasome in the immunostimulatory properties of aluminium adjuvants. *Nature* **453:**1122–1126.

Gans HA, Arvin AM, Galinus J, Logan L, DeHovitz R, Maldonado Y. 1998. Deficiency of the humoral immune response to measles vaccine in infants immunized at age 6 months. *JAMA* **280:**527–532.

Hassett DE, Zhang J, Slifka M, Whitton JL. 2000. Immune responses following neonatal DNA vaccination are long-lived, abundant, and qualitatively similar to those induced by conventional immunization. *J Virol* **74:**2620–2627.

Liao HX, Lynch R, Zhou T, Gao F, Alam SM, Boyd SD, Fire AZ, Roskin KM, Schramm CA, Zhang Z, Zhu J, Shapiro L, NISC Comparative Sequencing Program, Mullikin JC, Gnanakaran S, Hraber P, Wiehe K, Kelsoe G, Yang G, Xia SM, Montefiori DC, Parks R, Lloyd KE, Scearce RM, Soderberg KA, Cohen M, Kamanga G, Louder MK, Tran LM, Chen Y, Cai F, Chen S, Moquin S, Du X, Joyce MG, Srivatsan S, Zhang B, Zheng A, Shaw GM, Hahn BH, Kepler TB, Korber BT, Kwong PD, Mascola JR, Haynes BF. 2013. Co-evolution of a broadly neutralizing HIV-1 antibody and founder virus. *Nature* **496:**469–476.

Modelska A, Dietzschold B, Sleysh N, Fu ZF, Steplewski K, Hooper DC, Koprowski H, Yusibov V. 1998. Immunization against rabies with plant-derived antigen. *Proc Natl Acad Sci U S A* **95:**2481–2485.

Quinlivan ML, Gershon AA, Al Bassam MM, Steinberg SP, LaRussa P, Nichols RA, Breuer J. 2007. Natural selection for rash-forming genotypes of the varicella-zoster vaccine virus detected within immunized human hosts. *Proc Natl Acad Sci U S A* **104:**208–212.

Xu L, Sanchez A, Yang Z, Zaki SR, Nabel EG, Nichol ST, Nabel GJ. 1998. Immunization for Ebola virus infection. *Nat Med* **4:**37–42.

Zhou J, Sun XY, Stenzel DJ, Frazer IH. 1991. Expression of vaccinia recombinant HPV 16 L1 and L2 ORF proteins in epithelial cells is sufficient for assembly of HPV virion-like particles. *Virology* **185:**251–257.

9 Antiviral Drugs

Introduction

Historical Perspective

Discovering Antiviral Compounds

The Lexicon of Antiviral Discovery

Screening for Antiviral Compounds

Computational Approaches to Drug Discovery

The Difference between "R" and "D"

Examples of Some Antiviral Drugs

Approved Inhibitors of Viral Nucleic Acid Synthesis

Approved Drugs That Are Not Inhibitors of Nucleic Acid Synthesis

Expanding Target Options for Antiviral Drug Development

Entry and Uncoating Inhibitors

Viral Regulatory Proteins

Regulatory RNA Molecules

Proteases and Nucleic Acid Synthesis and Processing Enzymes

Two Success Stories: Human Immunodeficiency and Hepatitis C Viruses

Inhibitors of Human Immunodeficiency Virus and Hepatitis C Virus Polymerases

Human Immunodeficiency Virus and Hepatitis C Virus Protease Inhibitors

Human Immunodeficiency Virus Integrase Inhibitors

Hepatitis C Virus Multifunctional Protein NS5A

Inhibitors of Human Immunodeficiency Virus Fusion and Entry

Drug Resistance

Combination Therapy

Challenges Remaining

Perspectives

References

LINKS FOR CHAPTER 9

▶▶ *Video: Interview with Dr. Benhur Lee*
http://bit.ly/Virology_Lee

▶▶ *Draco's potion*
http://bit.ly/Virology_Twiv146

▶▶ *Hacking aphid behavior*
http://bit.ly/Virology_Twiv70

▶▶ *Combination antiviral therapy for hepatitis C*
http://bit.ly/Virology_10-14-14

Introduction

Public health measures and vaccines can control some viral infections effectively. For those that cannot, we must rely on antiviral drugs. Unfortunately, despite more than 50 years of research, our armamentarium of such drugs remains surprisingly small. As will be described in this chapter, this paucity reflects the many challenges that must be met in drug development. However, when available, antivirals can have a major impact on human health. Because of their medical importance, most of our antiviral drugs are directed against infections with human immunodeficiency virus and herpesviruses. In these cases, literally millions of lives have been saved by use of antiviral drugs.

One major limitation in antiviral drug development is the requirement for a high degree of **safety**. This restriction can be difficult to surmount because of the dependence of viruses on cellular functions: a compound that blocks a pathway that is critical for the virus can also have deleterious effects on the host cell. Another requirement is that antiviral compounds must be extremely **potent**: even modest reproduction in the presence of an inhibitor provides the opportunity for resistant mutants to prosper. Achieving sufficient potency to block viral reproduction completely is remarkably difficult. Other limitations can be imposed by the difficulty in propagating some medically important viruses in the laboratory

(e.g., hepatitis B virus and papillomaviruses) and the lack of small-animal models that faithfully reproduce infection in humans (such as measles and hepatitis C viruses). Lack of rapid diagnostic reagents has also hampered the development and marketing of antiviral drugs to treat many acute viral diseases, even when the effective therapies are available.

Historical Perspective

The first large-scale effort to find antiviral compounds began in the early 1950s with a search for inhibitors of smallpox virus reproduction. At that time, virology was in its infancy and smallpox was a worldwide scourge. Drug companies expanded efforts in the 1960s and 1970s, spurred on by increased knowledge and understanding of the viral etiology of common diseases, as well as by remarkable progress in the discovery of antibiotics to treat bacterial infections. The companies launched massive screening programs to find chemicals with antiviral activities. Despite much effort, there was relatively little success. One notable exception was amantadine (Symmetrel), approved in the late 1960s by the U.S. Food and Drug Administration (FDA) for treatment of influenza A virus infections. These antiviral discovery programs comprised **blind screening**, in which random chemicals and natural-product mixtures were tested for their ability to block the reproduction of a variety of viruses in cell culture systems. Candidate inhibitors were then tested in various cell and animal models for safety and efficacy. Promising molecules, called "leads," were modified systematically to reduce toxicity, increase solubility and bioavailability, or improve biological half-life. As a consequence, thousands of molecules

---

PRINCIPLES *Antiviral drugs*

- Antiviral compounds must be extremely potent to be effective: even modest viral reproduction in the presence of an inhibitor provides the opportunity for resistant mutants to prosper.

- Most antiviral compounds in clinical use target viral enzymes, such as proteases, and nucleic acid-synthesizing proteins.

- New drug design is focusing on blocking such viral functions as entry and uncoating, and the activities of viral regulatory proteins and RNA molecules.

- Whole-genome sequencing and methods to block gene expression make it possible to test for the requirement of every host gene in the reproduction of most viruses, further expanding the potential targets for antiviral drug design.

- Sophisticated computational methods have been developed to identify drug leads by "virtual screening," iterative docking of each chemical into a chosen site in a protein target.

- It is common for thousands of leads to yield but one promising drug candidate.

- Once a drug candidate is identified, clinical studies are needed to determine whether the compound gets to the right place in the body and at the appropriate concentration, persists in the body long enough to be effective, and is well tolerated and not toxic.

- More promising drug candidates are discarded because of toxicity and safety concerns than for any other reason.

- It often takes 5 to 10 years after identification of a candidate to get the drug to market.

- Emergence of drug-resistant mutants is of special concern during the extended period of therapy required for viruses that establish chronic infections. Combining two or more drugs with distinct targets circumvents the appearance of cells resistant to one treatment or the other.

- Potent drugs and drug combinations are now available to inhibit the reproduction of several human viruses, including herpesviruses, hepatitis B and C viruses, and human immunodeficiency virus.

were often made and screened before a specific antiviral compound was tested in humans. The mechanism by which these compounds inhibited the virus was often unknown. For example, the mechanism of action of amantadine, blocking the viral ion channel protein M2 and inhibiting uncoating, was not deduced until the early 1990s, almost 30 years after its discovery.

Discovering Antiviral Compounds

With the advent of modern molecular virology and recombinant DNA technology, the random, blind-screening procedures described above were all but discarded. Instead, viral genes essential for reproduction were cloned and expressed in genetically tractable organisms, and their products purified and analyzed in molecular and atomic detail. The life cycles of many viruses are known, allowing identification of numerous targets for intervention (Fig. 9.1). Inhibitors of critical processes can be found, even for viruses that cannot

be propagated in cultured cells. With the development of whole-genome sequencing and methods to block gene expression, it is now possible to test for the requirement of every host gene in the reproduction of most viruses, thereby expanding the potential targets for intervention. The practical challenge now is how to work from medical need to effective product.

The Lexicon of Antiviral Discovery

The antiviral discovery toolbox has expanded in the last decade to include many new devices and methods (Fig. 9.2). But lead compounds discovered using these approaches are only the starting point for the development of clinically useful drugs, i.e., compounds **approved** and licensed for use in humans. Many challenges remain after a drug candidate is identified. The developers must determine if the compound will get to the right place in the body and at the appropriate concentration (**bioavailability**), will persist in the body long enough to be effective (**pharmacokinetics**), and will be tolerated or toxic.

Figure 9.1 Knowledge of viral life cycles identifies general targets for antiviral drug discovery.

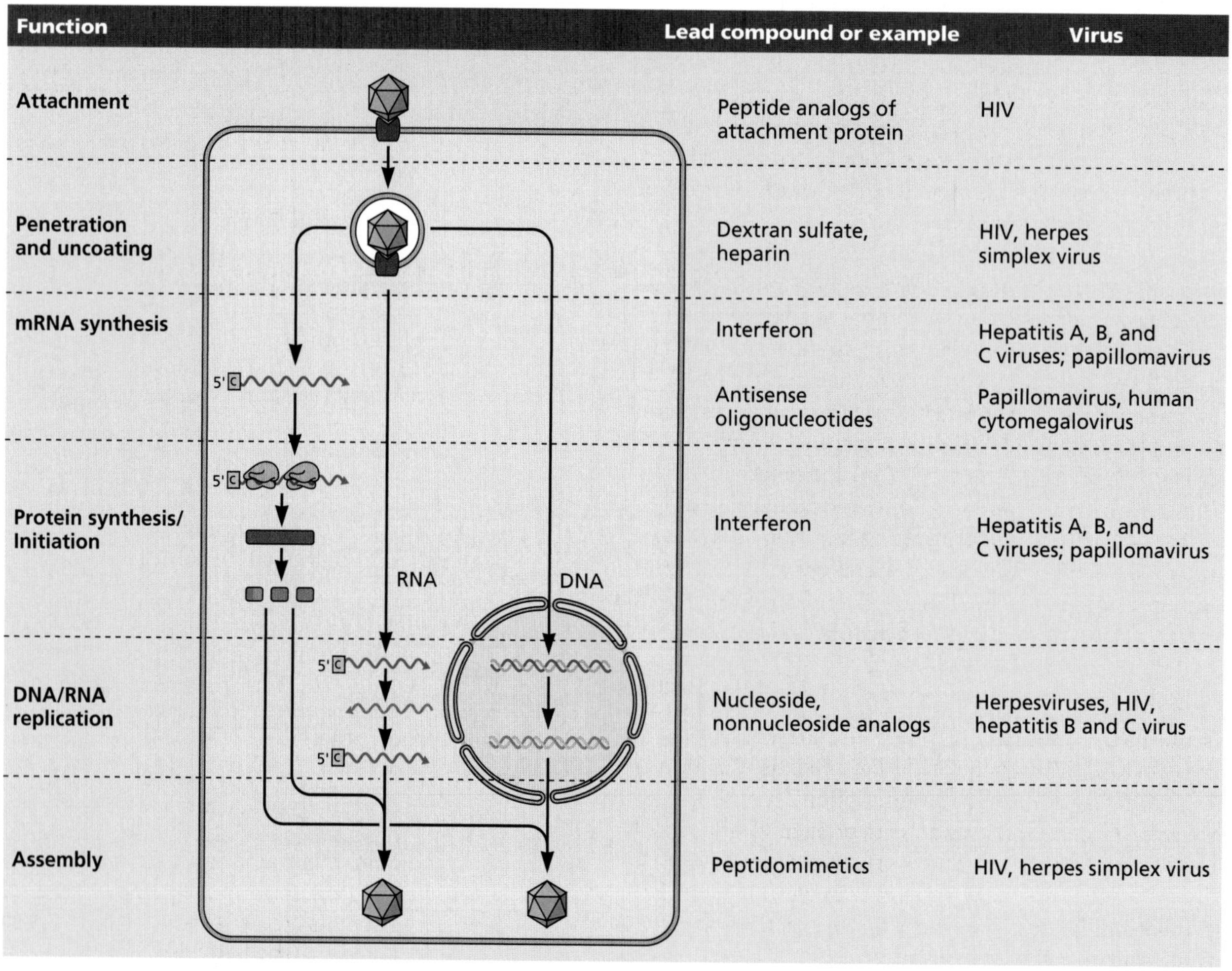

Function	Lead compound or example	Virus
Attachment	Peptide analogs of attachment protein	HIV
Penetration and uncoating	Dextran sulfate, heparin	HIV, herpes simplex virus
mRNA synthesis	Interferon	Hepatitis A, B, and C viruses; papillomavirus
	Antisense oligonucleotides	Papillomavirus, human cytomegalovirus
Protein synthesis/Initiation	Interferon	Hepatitis A, B, and C viruses; papillomavirus
DNA/RNA replication	Nucleoside, nonnucleoside analogs	Herpesviruses, HIV, hepatitis B and C virus
Assembly	Peptidomimetics	HIV, herpes simplex virus

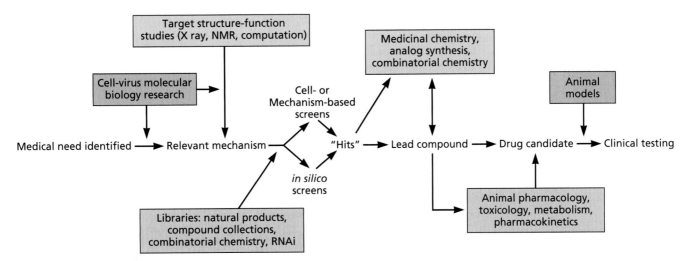

Figure 9.2 Path of drug discovery. The flow of information and action followed by modern drug discovery programs that ultimately yield compounds that can be tested clinically for efficacy is illustrated. NMR, nuclear magnetic resonance.

Screening for Antiviral Compounds

Genetics and Drug Discovery

Viral targets. Modern antiviral discovery methods have focused on genes known to be essential for viral reproduction and the use of mechanism-based screens (Fig. 9.1). Essential viral functions are defined by genetics and informed by our knowledge of viral genomes. Viral genomes can be manipulated to determine if a particular gene product is a valid target by construction and analysis of a mutant in which the gene of interest is inactivated or deleted.

Host targets. In the past, identifying host gene products that are required for viral reproduction was often impossible. Modern technology has changed this situation dramatically. As sequences of the genomes of humans and common laboratory animals are now available, host proteins that are essential for efficient virus reproduction and pathogenesis can be identified using RNA interference (RNAi) technology, genetic manipulation through transgenic and knockout approaches, and assays that detect protein-protein interactions. Maraviroc, an antagonist of the CCr5 cellular coreceptor for human immunodeficiency virus type 1, is one example of an approved drug that targets a host protein that enables viral infection.

Mechanism-Based Screens

As the name "mechanism-based" implies, this type of screen seeks to identify compounds that affect the function of a known viral target. Enzymes, transcriptional activators, cell surface receptors, and ion channels are popular targets. Often this screening is carried out with purified protein in formats that facilitate automated assay of many samples. One example of a mechanism-based screen designed to identify inhibitors of a viral protease is shown in Fig. 9.3.

Figure 9.3 Mechanism-based screen for inhibitors of a viral protease. The substrate is a short peptide encoding the protease cleavage site. A fluorogenic molecule is covalently joined to the N terminus of the peptide, and the entire complex is attached via its C terminus to a polystyrene bead. When the peptide-bead suspension is exposed to active protease, the peptide is cleaved such that the fluorogenic N terminus is released into the soluble fraction, which can be quickly and cleanly separated from the insoluble beads containing the nonfluorogenic product as well as the fluorogenic unreacted substrate. Protease activity is assayed by the appearance of soluble fluorescent peptide as a function of time as shown.

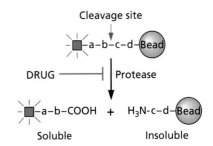

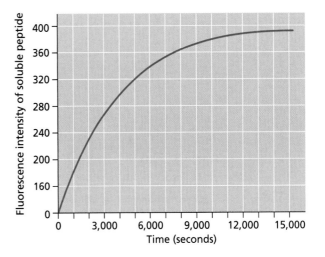

Cell-Based Screens

In cell-based assays, essential elements of the specific mechanism to be inhibited (e.g., a viral enzyme plus a readily assayable substrate) are engineered into an appropriate cell. An example of a bacteria-based screen with a convenient readout is shown in Fig. 9.4. Similar approaches work well in yeast or animal cells. In appropriate cell types, such assays can provide information not only about inhibition of the target reaction, but also about cytotoxicity and specificity. The use of several different reporter molecules may allow detection of more than one event at a time.

An interesting variation of the cell-based assay is the **minireplicon**. Certain viruses cannot be propagated readily in standard cultured cells (e.g., human papillomavirus) or are dangerous enough to require high biological containment (e.g., smallpox virus and the hemorrhagic fever viruses). The minireplicon system comprises a set of plasmids that separately carry genes that encode viral genome replication proteins, and an engineered viral genome segment marked with a reporter gene (i.e., the minireplicon); replication can be monitored by assaying for reporter gene expression. Inhibitors that block replication can be discovered, analyzed, and developed for therapeutic use. A subgenomic replicon system for human hepatitis C virus was used to great advantage for discovery of the first viral protease inhibitors capable of blocking reproduction of this virus.

High-Throughput Screens

High-throughput screens are mechanism- or cell-based screens that allow very large numbers of compounds to be tested in an automated fashion. It is not unusual for pharmaceutical companies to examine more than 10,000 compounds per assay per day, a rate inconceivable for early antiviral drug hunters. Compounds to be screened are typically arrayed in multiwell, plastic dishes. Robots then add samples of these compounds to other plastic dishes containing the cell-free or cell-based assay components, and after incubation, the signal created by the reporter gene (or other output) is read and recorded. Numerical data or images of cells or reactions can be captured, stored, and analyzed. Captured images are called **high-content screens** because they can examine more than one parameter simultaneously. For example, using antibodies, it is possible to monitor the import of transcriptional regulators from their site of cytoplasmic synthesis to their site

Figure 9.4 Cell-based screen for a viral protease inhibitor and a transcription regulatory protein. This cell-based assay for the protease of human immunodeficiency virus (HIV) uses the tetracycline resistance of genetically engineered bacteria as a readout. To facilitate uptake of small molecules that might inhibit the protease, a variety of *E. coli* strains that have reduced permeability barriers are available. It is important to include many controls and secondary assays to identify false-positive and false-negative results. This assay is described in more detail in R. H. Grafstrom et al., *Adv Exp Med Biol* **312:**25–40, 1992.

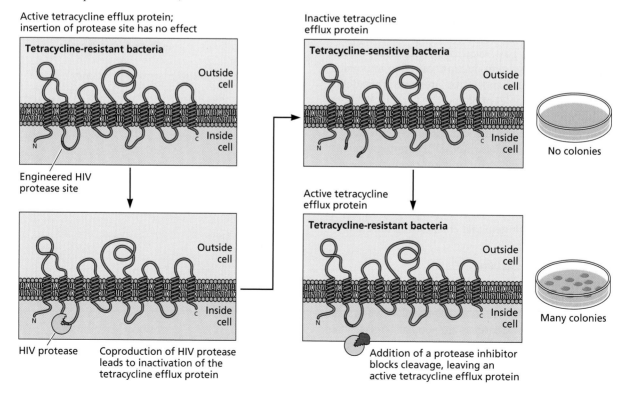

Active tetracycline efflux protein; insertion of protease site has no effect

Tetracycline-resistant bacteria

Outside cell

Inside cell

Engineered HIV protease site

HIV protease

Coproduction of HIV protease leads to inactivation of the tetracycline efflux protein

Inactive tetracycline efflux protein

Tetracycline-sensitive bacteria

Outside cell

Inside cell

No colonies

Active tetracycline efflux protein

Tetracycline-resistant bacteria

Outside cell

Inside cell

Many colonies

Addition of a protease inhibitor blocks cleavage, leaving an active tetracycline efflux protein of action in the nucleus at the same time that changes in cell morphology or protein production are visualized.

Sources of Chemical Compounds Used in Screening

Many pharmaceutical and chemical companies maintain large libraries of chemical compounds. Usually, a sample of every compound synthesized by the company for any project is archived, and its history is stored in a database. Chemical libraries of half a million or more distinct compounds are not unusual for a large company. Other kinds of libraries containing natural products collected from all over the world, including "broths" from microbial fermentations, extracts of plants and marine animals, and perfusions of soils containing mixtures of unknown compounds, can be searched for components that may have antiviral activities. A small-molecule repository of >200,000 compounds is maintained by the U.S. National Institutes of Health for use by the scientific community, and libraries of small bioactive molecules are available from a number of private entities.

Another type of chemical library may be produced by **combinatorial chemistry**, a technology that provides unprecedented numbers of small, synthetic molecules for screening (Fig. 9.5). Before implementation of this technology, a medicinal chemist could reliably synthesize and characterize only about 50 compounds a year. Combinatorial chemistry can provide all possible combinations of a basic set of modular components, often on uniquely tagged microbeads or other chemical supports, such that active compounds in the mixtures can be traced, purified, and identified with relative ease. Making thousands of compounds in days is now routine.

Computational Approaches to Drug Discovery

Structure-Assisted Drug Design

Structure-assisted design depends on knowing the atomic structure of the target molecule, usually obtained by X-ray crystallography. Computer programs, known and predicted mechanisms of enzyme action, fundamental chemistry, and personal insight all aid an investigator in the design of ligands that bind at a critical site and inhibit protein function. Currently, the atomic structures of tens of thousands of macromolecules, including important viral proteins, have been determined, many deposited in publicly accessible protein databases. Inhibitors of the human immunodeficiency virus protease may be the best example of successful antiviral agents that were designed from structure-assisted analyses (Fig. 9.6).

Figure 9.5 Combinatorial chemistry and the building-block approach to chemical libraries. Small organic molecules predicted to bind to different pockets on the surfaces of proteins can be grouped into subsets of distinctive chemical structures (different colored symbols). With automated procedures, these chemical entities can be joined together by various chemical linkers (lines) to produce a large but defined library of small compounds. For example, if assembled pairwise with 10 linkers, a collection of 10,000 small molecules yields a library of 1 billion new combinations. These defined chemical libraries allow a detailed exploration of the binding surfaces of complex proteins. Adapted from P. J. Hajduk et al., *Science* **278:**497–499, 1997, with permission.

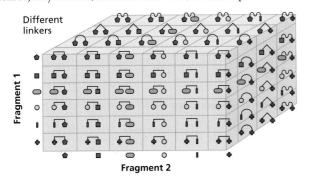

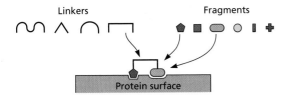

Figure 9.6 Structure of the human immunodeficiency virus (HIV) type 1 protease with the inhibitor saquinavir. The main chains of each monomer in the protease dimer are represented by blue and green ribbons inside the space-filling structure shown in pale blue mesh. A ball-and-stick model of saquinavir (Ligand) inside of a space-filling structure is shown in violet. The tight fit of the inhibitor in the active site of the protease is illustrated by the interlapping light blue and violet meshwork in the space-filling model. Two views, rotated by 90° (top and bottom), are shown. The images were prepared by J. Vondrasek and are reproduced from the protease database (http://xpdb.nist.gov/hivsdb).

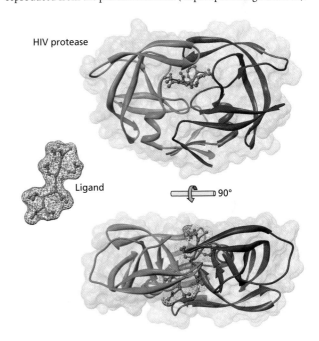

Genome Sequencing and Other Advances Expose New Targets for Antiviral Drugs

Standard approaches to antiviral drug development have focused on viral enzymes as primary targets (e.g., proteases, replicases, and reverse transcriptase), and most compounds in clinical use are such **direct-acting antivirals**. However, as virus reproduction is dependent on numerous host functions, host cell components and the interaction of viral proteins with them also represent potential targets for antiviral drug development. Because host genes have substantially lower mutation rates than do viral genes, drug resistance should be less of a problem with such compounds. A number of new, high-throughput methods now exist for determining the function of individual viral proteins in the host cell, their interactions with viral and cellular proteins, and the consequences of these interactions for both virus and host.

High-density arrays of DNA fragments on a DNA chip the size of a microscope slide have been used to assess the expression of thousands of genes in an infected cell in a single experiment (**microarray analyses**). This technology enabled scientists to identify the changes in mRNA concentration in response to viral infection in various cells and tissues, but has been largely superseded by the application of high-throughput, next-generation RNA sequencing methods (RNA-seq). RNA-seq provides information on the changes in all types of RNA in a cell in response to viral infection. Such data can be used to identify cellular genes and pathways that may be targets for antiviral drug development. With RNAi technology, it is possible to reduce the concentration of a potential target protein in cultured cells and observe the effects on virus reproduction. Modern methods for host gene editing using sequence-targeted nucleases have made it possible to change or delete a particular gene in both cultured cells and model organisms, and examine the contribution of that gene to virus reproduction or pathogenesis.

Advances in protein separation techniques, mass spectrometry, and bioinformatics have revolutionized our capacity determine the total protein repertoire (the **proteome**) of virus samples, and host cells and tissues, with great sensitivity. Proteomic analyses have also made it possible to assess protein interactions on global scales and to discover critical nodes and previously unknown connections. Genome-wide protein interaction maps are available for model organisms such as budding yeasts, and partial maps have been constructed for many others, using a number of *in vitro* or cell-based assays. An example of the use of such methods is a systematic survey of the binding of cellular proteins to each of the human immunodeficiency virus type 1 proteins, produced from bacterial expression plasmids in human cell lines (Fig. 9.7). The results identified not only previously documented interactions, but also new connections with cellular proteins and pathways that are potential targets for future pharmacological intervention.

In Silico Drug Discovery via Virtual Screening

With all the advances in the "-omics" (genomics, transcriptomics, proteomics, and metabolomics), an enormous number of potential targets for drug discovery have been identified. Furthermore, as a result of advances in chemistry, the number of chemical structures that can be tested as antiviral compounds has increased dramatically. In fact, one estimate suggests that as many as 10^{64} chemicals can be made to test for activity against human protein targets. Despite the paucity of lead antiviral compounds that engage such targets, no one would seriously think of urging medicinal chemists and biologists to do random screening with all possible compounds. However, many researchers now are using computers to take on this daunting task.

Structural biologists have provided atomic-resolution models for numerous viral and cellular proteins, and homologs of the vast majority of possible enzymatic active sites are present in current protein structure databases. Sophisticated computational methods have been developed to identify drug leads by "**virtual screening**," iterative docking of each chemical into a chosen site in a protein target. When a virtual small molecule "fits" into a pocket, the molecule is obtained or synthesized and then tested in a mechanism- or cell-based assay. Modifications to improve activity are made, and the computational analysis and testing are reiterated (Box 9.1).

As noted for the protease of human immunodeficiency virus type 1 (Fig. 9.6), computational methods are also used to optimize the binding strength or selectivity of lead compounds obtained by other screening methods. Further technological advances, such as improved systems for homology modeling from known structures, and development of algorithms for predicting protein structures *de novo* from coding sequences are expected to increase the future effectiveness of *in silico* approaches. The paradigm of virtual screening has been called "genome-to-drug-to-lead," and it has the potential to reduce the formidable human resource requirements for chemistry and biology.

The Difference between "R" and "D"

Antiviral Drugs Are Expensive To Discover, Develop, and Bring to the Market

Even with modern methods, it is common for thousands of leads to yield but one promising candidate for further drug development (Fig. 9.8). Research and lead identification, the "R" of "R&D," represent only the beginning of the process of producing a drug for clinical use. The "D" of "R&D" is development, comprising all the steps necessary to take an antiviral lead compound through safety testing, scale-up of synthesis, formulation, pharmacokinetic studies, and clinical trials.

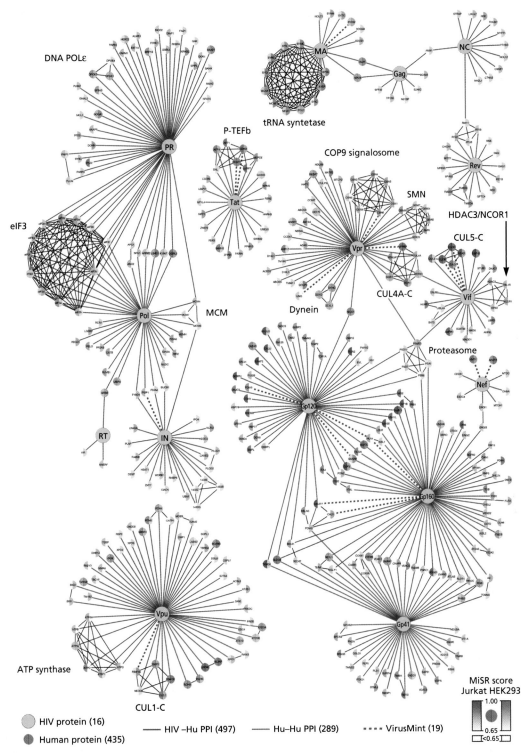

Figure 9.7 The human immunodeficiency virus (HIV) type 1 protein "interactome." The figure shows a network representation of 497 HIV-human interactions between 16 HIV proteins and 435 human proteins, connected by blue lines. Each node that represents a human protein is split into two colors; the intensity of each color represents a score for the relevant abundance of the protein and the reproducibility and specificity of the interaction in HEK293 (blue) or Jurkat (red) cultured cells. Black lines correspond to interactions between host proteins (289) that were obtained from publicly available databases; dashed lines correspond to links found in a virus interaction database. Reproduced from S. Jäger et al., *Nature* **481**:365–370, 2013, with permission.

EXPERIMENTS
An allosteric antiviral by in silico *design*

Retroviruses encode an enzyme, integrase, that catalyzes the specialized recombination reaction that joins viral and host DNAs, an essential step in the reproduction of these viruses (see Volume I, Chapter 6). The integrase protein of the human immunodeficiency virus type 1 is known to bind to the host cell transcriptional activator lens epithelium-derived growth factor (Ledgf/p75), which promotes this recombination reaction by tethering an integrase-viral DNA complex to host cell chromatin.

X-ray crystallographic analysis of the integrase-binding domain of Ledgf bound to the dimer interface of two isolated catalytic core domains (CCDs) of integrase revealed a well-defined binding pocket into which the end of an interhelical loop from the Ledgf domain was seen to extend (see panel A of the figure). On the basis of this structural information and other biochemical and genetic data, it was possible to define features (a pharmacophore) of a small molecule that would be optimal for binding in the CCD dimer pocket. Virtual, *in silico* screening of some 200,000 commercially available compounds that satisfied these features yielded several likely candidates. An *in vitro* assay for

inhibition of the Ledgf-integrase interaction by these candidates identified 2-(quinolin-3-yl) acetic acid derivatives as lead compounds for further development. Analyses of structure-activity relationships led to the synthesis, by various groups of investigators, of structurally related inhibitors with ever-increasing potency when tested for antiviral activity in cell cultures. Related compounds are now being tested for use in the clinic (see panels B and C).

The inhibitory compounds, called LEDGINs or ALLINIs (for Ledgf- or allosteric integrase inhibitors), were found to be dual-acting both *in vitro* and in cell culture experiments: they not only impeded integration by blocking Ledgf tethering, but also inhibited enzymatic activity by preventing critical conformational dynamics of the integrase protein. Most surprising, however, was the discovery that their antiviral potency was determined primarily by their ability to block proper virus particle maturation, not by their effects on enzymatic activity. The compounds promote integrase multimerization, a reaction that, for as yet unknown reasons, hinders formation of the normal electron-dense progeny viral cores. Most importantly, this research

demonstrated that allosteric sites on retroviral integrase are valid targets for antiviral drug discovery, and that *in silico* screening is a practical approach to identifying relevant lead compounds.

Christ F, Debyser Z. 2013. The LEDGF/p75 integrase interaction, a novel target for anti-HIV therapy. *Virology* **435:**102–109.

Christ F, Voet A, Marchand A, Nicolet S, Desimmie BA, Marchand D, Bardiot D, Van der Veken NJ, Van Remoortel B, Strelkov SV, De Maeyer M, Chaltin P, Debyser Z. 2010. Rational design of small-molecule inhibitors of the LEDGF/p75-integrase interaction and HIV replication. *Nat Chem Biol* **6:**442–448.

Jurado KA, Wang H, Slaughter A, Feng L, Kessl JJ, Koh Y, Wang W, Ballandras-Colas A, Patel PA, Fuchs JR, Kvaratskhelia M, Engelman A. 2013. Allosteric integrase inhibitor potency is determined through the inhibition of HIV-1 particle maturation. *Proc Natl Acad Sci U S A* **110:**8690–8695.

Kessl JJ, Jena N, Koh Y, Taskent-Sezgin H, Slaughter A, Feng L, de Silva S, Wu L, Le Grice SF, Engelman A, Fuchs JR, Kvaratskhelia M. 2012. Multimode, cooperative mechanism of action of allosteric HIV-1 integrase inhibitors. *J Biol Chem* **287:**16801–16811.

Tsiang M, Jones GS, Niedziela-Majka A, Kan E, Lansdon EB, Huang W, Hung M, Samuel D, Novikov N, Xu Y, Mitchell M, Guo H, Babaoglu K, Liu X, Geleziunas R, Sakowicz R. 2012. New class of HIV-1 integrase (IN) inhibitors with a dual mode of action. *J Biol Chem* **287:**21189–21203.

ALLINI structures and binding mechanisms. (A) X-ray cocrystal structure of the human immunodeficiency virus type 1 (HIV-1) integrase (IN)-Ledgf/p75 complex (left; PDB code 2B4J) shows one IN-binding domain (IBD) molecule (blue) bound at the interface of two IN CCD monomers (gold and silver), with IN active-site residues colored red and the interhelical loop of Ledgf/p75 penetrating into the cavity at the dimer interface. The side chains of Ledgf/p75 contact residues Ile365 and Asp366, and main chain atoms of Glu170 and His171 of IN are shown as sticks, with oxygen and nitrogen atoms colored red and blue, respectively, and H bonds drawn as dotted lines (right). **(B)** Chemical structures of LEDGIN-6, BI-1001, and GS-A, with the carboxyl group that mimics Ledgf/p75 hot spot residue Asp366 highlighted. **(C)** Binding of LEDGIN-6 (left; PDB code 3LPU), BI-1001 (middle; PDB code 4DMN), and GS-A (right; PDB code 4E1M) at the HIV-1 IN CCD-CCD interface. Red and blue colors identify oxygen and nitrogen atoms, respectively, in the compounds, and in Glu170, His171, and Thr174 (for simplicity, only main chain atoms of these residues are included). H bonds between the drugs and IN are dotted lines. Approximate values for 50% inhibition of HIV-1 reproduction in cell culture (EC$_{50}$) are indicated at the bottom. Adapted from L. Krishnan and A. Engelman, *J Biol Chem* 287:40858-40866, 2012, with permission.

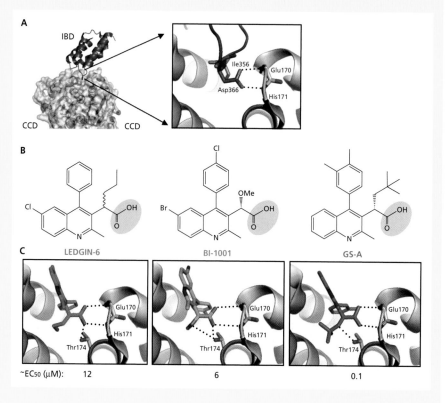

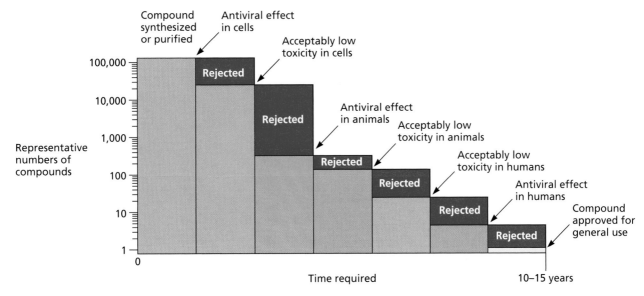

Figure 9.8 A descending staircase of drug discovery. Many compounds must be tested before a commercially viable antiviral drug will become available. The attrition rate is very high (red "rejected" label), as hundreds of thousands of chemicals are tested in multiple steps taking several years before one compound emerges as a drug. A few of the significant hurdles in the process and the extent of attrition at each step are illustrated.

With rare exceptions, it takes 5 to 10 years after the initial lead is identified to get a drug to the market. Decisions made by drug companies are influenced by these realities (Box 9.2).

According to an analysis in 2013 by *Forbes Magazine*, "a company hoping to get a single drug to market can expect to have spent $350 million before the medicine is available for sale … because so many drugs fail, large pharmaceutical companies that are working on dozens of drug projects at once spend an average of $5 billion per new medicine." Satisfying an unmet medical need, having no competitors, or being better than any competitive drug are all important issues for commercial development. The drug must also be relatively inexpensive to manufacture and easy to formulate and deliver (a pill to swallow is much preferred over injection, for example). Clinical testing must demonstrate that the drug is safe, effective, and has no serious side effects (Box 9.3). Finally, given the staggering costs noted above, the market for the drug must be large enough to ensure a profit.

Antiviral Drugs Must Be Safe

As in vaccine development, safety is the overriding concern of any company developing an antiviral drug. Toxicity to cultured cells and animals is the first indication that a compound may not be safe. More promising leads are discarded because of toxicity than for any other reason. Toxicity can be described in terms of the **cytotoxic index** (for cells) or the **therapeutic index** (for hosts). These indices are defined as the dose that inhibits virus reproduction divided by the dose that

---

BOX 9.2

DISCUSSION
New drugs, new mechanisms—no interest?

Two pharmaceutical companies independently discovered a new class of drug that inhibits herpes simplex virus reproduction. These compounds are targeted to the DNA helicase-primase, which is essential for viral genome replication. They represented the first new anti-herpes simplex virus drugs since acyclovir was developed in the 1970s. The helicase-primase inhibitors are more potent than acyclovir and its derivatives in animal models, and have remarkable potential.

However, neither company developed the inhibitors. The reason is that acyclovir is a safe, effective drug, and the expense of taking a new drug through clinical trials is enormous. Marketing strategy asserts that it is not cost-effective to compete with a proven drug. The reality is that companies must make choices about where to put their resources.

Crumpacker CS, Schaffer PA. 2002. New anti-HSV therapeutics target the helicase-primase complex. *Nat Med* **8:**327–328.

BOX 9.3

Clinical trials

Clinical trials are research studies in humans that test new ways to prevent, detect, diagnose, or treat a specific disease or condition. National and international regulations and policies have been developed to protect the rights, safety, and well-being of individuals who take part in clinical trials and to ensure that trials are conducted according to strict scientific and ethical principles. By taking part in clinical trials, participants can receive access to new treatments and help others by contributing to medical research.

Committees that are responsible for the protection of human subjects must approve all clinical trials. In the United States, this body is called the Institutional Review Board (IRB). Most IRBs are located in hospitals or other institutions in which the trial will be conducted, but approval by a central (independent/for profit) IRB may be acceptable for studies conducted at sites that do not have their own IRB.

Clinical trials are performed in a series of orderly, defined steps called "**phases**." The trials can vary in size and may involve a single site in one country or multiple sites in one or several countries, and it can take up to 10 or more years for a drug or treatment to be licensed for use in humans. The burden of paying for clinical trials (ultimately hundreds of millions of dollars) is borne by the sponsor, typically a pharmaceutical or biotech company, who designs the study in coordination with a panel of expert clinical investigators.

Phase I: The drug is tested in increasing doses with a small group of people (20 to 80). These trials are conducted mainly to evaluate the **safety** of chemical or biologic agents or other types of interventions. They help to determine the maximum dose that can be given safely (also known as the **maximum tolerated dose**), to assess whether an intervention causes harmful side effects, and to gain early evidence of effectiveness. Phase I trials may enroll people who have advanced disease that cannot be treated effectively with existing treatments or for which no treatment exists.

Phase II: These trials test the **effectiveness** of interventions in people who have a specific disease. They also continue to look at the safety of interventions. Phase II trials usually enroll fewer than 100 people but may include as many as 300. Although phase II trials can give an indication of whether or not an intervention works, they are almost never designed to show whether an intervention is better than standard therapy.

Phase III: These trials test the **effectiveness** of a new intervention or new use of an existing intervention. Phase III trials also examine how the **side effects** of the new intervention compare with those of the usual treatment. If the new intervention is more effective than the usual treatment and/or is easier to tolerate, it may become the new standard of care.

Phase III trials usually involve large groups of people (100 to several thousand), who are randomly assigned to one of two treatment groups, or "trial arms": a control group, in which everyone receives the usual treatment for their disease; or an investigational group, in which everyone receives the new intervention or new use of an existing intervention. Should a randomized clinical trial violate ethical standards (e.g., standard of care cannot be withdrawn), then an observational study of only a treated group is conducted.

Following successful completion of this phase, data are compiled in an application to the FDA for approval. If the drug is ruled "safe and effective," it is licensed for use in patients.

Phase IV: These trials further evaluate the effectiveness and long-term safety of drugs or other interventions. They usually take place after approval by the FDA for standard use, and are generally sponsored by drug companies. Several hundred to several thousand people may take part in a phase IV trial, which is also known as a postmarketing surveillance trial.

Approximate time frames from drug discovery to use in the clinic.

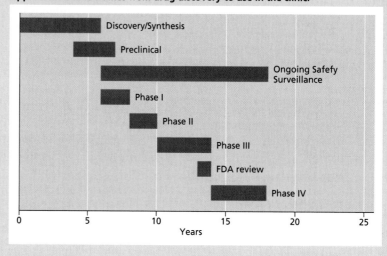

is toxic to cells or host. The lower the index, the better; indices of 1/1,000 or lower are preferred.

No human trial can be initiated without detailed safety studies in several animal species. Compounds that may be used in long-term treatment must be evaluated for toxicity, allergic effects, mutagenicity, and carcinogenicity. Safety overrides efficacy in most cases. On the other hand, when there are no other effective treatments, as in the early days of the acquired immunodeficiency syndrome (AIDS) pandemic, even drugs that caused some undesirable side effects can be licensed for human use (e.g., azidothymidine [AZT]).

Drug Formulation and Delivery

The science of formulation and delivery is an essential part of any antiviral discovery program. After administration, a drug must reach the proper place in a patient and remain at an effective concentration for a time sufficient to inhibit virus reproduction. A compound that cannot enter the bloodstream after ingestion is not likely to be effective. But satisfying this step may not be sufficient, as many compounds bind to albumin or other proteins in the blood and are thereby rendered ineffective. Other compounds may be inactivated as they pass through the liver and are cleared rapidly from the body. Such problems are generally discovered only by testing. Some insoluble compounds, or chemicals unable to enter the bloodstream after ingestion (poor bioavailability), can be modified by the addition of new side chains that may improve absorption from the intestine (Fig. 9.9). In addition, delivery vehicles

such as liposomes, minipumps, skin patches, or slow-release capsules may improve bioavailability. Other desirable features include stability and cost-effective synthesis in large quantities. Literally tons of precursor materials are needed to manufacture commercial quantities of an antiviral drug.

Examples of Some Antiviral Drugs

The antiviral drugs currently approved for general use are surprisingly few, belong to a limited number of chemical classes, and, as noted previously, mostly target viral proteins. While many are safe and effective, some are marginally efficacious or have side effects that limit their use to cases in which there are no alternatives. Some antiviral drugs that are of historic interest and still in general use are described below. In what follows, generic names for the drugs are used with the best-known brand/trade names in parentheses (Box 9.4).

Approved Inhibitors of Viral Nucleic Acid Synthesis

Most approved antiviral drugs are nucleoside or nucleotide analogs, directed against viral proteins that catalyze nucleic acid synthesis, or are nonnucleoside inhibitors of these proteins.

Acyclovir and Ganciclovir—Herpesvirus Infections

Acyclovir (Zovirax) is an example of a specific, nontoxic drug that is highly effective against herpes simplex virus (genital and oral herpes) and, to some extent, varicella-zoster virus (chickenpox and shingles). It was initially synthesized in 1974, but it was not until the mid-1980s that its full potential as an antiherpesviral drug was realized. Acyclovir is a nucleoside analog related to guanosine, containing an acyclic sugar group (Fig. 9.10). It is a **prodrug**, a precursor of the active antiviral compound. Conversion to the drug requires the sequential activities of three kinases that produce a triphosphate derivative, the actual antiviral compound (Fig. 9.11A). Herpes simplex virus and varicella-zoster virus genomes encode an enzyme that normally phosphorylates thymidine to form thymidine monophosphate, but this kinase can also accept a wide range of other substrates, including acyclovir. Cellular enzymes cannot perform this first reaction, but they can synthesize the di- and triphosphates, the latter of which is then used as a substrate by the viral polymerase for incorporation into viral DNA. As acyclovir lacks the $3'$-OH group of the sugar ring, the growing DNA chain is terminated upon its addition. The specificity of acyclovir for the herpesviruses depends therefore on the virally encoded thymidine kinase. Indeed, if this viral enzyme is synthesized in an uninfected cell and acyclovir is added, the cell will die because its DNA replication will also be blocked by the chain-terminating base analog. Such use of the viral enzyme is incorporated into several strategies for selective killing of tumor cells.

Figure 9.9 Valacyclovir (Valtrex), an l-valyl ester derivative of acyclovir with improved oral bioavailability. Acyclovir is not taken up efficiently after oral ingestion. However, a derivative of acyclovir, valacyclovir, has as much as 5-fold-higher oral bioavailability than acyclovir, as determined by the amount in serum relative to the dose of drug given. The addition of a new side group to acyclovir allows increased passage of drug from the digestive tract to the circulation. These acyclovir derivatives are prodrugs, which are converted to acyclovir by cellular enzymes that cleave off the valine side chain. Adapted from D. R. Harper, *Molecular Virology* (Bios Scientific Publishers, Ltd., Oxford, United Kingdom, 1994), with permission.

TERMINOLOGY
What's in a name?

Sorting out drug names can be mind-boggling. Many are tongue twisters and a challenge for the common mortal to pronounce. To make matters worse, every drug has more than one name. The definitions below are an attempt to provide some guidance to the novice.

Every drug has at least three names:

Chemical name: This is the scientific name, which based on the chemical composition and structure of the drug. It is the most specific and definitive name, but often too complicated to use for any but professional medicinal chemists or in scientific publications.

Generic name: The generic name, also known as the International Nonproprietary Name (INN), is assigned by a particular governing body, the FDA, in the United States. When the FDA approves a drug, it is given a generic name that is intended to be a shorthand derivative of the chemical name. Ergo: *N*-<u>acetyl</u>-*p*-<u>aminophen</u>ol is generic **acetaminophen**. [Although it is anyone's guess how (2*R*,3*S*, 4*R*,5*R*,8*R*,10*R*,11*R*,12*S*,13*S*,14*R*)-13-[(2,6-dideoxy-3-*C*-methyl-3-*O*-methyl-α-L-*ribo*-hexopyranosyl) oxy]-2-ethyl-3,4,10-trihydroxy-3,5,6,8,10,12,14-heptamethyl-11-[[3,4,6-trideoxy-3-(dimethylamino)-β-D-*xylo*-hexopyranosyl]oxy]-1-oxa-6-azacyclopentadecan-15-one becomes

generic **azithromycin**.] Letters are incorporated to refer to the action of the drug. For example, antiviral drugs all end in -vir, some monoclonal antibodies in -mab, and some antibiotics (as is azithromycin) end in -mycin.

Unfortunately, different countries may use different rules for their assignments, such that the same compound can have two or even more generic names. For example, generic acetaminophen in the United States is generic paracetamol in the United Kingdom. The latter name is used in other parts of Europe and Asia, which can be a source of confusion for world travelers.

Brand name: Also called a trade name, it is given to drugs by pharmaceutical companies. If more than one company markets a drug, the same chemical entity can have more than one name (at least 10 exist for acetaminophen worldwide), another potential source of confusion.

Brand naming is influenced strongly by the trademark system in the United States. Because almost all ordinary words are already taken, drug companies have to be quite creative in inventing entirely new names that can be identified with a registered trademark (®). These names are chosen with an eye to

both customer appeal and loyalty. Some names may have Latin roots, such as the *pax* (for "peace") in Paxil, an antidepressant; or *vir* ("man") in Viagra. Drugs for women often include soft letters such as S, M, and L, as in Sarafem and Vivelle; and the letters X, Z, K, and N are often chosen to denote cutting-edge science, as in Zantac, Nexium, and Protonix. The intention here is that even if generic acetaminophen is available, the customer will still reach for Tylenol in the drug store.

Koven S. July 14, 2012. How are drugs named? Boston.com. http://www.boston.com/lifestyle/health/blog/inpractice/2012/07/how_are_drugs_named.html

Hi! My name is "N-acetyl-p-aminophenol" or "Acetaminophen" for short. My friends call me "Tylenol."

Acyclovir remains the gold standard for treatment of herpes simplex and varicella-zoster virus infections, although other approved guanosine analogs, which show better oral absorption, have been approved for use in the clinic. **Ganciclovir (Cytovene)** is a derivative of acyclovir (Fig. 9.10) that was developed to treat human infections with the betaherpesvirus cytomegalovirus. The cytomegalovirus genome does not carry a thymidine kinase gene, but it does encode a protein kinase that can phosphorylate ganciclovir. Initial formulations of this drug given intravenously were quite toxic, and used only for life-threatening human cytomegalovirus infections in AIDS patients and immunosuppressed transplant recipients. Subsequently, an oral formulation that is much less toxic was developed and is effective for prophylaxis and long-term use for human cytomegalovirus infections.

AZT—Human Immunodeficiency Virus

Zidovudine; AZT (Retrovir) (Fig. 9.10), an analog of thymidine, was the first drug to be licensed for the treatment of AIDS. However, AZT was discovered initially in screens for antitumor cell compounds rather than for antiviral agents. The drug is phosphorylated by cellular enzymes and then incorporated into viral DNA, where, like acyclovir, it acts as a chain terminator (Fig. 9.11B). While phosphorylated AZT is not a good substrate for most cellular DNA polymerases, its selectivity depends mainly on the fact that retroviral reverse transcription takes place in the cytoplasm, where the drug appears first and in highest concentration. Because AZT monophosphate competes with TMP for the formation of nucleoside triphosphate, its presence causes depletion of the intracellular pool of TTP. Consequently, AZT has undesirable

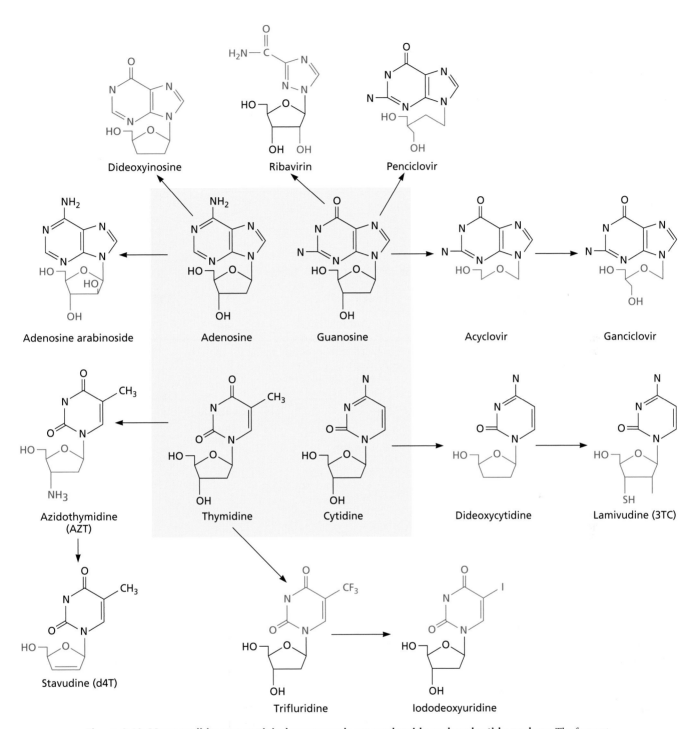

Figure 9.10 Many well-known antiviral compounds are nucleoside and nucleotide analogs. The four natural deoxynucleosides are highlighted in the yellow box. The chemical distinctions between the natural deoxynucleosides and antiviral drug analogs are highlighted in red. Arrows connect related drugs. Adapted from E. De Clercq, *Nat Rev Drug Discov* **1:**13–25, 2002, with permission.

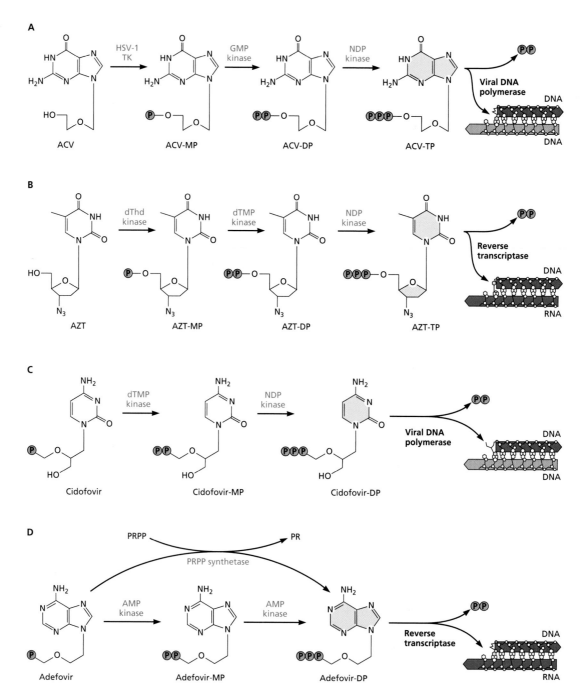

Figure 9.11 Chain termination by antiviral nucleos(t)ide analogs. (A) Acyclovir (ACV) is a prodrug that must be phosphorylated in the infected cell. The thymidine kinase of herpes simplex virus 1 (HSV-1 TK), but **not** the cellular kinase, adds one phosphate (orange circle labeled P) to the 5′ hydroxyl group of acyclovir. The monophosphate is a substrate for cellular enzymes that synthesize acyclovir triphosphate. The triphosphate compound is recognized by the viral DNA polymerase and incorporated into viral DNA. As acyclovir has no 3′ hydroxyl group, the growing DNA chain is terminated and viral genome replication ceases. **(B)** AZT targets the human immunodeficiency virus reverse transcriptase. The compound must be phosphorylated by cellular kinases in three steps to the triphosphate compound, which is incorporated into the viral DNA to block reverse transcription.

(C) Cidofovir [*S*-1-(3-hydroxy-2-phosphonylmethoxypropyl) cytosine] is an acyclic nucleotide analog. In contrast to acyclovir and AZT, cidofovir requires only two phosphorylations by cellular kinases to be converted to the active triphosphate chain terminator. **(D)** Adefovir [9-(2-phosphonylmethoxyethyl) adenine] is an acyclic nucleotide analog and also requires only two phosphorylations by cellular AMP kinases. Through the action of phosphoribosyl pyrophosphate (PRPP) synthetase, which forms the triphosphate from the monophosphate in one step, both cidofovir and adefovir bypass the nucleoside-kinase reaction that limits the activity of dideoxynucleoside analogs such as AZT. DP, diphosphate; dThd, (2′-deoxy)-thymidine; MP, monophosphate; NDP, nucleoside 5′-diphosphate; PR, 5-phosphoribose; TP, triphosphate. Adapted from E. De Clercq, *Nat Rev Drug Discov* **1:**13–25, 2002, with permission.

side effects when administered for long periods, including lactic acidosis (buildup of acid in the blood), liver problems, muscle weakness, and reduced numbers of red and white blood cells. The drug was used extensively in the early years of the AIDS pandemic, simply because there was nothing else available. Considerable effort has been devoted to discovering alternatives to AZT, and, as will be described in a later section, several nucleos(t)ide analogs that have better therapeutic value are now available and used for treatment of human immunodeficiency virus infections.

Lamivudine—Hepatitis B Virus

Lamivudine (Epivir) (Fig. 9.10), an orally delivered nucleoside analog, is also a prodrug that requires phosphorylation by cellular kinases to be incorporated into DNA, where it functions as a chain terminator. It is effective in blocking the reverse transcriptases of hepatitis B and human immunodeficiency viruses. Since the approval of lamivudine, a number of additional nucleoside analogs with improved properties have been developed for use in treatment of chronic hepatitis B virus infections. The latest of these, the human immunodeficiency virus reverse transcriptase nucleotide analog inhibitor tenofovir, was approved for treatment of hepatitis B in 2008, after its approval for treatment of AIDS.

Cidofovir—Broad-Spectrum Antiviral

Cidofovir (Vistide) (Fig. 9.11C) is an acyclic nucleoside phosphonate that can be considered a broad-spectrum drug, as it is active against various herpesvirus, papillomavirus, polyomavirus, adenovirus, and poxvirus infections. A prodrug, cidofovir, is converted to di- and triphosphate derivatives by host enzymes. The fully phosphorylated compound has a higher affinity for viral polymerases than host cell polymerases, a unique property of acyclic nucleotide analogs. For example, the binding affinity of cidofovir diphosphate to the DNA polymerase of human cytomegalovirus polymerase is 8 to 80 times higher than for any of the human DNA polymerases. Because the conversion of cidofovir to cidofovir monophosphate does not depend on a virus-induced thymidine kinase or a viral protein kinase, essentially all DNA viruses and retroviruses are susceptible. Related compounds are adefovir (Fig. 9.11D), approved for hepatitis B infections, and tenofovir, approved for HIV infections. These drugs have been formulated for intravenous, topical, and oral applications.

Ribavirin—RNA Virus Infections

Ribavirin (Virazole) (Fig. 9.10), one of the earliest antivirals, was synthesized in 1972 and purported to have broad-spectrum activity against many DNA and RNA viruses. However, the drug is relatively toxic, and its development and

indications for use have been controversial. In fact, ribavirin is not licensed for general use in many countries. Despite its long history, its primary mechanism of action is not clear. Ribavirin monophosphate is a competitive inhibitor of cellular inosine monophosphate dehydrogenase, and such inhibition leads to reduced GTP pools in the cell, which may adversely affect the replication of some viral genomes. The drug also blocks initiation and elongation by viral RNA-dependent RNA polymerases and interferes with capping of mRNA. Finally, ribavirin is an RNA virus mutagen: once incorporated into a template, it pairs with C or U with equal efficiency. In some studies, its antiviral activity correlates directly with its mutagenic activity. Even with its unknown mechanism and its toxicity, ribavirin is used as an aerosol for treatment of infants suffering from respiratory syncytial virus infection, as well as for treatment of Lassa fever virus and hantavirus infections. Viramidine and levovirin are analogs of ribavirin that are in clinical development for treatment of hepatitis C virus infections.

Foscarnet—Viral DNA Synthesis Blocker

Foscarnet (Foscavir) is the only nonnucleoside DNA replication inhibitor of herpesviruses (Fig. 9.12A). The drug is a noncompetitive inhibitor that binds to the pyrophosphate-binding site in the catalytic center of herpesvirus DNA polymerase. Its ability to chelate one of the active-site metals while making other electrostatic interactions is proposed to trap the enzyme in a closed, inactive configuration (see Volume I, Box 6.2). Foscarnet also inhibits hepatitis B virus polymerase and the reverse transcriptase of human immunodeficiency virus, at concentrations that do not affect human DNA polymerases. The drug must be administered intravenously. As it causes kidney and bone toxicity, its use is recommended only for life-threatening infections for which other antiviral drugs are no longer effective.

Figure 9.12 Two nonnucleoside/nucleotide antiviral compounds. (A) Foscarnet is an FDA-approved drug that inhibits herpesviral DNA polymerase by binding noncompetitively to the pyrophosphate-binding site in the catalytic center. The drug also inhibits the activities of the reverse transcriptases of hepatitis B virus and human immunodeficiency virus. **(B)** LJ-001 is an aryl methyldiene rhodanine lead compound that intercalates into the lipid bilayers of enveloped viruses and blocks virus-cell membrane fusion.

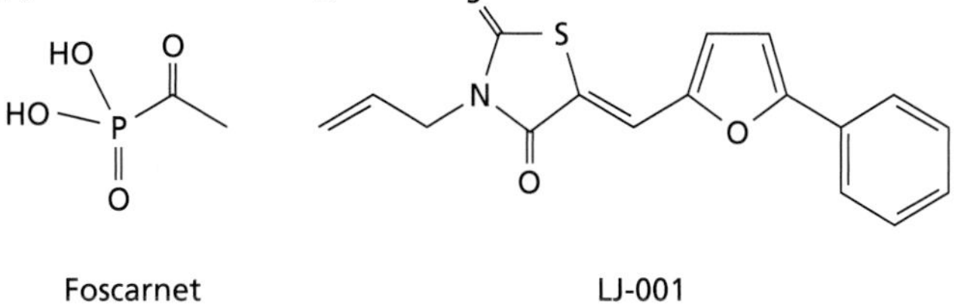

Foscarnet   LJ-001

Approved Drugs That Are Not Inhibitors of Nucleic Acid Synthesis

Amantadine—Influenza A Virus Uncoating Inhibitor

Amantadine (Symmetrel) is a three-ringed symmetric amine that was developed by DuPont chemists almost 50 years ago. It was the first highly specific, potent antiviral drug effective against any virus. The target of the drug was shown to be the influenza A virus M2 protein, a tetrameric, transmembrane ion channel that transports protons (Fig. 9.13). Amantadine has no effect on influenza B viruses, as their genomes do not encode an M2 protein. Influenza A virus mutants resistant to amantadine, which arise after therapy, all have amino acid changes in the M2 transmembrane

Figure 9.13 Interaction of amantadine with the transmembrane domain of the influenza A virus M2 ion channel. M2 protein is a tetramer with an aqueous pore in the middle of the four subunits. It is thought that at low concentrations (5 μM), amantadine (red) exerts its antiviral effect by blocking the M2 ion channel activity and preventing acidification of the virus particle. At higher concentrations, binding on the outside of the tetramer is thought to block proton entry allosterically. Acidification of the virus particle is required for release of viral nucleic acids (uncoating) in infected cells. For further detail see: J. R. Schnell and J. J. Chou, *Nature* **451:**591–595, 2008.

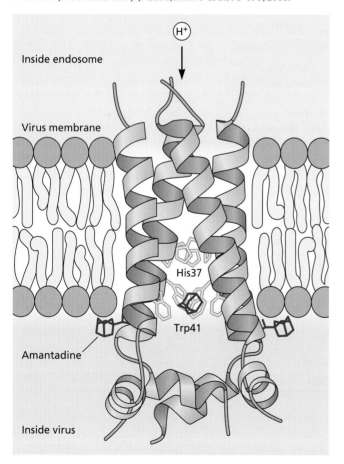

sequences predicted to form the ion channel. Amantadine blocks the channel, so that protons cannot enter the virion, effectively preventing the uncoating of influenza A virus (Volume I, Chapter 5). Unfortunately, during the 2008-2009 flu season, the U.S. Centers for Disease Control and Prevention found that 100% of seasonal H3N2 and the 2009 pandemic strain samples tested were resistant to amantadines, and these drugs are no longer recommended for the treatment of seasonal flu: the viral neuraminidase inhibitors oseltamivir and zanamivir are proposed as alternatives.

When treating infections with nonresistant strains, amantadine must be administered in the first 24 to 48 h and given at high doses for at least 10 days to have an impact on the clinical course of disease. The drug is most effective when susceptible patients are treated prophylactically, in anticipation of influenza virus infection. However, at high concentrations, side effects are common, particularly those affecting the central nervous system. On follow-up of these side effects, amantadine was found to be useful for relieving symptoms of Parkinson's disease in some patients. Today, more amantadine is sold for central nervous system disease than for antiviral treatment. Rimantadine, a methylated derivative, cannot cross the blood-brain barrier and therefore has fewer central nervous system side effects. For this reason, the drug often replaces amantadine in the treatment of influenza A virus infections.

Concentration dependence is an unusual property of amantadine. The drug has broad antiviral effects at high concentrations, but at low concentrations it is specific for influenza virus A. Analysis of resistant mutants provided insight into the apparently complex mechanism of action of amantadine. At concentrations of 100 mM or higher, the compound acts as a weak base and raises the pH of endosomes so that pH-dependent membrane fusion is blocked. Any virus with a pH-dependent fusion mechanism could be affected by high concentrations of amantadine. Resistant mutants of influenza A virus selected under these conditions in cultured cells harbor amino acid substitutions in hemagglutinin (HA) that destabilize the protein and enable fusion at higher pH. Influenza A virus mutants selected at concentrations of 5 mM or lower carried mutations in the M2 gene. These mutations affected amino acids in the membrane-spanning region of the M2 ion channel protein.

Zanamivir and Oseltamivir—Influenza Virus Neuraminidase Inhibitors

Zanamivir (Relenza) and **oseltamivir (Tamiflu)** are inhibitors of the neuraminidase enzyme synthesized by influenza A and B viruses (Box 9.5). Zanamivir is delivered via inhalation, while oseltamivir can be given orally. When used within 48 h of symptoms, the drugs reduce the median time to their alleviation by ~1 day compared to placebo. When used within 30 h of disease onset, the drugs reduce the duration of symptoms by ~3 days.

EXPERIMENTS
Inhibitors of influenza virus neuraminidase: development and impact

Influenza virus neuraminidase (NA) protein cleaves terminal sialic acid residues from glycoproteins, glycolipids, and oligosaccharides. It plays an important role in the spread of infection from cell to cell, because in cleaving sialic acid residues, the enzyme releases virus particles bound to the surfaces of infected cells and facilitates viral diffusion through respiratory tract mucus. Moreover, the enzyme can activate transforming growth factor β by removing sialic acid from the inactive protein. Because the activated growth factor can induce apoptosis, NA may influence the host response to viral infection.

NA is a tetramer of identical subunits, each of which consists of six four-stranded antiparallel sheets arranged like the blades of a propeller. The enzyme active site is a deep cavity lined by identical amino acids in all strains of influenza A and B viruses that have been characterized. Because of such invariance, compounds designed to fit in this cavity would be expected to inhibit the NA activity of all A and B strains of influenza virus, a highly desirable feature in an influenza antiviral drug. Moreover, as NA inhibitors are

predicted to block spread, they may be effective in reducing the transmission of infection to other individuals.

Sialic acid fits into the active-site cleft such that there is an empty pocket near the hydroxyl at the 4-position on its sugar ring. On the basis of computer-assisted analysis, investigators predicted, correctly, that replacement of this hydroxyl group with either an amino or a guanidinyl group would fill the empty pocket and therefore increase the binding affinity by contacting one or more neighboring glutamic acid residues. The resulting "designer drug," zanamivir (Relenza), was licensed to GlaxoSmithKline in 1990 and approved by the U.S. FDA in 1999. Following their lead, Gilead Sciences developed the better-known neuraminidase inhibitor oseltamivir (Tamiflu), which was licensed to Hoffmann-La Roche and approved by the FDA that same year.

Both drugs are inhibitors of influenza A and B viruses. They do not inhibit other viral neuraminidases, an important requirement for safety and lack of potential side effects. The United States reportedly stockpiled $1.5 billion

of oseltamivir prior to the global outbreak of H1N1 influenza in 2009, while the vaccine was being prepared. Many individuals established personal, preemptory "stockpiles" of the drug. However, given the limited reduction in duration of symptoms (8.4 to 25.1 h for adults; 12 to 47 h for children), the lack of reduction in hospitalizations and deaths, and the reported side effects (nausea, vomiting, headaches, and increased risk of renal and psychiatric syndromes), there remains uncertainty about whether the drug is worth taking by otherwise healthy flu patients.

Jefferson T, Jones M, Doshi P, Spencer EA, Onakpoya I, Heneghan CJ. 2014. Oseltamivir for influenza in adults and children: systematic review of clinical study reports and summary of regulatory comments *BMJ* **348:**g2545. doi:10.1136/bmj.g2545.

Varghese JN, Epa VC, Colman PM. 1995. Three-dimensional structure of the complex of 4-guanidino-Neu5Ac2en and influenza virus neuraminidase. *Protein Sci* **4:**1081–1087.

von Itzstein M, Wu WY, Kok GB, Pegg MS, Dyason JC, Jin B, Van Phan T, Smythe ML, White HF, Oliver SW, Colman PM, Varghese JN, Ryan DM, Woods JM, Bethell RC, Hotham VJ, Cameron JM, Penn CR. 1993. Rational design of potent sialidase-based inhibitors of influenza virus replication. *Nature* **363:**418–423.

Structure of influenza A virus NA and antiviral drugs. (A) Ribbon diagram of influenza A virus NA with α-sialic acid bound in the active site of the enzyme. The molecule is an N2 subtype from A/Tokyo/3/67. A monomer is viewed down the fourfold axis of an active tetramer. The C terminus is at the bottom right, near the subunit interface. The six β-sheets of the propeller fold are indicated in colors. Adapted from J. N. Varghese, p. 459–486, *in* P. Veerapandian (ed.), *Structure-Based Drug Design: Diseases, Targets, Techniques and Developments*, vol. 1 (Marcel Dekker, Inc., New York, NY, 1997), with permission. Courtesy of J. Varghese. **(B)** Chemical structures of two FDA-approved NA inhibitors.

A

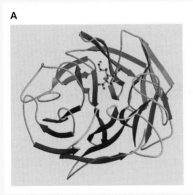

B

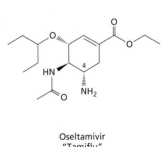

Zanamivir "Relenza"

Oseltamivir "Tamiflu"

Expanding Target Options for Antiviral Drug Development

As noted in an earlier section, technological advances have made it possible to identify both viral and host components that are required to complete the numerous steps in any virus life cycle successfully. Although viral nucleic acid synthesis inhibitors still predominate, it has now become practical to home in on other critical functions, viral protein activities, and viral host-interactions. (For additional comments on novel approaches see the interview with Dr. Benhur Lee: http://bit.ly/Virology_Lee)

Entry and Uncoating Inhibitors

The first step of the virus reproduction cycle has long been an attractive target, as virus-cell receptor interactions offer the promise of high specificity. Early enthusiasm for entry inhibitors came from experiments with monoclonal antibodies that blocked attachment or entry into cultured cells. Progress in the identification of potent or broadly neutralizing antibodies has promoted renewed interest in this approach. Passive immunization with these antibodies can often protect animals from challenge. One currently licensed antiviral monoclonal antibody, palivizumab, binds to the fusion protein of respiratory syncytial virus. This antibody is used to prevent infections of infants who are at high risk because of premature birth or medical problems such as congenital heart disease. There is considerable interest in developing neutralizing monoclonal antibodies against several other viruses including the human immunodeficiency virus, hepatitis C virus, and influenza viruses (Chapter 8). Monoclonal antibodies have also been valuable for identifying viral and host proteins required for entry and in elucidating their mechanisms of action. Moreover, their antiviral activities have suggested that small-molecule inhibitors of entry may be useful drug leads.

The binding sites of antibodies that block viral entry provide a starting point for screening chemical libraries or for design of small-molecule inhibitors. Optimally, the inhibitor should block viral entry but not interfere with the normal function of the cellular receptor. Because alternative receptors are available for some viruses (e.g., herpesviruses and human immunodeficiency virus), it may be necessary to block binding of a virus particle to more than one type of receptor for such treatment to be effective.

Membrane fusion, the usual process by which enveloped virus particles enter cells, is an attractive target for chemotherapeutic intervention because fusion mechanisms are conserved among enveloped viruses. To identify inhibitors of influenza virus fusion, a computer docking program was first used to predict which small molecules might bind into a pocket of the HA trimers and prevent a required low-pH-induced conformational change. From the molecules identified in this way, several benzoquinone- and hydroquinone-containing compounds were tested and found to prevent HA-mediated membrane fusion at low pH. One of these compounds inhibits influenza virus reproduction. Although it is not known how binding of the compound prevents fusion, it has been suggested that it blocks movements that juxtapose viral and cellular membranes (described in Volume I, Chapter 5). Several of these compounds are in clinical development.

The effectiveness of another type of virus-cell membrane fusion inhibitor depends on the fact that while cellular surface lipid bilayers are dynamic, viral envelopes are static. The inhibitor, a derivative of aryl methylene rhodanine (Fig. 9.12B), intercalates into both viral and cellular membranes, but can be removed rapidly from the cellular membrane by repair functions. As the viral envelope cannot be repaired, the inhibitor is retained and blocks the fusion step, most likely by affecting the fluidity/rigidity of the viral lipid bilayer. This broad-spectrum inhibitor was reported to block infection of cultured cells by numerous enveloped viruses but, consistent with its proposed mechanism of action, had no effect against nonenveloped viruses. Whether such inhibitors will be clinically useful awaits the results of testing in animal models and then in humans.

Microbicides

Considerable effort has been expended on development of **microbicides**, creams or ointments that contain compounds that either inactivate virus particles before they can attach and penetrate tissues, or enter cells and block virus reproduction. Particular attention has been focused on vaginal microbicides to prevent infection with sexually transmitted viruses. Formulations that incorporate acyclovir or tenofovir have been tested for prevention of transmission of herpes simplex viruses and human immunodeficiency virus, respectively. While investigations continue, there is as yet no approved microbicide for either of these viruses.

Viral Regulatory Proteins

Viral proteins that control transcription are often essential for virus reproduction and are prime targets for antiviral screens. Fomivirsen was the first licensed compound designed to inhibit the function of a viral regulatory protein. The drug is a phosphorothioate antisense oligonucleotide that is approved to treat retinitis caused by human cytomegalovirus, via direct injection into the vitreous of an infected eye. Inhibition depends on binding of the 21-nucleotide antisense molecule to the cytomegalovirus immediate-early 2 mRNA and hence preventing synthesis of this essential viral protein.

Regulatory RNA Molecules

Micro-RNAs (miRNAs), which induce degradation or inhibit translation of mRNA, are encoded in both host and viral genomes. One miRNA may regulate the expression of an entire network of genes. For example, the human miRNA miR-122, which is synthesized only in the liver, regulates expression of >400 genes, including those that participate in cholesterol metabolism. When expression of miR-122 is inhibited, not only do levels of cholesterol in the circulation drop, but the liver is also protected from hepatitis C virus infection. miR-122 protects the viral (+) strand RNA from degradation by binding to two complementary sequences in the 5′ untranslated region (Volume I, Box 10.12). Antagonists of miR-122, such as antisense oligonucleotides, block virus reproduction with no harmful effect in animal models. This finding suggests that small RNA molecules may be valid targets for antiviral compounds.

Proteases and Nucleic Acid Synthesis and Processing Enzymes

Viral proteases have become among the most attractive targets for antiviral drug discovery. These enzymes are responsible for cleaving protein precursors to form functional units or to release structural components during or following particle assembly (maturational proteases). The requirement for proteases in the life cycle of several viruses makes them an excellent target for drug development. All herpesviruses encode a serine protease that is required for formation of nucleocapsids (Volume I, Chapter 13). Many features of these enzymes and their substrates are conserved among the members of the family *Herpesviridae*, and the X-ray crystal structure of the human cytomegalovirus protease has been solved. Interest in this enzyme as a drug target is based on the unusual serine protease fold and the mechanism of catalysis. Some success has been reported with a small molecule that inhibits the dimerization of the enzyme that is required for its activity.

As already noted, most presently approved drugs are nucleos(t)ide analogs that block viral replication by acting as chain terminators. However, we know from genetic analyses that DNA polymerase accessory proteins, such as those that promote processivity or bind to viral origins, are essential for viral genome replication and, consequently, are also attractive targets. The RNA-dependent RNA polymerases of RNA viruses appear to be unique to the virus world. Their varied activities, which include synthesis of primers, cap snatching, and recognition of RNA secondary structure in viral genomes (Volume I, Chapter 6), can be exploited for drug discovery. The unique helicases encoded by many RNA viruses are also promising targets.

Newly replicated, concatemeric herpesviral DNA is cleaved by viral enzymes into monomeric units during packaging and assembly. These processes, which are essential for herpesviral reproduction, are carried out by specific viral enzymes, which represent promising targets for antiviral drugs. For example, the compound 5-bromo-5,6-dichloro-1-β-D-ribofuranosyl benzimidazole binds to the human cytomegalovirus UL89 gene product, which is a component of the terminase complex

responsible for cleaving and packaging replicated concatemeric DNA. Members of the UL89 gene family are highly conserved among all herpesviruses. This class of compounds may therefore be the basis for discovery of broad-based inhibitors of herpesvirus reproduction. Despite their promise, these compounds have not been developed for clinical use, primarily because the nucleoside analogs like acyclovir are so effective and safe.

Two Success Stories: Human Immunodeficiency and Hepatitis C Viruses

Recognition in the early 1980s that a retrovirus later named human immunodeficiency virus (HIV) was associated with the deadly AIDS disease galvanized the biomedical field. As the extent of the pandemic increased, the urgent need to develop effective therapies, together with pressure from effective political advocacy, led to unprecedented investment of both public and private resources in antiviral drug discovery. As of 2014, more than 20 antiretroviral drugs have been approved, with many more in clinical trials (Table 9.1). Although there are still more than **35 million** people living with HIV, the rate of new infections worldwide is declining as more people gain access to treatment. The impact of new drugs is especially striking in the United States and Europe, where for many patients AIDS has become not a death warrant, but rather a chronic disease.

A second virus of global impact, the (+) strand RNA flavivirus, hepatitis C virus (HCV), (Fig. 9.14), was discovered by tour de force screening of molecular clones from infected blood samples in 1989 (Box 11.6). An estimated **160 million to 185 million** people worldwide are infected with HCV, more than 4 times the number infected with HIV. Although HCV can be cleared in some individuals, the virus establishes a chronic infection in ~80% of those infected, and ~20% of these individuals develop liver cirrhosis within 20 to 30 years, with ~5% succumbing to fatal liver cancer. As with HIV, HCV is spread via exposure to virus-contaminated blood. Although progress in developing antivirals against HCV was impeded initially by the lack of a cell culture system for viral reproduction, substantial progress has been made in the last several years, based on lessons learned from experience with HIV drugs and the application of new technologies (Table 9.2). Drugs that can actually cure most patients entirely are already available, and many more are being developed.

With both HIV and HCV, the essential criteria for large-scale investment by the pharmaceutical industry are clearly satisfied: unmet medical needs; no (or insufficiently effective) existing/approved drugs; and, most assuredly, a market large enough to ensure a profit.

Knowledge of the single cell life cycles of these viruses has suggested many possible steps for antiviral drug intervention

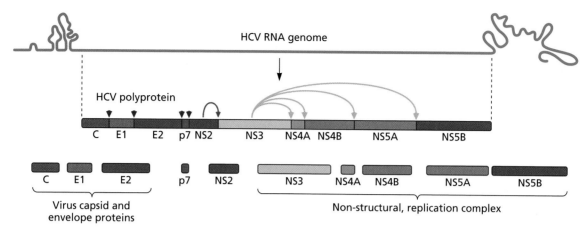

Figure 9.14 The hepatitis C virus polyprotein is cleaved by several proteases. Hepatitis C virus (HCV) is a human flavivirus with a (+) strand RNA genome. The viral proteins are encoded in one large open reading frame that is translated into a polyprotein. The polyprotein is processed by cellular and viral proteases to release the viral proteins. The cleavage sites for the viral proteases are indicated by arrows. The solid arrowheads show the cleavage sites for the host signal peptidase. The viral NS2 metalloprotease is an autoprotease comprising NS2 and the amino-terminal domain of NS3. The viral serine protease comprises the NS3 protein bound to an activator, NS4A. NS5B is the viral RNA polymerase. Viral proteins that are the targets of direct-acting antivirals in advanced clinical development are the protease NS3, the RNA-binding NS5A, and the polymerase NS5B (Fig. 9.15).

Table 9.1 Approved drugs targeted against HIV enzymes

Target	Generic name	Brandname	Manufacturer	Year
Reverse transcriptase Nucleos(t)ide inhibitors	Zidovudine (AZT)	Retrovir	GlaxoSmithKline	1987
	Didanosine (ddI)	Videx	Bristol-Myers Squibb	1991
	Zalcitabine (ddC)	Hivid	Hoffmann-La Roche	1992
	Stavudine (d4T)	Zerit	Bristol-Myers Squibb	1994
	Lamivudine (3TC)	Epivir	GlaxoSmithKline	1995
	Abacavir (ABC)	Ziagen	GlaxoSmithKline	1998
	Tenofovir (TDF)	Viread	Gilead Sciences	2001
	Emtricitabine (FTC)	Emtriva	Bristol-Myers Squibb	2003
Nonnucleoside inhibitors	Nevirapine (NVP)	Viramune	Roxane	1996
	Delavirdine (DLV)	Rescriptor	Pfizer	1997
	Efavirenz (EFV)	Sustiva	DuPont	1998
	Etravirine (ETR)	Intelence	Tibotec	2008
	Rilpivirine	Edurant	Tibotec	2011
Protease	Saquinavir (hard gel)	Invirase	Hoffmann-La Roche	1995
	Ritonavir	Norvir	Abbott	1996
	Indinavir	Crixivan	Merck	1996
	Nelfinavir	Viracept	Agouron	1997
	Amprenavir	Agenerase	GlaxoSmithKline	1999
	Lopinavir/ritonavir	Kaletra	Abbott	2000
	Atazanavir	Revataz	Bristol-Myers Squibb	2003
	Tipranavir	Aptivus	Boehringer Ingelheim	2005
	Darunavir	Prezista	Tibotec	2006
Integrase	Raltegravir	Isentress	Merck	2007
	Elvitegravir	Vitekta	Gilead Sciences	2012
	Dolutegravir	Tivicay	GlaxoSmithKline	2013
Combinations	TDF/FTC/EFV	Atripla	Bristol-Myers Squibb/ Gilead Sciences	2006
	TDF/FTC/rilpivirine	Complera	Gilead Sciences	2011
	TDF/FTC/elvitegravir + cobicistat	Stribild	Gilead Sciences	2012

Table 9.2 Examples of drugs targeted against HCV proteins

Target	Generic name	Brand name	Developer	Date approved/ Trial phase
Polymerase (NS5B)	Sofosbuvir	Sovaldi	Gilead Sciences	2013
Nucleoside	Mericitabine		Roche	II
Nonnucleoside	Deleobuvir		Boehringer Ingelheim	III
	ABT-333		Abbott	III
RNA binding (NS5A)	Ledipasvir		Gilead Sciences	III (filed)
	Daclatasvir		Bristol-Myers Squibb	III
	ABT-267		Abbott	III
Protease (NS3/4A)	Telaprevir	Incivek	Vertex/Johnson & Johnson	2011
	Boceprevir	Victrelis	Merck	2011
	Simeprevir	Olysio	Janssen/Tibotec/Medivir	2013
	Faldaprevir		Boehringer Ingelheim	III
	Vaniprevir		Merck	III
	Samatasvir		Idenix	II
Combinations	Sofosbuvir + ledipasvir		Gilead Sciences	III
	Faldaprevir + deleobuvir		Boehringer Ingelheim	III
	Simeprevir + samatasvir + TMC647055/r		Janssen	II
	ABT-450/r + ABT-267 and ABT-333		Abbott	II
	MK-8742 + MK-5172		Merck	II

(Fig. 9.15). However, while virus fusion and entry have been targeted successfully in the case of HIV, most drugs that are approved or close to approval are directed against the essential viral-encoded enzymes. In the following section, important examples are described individually, but in the clinic the drugs are administered mostly in combinations, as emergence of resistant mutants is a serious problem with both of these RNA viruses.

Inhibitors of Human Immunodeficiency Virus and Hepatitis C Virus Polymerases

Nucleoside and Nucleotide Analogs

Retroviruses are "defined" by their RNA- and DNA-dependent DNA polymerase, reverse transcriptase (Volume I, Chapter 7). The chain-terminating nucleoside AZT was the first inhibitor to be licensed, and was used extensively to treat AIDS before any other antivirals were available (Fig. 9.10). Since then, a number of nucleoside analog inhibitors with better pharmacological properties (such as once-a-day abacavir) have been approved for clinical use (Table 9.1). However, AZT is still used as a single drug (**monotherapy**) for prophylactic treatment of accidental needle sticks and, in some lower-income countries, for treatment of infected pregnant women, as it is relatively inexpensive and can reduce considerably the probability of delivering an HIV-infected baby.

The RNA polymerase of HCV (NS5B) presented several challenges for antiviral drug discovery: many nucleoside analogs initially appeared promising, demonstrating the feasibility of this approach, but it was not until December 2013 that the first inhibitor in this class, sofosbuvir, was approved for treatment of HCV infection, in combination with the standard-of-care regimen of interferon α and/or ribavirin. As with many of the nucleoside inhibitors, sofosbuvir is a prodrug that is converted in the cell to a chain-terminating substrate for HCV polymerase (Fig. 9.16). Treatment for 12 to 24 weeks with the approved regimen results in 60 to 90% cure rates, depending on the HCV genotype.

The success of this new drug generated great optimism but also some controversy, as its high price was likely to exclude access to many infected individuals who could benefit from it (Box 9.6). Agencies of the U.S. government have issued recommendations to test all individuals born between 1945 and 1965 for HCV infection: these "baby boomers" are 5 times more likely than others in the population to be infected because of increased drug use and sexual activity among young people during the 1960s and 1970s. It is estimated that detection and treatment of infected people in this group alone could save more than 120,000 lives.

Nonnucleoside Inhibitors

Nonnucleoside inhibitors of the HIV reverse transcriptase have been identified mainly by high-throughput screening.

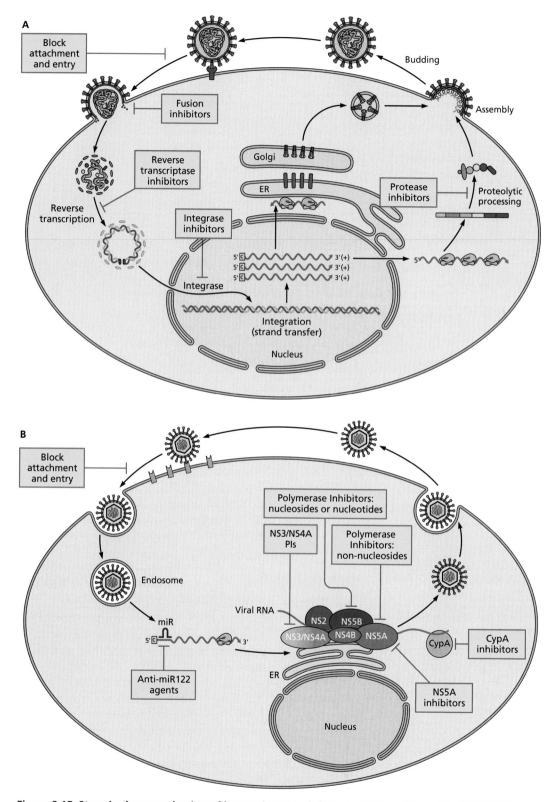

Figure 9.15 Steps in the reproduction of human immunodeficiency (HIV) and hepatitis C (HCV) viruses targeted by antiviral drugs. Steps in the life cycle of the viruses are illustrated, with important targets for antiviral development highlighted. **(A)** HIV; **(B)** HCV. ER, endoplasmic reticulum; PI, protease inhibitor.

Figure 9.16 The prodrug sofosbuvir: structure and activation.

Nucleoside → (Slow, Nucleoside kinase) → Nucleoside monophosphate → (Fast, Nucleotidyl and nucleoside diphosphate kinase) → Active triphosphate

Fast | Cellular enzymes

Sofosbuvir (a prodrug)

DISCUSSION
What price drugs?

In the United States, the FDA approves new drugs solely on the basis of safety and efficacy. Unlike similar agencies in some other countries (for example, the National Institute for Health and Care Excellence in the United Kingdom), there is no value assessment of the drug or treatment. Consequently, pharmaceutical companies in the United States are free to set their drug prices based mainly on what the market will bear. Nevertheless, even a jaded U.S. public reacted with "sticker shock" when Gilead Sciences announced that its just-approved anti-hepatitis C virus (HCV) drug sofosbuvir (Solvaldi) would be priced at $84,000 (plus the cost of necessary companion drugs) for a required 12-week course, and twice that for a 24-week course that would be needed to cure some patients.

According to a Gilead spokesman, the company considered the price to be fully justified: "We didn't really say, 'We want to charge $1,000 a pill.' . . . We're just looking at what we think was a fair price for the value that we're bringing into the health care system and to the patients."

Some medical specialists might agree, as it could cost up to $300,000 to treat patients with chronic HCV infection using less effective and less tolerable regimens. The potential benefit of a cure for patients with liver disease is clear, as the virus is the main reason that nearly 17,000 Americans are waiting for a liver transplant. The need for a well-tolerated, effective regimen is equally critical for people coinfected with human immunodeficiency virus (HIV) and HCV, because having both infections accelerates liver damage.

On the other hand, the high price will be a significant barrier to treatment access for others who could benefit, particularly those in limited- and fixed-budget programs, such as Medicare and Medicaid. Indeed, a panel of experts in San Francisco estimated that simply replacing current care of HCV-infected Californians with a Sovaldi-based regimen would raise drug expenditures in the state by $18 billion or more in a single year.

Gilead has agreed to help U.S. patients pay for Sovaldi if they can't afford it, or help patients look for drug coverage. In addition, the company will charge substantially less for a course of treatment in places such as India, Pakistan, Egypt, and China, where most people infected with HCV live. With deals announced early in 2014 of $2,000 for a 12-week course in India and $990 for the same in Egypt, U.S. citizens, government, and insurance companies may reasonably ask if they are being forced to subsidize the cost of the drug worldwide.

So what is a fair price for such a lifesaving drug? Gilead paid more than $11 billion in 2011 to acquire the smaller company that developed Sovaldi, and it is reasonable for it to seek to recoup that investment. On the other hand, Andrew Hill, of the Department of Pharmacology and Therapeutics at Liverpool University in the United Kingdom, and his colleagues have reported a conservative estimate of the manufacturing cost of a 12-week course of treatment with this drug to be on the order of $150 to $250 per person. Surely the answer to our question lies somewhere between this huge divide.

There are parallels between Sovaldi (and other new anti-HCV drugs in the pipeline) and the initially very pricey antivirals that were introduced ~20 years ago to treat HIV. In both cases, their use revolutionized the treatment of chronic, lethal infections that are major global health problems. But there are also important differences. Given the total number of people infected, HCV is actually a much larger public health threat than HIV. Furthermore, the new HCV antivirals can eliminate the virus completely, whereas anti-HIV drugs only suppress virus reproduction, so that they must be taken (and paid for) for life.

Callaway E. 2014. Hepatitis C drugs not reaching poor. *Nature* **508:**295–296.

The Editorial Board. March 15, 2014. How much should hepatitis C treatment cost? *New York Times.* http://nyti.ms/1fzwNQF.

Hill A, Khoo S, Fortunak J, Simmons B, Ford N. 2014. Minimum costs for producing hepatitis C direct-acting antivirals for use in large-scale treatment access programs in developing countries. *Clin Infect Dis* **58:**928–936.

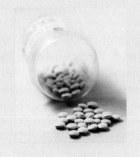

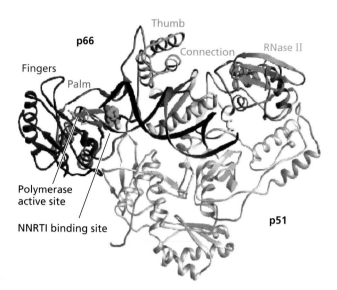

Figure 9.17 Structure of human immunodeficiency virus type 1 reverse transcriptase, highlighting the polymerase active site and the nonnucleoside reverse transcriptase inhibitor (NNRTI) binding site. The structure of the reverse transcriptase p66-p51 heterodimer is shown bound to a double-stranded DNA template-primer. Lines indicate the relative locations of the polymerase active site and the site at the bottom of the palm where NNRTIs are bound. Data from A. Jacobo-Molina et al., *Proc Natl Acad Sci U S A* **90:**6320–6324, 1993.

These compounds do not bind at the nucleotide-binding site of the enzyme, but in a hydrophobic pocket at the base of the palm subdomain (Fig. 9.17). Nevirapine was the first in this class of compounds, which are allosteric inhibitors of the enzyme. Although now used mainly in combination with other HIV antivirals, like AZT, nevirapine has proven to be valuable as a single-drug treatment for pregnant women before infant delivery. Indeed, a short course of the very affordable nevirapine is the preferred method for preventing transmission to newborns in lower-income countries.

Allosteric inhibitors of HCV polymerase have been identified in the same manner. The X-ray crystal structure of this protein has allowed optimization of drug binding to four distinct allosteric sites on the enzyme. Clinical trials of this class of compounds are ongoing, with several showing promise, especially for use in combination therapy.

Human Immunodeficiency Virus and Hepatitis C Virus Protease Inhibitors

HIV protease is encoded in the *pol* gene. During progeny particle maturation, the protease cleaves itself from the Gag-Pol precursor polyprotein and then cuts at seven additional sites in Gag-Pol to yield nine viral proteins, including the two other retroviral enzymes (Volume I, Appendix, Fig. 29). This small (only 99-amino-acid) aspartyl protease, which functions as a dimer, was the first HIV enzyme to be crystallized and studied at the atomic level (Fig. 9.6). The seven cleavage sites in Gag-Pol are similar but not identical. Mechanism-based screens benefited from the early discovery that the enzyme would cut short synthetic peptides that contained these sequences. Parameters of peptide binding and protease activity were determined by screening peptides that contained variations of the seven natural cleavage sites for their ability to be recognized and cut by the enzyme (Fig. 9.18). An additional boost to drug discovery was the similarity of this enzyme to another aspartyl protease, human renin, an enzyme implicated in hypertension. Indeed, the first inhibitor leads were peptide mimics (peptidomimetics) modeled after inhibitors of renin, and were developed into drugs such as saquinavir. Subsequent screens for mechanism-based and structure-based inhibitors designed *de novo* have yielded several powerful inhibitors of the protease and second-generation drugs, such as darunavir, for which many more viral mutations are needed to develop resistance.

The second HIV protease inhibitor to be approved, ritonavir, had an unexpected off-target effect that had important impact on the field: this compound was found to be an irreversible inhibitor of a detoxifying cytochrome P450 enzyme at only one-sixth the therapeutic dose for protease inhibition. Amazingly, this activity was found to improve the pharmacokinetic properties of other HIV protease inhibitors significantly. Ritonavir, or a derivative, is therefore included in many combination regimens at this lower dose as a "booster."

The HCV NS3/4A protein is a serine protease. Two HCV protease inhibitors, boceprevir and telaprevir (Fig. 9.19) were the first antivirals to be approved for treatment of chronic HCV infections. Although they are highly active against only one genotype, some efficacy against two others of the six known genotypes of HCV has been reported. Rapid development of resistance, which requires a substitution at only one site in the proteins, is a distinct disadvantage of these drugs. Two more recently developed HCV protease inhibitors, simeprevir and faldaprevir, have activity against several HCV genotypes and impose a higher genetic barrier for development of resistant mutants.

Human Immunodeficiency Virus Integrase Inhibitors

The HIV integrase is an excellent drug target, because it is a unique recombinase for which biochemical data and mechanism-based assays are available, thanks to earlier studies with avian retroviruses. Nevertheless, drug development lagged behind that targeted against the HIV reverse transcriptase and protease because of the difficulty in obtaining a crystal structure for an active tetramer of integrase, alone or bound to its DNA substrates. Despite this limitation,

A Natural substrate of the HIV-1 protease

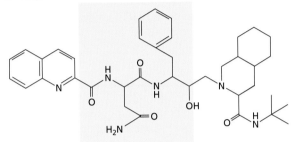

Valine   Serine   Glutamine Asparagine Tyrosine   Proline   Isoleucine   Valine

B Saquinavir

C Darunavir

Figure 9.18 Comparison of one natural cleavage site for the human immunodeficiency virus type 1 protease with a peptidomimetic inhibitor and a second-generation compound. (A) The chemical structure of eight amino acids comprising one of the cleavage sites in the Gag-Pol polyprotein. The cleavage site between tyrosine and proline is indicated by a red arrow. **(B)** The chemical structure of an inhibitory peptide mimic, saquinavir. **(C)** A second-generation protease inhibitor, darunavir, for which more viral mutations are needed to acquire resistance. Regions of similarity are highlighted in yellow. Adapted from D. R. Harper, *Molecular Virology* (Bios Scientific Publishers, Ltd., Oxford, United Kingdom, 1994), with permission.

Figure 9.19 Structure of the hepatitis C virus protease NS3/4A and with a bound inhibitor (A) Crystal structure of Telaprevir bound to NS3-4A protease crystal structure (PDB code: 3SV6). The NS3 protease domain is shown in a gold-colored ribbon depiction. The active site triad residues Ser139, His57 and Asp81 are highlighted as sticks (with carbons colored cyan, oxygen in red, nitrogen in blue). The NS4A cofactor peptide segment is shown as in green ribbon. Telaprevir is shown as a stick diagram (with carbons colored magenta, oxygen in red, nitrogen in blue). Figure courtesy of Dr. G.R. Bhisetti, Biogen, Cambridge MA **(B)** Chemical structure of the FDA-approved protease inhibitor, Telaprivir.

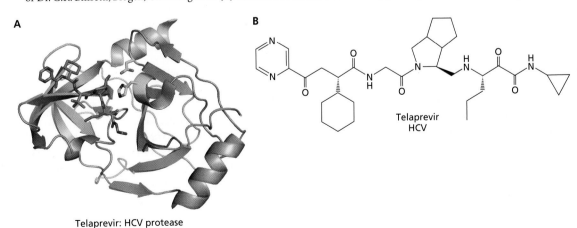

Telaprevir: HCV protease

successful application of a high-throughput assay that is specific for the second, joining step in the reaction (Volume I, Fig. 7.15 and Box 7.7) eventually led to the development of the first HIV integrase inhibitor, raltegravir, which was approved in 2007. Additional inhibitors of the same step in the reaction, elvitegravir and dolutegravir, were approved in 2012 and 2013, respectively. Solution of the crystal structure of the integrase of prototype foamy virus, with bound substrates and inhibitors, in 2010 provided the first clear picture of the mechanism of inhibition by these compounds. These drugs, called integrase strand transfer inhibitors, stabilize a viral DNA-protein intermediate while coordinating two catalytic magnesium ions bound to the three catalytic amino acids in the active site (Fig. 9.20). Allosteric inhibitors of HIV integrase, which bind at a specific dimer interface of the enzyme, are also being developed (Box 9.2).

A long-lasting analog of dolutegravir, which is suitable for monthly or quarterly clinical administration, has been shown to protect macaques from infection. These animal studies have suggested a promising approach to preexposure prophylaxis for humans at high risk for infection, such as partners of infected individuals.

Hepatitis C Virus Multifunctional Protein NS5A

The NS5A protein is an RNA-binding, phosphorylated protein that localizes to endoplasmic reticulum-derived membranes and functions in numerous steps of the virus life cycle, including genome replication and particle assembly. The mechanism by which the protein affects these processes, and causes hepatocyte apoptosis and carcinogenesis, is still unclear. Nevertheless, treatment with NS5A inhibitors, which are effective against all HCV genotypes, results in a rapid

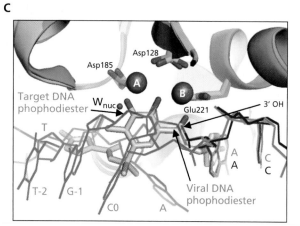

Figure 9.20 Strand transfer inhibitors of the human immunodeficiency virus integrase protein. (A) Chemical structures of the FDA-approved anti-HIV drugs raltegravir (RAL) and dolutegravir (DTG). **(B)** Comparison of crystal structures of the active sites of drug-free, RAL-bound, and DTG-bound prototype foamy virus integrase, which is also inhibited by these drugs. This integrase is shown in complex with a viral DNA end, and indicates how the terminal dA nucleotide and the reactive 3'-OH of the transferred DNA strand are displaced upon drug binding. Carbon, oxygen, and nitrogen atoms are colored magenta, red, and blue, respectively; the reactive 3'-OH is colored red and marked by arrows. **(C)** The integrase active site showing the relative positions of bound viral DNA ends, target DNA, and RAL. The unprocessed viral DNA end is green, whereas the processed viral DNA and bound target DNA strands are in dark blue and cyan, respectively. Mn²⁺ ions (labeled A and B), metal-chelating oxygens of RAL, bridging oxygen atoms of the scissile viral DNA and target DNA phosphodiester bonds (arrows), and 3'-processing (W_{nuc}, sphere) and DNA strand transfer (3'-OH, arrow) nucleophiles are colored red to highlight substrate mimicry of RAL during retroviral DNA integration. Adapted from L. Krishnan and A. Engelman, *J Biol Chem* 287:40858–40866, 2012, with permission.

decline in viral load. Other favorable properties include once-daily, low-dose efficacy and resistance profiles that do not overlap with other HCV antivirals.

Inhibitors of Human Immunodeficiency Virus Fusion and Entry

Some neutralizing antibodies block viral attachment of the viral envelope protein (Env) to the CD4 receptor on host cells by binding to the third variable domain (the so-called V3 loop) of the SU component of Env. A variety of natural and synthetic molecules also interfere with V3 loop-dependent host cell attachment. These compounds include specific antibodies (Volume II, Fig. 7.19) and polysulfated or polyanionic compounds such as dextran sulfate and suramin. Identified early in the search for antiviral agents, these compounds were subsequently discarded as antivirals because of intolerable side effects such as anticoagulant activity. Although considerable effort was expended to develop other inhibitors of the SU-CD4 interaction, including production of a "soluble CD4" that would act as a competitive inhibitor of infection, no effective antiviral agents have been found using this strategy. This lack of success can be attributed, in part, to the high concentration of SU on the virus particle, as well as to the existence of alternative mechanisms for spread in an infected individual.

As often happens, the early research with failed Env protein inhibitors provided much insight into how virus particles enter cells, and has focused attention on other targets in the process. For example, it was curious that mutants resistant to neutralizing antibodies have clustered substitutions in the V3 loop, yet virus-cell fusion is not affected. The implication was that CD4-V3 interactions did not contribute to entry. As described in Volume I, Chapter 5, entry is a multistep process requiring that the target cells synthesize not only CD4, but also any one of several chemokine receptors, such as CCr5 or CXCr4. Chemokine receptors are attractive targets, because individuals homozygous for mutations in one such receptor (CCr5) are partially resistant to infection. The first drug targeting an HIV chemokine receptor, maraviroc, was approved in 2007. The drug is an allosteric modulator of CCr5 function and blocks binding of HIV SU. As HIV can use other coreceptors for entry, the tropism of virus in an individual patient must first be determined to decide if treatment can be effective.

We now know that the interaction of the SU V3 loop with the chemokine receptor exposes previously buried SU sequences that are required for membrane fusion and that these transiently exposed surfaces can be targeted by antiviral agents. A 36-amino-acid synthetic peptide, termed T20, derived from the second heptad repeat of SU, binds to the exposed grooves on the surface of a transient triple-stranded coiled-coil and perturbs the transition of SU into the conformation active for fusion (Volume I, Chapter 5). T20 (enfuvirtide), the first drug with this mode of action, was approved in 2003. It is difficult to develop a peptide as a drug: large-scale synthesis is expensive, and patients must actually prepare a peptide solution for injection. Nevertheless, enfuvirtide is remarkably effective in reducing HIV titers in the blood.

Drug Resistance

Potent drugs are now available to inhibit the reproduction of several human viruses, including herpesviruses and hepatitis B virus, as well as HIV and HCV. However, because viral reproduction is so efficient and is accompanied by moderate to high mutation frequencies, resistance to any antiviral drug must be anticipated. The emergence of drug-resistant mutants is of special concern during the extended therapy required for viruses that establish chronic infections.

Mutations appear only when the viral genome is replicated. If replication is blocked completely by an inhibitor, no new drug-resistant mutants can arise. Consequently, if an individual harboring a small number of viral genomes with no relevant preexisting mutations is given a sufficient concentration of drug to block all viral replication, the infection will be held in check. When the viral genome numbers are small, the infection may be cleared by the host's immune system before resistant mutants take over. If the drug concentration is insufficient to block virus reproduction entirely or if the same antiviral drug is given after the viral population has expanded, genomes that harbor mutations will survive and will continue to replicate and evolve (Fig. 9.21).

If resistance to an antiviral drug requires multiple mutations, the chance that all mutations preexist in a single genome is much lower than if only a single mutation is required. Minimizing the differences between the natural ligand of the target protein and the antiviral drug will decrease the probability of emergence of mutants that are able to distinguish between ligand and drug. But when replication is allowed in the presence of the inhibitor, resistant mutants will accumulate. If there are no alternative antivirals, drug-resistant mutants can be devastating for patients and the population at large.

Although the development of drug resistance is a discouraging certainty with all direct-acting antivirals, genetic and biochemical analyses of the phenomenon can provide powerful insight into drug mechanisms and may identify new strategies to reduce or circumvent resistance. For example, acyclovir-resistant mutants of herpes simplex virus arise spontaneously during viral genome replication and are selected after exposure to the drug. As might be expected, the majority of mutations that confer resistance are in the viral thymidine kinase gene and inactivate kinase function. However, a subset of mutations leading to acyclovir resistance are not in

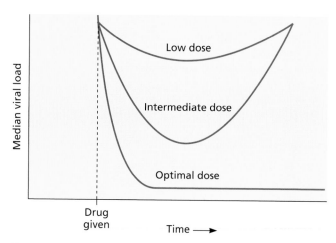

Median viral load

Low dose

Intermediate dose

Optimal dose

Drug given

Time ⟶

Figure 9.21 Viral load depends on the dose of antiviral drug. This relationship is illustrated by plotting median virus load in relative units on the *y* axis as a function of time after exposure to a drug on the *x* axis as indicated (Drug given). In the top curve (Low dose), the concentration of antiviral drug is insufficient to block virus reproduction, and the viral load is reduced only transiently, if at all. Virus mutants that are resistant to the drug may be enriched in this population following such treatment. In the middle curve (Intermediate dose), the concentration of antiviral drug is successful in lowering the viral load initially, indicating that some reproduction was blocked. In this example, the block was incomplete, and resistant mutants that can arise even during limited virus reproduction will eventually overwhelm the patient. In the bottom curve (Optimal dose), the concentration of the antiviral drug is high enough to block viral reproduction completely. As no new progeny viruses can be produced, the viral load drops dramatically, and the low level is maintained. Redrawn from J. H. Condra and E. A. Emini, *Sci Med* **4:**14–23, 1997, with permission.

this gene but, rather, in the viral DNA polymerase gene. The altered polymerases possess reduced ability to incorporate phosphorylated drug into DNA. Similar patterns of resistance have been reported when other nucleoside analog inhibitors are used against varicella-zoster virus.

Production of viral variants, drug-resistant mutants among them, is a hallmark of HIV infection, in which the initial, acute phase is followed by an asymptomatic period of clinical latency that lasts for years to decades (see Chapters 5 and 7). Although symptoms are kept in check by the patient's immune system, extensive viral reproduction and evolution continues throughout the asymptomatic period, until immune defenses finally break down and AIDS symptoms are manifested. Viral mutants that reproduced in the presence of AZT appeared almost immediately after the drug was approved for the treatment of AIDS. The genomes of the mutants were found to harbor single-base-pair changes at one of at least four sites in the reverse transcriptase gene. Reverse transcriptase enzymes bearing these substitutions no longer bound phosphorylated AZT, but they retained enzymatic activity. Mutants resistant to other nucleoside analogs, as well as to protease inhibitors, also

arose with disheartening frequency when these drugs were administered as monotherapy. These drug-resistant mutants were transmitted to new hosts and threatened to undermine the entire antiviral effort.

Combination Therapy

Combining two or more drugs with distinct targets circumvents the appearance of cells resistant to one treatment or the other. In theory, if resistance to one drug occurs once in every 10^3 genomes, and resistance to a second occurs once in every 10^4, then the likelihood that a genome carrying both mutations will arise is the product of the two probabilities, or one in every 10^7.

Mutants resistant to different nucleoside analogs were often found to carry different amino acid substitutions in the HIV reverse transcriptase. Furthermore, in some cases a mutation conferring resistance to one inhibitor suppressed resistance to another (Table 9.3). Consequently, combinations of nucleoside analogs were tested with the expectation that double-resistance mutants would be rare, perhaps nonviable, or at least severely crippled. While initially promising, many such combinations failed, with mutants resistant to both drugs appearing after less than a year of therapy. The frequency of resistance to many pairwise combinations of nucleoside and nonnucleoside inhibitors was lower than that for any single drug, but not low enough. Experience with protease substrate analog inhibitors was similar; resistance to two inhibitors emerged almost as quickly as resistance to either one alone. As current protease inhibitors are all peptide mimics that bind to the substrate pocket of the enzyme, a change in residues lining this pocket can affect the binding of more than one inhibitor. It became clear from these experiences that treatment of a patient with one antiviral drug, or in some cases even two, at a time is of limited clinical value. Consequently, inclusion of three or more antivirals has become standard practice in treating HIV infection.

Combination therapy can be demanding for physician and patient. For example, if other infections are being treated, as they almost always are in AIDS patients, then many pills a day may be required. Other problems arise because storing and keeping track of different medications are daunting tasks for someone who is ill. To compound the problems, every drug has side effects, and some are severe. For example, the gastrointestinal problems that accompany many protease inhibitors are particularly stressful. Some side effects, such as changes in fat distribution, may appear only after months of continuous use of current antiprotease drugs (Box 9.7). Because of these problems, some patients simply do not take their medication. The most insidious failure lies in wait when the patient begins to feel better and stops taking the medication. Viral replication resumes when the inhibitors are removed.

Table 9.3 Unpredicted drug resistance and susceptibility patterns

Compound	Substitution conferring resistance	Drug sensitivity phenotypes (amino acid substitution)
Zidovudine	T215F in reverse transcriptase	Didanosine resistance (L74V) restores zidovudine susceptibility
		Lamivudine resistance (M184V) restores zidovudine susceptibility
		Nevirapine and loviride resistance (Y181C) restores zidovudine susceptibility
		Foscarnet resistance (W88G) restores zidovudine susceptibility
Amprenavir	M46I + I47V + I50V in protease	Saquinavir resistance (G48V + I50V + I84L) restores amprenavir susceptibility
	VX-479 in protease	Indinavir resistance (V32I, A71V) restores amprenavir resistance
Delavirdine	P236L in reverse transcriptase	Increased susceptibility to nevirapine; R82913 (TIBO derivative); and L-697,661 (pyridinone)
Foscarnet	E89K + L92I + S56A + Q161L, H208Y in reverse transcriptase	Increased susceptibility to zidovudine, nevirapine, and R82150 (TIBO)

BOX 9.7

EXPERIMENTS
Highly specific, designed inhibitors may have unpredicted activities

The discovery and development of structure-based inhibitors of HIV protease have been pronounced a triumph of rational drug design. Structural biology and molecular virology came together to provide the protease inhibitors that anchor today's highly active antiretroviral therapy. However, patients receiving some protease inhibitors responded in unexpected ways. For example, one study showed that the protease inhibitor ritonavir inhibits the chymotrypsin-like activity of the proteasome. As a result, the drug blocks the formation and subsequent presentation of peptides to cytotoxic T lymphocytes (CTLs) by major histocompatibility complex class I proteins. In another study, the saquinavir protease inhibitor was found to inhibit Zmpste24, a protease required for conversion of farnesyl-prelamin A to lamin A, a structural component of the nuclear lamina.

The challenge is to determine if such secondary activities help or hinder AIDS therapy. As discussed in Chapter 4, CTLs not only kill virus-infected cells, but also are responsible for significant immunopathology in persistent infections. Perhaps a drug like ritonavir can block such immunopathology. On the other hand, reduction in immunosurveillance by CTLs potentiates persistent infections. In this case, the secondary activity of such a drug may presage long-term problems. We now know that the HIV protease inhibitors ritonavir and saquinavir interfere with proteasome activity, while indinavir and nelfinavir do not. The inhibition of Zmpste24 by the saquinavir class of compounds may contribute to the observed debilitating partial lipodystrophy side effect (redistribution of adipose tissue from the face, arms, and legs to the trunk). Genetic data indicate that individuals with missense mutations in *LmnA*, the gene encoding prelamin A and lamin C, have a significant loss of adipose tissues.

These experiences show that it is important to monitor lymphocyte functions and accumulation of prelamin A in patients under treatment with different protease inhibitors. Furthermore, tailoring HIV protease inhibitors to limit their action to the intended target is an important goal. As noted by the investigators who found these surprising activities, the human genome carries ~400 genes encoding proteases. About 70 of these proteases are targets for new drugs, and the unexpected side effects of antiviral protease inhibitors may be useful in finding new therapies.

André P, Groettrup M, Klenerman P, de Giuli R, Booth BL, Jr, Cerundolo V, Bonneville M, Jotereau F, Zinkernagel RM, Lotteau V. 1998. An inhibitor of HIV-1 protease modulates proteasome activity, antigen presentation, and T cell responses. *Proc Natl Acad Sci U S A* **95**:13120–13124.

Coffinier C, Hudon SE, Farber EA, Chang SY, Hrycyna CA, Young SG, Fong LG. 2007. HIV protease inhibitors block the zinc metalloproteinase ZMPSTE24 and lead to an accumulation of prelamin A in cells. *Proc Natl Acad Sci U S A* **104**:13432–13437.

Genome replication means mutation, and in such cases combination therapy may be ineffective if ever reinstated.

Because of these impediments to treatment compliance, there have been major efforts to develop drug combinations that can be taken less frequently and even together in a single pill for both HIV and HCV infections (Tables 9.1 and 9.2). A fixed-dose combination of two nucleos(t)ide inhibitors and a nonnucleoside inhibitor of HIV reverse transcriptase in a single pill (Atripla) that need be taken only once a day was approved in 2006. Development of this combination represented the first collaboration between two U.S. pharmaceutical companies to combine their patented anti-HIV drugs into one product. Atripla was followed in 2011 by Complera, comprising a similar cocktail, but with fewer side effects. Stribild, approved in 2012, contains four HIV inhibitors and is known as the "quad" pill. Stribild includes an integrase inhibitor, the two nucleos(t)ide analogs included in Atripla, and cobicistat. Cobicistat is a derivative of ritonavir that has no antiprotease activity but, like ritonavir, inhibits cytochrome P450, thereby increasing the effectiveness of the other three compounds. The clinical success of ever-improving combination therapy is truly remarkable and represents one of the high points in the battle against HIV. Similarly, once-a-day combinations are expected soon for the treatment of HCV.

Challenges Remaining

Infections with some viruses, such as HCV, can be cured. With aggressive use of potent antiviral drug combinations, reproduction of others, like hepatitis B virus and HIV, can be suppressed, but the infection **cannot** be cured. Even when HIV RNA has been undetectable in the bloodstream for years during drug therapy, virus appears again as soon as drug treatment is suspended. There have been isolated reports of cures, such as in one individual who received a bone marrow transplant from a donor carrying a mutation in the CCr5 cytokine receptor. However, there is no practical way at present to eliminate every last viral genome from the body of an HIV-infected individual. The current challenge is to devise workable strategies to rid an individual of all cells that contain proviruses or to remove established proviruses from all cells from which they may be activated. Some ideas that are being tested include activation of latent HIV proviruses with epigenetic drugs, followed by antiviral drug treatment, called "shock and kill." Cytokine therapy has yielded some promising results in inducing degradation of hepatitis B virus DNA in chronically infected hepatocytes.

Perspectives

The world's surprisingly small arsenal of antiviral drugs is directed against a subset of viral diseases. Few drugs are available for some of the most deadly established or emerging viral diseases, many of which are caused by RNA viruses. One formidable problem for delivery of antiviral drug therapy, even if available, is that many acute viral infections cannot be diagnosed accurately within sufficient time for effective intervention. Another arises from the fact that many debilitating viral infections affect people in the developing world, a population that lacks the means and possesses limited infrastructure for the delivery of antiviral drugs.

Persistent infections such as those caused by the human immunodeficiency virus, herpes simplex virus, and the hepatitis B virus present a special set of challenges. At present, these infections are controlled by drugs, but not cured. Often patients must take the drug, or more likely a combination of drugs, for the rest of their lives, a prospect that is both difficult and expensive. New approaches have been undertaken, and many promising lead compounds and therapies for treatment, and even cure, of persistent infections are being investigated. For example, in the future it may be possible to reduce viral load by antiviral drugs and then promote clearance of the remaining infection by treating with drugs that bolster immune responses. In most cases, however, selection of resistant mutants remains a problem for antiviral research and public health.

Despite the problems that remain, the successes in clinical development and distribution of increasingly effective antiviral drugs and drug combinations that target the human immunodeficiency virus and hepatitis C virus can be considered nothing less than a triumph, considering the millions of people whose lives have been saved not only in the United States and other high-income countries but also worldwide. A World Health Organization update in July 2013 reported an exponential increase in the number of people receiving anti-HIV therapy in lower- and middle-income countries since 2003 (Fig. 9.22). As a result, 4.2 million adult lives were saved and 800,000 infections of children were averted in these countries from 1996 to 2012. In addition, there were more than 700,000 fewer new HIV infections globally in 2011 than in 2001. Furthermore, prophylactic treatment against HIV infection for individuals at risk can be effective, and strategies for cures are being tested. Although much more needs to be done to end the AIDS pandemic, these results are most welcome news to the many millions worldwide who are infected with hepatitis C virus. In their cases, access to the new and expected antiviral therapies can promise a cure and elimination of the threat of fatal liver disease.

Based on experience gained from the successes with human immunodeficiency virus and hepatitis C virus, significant advances in technology, and increased understanding of virus biology and virus-host cell interactions, progress in developing drugs that block or even cure other viral infections should be more rapid in the future.

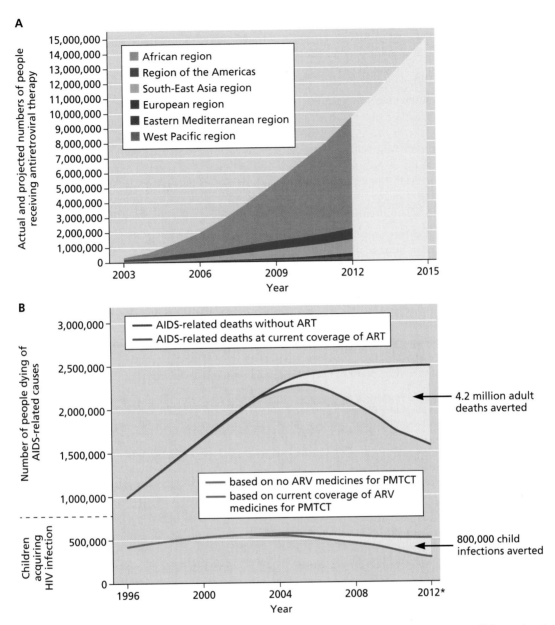

Figure 9.22 Anti-human immunodeficiency virus (HIV) therapy saves millions of lives. (A) Actual and projected numbers of people receiving antiretroviral therapy in lower- and middle-income countries. **(B)** Actual number of people dying from AIDS-related causes (top curves), and children acquiring HIV infection (bottom curves), in lower- and middle-income countries globally compared with a scenario of no antiretroviral therapy. ART, antiretroviral therapy; PMTCT, prevention of mother-to-child transmission; ARV, antiretroviral medication for prophylaxis in the third trimester of pregnancy (a regimen of twice-daily AZT, single-dose nevirapine at onset of labor, a combination of AZT + 3TC during delivery and 1 week postpartum, as well as infant prophylaxis for 1 week after birth). Adapted from World Health Organization, *Global Update on HIV Treatment 2013: Results, Impact and Opportunities* (World Health Organization, Geneva, Switzerland, 2013), with permission.

References

Books

Levy JA. 2007. *HIV and the Pathogenesis of AIDS*, 3rd ed. ASM Press, Washington, DC.

Veerapandian P (ed). 1997. *Structure-Based Drug Design: Diseases, Targets, Techniques and Developments*, vol 1. Marcel Dekker, Inc, New York, NY.

Reviews

Das K, Arnold E. 2013. HIV-1 reverse transcriptase and antiviral drug resistance. Part 1. *Curr Opin Virol* **3:**111–118.

Das K, Arnold E. 2013. HIV-1 reverse transcriptase and antiviral drug resistance. Part 2. *Curr Opin Virol* **3:**119–128.

De Clercq E. 2013. Antivirals: past, present and future. *Biochem Pharmacol* **85:**727–744.

Drews J. 2000. Drug discovery: a historical perspective. *Science* **287:**1960–1964.

Elion GB. 1986. History, mechanisms of action, spectrum and selectivity of nucleoside analogs, p 118–137. *In* Mills J, Corey L (ed), *Antiviral Chemotherapy: New Directions for Clinical Application and Research.* Elsevier Science Publishing Co, New York, NY.

Evans JS, Lock KP, Levine BA, Champness JN, Sanderson MR, Summers WC, McLeish PJ, Buchan A. 1998. Herpesviral thymidine kinases: laxity and resistance by design. *J Gen Virol* **79:**2083–2092.

Goff SP. 2008. Knockdown screens to knockout HIV-1. *Cell* **135:**417–420.

Halegoua-De Marzio D, Hann HW. 2014. Then and now: the progress in hepatitis B treatment over the past 20 years. *World J Gastroenterol* **14:** 401–413.

Harrison SC, Alberts B, Ehrenfeld E, Enquist L, Fineberg H, McKnight SL, Moss B, O'Donnell M, Ploegh H, Schmid SL, Walter KP, Theriot J. 2004. Discovery of antivirals against smallpox. *Proc Natl Acad Sci U S A* **101:** 11178–11192.

Manns MP, von Hahn T. 2013. Novel therapies for hepatitis C—one pill fits all? *Nat Rev Drug Discov* **12:**595–610.

Neutra MR, Kozlowski PA. 2006. Mucosal vaccines: the promise and the challenge. *Nat Rev Immunol* **6:**148–158.

Pang YP. 2007. *In silico* drug discovery: solving the "target-rich and lead-poor" imbalance using the genome-to-drug-lead paradigm. *Clin Pharmacol Ther* **81:**30–34.

Yu F, Lu L, Du L, Zhu X, Debnath AK, Jiang S. 2013. Approaches for identification of HIV-1 entry inhibitors targeting gp41 pocket. *Viruses* **5:**127–149.

Watkins WJ, Desai MC. 2013. HCV versus HIV drug discovery: déjà vu all over again? *Bioorg Med Chem Lett* **23:**2281–2287.

Zivin JA. 2000. Understanding clinical trials. *Sci Am* **282:**69–75.

Selected Papers

Bonhoeffer S, May RM, Shaw GM, Nowak MA. 1997. Virus dynamics and drug therapy. *Proc Natl Acad Sci U S A* **94:**6971–6976.

Coen DM, Schaffer PA. 1980. Two distinct loci confer resistance to acycloguanosine in herpes simplex virus type 1. *Proc Natl Acad Sci U S A* **77:**2265–2269.

Crotty S, Maag D, Arnold JJ, Zhong W, Lau JY, Hong Z, Andino R, Cameron CE. 2000. The broad-spectrum antiviral ribonucleoside ribavirin is an RNA virus mutagen. *Nat Med* **6:**1375–1379.

Crute JJ, Grygon CA, Hargrave KD, Simoneau B, Faucher AM, Bolger G, Kibler P, Liuzzi M, Cordingley MG. 2002. Herpes simplex virus helicase-primase inhibitors are active in animal models of human disease. *Nat Med* **8:**386–391.

Herold BC, Bourne N, Marcellino D, Kirkpatrick R, Strauss DM, Zaneveld LJ, Waller DP, Anderson RA, Chany CJ, Barham BJ, Stanberry LR, Cooper MD. 2000. Poly(sodium 4-styrene sulfonate): an effective candidate topical antimicrobial for the prevention of sexually transmitted diseases. *J Infect Dis* **181:**770–773.

Kumar P, Wu H, McBride JL, Jung KE, Kim MH, Davidson BL, Lee SK, Shankar P, Manjunath N. 2007. Transvascular delivery of small interfering RNA to the central nervous system. *Nature* **448:**39–43.

Perelson AS, Essunger P, Cao Y, Vesanen M, Hurley A, Saksela K, Markowitz M, Ho DD. 1997. Decay characteristics of HIV-1 infected compartments during combination therapy. *Nature* **387:**188–191.

Thomsen DR, Oien NL, Hopkins TA, Knechtel ML, Brideau RJ, Wathen MW, Homa FL. 2003. Amino acid changes within conserved region III of the herpes simplex virus and human cytomegalovirus DNA polymerases confer resistance to 4-oxo-dihydroquinolines, a novel class of herpesvirus antiviral agents. *J Virol* **77:**1868–1876.

10 Evolution

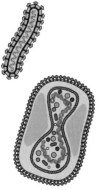

Virus Evolution

Classic Theory of Host-Parasite Interactions

How Do Virus Populations Evolve?
Two General Survival Strategies Can Be Distinguished
Large Numbers of Viral Progeny and Mutants Are Produced in Infected Cells
The Quasispecies Concept
Sequence Conservation in Changing Genomes
Genetic Shift and Genetic Drift
Fundamental Properties of Viruses That Constrain and Drive Evolution

The Origin of Viruses

Host-Virus Relationships Drive Evolution
DNA Virus Relationships
RNA Virus Relationships
The Protovirus Hypothesis for Retroviruses

Lessons from Paleovirology
Endogenous Retroviruses
DNA Fossils Derived from Other RNA Viral Genomes
Endogenous Sequences from DNA Viruses
The Host-Virus "Arms Race"

Perspectives

References

LINKS FOR CHAPTER 10

▶▶ *Video: Interview with Dr. Harmit Malik*
http://bit.ly/Virology_Malik

▶▶ *I want my MMTV*
http://bit.ly/Virology_Twiv242

▶▶ *Paleovirology with Michael Emerman*
http://bit.ly/Virology_Twiv237

▶▶ *Describing a viral quasispecies*
http://bit.ly/Virology_4-16-15

▶▶ *Viral genomes in 700 year old caribou scat*
http://bit.ly/Virology_12-8-14

▶▶ *A WORD on the constraints of influenza virus evolution*
http://bit.ly/Virology_5-23-14

Virus Evolution

The word "evolution" conjures up images of fossils, dusty rocks, and ancestral phylogenetic trees, covering eons. Thanks to the recent development of rapid sequencing methods, we can now discover fossils of ancient viruses, not in rocks, but in the DNA of living organisms. For currently circulating viruses, evolution is not only contemporary (and rapid), but also has profound effects on both viruses and their hosts: as host populations change or become resistant to infection, viruses that can overcome such changes are selected. Viral infections can also exert significant selective forces on the survival and evolution of host populations. In some ways, viral evolution can be thought of as the product of a continuing arms race in which both viral and host cell genes are selected in response to the pressures encountered during infection.

Classic Theory of Host-Parasite Interactions

Virus particles must spread from host to host to maintain a viable population. Spreading will occur if, on average, each infected host passes the agent to more than one new host before the host dies or clears the infection. The probability of such transmission is related to the size of the host population: infections can spread only if population density exceeds a minimal value.

These concepts have been incorporated into a comprehensive theory of host-parasite interactions that is well known in ecological circles, but not always appreciated among molecular virologists. This theory describes the parameters of viral infection in quantitative terms. The basic reproductive number for a virus population, R_0 (pronounced **R-naught**), is defined as the number of secondary infections that can arise in a large population of susceptible hosts from a single infected individual during its life span. If $R_0 < 1$, it is impossible to sustain an epidemic; in fact, it may be possible to eradicate the disease. If $R_0 > 1$, an epidemic is possible, but random fluctuations in the number of transmissions in the early stages of infection in a susceptible population can lead to either extinction or explosion of the infection. If R_0 is much greater than 1 (Table 10.1), an epidemic is almost certain. The proportion of the susceptible population that must be vaccinated to prevent virus spread is calculated as $1 - 1/R_0$.

In the simplest model, $R_0 = \tau \cdot c \cdot d$, where τ is the probability of infection, given contact between an infected and uninfected host; c is the average rate of contact between them; and d is the duration of infectivity. The original host-parasite theory assumed well-mixed, homogeneous host populations in which each individual host has the same probability of becoming infected. Although the general concepts remain valid, additional parameters and constraints have been added to the mathematical models as more has been learned about population diversity and the dynamics of viral infections (Box 10.1; see also Fig. 5.2). For example, immune-resistant viral mutants with differences in virulence

---

PRINCIPLES *Evolution*

- Virus evolution is the product of continuing interaction between viral and host cell genes and selection for the most fit.

- Diversity of a virus population allows adaptation to environmental changes; remove diversity, and the population suffers.

- Viral diversity is generated by mutation, recombination, and reassortment of viral genes.

- Virus populations can be sustained by production of many progeny, better competition for resources, or both.

- Virus populations exist as dynamic distributions of nonidentical but related and interactive replicons, called **quasispecies**.

- Genomes of RNA viruses are replicated close to the error thresholds, i.e., the number of mutations within populations at which viruses can no longer be propagated. Such replication contrasts with that of DNA viruses, which generally proceeds with higher fidelity, and well below the error threshold.

- Mutations accumulate at every viral replication cycle. Virus populations cannot survive unless genomes that are free from harmful mutations and conserve beneficial mutations can be produced by reassortment or recombination. This principle is illustrated by **Muller's ratchet**.

- The origins of viruses remain puzzling, and three non-mutually-exclusive hypotheses have been proposed: reduction by loss of genes from a cell, accumulation of **cellular** components that gained independent reproduction capacity, and **independent** coevolution with cells from the origin of life.

- Although the primordial history of viruses cannot be known, evidence of virus-derived sequences in host genomes provides important insights into the nature and consequences of viral and host interaction over evolutionary time.

- The discovery of viruses with extremely large genomes that include coding capacities equal to or surpassing those of some prokaryotes, and the diversity of viral genomes revealed by metagenomic analyses, indicate that there is much more to learn about virus history and evolution.

Table 10.1 Reproductive numbers for some viruses

Virus	$R_0{}^a$
Measles	12–18
Smallpox	5–7
Polio	5–7
Influenza	
2009 (H1N1)	1.47
1957, 1968 pandemics	1.8
1918 pandemic	2.4–5.4
Hemorrhagic fever (Ebola)	1.3–1.8[b]

[a]Values from Centers for Disease Control and Prevention website and literature.
[b]Source: G. Chowell et al., *J. Theor Biol* **229**:119–126, 2004.

and transmissibility can be selected, and some individuals (called super transmitters) can pass infection to others much more readily than the majority. We also now know that virus populations are more diverse than first imagined, and the constellation of possible host populations affects their evolution in ways not easily captured by mathematical equations. Consequently, although the calculations are useful indications of the thresholds that govern the spread of a virus in a population (i.e., they help to determine if a disease is likely to die out [$R_0 < 1$] or become endemic [$R_0 > 1$]), they cannot be used to compare possible outcomes in particular cases or for different diseases.

How Do Virus Populations Evolve?

A large, genetically variable host population dispersed in ever-changing environments may appear to present insurmountable barriers to the survival of viruses, yet viruses are plentiful and ubiquitous. The primary reason for this remarkable success is that virus populations display spectacular genetic diversity, manifested in the large collections of genome permutations that are present in a population at any given time. The sources of such diversity are **mutation**, **recombination**, and **reassortment** of viral genes. Virus evolution is driven by this diversity and selective pressures that promote survival of the most fit.

In most viral infections, thousands of progeny are produced after a single cycle of replication in one cell, and when genome copying is error prone, almost every new virus particle can differ from every other. Consequently, it is misleading to think of an individual particle as representing an average for that population. This great diversity of the virus population provides avenues for survival under varying conditions.

BOX 10.1

BACKGROUND
Virulence, selection, and evolution

Is virulence a positive or negative trait for selection? One idea is that increased virulence reduces transmissibility (hosts die faster, reducing exposure to uninfected hosts). Debilitating disease may actually reduce transmission because the infected individual may not interact with other susceptible hosts (see the figure).

If everything were simple, one might expect that all viruses would evolve to be maximally infectious and completely avirulent. A different view appears when real-life infections are studied. The interplay of contextual terms such as "severity of disease" and "transmissibility" is quite complicated. Indeed, for some diseases, and in some contexts, a strong case may be made that increased virulence actually increases R_0 and is strongly selected for in natural viral infections.

An instructive example of the complex relationship between virulence and transmissibility comes from the intentional release of myxoma virus in Australia to control the European rabbit population. Initially the virus was 99% lethal in rabbits, but within a few years both the virus and the rabbits evolved so that infection was 30% lethal. It is thought that moderate levels of virulence were selected because this was concomitant with production of sufficient amounts of virus particles to ensure spread in the rabbit population.

Boots M, Childs D, Reuman DC, Mealor M. 2009. Local interactions lead to pathogen-driven change to host population dynamics. *Curr Biol* **19**:1660–1664.

Fenner F. 1983. The Florey lecture, 1983. Biological control, as exemplified by smallpox eradication and myxomatosis. *Proc R Soc Lond B Biol Sci* **218**:259–285.

Weiss RA. 2002. Virulence and pathogenesis. *Trends Microbiol* **10**:314–317.

More virulent... ...less chance for transmission

Less virulent... ...greater chance for transmission

Every individual virion is a potential winner, and occasionally, the rarest genotype in a particular population will be the most common after a single selective event. As the famous biologist and science historian Stephen Jay Gould put it, the median is **not** the message when it comes to evolution.

Positive and negative selection of preexisting mutants in a population can occur at any step in a viral life cycle. The requirement to spread within an infected host, as well as between hosts, exposes virus particles to a variety of host antiviral defenses. In addition, host population density, social behavior, and health, represent but a few of the other forces that can affect the survival of virus populations.

Two General Survival Strategies Can Be Distinguished

Viral reproduction cycles typically produce large numbers of progeny particles, and in some cases, virus survival depends mainly on a very high reproductive output, the *r*-replication strategy. For other viruses, survival is compatible with a lower reproductive output but better competition for resources, the **K-replication strategy** (see Chapter 3). The notations *r* and *K* come from the following equation:

$$dN/dt = rN(1 - N/K)$$

where *r* is the intrinsic rate of increase (i.e., average rates of births minus deaths), *N* is the population size, and *K* is a measure of the resources available (see also Fig. 5.2).

The *r*-replication strategies are characterized by short reproductive cycles, and are effective when resources are in short supply. The *K*-replication strategies include the establishment of persistent or latent infections with little pathogenesis. In these cases, the viruses survive as long their hosts do. For some viruses (e.g., human immunodeficiency virus type 1), both strategies may be important for virus survival.

Large Numbers of Viral Progeny and Mutants Are Produced in Infected Cells

High reproduction rate (the *r*-replication strategy) is common among viruses. To illustrate the implications of this strategy, consider that a single cell infected with poliovirus yields ~10,000 virus particles in as little as 8 h. In theory, three or four cycles of reproduction at this rate could produce a sufficient number of particles to infect every cell in a human body. Such overreplication does not happen, for a variety of reasons, including a vigorous host defense and the fact that viruses can only reproduce in certain tissues or cell types. Nevertheless, this strategy is characteristic of many infections, including those of humans with human immunodeficiency virus type 1 and hepatitis B virus, in which high rates of particle production can continue for years. In the case of human immunodeficiency virus type 1, the time from release of thousands of virus particles from an infected T cell to infection and lysis of another T cell is estimated to be a mere 2.6 days during the later stages of infection (see Fig. 6.14). Mutations invariably accumulate during genome replication, although the *absolute* error rates for this process can be difficult to measure (Box 10.2).

RNA Virus Evolution

Most viral RNA genomes are replicated with considerably lower fidelity than those comprising DNA (see Volume I, Chapters 6 and 7). The average error frequencies reported for RNA genomes are about one misincorporation in 10^4 or 10^5 nucleotides polymerized, which is 1,000 to 10,000 times greater than the rate for a host genome. Given a typical RNA viral genome of 10 kb, a mutation frequency of 1 in 10^4 per template copied corresponds to an average of 1 mutation in *every* replicated genome. Not all viral RNA genomes have the same mutation rate: there is evidence that the replication machinery of viruses with larger RNA genomes operate with higher fidelity. One example is the 30,000-base human severe acute respiratory syndrome coronavirus; a proofreading exonuclease encoded in its genome may account for the lower mutation frequency. The rodent hantaviruses, members of the *Bunyaviridae*, appear to evolve very slowly, with mutation rates approaching those of double-stranded DNA viruses.

DNA Virus Evolution

The error rate for viral DNA replication is estimated to be from 10^{-6} to 10^{-8}, which is closer to the host rate than that for most RNA genomes described above. One reason for this difference is that many RNA polymerases lack error-correcting mechanisms, while most DNA polymerases can excise and replace misincorporated nucleotides (Volume I, Chapter 9). Experimental data indicate that replication of small, single-stranded DNA virus genomes (e.g., *Parvoviridae* and *Circoviridae*) is more error prone than is replication of the double-stranded DNA genomes of larger viruses.

Comparison of the number of mutations produced *per infected cell* shows an inverse relationship between genome size and mutation rate for both RNA and DNA viruses (Fig.10.1). These values can be somewhat higher than those estimated from the error rates of the respective polymerases as they include other sources of mutation, such as host enzyme-mediated base changes, additions or deletions (called RNA editing), or spontaneous damage of viral nucleic acids, e.g., via oxygen radicals or ionizing radiation. The lowest estimates determined for RNA viruses are close to the highest for DNA viruses. As illustrated in Fig. 10.1, the transition between them appears to be relatively smooth. The relationship between genome size and mutation rate suggests that extremes of mutation rate are selected against.

BOX **10.2**

DISCUSSION
Error rates are difficult to quantify

Estimates of mutation rates must be viewed with caution. Determining the absolute error rate (measured as the number of misincorporations per nucleotide polymerized) for any nucleic acid polymerase is difficult, if not impossible. Estimates can vary substantially, depending on the experimental method by which they are assessed. For example, PCR technology is commonly used to sample viral genomes, but the polymerase used may itself introduce copying errors that must be factored into the analysis. Cloning viral sequences from a given population in a plasmid, followed by sequencing, can reduce this problem, but depending on the conditions, lethal mutations may be selected against.

A popular method makes use of reporter genes (e.g., the *lacZ* gene, which encodes β-galactosidase). The reporter gene can be inserted into a viral genome such that enzyme-inactivating mutations can be scored simply by inspection of virus plaques. The mutation frequency for the viral genome is then extrapolated from that determined for the reporter gene. While this method is relatively simple, it can yield misleading data, because assay conditions and host editing functions can affect the results. Furthermore, errors of incorporation are not uniformly distributed as each genome is copied, and can be under- or over-estimated depending on the particular polymerase and sequence analyzed. For example, when measured *in vitro* using the *lacZ* system, error rates for the human immunodeficiency virus type 1 reverse transcriptase are both extremely high and variable. However, despite the fact that both reverse transcriptase and the host cell RNA polymerase II can contribute to nucleotide misincorporations *in vivo*, the rate calculated with the same reporter in a single replication cycle of this virus is substantially lower and, at 1.4×10^{-5}, similar to the average rate measured for other retroviruses. Analysis of the recovered reporter genes shows that certain nucleotides are hot spots for mutation, as would be predicted for some loci in the viral genome.

Abram ME, Ferris AL, Shao W, Alvord WG, Hughes SH. 2010. Nature, position, and frequency of mutations made in a single cycle of HIV-1 replication. *J Virol* **84:**9864–9878.

Rezende LF, Prasad VR. 2004. Nucleotide-analog resistance mutations in HIV-1 reverse transcriptase and their influence on polymerase fidelity and viral mutation rates. *Int J Biochem Cell Biol* **36:**1716–1734.

Svarovskaia ES, Cheslock SR, Zhang WH, Hu WS, Pathak VK. 2003. Retroviral mutation rates and reverse transcriptase fidelity. *Front Biosci* **8:**d117–d134.

Single nucleotide substitutions detected in a human immunodeficiency virus type 1-encoded *lacZα* reporter. Numbers, types, and locations of the independent substitution errors are shown for both the forward orientation [(+) strand nucleotide sequence] and the reverse orientation [(−) strand nucleotide sequence] of the *lacZα* reporter. Opposing directional arrows indicate the actual sequence context and direction of minus-strand DNA synthesis during reverse transcription. The total length of the *lacZα* target sequence was defined as 174 nucleotides, representing codons 6 to 63 from GGA to the first TAA termination codon. Only the first 120 nucleotides, which incurred most of the substitutions, are shown here. Single substitution errors are shown as letters above the original wild-type template sequence, limited to 11 per position, with additional errors indicated by +n. Runs of three or more identical nucleotides are underlined. Misalignment/slippage of the primer or template strand that could result in a substitution error is highlighted in orange or green, respectively. Mutational hot spots for which there are significant differences in the forward and reverse *lacZα* orientations are indicated below the sequence by asterisks. Reprinted from M. E. Abram et al., *J Virol* **84:**9864–9878, 2010, with permission.

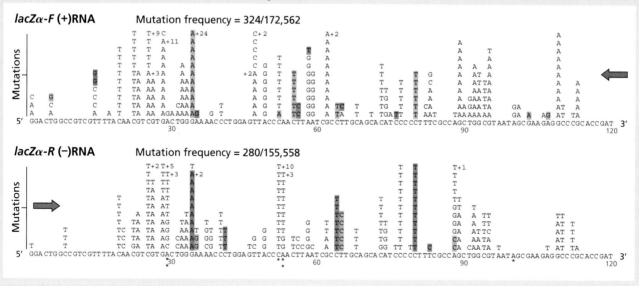

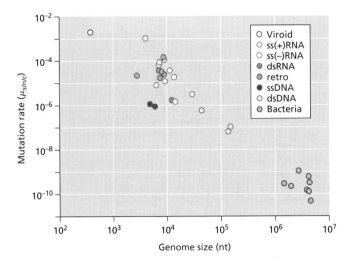

Figure 10.1 Relationship between mutation rate and genome size. Mutation rate is expressed as the number of substitutions per nucleotide per generation, defined as a single cell infection ($\mu_{s/n/c}$). ss, single-stranded; ds, double-stranded; retro, retroviruses; Bacteria includes points for *Bacillus*, *Deinococcus*, *Helicobacter*, *Sulfolobus*, enterobacteria, and mycobacteria. The white circle is a single reported rate for a viroid, expressed in substitutions per strand copied. Adapted from R. Sanjuán et al., *J Virol* **84**:9733–9748, 2010, with permission.

The Quasispecies Concept

A 1978 paper, which described a detailed analysis of an RNA bacteriophage population (phage Qβ), made the following, startling conclusion:

> A Qβ phage population is in a dynamic equilibrium, with viral mutants arising at a high rate on the one hand, and being strongly selected against on the other. The genome of Qβ phage cannot be described as a defined unique structure, but rather as a weighted average of a large number of different individual sequences.
>
> E. Domingo, D. Sabo, T. Taniguchi, and
> C. Weissmann, *Cell* **13**:735–744, 1978

This conclusion has since been validated for many virus populations. Indeed, we now understand that virus populations exist as dynamic distributions of nonidentical but related replicons, often called **quasispecies**, a concept developed by Manfred Eigen. A steady-state, equilibrium population of a given viral quasispecies must comprise vast numbers of particles. Indeed, such equilibria cannot be attained in the small populations typically found after isolated infections in nature or in the laboratory. In these cases, extreme fluctuations in genotype and phenotype are possible.

For a given RNA virus population, the genome sequences cluster around a consensus or average sequence, but virtually every genome can be different from every other. A rare genome with a particular mutation may survive a specific selection event, and this mutation will be found in all progeny genomes. However, any linked but unselected mutations in

that genome will also be retained. Consequently, the product of selection after replication is a new, diverse population of genomes that share only the selected and closely linked mutations (Fig. 10.2).

The quasispecies theory predicts that a viral quasispecies is not simply a collection of diverse mutants, but rather a group of *interactive* variants that characterize the particular population. Diversity of the population, therefore, is critical for survival. It has been possible to test the idea that virus populations, **not** individual mutants, are the targets of selection by limiting diversity. Certain spontaneous mutants of human immunodeficiency virus type 1 that are resistant to the reverse transcriptase inhibitor lamivudine exhibit a 3.2-fold reduction in error frequency. However, this seemingly modest increase in fidelity was found to be associated with a significant reproductive **disadvantage** in infected individuals. As another example, poliovirus replication is notoriously error prone, producing a remarkably diverse population. Certain ribavirin-resistant poliovirus mutants have increased fidelity of ~6-fold, but such mutants were found to be **much less** pathogenic in animals than was the wild-type virus, and the reduced diversity led to attenuation and loss of neurotropism. Further studies showed that in a genetically diverse population, isolated viral mutants complement each other, consistent with the idea that it is the population, not the individual, that is evolving. In any case, as virus populations have maintained high mutation rates, we can infer that lower rates are neither advantageous nor selected in nature.

Another outcome of quasispecies dynamics is that viral mutants with low fitness can sometimes outcompete viruses with higher fitness if the low-fitness mutations are surrounded by beneficial ones. In other words, a population whose mutants have a similar mean fitness can outcompete a population with a lower average fitness that contains mutants with higher fitness. This situation has been called the **quasispecies effect**, or **survival of the flattest**. In contrast, in classic population genetics models, individual high-fitness variants are favored, a situation known as **survival of the fittest**.

Sequence Conservation in Changing Genomes

Despite high mutation rates, not all is in flux during viral genome replication. For example, the *cis*-acting sequences of RNA viruses change very little during propagation. These sequences include signals that are required for genome replication, messenger RNA (mRNA) synthesis, and genome packaging. They are often the binding sites for one or more viral or cellular proteins. Any genome with mutations in such sequences, or in the gene that encodes the corresponding viral binding protein, is likely to be less fit, or may not be replicated at all. Changes must occur in both interacting components for restoration of function. The tight, functional coupling of binding protein and target sequence is a marked constraint

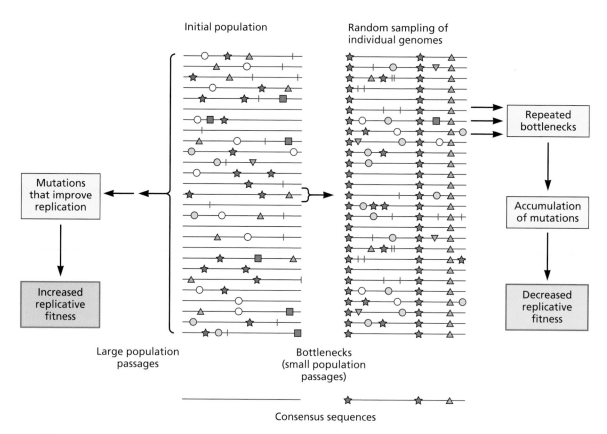

Initial population

Random sampling of individual genomes

Mutations that improve replication

Increased replicative fitness

Large population passages

Bottlenecks (small population passages)

Repeated bottlenecks

Accumulation of mutations

Decreased replicative fitness

Consensus sequences

Figure 10.2 Viral quasispecies, population size, bottlenecks, and fitness. Genomes are indicated by the horizontal lines. Mutations are indicated by different symbols. When a viral genome is replicated, mutations accumulate in the progeny genomes. While every viral population exhibits genomic diversity, the extent of diversity depends on many parameters, including the frequency of mutation. (Left) A hypothetical population of genomes, in which each member contains a characteristic set of mutations. The consensus sequence for this population is shown as a single line at the bottom. Note that there are **no** mutations in the consensus sequence, despite their presence in most genomes in the population (almost every genome is different). (Right) A population of genomes that emerges after passage of one genome through one bottleneck. The consensus sequence for this population is shown as a single line at the bottom. Note that in this example, three mutations selected to survive the bottleneck are found in every member of the population, and these appear in the consensus sequence. If the large population is propagated without passage through bottlenecks (situation on the left), repeated passage enriches for mutant genomes that **improve replication and increase the fitness** of the population. Recent studies with plaque purified virus (a bottleneck; see text and Table 10.2), and using techniques that eliminate some artifacts of sequencing, found that for some RNA viruses (e.g., poliovirus), almost half of the particles at an early passage at low MOI had genomes with no mutations. As the population was passaged serially, the proportion of genomes with no mutations decreased and those containing multiple changes increased. If a viral population continues to be propagated through serial bottlenecks, mutations that result in **reduced fitness will accumulate**. Adapted from E. Domingo et al., p. 144, *in* E. Domingo et al. (ed.), *Origin and Evolution of Viruses* (Academic Press, Inc., San Diego, CA, 1999), with permission.

for evolution. In some instances, these sequences are stable enough to represent lineage markers for molecular phylogeny.

The Error Threshold

The capacity to sustain prodigious numbers of mutations is a powerful advantage for virus populations. Yet, at some point, selection and survival must balance genetic fidelity and mutation rate. Many mutations are detrimental, and if the mutation rate is high, their accumulation can lead to a phenomenon called **lethal mutagenesis** when the population is driven to extinction. Intuitively, mutation rates higher than

one error in 1,000 incorporated nucleotides must challenge the very existence of the viral genome, as essential genetic information can be lost irreversibly. The **error threshold** is a mathematical parameter that measures the complexity of the information that must be maintained to ensure survival of the population. RNA viruses tend to evolve close to their error threshold, while DNA viruses have evolved to exist far below it. We can infer these properties from experiments with mutagens. After treatment of cells infected by an RNA virus (such as vesicular stomatitis virus or poliovirus) with a nucleoside analog such as 5-azacytidine, virus titers drop dramatically, but the error frequency per surviving genome increases by only 2- to 3-fold at most. In contrast, a similar experiment performed with a DNA virus, such as herpes simplex virus or simian virus 40, results in a less precipitous drop in survival, but an increase of several orders of magnitude in single-site mutations among the survivors.

It is clear that many important biological parameters contribute to virus survival, including a complex property called **fitness**, the adaptability of an organism to its environment. Fitness depends on the context in which reproduction is examined and what outcome is measured. In the laboratory setting, it may be measured simply by comparison of reproduction rates or virus yields. However, fitness is far more difficult to measure under natural conditions, such as infection of organisms that live in large, interacting populations. Another essential parameter, equally difficult to measure, is the stability or predictability of the environment as it affects propagation of a virus. Host population dynamics and seasonal variation are but two examples of such complicated environmental parameters. Finally, given the diversity of any viral population, determining the fitness of one population versus that of another requires application of the mathematics of population genetics, a subject beyond the scope of this text.

Genetic Bottlenecks

Unlike lethal mutagenesis, which can lead to the extinction of large populations, the **genetic bottleneck** represents extreme selective pressure on small populations, which results in loss of diversity, accumulation of nonselected mutations, or both (Fig. 10.2). A simple experiment illustrating this principle can be performed readily in the laboratory (Table 10.2). Virus particles are isolated from a single, isolated plaque and used to infect fresh cells; the process is then repeated many times (**serial passage**). The perhaps surprising result is that, after about 20 or 30 cycles of single-plaque selection, the virus populations are barely able to propagate; they are markedly less fit than the original populations. The bottleneck is the consequence of restricting further viral reproduction to the progeny found in a single plaque, which contains only a few thousand virions—all derived from a single infected cell.

Table 10.2 Fitness decline compared to initial virus clone after passage through a bottleneck[a]

Virus	No. of bottleneck passages	% Decrease in fitness (avg)
φ6 (bacteriophage)	40	22
Vesicular stomatitis virus	20	18
Foot-and-mouth disease virus	30	60
Human immunodeficiency virus	15	94
MS2 (bacteriophage)	20	17

[a]Source: A. Moya et al., *Proc Natl Acad Sci U S A* **97**:6967–6973, 2000.

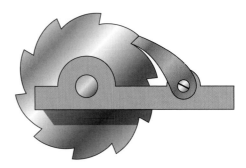

Figure 10.3 Muller's ratchet turns in only one direction. Mutations occur and accumulate at every replication cycle. Without recombination to recreate a genome that is free from harmful mutations, and conserve beneficial mutations, the population cannot survive.

The environment is constant, and the only apparent requirement is that the population of viruses obtained from a single plaque must be able to reproduce. Why does fitness plummet?

The answer lies in **Muller's ratchet** model, which explains why small, asexually reproducing populations decline in fitness over time if the mutation rate is high. The ratchet metaphor is fitting: a ratchet on a gear allows the gear to move forward, but not backward (Fig. 10.3). After each round of genome replication, mutations accumulate but are not removed. Each round of error-prone replication works like a ratchet, "clicking" relentlessly as mutations accumulate. Each mutation has the potential to erode the fitness of subsequent, limited populations. We have noted that the genomes of replicating RNA viruses accumulate many mutations and survive close to their error threshold. By restricting population growth to serial single founders (serial bottlenecks), so many mutations accumulate in **all** of the progeny that fitness decreases.

Simple studies such as the serial plaque transfer experiment show that Muller's ratchet can be avoided if a more diverse viral population is subjected to serial passage. One such study showed that pools of virus from 30 individual plaques were required in serial transfer to maintain the population's original fitness. This observation can be explained as follows: greater diversity in the population facilitates the construction of mutation-free genomes by recombination or reassortment, and hence removes or compensates for deleterious mutations. Such recombination or reassortment may be quite rare, but it imparts a powerful selective advantage in this experimental paradigm. Indeed, the progeny of such a rare virus will ultimately predominate in the population. The message is simple but powerful: the diversity of a virus population is important for its survival; remove diversity, and the population suffers.

As the particular bottleneck of single-plaque passage is obviously artificial, it is reasonable to ask: Does Muller's ratchet ever occur in nature? Infections by small virus populations include the tiny droplets of suspended virus particles during transmission as an aerosol, activation of a latent virus from a limited population of cells, and the small volumes of inoculum introduced by insect bites. How virus populations pass through such natural bottlenecks and survive is not yet clear.

Genetic Shift and Genetic Drift

Selection of mutants that are resistant to elimination by antibodies or cytotoxic T lymphocytes is inevitable when successful virus reproduction occurs in an immunocompetent individual. This process of antigenic variation and its contribution to modulating the immune response is discussed in Chapter 5. The terms **genetic drift** and **genetic shift** describe distinct mechanisms for generation of diversity. Diversity that arises from genome replication errors and immune selection of single-site mutants (drift) is contrasted with diversity that results from recombination among genomes, or reassortment of genome segments (shift). Drift is possible every time a genome replicates, but shift is relatively rare. The episodic pandemics of influenza (Fig. 10.4) provided strong evidence for this conclusion. For example, there are only six established instances of genetic shift for the influenza virus hemagglutinin gene since 1889. However, the combination of frequent drift and infrequent shift, together with the availability of intermediate host species, contributes significantly to diversity in influenza virus populations (Box 10.3). When retrovirus infection results in integration of multiple proviral genomes in a single cell, genetic shift can occur via recombination if two different viral genomes are packaged in a progeny virus particle.

Exchange of Genetic Information

Genetic information is exchanged by recombination or by reassortment of genome segments (Volume I, Chapters 6, 7, and 9). In a single step, recombination can create new linkages of many mutations that may be essential for survival under selective pressures. As discussed above, this process allows the construction of viable genomes from debilitated ones. Recombination occurs when the polymerase that copies a viral genome changes templates (copy choice) during replication or when nucleic acid segments are broken and rejoined. The former mechanism is common among RNA viruses, whereas the latter is more typical of double-stranded DNA viruses. Reassortment of genomic segments takes place when cells are coinfected with segmented RNA viruses. This can be an important source of variation, as exemplified by reoviruses and orthomyxoviruses (Fig. 10.4 and Box 10.3).

Insertion of nonviral nucleic acid into a viral genome is also a well-documented phenomenon that can contribute to virus evolution. Incorporation of cellular sequences can lead to defective genomes, or to more-pathogenic viruses. Examples of such recombination include the appearance of a cytopathic

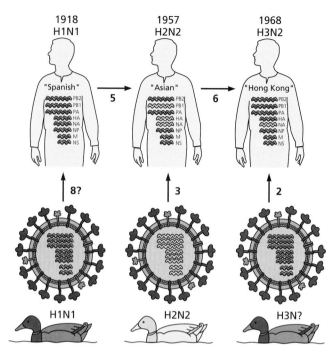

Figure 10.4 Appearance and transmission of distinct serotypes of influenza A virus in human pandemics in the 20th century. The major influenza pandemics are characterized by viral reassortants. The reassortants carried HA (H) and NA (N) genes that had not been in circulation in humans for some time, and consequently, immunity was low or nonexistent. With the introduction of each subtype, the world experienced an influenza pandemic characterized by a new combination of H and N. The viral genome segments are illustrated in three different colors, with each representing a particular viral genotype. Segments and gene products of the pandemic strains are indicated in each human silhouette. The numbers next to the arrows indicate how many segments of the viral genome are known to have been transmitted in each episode. Adapted from R. G. Webster and Y. Kawaoka, *Semin Virol* **5**:103–111, 1994, with permission.

virus in an otherwise nonpathogenic infection by the pestivirus bovine viral diarrhea virus (see Volume I, Chapter 6) or the sudden appearance of oncogenic retroviruses following infection with nononcogenic strains. The latter results from the acquisition of oncogenes from the host cell genome, and is characteristic of acutely transforming retroviruses such as Rous sarcoma virus (Chapter 7). Poxvirus and gammaherpesvirus genomes carry virulence genes with sequence homology to host immune defense genes, which must also have been acquired via genetic recombination. These genes are usually found near the ends of the genome. One explanation for this location is that the process of DNA packaging (gammaherpesviruses) or initiation of DNA replication (poxviruses) stimulates virus-host recombination when viral DNA is cleaved.

Information can be exchanged in a variety of unexpected ways during viral infections. For example, a host can be infected or coinfected by many different viruses during

BOX 10.3

BACKGROUND
Reassortment of influenza virus genome segments

Pandemic influenza strains result from shifts in H and N serotypes via exchange of the genome segments of mammalian and avian influenza viruses. Virologists have demonstrated that certain combinations of H and N are selected more strongly in avian hosts than in humans. An important observation was that both avian and human viruses replicate well in certain species such as pigs, no matter what the H-N composition. Indeed, the lining of the throats of pigs contains receptors for both human and avian influenza viruses, providing an environment in which both can flourish. As a result, the pig is a good host for mixed infection of avian and human viruses, in which reassortment of H and N segments can occur, creating new viruses that can reinfect the human population.

One might think that this combination of human, bird, and pig infections must be extremely rare. However, the dense human populations in Southeast Asia that come in daily contact with domesticated pigs, ducks, and fowl create conditions in which these interactions are likely to be frequent. Indeed, epidemiologists and virologists can show that the 1957 and 1968 pandemic influenza A virus strains (Fig. 10.4) originated in China and that the human H and N serotypes are circulating in wildfowl populations.

Studies of Italian pigs also provide evidence for reassortment between avian and human influenza viruses. The figure shows how the avian H1N1 viruses in European pigs reassorted with H3N2 human viruses. The color of the segments of the influenza genome indicates the origin of the avian and human viruses. The host of origin of the influenza virus genes was determined by sequencing and phylogenetic analysis. Results from these studies show that pigs can serve as an intermediate host in the emergence of new pandemic influenza viruses.

Reassortment among human, avian, and swine influenza virus strains led to the emergence of the 2009 H1N1 influenza pandemic. The U.S. Centers for Disease Control and Prevention has estimated that 43 million to 89 million people contracted H1N1 between April 2009 and April 2010, with 8,870 to 18,300 related deaths.

Peiris JS, Guan T, Markwell D, Ghose P, Webster RG, Shortridge KF. 2001. Cocirculation of avian H9N2 and contemporary "human" H3N2 influenza A viruses in pigs in southeastern China: potential for genetic reassortment? *J Virol* **75**:9679–9686.

New influenza A strains can emerge following reassortment of human and avian influenza viruses in pigs. Adapted from R. G. Webster and Y. Kawaoka, *Semin Virol* **5**:103–111, 1994, with permission.

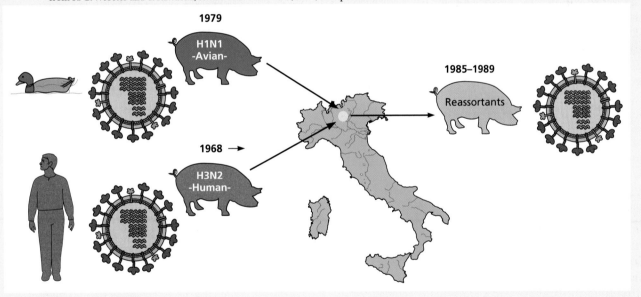

its lifetime. In fact, serial and concurrent infections are commonplace and can have a major effect on virus evolution. In the simplest case, propagation of a viral quasispecies in an infected individual allows coinfection of single cells, phenotypic mixing, and genetic complementation. As a result, recessive mutations are not eliminated immediately, despite the haploid nature of most viral genomes. Of course, such coinfection also provides an opportunity for the physical exchange of genetic information (Box 10.4).

Two General Pathways for Virus Evolution

Because viruses are absolutely dependent on their hosts for their reproduction, viral evolution tends to take one of two general pathways. In one, viral populations coevolve with their hosts so that they share a common fate: as the host prospers, so does the viral population. However, given no other host, a serious roadblock exists: the entire virus population can be eliminated with potent antiviral measures (e.g., smallpox virus) or by extinction of the host. In the other pathway,

BACKGROUND

Evolution by nonhomologous recombination and horizontal gene transfer

In the early 1970s, scientists working with the bacteriophage lambda and related bacteriophages found that various pairs of viral DNA formed heteroduplexes when visualized in the electron microscope; as illustrated in the figure, homologous, double-stranded stretches were seen connected to single-stranded bubbles corresponding to nonhomologous stretches that cannot form base pairs. The images were striking, and showed that the genomes of this group of lambdoid phages were **mosaics**; that is, they contained blocks of genes that were shuffled by recombination during evolution. Further analyses of bacteriophages that had picked up host genes by nonhomologous recombination established that **horizontal gene transfer** among bacteria by bacteriophages was a central feature in the evolution of both. With large-scale genome sequencing, we now know that bacteriophage genomes have ancestral connections to viruses of the *Eukarya* and *Archaea*.

Murray NE, Gann A. 2007. What has phage lambda ever done for us? *Curr Biol* 17:R305–R312.

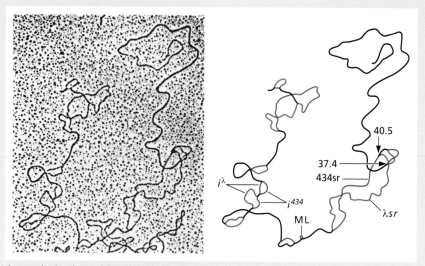

A heteroduplex formed from DNA of bacteriophages lambda and 434. In the explanatory tracing, homologous double-stranded regions are shown as solid black lines and nonhomologous single-stranded regions are red (phage lambda) or blue (phage 434) lines. The numbers 37.4 and 40.5 mark the left termini of the first and second nonhomology loops, starting from the left end of the genetic maps. ML is a minute deletion. *sr*, silent genetic regions; *i*, immunity regions. Reprinted from M. Simon et al., p. 315, in A. D. Hershey (ed.), *The Bacteriophage Lambda* (Cold Spring Harbor Laboratory Press, Cold Spring Harbor, NY, 1971), with permission.

virus populations occupy broader niches and infect multiple host species. When one species is compromised, the virus population can thrive in another. As discussed below, the first pathway is generally typical of DNA viruses, whereas the second is common for RNA viruses.

Fundamental Properties of Viruses That Constrain and Drive Evolution

The very characteristics that enable us to define and classify viruses are the primary barriers to major genetic change. Once a genome replication and expression strategy has evolved, there can be no turning back. For example, genomes that have suffered extreme alterations in the consensus do not survive selection. The nature of the genome is also fixed: DNA genomes cannot become RNA genomes, and vice versa. Furthermore, the solutions to replication or the decoding of viral information are limited, and because every step in viral reproduction requires interactions with host cell machinery, any change in a viral component without a compensating change in the cellular machinery may compromise viral propagation. Similarly, inappropriate synthesis, concentration, or location of a viral component is likely to be detrimental.

A second common constraint is the physical nature of the capsid required for transmission of the genome. Closed capsids have defined internal volumes that establish a limit on the size of the nucleic acids that may be packaged. Once the genes that encode the structural proteins of such capsids are selected, genome size is essentially fixed; only small duplications or acquisitions of sequences are possible without compensating deletion of other sequences. A final constraint is the requirement for balance (Box 10.5). All viral genomes encode products capable of modulating a broad spectrum of host defenses, including physical barriers to viral access and the vertebrate immune system. A mutant that is too efficient in bypassing host defenses will kill its host and suffer the same fate as one that does not reproduce efficiently enough: it will be eliminated. These general constraints define the viruses that we see today, as well as the further evolution of new ones.

Finite Strategies for Replication and Expression of Viral Genomes

We have described the seven viral genome replication strategies that are likely to represent all possible solutions, as well as the small number of strategies for expression of these

BOX 10.5

DISCUSSION

An unexpected constraint on evolution: selection for transmission and survival inside a host

The human immunodeficiency virus type 1 particles that initiate infection of their human hosts appear to be underglycosylated and are characterized by use of the CCr5 coreceptor and requirement for large amounts of the CD4 receptor on their target cells, suggesting selection for reproduction in T cells. At the late stage of disease, the infected individual produces billions of virus particles that survive in the face of host defenses and antiviral therapy. Invariably these late-stage virions can infect an expanded range of targets that include not only mature T cells, but also CCr5-positive macrophages that display small amounts of the CD4 receptor, and naïve T cells that produce very little CCr5 but large amounts of the CXCr4 coreceptor. Importantly, diversity in this final virus population is a result of the evolution of viral envelope receptor determinants inside a single infected individual.

When virus particles are transmitted from individuals with late-stage disease to new hosts, the first viruses that can be detected in the new hosts have the same characteristics as those that initiated the infections in the donors, indicating that only a few of the diverse variants present at the late stage in the donor are passed on. The processes that select these variants from the infecting virus population are not well understood. However, one conclusion is clear: the virus particles that ultimately devastate the immune system after years of reproduction and selection within a host are not those most fit for infection of new hosts.

Joseph SB, Arrildt KT, Swanstrom AE, Schnell G, Lee B, Hoxie JA, Swanstrom R. 2014. Quantification of entry phenotypes of macrophage-tropic HIV-1 across a wide range of CD4 densities. *J Virol* 88:1858–1869.

Ping LH, Joseph SB, Anderson JA, Abrahams MR, Salazar-Gonzalez JF, Kincer LP, Treurnicht FK, Arney L, Ojeda S, Zhang M, Keys J, Potter EL, Chu H, Moore P, Salazar MG, Iyer S, Jabara C, Kirchherr J, Mapanje C, Ngandu N, Seoighe C, Hoffman I, Gao F, Tang Y, Labranche C, Lee B, Saville A, Vermeulen M,

Fiscus S, Morris L, Karim SA, Haynes BF, Shaw GM, Korber BT, Hahn BH, Cohen MS, Montefiori D, Williamson C, Swanstrom R, CAPRISA Acute Infection Study and the Center for HIV-AIDS Vaccine Immunology Consortium. 2013. Comparison of viral Env proteins from acute and chronic infections with subtype C human immunodeficiency virus type 1 identifies differences in glycosylation and CCR5 utilization and suggests a new strategy for immunogen design. *J Virol* 87:7218–7233.

Bottleneck for transmission of human immunodeficiency virus type 1.

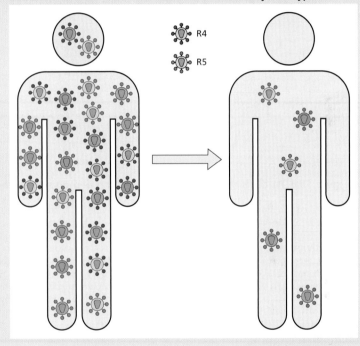

genomes (Volume I, Chapter 3). That the provenance of all viruses can be described by such a short list is remarkable. We have no clear understanding of how these replication and expression strategies have evolved, but some observations have been thought-provoking. For example, the replication complexes of different RNA virus families exhibit some fundamental similarities. Localization of genomes to membrane sites or to assembling capsids leads to the precise temporal and spatial organization of viral compartments, a property that is important for gene expression, replication, and particle assembly (Fig. 10.5). Are these overtly similar mechanisms products of convergent evolution and coincidence, or do they imply a common evolutionary origin for this abundant group of viral genomes? One notion is that similar mechanisms

were selected because they sequestered viral nucleic acid from intrinsic defense proteins present in the cytoplasm, such as RigI/Mda5, Pkr, and Tlrs.

The Origin of Viruses

One cannot help but conclude that nothing looks quite like the world of viral genomes (the virosphere). Soon after their discovery, many speculated that viral genomes might be very ancient, even predecessors to cellular microbes. Consistent with this hypothesis, the genomes of viruses that infect hosts in all three domains of life (*Archaea*, *Bacteria*, and *Eukarya*) share structural and coding features. Hypotheses about the origins of viruses center around three nonexclusive ideas. One is that viruses originated **before cells**, >3.6 billion years ago,

Retrovirus

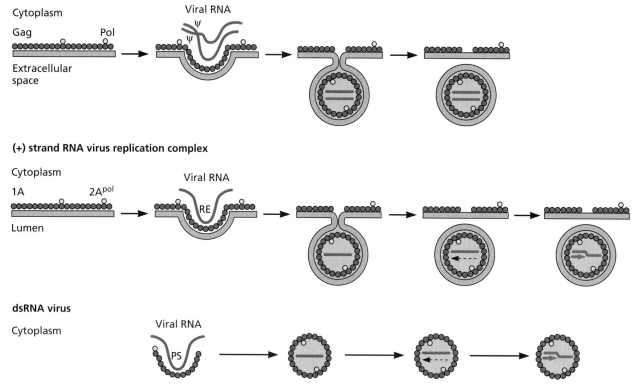

Figure 10.5 Parallels in replication of (+) strand and double-stranded RNA genomes. The mRNA templates of viruses with double-stranded RNA (dsRNA) and single (+) strand RNA genomes (including retroviruses) are sequestered in a multisubunit protein core that directs synthesis of the RNA or DNA intermediate from which more viral mRNA is made. Similarities in how the mRNA template and core proteins are assembled suggest that, despite a complete lack of genome sequence homology, all three virus groups may share evolutionary history. It is possible that this ancient replicative strategy provides RNA genomes with increased template specificity and retention of (−) strand products in the core or vesicle for template use. In addition, by sequestering RNA in vesicles or capsids, host defenses such as RNA interference, dsRNA-activated protein kinase, and RNase L are avoided. (Top) Retroviruses: specific sequences in the RNA genome (ψ) bind to Gag proteins that define the budding site. Gag proteins encapsidate two viral RNA genomes and reverse transcriptase within the plasma membrane. The final reaction for retroviruses is the release of an enveloped particle with two copies of the RNA genome and polymerase. (Middle) (+) strand RNA genomes: intracellular membrane vesicles form in response to a viral protein that binds to membranes and polymerase complexes, and viral RNA templates are recruited to these vesicles. In the case of (+) strand RNA viruses, the product is not an enveloped virion, but rather an involuted vesicle or the surface of a membrane vesicle where mRNA synthesis, (−) strand genome template synthesis, and (+) strand genome synthesis occur. (Bottom) dsRNA genomes: replication occurs in compartments formed by assembling capsid proteins that sequester single-stranded genome templates via specific protein-RNA interactions. In this case, the product is a capsid compartment within which mRNA synthesis and complementary strand genome replication occur. Blue circles are Gag proteins (retrovirus), 1A protein [(+) strand RNA virus], and the inner capsid protein (dsRNA virus). The polymerase proteins (Pol or 2A^{pol}; yellow circles) are incorporated with Gag or 1A, respectively. The polymerase is part of the assembling capsid of dsRNA viruses. RNA genomes are indicated in green, and the binding sites for interaction with Gag, 1A, or capsid protein are ψ, RE, or PS, as indicated for each virus. Adapted from M. Schwartz et al., *Mol Cell* **9:**505–514, 2002, with permission.

and might have contributed to the structure of the first cells. A second hypothesis is that viruses arose **after the first cellular organisms**, acquiring the ability to replicate and become packaged in particles. The third hypothesis is that viruses are **derived from intracellular parasites** that have lost all but the most essential genes, those encoding products required for genome replication and maintenance. These hypotheses are certainly not mutually exclusive; for example, some viruses predate cells, while others arose after that time.

It is widely believed that the first genomes and enzymes comprised RNA, with the transition from RNA to DNA genomes made possible by the evolution of reverse transcriptase. However, this transition would have had to wait for the evolution of genes that encoded the machinery for synthesis of deoxyribonucleotides (e.g., ribonucleotide reductase and thymidylate synthases). Some have hypothesized that DNA genomes were a viral invention that was shared later with cells harboring RNA genomes (Box 10.6).

BOX 10.6

DISCUSSION
Hypothetical origins of cells and viruses

The coexistence of two distinct genomes in a common cell is thought to have driven major evolutionary leaps such as the acquisition of mitochondria by eukaryotic cells. Some evolutionary biologists have proposed that eukaryotic nuclei arose following fusion of the cells of primordial bacteria and archaea, as genes derived from both have been identified in eukaryotic genomes. Others have proposed that eukaryotic cells existed before bacteria and archaea, noting that nucleus-like structures are present in certain unusual soil and freshwater bacteria, the planctomycetes. In this scenario, a nucleus was retained in eukaryotes, but lost in most bacteria and archaea over evolutionary time. Both of these hypotheses are controversial, as many gaps in

our knowledge of evolutionary relationships remain to be filled.

Because it is commonly believed that the very first genomes (and enzymes) comprised RNA, yet a third group of scientists have proposed that cellular DNA genomes arose from the infection of a primordial cell with an RNA genome by a DNA virus. In this scenario, it is hypothesized that viral DNA genomes arose via reverse transcription from a viral RNA genome, following the emergence of reverse transcriptase and enzymes required for the synthesis of desoxyribonucleotides, including ribonucleotide reductase and thymidylate synthetase. It is noted that many viruses with large DNA genomes encode both of the latter enzymes. However, how these enzymes might

have evolved in viral genomes is unclear, and the possibility that they were acquired from some ancestral host cell cannot be excluded. Consequently, while the proposal of a viral origin of cellular genomes has promoted much lively discussion, this idea remains controversial. Ideas about the origin of viruses with RNA genomes are even more speculative. They could be escapees from an ancient RNA world, but their distribution among the archaea, eukaryotes, and bacteria is decidedly nonuniform.

Claverie JM. 2006. Viruses take center stage in cellular evolution. *Genome Biol* **7**:110. doi:10.1186/gb-2006-7-6-110.

Forterre P. 2005. The two ages of the RNA world, and the transition to the DNA world: a story of viruses and cells. *Biochimie* **87**:793–803.

Hypothetical origin of the eukaryotic cell nucleus. (A) An early DNA virus (perhaps a bacteriophage ancestor) infects a cell with an RNA genome. **(B)** The DNA virus is sequestered within a vesicle in the "cytoplasm" and replicates in this compartment. Cellular genes are recruited to the enlarging nucleus; new DNA chemistry provides selective advantages. **(C)** This unstable situation may produce altered viruses better adapted to infection of and cytoplasmic replication in cells with RNA genomes, as well as **(D)** the evolution of a stable eukaryotic cell with a nucleus and DNA replication machinery. Adapted from J.-M. Claverie, *Genome Biol* **7**:110–114, 2006, with permission.

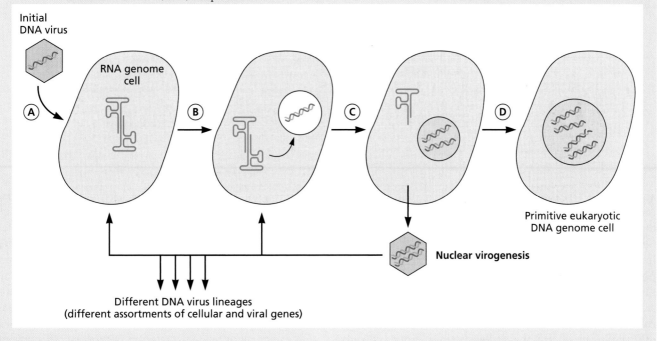

Speculation about the origin of viral genomes was provoked anew by the discovery of giant DNA viruses with genomes of up to 2.5 Mbp, larger than some parasitic bacteria. Based on analyses of their DNA polymerases, these giants have been assigned to a group known as the "nucleocytoplasmic large DNA viruses," which includes herpesviruses and

poxviruses (Fig. 10. 6). The genomes of all families in this group include genes normally found in host cells, for example, genes for proteins of the glycosylation machinery, nucleotide processing, DNA transcription, and even some translation components. The largest members in the group of nucleocytoplasmic DNA viruses, the megaviruses, pandoraviruses,

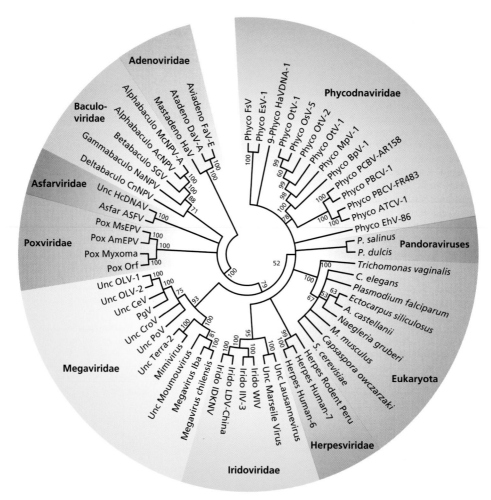

Figure 10.6 Phylogenetic analysis of the B-family DNA polymerase of nucleocytoplasmic large DNA viruses. A multiple alignment of 9 eukaryote and 50 viral DNA polymerase B sequences. Adapted from N. Philippe et al., *Science* **341**:281–286, 2013, with permission.

and an ancient iridovirus-related species, have unusual structural features or architecture (Box 10.7). A puzzling fact is that the majority of genes in the largest genomes have no obvious homology to sequences in the databases, a feature that is hard to reconcile with a gene reduction model. Some have proposed that these genes originated in a fourth domain of life that is either extinct or not yet discovered.

Genomes of many of the large DNA viruses have **sequence coherence**: they do not appear to be mosaics (i.e., the products of multiple recombination reactions between groups of genes [Box 10.4]). The homogeneity of genomes within a family and the lack of any obvious homology among families are difficult to explain using the model that they arose by the sequential acquisition of exogenous genes by a single, primordial precursor viral genome. However, technological advances in nucleic acid chemistry, sequencing, proteomics, and bioinformatic analyses have yielded data that are consistent with the

idea that members of these families coexisted with the first primordial cells (Boxes 10.7 and 10.8).

We currently know the sequences of thousands of viral genomes (roughly equal numbers of RNA and DNA genomes are in the databases), and more are being added all the time. Furthermore, it is now possible to sample the ecosystem and determine the nature and diversity of viral genomes without having to propagate the viruses in the laboratory, for example by purifying particles from water samples and then sequencing all of the DNA released from them. This type of unbiased survey, called **metagenomic analysis**, has revealed remarkable diversity among known virus families. Even more amazing is the fact that the vast majority of viral sequences determined so far by these technologies represent previously **unknown** viral genomes. It may be that future analyses of this type will stimulate new or refined hypotheses about the origin of viruses and their place in the biosphere.

BOX 10.7

BACKGROUND

Discovery of virus giants: the largest known viral particles and genomes

An outbreak of pneumonia in Bradford, England, led to the isolation in 1992 of what was then the world's largest virus. Investigators attempted to isolate *Legionella*-like pathogens of amoebae from hospital cooling towers and recovered what appeared to be a small, Gram-positive bacterium. All attempts to identify it using universal bacterial 16S ribosomal RNA PCR amplification failed. Transmission electron microscopy of *Acanthamoeba polyphaga* infected with this agent revealed 400-nm icosahedral virus particles in the cytoplasm. Mature particles were surrounded by a profusion of fibers, the bases of which form an external protein capsid layer. The virus was named "mimivirus" because it mimicked a microbe.

Approximately 10 years later, this giant was eclipsed by discovery of two other pathogens of amoebae, one in marine sediment off the coast of central Chile and a second from a freshwater pond near Melbourne, Australia. These new giant virus particles exhibit no morphological or genomic resemblance to any previously defined virus families, and they have been proposed as the first members of a new genus, *Pandoravirus*. A third giant virus was discovered in 2014 in a sample from the frozen permafrost of Siberia, estimated to be >30,000 years old from carbon dating of associated late Pleistocene sediments. Amazingly, the virus was nevertheless still able to infect a cultured amoeba host. This ancient virus looks somewhat like pandoraviruses, but the replication cycle and genomic features are more like icosahedral DNA viruses. Another unique feature is the large fraction (21.2%) of the genome comprising multiple, regularly interspersed copies of 2-kb-long tandem arrays of a conserved 150-bp palindrome. A movie of the reproduction of this virus, called *Pithovirus sibericum*, in *Acanthamoeba castellanii*-infected cells can be found at http://www.pnas.org/content/suppl/2014/02/26/1320670111.DCSupplemental/sm01.mp4.

General features of these three giant viruses of amoebae are summarized below:

	Mimivirus	Pandoravirus	Pithovirus
Virion size	0.75 µm (diameter)	~1.0 µm (length)	1.5 µm (length)
Capsid shape	Icosahedral	Ovoid	Ovoid
Genome composition	AT-rich (>70%)	GC-rich (>61%)	AT-rich (64%)
Genome size (bp)	1.2×10^6	$1.9 \times 10^6 - 2.5 \times 10^6$	0.6×10^6
Genes (protein-coding)	911	2,556	467

Despite having a number of genes predicted to specify components of protein synthesis machinery, the giant viral genomes do not encode a complete translation system. As is typical for all viruses, they also undergo an eclipse period after entering their host cells. Similarities among the genomes of these viruses and some parasitic bacteria have prompted the suggestion that the viruses arose from a common cellular ancestor by gene loss. However, the presence of a large number of coding sequences with no homology to any genes in the current database (>50% of the total) implies that the putative cellular ancestor is now extinct or has not yet been discovered.

Legendre M, Bartoli J, Shmakova L, Jeudy S, Labadie K, Adrait A, Lescot M, Poirot O, Bertaux L, Bruley C, Couté Y, Rivkina E, Abergel C, Claverie JM. 2014. Thirty-thousand-year-old distant relative of giant icosahedral DNA viruses with a pandoravirus morphology. *Proc Natl Acad Sci U S A* **111**:4274–4279.

Philippe N, Legendre M, Doutre G, Couté Y, Poirot O, Lescot M, Arslan D, Seltzer V, Bertaux L, Bruley C, Garin J, Claverie JM, Abergel C. 2013. Pandoraviruses: amoeba viruses with genomes up to 2.5 Mb reaching that of parasitic eukaryotes. *Science* **341**:281–286.

Raoult D, Audic S, Robert C, Abergel C, Renesto P, Ogata H, La Scola B, Suzan M, Claverie JM. 2004. The 1.2-megabase genome sequence of Mimivirus. *Science* **306**:1344–1350.

Zauberman N, Mutsafi Y, Halevy DB, Shimoni E, Klein E, Xiao C, Sun S, Minsky A. 2008. Distinct DNA exit and packaging portals in the virus *Acanthamoeba polyphaga mimivirus*. *PLoS Biol* **6**:e114. doi:10.1371/journal.pbio.0060114.

Mimivirus, pandoravirus, and pithovirus particles. (Left) Cryo-electron micrograph showing the icosahedral capsid with copious attached fibers of mimivirus. Arrows mark the position of "stargate," the structure that allows release of the viral DNA following infection of a new host cell. Scale bar, 600 nm. Reprinted from Xiao C et al. *J Mol Biol* **353**:493–496, 2005, with permission. **(Center)** Electron micrograph of *Pandoravirus salinus*, the larger, Chilean isolate. Like mimivirus, pandoravirus particles are internalized via phagocytic vacuoles in host cells and the viral genome is later emptied into the cytoplasm through an apical pore (top, right corner). Scale bar, 200 nm. **(Right)** Electron micrograph of *Pithovirus sibericum*. Shaped like the pandoravirus, this particle has a cork structure (top, left corner) that opens to allow viral contents to be emptied into the cytoplasm after fusion of an inner membrane with that of the phagocytic vesicles. Scale bar, 200 nm. *Pandoravirus* and *Pithovirus* images were kindly provided by Dr. Chantal Abergel, CNRS, Aix-Marseille Université.

Mimivirus

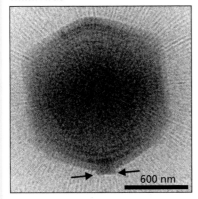

Pandoravirus

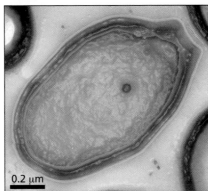

Pithovirus

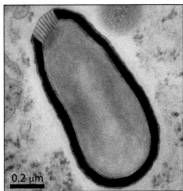

BOX **10.8**

DISCUSSION
Protein analyses point to primordial origins for the nucleocytoplasmic large DNA viruses

Protein domains that exhibit high amino acid sequence conservation (>30% identity) and the same major secondary structure (folds) can be grouped into families. Fold families that include members with similar function and are likely of common origin are defined as fold superfamilies (FSFs). The number of FSFs (~2,000) identified in the Structural Classification of Proteins website (version 1.75: http://scop.mrc-lmb.cam.ac.uk/scop/) is small in comparison to the approximately half-million protein sequence entries in the databases, indicating that a limited number of protein design units exist in nature. In multidomain proteins, such units are gained, lost, or rearranged during evolution.

Phylogenomic trees can be derived by assuming that the most abundant FSFs appeared earliest in evolution and have been reused as gene duplications and rearrangements increased proteomic repertoires. An analysis of FSFs in the proteomes of *Archaea*, *Bacteria*, *Eukarya*, and nucleocytoplasmic large DNA viruses showed that each forms a distinct group, with the viruses the most ancient and *Archaea* the second oldest. These large DNA viruses were also found to encode a number of FSFs with no current cellular representatives. It was therefore proposed that their primordial host predated or coexisted with the first living cells. A plot of abundance versus diversity of FSFs showed that the viruses

have the simplest proteomes, and that structural diversity and organismal complexity are congruent. These and other results are consistent with the idea that the lack of diversity in these viral proteomes can be explained by a reductive evolutionary history in which, as with parasitic bacteria, nonessential genes have been lost.

Caetano-Anollés G, Nasir A. 2012. Benefits of using molecular structure and abundance in phylogenomic analysis. *Front Genet* **3**:172. doi:10.3389/fgene.2012.00172.

Nasir A, Kim KM, Caetano-Anolles G. 2012. Giant viruses coexisted with the cellular ancestors and represent a distinct supergroup along with superkingdoms Archaea, Bacteria and Eukarya. *BMC Evol Biol* **12**:156. doi:10.1186/1471-2148-12-156.

The universal tree of life and proteomic diversity. (A) An optimal most parsimonious phylogenomic tree describing the evolution of 200 proteomes (50 each from *Archaea*, *Bacteria*, *Eukarya*, and viruses) generated using the census of abundance of 1,739 protein domain FSFs. Terminal leaves of viruses, *Archaea*, *Eukarya*, and *Bacteria* are labeled in red, blue, black, and green, respectively. Numbers on the branches indicate bootstrap values. **(B)** FSF diversity (number of distinct FSFs in a proteome) plotted against FSF abundance (total number of FSFs that are encoded) for 200 proteomes. Major families/phyla/kingdoms are labeled. Reprinted from A. Nasir et al., *BMC Evol Biol* **12**:156, 2012, with permission.

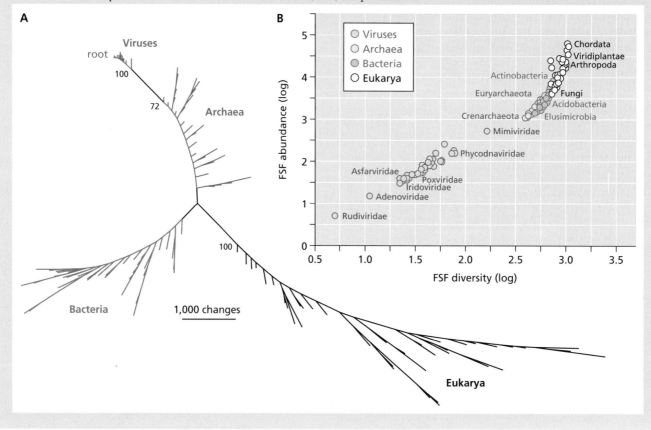

Host-Virus Relationships Drive Evolution

Although the primordial history of viruses cannot be known, nucleic acid sequence analyses have identified many relationships among contemporary viruses and their host species, providing considerable insight into viral evolution.

DNA Virus Relationships

Papillomaviruses and Polyomaviruses

Coevolution with a host is a characteristic of small DNA viruses, the parvoviruses, polyomaviruses, and papillomaviruses. The evidence for coevolution comes from finding close association of a given viral DNA sequence with a particular host group. The linkage of host to virus was particularly striking when human papillomavirus types 16 and 18 were compared: the distribution of these viral genomes is congruent with the racial and geographic distribution of the human population. Another example of the same phenomenon is provided by JC virus, a ubiquitous human polyomavirus associated with a rare, fatal brain infection of oligodendrocytes. This virus exists as five or more genotypes identified in the United States, Africa, and parts of Europe and Asia. Recent polymerase chain reaction analyses of subtypes of JC and BK polyomaviruses indicate that they not only coevolved with humans, but also did so within specific human subgroups. Some studies have indicated that these viruses may provide convenient markers for human migrations in Asia and the Americas in prehistoric and modern times.

How can virus evolution be linked with specific human populations in a manner akin to vertical transmission of a host gene? We can begin to appreciate this perhaps counterintuitive phenomenon from the unusual biology of human papillomaviruses. Infection of the basal keratinocytes of adult skin leads to viral reproduction that is coordinated with cellular differentiation. The final step, assembly of progeny virus particles, occurs only as cells undergo terminal differentiation near the skin surface. Mothers infect newborns with high efficiency, because of close contact or reactivation of persistent virus during pregnancy or birth. The infection therefore appears to spread vertically, in preference to the more standard horizontal spread between hosts. This mode of transmission is the predominant mechanism for papillomaviruses and polyomaviruses. It stands in contrast to that observed for most acutely infecting viruses of humans, which are spread by aerosols, contaminated water, or food.

Herpesviruses

The three main subfamilies of the family *Herpesviridae* (*Alphaherpesvirinae*, *Betaherpesvirinae*, and *Gammaherpesvirinae*) are readily distinguished by genome sequence analysis even though the original taxonomic separation of these families was based on general, often arbitrary, biological properties. Researchers have related the timescale of herpesviral genome evolution to that of the hosts. For most of these viruses, points of sequence divergence coincide with well-established points of host divergence. The conclusion is that an early herpesvirus infected an ancient host progenitor, and subsequent viruses developed by coevolution with their hosts. Consistent with this conclusion, the genomes of **all** members of the three major subfamilies that have been sequenced contain a core block of genes, often organized in similar clusters in the genome.

Our current best estimate is that the three major groups of herpesviruses arose ~180 million to 220 million years ago. This implies that the three subfamilies must have been in existence before mammals spread over the earth 60 million to 80 million years ago. Fish, oyster, and amphibian herpesviruses have virtually identical architecture but little or no sequence homology to the major subfamilies, and therefore must represent a very early branch of this ancient family.

RNA Virus Relationships

Relationships among RNA viruses can also be deduced from nucleotide sequence analyses, but the high rates at which mutations accumulate in these genomes impose some difficulties. Moreover, genomes of RNA viruses are often small and, in contrast to those of the large DNA viruses, contain few, if any, genes in common with a host that might be used to correlate virus and host evolution. Nevertheless, when nucleotide sequences of many (+) and (−) strand RNA viral genomes are compared, blocks of genes that encode proteins with similar functions can be identified. Common coding strategies can also be inferred. These groups are often called "supergroups" because the similarities suggest a common ancestry.

An obvious feature common to the sequences of many (−) strand RNA genomes is the limited number of genes that encode proteins (as few as 4 and not more than 13). These proteins can be placed in one of three functional classes: core proteins that interact with the RNA genome, envelope glycoproteins that are required for attachment and entry of virus particles, and a polymerase required for replication and mRNA synthesis (Fig. 10.7).

The (+) strand RNA viruses (excluding the retroviruses) are the largest and most diverse subdivision. The number of genes that encode proteins in their genomes ranges from 3 to more than 12 and, as with the (−) strand RNA viruses, the proteins can be divided into the same three groups by function, although in this case their organization is not necessarily colinear (Fig. 10.8). A unifying feature is that the RNA polymerase gene appears to be the most highly conserved, implying that it arose once in the evolution of these viruses. Consequently, these viruses are organized into three virus supergroups, based on similarities in their polymerases. As each of the supergroups contains members that infect a broad variety of animals and plants, an ancestor present before the

Nonsegmented

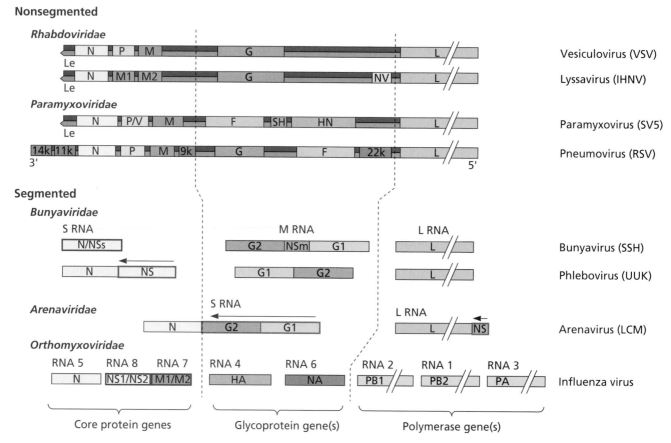

Figure 10.7 The genetic maps of selected (−) strand RNA viral genomes. Maps of the genes of *Rhabdoviridae,* *Paramyxoviridae, Bunyaviridae, Arenaviridae,* and *Orthomyxoviridae* are aligned to illustrate the similarity of gene products. The individual gene segments of the *Orthomyxoviridae* are arranged according to functional similarity to the two other groups of segmented viruses. Within a given genome, the genes are approximately to scale. For segmented genomes, blue-outlined genes are those that encode multiple proteins from different open reading frames. Red-outlined genes are expressed by the ambisense strategy, as indicated by the arrow. Virus abbreviations: VSV, vesicular stomatitis virus; IHNV, infectious hematopoietic necrosis virus; SV5, simian virus 5; RSV, respiratory syncytial virus; SSH, snowshoe hare virus; UUK, Uukuniemi virus; LCM, lymphocytic choriomeningitis virus. Le is a nontranslated leader sequence. Gene product abbreviations: N, nucleoprotein; P, phosphoprotein; M (M1 and M2), matrix proteins; G (G1 and G2), membrane glycoproteins; F, fusion glycoprotein; HN, hemagglutinin/neuraminidase glycoprotein; L, replicase; NA, neuraminidase glycoprotein; HA, hemagglutinin glycoprotein; NS (NV, SH, NSs, and NSm), nonstructural proteins; PB1, PB2, and PA, components of the influenza virus replicase. Figure derived from J. H. Strauss and E. G. Strauss, *Microbiol Rev* **58:**491–562, 1994, with permission.

separation of these kingdoms might have provided the primordial RNA polymerase gene. Alternatively, the ancestral (+) strand virus could have infected hosts in all branches of the tree of life.

As discussed above for herpesviruses, it has been possible to determine the timescale of RNA virus evolution by using host divergence times. A good example is the retrovirus simian foamy virus, for which there is a 30-million-year match between host and virus phylogenetic trees. The use of endogenous viral sequences to determine the timescale of viral evolution places foamy viruses in mammals for the past 100 million years (see "Lessons from Paleovirology" below).

Influenza Virus

Study of the ecology and biology of influenza has shown that the same virus population can infect many different species, and each host species imposes new selections for reproduction and spread of the infection. As a result, the influenza virus gene pool is immense, with a dynamic ebb and flow of genetic information as the virus is transmitted among many different animals. Large-scale sequencing has provided a view of the state of the viral gene pool at various points in time and space, during transmission from human to animal, animal to human, and human to human. In one analysis alone, a consortium of scientists sequenced >200 human influenza virus

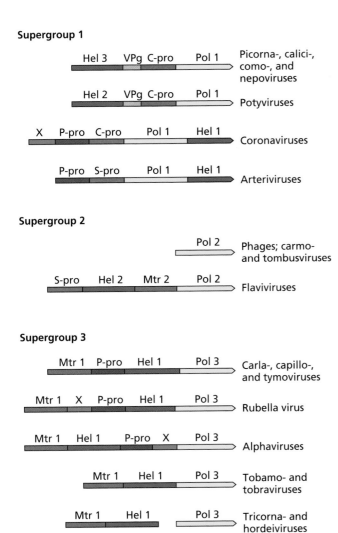

Supergroup 1

Hel 3 VPg C-pro   Pol 1   Picorna-, calici-,
                              como-, and
                              nepoviruses

Hel 2 VPg C-pro   Pol 1   Potyviruses

X P-pro C-pro   Pol 1   Hel 1   Coronaviruses

P-pro S-pro   Pol 1   Hel 1   Arteriviruses

Supergroup 2

Pol 2   Phages; carmo-
         and tombusviruses

S-pro   Hel 2   Mtr 2   Pol 2   Flaviviruses

Supergroup 3

Mtr 1 P-pro   Hel 1   Pol 3   Carla-, capillo-,
                                 and tymoviruses

Mtr 1 X P-pro   Hel 1   Pol 3   Rubella virus

Mtr 1 Hel 1   P-pro   X   Pol 3   Alphaviruses

Mtr 1   Hel 1   Pol 3   Tobamo- and
                          tobraviruses

Mtr 1   Hel 1   Pol 3   Tricorna- and
                          hordeiviruses

Figure 10.8 RNA virus genomes and evolution. Organization of (+) strand RNA genomes. The genomes of (+) strand RNA viruses comprise several genes for replicative functions that have been mixed and matched in selected combinations over time. These functions include a helicase (Hel, purple), a genome-linked protein (VPg, orange), a chymotrypsin-like protease (C- or S-pro, red), a polymerase (Pol, yellow), a papain-like protease (P-pro, brown), a methyltransferase (Mtr, dark blue), and a region of unknown function (X, green). Differences in the polymerase gene define the three supergroups. In this figure, the genes are not shown to scale and the structural proteins have been omitted for clarity. Derived from J. H. Strauss and E. G. Strauss, *Microbiol Rev* **58:**491–562, 1994, with permission.

genomes and collected almost 3 million bases of sequence. One salient finding was that a given influenza virus population in circulation contains multiple lineages at any time. In addition, alternative minor lineages exchange information with the dominant one. As selection pressures change, the numbers of distinct immune escape mutants rise and fall, as do the numbers of mutants with alterations in receptor-binding affinity.

Important clues to the epidemiology of influenza virus came from the sequencing and analysis of the genomes of >1,300 influenza A virus isolates from various geographic locations. It was clear that the viral genome changes by frequent gene reassortment and occasional bottlenecks of strong selection. More importantly, the study suggests that new antigenic subtypes have different dynamics but that all follow a classical "sink-source" model of viral ecology: in this model, antigenic variants emerge at intervals from a persisting reservoir in the tropics (the source) and spread to temperate regions, where they have only a transient existence before disappearing (the sink) (Fig. 10.9).

The Protovirus Hypothesis for Retroviruses

Hypotheses about the origin of retroviruses and their evolution may be easier to propose than for any other virus group, because of their distinctive association with their hosts. Upon infection, accompanying reverse transcriptase and integrase enzymes convert their RNA genomes into a DNA copy, which is then inserted into the host DNA (the provirus). Howard Temin, who shared the Nobel Prize for the discovery of reverse transcriptase, first proposed the "protovirus theory" for the origin of this virus family. This theory posits that a cellular reverse transcriptase-like enzyme copied segments of cellular RNA into DNA molecules that were then inserted into the genome to form retroelements. These DNA segments in turn acquired more sequences, including those encoding RNase H, integrase, regulatory sequences, and structural genes. This hypothesis predicts that evidence of such sequential acquisitions might exist in the genomes of mammals and other species. Indeed, many of the proposed intermediates are found in abundance, including long interspersed nuclear elements (LINEs), retrotransposons, and a variety of endogenous retroviruses (see Volume I, Chapter 7). As reverse transcriptase-encoding, mobile elements are present in bacterial, archaeal, and bacteriophage genomes, this function is likely to have arisen quite early in evolutionary history. Recent genomic and computational analyses indicate that the retroviral reverse transcriptase was likely derived from a retrotransposon precursor related to LINEs; RNase H, integrase, and the regulatory long terminal repeats (LTRs) were added later during evolution of retroviral genomes.

Lessons from Paleovirology

Traces of virus-derived sequences are present in all living species but historically have been considered to be mostly genomic "junk." In the last decade or so, the rapid expansion of nucleic acid sequence databases and new methods of genomic analyses have altered that perception, providing unique insights into the nature and consequences of viral and host interactions over evolutionary time.

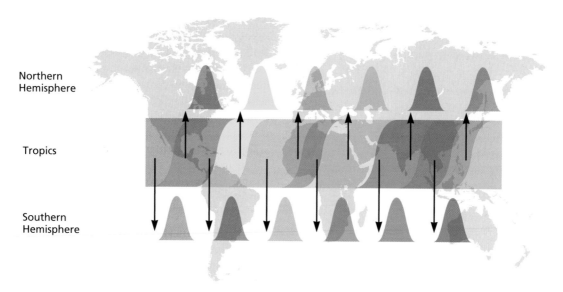

Figure 10.9 The genomic and epidemiological dynamics of human influenza A virus. Viral genetic and antigenic diversity (shown by different colors) is continuously generated in a reservoir, or "source" population, perhaps represented by the tropics, before being exported to "sink" populations in the Northern and Southern Hemispheres, as shown by the arrows. The continuous transmission of influenza A virus in the source population, and hence its larger effective population size, allows natural selection for antigenic diversity to proceed more efficiently than in the sink populations that are afflicted by major seasonal bottlenecks. Reprinted from A. Rambaut et al., *Nature* **453**:615–619, 2008, with permission.

Endogenous Retroviruses

We now know that endogenous retroviruses are abundant in vertebrate genomes; they account for 6 to 14% of all genomes analyzed to date. In humans, endogenous retroviruses comprise ~8% of genomic DNA, almost an order of magnitude greater than that encoding all of our proteins! Endogenous retroviral sequences originate from proviral DNA integrations into the genomes of the host's germ line cells that are passed on through subsequent generations. Consequently, a comparison of the insertion sites of endogenous retroviruses in present-day vertebrate species, and knowledge of the species' evolutionary relationships, can allow one to estimate when shared endogenous sequences were inserted into an ancestral host germ line. Such analyses have revealed that most human endogenous retroviruses (HERVs) are at least 10 million to 50 million years old and were circulating on the earth long before the emergence of *Homo sapiens*.

Most endogenous retroviral proviruses are defective, but some do retain an ability to replicate, and if their transcription is not repressed, they can reemerge as infectious agents. For example, the high incidence of spontaneous leukemogenesis in the well-studied AKR strain of mice has been traced to the production of replication-competent leukemic viruses that arise via recombination between the genomes of three different endogenous retroviruses. Of course, endogenous retroviruses can also serve as genetic reservoirs for recombination with exogenously infecting viruses. Although all of the endogenous retroviruses in the human genome are defective, scientists have managed to regenerate one of the "youngest" of these ancient viruses, which was circulating in ancestral species ~1 million years ago. They accomplished this feat by deriving a nondefective consensus sequence from several endogenous members of the family. Shades of *Frankenstein* (or maybe *Jurassic Park*), the resulting proviral clone, called HERV-K$_{con}$, forms ("raised from the dead") infectious particles when introduced into human cells!

Endogenous retroviruses can have profound effects on the evolution and function of their host's genome. For example, recombination between endogenous retroviral sequences integrated at different loci can account for several large-scale deletions, duplications, and other types of chromosomal reshuffling that occurred during the evolution of primate genomes. In humans, such recombination has contributed to the extensive duplication of gene blocks that comprise the major histocompatibility complex class I locus. The diversity that then arose in such duplications, as well as heterozygosity at this locus, confers a strong selective advantage on human populations in the battle against pathogenic agents. In addition, transcription factor-binding sites and regulatory sequences in the LTRs that flank endogenous retroviruses can modify the expression of neighboring host genes by providing alternate promoters and enhancers. In some cases, inserted LTRs may be responsive to tissue-specific regulators: expression of the human salivary amylase is controlled by an endogenous LTR and may have enabled our ancestors to thrive on a starch diet. Finally, some endogenous retroviral proteins have been repurposed by their hosts to serve new functions. One example is the resistance to retroviral infection of certain mouse strains, which is conferred by the expression of endogenous sequences related to a retroviral capsid gene (called *Fv-1*). Similarly, retroviral Env proteins called syncytins have been coopted independently by various host species at least seven times during evolution. These fusogenic proteins are essential for formation of placental syncytiotrophoblasts, which allow passage of essential nutrients and may protect a fetus from the maternal immune system. Indeed, selection and fixation of this retroviral protein was a pivotal step in the evolution of placental mammals.

DNA Fossils Derived from Other RNA Viral Genomes

Although the existence of endogenous retroviruses in ancient evolutionary time has long been appreciated, until quite recently the ages of other RNA viruses remained a mystery. Approximations could be made only by comparison of currently circulating representatives and measurements of their rates of genetic change. It was not until databases for numerous viruses and vertebrate species became available that sequences related to other RNA viruses were discovered in host genomes. The first hints of such unexpected inheritance came from reports that sequences related to the flaviviruses and picornaviruses were incorporated into the genomes of plants and insects. These findings stimulated comprehensive bioinformatic analyses in which the sequences of all currently known viruses with RNA genomes were matched against the now extensive library of vertebrate genomes. The results revealed that as long as 30 million to 40 million years ago almost half of the vertebrate species analyzed had acquired sequences related to genes in two currently circulating single (−) strand RNA virus families, the filoviruses and the bornaviruses, deadly pathogens that cause lethal hemorrhagic fever and neurological disease, respectively (Fig. 10.10, left). Some of these sequence fossils could be traced back ~90 million years. Because host genomes are also subject to genetic drift, and sequences can change beyond recognition, this is close to the limit of such analyses. The conservation and current-day expression of some of these endogenous sequences suggest that they may have afforded a selective advantage in vertebrate populations at some time. Integration of sequences related to other single-strand RNA viruses was also found in the genomes of various insect species. Analysis of the nucleotides that flanked many of these viral fossils indicates that they were likely derived from viral mRNAs that were reverse-transcribed and integrated into the host genome by mobile LINEs (see Volume I, Fig 7.12). It is noteworthy that LINEs tend to be active in germ line cells, a necessary requirement for fixation.

The error rate for the replication of host DNA is much lower than the rates of the RNA-dependent RNA synthesis of currently circulating viruses. Consequently, the ability to detect ancient sequences that are related to particular currently circulating RNA virus families may be due partly to the fact that their genome sequences are relatively stable, as appears to be the case for the bornaviruses. It seems probable that LINEs have inserted DNA copies of the mRNAs of many other viruses into their host genomes over evolutionary time, but our ability to recognize such sequences has been clouded by the high rate of mutation in most present-day relatives.

Endogenous Sequences from DNA Viruses

A survey of vertebrate genomes for sequences related to DNA viruses revealed numerous examples of the integration of sequences related to members of two single (−) strand DNA viral families, the circoviruses and the parvoviruses. These were found broadly distributed among ~70% of the vertebrate species tested. Some of these insertions are >50 million years old, but others have occurred more recently (Fig. 10.10, right). Both of these virus families have tiny genomes that encode only two proteins, replicase and capsid. Host enzymes are recruited by the replicase to hairpin regions in the viral genomes where self-primed viral DNA synthesis is initiated. The parvovirus, adenovirus-associated virus is known to be able to insert its DNA into its host genome at sequences that are recognized by the viral replicase protein (Volume I, Chapter 9). It seems likely, therefore, that the germ line insertions that gave rise to the endogenous sequences result from the occasional copying of circovirus and parvovirus DNA at loci in their host genomes that resemble viral replication hairpin regions.

As contemporary hepadnavirus genomes are known to be incorporated randomly into host genomes via nonhomologous recombination, and a member of this virus family infects birds, it is not so surprising that sequences related to this double-strand DNA hepadnavirus genome have been detected in the zebra finch genome.

The Host-Virus "Arms Race"

The discovery of ancient viral fossil sequences paints a picture in which all current living forms have evolved in a virtual sea of viruses that are capable of rapid evolution. We can assume that host genes associated with antiviral defense evolved by mutation, with individuals who encoded ineffective alleles dying from infection and thereby being eliminated from the population. Virus populations with compensatory mutations would then emerge, exerting selective pressure back on the host in an ever-escalating, molecular "arms race." (For more insight into the arms race concept see the comments of Dr. Harmit Malik: http://bit.ly/Virology_Malik.) Constitutively expressed host cell genes encode antiviral proteins that function in a cell-autonomous manner. These proteins

History of ssRNA virus integrations

History of ssDNA virus integrations

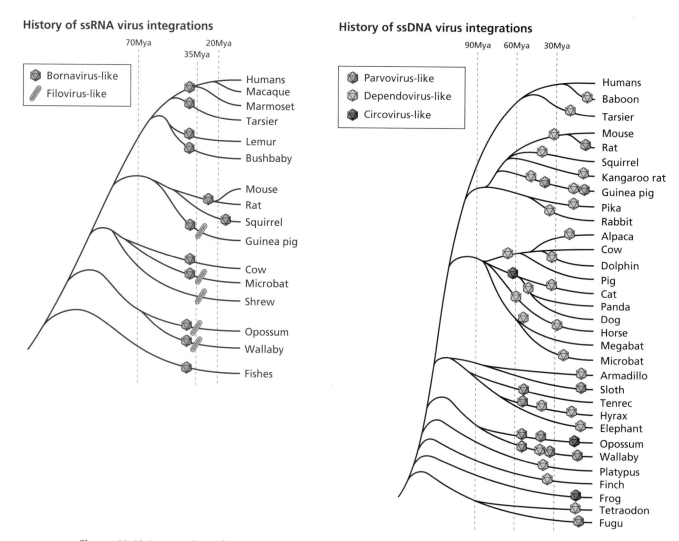

Figure 10.10 Integration of nonretroviral sequences into vertebrate genomes. (Left) Comparisons of sequences representing all known single-stranded (nonretroviral) RNA (ssRNA) genomes with the genomes of vertebrate species revealed that almost all of the nearly 80 integrations identified were related to only two viral families, the filoviruses and the bornaviruses. Based on signature landmarks, some, and perhaps all, of the endogenous virus-like DNA sequences appear to be LINE-facilitated integrations of viral mRNAs. The integrations occurred in a limited time frame, ~40 million years ago, coincident with the rapid evolutionary diversification of mammalian species. It is noteworthy that an analysis of pseudogene formation suggests that the predominant mammalian LINE-1 underwent a peak of retrotransposition activity around this time. The reason for finding mRNA copies of genes from only these viral families is unknown. (Right) Similar comparisons with ssDNA viral genomes also revealed a preference for integration of sequences from certain virus families, in this case the parvoviruses and circoviruses. However, these integrations occurred both throughout evolution, stretching back to >90 million years ago, and in the present time. It has been suggested that shared features of the replication strategies of these viruses may explain the broad history and high incidence of their integration. Times of the viral gene integrations are approximate. Adapted from V. A. Belyi et al., *PLoS Pathog* 6:e1001030, 2010, and V. A. Belyi et al., *J Virol* 84:12458–12462, 2010, with permission.

contribute to intrinsic cellular defenses by interacting directly with viral components and inhibiting the reproduction of infecting viruses, at various stages and by a variety of mechanisms (Chapter 3). They include, among others, the Apobec3 family of cytidine deaminases, which create mutations in viral DNAs; the tripartite motif (Trim) proteins that interact with the capsids of infecting retroviruses and block uncoating; and the membrane protein tetherin, which inhibits budding of a number of enveloped viruses from the cell surface. These proteins have been studied most intensely in the context of the primate immunodeficiency viruses that have altered capsid proteins, and the viral accessory protein Vpr, respectively, each of which provides a countering function (described in Chapter 6). Points of contact between the viral and host

BOX 10.9

EXPERIMENTS
Host-virus arms race and the transferrin receptor

Located at the cell surface, the dimeric transferrin receptor protein (TfR1) controls the uptake of iron—a "housekeeping" function essential for all living cells. The observation that this protein is also the receptor for a variety of viruses prompted an investigation into how opposing selective pressures, avoiding infection and maintaining iron uptake, might be balanced during evolution of the *TFR1* gene. By analysis of this gene in a number of evolutionarily related rodent species, it was found that, while most of the amino acids in the encoded proteins were conserved, several residues were quite variable, with dN/dS > 1, indicating a high probability of positive selection. Using the crystal structure of the ectodomain of human TfR1 as a model, it was noted that these residues are included in the known overlapping binding sites for arenaviruses and the retrovirus mouse mammary tumor virus, but separate from the transferrin-binding site. Furthermore, experiments showed that naturally occurring substitutions of these residues block virus entry while preserving iron uptake by both rodent and human TfR1.

The TfR1 protein is also the receptor for parvoviruses that infect dogs and related species of carnivores. Residues that show evidence for positive selection in these receptors have also been mapped to the ectodomain in a region close to the mouse mammary tumor virus-binding site and distinct from the transferrin-binding site. These results indicate that by the repeated changing of codons in a few key places, in a cyclic game of rock-paper-scissors between the *TFR1* gene and the viral receptor-binding genes, virus entry can be blocked while iron uptake function is maintained in host cells.

Demogines A, Abraham J, Choe H, Farzan M, Sawyer SL. 2013. Dual host-virus arms races shape an essential housekeeping protein. *PLoS Biol* **11**:e1001571. doi:10.1371/journal.pbio.1001571.

Kaelber JT, Demogines A, Harbison CE, Allison AB, Goodman LB, Ortega AN, Sawyer SL, Parrish CR. 2012. Evolutionary reconstructions of the transferrin receptor of Caniforms supports canine parvovirus being a re-emerged and not a novel pathogen in dogs. *PLoS Pathog* **8**:e1002666. doi:10.1371/journal.ppat.1002666.

Meyerson NR, Sawyer SL. 2011. Two-stepping through time: mammals and viruses. *Trends Microbiol* **19**:286–294.

Identification of residues in the TfR1 protein that are under positive selection and in binding sites for two viral families. (A) Red stars represent the six rapidly evolving codon positions identified in rodent TfR1, mapped to a linear schematic of the TfR1 ectodomain. The amino acid encoded by human TfR1 at each of these positions is indicated. **(B)** Residue positions under positive selection are indicated in red on the structure of human TfR1 (PDB 1CX8). TfR1 is a homodimer, and the six sites of positive selection are indicated on the outer edge of each monomer. Known binding regions on TfR1 for the arenavirus Machupo virus GP protein and the retrovirus mouse mammary tumor virus (MMTV) Env protein are indicated in gray and blue, respectively, and the small region where they overlap is indicated with cross-hatching. The binding region for transferrin is indicated with a black arrow. Adapted from A. Demogines et al., *PLoS Biol* **11**:e1001571, 2013, with permission.

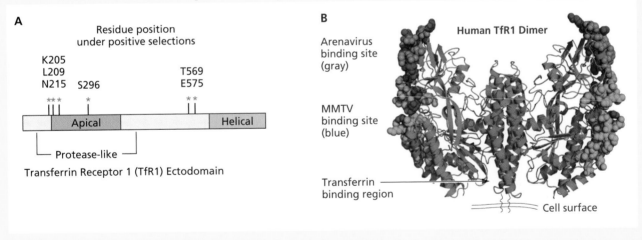

proteins that participate in such arms races can be identified by surveying orthologous genes in related host species for codons that exhibit a higher proportion of nucleotide substitutions that change an amino acid (called a nonsynonymous substitution, dN) than those that are silent (called synonymous, dS). A ratio of dN/dS > 1 indicates a high probability of positive selection. An example of the power of this approach, applied to viral-host receptor interactions in nonprimate species, is described in Box 10.9.

Perspectives

The relationships of viruses and their hosts are in constant flux. The combined perspectives of evolutionary biologists, ecologists, and epidemiologists are needed to decipher both the nature and consequences of these relationships. At present, the extent and significance of interplay between environment and genes, as well as the interactions of virus and host populations, are largely unknown. For viruses, rapid production of large numbers of progeny, tolerance to changes in host populations, and a capacity to produce enormous genetic diversity provide the adaptive palette that ensures their survival (Box 10.10). Host survival in this competition has depended on the evolution of intrinsic, innate, and adaptive immune defense systems, which are capable of recognizing and then blocking reproduction of or destroying invading viruses.

Present-day hosts represent progeny of survivors of ancient infections. Until quite recently, we have been limited to considering virus and host evolution mainly in the context of the currently circulating populations, to which we have access.

With the exception of the retroviruses, there had been no lasting record of other viral families to estimate how old they might be or how they may have changed over eons. As is now clear from the constantly expanding DNA fossil record, experience with numerous other viral families over vast stretches of evolutionary time is also recorded in the gene pools of host survivors.

Viral infections have far-reaching effects, ranging from shaping the host immune system in survivors to eliminating entire populations. However, given the ever-changing viral populations and the drastic modifications of the ecosystem that have accompanied the current human population explosion, we are hard-pressed to predict the future. The trajectory of evolution has long been a subject of deliberation, and both scientists and philosophers have pondered the parameters that may determine its paths. Virology provides a productive area for research in this area, with the caveat that outcomes cannot be judged as "good" or "bad." From the first principle that there is no goal but survival, we can deduce that evolution does not move a viral genome from "simple" to "complex" or along some trajectory aimed at "perfection." Change is effected by elimination of the ill-adapted of the moment, not by the prospect of building something better for some unknown future.

References

Books

Diamond J. 1997. *Guns, Germs and Steel: The Fates of Human Societies.* W. W. Norton and Company, New York, NY.

Domingo E, Webster RW, Holland J (ed). 1999. *Origin and Evolution of Viruses.* Academic Press, Inc, San Diego, CA.

Ewald PW. 1994. *Evolution of Infectious Disease.* Oxford University Press, Oxford, United Kingdom.

Harvey PH, Leigh Brown AJ, Maynard Smith J, Nee S (ed). 1996. *New Uses for New Phylogenies.* Oxford University Press, Oxford, United Kingdom.

Ryan F. 2009. *Virolution.* HarperCollins Publishers Ltd, London, United Kingdom.

Review Articles

Ahlquist P. 2006. Parallels among positive-strand RNA viruses, reverse-transcribing viruses and double-stranded RNA viruses. *Nat Rev Microbiol* **4:**371–382.

Daugherty MD, Malik HS. 2012. Rules of engagement: molecular insights from host-virus arms races. *Annu Rev Genet* **46:**677–700.

Domingo E, Holland JJ. 1997. RNA virus mutations and fitness for survival. *Annu Rev Microbiol* **51:**151–178.

Duffy S, Shackelton LA, Holmes EC. 2008. Rate of evolutionary change in viruses: patterns and determinants. *Nat Rev Genet* **9:**267–276.

Holmes EC. 2010. The RNA virus quasispecies: fact or fiction? *J Mol Biol* **400:**271–273.

Malim MH, Emerman M. 2001. HIV-1 sequence variation: drift, shift, and attenuation. *Cell* **104:**469–472.

Moya A, Holmes EC, González-Candelas F. 2004. The population genetics and evolutionary epidemiology of RNA viruses. *Nat Rev Microbiol* **2:**279–288.

Palumbi SR. 2001. Humans as the world's greatest evolutionary force. *Science* **293:**1786–1790.

Solomon T, Ni H, Beasley DWC, Ekkelenkamp M, Cardosa MJ, Barrett ADT. 2003. Origin and evolution of Japanese encephalitis virus in Southeast Asia. *J Virol* **5:**3091–3098.

Stoye JP. 2012. Studies of endogenous retroviruses reveal a continuing evolutionary saga. *Nat Rev Microbiol* **10:**395–406.

BOX 10.10

BACKGROUND
The world's supply of human immunodeficiency virus genomes provides remarkable opportunity for selection

Tens of millions of humans are infected by human immunodeficiency virus. Before the end stage of disease, each infected individual produces billions of viral genomes per day. As a result, $>10^{16}$ genomes are produced each day on the planet. Almost every genome has a mutation, and every infected human harbors viral genomes with multiple changes resulting from recombination and selection. Practically speaking, these large numbers provide an amazing pool of diversity. For example, thousands of times each day simply by chance, mutants arise that are resistant to **every combination** of the current >20 Food and Drug Administration-approved antiviral drugs or any others that might be developed in the future to treat AIDS.

Global view of HIV infection. More than 30 million people are living with HIV.

Sturtevant AH. 1937. Essays on evolution. I. On the effects of selection on mutation rate. *Q Rev Biol* **12**:464–467.

Suttle CA. 2007. Marine viruses—major players in the global ecosystem. *Nat Rev Microbiol* **5**:801–812.

Van Doorslaer K. 2013. Evolution of the *Papillomaviridae. Virology* **445**:11–20.

Van Etten JL, Lane LC, Dunigan DD. 2010. DNA viruses: the really big ones (giruses). *Annu Rev Microbiol* **64**:83–99.

Van Regenmortel MH, Bishop DH, Fauquet CM, Mayo MA, Maniloff J, Calisher CH. 1997. Guidelines to the demarcation of virus species. *Arch Virol* **142**:1505–1518.

Vignuzzi M, Stone JK, Arnold JJ, Cameron CE, Andino R. 2006. Quasispecies diversity determines pathogenesis through cooperative interactions in a viral population. *Nature* **439**:344–348.

Wasik BR, Turner PE. 2013. On the biological success of viruses. *Annu Rev Microbiol* **67**:519–541.

Wolfe ND, Dunavan CP, Diamond J. 2007. Origins of major human infectious diseases. *Nature* **447**:279–283.

Papers of Special Interest

Agostini HT, Yanagihara R, Davis V, Ryschkewitsch CF, Stoner GL. 1997. Asian genotypes of JC virus in Native Americans and in a Pacific Island population: markers of viral evolution and human migration. *Proc Natl Acad Sci U S A* **94**:14542–14546.

Baric RS, Yount B, Hensley L, Peel SA, Chen W. 1997. Episodic evolution mediates interspecies transfer of a murine coronavirus. *J Virol* **71**:1946–1955.

Benachenhou F, Sperber GO, Bongcam-Rudloff E, Andersson G, Boeke JD, Blomberg J. 2013. Conserved structure and inferred evolutionary history of long terminal repeats (LTRs). *Mob DNA* **4**:5. doi:10.1186/1759-8753-4-5.

Bonhoeffer S, Nowak MA. 1994. Intra-host versus inter-host selection: viral strategies of immune function impairment. *Proc Natl Acad Sci U S A* **91**:8062–8066.

Chao L. 1990. Fitness of RNA virus decreased by Muller's ratchet. *Nature* **348**:454–455.

Chao L. 1997. Evolution of sex and the molecular clock in RNA viruses. *Gene* **205**:301–308.

Clarke DK, Duarte EA, Elena SF, Moya A, Domingo E, Holland J. 1994. The Red Queen reigns in the kingdom of RNA viruses. *Proc Natl Acad Sci U S A* **91**:4821–4824.

Clarke DK, Duarte EA, Moya A, Elena SF, Domingo E, Holland J. 1993. Genetic bottlenecks and population passages cause profound fitness differences in RNA viruses. *J Virol* **67**:222–228.

Domingo E, Sabo D, Taniguchi T, Weissmann C. 1978. Nucleotide sequence heterogeneity of an RNA phage population. *Cell* **13**:735–744.

Drake JW, Allen EF. 1968. Antimutagenic DNA polymerases of bacteriophage T4. *Cold Spring Harb Symp Quant Biol* **33**:339–344.

Eigen M. 1971. Selforganization of matter and the evolution of biological macromolecules. *Naturwissenschaften* **58**:465–523.

Eigen M. 1996. On the nature of virus quasispecies. *Trends Microbiol* **4**:216–218.

Elena SF, Sanjuán R. 2005. Adaptive value of high mutation rates of RNA viruses: separating causes from consequences. *J Virol* **79**:11555–11558.

Esposito JJ, Sammons SA, Frace AM, Osborne JD, Olsen-Rasmussen M, Zhang M, Govil D, Damon IK, Kline R, Laker M, Li Y, Smith GL, Meyer H, Leduc JW, Wohlhueter RM. 2006. Genome sequence diversity and clues to the evolution of variola (smallpox) virus. *Science* **313**:807–812.

Ghedin E, Sengamalay NA, Shumway M, Zaborsky J, Feldblyum T, Subbu V, Spiro DJ, Sitz J, Koo H, Bolotov P, Dernovoy D, Tatusova T, Bao Y, St George K, Taylor J, Lipman DJ, Fraser CM, Taubenberger JK, Salzberg SL. 2005. Large-scale sequencing of human influenza reveals the dynamic nature of viral genome evolution. *Nature* **437**:1162–1166.

Guidotti LG, Borrow P, Hobbs MV, Matzke B, Gresser I, Oldstone MB, Chisari FV. 1996. Viral cross talk: intracellular inactivation of the hepatitis B virus during an unrelated viral infection of the liver. *Proc Natl Acad Sci U S A* **93**:4589–4594.

Hall JD, Coen DM, Fisher BL, Weisslitz M, Randall S, Almy RE, Gelep PT, Schaffer PA. 1984. Generation of genetic diversity in herpes simplex virus: an antimutator phenotype maps to the DNA polymerase locus. *Virology* **132**:26–37.

Iyer LM, Aravind L, Koonin EV. 2001. Common origin of four diverse families of large eukaryotic DNA viruses. *J Virol* **75**:11720–11734.

Katzourakis A, Gifford RJ. 2010. Endogenous viral elements in animal genomes. *PLoS Genet* **6**:e1001191. doi:10.1371/journal.pgen.1001191.

Krisch HM. 2003. The view from Les Treilles on the origins, evolution and diversity of viruses. *Res Microbiol* **154**:227–229.

Mayr E. 1997. The objects of selection. *Proc Natl Acad Sci U S A* **94**:2091–2094.

McGeoch DJ, Cook S, Dolan A, Jamieson FE, Telford EA. 1995. Molecular phylogeny and evolutionary timescale for the family of mammalian herpesviruses. *J Mol Biol* **247**:443–458.

Mindell DP, Villarreal LP. 2003. Don't forget about viruses. *Science* **302**:1677.

Muller HJ. 1964. The relation of recombination to mutational advance. *Mutat Res* **106**:2–9.

Nichol ST, Rowe JE, Fitch WM. 1993. Punctuated equilibrium and positive Darwinian evolution in vesicular stomatitis virus. *Proc Natl Acad Sci U S A* **90**:10424–10428.

Ohshima K, Hattori M, Yada T, Gojobori T, Sakaki Y, Okada N. 2003. Whole-genome screening indicated a possible burst of formation of processed pseudogenes and Alu repeats by particular L1 subfamilies in ancestral primates. *Genome Biol* **4**:R74. doi:10.1186/gb-2003-4-11-r74.

Oude Essink BB, Back NK, Berkhout B. 1997. Increased polymerase fidelity of the 3TC-resistant variants of HIV-1 reverse transcriptase. *Nucleic Acids Res* **25**:3212–3217.

Powers AM, Brault AC, Shirako Y, Strauss EG, Kang W, Strauss JH, Weaver SC. 2001. Evolutionary relationships and systematics of the alphaviruses. *J Virol* **75**:10118–10131.

Prangishvili D, Forterre P, Garrett RA. 2006. Viruses of the Archaea: a unifying view. *Nat Rev Microbiol* **4**:837–848.

Preston BD, Poiesz BJ, Loeb LA. 1988. Fidelity of HIV-1 reverse transcriptase. *Science* **242**:1168–1171.

Rambaut A, Pybus OG, Nelson MI, Viboud C, Taubenberger JK, Holmes EC. 2008. The genomic and epidemiological dynamics of human influenza A virus. *Nature* **453**:615–619.

Roossinck M. 2005. Symbiosis versus competition in plant virus evolution. *Nat Rev Microbiol* **3**:917–924.

Sáiz JC, Domingo E. 1996. Virulence as a positive trait in viral persistence. *J Virol* **70**:6410–6413.

Schrag SJ, Rota PA, Bellini WJ. 1999. Spontaneous mutation rate of measles virus: direct estimation based on mutations conferring monoclonal antibody resistance. *J Virol* **73**:51–54.

Simmonds P, Smith DB. 1999. Structural constraints on RNA virus evolution. *J Virol* **73**:5787–5794.

Spiegelman S, Pace NR, Mills DR, Levisohn R, Eikhom TS, Taylor MM, Peterson RL, Bishop DH. 1968. The mechanism of RNA replication. *Cold Spring Harb Symp. Quant Biol* **33**:101–124.

Summers J, Litwin S. 2006. Examining the theory of error catastrophe. *J Virol* **80**:20–26.

Takemura M. 2001. Poxviruses and the orgin of the eukaryotic nucleus. *J Mol Evol* **52**:419–425.

Wagenaar TR, Chow VT, Buranathai C, Thawatsupha P, Grose C. 2003. The out of Africa model of varicella-zoster virus evolution: single nucleotide polymorphisms and private alleles distinguish Asian clades from European/North American clades. *Vaccine* **21**:1072–1081.

Webster RG, Laver WG, Air GM, Schild GC. 1982. Molecular mechanisms of variation in influenza viruses. *Nature* **296**:115–121.

Zhu T, Korber BT, Nahmias AJ, Hooper E, Sharp PM, Ho DD. 1998. An African HIV-1 sequence from 1959 and implications for the origin of the epidemic. *Nature* **391**:594–597.

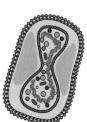

11 Emergence

The Spectrum of Host-Virus Interactions
Stable Interactions
The Evolving Host-Virus Interaction
The Dead-End Interaction
Common Sources of Animal-to-Human Transmission
The Resistant Host

Encountering New Hosts: Ecological Parameters
Successful Encounters Require Access to Susceptible and Permissive Cells
Population Density, Age, and Health Are Important Factors
Experimental Analysis of Host-Virus Interactions
Learning from Accidental Infections

Expanding Viral Niches: Some Well-Documented Examples
Poliomyelitis: Unexpected Consequences of Modern Sanitation
Smallpox and Measles: Exploration and Colonization

Notable Zoonoses
Hantavirus Pulmonary Syndrome: Changing Climate and Animal Populations
Severe Acute and Middle East Respiratory Syndromes (SARS and MERS): Two New Zoonotic Coronavirus Infections

Acquired Immunodeficiency Syndrome (AIDS): Pandemic from a Zoonotic Infection

Host Range Can Be Expanded by Mutation, Recombination, or Reassortment
Canine Parvoviruses: Cat-to-Dog Host Range Change by Two Mutations
Influenza Epidemics and Pandemics: Escaping the Immune Response by Reassortment

New Technologies Uncover Hitherto Unrecognized Viruses
Hepatitis Viruses in the Human Blood Supply

A Revolution in Virus Discovery

Perceptions and Possibilities
Virus Names Can Be Misleading
All Viruses Are Important

What Next?
Can We Predict the Next Viral Pandemic?
Emerging Viral Infections Illuminate Immediate Problems and Issues
Humans Constantly Provide New Venues for Infection
Preventing Emerging Virus Infections

Perspectives

References

LINKS FOR CHAPTER 11

▶▶ *Video: Interview with Dr. Ian Lipkin*
http://bit.ly/Virology_Lipkin

▶▶ *The real Batman, Linfa Wang*
http://bit.ly/Virology_Twiv296

▶▶ *Heartland virus disease*
http://bit.ly/Virology_3-28-14

▶▶ *MERS-coronavirus in dromedary camels*
http://bit.ly/Virology_2-26-14

▶▶ *Influenza A viruses in bats*
http://bit.ly/Virology_11-12-13

▶▶ *An epidemic of porcine diarrhea in North America*
http://bit.ly/Virology_1-28-14

Humans have suffered for millions of years from infectious diseases. However, since the rise of agriculture (the past 11,000 years), new infectious agents have invaded human populations, primarily because these infections (e.g., measles and smallpox) can be sustained only in large, dense populations that were unknown before agriculture and commerce. The source of such emerging infectious agents is a popular topic of research, debate, and concern.

We define an **emerging virus** as the causative agent of a new or hitherto unrecognized infection. Occasionally, emerging infections are manifestations of expanded host range with an increase in disease that was not previously obvious. More generally, emerging infections of humans reflect transmission of a virus from a wild or domesticated animal, with attendant human disease (**zoonotic infections**). Occasionally, such cross-species infection will establish a new virus in a population (e.g., human immunodeficiency viruses moving from chimpanzees to humans). On the other hand, some cross-species infections, although not without consequence, cannot be sustained (e.g., Ebola and Marburg viruses moving from bats to humans).

While the term "emerging virus" became common in the popular press in the 1990s (usually with dire implications ["killer viruses on the loose"]), such infections are not new to virologists, epidemiologists, or public health officials but have long been recognized as an important manifestation of virus evolution. Parameters that drive such evolution include changes brought about by unprecedented human population growth and large-scale disturbances of ecosystems that result from human occupation of almost every corner of the planet (Fig. 11.1). In recent years, emerging infections have been detected with increasing frequency, thanks to advances in technology and better communication about disease outbreaks. Indeed, global communication has brought some emerging viral infections to center stage on the local news. Anyone with access to television, radio, the Internet, or newspapers has heard something about the AIDS virus, human immunodeficiency virus type 1, Ebola virus, and certainly H5N1 avian influenza virus. Examples of zoonotic infections and conditions that contributed to the emergence of particular viruses are provided in Table 11.1. Despite the variety of virus families and different geographic locations of these outbreaks, some common parameters do exist. These parameters define the rules of engagement for viruses and their potential hosts.

The Spectrum of Host-Virus Interactions

Four general types of interactions between hosts and viruses can be recognized: **stable**, **evolving**, **dead-end**, and **resistant** (Fig. 11.2). These four categories identify the extremes of

---

PRINCIPLES *Emergence*

- An emerging virus is defined as the causative agent of a new or previously unrecognized virus infection in a population.

- Zoonoses are infections of humans by viruses that preexist in stable relationships with nonhuman hosts. Most emerging viruses come from zoonotic infections.

- There are four general types of interaction between a virus and its host: stable, evolving, dead-end, and resistant.

  - Stable host-virus interactions are those in which both participants survive and reproduce.

  - The hallmarks of the evolving host-virus interaction are instability and unpredictability.

  - The dead-end interaction, in which the virus is not transmitted to other members of the new host species, is a frequent outcome of cross-species infection.

  - The **resistant host** interaction represents situations in which the host blocks infection completely.

- The vast majority of human encounters with viruses are uneventful because host cells are not susceptible or the body's defenses are so strong that potential invaders cannot initiate an infection.

- Outcomes of a virus-host interaction depend on many factors, including ecological, host, and viral parameters.

- The predominant parameters for spread of infection are the population density and the age and health of individuals in that population.

- Even when large numbers of susceptible individuals are inoculated with an equivalent quantity of a virulent virus, the results can be quite variable, and not everyone succumbs to disease.

- The rate of virus discovery has risen with the development of new technologies; it is now possible to detect and characterize unknown viruses with comparative ease and speed.

- Virus names are often misleading.

- Despite much new knowledge and advances in technologies to detect viruses, we cannot predict the next pandemic.

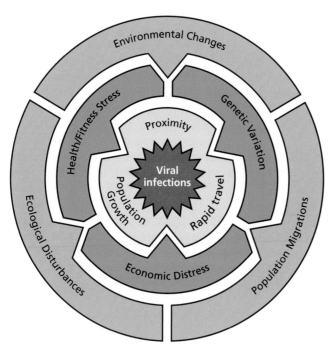

Figure 11.1 Multiple parameters converge to promote emerging viral infections. In the ecology of virus-host interactions, many interlocking and interconnected parameters are in play.

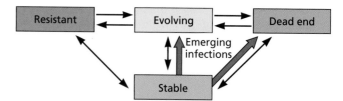

Figure 11.2 General categories of interactions between hosts and viruses. Four categories of host-virus interactions are indicated in the boxes. The **stable** interaction maintains the virus in the ecosystem. The **evolving** interaction describes the passage of a virus from "experienced" populations to naive populations in the same or other host species. The **dead-end** interaction represents one-way passage of a virus to different species. The host usually dies, or if it survives, the virus is not transmitted efficiently to the new host species. The **resistant host** interaction represents situations in which the host blocks infection completely. The arrows indicate possible transition from one category to another or the possible transformation of one into another. The red, filled arrows indicate the major pathways of zoonotic or other emerging viruses.

relationships that preexist in nonhuman hosts, which serve as **reservoirs** for particular viruses.

Stable Interactions

Stable host-virus interactions are those in which both participants survive and reproduce. Such relationships are essential for the continued existence of the virus and may influence host survival as well. This state is optimal for a host-parasite interaction, but need be neither benign nor permanent in an outbred population. Infected individuals can become ill, recover, develop immunity, or die, yet in the long run, both virus and host populations survive. While this situation is often described as an equilibrium, the term is misleading because the interactions are dynamic and fragile, and are rarely reversible. Viral populations may become more or less

dynamic host-virus interactions, and their names emphasize the defining feature of each. The figure shows how relationships can shift from one category to another and illustrates the continuity of viral interactions in nature. It is important to note that these categories are meant to describe interactions among large populations and **not** single virus-host interactions. In this framework, emerging viral infections are defined as human infections that derive from stable host-virus

Table 11.1 Examples of viruses that cause zoonotic infections

Virus	Family	Drivers of emergence
Dengue virus	*Flaviviridae*	Urban population density; open water storage favors mosquito breeding (e.g., millions of used tires)
Ebola virus	*Filoviridae*	Human contact with unknown natural host (Africa); importation of monkeys in Europe and the United States
Hantaan virus	*Bunyaviridae*	Agriculture techniques: human-rodent contact
Hendra virus	*Paramyxoviridae*	Domestic horses to stable workers
Human immunodeficiency virus	*Retroviridae*	Hunting and butchering of infected primates (bushmeat trade)
Influenza virus	*Orthomyxoviridae*	Integrated pig-duck agriculture; mobile population
Junin virus	*Arenaviridae*	Agriculture techniques: human-rodent contact
Machupo virus	*Arenaviridae*	Agriculture techniques: human-rodent contact
Nipah virus	*Paramyxoviridae*	Proximity of fruit bats, the natural reservoir, favors transmission to pigs and then to humans
Rift Valley virus	*Bunyaviridae*	Dams, irrigation
Sin Nombre virus	*Bunyaviridae*	Natural increase of deer mice and subsequent human-rodent contact
West Nile virus	*Flaviviridae*	Birds ↔ mosquitoes ↔ humans; recent introduction into United States

virulent, if such a change enables them to be maintained in the population, while host mechanisms that attenuate the more debilitating effects of the viruses may be selected.

Some stable interactions are effectively permanent. This is the case when there is only one natural host, because a stable relationship is required for the survival of the virus population. Examples include measles virus, herpes simplex virus, human cytomegalovirus, and smallpox virus in humans, and simian cytomegalovirus, monkeypox virus, and simian immunodeficiency virus, which infect only certain species of monkeys. Stable interactions can also be sustained by infection of more than one host species with the same virus: influenza A virus, flaviviruses, and togaviruses are capable of propagating in a variety of species. Indeed, many members of these virus families replicate efficiently in some insects as well as in mammals and birds.

Establishment of a stable host-virus interaction is not necessarily the optimal solution for survival. The trajectory of evolution is unpredictable: what is successful today may be lethal at another time. Furthermore, if a virus population becomes completely dependent on one, and only one, host, it will have entered a potential bottleneck that may constrain its further evolution. If the host becomes extinct for whatever reason, so will the virus. For example, if humans disappeared, many virus populations, including poliovirus, measles virus, and several herpesviruses, would cease to exist. Eradication of natural smallpox virus was possible because humans are the only hosts and worldwide immunization was achieved.

The Evolving Host-Virus Interaction

The hallmarks of the evolving host-virus interaction (Fig. 11.2) are instability and unpredictability. These properties are to be expected, as selective forces are applied to both host and virus, and are magnified when host populations are small. An example of such an interaction is the introduction of measles viruses to natives of the Americas by Old World colonists and slave traders. Measles infections devastated the native populations. While less affected at the time, European populations had experienced the same horror when these viruses first spread from Asia, and only developed some resistance and immunity over time. Other opportunities to enter the evolving host-virus interaction may arise if the virus in a stable relationship acquires a new property that increases its virulence or spread, or if the host population suffers a far-reaching catastrophe that reduces resistance (e.g., famine or mass population changes during wars). The introduction of West Nile virus into the Western Hemisphere in 1999 provides a contemporary example of an evolving host-virus interaction in which a virus was introduced **accidentally** into a new geographic location (Box 11.1). A classic case of the **deliberate** release of a virus in a new geographic location is the attempt to use poxvirus infection to rid Australia of rabbits.

The consequences of this "experiment" provide another example of an evolving virus-host interaction (Box 11.2).

The Dead-End Interaction

In a dead-end interaction (Fig. 11.2), the virus is not transmitted to other members of the new host species. Like the evolving host-virus interaction, it represents a departure from a stable relationship, often with lethal consequences. A dead-end interaction is a frequent outcome of cross-species infection. In many cases of such infection, the host is killed so quickly that there is little or no subsequent transmission of the virus to others. In other cases, the virus cannot be transmitted to other individuals of the same species.

The dead-end interaction is often observed with viruses carried by arthropods, such as ticks and mosquitoes, which cycle in the wild in a stable relationship with a vertebrate host. Occasionally the infected insect bites a new species (e.g., humans) and transmits the virus (Fig. 11.3 and 11.4). Even though the consequences to the infected individual may be severe, because the human is not part of the natural, stable host-virus relationship, these interactions have little, if any, effect on the evolution of the virus and its natural host. While generally accurate, such a view may be too simplistic in some cases; infection by a less virulent mutant or infection of a more resistant individual may be the first step in establishing a new host-virus interaction.

Infection of humans by the filoviruses, such as marburgviruses or ebolaviruses, is an example of one type of dead-end virus-host interaction. In these cases, disease onset is sudden, with 25 to 90% case-fatality ratios reported (Appendix, Fig. 5). Virus disseminates through the blood and reproduces in many organs, causing focal necrosis of the liver, kidneys, lymphatic organs, ovaries, and testes. Capillary leakage with massive hemorrhaging, shock, and acute respiratory disorders are observed in fatal cases. Patients usually die rapidly of intractable shock without evidence of an effective immune response. Even when recovery is under way, survivors do not have detectable neutralizing antibodies. The infection clearly overwhelms a particular individual, but does not spread widely because these viruses can be transmitted to other humans only by contact with infected blood and tissue. Consequently, human infections tend to cluster in local areas. Current evidence points to fruit bats as the reservoir species for Ebola virus in Central Africa. Humans are not the only dead-end hosts: gorillas are susceptible, and large numbers of these primates have died from Ebola virus infection.

Many animal models of disease might also be considered examples of dead-end interactions. Herpes simplex virus is a human virus, but when it is introduced into mice, rabbits, or guinea pigs in the laboratory, these animals become infected and show pathogenic effects that mimic some aspects of the human disease. However, in their natural environment, these

BOX **11.1**

DISCUSSION
An evolving virus infection: the West Nile virus outbreak

In August 1999, six people were admitted to Flushing Hospital in Queens, NY, with similar symptoms of high fever, altered mental status, and headache. These people were discovered subsequently to be infected with West Nile virus. This Old World flavivirus was discovered in 1937 in the West Nile district of Uganda and had never before been isolated in the Western Hemisphere. The virus has now spread in the United States from the Atlantic to the Pacific, as well as both south and north as far as the Canadian provinces and territories.

The New York isolate of West Nile virus is nearly identical to a virus isolated in 1998 from a domestic goose in Israel during an outbreak of the disease. The close relationship between these two isolates suggests that the virus was brought to New York City from Israel in the summer of 1999. How it crossed the Atlantic will probably never be known for sure, but it might have been via an infected bird, mosquito, human, horse, or other vertebrate host. These events mark the first introduction in recent history of an Old World flavivirus into the New World. A fascinating, and yet unanswered, question is how the epidemic got started in New York City. The summer

of 1999 was particularly hot and dry. Similar conditions spawn outbreaks of West Nile virus encephalitis in Africa, the Middle East, and the Mediterranean basin of Europe.

West Nile virus or virus-specific antibodies have been found in numerous species of birds. Crows and jays appear to be particularly sensitive, but many zoos have reported deaths of their exotic birds from West Nile virus infections. Humans and other animals acquire the virus from mosquito bites after the insect has fed on infected birds. Infected horses develop lethal encephalitis, of which hundreds of cases have been reported.

About 20% of infected humans experience flu-like symptoms when infected by West Nile virus; <1% of these individuals develop life-threatening neuroinvasive disease with meningitis-, encephalitis-, or poliomyelitis-like symptoms. However, the risk is higher in immune-compromised and elderly individuals. In the summers of 2002 and 2003, human infection reached epidemic status, causing encephalitis in hundreds of individuals. As the majority of infected people do not develop symptoms, the importance of screening the blood supply was appreciated

early on. Since tests were developed in 2003, thousands of contaminated blood donations were removed from the blood supply, and infections attributable to blood transfusions are now quite rare.

By 2007, the North American epidemic seemed to be resolving, but the numbers reported to the U.S. Centers for Disease Control and Prevention (CDC) rose again in 2012 and 2013. Warmer winter temperatures that allow survival of more infected mosquitoes is a likely contributing factor. Sequencing of isolates did not reveal anything unique about the circulating 2012 strains relative to previous strains. Unfortunately, there is no treatment for West Nile disease beyond supportive care, and as yet, no vaccine exists for humans. Until we understand the complex ecology of this viral infection, the consequences for public health are difficult to predict.

FAQ: West Nile Virus, a publication of the American Academy of Microbiology, is available online: http://academy.asm.org/index.php/faq-series/793-faq-west-nile-virus-july-2013. For timely updates on West Nile virus, see http://www.cdc.gov/westnile/index.html

Data on West Nile virus cases, derived from CDC statistics, 1999 to 2013.

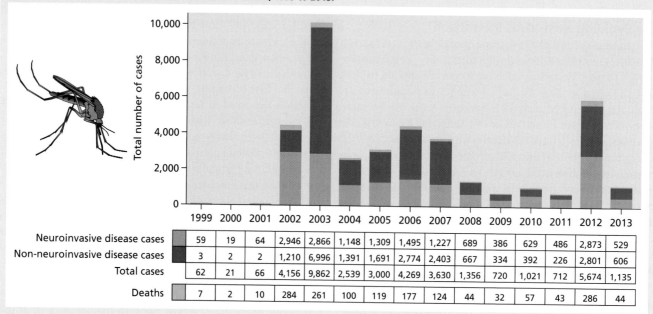

	1999	2000	2001	2002	2003	2004	2005	2006	2007	2008	2009	2010	2011	2012	2013
Neuroinvasive disease cases	59	19	64	2,946	2,866	1,148	1,309	1,495	1,227	689	386	629	486	2,873	529
Non-neuroinvasive disease cases	3	2	2	1,210	6,996	1,391	1,691	2,774	2,403	667	334	392	226	2,801	606
Total cases	62	21	66	4,156	9,862	2,539	3,000	4,269	3,630	1,356	720	1,021	712	5,674	1,135
Deaths	7	2	10	284	261	100	119	177	124	44	32	57	43	286	44

EXPERIMENTS

A classic experiment in virus evolution: deliberate release of rabbitpox virus in Australia

In 1859, 24 European rabbits were introduced into Australia for sport, and lacking natural predators, the amorous bunnies went on to reproduce in plague proportions. In 1907, the longest unbroken fence in the world (1,139 miles long) was built to protect portions of the country from invading rabbits, who consumed all vegetation in their paths. Such desperate actions were to no avail. As a last resort, the rabbitpox virus, myxoma virus was released in Australia in the 1950s in an attempt to rid the continent of these pests. The natural hosts of myxoma virus are the cottontail rabbit, the brush rabbit of California, and the tropical forest rabbit of Central and South America. The virus is spread by mosquito vectors, and the natural hosts develop superficial warts on their ears. However, European rabbits, a distinct species, are killed rapidly by myxoma virus. In fact, infection is 90 to 99% fatal in these hosts!

In the first year, the virus was efficient in killing rabbits, with a 99.8% mortality rate. However, by the second year, the mortality dropped dramatically to 25%. In subsequent years, the rate of killing was lower than the reproductive rate of the rabbits, and hopes for 100% eradication were dashed. Careful epidemiological analysis of this artificial epidemic provided important information about the evolution of viruses and their hosts.

As expected, the infection spread rapidly during spring and summer, when mosquitoes are abundant, but slowly in winter. Given the large numbers of rabbits and virus particles, and the almost 100% lethal nature of the infection, attenuating mutations were selected

quickly; within 3 years, less-virulent viruses appeared, and some infected rabbits were able to survive over the winter. The host-virus interaction observed was that predicted for an evolving host, coming to an equilibrium with the pathogen. A balance is struck: some infected rabbits die, but many survive.

The most obvious lesson from this experience was that the original idea to eliminate rabbits with a lethal viral infection was flawed. Powerful selective forces that could not be controlled or anticipated were at work. Surprisingly, more experiments in the virological control of rabbits are under way in Australia.

One approach used a lethal rabbit calicivirus. As resistance has developed in the rabbit population (as might have been predicted), more-lethal strains are being tested for possible use. Another approach employs a genetically engineered myxoma virus designed to sterilize, but not kill, rabbits. The latter viruses encode a rabbit zona pellucida protein, and infected animals synthesize antibodies against their own eggs (so-called immunocontraception). Stay tuned ...

Cooke, B. 7 March 2012. Controlling rabbits: let's not get addicted to viral solutions. The Conversation. http://theconversation.com/controlling-rabbits-lets-not-get-addicted-to-viral-solutions-5701,

Rabbits and ever more rabbits!

animals contribute nothing to transmission or maintenance of the virus.

Common Sources of Animal-to-Human Transmission

Rodents play critical roles in the introduction of new viruses into human populations in areas where these animals are abundant. Most hemorrhagic disease viruses, including Lassa, Junin, and the Sin Nombre virus, are endemic in rodents, their natural hosts. The viruses establish a persistent infection, and the rodents show few, if any, ill effects. However, substantial numbers of virus particles are excreted in urine, saliva, and feces to maintain the virus in the rodent population. Humans become infected when they happen to come in contact with rodent excretions that contain

infectious virus particles. Unfortunately, infection by such rodent viruses can cause lethal outbreaks in humans as dead-end hosts.

Bats are the natural hosts of several viruses that cause dead-end, zoonotic infections. Hendra virus and Nipah virus are known to have entered the human population via bats. The Old World fruit bats (genus *Pteropus*), commonly called flying foxes, are widely distributed in Southeast Asia, Australia, and the Indian subcontinent. Despite having high antibody titers against the deadly viruses, the animals exhibit no obvious disease (Box 11.3). High virulence in humans and our complete lack of therapeutic interventions require that these viruses be studied only under the highest biological and physical containment (biosafety level 4 [BSL4]). Accordingly, we know very little about their biology, ecology, and pathogenesis.

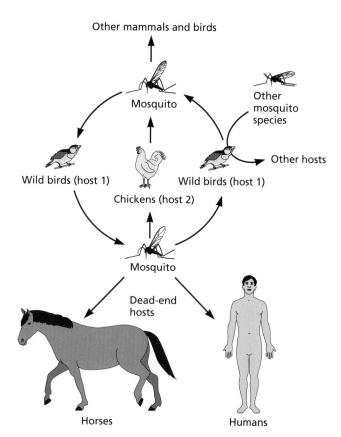

Figure 11.3 The dead-end host scenario as illustrated by a complex host-virus relationship. This illustration summarizes how multiple host species can maintain and transmit a virus (e.g., arthropod-transmitted dengue yellow fever viruses). In this example, the virus population is maintained in two different hosts (wild birds and domestic chickens) and is spread among individuals by a mosquito vector. The virus reproduces both in species of bird and in the mosquito. Disease is likely to be nonexistent or mild in these species, as these hosts have adapted to the infection. A third host (in this example, horses or humans) occasionally is infected when bitten by a mosquito that previously fed on an infected bird. Horses and humans are dead-end hosts and contribute little to the spread of the natural infection, but they may suffer from serious, life-threatening disease. Occasionally another species of biting insect (e.g., other mosquito species) can feed on an infected individual (bird, horse, or human) and then transmit the infection to another species not targeted by the original mosquito vector.

The Resistant Host

All living things are exposed continuously to viruses of all types, yet the vast majority of these interactions are uneventful (Fig. 11.2). In some cases, there is no infection because host cells are not susceptible, not permissive, or the primary physical, intrinsic, and innate defenses are so strong that most potential invaders are diverted or destroyed upon contact. In other cases, organisms may become infected and produce some virus particles, but the virus is cleared rapidly without activation of the host's acquired immune system (Chapter 5). This outcome contrasts with an inapparent infection, in which

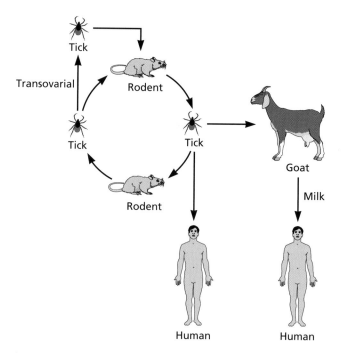

Figure 11.4 Replicative cycle of the central European tick-borne flavivirus may involve zoonotic infections. The European tick-borne flavivirus is maintained and spread by multiple host infections. Congenital transmission in the tick maintains the virus in the tick population as they feed upon rodents. The infected newborn ticks have adapted to the virus and thrive. The virus is transmitted from the tick to a variety of animals, including cows, goats, and humans. Humans can also be infected by drinking milk from an infected goat, sheep, or cow. As the virus cannot be transmitted from human to human, this zoonotic infection is another example of the dead-end host interaction.

an immune response is mounted but the individual exhibits no signs of disease.

The obstacles that limit the ability of a host to support viral reproduction need not be insurmountable. This is certainly the case in the practice of **xenotransplantation** (the use of animal organs in humans). Not only does such transplantation bypass physical and innate defenses by surgery, but the drugs used to block transplant rejection also suppress the immune response. Consequently, any virus particles or genomes in xenografts would have direct access to the once-resistant host in the absence of crucial antiviral defenses. As many of these viruses can infect human cells or have close, human-adapted relatives, the xenotransplantation patient represents a potential source of new viral diversity. Clearly this practice raises many medical and ethical issues.

Encountering New Hosts: Ecological Parameters

Establishment of all virus-host relationships depends on the concentrations of the participants and the probability of

BOX 11.3

What is it about bats?

Bats comprise a quarter of all mammalian species. There are almost 1,000 recognized bat species, which are grouped into two suborders: *Microchiroptera* (microbats, mostly insect eaters) and *Megachiroptera* (megabats, mostly fruit eaters). Bats are the only mammals capable of sustained flight and have much longer life spans than any other mammals of similar size; for example, ~2 years for a mouse versus almost 40 years for the little brown bat common to North America. These amazing animals have been around for more than 65 million years, since long before the appearance of *Homo sapiens*.

Bats are gregarious, social creatures, some existing in extremely large populations, likely to facilitate the transmission of viruses among individuals. We now know that bats are natural reservoirs that can sustain many viruses that are extremely pathogenic to humans, including Nipah, Hendra, rabies, Ebola, and SARS coronavirus. As illustrated in Fig. 10.10 such associations are likely to have occurred throughout evolutionary times, as sequences related to filoviruses and another deadly pathogen, Bornavirus, were integrated into the bat genome more than 40 million years ago.

Recent comparative analyses of their genomes have provided clues both to the long life span of bats and their ability to interact with

numerous deadly pathogens without ill effect. In these studies, a large number of genes in the DNA damage checkpoint-DNA repair pathway of bats showed evidence of positive selection (dN/dS > 1). It was hypothesized that such selection may have been driven by the need to counteract a high production of DNA-damaging reactive oxygen species associated with adaptation to flight. A more efficient DNA damage repair system might contribute to longevity.

These genomic studies also revealed genetic changes that affect components shared by the DNA damage pathway and the innate immune system (i.e., Nf-κb signaling), as well as an absence of the genes that encode certain natural killer cell receptors and proteins required for formation of inflammasomes. These findings indicate that there are likely to be significant differences in immune function among bats and other mammals. Such differences might explain how bats can provide a stable reservoir for viruses that wreak havoc on other mammalian hosts.

Zhang G, Cowled C, Shi Z, Huang Z, Bishop-Lilly KA, Fang X, Wynne JW, Xiong Z, Baker ML, Zhao W, Tachedjian M, Zhu Y, Zhou P, Jiang X, Ng J, Yang L, Wu L, Xiao J, Feng Y, Chen Y, Sun X, Zhang Y, Marsh GA, Crameri G, Broder CC, Frey KG, Wang LF, Wang J. 2013. Comparative analysis of bat genomes provides insight into the evolution of flight and immunity. *Science* **339**:456–460.

All Pteropus species of megabats are considered flying foxes. *Pteropus vampyrus* is the Malayan flying fox (found in peninsular Malaysia), and is one of the species that carries Nipah virus. Photograph courtesy of Juliet Pulliam, Princeton University.

---

productive encounters. Many ecological and social parameters affect the transmission of infection to new hosts in natural populations (Tables 11.2 and 11.3). Living together and sharing resources facilitates inter- and intraspecies transmission. Droughts concentrate many species at water holes, which increases the probability for transmission; destruction of habitat forces new species interactions. Predators eat their prey and become unwitting "test tubes" for cross-species infection by viruses found in tissues of the prey.

In contrast, rare chance encounters of viruses with new hosts may give rise to infections that are never seen, or at least never appreciated. These rare single-host infections may not be transmitted among humans for any number of reasons, including insufficient quantity of progeny virus shed, limited duration of shedding, and small numbers of new human hosts exposed to the infected individual. In addition, the progeny virus produced in the new host may not have the genetic repertoire to facilitate high levels of reproduction and transmission to other hosts.

Successful Encounters Require Access to Susceptible and Permissive Cells

Potential new hosts must have cells with accessible receptors that can engage ligands on virus particles. The influenza virus hemagglutinin protein has a high affinity for sialosaccharides found on the cell surfaces of many different host species. The linkage of the terminal sialic acid/galactose residues is an important determinant of tropism. Avian influenza virions bind sialic acid α(2,3)-galactose-terminated oligosaccharides, whereas the human influenza virus hemagglutinin proteins bind tightly to oligosaccharides carrying a terminal α(2,6)-linked sugar. Cells of the human respiratory tract do display α(2,3)-galactose-terminated oligosaccharides, but such cells lie deep in respiratory tissues. Conversely, sialic acid with terminal α(2,6) linkages is abundant in the more accessible regions of the upper respiratory tract. This anatomical fact appears to be a prime reason why humans cannot be infected easily with avian influenza viruses.

Nipah virus was first identified during an outbreak in swine and humans in Malaysia in 1998 and 1999. While Nipah virus

Table 11.2

Human actions

Dams and water impoundments

Irrigation

Massive deforestation

Rerouting of wildlife migration patterns

Wildlife parks

Long distance transport of livestock and birds

Air travel

Uncontrolled urbanization

Day care centers

Hot tubs

Air conditioning

Millions of used tires

Blood transfusion

Xenotransplantation

Societal changes with regard to drug abuse and sex

infection of bats is apparently nonpathogenic, large quantities of virus particles are excreted in bat urine and feces. Two facts are relevant to the Malaysian outbreak: (i) pig farmers often plant mango and durian trees next to pig pens, and (ii) fruit bats are messy eaters. When pigs come in contact with partially eaten contaminated fruit, they suffer a respiratory disease, and spread virus particles into the environment efficiently by sneezing and from mucous secretions.

Table 11.3 Ecological and social parameters facilitate transmission of infection to new hosts

Transmission parameter	**Action or example**
Contact with bodily fluids of infected hosts	Hunting and consumption of wild game; intimate contact with infected animals in the wild, in farms, at zoos, or in the home
Sharing a resource with different species	Infected fruit bats, pigs, rodents, and humans share food or inhabit the same or nearby space
Being host to the same insect vector	Japanese encephalitis virus infection is spread by mosquitoes that feed on herons, people, and pigs
Encroachment by one species into the habitat of another	Humans enter the jungle and are bitten by mosquitoes that are part of an established host-virus interaction or cycle

Bats and pigs establish a one-way conduit for a zoonotic infection of humans in rural Indonesian communities, where humans often share accommodations with domestic swine. In addition, slaughterhouse workers are exposed to infected pigs. Remarkably, when Nipah virus infects humans, it causes encephalitis, not respiratory infection. While often lethal for the infected human (Nipah virus killed 105 of 265 infected people in the Malaysian outbreak mentioned above), the infection is contained in infected brain tissue and does not spread. Nipah virus can also infect the human upper respiratory tract, and these infections are spread among humans in close contact.

Population Density, Age, and Health Are Important Factors

The predominant parameters influencing the spread of infection are the population density and the age and health of individuals in that population. The importance of population density is illustrated by the fact that at least half a million people in a more or less confined urban setting are required to ensure a large enough annual supply of susceptible hosts to maintain measles virus in a human population (Chapter 1). When this large population of interacting hosts is not available, measles virus cannot be propagated.

Variables such as duration of immunity and the quantity of virus particles that are produced and shed from each individual have marked effects on spread of infection, as do opportunities for direct (e.g., via sexual contact) or indirect (e.g., contaminated water) exposure. The age distribution of any potential host population is also an important determinant for the spread of infection. For example, the very young and the very old are commonly more susceptible to a given virus than is the general population and, consequently, serve as sources of transmission. Predictably, prevention of infection in these groups tends to reduce the overall infection rate in the population at large. The distribution of poor and wealthy individuals in a population can also influence infection rates.

Experimental Analysis of Host-Virus Interactions

Consideration of the different types of host-virus interactions listed above highlights some of the difficulties in identifying the parameters that affect virus evolution and emergence in natural populations. Mathematical models are being developed to analyze the dynamics of this process and to predict the critical population size necessary to support the continual transmission of viruses. The goals of such modeling are to increase our understanding of the conditions that lead to the persistence of a virus population in its reservoir species and its spread to other hosts, and to test hypotheses for effective methods to prevent and control viral diseases. However, such models can be hard to evaluate simply because it is difficult to perform controlled experiments in nature.

Learning from Accidental Infections

Our understanding of the dynamics of a viral infection in a large outbred human population is rudimentary. But some insight has been gained from a limited number of accidental "experiments." Two classic examples are provided by hepatitis B virus and poliovirus infections. During World War II, large doses of infectious hepatitis B virus were accidentally introduced into ~45,000 soldiers when they were injected with a contaminated yellow fever vaccine. Only 900 (2%) developed clinical hepatitis, and <36 developed severe disease. Similarly, in 1955, 120,000 school-aged children were vaccinated with an improperly inactivated poliovirus vaccine. About half were protected by preexisting antibodies to poliovirus as a result of inapparent infections. Of the remainder, ~10 to 25% were infected by the vaccine virus, as determined by the appearance of antibodies. More than 60 cases of paralytic poliomyelitis were documented among these infected children, but the remainder escaped disease. These two experiments tell us that even when large numbers of individuals are inoculated with an equivalent amount of a virulent virus, the outcomes can be quite variable, and not everyone succumbs to disease. The finding that infection was relatively rare even in these cases, where large amounts of virus particles were administered to numerous individuals, suggests that special circumstances or conditions are required for natural cross-species infections to emerge.

Expanding Viral Niches: Some Well-Documented Examples

Poliomyelitis: Unexpected Consequences of Modern Sanitation

Host populations change with time, and each change can have unpredictable effects on virus evolution. An example is poliomyelitis, a disease caused by poliovirus infection. The disease is ancient, postulated by some to be present >4,000 years ago (see Volume I, Chapter 1). For centuries, the host-virus relationship was stable, and infection was endemic in the human population. Poliomyelitis epidemics were not reported, but we imagine that occasional outbreaks of disease occurred in scattered areas. This state of affairs changed radically in the first half of the 20th century, when large annual outbreaks of poliomyelitis were seen in Europe, North America, and Australia (Fig. 11.5). Retrospective analysis established that these outbreaks were not correlated with any substantial change in the viral genome.

Emergence of epidemic poliomyelitis can be explained by a change in human lifestyle: unprecedented urbanization and improvement in sanitation. Poliomyelitis is caused by an enteric virus that is spread by oral-fecal contact. As a consequence, endemic disease was characteristic of life in rural communities, which generally had poor sanitation and small populations. Because the virus circulated freely, most children were infected at an early age and developed antibodies to at least one of the serotypes. Maternal antibodies, which protect

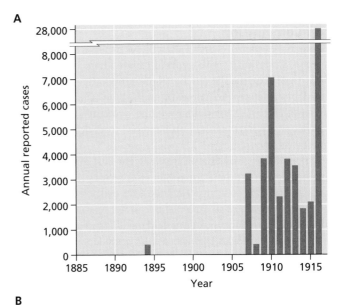

Figure 11.5 Poliovirus in the early 20th century. (A) The emergence of paralytic poliomyelitis in the United States, 1885 to 1915. From N. Nathanson, *ASM News* **63**:83–88, 1997, with permission. **(B)** Board of Health quarantine notice, San Francisco, CA, circa 1910.

newborns, were also prevalent, as most mothers had experienced a poliovirus infection at least once. A salient point is that most infected children do not develop paralysis, the most visible symptom of poliomyelitis. Paralysis is a more frequent result when older individuals are infected. Even the most virulent strains of poliovirus cause 100 to 200 subclinical infections for every case of poliomyelitis. These inapparent infections in children provided a form of natural vaccination. As childhood disease and congenital malformations were not uncommon in rural populations, the few individuals who developed poliomyelitis were not seen as out of the ordinary. No one noticed endemic poliovirus.

During the 19th and 20th centuries, industrialization and urbanization changed the pattern of poliovirus transmission. Improved sanitation broke the normal pattern and effectively stopped natural vaccination, thereby increasing the pools of susceptible children. As a result, children tended to encounter the virus for the first time at a later age, without the protection of maternal antibodies, and were therefore at far greater risk for developing paralytic disease. Consequently, epidemic poliovirus infections emerged time and time again in communities across the world.

Smallpox and Measles: Exploration and Colonization

Explosive epidemic spread may occur when a virus enters a naive population (**the evolving host-virus interaction**) (Fig. 11.2). The results can be frightening, often devastating, as infections appear to "come out of the blue." Charles Darwin was aware of this phenomenon, as he wrote in *The Voyage of the Beagle*: "Wherever the European has trod, death seems to pursue the aboriginal."

Smallpox reached Europe from the Far East in AD 710 and attained epidemic proportions in the 18th century as populations grew and became concentrated. The effects on society are hard to imagine today, but as an example, at least five reigning monarchs died of smallpox. Smallpox virus continued its spread around the world when European colonists and slave traders moved to the Americas and Australia. This viral infection certainly changed the balance of human populations in the New World. The first recorded outbreak of smallpox in the Americas occurred among African slaves on the island of Hispaniola in 1518, and the virus spread rapidly through the Caribbean islands. This toehold of smallpox in the New World enabled the conquest of the Aztecs by European colonists. In 1520, smallpox reached the American mainland from Cuba. Within 2 years, 3.5 million Aztecs were dead, far more than could be accounted for by the bullets and swords of Hernán Cortez's small band of conquistadors. Smallpox spread like wildfire in the native population, which was highly interactive and of sufficient density for efficient virus transmission. Infection reached as far as the Incas in Peru before Francisco Pizarro made his initial invasion in 1533.

As is true in most smallpox epidemics, some Aztecs and Incas survived, but those who did were then devastated by measles virus, probably brought in by Cortez's and Pizarro's men. Conquest was accomplished by a one-two virological punch rather than by military prowess. Slave traders (who were most likely immune to infection) were populating Brazil with their infected human cargo at approximately the same time, with the same horrible result.

The devastation of indigenous peoples by these viruses was recapitulated in the colonization of North America and continued into the 20th century as contaminated explorers infected isolated groups of Alaskan Inuit and native populations in New Guinea, Africa, South America, and Australia.

Notable Zoonoses

Hantavirus Pulmonary Syndrome: Changing Climate and Animal Populations

A small but alarming epidemic of a highly lethal infectious disease appeared in the Four Corners area of New Mexico in the United States in 1993. Individuals who were in excellent health developed flu-like symptoms that were followed quickly by a variety of pulmonary disorders, including massive accumulation of fluid in the lungs, and death. Rapid action by local health officials and a prompt response by the Centers for Disease Control and Prevention were instrumental in discovering that these patients had low-level, cross-reacting antibodies to previously identified hantaviruses. These members of the family *Bunyaviridae* had been associated with renal diseases in Europe and Asia and were well known to be associated with viral hemorrhagic fever during the Korean War. Hantaviruses commonly infect rodents and are endemic in these populations around the world. PCR technology was used to determine that the patients were infected with a new hantavirus. Subsequently, field biologists found this virus in a rodent called the deer mouse (*Peromyscus maniculatus*), which is common in New Mexico. The virus, which was given the name Sin Nombre virus (no-name virus), is an example of an emerging virus, endemic in rodents, that causes severe problems when it crosses the species barrier and infects humans. Sin Nombre virus has since caused a few additional isolated outbreaks in North America.

Humans became infected with Sin Nombre virus most likely because of a dramatic increase in the deer mouse population. A higher-than-normal rainfall resulted in a bumper crop of piñon nuts, a favorite food for deer mice and local humans. Mouse populations increased in response, and contacts with humans inevitably increased as well. Hantavirus infection is asymptomatic in mice, but virus particles are excreted in large quantities in urine and droppings, where they are quite stable. Human contact with contaminated blankets or dust from floors or food storage areas provided ample opportunities for infection. Hantavirus syndrome is rare because humans are not the natural host, and apparently are not efficient vehicles for virus spread.

Severe Acute and Middle East Respiratory Syndromes (SARS and MERS): Two New Zoonotic Coronavirus Infections

A new human viral disease called severe acute respiratory syndrome (SARS) first appeared in humans in Guangdong Province in China in the fall of 2002. A doctor who treated these patients traveled to Hong Kong on February 21, 2003, and checked into a hotel. He became ill and died in the hospital the very next day. During his stay in the hotel, the virus was transmitted to 10 other residents, who subsequently flew to Singapore, Vietnam, Canada, and the United States before symptoms were evident. A major viral epidemic was spread by air travel. This small number of infected people efficiently transmitted the new SARS coronavirus to other individuals around the world, such that ~8,000 people in 29 countries became infected in less than a year. The case-fatality ratio was almost 1 in 10, a chilling statistic that activated health organizations worldwide. The scientific community mobilized with unprecedented speed and cooperation, and the causative agent was identified within only a few months.

It is now generally accepted that bats serve as the natural reservoir for the SARS coronavirus. Whether the virus was transmitted directly from bats to humans or though an intermediate host is not yet known (Box 11.4). Amazingly, the epidemic never reached pandemic proportions, despite the existence of billions of susceptible hosts and widespread seeding of infected people around the world. After a frightening few months, SARS all but disappeared from the human population, although a few cases were reported subsequently.

Ten years later, reports from the Arabian Peninsula described the emergence of a new virus that caused severe pneumonia in infected humans, with an ~50% case-fatality ratio. Rapid sequencing of isolates identified the agent as another member of the family *Coronaviridae*, and the virus was given the name Middle East respiratory syndrome (MERS) coronavirus. As of this writing, the reservoir of the virus is believed to be dromedary camels. Although most cases of this disease have occurred in Saudi Arabia and countries nearby, some have been reported in different parts of Europe in people who had visited the Middle East. The total number of cases is small, but the virus has been seen to spread through close contact from infected individuals to others. This situation appears to be a contemporary example of an evolving host-virus interaction. (For more on such emerging viruses, check out the video interview with Dr. Ian Lipkin: http://bit.ly/Virology_Lipkin).

Acquired Immunodeficiency Syndrome (AIDS): Pandemic from a Zoonotic Infection

The emergence of human immunodeficiency virus type 1 can be traced to transmission from a chimpanzee to a human in West Central Africa (Box 1.2). There is strong evidence that humans are exposed to many zoonotic infections by the bushmeat trade in West Africa, which involves killing and consumption of wild animals, including chimpanzees, gorillas, other primates, and rodents. It is easy to imagine how

---

BOX 11.4

DISCUSSION
A SARS-like coronavirus from bats can infect human cells

The SARS pandemic of 2002–2003 is believed to have been caused by a bat coronavirus that first infected a civet and was then passed on to humans. The isolation of a new SARS-like coronavirus from bats suggests that the SARS coronavirus could have infected humans directly from bats.

A single colony of horseshoe bats (*Rhinolophus sinicus*) in Kunming, Yunnan Province, China, was sampled for coronavirus sequences over a 1-year period. Of a total of 117 anal swabs or fecal samples collected, 27 (23%) were positive. Using PCR amplification, seven different SARS-like coronavirus sequences were identified, including two new ones. The complete genome sequences were determined for the new coronaviruses, which showed a higher nucleotide sequence identity (95%) with the SARS coronavirus than had been previously observed among bat viruses.

One of the new viruses was recovered by infecting monkey cell cultures with one of the positive bat samples. The recovered virus was able to infect human cells and for entry utilize human angiotensin-converting enzyme 2 (ACE2), which is the receptor for SARS coronavirus. Furthermore, infectivity of the recovered virus could be neutralized with sera collected from seven SARS patients.

The spike glycoprotein of SARS-like coronaviruses previously isolated from bats does not recognize the ACE2 receptor, and thus these viruses are unable to infect human cells. Because the SARS-like coronaviruses isolated from palm civets during the 2002–2003 outbreak have amino acid changes in the viral spike glycoprotein that improve its interaction with ACE2, these animals were believed to be an intermediate host for adaptation of the SARS virus to humans. However, the properties of the new bat SARS-like coronavirus suggest that humans might have been infected directly from bats.

This finding has important implications for public health: if SARS-like coronaviruses that can infect human cells are currently circulating in bats, they have the potential to cause another outbreak of disease. The investigators believe that the diversity of bat coronaviruses is probably greater than previously suspected. They speculate that further surveillance may reveal a broad diversity of bat SARS-like coronaviruses that are able to use ACE2, some of which may have closer homology to the SARS virus than even their latest isolate.

Ge XY, Li JL, Yang XL, Chmura AA, Zhu G, Epstein JH, Mazet JK, Hu B, Zhang W, Peng C, Zhang YJ, Luo CM, Tan B, Wang N, Zhu Y, Crameri G, Zhang SY, Wang LF, Daszak P, Shi ZL. 2013. Isolation and characterization of a bat SARS-like coronavirus that uses the ACE2 receptor. *Nature* **28**:535–538.

Deciphering the source of human SARS coronavirus (SARS-CoV).

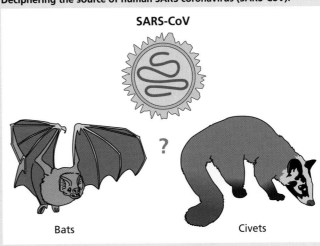

SARS-CoV

Bats   ?   Civets

butchering an infected chimp could facilitate the transfer of a retrovirus particle from an animal's blood. It is sobering to consider that the progenitor of human immunodeficiency virus type 1 was on the path to extinction, because the population of wild chimpanzees had dropped to about 150,000 animals living in isolated troops. The new human host exceeds 7 billion individuals. A second, related human immunodeficiency virus subtype, called type 2, was acquired by zoonotic infection from another primate, the sooty mangabey.

Host Range Can Be Expanded by Mutation, Recombination, or Reassortment

Canine Parvoviruses: Cat-to-Dog Host Range Change by Two Mutations

Canine parvovirus was identified in several countries in 1978 as the cause of a new enteric and myocardial disease in dogs. Canine parvovirus apparently evolved from the feline panleukopenia virus that infects cats, mink, and raccoons, but not dogs. However, the new canine virus did not reproduce in cats. Because canine parvovirus appeared less than 40 years ago, it has been possible to analyze dog and cat tissue collected in Europe in the early 1970s to search for the progenitor canine parvovirus. The ancestor of canine parvovirus began infecting dogs in Europe during the early 1970s, and within 8 years, it had spread to several other continents. The stability

of the new virus and its efficient fecal-oral transmission were important factors in its emergence.

It is now clear that only two amino acid substitutions in the VP2 capsid protein were necessary to change the tropism from cats to dogs (Fig. 11.6). These critical amino acids are located on a raised region of the capsid that binds the host transferrin receptor, the protein used to establish infection. Feline panleukopenia virus particles bind only to the feline transferrin receptor, but the substitutions in VP2 allow the canine parvovirus particles to bind the canine transferrin receptors. The emergence of the canine parvovirus group provided an extraordinary opportunity to study virus-host adaptation and host-range shifts in the field.

Influenza Epidemics and Pandemics: Escaping the Immune Response by Reassortment

Influenza serves as the paradigm for the situation in which continued evolution of the virus in several host species is essential for its maintenance. The life cycle of influenza virus, while comparatively well understood at the molecular level, is remarkable for its complexity in nature (Box 10.3; Fig. 11.7). Sequencing data indicate that the H1N1 virus, which claimed >25 million human lives in the pandemic of 1918 (Table 11.4), is likely to be the ancestor of all current human influenza viruses, as well as some that are circulating in the world's swine populations. However, new influenza viruses constantly

Figure 11.6 The transferrin receptor mediates canine and feline parvovirus host range. The transferrin receptors for feline and canine parvoviruses have a large extracellular domain (ectodomain) that is a homodimer of a single protein (see Box 10.9). The binding of the canine parvovirus virion to the ectodomain is determined by combinations of amino acid residues on the surface of the capsid. **(A)** Cryo-electron microscopy was used to determine the structure of the purified ectodomain of the feline transferrin receptor bound to the canine parvovirus capsid. Only a small number of transferrin receptors bind to each capsid, and in the model, only one is shown. **(B)** The binding site (footprint) of the transferrin receptor, in a representation of the surface-exposed amino acids of the capsid, is colored green. One of the 60 asymmetric units of the icosahedral capsid is outlined. Residues that are known to affect binding of the canine transferrin receptor or the host range are indicated in yellow. Figures prepared by Susan Hafenstein and Colin Parrish, Cornell University. See S. Hafenstein et al., *Proc Natl Acad Sci U S A* **104:** 6585–6589, 2007.

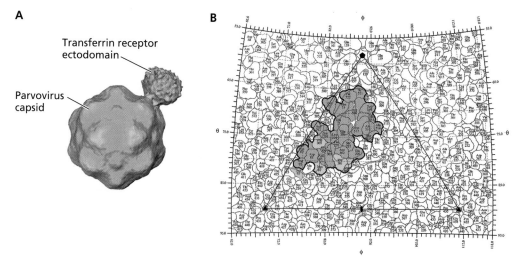

A

Transferrin receptor ectodomain

Parvovirus capsid

B

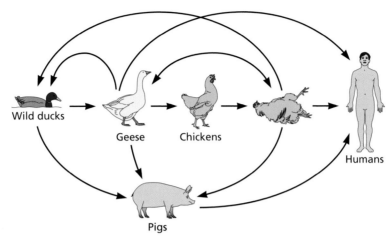

Figure 11.7 Emergence and transmission of H5N1 influenza virus. H5N1 influenza virus has its origins in wild waterfowl, where it is relatively nonpathogenic. Infection is thought to have spread to domestic ducks and chickens, and the virus evolved to be highly pathogenic in chickens. Transmitted back to ducks and geese, the viral genome underwent reassortment with the genomes of other influenza viruses of aquatic birds, resulting in a virus that could be transmitted directly to domestic chickens, humans, and pigs. Transmission was facilitated by mutations or changes in the viral PB2, HA, NA, and NS genes, which made the virus more pathogenic to domestic and wild waterfowl and humans. Spread to humans without need for an intervening "mixing host" is a particularly worrisome feature of this virus.

emerge from migratory populations of aquatic birds to infect humans, pigs, horses, domestic poultry, and aquatic mammals. In birds, influenza virus reproduces in the gastrointestinal tract and particles are excreted in large quantities, a most efficient virus distribution system. The widespread dispersal of virus particles in water, the facile changing of hosts, and the ease of genetic reassortment form a powerful engine for creation of new pathogenic strains.

Outbreaks of swine and avian influenza periodically devastate agricultural operations that produce these animals for food. Despite large-scale immunization programs, virulent strains of swine influenza virus continue to emerge in these animal hosts. The lethal consequences of direct transfer of a virulent avian H5N1 virus to humans were first documented in 1997. The World Health Organization recognizes the avian influenza virus subtypes H5, H7, and H9 as potential pandemic strains, because humans have no immunity to them. Attempts to predict critical changes that can affect the transmissibility of some of these subtypes to humans, using laboratory animal models, have generated considerable controversy worldwide (Box 11.5).

One surprising finding is that, in contrast to the genomes of human and other nonavian influenza viruses, the genome of the avian influenza virus has not changed much in more than 60 years. Although the avian viral genomes exhibit mutation and reassortment rates as high as those of human and swine influenza viruses, only sequences with neutral mutations are selected and maintained in the bird population.

While virulent mutants do arise occasionally, birds infected with avian influenza viruses generally experience no overt pathogenesis. These properties indicate that influenza virus is in evolutionary stasis in birds. The avian hosts provide the stable reservoir for influenza virus gene sequences that emerge as recombinants capable of transspecies infection.

New Technologies Uncover Hitherto Unrecognized Viruses

In addition to being acquired from animals, emerging viral infections may simply be caused by previously unknown agents. The rate of virus discovery has risen with

Table 11.4 The 1918–1919 influenza pandemic: one of history's most deadly events[a]

Event	Estimated no. of deaths (millions)
Influenza pandemics (1918–1919)	20–50
Black Death (1348–1350)	20–25
AIDS pandemic (through 2013)	35
World War II (1937–1945)	
Military	15
Civilian	45
World War I (1914–1918)	
Military	10
Civilian	7

[a]Data from *The New York Times*, August 21, 1998, and from the World Health Organization.

BOX **11.5**

DISCUSSION
Avian influenza viruses: scientific and societal implications of transmissibility experiments using animal models

Highly pathogenic variants of the H5N1 avian influenza virus, known commonly as "bird flu," moved from Asia to India to Europe, and also to Africa, in the space of only 10 years. Variants of H5N1 continue to evolve, and are moving around the world in wild birds, notably waterfowl, which are considered the natural reservoirs and sources of infection for other species. Highly pathogenic avian influenza viruses also become established in domestic fowl, which are kept at very high densities in farms and markets. Efforts to control these infections have led to the culling of millions of domestic birds in Asia and Europe. Humans have helped to spread the virus by transporting infected "exotic" birds by car, truck, and railroad. Although the virus has not yet reached the Americas, it is sobering to note that, after narcotics, live birds for the pet trade are the next most commonly smuggled items brought into the United States.

Transmission of this virus to humans is rare, requiring close contact with bodily fluids of infected birds. As of July 2013, the World Health Organization reported a total of only 630 confirmed human cases of H5N1 infection since 2003, with a 60% case-fatality ratio. Several species of cats can be infected with and transmit the H5N1 virus to other cats. Transmission of virus from humans to humans or from cats to humans has not been demonstrated.

In 2011, two groups of scientists, one in the Netherlands and one in the United States, reported results from experiments that used a combination of genetic engineering and animal models to show that a very few mutations, 4 in the hemagglutinin (HA) gene and 1 in the PB2 gene, are sufficient to create a variant of H5N1 that could spread among ferrets in the laboratory, without direct contact and loss of virulence. As ferrets are susceptible to many of the same viruses as humans, these so-called gain-of-function studies sparked international concern and heated controversy about the potential for the creation and intentional or unintentional release of a pandemic agent, or whether transmission in ferrets is even a valid model for human transmission.

Some scientists called for censoring experimental details in the publications from the two groups. The public concerns eventually prompted 39 leading influenza researchers, including the 2 who led the controversial studies, to impose a voluntary moratorium

■ Countries with humans, poultry and wild birds killed by H5N1.

■ Countries with poultry or wild birds killed by H5N1 and reported human cases.

□ Countries with poultry or wild birds killed by H5N1.

Map of the global spread of H5N1 influenza virus.

on research designed to increase the transmissibility of H5N1 viruses in mammals until the development of government policies on biosafety and biosecurity for such research. Following two separate reviews by the U.S. National Science Advisory Board for Biosecurity, which advises the Department of Health and Human Services, it was decided that the full versions of both studies should be published. Finally, in March 2012, the United States government issued a Policy for Institutional Oversight of Life Sciences Dual Use Research of Concern, defined as "research that is intended for benefit, but might easily be misapplied to do harm." International bodies, including the World Health Organization, are also expected to propose international guidelines for such research.

In an unprecedented action, 22 influenza researchers from the United States,

United Kingdom, and China, including the 2 leaders of the H5N1 gain-of-function experiments, published a letter in August 2013 announcing, and justifying, their intentions to conduct similar experiments (under appropriate biosafety conditions) with a second avian influenza virus, H7N9, which began infecting humans in eastern China earlier that year. The H7N9 virus can infect ferrets and can be transmitted between them by close contact. It can also infect pigs, a common intermediate host for the evolution of viruses that can then circulate in humans, but is not transmitted between pigs. This virus is highly pathogenic in humans, but in contrast to H5N1, infection of domestic fowl is asymptomatic, so that virus spread is more difficult to track.

Supporters of research on influenza virus transmissibility note that there are both scientific and practical needs to increase our understanding of the parameters of immu-nogenicity, virulence, and transmissibility of these viruses. They suggest further that the results could allow public health workers to monitor wild viruses for critical mutations. Health agencies could then advise manufac-turers of drugs and vaccines to increase pro-duction appropriately, or they may impose stricter public health measures to prevent transmission. Detractors of such research note that it is unclear if laboratory experiments on transmissibility can ever recapitulate events that occur in nature, and therefore such exper-iments are unlikely to yield results of relevance to natural human pandemics. How the labora-tory results relate to evolution in nature is not clear, as the fitness of a virus with the identi-fied mutations, singularly or in combinations,

is unknown. Other scientists, and some groups in the public sector, contend that the poten-tial risks of such experiments outweigh any possible benefits. Given the yearly death toll of seasonal influenza and the potential cata-strophic consequences of a future pandemic, the current consensus is that such work should continue, albeit with appropriate oversight and safeguards.

Fouchier RA, Kawaoka Y, Cardona C, Compans RW, García-Sastre A, Govorkova EA, Guan Y, Herfst S, Orenstein WA, Peiris JS, Perez DR, Richt JA, Russell C, Schultz-Cherry SL, Smith DJ, Steel J, Tompkins SM, Topham DJ, Treanor JJ, Tripp RA, Webby RJ, Webster RG. 2013. Gain-of-function experiments on H7N9. *Science* **341**:612–613.

Herfst S, Schrauwen EJ, Linster M, Chutinimitkul S, de Wit E, Munster VJ, Sorrell EM, Bestebroer TM, Burke DF, Smith DJ, Rimmelzwaan GF,

Osterhaus AD, Fouchier RA. 2012. Airborne transmission of influenza A/H5N1 virus between ferrets. *Science* **336**:1534–1541.

Imai M, Watanabe T, Hatta M, Das SC, Ozawa M, Shinya K, Zhong G, Hanson A, Katsura H, Watanabe S, Li C, Kawakami E, Yamada S, Kiso M, Suzuki Y, Maher EA, Neumann G, Kawaoka Y. 2012. Experi-mental adaptation of an influenza H5 HA confers respiratory droplet transmission to a reassortant H5 HA/H1N1 virus in ferrets. *Nature* **486**:420–428.

Patterson AP, Tabak LA, Fauci AS, Collins FS, Howard S. 2013. Research funding. A framework for decisions about research with HPAI H5N1 viruses. *Science* **339**:1036–1037.

Russell CA, Fonville JM, Brown AE, Burke DF, Smith DL, James SL, Herfst S, van Boheemen S, Linster M, Schrauwen EJ, Katzelnick L, Mosterín A, Kuiken T, Maher E, Neumann G, Osterhaus AD, Kawaoka Y, Fouchier RA, Smith DJ. 2012. The potential for respiratory droplet-transmissible A/H5N1 infuenza virus to evolve in a mammalian host. *Science* **336**:1541–1547.

the development of new technologies, as illustrated for those that infect humans (Fig. 11.8). It is now possible to detect and characterize unknown viruses with comparative ease.

Hepatitis Viruses in the Human Blood Supply

One of the first examples of the power of the new technologies was the recognition of hepatitis C virus. With the development of specific diagnostic tests for hepatitis A and B viruses in the 1970s, it became clear that most cases of hepatitis that occur after blood transfusion are caused by other agents. Recombinant DNA technology was used in the late 1980s to identify one of the non-A, non-B hepati-tis (NANBH) agents as a new virus, named hepatitis C virus (Box 11.6). The availability of the hepatitis C virus genome sequence made possible the development of diagnostic reagents that effectively eliminated the virus from the U.S. blood supply, substantially reducing the incidence of trans-fusion-derived NANBH.

Hepatitis C virus was not the only previously hidden virus in the human blood supply; the same recombinant DNA technology permitted the discovery of several other virus-es, including TT virus, a ubiquitous human circovirus of no known consequence.

A Revolution in Virus Discovery

Newly discovered nucleic acid sequences can now be associ-ated with diseases and characterized in the absence of stan-dard virological techniques, in time frames measured in days rather than month or years. The etiological agent of SARS was identified after PCR amplification using a "virochip" that contained oligonucleotides from each of the known viral genomes. However, because of high sensitivity and the poten-tial for contamination by adventitious viruses, special caution

Figure 11.8 Major developments in methods for virus dis-covery drive the identification of viruses that infect humans. **(A)** Discovery by species of virus. Vertical lines indicate significant upward breakpoints. **(B)** Discovery by virus family. **(C)** Technologi-cal advances through the 20th century to the present. Reprinted from M. Woolhouse et al., *Philos Trans R Soc Lond B Biol Sci* **367**:2864–2871, 2012, with permission.

A

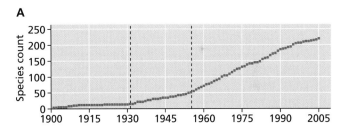

B

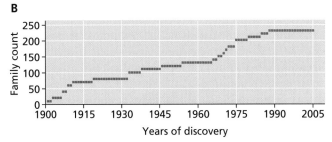

Years of discovery

C

Year	Technology
1890s	Filtration
1929	Complement fixation
1948	Tissue culture
1970s	Monoclonal antibodies
1985	Polymerase chain reaction (PCR)
2000s	High throughput sequencing

BOX 11.6

BACKGROUND
Discovery of hepatitis C virus, a triumph of persistence

The virus called non-A, non-B hepatitis (NANBH) virus was known to be contracted via blood transfusion. As it was refractory to laboratory culture, the identity of this virus remained elusive until Chiron Corporation, a California biotechnology company, isolated a DNA copy of a fragment of what was later identified as the hepatitis C virus genome from a chimpanzee with NANBH. Following nearly 6 years of intensive investigations, this remarkable feat was accomplished by screening a library of ~1,000,000 randomly primed complementary DNA clones made from RNA in the plasma of the infected chimp. The researchers tested the clones for their ability to produce proteins that were recognized by serum from a patient with chronic NANBH. A single small cDNA clone was positive in this assay and was found to be derived from an RNA molecule comprising a (+) strand genome of ~10,000 nucleotides with high homology to those of known flaviviruses.

Choo QL, Kuo G, Weiner AJ, Overby LR, Bradley DW, Houghton M. 1989. Isolation of a cDNA clone derived from a blood-borne non-A, non-B viral hepatitis genome. *Science* **244:**359–362.

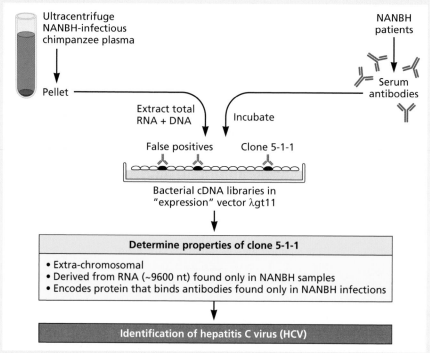

Determine properties of clone 5-1-1
- Extra-chromosomal
- Derived from RNA (~9600 nt) found only in NANBH samples
- Encodes protein that binds antibodies found only in NANBH infections

Identification of hepatitis C virus (HCV)

Schematic for molecular isolation and identification of hepatitis C virus. Adapted from Q.-L. Choo et al., *Science* **244:**359–362, 1989, with permission.

---

is necessary when using any PCR-based or hybridization method. Without proper controls, these techniques have the potential to associate a particular virus with a disease incorrectly, confounding the deduction of etiology. One example was the misidentification of what later proved to be a contaminating recombinant mouse retrovirus (XMRV) as a "novel" human virus associated with prostate cancer and chronic fatigue syndrome. Similarly, a variety of studies have been published in which PCR identified human herpesviruses in the brains of patients who had died of Alzheimer's disease, implying a causal link between the disease and the virus. The latter conclusions are called into question because the same viruses can be found in the tissues of similarly aged patients who died of other causes. These examples illustrate the fact that availability of the new technologies does not circumvent the need to satisfy the modified Koch's postulates (Volume I, Box 1.4).

For virus discovery, virochip technology was superseded rapidly by the use of methods that apply high-throughput, next-generation sequencing technology and associated computational tools. For example, a new virus of major agricultural importance was discovered in 2011 by the use of metagenomic analysis of sequences in tissues obtained from diseased animals. The agent was named Schmallenburg virus after the German town from which the first positive tissue samples were obtained. The virus is transmitted by biting midges and has spread rapidly to farms across Europe and the United Kingdom. Infection of cattle, goats, or sheep is associated with fairly mild disease symptoms, but the fetuses of these animals are stillborn or malformed. The Schmallenburg virus has been identified as a member of the family *Bunyaviridae*. Following Koch's postulates, it has been shown that animals injected with purified virus particles do, indeed, develop the disease. Schmallenburg virus dissemination is now being monitored by PCR.

Perceptions and Possibilities

While emerging virus infections are well known to virologists, in recent years they have become the subject of widespread public interest and concern. Less than 40 years ago, many people were ready to close the book on infectious diseases. The public perception was that wonder drugs and vaccines had microbes fully under control. This optimistic view has now changed dramatically. Announcements of new and destructive viruses and bacteria appear with increasing frequency. The reality of the human immunodeficiency virus pandemic and its effects at every level of society have attracted worldwide attention, while exotic viruses like Ebola virus capture front-page headlines. Movies and books bring viruses to the public consciousness more effectively and dramatically than ever before. After the events of September 11, 2001, concern that terrorists might use infectious agents was widespread (Box 11.7).

Virus Names Can Be Misleading

Much information is implied (inappropriately in some cases) when naming a virus by the host from which it was isolated. By using the name **human** immunodeficiency virus for the virus that causes AIDS, we give short shrift to its nonhuman origins. Canine parvovirus is clearly a feline virus that recently switched hosts. Similarly, canine distemper virus is not confined to dogs, but can cause disease in lions, seals, and dolphins. Well-known viruses can cause new diseases when they change hosts. Much is implied, and more is ignored, about the host-virus interaction when the virus is given a host-specific name.

All Viruses Are Important

It is not uncommon to consider disease-causing viruses important while deeming nonpathogenic viruses uninteresting and irrelevant. But as we have seen, a virus that is stable in one host may have devastating effects when it enters a different species. Conversely, a virus may be pathogenic in one species but not in another. Some viruses are even beneficial to their hosts or the environment (Volume I, Chapter 1). Misconceptions often arise from human-centered thinking, a belief that viruses causing human diseases are more important than those that infect mammals, birds, fish, or other hosts, forgetting that all life forms are interconnected. By focusing solely on viruses that can infect humans, we are blinded to the intertwined networks of interactions that comprise host-virus relationships.

What Next?

Can We Predict the Next Viral Pandemic?

It is now clear that some of the most serious threats to the human population come not come from the popularized, highly lethal filoviruses (e.g., Ebola virus), the hemorrhagic disease viruses (e.g., Lassa virus), or even some undiscovered virus lurking in the wild. Rather, the most dangerous viruses are likely to be the well-adapted, multihost, evolving viruses already in the human population. Influenza virus fits this description perfectly. Its yearly visits show no signs of diminishing; genes promoting pandemic spread and virulence are already circulating in the virus population, and the world is ever more prone to its dissemination. A pandemic of influenza on the scale of, or even greater than, that of the 1918–1919 outbreak is thought by many to be the next emerging disease most likely to affect humans (Table 11.4).

Traditional monitoring tools are used to detect the onset and gauge the severity of the yearly influenza outbreaks in various countries. Such monitoring depends, in large part, on networks of physicians who report cases of patients with flu-like symptoms. Traditional monitoring might also detect the emergence of some new infection, especially if suspicious symptoms appear at an unusual time of the year. Other methods attempt to predict the onset of influenza outbreaks from Internet queries and communication on social media. For example, Google-flu relies on data-mining records of flu-related search terms entered into its search engine, combined with computer modeling. In the United States, Google-flu estimates have matched those obtained from traditional methods by the Centers for Disease Control and Prevention, but Google can deliver the information several days faster. Another computational approach, intended to

---

BOX 11.7

DISCUSSION
Viral infections as agents of war and terror

Infectious agents have a documented capacity to cause harm, and can cause epidemics as well as pandemics. Well-known deadly viruses range from agents of universal scourges, such as the recently eradicated smallpox virus and the influenza viruses, to the less widely distributed, but no less deadly, hemorrhagic fever viruses. Any viral infection that can kill, maim, or debilitate humans, their crops, or their domesticated animals has the potential to be used as a biological weapon. Obviously, a biological attack need not cause mass destruction to be an instrument of terror, as was demonstrated by the far-reaching effects of the introduction of bacteria causing anthrax into the United States mail system. Society has only a limited set of responses to frightening outbreaks: vaccination, quarantine, and antimicrobial drugs. One example is the unintentional 1947 outbreak of smallpox in New York City, which originated from a single businessman who had acquired the disease in his travels. He died after infecting 12 others; to stop the epidemic, >6 million people were vaccinated within a month.

As with such natural outbreaks, potential bioterrorism threats pose serious problems with few clear solutions. Some argue that the resources devoted by governments to counterterrorism and research on category A pathogens would be better applied to research on public health or naturally occurring common diseases. Others are concerned that publication of research data on such pathogens could aid terrorists, and maintain that measures to control the publication of such information must be considered (Box 11.5). Practically speaking, public health officials view bioterrorism as a low-probability but high-impact event, much like major hurricanes or tsunamis. When such events occur, they are devastating. However, the hallmarks of these calamities are that they cannot be predicted with accuracy or easily prevented. Societies can only prepare by ensuring that remedial actions can be taken as quickly as possible.

inform influenza vaccine production, has been to incorporate genetic data on beneficial and deleterious mutations in the antibody-binding domains on the hemagglutinin proteins of circulating strains from previous years into a fitness-based model that will predict the frequency of descendent strains in the following year. A retrospective analysis indicates that the model can successfully predict the year-to-year evolution of individual influenza clades. Although not yet perfect, these nontraditional approaches represent a promising complementary strategy for the future, at least for the short-term prediction of potential epidemics.

Emerging Viral Infections Illuminate Immediate Problems and Issues

The AIDS pandemic and our experience with SARS illustrate the ease with which a new virus can enter the human population. The secondary infections that accompany AIDS have underscored the fragility of both the world's health care systems and the infrastructure of developing countries. Outbreaks of new exotic viruses, zoonoses, or well-known viruses that have invaded new geographic niches highlight the need for standardized methods of diagnosis, epidemiology, treatment, and control.

Humans Constantly Provide New Venues for Infection

Past experiences with poliovirus, measles virus, and smallpox virus demonstrate that viruses can cause illness and death on a catastrophic scale following a change in human behavior. Current technological advances and changing environmental and social behaviors continue to influence the spread of viruses (Table 11.2). Most of the contemporary opportunities for interaction between humans and viruses did not exist 50 years ago. Furthermore, the highly connected human population is larger than it has ever been, and is still growing (Fig. 11.9). As a consequence, humans are interacting among themselves and with the environment on a scale unprecedented in history.

Many human activities have major effects upon the transmission of viruses by vectors, such as insects and rodents (Table 11.2). Population movements, the transport of livestock and birds, the construction and use of irrigation systems, and deforestation provide not only new contacts with mosquito and tick vectors, but also mechanisms for transport of infected hosts to new geographic areas.

Humans can also provide new habitats for viruses, as demonstrated with used tires. Several species of tropical mosquitoes (e.g., *Aedes* species) prefer to breed in small pockets of water that accumulate in tree trunks and flowers in the tropics. The used tire has provided a perfect mimic of this breeding ground, and as a consequence, the millions of used tires (almost all carrying a little puddle of water inside) accumulating around the world provide a new habitat for mosquitoes and their viruses. Furthermore, used tires are shipped all around the world for recycling, effectively transmitting mosquito larvae along with them. Insect vectors for viruses may be given the chance to establish new ranges thanks to such shipments.

Another contemporary example of the consequences of humans moving viruses to new hosts comes from transport of livestock. African swine fever virus, a member of the family *Iridoviridae*, causes a serious viral disease that is threatening the swine industries of both developing and industrialized countries. The African swine fever virus was spread from Africa to Portugal in 1957, to Spain in 1960, and to the Caribbean and South America in the 1960s and 1970s, via long-distance transport of livestock and their resident infected arthropods.

The construction of dams and irrigation systems can influence host-virus interactions through creation of vast areas of standing water. The 1987 outbreak of Rift Valley fever along

Figure 11.9 World population growth over the last 2 millennia. (A) The world population grew from 1.6 billion to 6.1 billion in the 20th century, and is expected to increase to 9.2 billion by 2050, with growth almost entirely in less-developed regions. Graph from the United Nations Population Division, used with permission. **(B)** A connected world: one day in global air traffic. Each dot represents one plane.

A

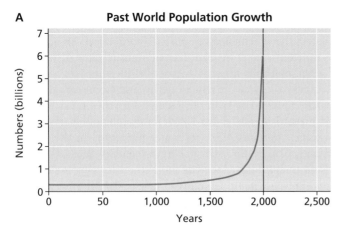

Past World Population Growth

B

the Senegal River was associated with the new Diama Dam, which provided conditions ideal for mosquito propagation. Not only do water impoundments affect insects, but they also alter the population and migration patterns of waterfowl and other animals, including the viruses they carry, bringing together previously separated viruses and potential new hosts.

In industrialized countries, the increasing need for day care centers has led to new opportunities for viral transmission. In the United States, many millions of children are in day care centers for several hours a day, and the vast majority are under 3 years of age. As most parents can testify, respiratory and enteric infections are common, and these infections spread easily among other children, day care workers, and the family at home.

Finally, among the most important human activities likely to affect the emergence of viral disease are those that are causing climate change. Global warming is already having an impact on all living things; viruses are no exception. Warming temperatures and increased rainfall in certain areas have led to an upsurge in the incidence of insect vector-borne infections; the spread of dengue virus from the equatorial areas to which it was previously confined is one clear example. Indirect effects of climate change such as flooding, disruption of human and wildlife populations, reduced food supply, and economic distress can all increase opportunities for new virus-host relationships to be established. As global warming continues, new reports of emerging viruses can be expected.

Preventing Emerging Virus Infections

The modernization of society and the expanding human population have facilitated the spread of infection, selection of virus variants, and virus emergence. We cannot turn back the clock, but experience and acquired knowledge can provide some guidance for ameliorating actions in the future. In some cases, viral emergence can be blocked quite effectively, as illustrated for infections of humans by the highly pathogenic influenza strain H7N9—once wild bird markets were identified as the major route of infection (Fig. 11.10). Knowledge that camels are a likely source of the equally pathogenic MERS coronavirus should lead to methods for preventing future human infections.

Modern diagnostic techniques have made it possible, although not yet entirely practical, to estimate the total viral diversity in any, or all, animal species. For example, one study of almost 2,000 samples collected from a particular species of bat, the Indian flying fox (*Pteropus giganteus*), showed that these animals collectively harbor 58 different viruses from 7 known families, the majority of which had not been identified previously. Extrapolation from these results to all mammalian species suggests that at least ca. 320,000 different viruses are waiting to be discovered. Although that number seems daunting, it would be possible to screen viruses identified in common or suspected reservoir species for their

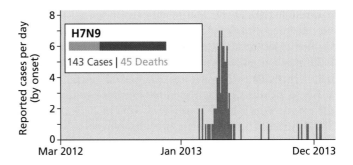

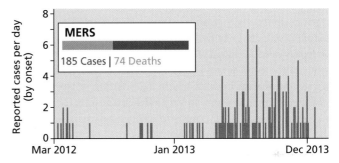

Figure 11.10 Knowledge of the reservoir species affects the outcome of infection of humans by emerging viruses. Adapted from Andrew Rambaut/WHO, with permission.

potential to bind to receptors on human cells (Box 11.4). Diagnostic reagents for such viruses could then be applied in disease cases of unknown etiology. Identification of the particular virus and animal reservoir could help to "nip" nascent zoonoses "in the bud" before they become more widespread public health problems.

Because of the potential for rapid spread of viruses via air travel and urban development, a system of global surveillance and early warning is required to alert primary care physicians and health care workers. Such efforts have begun, as illustrated in the recent SARS epidemic and MERS outbreaks. With modern technology, it is conceivable that all viral pathogens circulating in humans could be monitored. When a new (or old) viral disease is suspected, the agent could be identified and characterized and the information could be shared widely. These responses would be facilitated with the development of methods for rapid diagnosis in the field and ready access to early-warning databases for primary health care workers.

Perspectives

It has been estimated that upwards of 60% of human infectious diseases originate from animal reservoirs, and many of these zoonoses are caused by viruses. The relationships between viruses and their hosts are in constant flux, and numerous factors, most related to the modern human population explosion, have led to an increase in the adoption of new ecological niches or geographic zones by emerging viruses.

Given the ever-changing viral populations and drastic modifications of the earth's ecosystem, we are hard-pressed to predict the future. Mathematical modeling, powerful new diagnostic tools, and increased efforts at surveillance in various governmental and international agencies, particularly of animal species known to present the greatest risk, should help to provide early warnings of potential emergence. Our ability to identify viral pathogens has increased enormously in the past decade. While it took a few weeks in 2003 to identify SARS coronavirus by nucleic hybridization on a "virochip," only a few days were needed in 2012 to identify the MERS coronavirus by next-generation sequencing and bioinformatic tools. A deeper understanding of the diversity of viruses in various species will point to areas or situations in which particular vigilance may be warranted for possible cross-species infections. In addition, computer simulations are improving our ability to track the potential spread of emerging viruses, a great advantage in cases where the isolation or treatment of infected individuals can prevent further transmission.

Open avenues of communication between scientists, health care workers, and veterinarians in all parts of the globe can help to minimize the spread of infections and enable the development of strategies to cope with the consequences. Experience tells us that future incidences of infection by newly emerging viruses are inevitable.

References

Books

Dieckmann U, Metz J, Sabelis M, Sigmund K. 2002. *Adaptive Dynamics of Infectious Diseases.* Cambridge University Press, Cambridge, United Kingdom.

Koprowski H, Oldstone MBA (ed). 1996. *Microbe Hunters, Then and Now.* Medi-Ed Press, Bloomington, IL.

Krauss H, Weber A, Appel M, Enders B, Isenberg HD, Schiefer HG, Slenczka W, von Graevenitz A, Zahner H. 2003. *Zoonoses: Infectious Diseases Transmissible from Animals to Humans,* 3rd ed. ASM Press, Washington, DC.

Scheld WM, Armstrong D, Hughes JM (ed). 1997. *Emerging Infections 1.* ASM Press, Washington, DC.

Scheld WM, Craig WA, Hughes JM (ed). 1998. *Emerging Infections 2.* ASM Press, Washington, DC.

Reviews

Emerging Viruses

American Association for the Advancement of Science. 2012. Special edition: H5N1. *Science* **336:**1521–1547.

Brinton MA. 2002. The molecular biology of West Nile virus: a new invader of the Western hemisphere. *Annu Rev Microbiol* **56:**371–402.

Brody H. 2011. Influenza. *Nature* **480:**S1. doi:10.1038/480S1a.

Eaton BT, Broder CC, Middleton D, Wang LF. 2006. Hendra and Nipah viruses: different and dangerous. *Nat Rev Microbiol* **4:**23–35.

Holmes EC. 2013. What can we predict about viral evolution and emergence? *Curr Opin Virol* **3:**180–184.

Hui EK. 2006. Reasons for the increase in emerging and re-emerging viral infectious diseases. *Microbes Infect* **8:**905–916.

Mackenzie JS, Jeggo M. 2013. Reservoirs and vectors of emerging viruses. *Curr Opin Virol* **3:**170–179.

Smith I, Wang LF. 2013. Bats and their virome: an important source of emerging viruses capable of infecting humans. *Curr Opin Virol* **3:**84–91.

Taubenberger JK, Morens DM. 2006. 1918 influenza: the mother of all pandemics. *Emerg Infect Dis* **12:**15–22.

Weiss RA. 1998. Transgenic pigs and virus adaptation. *Nature* **391:**327–328.

Woolhouse M, Scott F, Hudson Z, Howey R, Chase-Topping M. 2012. Human viruses: discovery and emergence. *Philos Trans R Soc Lond B Biol Sci* **367:**2864–2871.

Theory of Host-Parasite Interactions

Boots M, Mealor M. 2007. Local interactions select for lower pathogen infectivity. *Science* **315:**1284–1286.

Levin S, Pimental D. 1981. Selection of intermediate rates of increase in parasite-host systems. *Am Nat* **117:**308–315.

Matthews L, Woolhouse M. 2005. New approaches to quantifying the spread of infection. *Nat Rev Microbiol* **3:**529–536.

Rambaut A, Pybus OG, Nelson MI, Viboud C, Taubenberger JK, Holmes EC. 2008. The genomic and epidemiological dynamics of human influenza A virus. *Nature* **453:**615–619.

Tuckwell HC, Toubiana L. 2007. Dynamical modeling of viral spread in spatially distributed populations: stochastic origins of oscillations and density dependence. *Biosystems* **90:**546–559.

Selected Papers

Anderson NG, Gerin JL, Anderson NL. 2003. Global screening for human viral pathogens. *Emerg Infect Dis* **9:**768–773.

Henige D. 1986. When did smallpox reach the New World (and why does it matter)? p 11–26. *In* Lovejoy PE (ed), *Africans in Bondage.* University of Wisconsin Press, Madison, WI.

Jin L, Brown DW, Ramsay ME, Rota PA, Bellini WJ. 1997. The diversity of measles virus in the United Kingdom, 1992–1995. *J Gen Virol* **78:**1287–1294.

Jones KE, Patel NG, Levy MA, Storeygard A, Balk D, Gittleman JL, Daszak P. 2008. Global trends in emerging infectious diseases. *Nature* **451:**990–993.

Morimoto K, Patel M, Corisdeo S, Hooper DC, Fu ZF, Rupprecht CE, Koprowski H, Dietzschold B. 1996. Characterization of a unique variant of bat rabies virus responsible for newly emerging human cases in North America. *Proc Natl Acad Sci U S A* **93:**5653–5658.

Luksza M, Lässig M. 2014. A predictive fitness model for influenza. *Nature* **507:**57–61.

Petsko GA. 2007. They fought the law and the law won. *Genome Biol* **8:**111. doi:10.1186/gb-2007-8-10-111.

Shackelton LA, Parrish CR, Truyen U, Holmes EC. 2005. High rate of viral evolution associated with emergence of carnivore parvovirus. *Proc Natl Acad Sci U S A* **102:**379–384.

Sharp GB, Kawaoka Y, Jones DJ, Bean WJ, Pryor SP, Hinshaw V, Webster RG. 1997. Coinfection of wild ducks by influenza A viruses: distribution patterns and biological significance. *J Virol* **71:**6128–6135.

Sharp PM, Bailes E, Chaudhuri RR, Rodenburg CM, Santiago MO, Hahn BH. 2001. The origins of acquired immune deficiency syndrome viruses: where and when? *Philos Trans R Soc Lond B Biol Sci* **356:**867–876.

Smith AW, Skilling DE, Cherry N, Mead JH, Matson DO. 1998. Calicivirus emergence from ocean reservoirs: zoonotic and interspecies movements. *Emerg Infect Dis* **4:**13–20.

Tumpey TM, Basler CF, Aguilar PV, Zeng H, Solórzano A, Swayne DE, Cox NJ, Katz JM, Taubenberger JK, Palese P, García-Sastre A. 2005. Characterization of the reconstructed 1918 Spanish influenza pandemic virus. *Science* **310:**77–80.

Webster RG, Yi G, Peiris M, Chen H. 2006. H5N1 influenza continues to circulate and change. *Microbe* **1:**559–565.

Wolfe ND, Switzer WM, Carr JK, Bhullar VB, Shanmugam V, Tamoufe U, Prosser AT, Torimiro JN, Wright A, Mpoudi-Ngole E, McCutchan FE, Birx DL, Folks TM, Burke DS, Heneine W. 2004. Naturally acquired simian retrovirus infections in central African hunters. *Lancet* **363:**932–937.

12 Unusual Infectious Agents

Introduction

Viroids
  Replication
  Sequence Diversity
  Movement
  Pathogenesis

Satellites
  Replication
  Pathogenesis
  Virophages or Satellites?
  Hepatitis Delta Satellite Virus

Prions and Transmissible Spongiform Encephalopathies
  Scrapie
  Physical Nature of the Scrapie Agent
  Human TSEs
  Hallmarks of TSE Pathogenesis
  Prions and the *prnp* Gene
  Prion Strains
  Bovine Spongiform Encephalopathy
  Chronic Wasting Disease
  Treatment of Prion Diseases

Perspectives

References

LINKS FOR CHAPTER 12

▶▶▎ *Darwin gets weird*
http://bit.ly/Virology_Twiv78

▶▶▎ *Wasting deer and the Hulk rabbit*
http://bit.ly/Virology_Twiv67

▶▶▎ *Is chronic wasting disease a threat to humans?*
http://bit.ly/Virology_3-11-15

> *Simple solutions seldom are.*
> ALFRED NORTH WHITEHEAD

Introduction

Genomes of nondefective viruses range in size from 2,400,000 bp of double-stranded DNA (*Pandoravirus salinus*) to 1,759 bp of single-stranded DNA (porcine circovirus). Are even smaller viral genomes possible? **Viroids, satellites,** and **prions** provide answers to these questions. The adjective "subviral" was coined, in part, because these agents did not fit into the standard taxonomy schemes for viruses. The Subviral RNA Database (http://subviral.med.uottawa.ca/cgi-bin/home.cgi) lists 2,923 nucleotide sequences for viroids and satellites. No prion sequences will be found in this database, because these infectious agents are, remarkably, devoid of nucleic acid.

Viroids

Viroids, the smallest known pathogens, are unencapsidated, circular, single-stranded RNA molecules that do not encode protein yet replicate autonomously when introduced into host plants. Potato spindle tuber viroid, discovered in 1971, is the prototype (Fig. 12.1); 29 other viroids ranging in length from 120 to 475 nucleotides have since been discovered. Viroids are known to infect only plants (Box 12.1); some cause economically important diseases of crop plants, while others appear to be benign, despite their widespread presence in the plant world. Two examples of economically important viroids are coconut cadang-cadang viroid (which causes a lethal infection of coconut palms) and apple scar skin viroid (which causes an infection that results in visually unappealing apples).

The 30 known viroids have been classified in two families. Members of the *Pospiviroidae*, named for potato spindle tuber viroid, have a rodlike secondary structure with small single-stranded regions, and a central conserved region (Fig. 12.1A), and replicate in the nucleus. The *Avsunviroidae*, named for avocado sunblotch viroid, have both rodlike and branched regions, but lack a central conserved region (Fig. 12.1B) and replicate in chloroplasts. In contrast to the *Pospiviroidae*, the latter RNA molecules are functional ribozymes, and this activity is essential for replication.

Viroids typically have a narrow host range that is implied by their names. However, host range expansion of some viroids has been observed. For example, potato spindle tuber viroid can infect avocado and tomato, and the weeds found in potato and hop fields can support the replication of both this viroid and hop stunt viroid.

After introduction into a plant, all viroids reproduce according to the following steps: import into a cellular organelle, replication, export out of the organelle, trafficking to adjacent cells, entry into the phloem (the plant vascular system that conducts nutrients downward from the leaves), long-distance movement to leaves and roots, and exit from phloem into new cells to repeat the cycle.

Replication

There is no evidence that viroids encode proteins or mRNA. Unlike viruses, which are parasites of the host translation machinery, **viroids are parasites of cellular transcription proteins:** they depend on cellular RNA polymerases for replication. Such polymerases normally recognize DNA templates, but can copy viroid RNAs.

In plants infected with members of the *Pospiviroidae*, viroid RNA is imported into the nucleus, probably by the nuclear import machinery. Plant DNA-dependent RNA polymerase II binds to the left terminal domain of potato spindle tuber viroid, suggesting that this structure serves as an origin of replication. In the nucleus the viroid is copied by a rolling circle mechanism that produces complementary linear, concatemeric, RNAs (Fig. 12.2). These products are copied again to produce concatemeric, linear RNA molecules, which are cleaved by RNase III. The linear, monomeric RNA molecules produced by cleavage of multimers have 5′-monophosphate and 3′-hydroxyl at their termini, groups required for ligation by DNA ligase I.

In plants infected with members of the *Avsunviroidae*, viroid RNA is imported into the chloroplast by an unknown import pathway, and complementary RNAs are produced by chloroplast DNA-dependent RNA polymerase. The circular RNA is then copied into a linear concatemeric RNA. After self-cleavage by ribozymes and ligation, the RNAs serve as templates for a round of concatemeric RNA synthesis,

---

PRINCIPLES *Unusual infectious agents*

- Viroids and prions are the smallest known pathogenic agents.
- Viroids comprise only noncoding RNA that is replicated by enzymes of plant host cells.
- Satellites depend on helper viruses for their reproduction.
- Diseases caused by viroids and satellites appear to be the result of silencing of expression of host genes.
- Hepatitis delta satellite virus, which exacerbates the pathogenesis of hepatitis B virus, is a unique hybrid of a viroid and a satellite.

- Prions are infectious proteins that cause neurological diseases of protein misfolding (transmissible spongiform encephalopathies, TSEs).
- There are three ways to contract a TSE: sporadic, infectious, and familial.
- Humans have increased the prevalence of TSEs by feeding cattle the remains of diseased animals.
- TSEs are surprisingly prevalent in wild deer and elk populations in North America, and represent a potential source of transmission to hunters and agricultural animals.

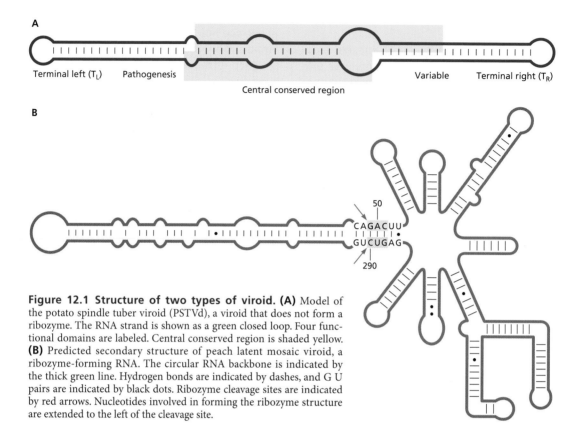

Figure 12.1 Structure of two types of viroid. (A) Model of the potato spindle tuber viroid (PSTVd), a viroid that does not form a ribozyme. The RNA strand is shown as a green closed loop. Four functional domains are labeled. Central conserved region is shaded yellow. **(B)** Predicted secondary structure of peach latent mosaic viroid, a ribozyme-forming RNA. The circular RNA backbone is indicated by the thick green line. Hydrogen bonds are indicated by dashes, and G U pairs are indicated by black dots. Ribozyme cleavage sites are indicated by red arrows. Nucleotides involved in forming the ribozyme structure are extended to the left of the cleavage site.

DISCUSSION
Why do viroids only infect plants?

Viroids infect plants cells after mechanical damage of the plant cell wall. Unlike plant viruses, there are no known animal vectors that transmit viroids from plant to plant. After genome replication in the plant cell, viroid RNA moves to the next cell by passage through plasmodesmata, the microchannels that connect neighboring plant cells (Chapter 12, Box 12.2). Animal cells do not have such connections, requiring that viruses travel from cell to cell either after release in the extracellular fluids, or by direct fusion with the membrane of a neighboring cell. Viruses may also travel from host to host in many ways, including aerosols and vectors, that are not available to viroids. Plants therefore seem well suited to serve as hosts for small, circular naked RNA molecules.

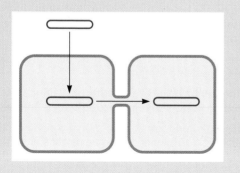

followed by cleavage and ligation to produce mature viroids. The self-ligating activity of viroids in the family *Avsunviroidae* is enhanced by a chloroplast tRNA ligase.

Replication of viroids therefore requires three enzymatic activities: RNA polymerase, RNase, and RNA ligase, and, for the former and the latter cases, the unusual situation of DNA enzymes working on RNA templates.

Sequence Diversity

Members of the family *Pospiviroidae* display little sequence diversity: the consensus sequence of different strains does not differ substantially. In contrast, members of the *Avsunviroidae* vary considerably. The difference is likely a consequence of the fidelity of the two RNA polymerases that carry out viroid replication. Two viroids that infect the same plant, the nuclear Chrysanthemum stunt viroid and the chloroplastic Chrysanthemum chlorotic mottle viroid have widely different mutation frequencies and provided the first experimental support of the "survival of the flattest" model of evolution (Box 12.2).

Movement

After replication, viroid progeny leave the nucleus or chloroplast and move to adjacent cells through plasmodesmata, and can travel systemically via the phloem. The results of mutational analyses demonstrate the requirement of specific

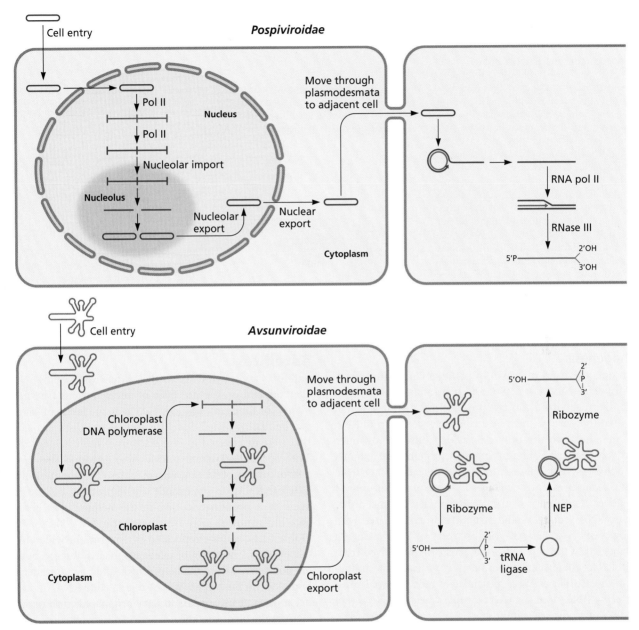

Figure 12.2 Replication of two different types of viroid in plants. (Top) Replication of *Pospiviroidae* in the nucleus. After entering the cell, circular viroid RNA is imported into the nucleus and copied by RNA polymerase II to form concatemeric RNAs. These are imported into the nucleolus where they are processed by cellular enzymes to genome-length RNAs and circularized before export and movement to the next cell through plasmodesmata (Pd). Right panel shows the details of viroid rolling-circle replication without showing cellular compartments. Complementary RNAs are red. **(Bottom)** Replication of *Avsunviroidae* in the chloroplast. After entering the cell, circular viroid RNA is imported into the chloroplast copied to form concatemeric RNAs. These are processed by the ribosome activity of the viroid. Right panel shows the details of rolling-circle replication without showing cellular compartments.

RNA loops and bulges for systemic transport within plants. Furthermore, a variety of host proteins that bind viroid RNA have been identified. These results have led to the hypothesis that trafficking of viroids within plants, from cell to cell or over longer distances, depends upon specific RNA sequences and structures that interact with cell proteins, including those that participate in movement. Some of these movement proteins also move viruses within plants (Volume I, Chapter 13). Viroids enter the pollen and ovule, from where they are transmitted to the seed. When the seed germinates, the new plant becomes infected. Viroids can also be transmitted among plants by contaminated farm machinery and insects.

DISCUSSION
Viroids and mutation rates

The quasispecies model of evolution predicts that, at a high mutation rate, populations with higher mutational robustness can displace those with a higher replicative capacity. This phenomenon has been called "survival of the flattest." The fitness associated with a sequence depends on the average fitness of its neighbors. In the survival of the flattest effect, a population in an area with neutral neighbors can out-compete another population with a higher fitness peak but surrounded by more deleterious neighbors (Figure). As predicted by the "survival of the fittest" paradigm, Chrysanthemum stunt viroid out-competed Chrysanthemum chlorotic mottle viroid because it is faster replicating and more genetically homogeneous.

However, when mutation rates were increased by ultraviolet irradiation of infected plants, the opposite effect was observed: Chrysanthemum chlorotic mottle viroid won the race.

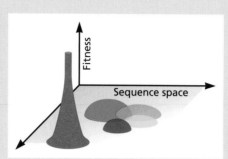

Codoñer FM, Daros JA, Solé RV, Elena SF. 2006. The fittest versus the flattest: experimental confirmation of the quasispecies effect with subviral pathogens. *PLoS Pathog* 2:e136.

Illustration of a population with high fitness surrounded by deleterious neighbors (red) and one with lower fitness and neutral neighbors (multiple colors).

Pathogenesis

Symptoms of viroid infection in plants include stunting of growth, deformation of leaves and fruit, stem necrosis, and death. Because viroids do not produce mRNAs, it was first proposed that disease must be a consequence of viroid RNA binding to host proteins or nucleic acids. Consequently much effort was directed at identifying host proteins that interact with viroid RNAs. These studies also revealed that viroid infection causes extensive changes in the expression of many host genes. The results of mutational analyses showed that specific RNA sequences and structures are associated with pathogenesis (Fig. 12.1A). The discovery of RNA silencing in plants led to the hypothesis that small RNAs derived from viroid RNAs guide silencing of host genes, leading to induction of disease. For example, peach latent mosaic viroid small interfering RNAs that silence chloroplast heat shock protein 90 and lead to disease symptoms have been identified. The two small interfering RNAs that target heat shock protein 90 mRNA are derived from the less abundant complementary strand. These observations suggest that different disease patterns caused by viroids in their hosts might all have in common an origin in RNA silencing.

Our current understanding is that the disease-causing viroids were transferred from wild plants used for breeding modern crops. The widespread prevalence of these agents can be traced to the use of genetically identical plants (monoculture), worldwide distribution of breeding lines, and mechanical transmission by contaminated farm machinery. As a consequence, these unusual pathogens now occupy niches around the planet that never before were available to them.

Satellites

Satellites are subviral agents that differ from viroids because they depend on the presence of another virus (the **helper virus**) for their propagation. Two general classes of satellites can be distinguished. **Satellite viruses** are distinct particles that were discovered in preparations of their helper viruses. These particles contain nucleic acid genomes that encode a structural protein that encapsidates the satellite genome. **Satellite RNAs** do not encode capsid proteins, but are packaged by a protein encoded in the helper virus genome. Satellite genomes may be single-stranded RNA or DNA (Table 12.1), and are replicated by enzymes provided by the helper virus. The origin of satellites remains obscure, but they are **not** derived from the helper virus: their genomes have no homology to the helper. At least one satellite RNA, of cucumber mosaic virus, appears to have originated from repetitive DNA in the plant genome.

Table 12.1 Viroids and satellites

Property	Viroids		Satellites
	Avsunviroidae	*Pospiviroidae*	
Requires coinfection with helper virus	No	No	Yes
Encodes protein	No	No	Yes
Replication	By host RNA polymerase and viroid ribozyme	By host RNA polymerase and host RNase	By helper virus replication proteins

Satellite viruses may infect plants, animals, or bacteria. An example of a satellite virus is satellite tobacco necrosis virus, which encodes a capsid protein that forms a $t = 1$ icosahedral capsid that selectively packages the 1,260-nucleotide satellite RNA. The helper virus, tobacco necrosis virus, encodes an RNA polymerase that replicates its genome and that of the satellite. The tobacco necrosis virus 5 kb RNA genome is packaged in a $t = 3$ icosahedral capsid made only from viral subunits. The structures of satellite virus and helper virus need not always be similar: the capsid of satellite tobacco mosaic virus is icosahedral, while virus particles of its helper virus are helical rods.

Satellite RNAs do not encode a capsid protein and therefore require helper virus proteins for both genome encapsidation and replication. Satellite RNA genomes range in length from 220 to 1,500 nucleotides, and have been placed into one of three classes. Class 1 satellite RNAs are 800- to 1,500-nucleotide linear molecules with a single open reading frame encoding at least one nonstructural protein. Class 2 satellite RNAs are also linear, but less than 700 nucleotides long and do not encode protein. Class 3 satellite RNAs are 350- to 400-nucleotide-long circles without an open reading frame.

Replication

The linear genomes of satellite viruses and satellite RNAs are copied by helper virus enzymes, and the mechanisms of replication are presumably similar (Volume I, chapters 6, 9). Satellite viruses typically impair production of the helper virus. For example, satellite tobacco necrosis virus reduces the yield of tobacco necrosis virus to undectable levels, while adenovirus-associated virus decreases the yield of adenovirus by >90%. How helper virus RNA polymerases recognize satellite RNA genomes is not known, because they have no sequence or structural similarity with the genome of the helper virus. Low-level replication in the absence of a helper virus has been demonstrated for a satellite RNA of cucumber mosaic virus. Such replication, which occurs in the cell nucleus and likely requires a host enzyme, may be a mechanism for persistence of satellite RNA.

Circular satellite RNA genomes are replicated by a rolling-circle mechanism like that of viroids (Fig. 12.2), except that replication by the helper virus RNA polymerase takes place in the cytoplasm. This enzyme recognizes a sequence on the satellite encapsidated RNA genome and produces complementary concatemers (Fig. 12.3). Depending on the satellite RNA, this product may be the template for the synthesis of multimeric copies of the encapsidated strand, or it may be cleaved and circularized by a ribozyme, prior to production of multimeric copies. The latter are then cleaved and ligated by a ribozyme followed by packaging. In some cases, linear strands may be packaged, but these are circularized upon infection of a new cell.

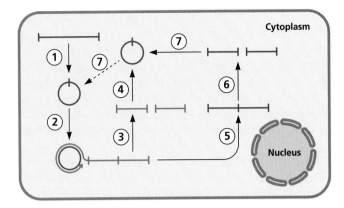

Figure 12.3 Replication of satellite RNA. Satellite RNA enters the plant cell and, if linear, is converted to circular RNA (1) which is then copied by a rolling-circle mechanism to produce concatemeric complementary copies (2). These are cleaved to form genome-length RNAs, possibly by ribozymes (3). Alternatively, the concatemeric RNAs might be copied (5) and then cleaved (6). Newly synthesized RNAs could then reenter the replicative pathway after circularization by a ribozyme or host enzyme.

Pathogenesis

In plants, satellites and satellite viruses may attenuate or exacerbate disease caused by the helper virus. Examples of disease include necrosis and systemic chlorosis, or reduced chlorophyll production leading to leaves that are pale, yellow, or yellow-white. Most satellite RNAs reduce the reproduction and yield of helper viruses, which leads to milder disease. The same satellite RNA may attenuate symptoms in one host, and cause greater disease in another. For example, disease caused by cucumber mosaic virus in tomatoes is much more severe in the presence of satellite cucumber mosaic virus RNA. However, in other hosts, the presence of the satellite RNA attenuates signs of disease symptoms. This effect may be due to the induction of a strong plant antiviral response mediated by RNA interference, resulting from high concentrations of satellite RNA.

The symptoms induced by satellite RNAs are thought to be a consequence of silencing of expression of host genes. For example, the Y-satellite RNA of cucumber mosaic virus causes systemic chlorosis in tobacco. This syndrome is caused by production of a small RNA from the Y-satellite RNA that has homology to a gene needed for chlorophyll biosynthesis. Production of this small RNA leads to degradation of the corresponding mRNA, resulting in bright yellow leaves. Consistent with this hypothesis, production of a potyvirus suppressor of silencing in tobacco plants reduces the severe yellowing caused by the cucumber mosaic virus and its satellite RNA.

Virophages or Satellites?

The giant DNA viruses including *Acanthamoeba polyphaga* mimivirus, *Cafeteria roenbergensis* virus, and others are associated with much smaller viruses (sputnik and mavirus, respectively) that depend upon the larger viruses for reproduction. For example, sputnik virus can only replicate in cells infected with mimivirus, and does so within viral factories. Whether these are satellite viruses or something new (they have been called virophages) has been a matter of controversy.

Sputnik and others have similar relationships with their helper viruses as satellite viruses: they require their helper for their propagation, but their genomes are not derived from the helper, and they negatively impact helper reproduction. Others argue that the definition of satellite viruses as subviral agents cannot apply to these very large viruses. For example, sputnik virophage contains a circular double-stranded DNA genome of 18,343 bp encoding 21 proteins encased in a 75-nm *t* = 27 icosahedral capsid. Sputnik is dependent upon mimivirus not for DNA polymerase (it encodes its own), but probably for the transcriptional machinery of the helper virus. Those who favor the name virophage argue that dependence upon the cellular transcriptional machinery is a property of many autonomous viruses, the only difference is that Sputnik depends upon that provided by another virus. For example, the replication-defective adenovirus-associated viruses, which require adenovirus as a helper virus, are classified by the International Committee on the Taxonomy of Viruses both in the parvovirus family and as a satellite. It seems likely that a redefinition of what constitutes a satellite virus will be required to solve this disagreement.

Hepatitis Delta Satellite Virus

Most known satellites are associated with plant viruses, but hepatitis delta satellite virus is associated with a human helper virus, hepatitis B virus. This satellite virus was discovered in 1977 in the nucleus of hepatocytes from patients with severe hepatitis. It was thought to be another marker of hepatitis B virus infection and was therefore called delta antigen. Later, the antigen was found to be encoded in the genome of a separate, defective virus. The genome is 1.7 kb (the smallest of any known animal virus) of circular single-stranded RNA that is 70% base paired and folds upon itself in a tight rodlike structure (Fig. 12.4A). The RNA molecule is replicated by cellular RNA polymerase II, a process that requires the self-cleaving activity of a ribozyme that is formed by a part of the delta satellite virus RNA (Chapter 6, Fig. 6.24). These properties resemble those of viroid genomes. On the other hand, the genome encodes a protein (delta) that encapsidates the RNA, a property shared with satellite nucleic acids. The hepatitis delta satellite virus particle comprises the satellite nucleocapsid packaged within an envelope that contains the surface protein of the helper, hepatitis B virus (Fig. 12.4B).

Upon entry into a cell, the hepatitis delta satellite virus RNA moves to the nucleus where antigenomic RNA is produced. This molecule in turn serves as template for the synthesis of an mRNA that encodes delta protein. Two functionally

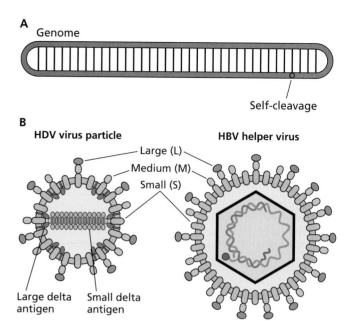

Figure 12.4 Genome and virus particle of hepatitis delta satellite virus. (A) Schematic of the circular (−) strand hepatitis delta virus RNA. Red dot indicates the ribozyme cleavage site. **(B)** Schematic of hepatitis delta satellite virus particle (left) and its helper virus, hepatitis B virus. The hepatitis delta RNA genome is encapsidated with the small and large delta antigens. The lipid envelope, derived from the host cell, contains the hepatitis B virus glycoproteins, comprising the large, medium, and small antigens.

distinct forms of the delta protein are made as a result of RNA editing (Fig. 6.24). Cellular adenosine deaminase changes an A in the delta antigen stop codon to inosine. Consequently, the stop codon becomes a tryptophan codon, extending the open reading frame by 19 amino acids. Small delta antigen is essential for viral RNA synthesis, regulating RNA editing, and is required for the accumulation of processed viral RNAs. It is also present in the virus particle, and assists in transport of the viral genome into the cell nucleus via an RNA binding domain and a nuclear localization signal. Large delta antigen is an inhibitor of viral replication, contains a nuclear export signal for transport of viral ribonucleoprotein from the nucleus to the cytoplasm, and is essential for the assembly of new virus particles. The latter activity is made possible by a cysteine that is four amino acids from the C terminus, which becomes farnesylated, allowing interaction of the proteins with membranes. It is remarkable that such a short C-terminal extension of the protein endows it with numerous new functions.

Infection with hepatitis delta satellite virus occurs only in individuals infected with hepatitis B virus: it is globally distributed, present in approximately 5% of the 350 million carriers of hepatitis B virus. Acute coinfections of the two viruses can be more severe than infection with hepatitis B virus alone, increasing the rate of liver failure. In chronic hepatitis B virus infections, hepatitis delta satellite virus aggravates preexisting liver disease, and may lead to more rapid progression to cirrhosis and death than monoinfections. Why coinfection with both viruses leads to more serious outcomes is not known.

Prions and Transmissible Spongiform Encephalopathies

The question of whether infectious agents exist without genomes arose with the discovery and characterization of infectious agents associated with a group of diseases called **transmissible spongiform encephalopathies (TSEs)**. These diseases are rare, but always fatal, neurodegenerative disorders that afflict humans and other mammals (Table 12.2). They are characterized by long incubation periods, spongiform changes in the brain associated with loss of neurons, and the absence of host responses. We now know that TSEs are caused by infectious proteins called **prions**.

Scrapie

The first TSE recognized was scrapie, so called because infected sheep tend to scrape their bodies on fences so much that they rub themselves raw. A second characteristic symptom, skin tremors over the flanks, led to the French name for the disease, tremblant du mouton. Motor disturbances then manifest as a wavering gait, staring eyes, and paralysis of the hindquarters. There is no fever, but infected sheep lose weight and die, usually within 4 to 6 weeks of the first appearance of symptoms. Scrapie has been recognized as a disease of European sheep for more than 250 years. It is endemic in some countries, for example, the United Kingdom, where it affects 0.5 to 1% of the sheep population each year.

Physical Nature of the Scrapie Agent

Sheep farmers discovered that animals from affected herds could pass the disease to a scrapie-free herd, implicating an infectious agent. Infectivity from extracts of scrapie-affected sheep brains was shown to pass through filters with pores small enough to retain everything but viruses. As early as 1966, scrapie infectivity was shown to be considerably more resistant than that of most viruses to ultraviolet (UV) and

Table 12.2 Some transmissible spongiform encephalopathies

TSE diseases of animals
Bovine spongiform encephalopathy (mad cow disease)
Chronic wasting disease (deer, elk)
Scrapie in sheep and goats
TSE diseases of humans
Creutzfeldt-Jakob disease
Variant Creutzfeldt-Jakob disease
Fatal familial insomnia
Gerstmann-Straussler-Scheinker syndrome
Kuru

ionizing radiation. Other TSE agents exhibit similar UV resistance. On the basis of this relative resistance to UV irradiation, some investigators argued that TSE agents are viruses well shielded from irradiation, whereas others claimed that TSE agents have little or no nucleic acids.

The infectivity of scrapie agents is also more resistant to chemicals, such as the combination of 3.7% formaldehyde and autoclaving routinely used to inactivate virus particles. While it is possible to reduce infectivity by 90 to 95% after several hours of such treatment, complete elimination is exceedingly difficult. This property has led to unfortunate human infections caused by sterilized surgical instruments.

Human TSEs

Several lines of evidence indicated that human spongiform encephalopathies might be caused by an infectious agent. Carleton Gajdusek and colleagues studied the disease **kuru**, found in the Fore people of New Guinea. This disease is characterized by cerebellar ataxia (defective motion or gait) without loss of cognitive functions. Kuru spread among women and children as a result of ritual cannibalism of the brains of deceased relatives. When cannibalism stopped in the late 1950s, kuru disappeared. Others observed that lesions in the brains of humans with kuru were similar to lesions in the brains of animals with scrapie. It was soon demonstrated that kuru and other human TSEs can be transmitted to chimpanzees and small laboratory animals.

Human spongiform encephalopathies are placed into three groups: infectious, familial or genetic, and sporadic, distinguished by how the disease is acquired initially. An infectious (or transmissible) spongiform encephalopathy is exemplified by kuru and the iatrogenic spread of disease to healthy individuals by transplantation of infected corneas, the use of purified hormones, or transfusion with blood from patients with the TSE Creutzfeldt-Jakob disease (CJD). Over 400 cases of iatrogenic Creutzfeldt-Jakob disease have been reported worldwide. The epidemic spread of bovine spongiform encephalopathy (mad cow disease, see below) among cattle in Britain can be ascribed to the practice of feeding processed animal by-products to cattle as a protein supplement. Similarly, the new human disease, variant CJD, arose after consumption of beef from diseased cattle. Sporadic CJD is a disease affecting one to five per million annually, usually late in life (with a peak at 68 years). As the name indicates, the disease appears with no warning or epidemiological indications. Kuru may have been originally established in the small population of Fore people in New Guinea when the brain of an individual with sporadic CJD was eaten. Familial spongiform encephalopathy is associated with an autosomal dominant mutation in the *prnp* gene (see below). Together familial and sporadic forms of prion disease account for ~99% of all cases. Diseases of all three classes can usually be transmitted experimentally or naturally by inoculation or ingestion of diseased tissue.

Hallmarks of TSE Pathogenesis

Clinical signs of infection commonly include cerebellar ataxia, memory loss, visual changes, dementia, and akinetic mutism, with death occurring after months or years. The infectious agent first accumulates in the lymphoreticular and secretory organs and then spreads to the nervous system. In model systems, spread of the disease from the site of inoculation to other organs and the brain requires dendritic and B cells. The disease agent then invades the peripheral nervous system and spreads from there to the spinal cord and brain. Once the infectious agent is in the central nervous system, the characteristic pathology includes severe astrocytosis, vacuolization (hence, the term spongiform), and loss of neurons. Occasionally, dense fibrils or aggregates (sometimes called plaques) can be detected in brain tissue at autopsy. There are no inflammatory, antibody, or cellular immune responses. The time course, degree, and site of cytopathology within the central nervous system are dependent upon the particular TSE agent and the genetic makeup of the host.

Prions and the *prnp* Gene

The unconventional physical attributes and slow infection pattern originally prompted many to argue that TSE agents are not viruses at all. In 1967, it was suggested that scrapie could

be caused by a host protein, not by a nucleic acid-carrying virus. These ideas were among the first of the protein-only hypotheses to explain TSE.

An important breakthrough occurred in 1981, when characteristic fibrillar protein aggregates were visualized in infected brains. These aggregates could be concentrated by centrifugation and remained infectious. Stanley Prusiner and colleagues developed an improved bioassay, as well as a fractionation procedure that allowed the isolation of a protein with unusual properties from scrapie-infected tissue. This protein is insoluble and relatively resistant to proteases. He named the scrapie infectious agent a **prion**, from the words protein and infectious.

Prusiner's unconventional proposal was that an altered form of a normal cellular protein, called PrP^C, causes the fatal encephalopathy characteristic of scrapie. This controversial protein-only hypothesis caused a firestorm among those who study infectious disease. The hypothesis was that the essential pathogenic component **is** the host-encoded PrP^C protein with an altered conformation, called PrP^{Sc} ("PrP-scrapie"). Furthermore, PrP^{Sc} was proposed to have the property of converting normal PrP^C protein into more copies of the pathogenic form (Fig. 12.5). PrP^C and PrP^{Sc} can be differentiated by sensitivity to protease digestion: PrP^C is completely degraded

Figure 12.5 The conversion of nonpathogenic, α-helix-rich PrP^C protein to the β-sheet-rich conformation of PrP^{Sc}, the pathogenic prion. (A) PrP^C is the mature normal cellular protein. The precursor is 254 amino acids long with a signal sequence that is removed. Twenty-three amino acids of the carboxy terminus also are removed as the glycosylphosphatidylinositol (GPI) anchor is added. PrP^{Sc} is the β-sheet-rich, pathogenic prion. This conformation is relatively resistant to protease K digestion, in contrast

to PrP^C, as indicated. This protease K-resistant PrP fragment of PrP^{Sc} is diagnostic of the prion protein. H1, H2, and H3 are helical regions of PrP^C. The yellow boxes indicate repeats of 8 amino acids [P(Q/H)GGGWGQ]. CHO indicates two N-linked carbohydrate chains. S–S indicates disulfide bonds. **(B)** Ribbon diagram of the PrP^C and PrP^{Sc} protein backbones with α-helices in red and β-sheets in blue. From P. Chien, J. Weissman, and A. DePace, *Annu Rev Biochem* **73**:617–656, 2004, with permission.

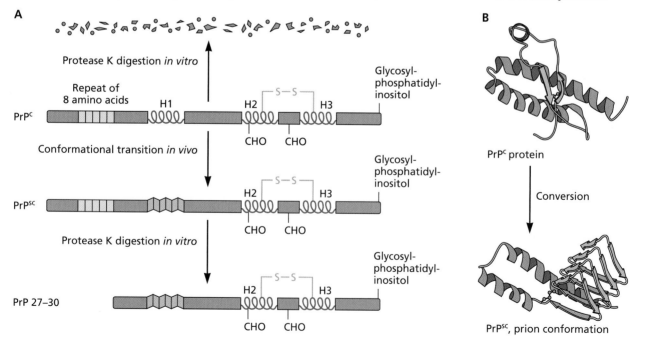

by proteinase K, while digestion of PrP^{Sc} produces a 27- to 30-kDa fragment. PrP^{C} has little β-sheet structure and high α-helical content, whereas PrP^{Sc} has high β-sheet structure and low α-helical content. In recognition of his work on prions, Prusiner was awarded the Nobel Prize in physiology or medicine in 1997.

Sequence analysis of this protein led to the identification of the *prnp* gene, which is highly conserved in the genomes of many mammals, including humans. Expression of this gene is now known to be essential for the pathogenesis of TSEs. The *prnp* gene encodes a 35-kDa membrane-associated neuronal glycoprotein, PrP^{C}. The function of this protein has been difficult to determine, because mice lacking both copies of the *prnp* gene develop normally and have few obvious defects. However, these mice are resistant to TSE infection, showing that PrP^{C} is essential for prion propagation. When *prnp$^{-/-}$* mice are inoculated with PrP^{Sc} they develop antibodies that also recognize PrP^{C}, showing that the two forms of the protein share epitopes.

The discovery of the *prnp* gene has helped explain the basis of familial TSE diseases such as Creutzfeldt-Jakob disease, Gerstmann-Straussler-Scheinker disease, and fatal familial insomnia. Gerstmann-Straussler-Scheinker disease is associated with the change of PrP^{C} amino acid 102 from proline to leucine. Introduction of this amino acid change into mice gives rise to a spontaneous neurodegenerative disease characteristic of a TSE. Familial Creutzfeldt-Jakob disease may be associated with an insertion of 144 bp at codon 53, or changes at amino acids 129, 178, or 200. In fatal familial insomnia, adults develop a progressive sleep disorder and typically die within one year. In this disease, PrP^{Sc} is found only in the anteroventral and dorsal medial nuclei of the thalamus. Development of the disease is strongly linked to the D178N amino acid change and V129. When D178N is present with V129, the patients develop familial Creutzfeldt-Jakob disease, which is characterized by dementia; in this case, PrP^{Sc} is found throughout the brain. How the sequence of the protein affects pathogenesis is not known. Over 40 different mutations of the PrP gene have been identified. The familial prion diseases are fully penetrant. Both infectious and sporadic TSEs develop in the absence of mutations in the wild-type *prnp* gene (Fig. 12.6).

Although altered PrP proteins are produced early in human development, progress of neurological disease is generally delayed for decades. This observation has led to the suggestion that an event associated with aging is required for producing TSE. However, there is no evidence that any age-dependent process, such as mitochondrial DNA mutations, oxidative modifications of DNA and proteins, or proteasome malfunction, is responsible for TSE. Another possibility is that the accumulation of PrP^{Sc} in numbers sufficient to be self-sustaining is a time-consuming process. Not until

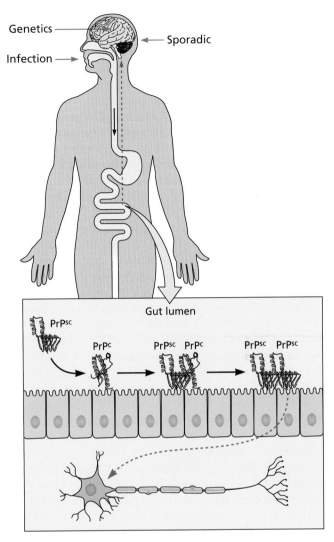

Figure 12.6 Human prion infections. Human prion diseases are spontaneous, genetic, or acquired by exposure to infectious materials. When prions are ingested, they move to the intestine and are taken up by mucosa and Peyer's patches. It is thought that ingested infectious prions convert PrP^{C} to PrP^{Sc}, which spreads to the central nervous system via enteric nerves (red line).

the number of prions being produced reaches a threshold would infection continue unchecked, at which point neurological dysfunction could occur.

The rate of prion formation in an inoculated animal is influenced by many parameters. It is inversely related to the incubation time, and proportional to the quantity of PrP^{C} in the brain and in the inoculum. The PrP sequence also matters: prion propagation is faster when PrP^{C} and PrP^{Sc} are identical in sequence. The incubation time until development of disease in mice inoculated with Syrian hamster prions is >500 days. In contrast, the incubation time of mice transgenic for the Syrian hamster *prnp* gene is 70 to 75 days, the same for hamsters inoculated with hamster PrP^{Sc}. Transgenic mice inoculated with Syrian hamster prions produce Syrian hamster, and not mouse, prions: in other words, they will only infect hamsters. When the same mice are inoculated with mouse prions, only mouse prions are produced.

Some prions have a distinct host range. For example, mouse-adapted scrapie prions (produced by serial passage of scrapie prions in mice) cannot propagate in hamsters, but hamster-adapted scrapie prions can propagate in mice. A single amino acid substitution in the hamster protein enables it to be converted efficiently by mouse PrP^{Sc} into hamster PrP^{Sc}. The barrier to interspecies transmission is therefore in the sequence of the PrP protein: the infecting PrP^{Sc} must match the PrP^{C} of the host. Bovine spongiform encephalopathy prions have an unusually broad host range, infecting a number of meat-eating animals, including domestic cats, wild cats, and humans, but not mice. The latter can be infected with bovine prions if they are made transgenic for the bovine *prnp* gene.

How prions propagate after PrP^{Sc} enters a cell remains obscure. PrP^{Sc} may act as a seed or template, recruiting PrP^{C} monomers into ordered polymers and altering their conformation. Despite widespread presence of PrP^{C}, formation of PrP^{Sc} is restricted to a few cell types (neurons, cornea, myocytes, follicular dendritic cells), suggesting that auxiliary molecules participate in the formation of PrP^{Sc}. Conversion of PrP^{C} to PrP^{Sc} has been achieved *in vitro* using purified proteins, but the efficiency of conversion is very low. Addition of glycosaminoglycans increases the conversion frequency, providing evidence for the role of auxiliary molecules in the formation of PrP^{Sc}. Increased efficiency of conversion of PrP^{C} to PrP^{Sc} has been achieved using a technique called protein misfolding cyclic amplification. In this method, crude brain homogenates containing PrP^{Sc} and PrP^{C} are mixed and incubated for 1 to 3 days with intermittent sonication, resulting in amplification of PrP^{Sc}. The cyclic sonications are thought to speed the polymerization reaction by fragmenting PrP^{Sc} polymers, increasing the concentration of seeds. By depleting different molecules from brain extracts it was found that RNA or phosphatidylethanolamine can facilitate the conversion of PrP^{C} to PrP^{Sc}. It seems likely that other cofactor molecules participate in this process. Addition of phosphatidylethanolamine to mouse PrP^{C} produced in *E. coli*, in the absence of PrP^{Sc}, leads to the production of infectious prions. This finding will facilitate studies on the mechanism of conversion, and could enable development of therapeutics or diagnostic tests.

Until recently, the presence of a prion could be detected definitively only by injection of organ homogenates into susceptible recipient species or by proteinase K digestion, procedures that cannot be done with living patients. This challenge has been overcome by using amplification procedures to detect Creutzfeldt-Jakob prions in nasal brushings and in urine (Box 12.3).

Prion Strains

Serial infections of mice and hamsters with infected sheep brain homogenates have led to the production of distinct strains of scrapie prion. Strains are distinguished by length of incubation time before the appearance of symptoms, brain pathologies, relative abundance of various glycoforms of PrP^{C}, and electrophoretic profiles of protease-resistant PrP^{Sc}. A striking finding is that different scrapie strains can be propagated in the same inbred line of mice, yet maintain their original phenotypes.

Prion strains do not differ in amino acid sequence, but rather in their glycosylation patterns, protease resistance, and conformation. Each of the distinctive pathogenic conformations is postulated to convert the normal PrP protein into a conformational image of itself. Recent evidence suggests that strain diversity may be a function of different compounds present during the formation of PrP^{Sc} molecules. In support of this hypothesis, it was found that when three different PrP^{Sc} strains are propagated *in vitro* in the presence of phosphatidylethanolamine, they were converted into a single novel strain.

These observations demonstrate that the properties of PrP^{Sc}, including its conformation, can override sequence differences between the infecting prion and host PrP^{Sc}. A mechanistic understanding of this process will require determining the structures of different PrP^{Sc} strains. Unfortunately to date no structure of PrP^{Sc} has been determined because the protein is insoluble.

Bovine Spongiform Encephalopathy

In the mid-1980s, a new disease appeared in cows in the United Kingdom: bovine spongiform encephalopathy, also called mad cow disease (Fig. 12.7). It is believed to have been transmitted to cows by feeding them meat and bone meal, a high-protein supplement prepared from the offal of sheep, cattle, pigs, and chicken. In the late 1970s the method of preparation of meat and bone meal was changed, resulting in material with a higher fat content. It is believed that this change allowed prions, from either a diseased sheep or cow, to retain infectivity and pass on to cattle. Before the disease was recognized in 1985, it was amplified by feeding cows the remains of infected bovine tissues. The incubation period for bovine spongiform encephalopathy is 5 years, but disease was not observed because most cattle are slaughtered between 2 and 3 years of age. Three years later, as the number of cases of mad cow disease increased, a ban on the use of meat and bone meal was put in place, a practice that together with culling infected cattle stopped the epidemic. Over 180,000 cattle, mostly dairy cows, died of bovine spongiform encephalopathy from 1986 to 2000.

Cases of variant Creutzfeldt-Jakob disease, a new TSE of humans, began to appear in 1994 in the United Kingdom. These were characterized by a lower mean age of the

BOX **12.3**

EXPERIMENTS
Detection of Creutzfeldt-Jakob prions in nasal brushings and urine

The human prion disease, Creutzfeldt-Jakob, is diagnosed by a variety of criteria, including clinical features, electroencephalograms, and magnetic resonance imaging. Until recently, there was no noninvasive assay to detect PrP^{Sc}, the only specific marker for the disease. New diagnostic tests using nasal brushings or urine seem to fill this need.

These assays utilize two different methods for amplifying the quantity of prions *in vitro*. In real-time quaking-induced conversion, PrP^{C} (produced in *E. coli*) is mixed with a small quantity of PrP^{Sc}. The mixtures are subjected to cycles of shaking and rest at 42°C for 55 to 90 hours, a procedure that leads to the formation of amyloid fibrils that can be detected by fluorescence. The assay can detect femtograms of PrP^{Sc} in brain homogenates from humans with Creutzfeldt-Jakob disease. In protein misfolding cyclic amplification, samples are incubated for 30 minutes at 37 to 40°C, subjected to a pulse of sonication, and the cycle is then repeated 96 times. Prions are detected by Western blot analysis after treatment with proteinase K. This process can detect a single oligomeric PrP^{Sc}. Although done with protein, the assay resembles PCR in the use of templates to provide amplification of PrP^{Sc}.

Two noninvasive assays using these amplification approaches were developed. The first is a nasal-brushing procedure to sample the olfactory epithelium, where PrP^{Sc} is known to accumulate in patients with the disease. The real-time quaking-induced conversion assay was positive in 30 of 31 patients with Creutzfeldt-Jakob disease, and negative in 43 of 43 healthy controls (a sensitivity of 97%). Furthermore, nasal brushings gave stronger and faster positive results than cerebrospinal fluid in this assay. The high concentrations of PrP^{Sc} detected in nasal brushings suggest that prions can contaminate nasal discharge of patients with the disease, a possible source of iatrogenic transmission that has implications for infection control.

Protein misfolding cyclic amplification was used to assay for the presence of PrP^{Sc} in the urine of patients with variant Creutzfeldt-Jakob disease, which had been previously shown to contain prions. PrP^{Sc} was detected in 13 of 14 urine samples from patients with the disease, but not in 224 urine samples from healthy controls and patients with other neurological diseases, including other TSEs. The estimated concentration of PrP^{Sc} in urine was 40 to 100 oligomeric particles per ml.

Because Creutzfeldt-Jakob disease is so rare, any assay for the disease must have near-perfect specificity. A problem with both cycling assays is that PrP^{C} converts into oligomers and fibrils in the absence of PrP^{Sc}. Additional work is needed to address this problem. Nevertheless it is possible that these assays could one day lead to earlier diagnosis and treatments, if the latter become available.

Moda F, Gambetti P, Notari S, Concha-Marambio L, Catania M, Park KW, Maderna E, Suardi S, Haïk S, Brandel JP, Ironside J, Knight R, Tagliavini F, Soto C. 2014. Prions in the urine of patients with variant Creutzfeldt-Jakob disease. *N Engl J Med* **371**:530–539.

Orrú CD, Bongianni M, Tonoli G, Ferrari S, Hughson AG, Groveman BR, Fiorini M, Pocchiari M, Monaco S, Caughey B, Zanusso G. 2014. A test for Creutzfeldt-Jakob disease using nasal brushings. *N Engl J Med* **371**:519–529.

Detection of PrP^{Sc} by protein misfolding cyclic amplification. Inset, sampling of olfactory neuroepithelium. Samples (labeled oligomeric seeds) are mixed with PrP^{C} monomers and incubated to allow growth of the polymers. The mixture is sonicated to fragment the aggregates and increase the number of nuclei for prion replication. After 96 cycles, additional PrP^{C} substrate is added, and the samples are subjected to another cycle.

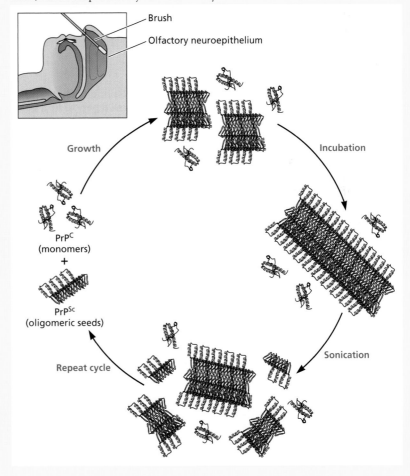

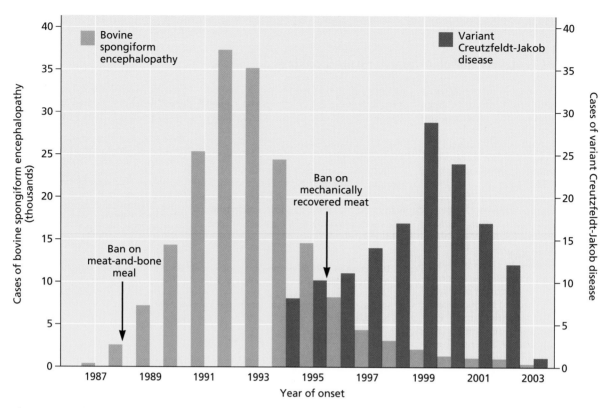

Figure 12.7 Time course of the reporting of bovine spongiform encephalopathy in cattle and variant Creutzfeldt-Jakob disease in humans in the United Kingdom over a period of 9 years. The peak of the bovine epidemic was in 1992, and the peak of the human disease was in 1999. The incidence of both is now rare. Data obtained from http://webarchive.national archives.gov.uk/20090505194948/http://www.bseinquiry.gov.uk/report/index.htm

patients (26 years), longer duration of illness, and differences in other clinical and pathological characteristics. The results of epidemiological and experimental studies indicate that variant Creutzfeldt-Jakob disease is caused by prions transmitted by the consumption of cattle with bovine spongiform encephalopathy. As of 2011 there had been 177 cases of variant Creutzfeldt-Jakob disease in the United Kingdom, and 229 globally.

Bovine spongiform encephalopathy continues to be detected in cattle. As of April 2012, 4 cases had been identified in the United States and 19 in Canada. These cases may arise sporadically, or through consumption of contaminated feed. Because cattle are slaughtered before disease symptoms are evident, there is concern that variant Creutzfeldt-Jakob might increase as contaminated meat enters the food supply. These concerns are being addressed by imposing bans on animal protein-containing feed, and increased surveillance of cows for the disease, for which diagnostic tests are being developed.

Chronic Wasting Disease

Chronic wasting disease is a transmissible spongiform encephalopathy of cervids such as deer, elk, and moose. It is the only TSE known to occur in free-ranging animals.

The disease has been reported in the United States, Canada, and South Korea (Fig. 12.8). In captive herds in the United States and Canada, up to 90% of mule deer and 60% of elk are infected, and the incidence in wild cervids is as high as 15%.

Mice have been used to understand whether chronic wasting disease prions might be transmitted to humans. Mice are not efficiently infected with chronic wasting disease prions unless they are made transgenic for the cervid *prnp* gene. Four different research groups have found that mice transgenic for the human *prnp* gene are not infected by chronic wasting disease prions. These findings suggest that such prions are not likely to be transmitted directly to humans. However, changing four amino acids in human *prnp* to the cervid sequence allows efficient infection of transgenic mice with cervid prions.

Another concern is that prions of chronic wasting disease could be transmitted to cows grazing in pastures contaminated by cervids. It is not known how the disease is spread among cervids, but transmission by grass contaminated with saliva and feces is one possibility. When deer are fed prions they excrete them in the feces before developing signs of infection, and prions can also be detected in deer saliva. In the laboratory, brain homogenates from infected deer advanced Creutzfeldt-Jakob disease showed that the drug is not effective. This failure was suggested to be a consequence of poor penetration of the drug into the central nervous system. When quinacrine was given to genetically altered mice in which drugs can more easily penetrate the brain, PrP^{Sc} levels were depressed transiently but disease was not prevented. Monoclonal antibodies specific for PrP inhibit scrapie prion propagation in mice and delay the development of prion disease. Although delivery of antibodies into the central nervous system is not efficient, clinical trials to evaluate the efficacy of these molecules in treating prion disease are planned. Small molecules that bind to prion proteins, enhance their clearance, or cause dominant-negative inhibition of prion propagation have been identified, but none have yet been tested in humans.

Perspectives

We have discussed viroids, satellites, and prions together in this chapter because they are not accommodated by the classification schemes for viruses. The origin of satellites and viroids remains an enigma, but it has been proposed that they are relics from the RNA world, which is thought to have been populated only by noncoding RNA molecules that catalyzed their own synthesis. Both types of infectious agents have properties that make them candidates for survivors of the RNA world: small genome size (to avoid error catastrophe caused by error-prone replication), high G+C content (greater thermodynamic stability), circular genomes (to avoid the need for mechanisms to prevent loss of information at the ends of linear genomes), no protein content, and the presence of a ribozyme, a fingerprint of the RNA world. Today's viroids can no longer self-replicate, possibly having lost that function when they became parasites of plants. What began as a search for viruslike agents that cause disease in plants has led to new insights into the evolution of life.

Many intriguing questions about viroids and satellites remain, including the nature of plant defense mechanisms against these elements, how they enter and exit cellular organelles, the precise mode of their spread within plants, the role of host proteins in reproduction, and the mechanisms of pathogenesis. Perhaps the thorniest question is why satellites impair the reproduction of their helper viruses, rather than being mutually beneficial as might be anticipated. The intriguing possibility that satellite viruses provide a function to helper viruses remains unaddressed. Even more enigmatic is hepatitis delta virus, a hybrid of viroid and satellite with a mammalian helper virus. This satellite virus likely arose in the liver of a patient infected with hepatitis B virus, but the source might have been a plant viroid or satellite passing through the host's intestine. Given the very high rate of discovery of small RNAs, it seems likely that other elements similar to hepatitis delta virus will be identified.

Figure 12.8 Chronic wasting disease in North America. CWD, chronic wasting disease. Figure courtesy of the Chronic Wasting Disease Alliance (www.cwd-info.org).

Infected cervid populations

CWD in captive populations

can transmit the disease to cows. Therefore, it is possible that contamination of grass could pass the agent on to cows, from where it could then enter the human food chain.

A further worry is that bovine spongiform encephalopathy (BSE) prions shed by cows in pastures might infect cervids, which would then become a reservoir of the agent. BSE prions do not infect mice that are transgenic for the cervid *prnp* gene. However, intracerebral inoculation of deer with BSE prions causes neurological disease, and the prions from these animals can infect mice that are transgenic for the cervid *prnp* gene. Therefore, caution must be used when using transgenic mice to predict the abilities of prions to cross species barriers.

No case of transmission of chronic wasting disease prions to deer hunters has yet been reported. Although the risk of human infection with chronic wasting disease prions appears to be low, hunters are advised not to shoot or consume an elk or deer that is acting abnormally or appears to be sick, to avoid the brain and spinal cord when field dressing game, and not to consume brain, spinal cord, eyes, spleen, or lymph nodes.

Treatment of Prion Diseases

There are currently no therapeutics available to slow or stop the neurodegeneration characteristic of transmissible spongiform encephalopathies, although symptoms may be mitigated by the drug L-dopa. A potential breakthrough came when researchers discovered that the antimalarial drug quinacrine blocked accumulation of infectious prions in cultured cells. Unfortunately human trials of quinacrine in patients with

If the reader does not believe that viroids and satellites are distinctive, then surely prions, infectious agents composed only of protein, must impress. Precisely how prions are formed from normal cell proteins, and how their structures provide strain differences are just two of many important unanswered questions. While TSEs are rare, they are uniformly fatal, and better methods of diagnosis and treatment are needed. Since prions were discovered it has become clear that protein misfolding is involved in a wide spectrum of neurodegenerative diseases. For example, the amyloid fibrils in Alzheimer's disease contain the amyloid-β peptide that is processed from the amyloid precursor protein; familial disease is caused by mutations in the gene for this protein. Mutations in the tau gene are responsible for heritable tauropathies including familial frontotemporal dementia and inherited progressive supranuclear palsy. Self-propagating tau aggregates pass from cell to cell. The prionlike spread of misfolded α-synuclein is believed to be associated with Parkinson's disease. In these cases there is good evidence that the causative protein, like PrP^{Sc}, adopts a conformation that becomes self-propagating.

Despite the contribution of prions in human neurological diseases, in other organisms such proteins are not pathogenic but rather impart diverse functions through templated conformational change of a normal cellular protein. Such prions have been described in fungi where they do not form infectious particles and do not spread from cell to cell. These proteins change conformation in response to an environmental stimulus and acquire a new, beneficial function. An example is the *Saccharomyces cerevisiae* Ure2p protein, which is a nitrogen catabolite repressor when cells are grown in the presence of a rich source of nitrogen. In the aggregated prion state, called [URE3], the protein allows growth on poor nitrogen sources. These findings prompt the question of whether the conversion of PrP^C to PrP^{Sc} once had a beneficial function that became pathogenic. If so, identifying that function, and how it was usurped, will be important for understanding the pathogenesis of transmissible spongiform encephalopathies.

References

Reviews

Alves C, Branco C, Cunha C. 2013. Hepatitis delta virus: a peculiar virus. *Adv Virol* **2013**:560105.

Diener TO. 2003. Discovering viroids—a personal perspective. *Nat Rev Microbiol* **1**:75–80.

Ding B. 2009. The biology of host-viroid interactions. *Annu Rev Phytopathol* **47**:105–131.

Fischer MG. 2011. Sputnik and Mavirus: more than just satellite viruses. *Nat Rev Microbiol* **9**:762–763.

Fischer MG, Suttle CA. 2011. A virophage at the origin of large DNA transposons. *Science* **332**:231–234.

Flores R, Gago-Zachert S, Serra P, Sanjuan R, Elena SF. 2014. Viroids: survivors from the RNA world? *Annu Rev Microbiol* **68**:395–414.

Fraser PE. 2014. Prions and prion-like proteins. *J Biol Chem* **289**:19839–19840.

Hope J. 2012. Bovine spongiform encephalopathy: a tipping point in one health and food safety. *Curr Top Microbiol Immunol* **366**:37–47.

Kraus A, Groveman BR, Caughey B. 2013. Prions and the potential transmissibility of protein misfolding diseases. *Annu Rev Microbiol* **67**:543–564.

Krupovic M, Cvirkaite-Krupovic V. 2012. Towards a more comprehensive classification of satellite viruses. *Nat Rev Microbiol* **10**:234.

Navarro B, Gisel A, Rodio M-E, Delgado S, Flores R, Di Serio F. 2012. Viroids: how to infect a host and cause disease without encoding proteins. *Biochimie* **94**:1474–1480.

Palukaitis P. 2014. What has been happening with viroids? *Virus Genes* **49**:175–184.

Panegyres PK, Armari E. 2013. Therapies for human prion diseases. *Am J Neurodegener Dis* **2**:176–186.

Prusiner SB. 2013. Biology and genetics of prions causing neurodegeneration. *Annu Rev Genet* **47**:601–623.

Rao ALN, Kalantidis K. 2015. Virus-associated small satellite RNAs and viroids display similarities in their replication strategies. *Virology* **479–480**:627–636.

Raoult D. 2014. How the virophage compels the need to readdress the classification of microbes. *Virology* **477**:119–124.

Saunders SE, Bartelt-Hunt SL, Bartz JC. 2012. Occurrence, transmission, and zoonotic potential of chronic wasting disease. *Emerg Infect Dis* **18**:369–376.

Watts JC, Prusiner SB. 2014. Mouse models for studying the formation and propagation of prions. *J Biol Chem* **289**:19841–19849.

Papers of special interest

Haley NJ, Seelig DM, Zabel MD, Telling GC, Hoover EA. 2009. Detection of CWD prions in urine and saliva of deer by transgenic mouse bioassay. *PLoS One* **4**:e4848.

Kurt TD, Jiang L, Fernández-Borges N, Bett C, Liu J, Yang T, Spraker TR, Castilla J, Eisenberg D, Kong Q, Sigurdson CJ. 2015. Human prion protein sequence elements impede cross-species chronic wasting disease transmission. *J Clin Invest* **125**:1485–1496.

Moda F, Gambetti P, Notari S, Concha-Marambio L, Catania M, Park KW, Maderna E, Suardi S, Haïk S, Brandel JP, Ironside J, Knight R, Tagliavini F, Soto C. 2014. Prions in the urine of patients with variant Creutzfeldt-Jakob disease. *N Engl J Med* **371**:530–539.

Orru CD, Bongianni M, Tonoli G, Ferrari S, Hughson AG, Groveman BR, Fiorini M, Pocchiari M, Monaco S, Caughey B, Zanusso G. 2014. A test for Creutzfeldt-Jakob disease using nasal brushings. *N Engl J Med* **371**:519–529.

Shimura H, Pantaleo V, Ishihara T, Myojo N, Inaba J-I, Sueda K, Burgyan J, Masuta C. 2011. A viral satellite RNA induces yellow symptoms on tobacco by targeting a gene involved in chlorophyll biosynthesis using the RNA silencing machinery. *PLoS Pathog* **7**:e1002021.

Supattapone S. 2014. Synthesis of high titer infectious prions with cofactor molecules. *J Biol Chem* **289**:19850–19854.

Vickery CM, Lockey R, Holder TM, Thorne L, Beck KE, Wilson C, Denyer M, Sheehan J, Marsh S, Webb PR, Dexter I, Norman A, Popescu E, Schneider A, Holden P, Griffiths PC, Plater JM, Dagleish MP, Martin S, Telling GC, Simmons MM, Spiropoulos J. 2013. Assessing the susceptibility of transgenic mice overexpressing deer prion protein to bovine spongiform encephalopathy. *J Virol* **88**:1830–1833.

Zahid K, Zhao J-H, Smith NA, Schumann U, Fang Y-Y, Dennis ES, Zhang R, Guo H-S, Wang M-B. 2014. *Nicotiana* small RNA sequences support a host genome origin of Cucumber mosaic virus satellite RNA. *PLoS Genet* **11**:e1004906.

Website

http://www.prion.ucl.ac.uk/clinic-services/research/drug-treatments/
Prion therapies

APPENDIX

Diseases, Epidemiology, and Disease Mechanisms of Selected Animal Viruses Discussed in This Book

This appendix presents key facts about the pathogenesis of selected animal viruses that cause human disease. Information about each virus or virus group is presented in four sections. In the first section, the selected viruses and associated diseases are listed. The second section, "Epidemiology," outlines virus transmission, worldwide distribution, those at risk or risk factors, and vaccines or antiviral drugs currently available. The third section, "Pathogenesis," provides simple images that will enable the reader to visualize infection and the resulting pathogenesis, with green arrows indicating portals of viral entry and red arrows indicating sites of virus egress from the human body. Finally, in a section entitled "Human Infections," we provide an indication of the impact of the infection on the human population. These pages were designed to be made into slides for lectures or teaching, providing "snapshots" of the pathogenesis of specific viruses.

Adenoviruses

Virus

57 adenovirus serotypes that infect humans, classified into 6 subgroups

Disease

Respiratory diseases
- Upper tract infection
- Pharyngoconjunctival fever
- Pertussis-like disease
- Pneumonia

Other diseases
- Acute hemorrhagic cystitis
- Epidemic keratoconjunctivitis
- Gastroenteritis
- Myocarditis

Epidemiology

Transmission
- Aerosol, fecal matter, fomites
- Poorly sanitized swimming pools
- Ophthalmologic instruments (eye infections)

At risk or risk factors
- Children aged <14 years
- Day care centers, military camps, swimming clubs
- Immunosuppression

Distribution
- Worldwide
- No seasonal incidence

Vaccines or antiviral drugs
- Attenuated vaccine for serotypes 4 and 7 has been produced for the military
- Cidofovir

Pathogenesis

Virus infects mucoepithelial cells of respiratory and gastrointestinal tract, conjunctivae

Virus can persist in lymphoid tissue (tonsils, adenoids, and Peyer's patches)

Human Infections

>80% of humans are seropositive

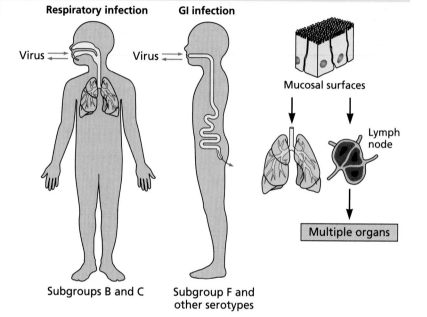

Respiratory infection — Virus — Subgroups B and C

GI infection — Virus — Subgroup F and other serotypes

Mucosal surfaces → Lymph node → Multiple organs

Figure 1

Arenaviruses

Virus	Disease
Lymphocytic choriomeningitis virus	Fever, muscle pain, meningitis
Lassa virus	Lassa hemorrhagic fever (severe systemic illness, increased vascular permeability, shock)
Junin virus	Argentine hemorrhagic fever similar to Lassa fever but more extensive bleeding

Epidemiology

Transmission
- Contact with infected rodents or their excreta

At risk or risk factors
- Lymphocytic chorio-meningitis virus: contact with pet hamsters, areas with rodent infestation
- Other arenaviruses: proximity to rodents

Distribution
- Lymphocytic choriomeningitis virus: hamsters and house mice in Europe, Americas, Australia, possibly Asia
- Other arenaviruses: Africa, South America, United States
- No seasonal incidence

Vaccines or antiviral drugs
- No vaccines
- Antiviral drug: ribavirin

Pathogenesis

Persistent infection of rodents caused by neonatal infection and induction of immune tolerance

Viruses infect macrophages, which release mediators of cell and vascular damage

Tissue destruction caused by T cell immunopathology

Human Infections

Lassa fever: several thousand cases per year worldwide; none in U.S.

~50% seropositive in Africa

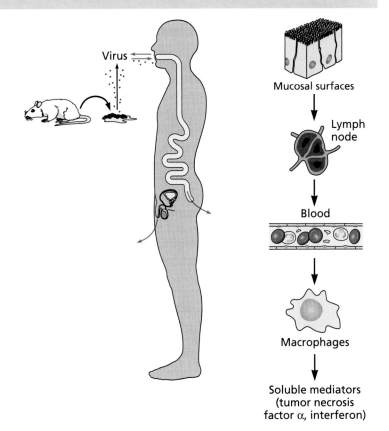

Figure 2

Bunyaviruses

Virus	Vector	Disease
Bunyavirus (>350 species) • Bunyamwera virus • California encephalitis virus • La Crosse virus • Oropouche virus	Mosquito	Febrile illness, encephalitis, rash
Hantavirus (22 species) • Hantaan virus	None	Hemorrhagic fever with renal syndrome, adult respiratory distress syndrome
• Sin Nombre virus	None	Hantavirus pulmonary syndrome, shock, pulmonary edema
Nairovirus (6 species) • Crimean-Congo hemorrhagic fever virus	Tick	Sandfly fever, hemorrhagic fever, encephalitis, conjunctivitis, myositis
Phlebovirus (9 species) • Rift Valley fever virus • Sandfly fever virus	Fly	Hemorrhagic fever

Epidemiology

Transmission
- Arthropod bite
- Contact with rodent excreta

Distribution of virus
- Depends on distribution of vector or rodents
- Disease more common in summer

At risk or risk factors
- People in area of vector, e.g., campers, forest rangers, woodspeople

Vaccines or antiviral drugs
- No vaccines
- No antivirals

Pathogenesis

Primary viremia, then secondary viremia leads to virus spread to target tissues, including central nervous system, various organs, and vascular endothelium

Human Infections

Hantavirus pulmonary syndrome in U.S.: ~650 cases (through 2013)

Sporadic, limited outbreaks worldwide, especially South America

Case fatality: 20–50%

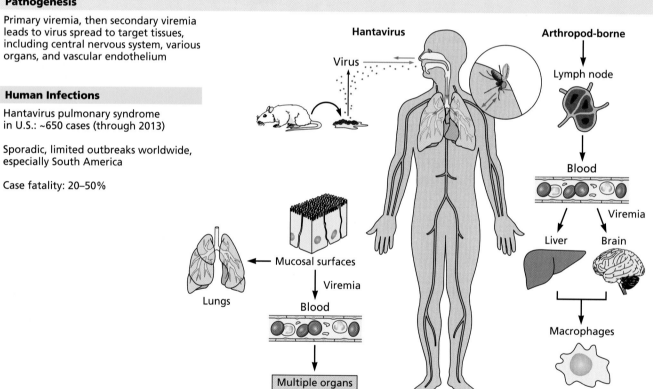

Figure 3

Caliciviruses

Virus	Disease
Logovirus	Gastroenteritis
Norovirus	
• Norwalk virus	
Sapovirus	Gastroenteritis
• Sapporo virus	

Epidemiology

Transmission
- Fecal-oral route from contaminated water and food
- Virus particles are resistant to detergents, drying, and acid

At risk or risk factors
- Children in day care centers and schools
- Resorts, hospitals, nursing homes, restaurants, cruise ships

Distribution
- Worldwide
- No seasonal incidence

Vaccines or antiviral drugs
- None

Pathogenesis

Infection of intestinal brush border, prevents proper absorption of water and nutrients

Cause diarrhea, vomiting, abdominal cramps, nausea, headache, malaise, and fever

May cause persistent infection, but usually resolved

Human Infections

Noroviruses: Of 348 outbreaks during Jan. 1996–Nov. 2000, 39% occurred in restaurants; 29% occurred in nursing homes and hospitals; 12% in schools and day care centers; 10% in vacation settings, including cruise ships; and 9% in other settings

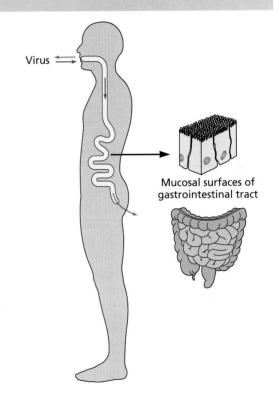

Virus

Mucosal surfaces of gastrointestinal tract

Figure 4

Filoviruses

Virus	Disease
Marburgvirus • Lake Victoria	Hemorrhagic fever
Ebolavirus • Reston • Sudan • Zaire • Bundibugyo • Tai Forest	Hemorrhagic fever

Epidemiology

Transmission
- Fruit bats are reservoirs
- Contact with infected fruit bats, monkeys, or their tissues, secretions, or body fluids
- Contact with infected humans or body fluids
- Accidental injection, contaminated syringes

Distribution
- Africa, Philippines
- No seasonal incidence

At risk or risk factors
- Bushmeat hunting and preparation
- Family members of the sick
- Burial preparation
- Health care workers attending sick persons

Vaccines or antiviral drugs
- In clinical trials

Pathogenesis

Virus reproduction causes necrosis in liver, spleen, lymph nodes, and lungs

Hemorrhage causes edema and shock

Death from dehydration

Human Infections

Case fatality: 40–90%

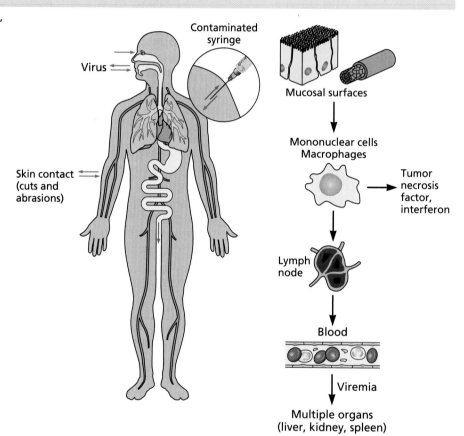

Figure 5

Flaviviruses

Virus	Vector	Disease
Flavivirus		
• Yellow fever virus	*Aedes* mosquitoes	Hepatitis, hemorrhagic fever
• Powassan virus	*Ixodes* ticks	Encephalitis
• Dengue virus	*Aedes* mosquitoes	Breakbone fever, dengue hemorrhagic fever, dengue shock syndrome
• Japanese encephalitis virus	*Culex* mosquitoes	Encephalitis
• St. Louis encephalitis virus	*Culex* mosquitoes	Encephalitis
• West Nile virus	*Culex* mosquitoes	Fever, encephalitis, hepatitis

Epidemiology

Transmission
• Mosquito or tick vectors

Distribution
• Determined by habitat of vector
  Aedes mosquito (urban areas)
  Culex mosquito (forest, urban areas)
• More common in summer

At risk or risk factors
• Proximity to vector

Vaccines or antiviral drugs
• Attenuated vaccines for yellow fever and Japanese encephalitis
• No antiviral drugs

Pathogenesis

Viruses cause viremia and systemic infection

Nonneutralizing antibodies can facilitate infection of monocytes/macrophages via Fc receptors

Human Infections

Dengue cases worldwide/year: 50 million–100 million (estimated); 419 dengue cases in U.S. since its identification

Yellow fever cases worldwide: 200,000 (90% in Africa)

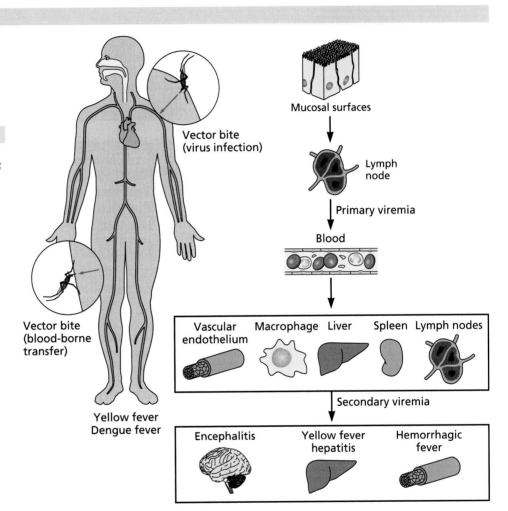

Figure 6

Flaviviruses (Hepatitis C)

Virus	Vector	Disease
Hepacivirus • Hepatitis C virus	None	Hepatitis

Epidemiology

Transmission
• Blood
• Sex

Distribution
• Worldwide
• No seasonal incidence

At risk or risk factors
• IV drug users
• Health care workers

Vaccines or antiviral drugs
• Currently 7 FDA-approved antivirals
• No vaccines

Pathogenesis

Viruses are noncytolytic and chronic

Disease caused by ongoing immune response

Liver cancer can result from chronic cirrhosis

Human Infections

Persons living with hepatitis C
Worldwide: 130 million–150 million
In United States: 3.2 million

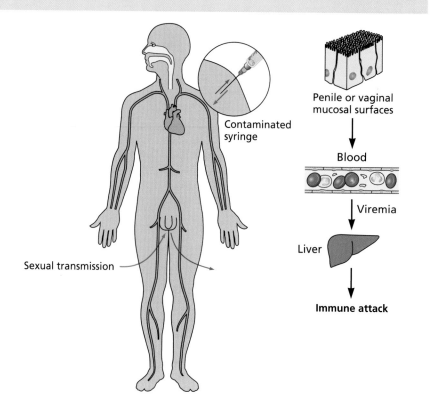

Contaminated syringe

Sexual transmission

Penile or vaginal mucosal surfaces

Blood

Viremia

Liver

Immune attack

Figure 7

Hepadnaviruses

Virus	Disease
Hepatitis B virus	Hepatitis, primary hepatocellular carcinoma, cirrhosis • Children (mild, chronic infection) • Adults (insidious hepatitis) • Adults with chronic hepatitis

Epidemiology

Transmission
• Blood, semen, vaginal secretions, transfusions, sex, breast-feeding, saliva, contaminated needles

At risk or risk factors
• IV drug use
• Promiscuity

Distribution
• Worldwide
• No seasonal incidence

Vaccines or antiviral drugs
• Currently 6 FDA-approved antiviral drugs
• Subunit vaccine

Pathogenesis

Primary reproduction in hepatocytes followed by viremia

Tissue damage caused by cell-mediated immune response

Acute infections

Chronic infections that can lead to hepatocellular carcinoma

Human Infections

Worldwide incidence of hepatitis B: 240 million

Annual number of deaths: ~1 million

Hepatitis B virus vaccine efficacy: >95%

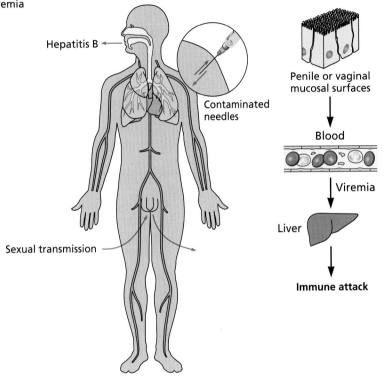

Figure 8

Herpesviruses (Herpes simplex virus)

Virus

Alphaherpesviruses
- Herpes simplex virus types 1 and 2

Disease

Mucosal lesions, encephalitis

- Children (HSV-1)
- Sexually active people (HSV-2)

Epidemiology

Transmission
- Saliva, vaginal secretions, secretions from blisters in oral and anogenital tracts
- Eyes; skin lesions
- Herpes simplex virus type 1: mainly oral; herpes simplex virus type 2: mainly sexual

At risk or risk factors
- Promiscuity

Distribution
- Worldwide
- No seasonal incidence

Vaccines or antiviral drugs
- Antivirals: acyclovir, ganciclovir

Pathogenesis

Cell-to-cell spread, not neutralized by antibody

Establishes latency in neurons

Reactivated from latency by stress or immune suppression

Human Infections

HSV-1 seroprevalence: >65%

HSV-2 seroprevalence: 16.2%

Incidence increases with number of sex partners

Demographic	(Females)	(Males)
1 Lifetime sex partner	5.4%	1.7%
2–4 Lifetime sex partners	18.8%	7.3%
5–9 Lifetime sex partners	21.8%	10.1%
≥10 Lifetime sex partners	37.1%	19.1%

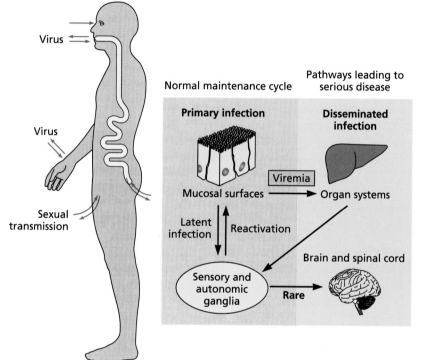

Figure 9

Herpesviruses (Varicella-zoster virus)

Virus	Disease
Alphaherpesviruses • Varicella-zoster virus	Chickenpox, shingles • Children (ages 5–9 years) (mild disease) • Teenagers and adults (more severe disease, possibly pneumonia) • Elderly, immunocompromised (recurrent zoster)

Epidemiology

Transmission
• Aerosol

At risk or risk factors
• Immunosuppression
• Shingles: previous exposure

Distribution
• Worldwide
• No seasonal incidence

Vaccines or antiviral drugs
• Attenuated vaccine
• Antiviral drugs: acyclovir, foscarnet

Pathogenesis

Infects epithelial cells and fibroblasts, spread by viremia to skin, causes lesions of chickenpox

Latent infection in neurons

Reactivation by immune suppression, stress, or other infections leads to zoster or shingles, formation of lesions over entire dermatome

Human Infections

99.5% of people are seropositive

30% chance of shingles following primary infection

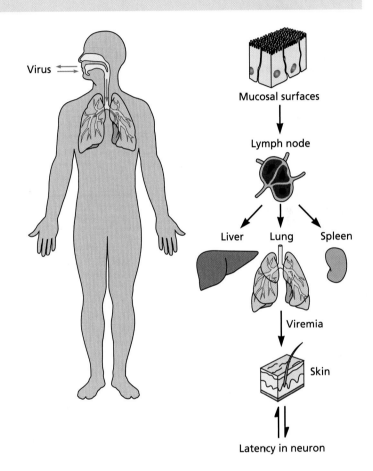

Figure 10

Herpesviruses (Cytomegalovirus)

Virus	Disease
Betaherpesviruses	
• Cytomegalovirus	Congenital defects (cerebral palsy)
	Disseminated disease in immunosuppressed patients

Epidemiology

Transmission
- Blood, tissue, and body secretions (urine, saliva, semen, cervical secretions, breast milk, tears)

At risk or risk factors
- Babies whose mothers become infected during pregnancy
- Sexual activity
- Transplant recipients
- Burn victims
- Immunosuppression

Distribution
- Worldwide
- No seasonal incidence

Vaccines or antiviral drugs
- No vaccines
- Antiviral drugs: acyclovir, ganciclovir

Pathogenesis

Infects epithelial and other cells

Mainly causes subclinical infections

Latent infection in CD34$^+$ bone marrow progenitor cells, macrophages, other cells

Immunosuppression leads to recurrence and severe disease

Human Infections

60–70% seropositive in resource-rich countries

100% seropositive in resource-poor countries

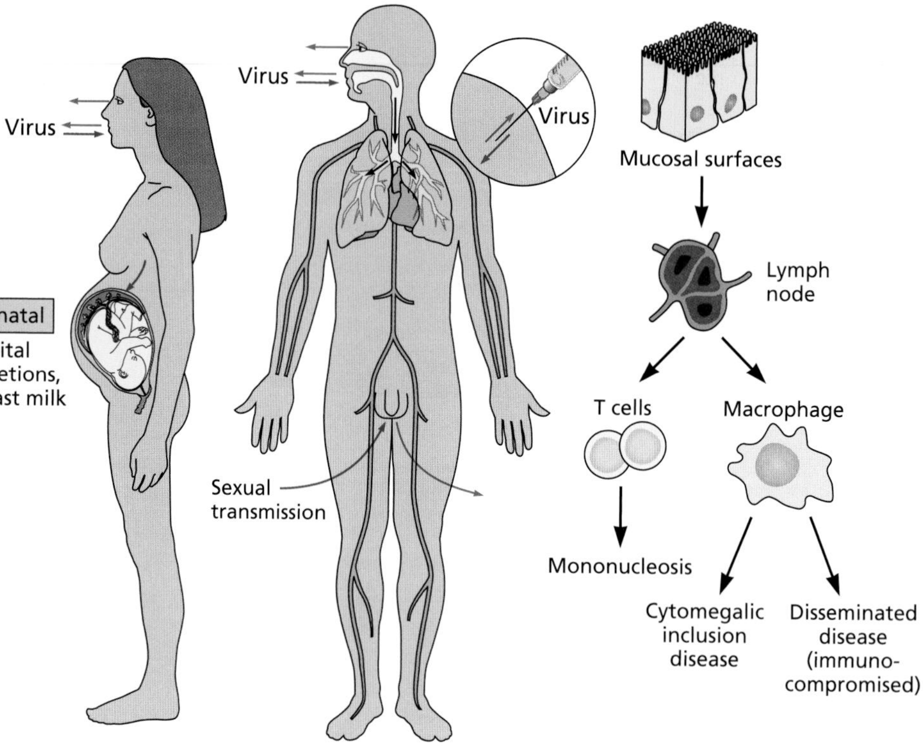

Virus

Virus

Virus

Perinatal
Genital secretions, breast milk

Sexual transmission

Mucosal surfaces

Lymph node

T cells

Macrophage

Mononucleosis

Cytomegalic inclusion disease

Disseminated disease (immuno-compromised)

Figure 11

Herpesviruses (Gammaherpesviruses)

Virus	Disease
Gammaherpesviruses • Epstein-Barr virus (EBV)	Infectious mononucleosis; associated with a variety of lymphomas Nasopharyngeal carcinoma
• Human herpesvirus 8 (HHV-8; Kaposi's sarcoma-associated herpesvirus [KSHV])	Kaposi's sarcoma, rare B cell lymphoma

Epidemiology

Transmission
• Saliva, close oral contact, or shared items (cup or toothbrush)

At risk or risk factors
• Immunosuppression
• Malaria (Burkitt's lymphoma)
• Kissing

Distribution
• EBV: Worldwide
• KSHV: Mediterranean basin
• Burkitt's lymphoma in malaria belt
• No seasonal incidence

Vaccines or antiviral drugs
• None

Pathogenesis

EBV:
Infects oral epithelial cells, B cells

Immortalizes B cells

HHV-8:
Characteristic skin rash

Human Infections

In the U.S.: 90–95% of adults show evidence of prior EBV infection

HHV-8: Prevalence highly variable:
Japan: 0.2%
Africa: >50%

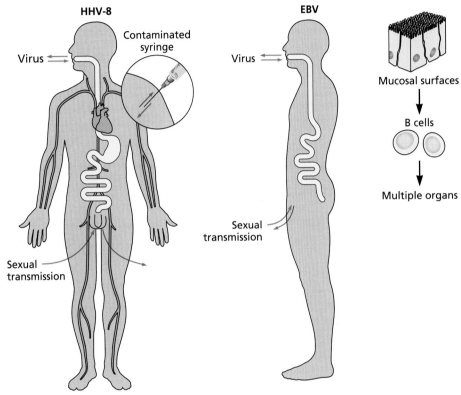

Figure 12

391

Orthomyxoviruses

Virus	Disease
Influenza A, B, and C viruses	Acute febrile respiratory tract infection
	Children may also have abdominal pain, vomiting, otitis media, muscle inflammation, croup
	Primary viral pneumonia
	Reye's syndrome

Epidemiology

Transmission
- Aerosols; fomites

At risk or risk factors
- Age (very young and elderly)
- Immunosuppression
- Pregnancy

Distribution
- Worldwide
- More common in winter

Vaccines or antiviral drugs
- Inactivated vaccine against annual strains of influenza A and B viruses; Flublok subunit vaccine
- Attenuated influenza A and B vaccine (nasal spray)
- Antiviral drugs: amantadine, oseltamivir (Tamiflu)

Pathogenesis

Infects upper and lower respiratory tract

Pronounced systemic symptoms caused by cytokine response to infection

Susceptibility to bacterial superinfection because of compromised natural epithelial barriers

Human Infections

Worldwide deaths due to flu epidemics:
1918: >50 million
2009: 151,000–575,000

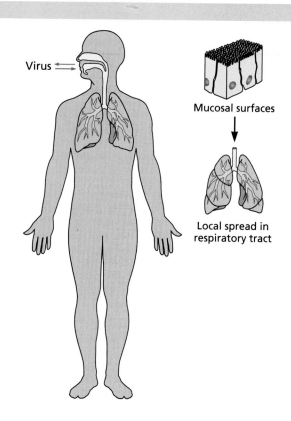

Virus

Mucosal surfaces

Local spread in respiratory tract

Figure 13

Papillomaviruses

Virus	Disease
Papillomavirus (120 genotypes)	Skin warts: plantar, common, and flat warts; epidermodysplasia verruciformis
	Head and neck tumors: laryngeal, oral, and conjunctival papillomas
	Anogenital warts: condyloma acuminatum, cervical intra-epithelial neoplasia, cancer

Epidemiology

Transmission
- Direct contact, sexual contact
- During birth, from infected birth canal

At risk or risk factors
- Genital and oral sex

Distribution
- Worldwide
- More common in winter

Vaccines or antiviral drugs
- Vaccine against types 6, 11, 16, and 18 (e.g., Gardasil)

Pathogenesis

Infect epithelial cells of skin, mucous membranes

Reproduction depends on stage of epithelial cell differentiation

Cause benign outgrowth of cells into warts

Some types are associated with dysplasia that may become cancerous

Human Infections

Head and neck cancers in U.S. annually:
Males: 10,000
Females: 2,500

Cervical cancers: 12,000

Penile cancers: 1,000

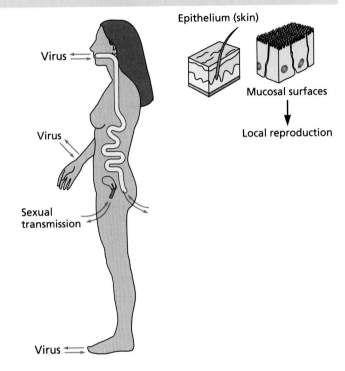

Figure 14

Paramyxoviruses (Measles virus)

Virus	Disease
Morbilliviruses • Measles virus	Ear infections, croup, bronchopneumonia, encephalitis, congestion, immunosuppression, skin rash

Epidemiology

Transmission
• Aerosols
• Highly contagious

At risk or risk factors
• Malnutrition
• Unvaccinated
• Immunosuppression

Distribution
• Worldwide
• Endemic from autumn to spring

Vaccines or antiviral drugs
• Attenuated vaccine
• No antiviral drugs

Pathogenesis

Infects epithelial cells of respiratory tract, spreads in lymphocytes and by viremia

Reproduces in conjunctivae, respiratory tract, urinary tract, lymphatic system, blood vessels, and central nervous system

T cell response to virus-infected capillary endothelial cells causes rash

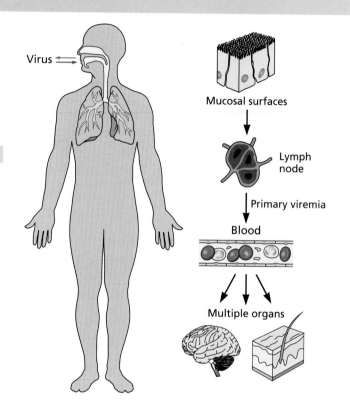

Human Infections

Increased incidence of measles with decreased herd immunity

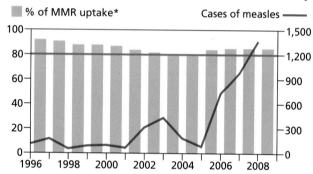

*Figures relate to financial years 1996/97, 1997/98, etc.

Figure 15

Paramyxoviruses (Respiratory syncytial virus)

Virus	Disease
Pneumoviruses • Respiratory syncytial virus	Bronchiolitis, pneumonia, febrile rhinitis, pharyngitis, common cold

Epidemiology

Transmission
• Aerosols

Distribution
• Worldwide
• Winter and spring

At risk or risk factors
• Age (<6 months)
• Immunosuppression
• Adults with other respiratory problems (chronic obstructive pulmonary disease)

Vaccines or antiviral drugs
• Monoclonal antibody for treatment
• Antiviral drug: ribavirin for infants

Pathogenesis

Infects the respiratory tract, does not spread systemically

In newborns, the infection may be fatal because narrow airways are blocked by virus-induced pathology

Infants are not protected from infection by maternal antibody

Reinfection may occur after a natural infection

Human Infections

Most common cause of bronchiolitis and pneumonia in children <1 year of age in the U.S.

Almost all children will have had respiratory syncytial virus infection before 2 years of age, though symptoms vary from severe respiratory disease to mild

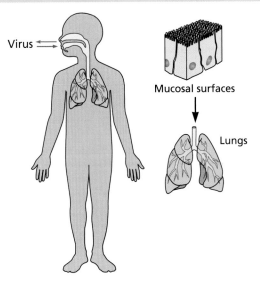

Virus

Mucosal surfaces

Lungs

Figure 16

Picornaviruses (Poliovirus and Hepatitis A)

Virus	Disease
Polioviruses	
• Poliovirus types 1–3	Paralytic disease, encephalitis, meningitis, respiratory tract infections, undifferentiated fever, disease in immunodeficient patients
Hepatitis A virus	Hepatitis

Epidemiology

Transmission
• Fecal-oral

At risk or risk factors
• **Polio:** Poor sanitation
• **Hepatitis A:** Intravenous drugs, sex, contaminated food supply

Distribution
• Nearly eradicated (polio)
• Worldwide hepatitis A

Vaccines or antiviral drugs
• Attenuated oral or inactivated polio vaccines
• No licensed antiviral drugs
• Hepatitis A virus: inactivated vaccine

Pathogenesis

99% of infections are mild or asymptomatic

Only 1% associated with paralysis

Reversion of attenuated vaccine may lead to vaccine-associated paralysis

Human Infections

Polio continues to be endemic in Pakistan, Afghanistan, Nigeria

Only 350–400 cases in recent years

Hepatitis A: 1.4 million cases annually

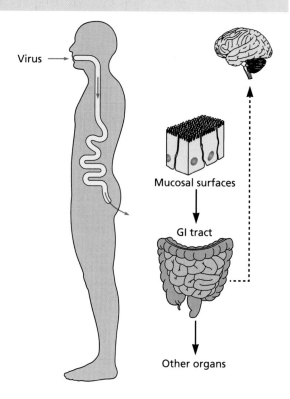

Figure 17

Picornaviruses (Rhinoviruses)

Virus	Disease
Rhinoviruses • A, B, and C (>150 genotypes)	Respiratory tract infections Major cause of common cold

Epidemiology

Transmission
• Aerosols
• Contact with contaminated hands

At risk or risk factors
• Preexisting respiratory conditions

Distribution
• Worldwide
• Disease most common in early autumn, late spring

Vaccines or antiviral drugs
• No vaccines
• No licensed antiviral drugs

Pathogenesis

Enter upper respiratory tract and may remain localized or spread to lower respiratory tract

Major factor in asthma exacerbations

Human Infections

Millions of cases per year in U.S. alone

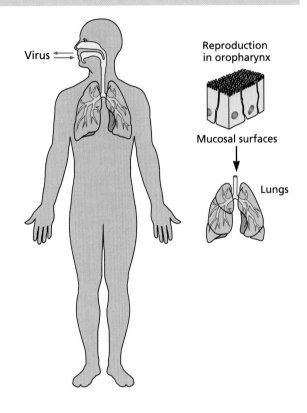

Virus ⇌

Reproduction in oropharynx

Mucosal surfaces

Lungs

Figure 18

Picornaviruses (Enteroviruses)

Virus	Disease
Enteroviruses	
• Enterovirus 68	Severe respiratory disease
• Enterovirus 70	Paralytic disease, acute hemorrhagic conjunctivitis
• Enterovirus 71	Paralytic disease, encephalitis, meningitis, hand-foot-and-mouth disease

Epidemiology

Transmission
• Fecal-oral

At risk or risk factors
• Poor sanitation
• Age (newborns and neonates)

Distribution
• Worldwide
• Disease most common in summer

Vaccines or antiviral drugs
• No vaccines
• No licensed antiviral drugs

Pathogenesis

Enter oropharyngeal or intestinal mucosa

Serum antibody blocks spread

Virus may be shed in feces

High asymptomatic infection rate

Human Infections

10 million–15 million enterovirus infections in the U.S. each year

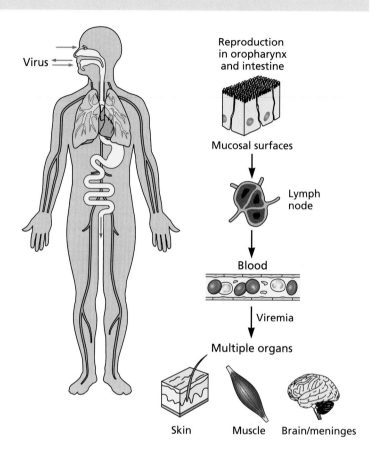

Figure 19

Polyomaviruses

Virus	Disease
Polyomavirus	
• BK virus	Renal disease in immunosuppressed patients
• JC virus	Progressive multifocal leukoencephalopathy (PML) in immunosuppressed patients

Epidemiology	
Transmission	**Distribution**
• Aerosols	• Worldwide
	• No seasonal incidence
At risk or risk factors	**Vaccines or antiviral drugs**
• Immunocompromised persons (HIV infection)	• None

Pathogenesis

Acquired through the respiratory route, spread by viremia to kidneys early in life

Infections are usually asymptomatic

Virus establishes persistent and latent infection in organs such as the kidneys and lungs

In immunocompromised people, JC virus is activated, spreads to the brain, and causes progressive multifocal leukoencephalopathy; oligodendrocytes are killed, causing demyelination

Human Infections

Ubiquitous in human population

PML rare, even in immunosuppressed patients

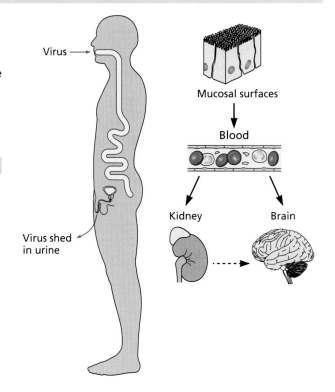

Virus

Mucosal surfaces

Blood

Virus shed in urine

Kidney   Brain

Figure 20

Poxviruses

Virus	Disease
Variola virus	Smallpox
Vaccinia virus (smallpox vaccine)	Encephalitis and vaccinia necrosum (complications of vaccination)
Cowpox virus	Localized lesion
Monkeypox virus	Generalized disease
Molluscum contagiosum virus	Disseminated skin lesions

Epidemiology

Transmission
- Smallpox: respiratory droplets, contact with virus on fomites
- Other poxviruses: direct contact or fomites

At risk or risk factors
- Molluscum contagiosum: sexual contact, wrestling
- Pet owners, animal handlers (contact with lesion)

Distribution
- Worldwide
- No seasonal incidence

Vaccines or antiviral drugs
- Attenuated vaccine against smallpox (vaccinia virus)

Pathogenesis

Infects respiratory tract, spreads through lymphatics and blood

Sequential infection of multiple organs

Skin lesions are prominent

Human Infections

Smallpox is the only human virus to be eradicated

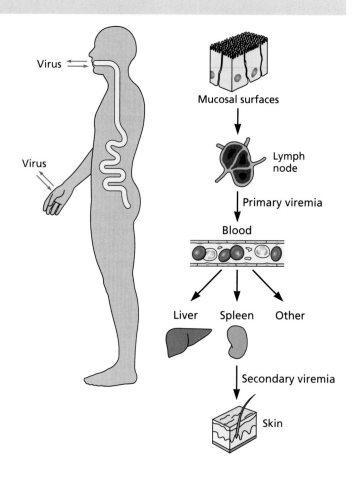

Figure 21

Reoviruses

Virus	Disease
Orthoreovirus	Mild upper respiratory tract disease, gastroenteritis, biliary atresia
Coltivirus	Colorado tick fever: febrile disease, headache, myalgia
Rotavirus	Gastroenteritis

Epidemiology

Transmission
- Fecal-oral route
- Arthropod

At risk or risk factors
- Malnourishment
- Age (<2 years)
- Proximity to vectors

Distribution
- Worldwide; type B prevalent in China
- Less common in summer

Vaccines or antiviral drugs
- Attenuated, oral vaccines available

Pathogenesis

Rotavirus: Large quantities of virus particles released in diarrhea

Colorado tick: persists in red blood cells up to 120 days

Human infections

Rotavirus:
Estimated worldwide cases annually: 111 million
Estimated deaths annually: 440,000
82% of deaths occur in the poorest countries from dehydration

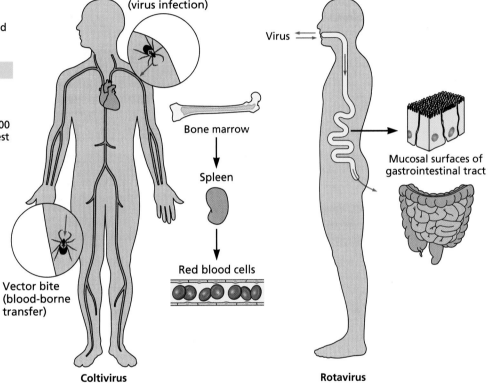

Vector bite (virus infection)

Bone marrow

Spleen

Red blood cells

Vector bite (blood-borne transfer)

Coltivirus

Virus

Mucosal surfaces of gastrointestinal tract

Rotavirus

Figure 22

Retroviruses (Human T-lymphotropic virus type 1)

Virus	Disease
Deltaretrovirus	
• Human T-lymphotropic virus type 1	Adult T-cell leukemia, tropical spastic paraparesis
• Human T-lymphotropic virus type 2	Hairy-cell leukemia
• Human T-lymphotropic virus type 5	Malignant cutaneous lymphoma

Epidemiology

Transmission
- Transfusions, needle sharing among drug users
- Virus in semen
  Anal and vaginal intercourse
- Perinatal transmission

At risk or risk factors
- Intravenous drug users
- Homosexuals and heterosexuals with many partners
- Newborns of virus-positive mothers

Distribution
- Worldwide
- No seasonal incidence

Vaccines or antiviral drugs
- No vaccines
- Some antivirals may be useful

Pathogenesis

Infects T lymphocytes

Remains latent or reproduces slowly, induces clonal outgrowth of T cell clones

Long latency period (30 years) before onset of leukemia

Infection leads to immunosuppression

Human Infections

Highest incidence in Japan and islands in the Caribbean

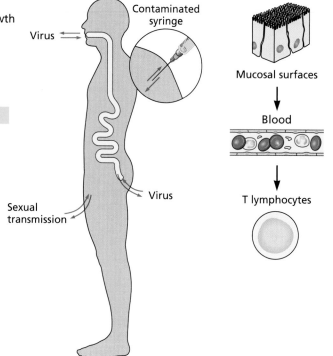

Figure 23

Retroviruses (Human immunodeficiency virus types 1 and 2)

Virus	Disease
Lentivirus • Human immunodeficiency virus types 1 and 2	Acquired immune deficiency syndrome (AIDS)

Epidemiology

Transmission
- Transfusions, needle sharing among drug users
- Virus in semen
    Anal and vaginal intercourse
- Perinatal transmission

At risk or risk factors
- Intravenous drug users
- Homosexuals and heterosexuals with many partners
- Prostitutes
- Newborns of virus-positive mothers

Distribution
- Worldwide
- No seasonal incidence

Vaccines or antiviral drugs
- No vaccines
- Antiviral drugs
    Nucleoside analog reverse transcriptase inhibitors (e.g., azidothymidine, dideoxycytidine)
    Nonnucleoside reverse transcriptase inhibitors (e.g., nevirapine, delavirdine)
    Protease inhibitors (e.g., saquinavir, ritonavir)
    Integrase inhibitors (e.g., raltegravir, elvitegravir)
    Fusion inhibitors (e.g., enfuvirtide, maraviroc)

Pathogenesis

Infects mainly CD4$^+$ T cells and macrophages

Lyses CD4$^+$ T cells, persistently infects macrophages

Infection alters T cell and macrophage function; immunosuppression leads to secondary infection and death

Infects long-lived cells, establishing reservoir for persistent infection

Infected monocytes spread to brain, causing dementia

Human Infections

Percentage of adults (ages 15-29 years) infected worldwide (2011)

Afghanistan	<0.1
Argentina	0.5
Belgium	0.3
Botswana	23.4
Canada	0.3
Ethiopia	1.4
India	0.3
Philippines	>0.1
Russia	1.1
South Africa	17.3
Swaziland	26.5
United States	0.6

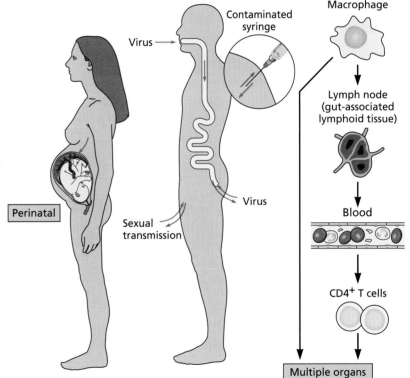

Figure 24

Rhabdoviruses (Rabies virus)

Virus	Disease
Lyssavirus	
• Rabies virus	Rabies
Vesiculovirus	
• Vesicular stomatitis virus	Flu-like illness

Epidemiology

Transmission
- Bites of wild animals and unvaccinated dogs and cats

At risk or risk factors
- Animal handlers, veterinarians
- Those in countries with no pet vaccinations or quarantine

Distribution
- Worldwide, except certain islands and U.K.
- No seasonal incidence

Vaccines or antiviral drugs
- Vaccines for pets and wild animals
- Inactivated virus vaccine for at-risk personnel, postexposure prophylaxis
- No antiviral drugs

Pathogenesis

Reproduces in muscle at bite site

Incubation period of weeks to months, depending on inoculum and distance of bite from central nervous system

Infects peripheral nerves and travels to brain

Reproduction in brain causes hydrophobia, seizures, hallucinations, paralysis, coma, and death

Spreads to salivary glands of nonhuman animals, from which it is transmitted

Postexposure immunization can prevent disease due to long incubation period

Human Infections

Rabies deaths estimated 20,000 per year worldwide

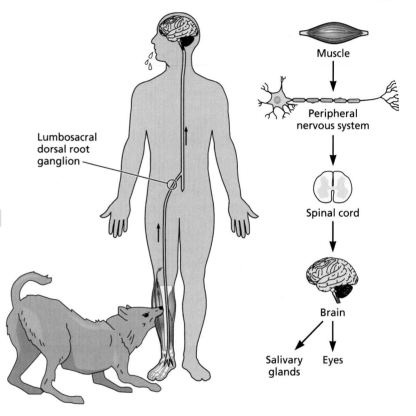

Lumbosacral dorsal root ganglion

Muscle

Peripheral nervous system

Spinal cord

Brain

Salivary glands

Eyes

Figure 25

Togaviruses (Alphaviruses)

Virus	Vector	Disease
Alphaviruses		
• Venezuelan equine encephalitis virus, Eastern equine encephalitis virus, Western equine encephalitis virus	*Aedes, Culex, Culiseta* mosquitoes	Mild systemic; severe encephalitis
• Chikungunya virus	*Aedes* mosquitoes	Fever, arthralgia, arthritis

Epidemiology

Transmission
• Mosquito vectors

At risk or risk factors
• Proximity to vector

Distribution
• Range determined by habitat of vector
• Most common in summer

Vaccines or antiviral drugs
• None

Pathogenesis

Antibodies limit virus spread by viremia (e.g., to fetus in pregnant host)

Cell-mediated immunity important to resolve infection

Human Infections

Chikungunya endemic in >40 countries and associated with sporadic epidemics

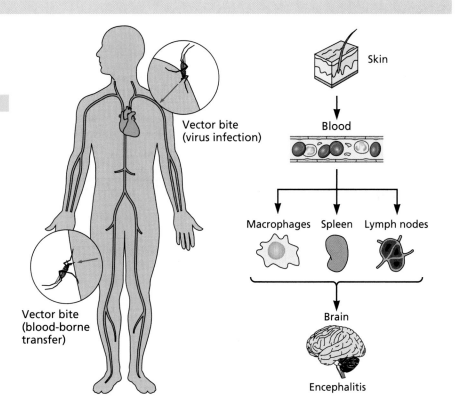

Vector bite (virus infection)

Vector bite (blood-borne transfer)

Skin

Blood

Macrophages Spleen Lymph nodes

Brain

Encephalitis

Figure 26

Togaviruses (Rubella virus)

Virus	Disease
Rubella virus	Rubella Cerebral palsy

Epidemiology

Transmission
- Aerosols

Distribution
- Worldwide
- No seasonal incidence

At risk or risk factors
- Age (<20 weeks)
- Lack of vaccination

Vaccines or antiviral drugs
- Attenuated vaccine

Pathogenesis

Antibodies limit virus spread by viremia (e.g., to fetus in pregnant host)

Cell-mediated immunity important to resolve infection

Human Infections

U.S. rubella epidemic 1964–65 (pre-vaccine):
12.5 million cases
2,000 cases of encephalitis
2,100 deaths
11,200 abortions

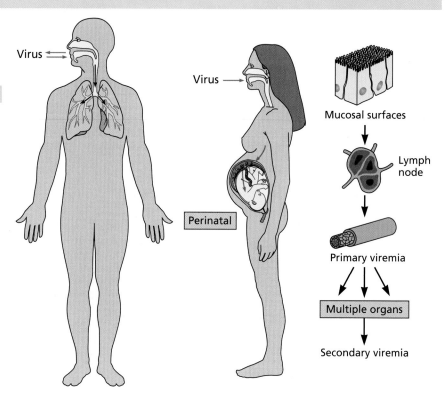

Figure 27

Glossary

Abortive infection An incomplete infectious cycle; virions infect a susceptible cell or host but do not complete reproduction, usually because an essential viral or cellular gene is not expressed. *(Chapter 5)*

Accessibility An attribute that describes the physical availability of cells to virus particles at the site of infection. *(Chapter 2)*

Active immunization The process of inducing an immune response by exposure to a vaccine; contrasts with passive immunization. *(Chapter 8)*

Acute infection A common pattern of infection in which virus particles are produced rapidly, and the infection is resolved quickly by the immune system; survivors are usually immune to subsequent infection. *(Chapter 5)*

Adaptive response The immune response consisting of antibody (humoral) and T lymphocyte-mediated responses; unlike the innate response, the adaptive response is tailored to the particular foreign invader; the adaptive response has memory: subsequent infections by the same agent are met with a robust and highly specific response. Also known as the acquired immune response. *(Chapters 2 and 4)*

Adjuvant A compound or mixture that stimulates immune responses to an antigen *(Chapter 8)*

Adoptive transfer The transfer of cells, usually lymphocytes, from an immunized donor to a nonimmune recipient. *(Chapters 4 and 8)*

Alternative pathway One of three pathways in the complement system; activates the C3 and C5 convertases without going through the C1-C2-C4 complex. *(Chapter 3)*

Anchorage independence The ability of some cells to grow in the absence of a surface on which to adhere; often detected by the ability to form colonies in semisolid media. *(Chapter 6)*

Antibody-dependent cell-mediated cytotoxicity The process in which binding of an anti-viral IgG antibody to Fc receptors on macrophages and some NK cells targets these cells to kill infected cells that carry on their surfaces the antigen recognized by the antibody; also known as ADCC. *(Chapter 4)*

Antigen Protein, DNA, lipid or polysaccharide that induces an immune response. *(Chapter 4)*

Antigenic drift The appearance of virus particles with a slightly altered surface protein (antigen) structure as a result of the accumulation of point mutations following passage and immune selection in the natural host. *(Chapter 5)*

Antigenic shift A major change in one or more surface proteins of a virus particle when genes encoding markedly different surface proteins are acquired during infection; this process occurs when viruses with segmented genomes exchange segments, or when nonsegmented viral genomes recombine after coinfection. *(Chapter 5)*

Antigenic variation The display by virus particles or infected cells of new protein sequences that are not recognized by antibodies or T cells that responded to previous infections. *(Chapter 5)*

Antiviral state A condition in which cells cannot support reproduction of viruses as a result of binding and responding to interferon. *(Chapter 3)*

Apoptosis Cell death following a sequence of tightly regulated reactions induced by external or internal stimuli that signal DNA damage or other forms of stress; characterized by chromosome degradation, nuclear degeneration and cell lysis; a natural process in development and the immune system, but also an intrinsic defense of cells to viral infection. Also called programmed cell death. *(Chapters 2 and 3)*

Attenuated Having mild or inconsequential instead of normally severe symptoms or pathology as an outcome of infection; having a state of reduced virulence. *(Chapter 5)*

Autocrine growth stimulation Stimulation of cell growth by proteins produced and sensed by the same cell. *(Chapter 6)*

Autophagy A process in which cells are induced to degrade the bulk of their cellular contents for recycling within specialized membrane-bounded compartments called autophagolysosomes. *(Chapter 3)*

Blind screening Screening for antiviral compounds without regard to a specific mechanism. *(Chapter 9)*

Case fatality ratio The number of deaths divided by the number of clinically confirmed infections. *(Chapter 1)*

Caspases Crirical proteases in apoptosis; members of a family of cysteine proteases that specifically cleave after aspartate residues. *(Chapter 3)*

CD markers *See* Cluster-of-differentiation markers.

CD4⁺ T cells T lymphocytes that carry the coreceptor protein CD4 on their surfaces. *(Chapter 4)*

CD8⁺ T cells T lymphocytes that carry the coreceptor CD8 on their surfaces. *(Chapter 4)*

Cell cycle The orderly and reproducible sequence in which cells increase in size, duplicate the genome, segregate duplicated chromosomes, and divide. *(Chapter 6)*

Cell-mediated response The arm of the adaptive immune response consisting of helper and effector T lymphocytes. *(Chapter 4)*

Central memory T cells Self-renewing memory T cells that are abundant in lymph nodes and other lymphoid tissues. *(Chapter 4)*

Chemokines Small proteins that attract and stimulate cells of the immune defense system; produced by many cells in response to infection. *(Chapter 3)*

Circadian rhythm The cycle (roughly 24 hours in humans) that regulates many physiological processes, such as sleep-wake cycles. *(Chapter 1)*

Clades Subtypes of human immunodeficiency virus that are prevalent in different geographic areas. *(Chapter 7)*

Classical pathway One of three complement pathways that lead to activation of C3-C5 convertases; activation occurs by direct interaction of C1q or C3b proteins with a viral protein/antibody complex on the surface of an infected cell or a virus particle. *(Chapter 3)*

Clinical latency A state of persistent viral infection in which no clinical symptoms are manifested. *(Chapter 7)*

Cluster-of-differentiation markers Distinct surface proteins that are recognized by specific monoclonal antibodies; these antibodies bind to various cluster-of-differentiation markers and are used to distinguish different cell types (e.g., CD4 on helper T cells). Also called CD markers. *(Chapter 4)*

Cold chain A supply chain in which the low temperatures required to preserve vaccines (and other biological agents) are maintained continuously. *(Chapter 8)*

Complement A general term referring to all the components of the complement system. *(Chapter 3)*

Complement system A set of blood plasma proteins that act in a concerted fashion to destroy extracellular pathogens and infected cells; originally defined as a heat-labile activity that lysed bacteria in the presence of antibody (it "complemented" antibody action); the activated complement pathway also stimulates phagocytosis, chemotaxis, and inflammation. *(Chapter 4)*

c-Oncogene A cancer-causing gene encoded in cellular genomes: may be formed via mutagenesis of a gene that does not cause cancer, known therefore as a proto-oncogene. *(Chapter 7)*

Contact inhibition Cessation of cell division when cells make physical contact, as occurs at high density in a culture dish. *(Chapter 6)*

Control group A group of experimental subjects, such as patients in a clinical trail, who receive no treatment or a placebo. *(Chapter 9)*

Cutaneous immune system The lymphocytes and scavenger antigen-presenting cells (Langerhans cells) that comprise the skin-associated lymphoid tissue. *(Chapter 4)*

Cytokines Soluble proteins produced by cells in response to various stimuli, including virus infection; they affect the behavior of other cells both locally and at a distance, by binding to specific cytokine receptors. *(Chapters 3 and 4)*

Cytokine storm See systemic inflammatory response syndrome *(Chapters 1 and 5)*

Cytopathic effect Deleterious morphological changes induced in cells by viral infection. *(Chapter 3)*

Cytopathic virus A virus that causes characteristic visible cell damage and death upon infection of cells in culture. *(Chapter 5)*

Damage-associated molecular patterns Host components that are released upon cellular damage. *(Chapter 3)*

Defensins Small (29- to 51- residue), cysteine-rich, cationic proteins produced by lymphocytes and epithelial cells that are active against bacteria, fungi, and enveloped viruses; usually found in the gut. *(Chapter 2)*

Delayed-type hypersensitivity A reaction caused by CD4$^+$ T cells that recognizes antigens in the skin; the reaction typically occurs hours to days after antigen is injected, hence its name; it is partially responsible for characteristic local responses to virus infections, such as rashes. *(Chapter 4)*

Diapedesis The process by which viruses cross the vascular endothelium, while being carried within monocytes or lymphocytes. *(Chapter 2)*

Direct-acting antivirals Drugs that inhibit viral enzymes. *(Chapter 9)*

Disseminated infection An infection that spreads beyond the primary site; often includes viremia and infection of major organs such as the liver, lungs, and kidneys. *(Chapter 2)*

DNA synthesis phase *See* S phase.

DNA vaccine A preparation of DNA containing the genes for one or more antigenic proteins; when the pure DNA preparation is injected into a test subject and enters cells, the proteins are synthesized, and an immune response to those proteins is elicited. *(Chapter 8)*

Double blind A trial in which neither the investigators nor the patients know which patients belong to treated and control groups. *(Chapter 1)*

Effector memory T cells Memory T cells that produce cytokines rapidly upon re-encountering a viral antigen, and are generally present in the circulatory system. *(Chapter 4)*

Emerging virus A viral population responsible for a marked increase in disease incidence, usually as result of changed societal, environmental, or population factors. *(Chapter 11)*

Endemic A disease or condition typical of a particular population or geographic area; persisting in a population for a long period without reintroduction of the causative agent from outside sources. *(Chapter 1)*

Endogenous antigen presentation The cellular process by which viral proteins are degraded inside the infected cell, and the resulting peptides are loaded onto major histocompatibility complex class I molecules that move to the cell surface. *(Chapter 4)*

Enhancing antibodies Antibodies that can facilitate viral infection by allowing virus particles to which they bind to enter susceptible cells. *(Chapter 7)*

Epidemic A pattern of disease characterized by rapid and sudden appearance of cases spreading over a wide area. *(Chapter 1)*

Epidemiology The study of the incidence, distribution, and spread of infectious disease in populations with particular regard to identification and subsequent control. *(Chapter 1)*

Epitope The parts of an antigen that are bound by an antibody or that are recognized by a T cell receptor in the context of major histocompatibility proteins. *(Chapter 4)*

Error threshold A mathematical parameter that measures the complexity of the information that must be maintained to ensure survival of a population. *(Chapter 10)*

Etiology The cause or causes of a disease. *(Chapter 1)*

Exogenous antigen presentation The cellular process in which viral proteins are engulfed from the outside of the cell, degraded, and the resulting peptides loaded onto major histocompatibility complex class II molecules that then move to the cell surface for presentation to T cells. *(Chapter 4)*

Extrinsic pathway Pathway by which apoptosis is induced when a proapoptotic ligand binds to its cell surface receptor. *(Chapter 3)*

Fitness The degree to which an organism is able to reproduce its environment. *(Chapter 10)*

Foci Clusters of cells that are derived from a single progenitor and share properties, such as unregulated growth, that cause them to pile up on one another. A single such cluster is called a **focus** *(Chapter 6)*

Fomites Inanimate objects that may be contaminated with microorganisms and become vehicles for transmission. *(Chapter 1)*

Gap phases (G$_1$ and G$_2$) Phases in the cell cycle between the mitosis (M) and DNA synthesis (S) phases. *(Chapter 6)*

Genetic bottleneck A descriptive term evoking the extreme selective pressure on small populations that results in loss of diversity, accumulation of selected mutations, or both. *(Chapter 10)*

Genetic drift Diversity in viral genomes that arises as a result of errors during genome replication and immune selection. *(Chapter 10)*

Genetic shift Diversity in viral genomes that arises as a result of re-assortment of genome segments or recombination between genomes. *(Chapter 10)*

Granzymes Members of a family of serine proteases that are released from activated cytotoxic T cells and induce apoptosis of target cells. *(Chapter 4)*

G$_0$ *See* Resting state.

Helper virus A virus that provides viral proteins needed for the reproduction of a coinfecting defective virus or subviral agents. *(Chapter 12)*.

Hematogenous spread Spread of virus particles through the bloodstream. *(Chapter 2)*

Hepatitis Inflammation of the liver. *(Chapters 2 and 5)*

Herd immunity The immune status of a population, rather than an individual. *(Chapter 8)*

Heterologous T cell immunity A secondary T cell response to antigen that is related but not identical to the immunodominant antigens that elicited the primary T cell response. *(Chapter 5)*

Humoral response The arm of the adaptive immune response that produces antibodies. *(Chapter 4)*

Immortality The capacity of cells to grow and divide indefinitely. *(Chapter 6)*

Immune memory A property provided by specialized B and T lymphocytes (memory B and T cells) that respond rapidly upon reexposure to an antigen. *(Chapter 8)*

Immunodominant Having the property of being recognized most efficiently by cytotoxic T lymphocytes and antibodies; said of peptides and epitopes. *(Chapter 5)*

Immunological synapse A specialized, organized structure formed upon aggregation of the T cell receptors of a cytotoxic T cell bound to peptide presented by MHC on the target cells; this structure allow prolonged signaling from the engaged T cell receptors and associated co-receptors, and facilitates polarization of the T cell secretion machinery. *(Chapter 4)*

Immunopathology Pathological changes caused partly or entirely by the immune response. *(Chapters 1 and 5)*

Immunotherapy A treatment that provides an infected host with exogenous antiviral cytokines, other immunoregulatory agents, antibodies, or lymphocytes in order to reduce viral pathogenesis. *(Chapter 8)*

Inactivated vaccine A vaccine made by taking a disease-causing virus and treating it (e.g., with chemicals) to reduce infectivity to undetectable levels. *(Chapter 8)*

Incidence The frequency with which a disease appears in a particular population or area (e.g., the number of newly diagnosed cases during a specific period); distinct from the prevalence (i.e., the number of cases in a population on a certain date). *(Chapter 1)*

Incubation period The period before symptoms of disease appear after an infection. *(Chapter 5)*

Index case The human or other animal originally infected in an epidemic. *(Chapter 1)*

Infectious mononucleosis An infectious disease caused by Epstein-Barr virus; characterized by an increase in the number of lymphocytes with a single nucleus. *(Chapter 5)*

Inflammation A general term for the elaborate response that leads to local accumulation of white blood cells and fluid; initiated by local infection or tissue damage; many different forms of this response, characterized by the degrees of tissue damage, capillary leakage, and immune cell infiltration, occur after infection with pathogens. *(Chapter 3)*

Innate response The first line of immune defense; able to function continually in the host without prior exposure to the invading pathogen. This elaborate system includes cytokines, sentinel cells, complement, and natural killer cells. *(Chapters 2 and 3)*

Insertional activation The mechanism of oncogenesis by nontransducing retroviruses; integration of a proviral promoter or enhancer in the vicinity of a proto-oncogene results in inappropriate transcription of that gene, making it a cellular oncogene (c-oncogene) *(Chapter 6)*

Interfering antibodies Antibodies that can bind to virus particles or infected cells and block interaction with neutralizing antibodies. *(Chapter 7)*

Interferons Cytokines that activate antiviral programs. *(Chapter 3)*

Interleukins Secreted cytokines that allow communication among leukocytes. *(Chapter 4)*

Intrinsic cellular defenses The conserved cellular programs that respond to various stresses, such as starvation, irradiation, and infection; intrinsic defenses include apoptosis, autophagy, and RNA interference. *(Chapters 2 and 3)*

Intrinsic pathway Pathway of apoptosis in which cell death is induced in response to indicators of *internal* stress, such as DNA damage. *(Chapter 3)*

Koch's postulates Criteria developed by the German physician Robert Koch in the late 1800s to determine whether a given agent is the cause of a specific disease. *(Chapter 1)*

Koplik's spots Small spots inside the mouth that are hallmarks of measles virus infection. *(Chapter 1)*

Kupffer cells Macrophages of the liver that are part of the reticuloendothelial system. *(Chapter 2)*

Latency-associated transcript RNA produced specifically during a latent infection by herpes simplex virus. *(Chapter 5)*

Latent infection A class of persistent infection that lasts the life of the host; few or no virus particles can be detected, despite continuous presence of the viral genome. *(Chapter 5)*

Lectin pathway One of three complement pathways that lead to activation of C3-C5 convertases; mannose-binding, lectin-associated proteases cleave the C2 and C4 proteins. *(Chapter 3)*

Lethal mutagenesis The elevation of mutation rates by exposure to a mutagen or an error-prone polymerase to the point at which the resulting population of genomes has lost fitness and is incapable of propagating. *(Chapter 10)*

Long-latency retrovirus A retrovirus that causes cancer in a host many years after infection; the viral genome does not encode cellular oncogenes, nor does it cause cancer by perturbing the expression of cellular oncogenes. *(Chapter 6)*

Maximum tolerated dose The highest dose of a drug or other treatment that does not cause unacceptable side effects. *(Chapter 9)*

M cell Microfold or membranous epithelial cell; cells of mucosal surfaces specialized for delivery of antigens to underlying lymphoid tissues. *(Chapters 2 and 4)*

Memory cells A subset of B and T lymphocytes maintained after each encounter with a foreign antigen; these cells survive for years and are ready to respond and proliferate upon subsequent encounter with the same antigen. *(Chapter 4)*

Metagenomic analysis Nucleic acid sequencing of samples recovered directly from the environment, and comprising multiple genomes. *(Chapter 10)*

Metastases Secondary tumors, often at distant sites, that arise from the cells of a malignant tumor. *(Chapter 6)*

MHC restriction The recognition of an antigen by T cells only when it is presented by MHC of the haplotype identical to that of the T cells. *(Chapter 4)*

Microbicides Creams or ointments that inactivate or block virus particles before they can attach and penetrate tissues. *(Chapter 9)*

Mitogens Extracellular signaling molecules that induce cell proliferation. *(Chapter 6)*

Mitosis The phase of the cell cycle in which newly duplicated chromosomes are distributed to two new daughter cells as a result of cell division. Also called M phase. *(Chapter 6)*

Molecular mimicry Sequence similarities between viral peptides and self-peptides that result in the cross-activation of autoreactive T or B cells by virus-derived peptides. *(Chapters 5 and 8)*

Monoclonal antibody-resistant mutants Viral mutants selected to propagate in the presence of neutralizing monoclonal antibodies; often carry mutations in viral genes encoding structural proteins. *(Chapter 4)*

Morbidity The percentage of individuals in a specified population who show symptoms of infection in a given period. *(Chapter 1)*

Mortality The percentage of deaths in a specified population of infected individuals. *(Chapter 1)*

M phase *See* Mitosis.

Muller's ratchet A model positing how small, asexual populations decline in fitness over time if the mutation rate is high. *(Chapter 10)*

Natural killer cells An abundant lymphocyte population that comprises large, granular cells; distinguished from other lymphocytes by the absence of B and T cell antigen receptors; these cells are part of the innate defense system. Also called NK cells. *(Chapter 3)*

Negative selection (re: T cells) Elimination of T cells that recognize target cells that display "self" peptides on their surfaces. *(Chapter 4)*

Neuroinvasive virus A virus that can enter the central nervous system (spinal cord and brain) after infection of a peripheral site. *(Chapter 2)*

Neurotropic virus A virus that can infect neurons. *(Chapter 2)*

Neurovirulent virus A virus that can cause disease in nervous tissue, manifested by neurological symptoms and often death. *(Chapter 2)*

Neutralizing antibodies Antibodies that block the infectivity of virus particles. *(Chapter 7)*

NK cells *See* Natural killer cells.

Noncytopathic virus A virus that produces no visible signs of infection in cells. *(Chapter 5)*

Nontransducing oncogenic retroviruses Retroviruses that do not encode cell-derived oncogene sequences but can cause cancer (at low efficiency) when their DNA becomes integrated in the vicinity of a cellular oncogene, thereby perturbing its expression. *(Chapter 6)*

Oncogene A gene encoding a protein that causes cellular transformation or tumorigenesis. *(Chapter 6)*

Oncogenesis The processes leading to cancer. *(Chapter 6)*

Original antigenic sin A secondary immune response to an antigen that is related, but not identical to, the antigen that elicited the primary response. *(Chapter 5)*

Pandemic A worldwide epidemic. *(Chapter 1)*

Pantropic virus A virus that replicates in many tissues and cell types. *(Chapter 2)*

Passive immunization Direct administration of the products of the immune response (e.g., antibodies or stimulated immune cells) obtained from an appropriate donor(s) to a patient; contrasts with active immunization. *(Chapter 8)*

Pathogen A disease-causing virus or other microorganism. *(Chapters 1 and 4)*

Pathogen-associated molecular patterns Molecules or molecular features unique to pathogens. *(Chapter 3)*

Pattern recognition receptors Protein receptors of the innate immune system that bind definitive molecular features of pathogens; present in sentinel cells, such as immature dendritic cells and macrophages. *(Chapter 3)*

Permissive Able to support virus reproduction when the viral genome is introduced; refers to cells. *(Chapter 2)*

Permissivity A cellular environment that provides all cellular components required for viral reproduction. *(Chapter 2)*

Persistent infection A viral infection that is not cleared by the combined actions of the innate and adaptive immune response. *(Chapter 5)*

Phagocytosis Engulfment of dying cellular debris and virus particles by myeloid cells, including dendritic cells. *(Chapter 4)*

Plasma cells Mature B cells that synthesize secreted antibodies. *(Chapter 4)*

Polymorphic gene A gene that has many allelic forms in outbred populations. *(Chapter 4)*

Positive selection (re: T cells) The process in which only T cells with T cell receptors that can bind to MHC proteins are retained during T cell differentiation. *(Chapter 4)*

Power The probability that a meaningful difference or effect can be detected, if one were to occur. *(Chapter 1)*

Prevalence The proportion of individuals in a population having a disease; the number of cases of a disease present in a particular population at a given time. *(Chapter 1)*

Primary antibody response The response of B cells following first exposed to a pathogen. *(Chapter 4)*

Primary viremia Progeny virus particles released into the blood after initial virus propagation at the site of entry. *(Chapter 2)*

Prodrug An inactive precursor to an active antiviral compound. *(Chapter 9)*

Prions Infectious agents comprising an abnormal isoform of a normal cellular protein but no nucleic acid; implicated as the causative agents of transmissible spongiform encephalopathies. *(Chapter 12)*

Professional antigen-presenting cells Dendritic cells, macrophages, and B cells; defined by their ability to take up antigens and present them to naïve T lymphocytes in the groove of an major histocompatibility complex class II molecule. *(Chapter 4)*

Prospective studies Studies in which cohorts of subjects with and without the condition(s) or treatment(s) of interest are examined for a specified period. *(Chapter 1)*

Proteome The total protein repertoire of a sample, such as a preparation of virus particles or a type of host cell. *(Chapter 9)*

Proto-oncogene A normal cellular gene that, when altered by mutation or misregulation, can contribute to cancer; thereafter called a cellular oncogene (c-oncogene). *(Chapter 6)*

Quasispecies Virus populations that exist as dynamic distributions of nonidentical but related replicons. *(Chapter 10)*

Recombinant vaccine A vaccine produced by recombinant DNA technology. *(Chapter 8)*

Replication-competent, attenuated vaccine A vaccine made from viral mutants that have reduced virulence but can reproduce; they often also have reduced capacity for transmission. *(Chapter 8)*

Reservoir The host population in which a viral population is maintained. *(Chapters 1 and 11)*

Resting state A state in which the cell has ceased to grow and divide and has withdrawn from the cell cycle. Also called G_0. *(Chapter 6)*

Reticuloendothelial system Macrophages that line sinusoids present in organs such as liver, spleen, bone marrow, and adrenal glands. *(Chapter 2)*

Retrospective study A study that looks backwards in time and examines exposures to suspected risk or protection factors in relation to a particular outcome. *(Chapter 1)*

Satellites Small, single-stranded RNA molecules that lack genes required for their replication, but are replicated in the presence of another virus that can supply the required proteins (the **helper virus**). *(Chapter 12)*

Satellite RNA An RNA that does not encode capsid proteins, and is packaged by a protein(s) encoded in a helper virus genome. *(Chapter 12)*

Satellite virus A satellite with a genome that encodes one or two proteins. *(Chapter 12)*

Secondary antibody response The antibody response produced after a subsequent infection or challenge with the same antigen or virus. *(Chapter 4)*

Secondary viremia Delayed appearance of a high concentration of infectious virus in the blood as a consequence of disseminated infections. *(Chapter 2)*

Sentinel cells Dendritic cells and macrophages; migratory cells that are found in the periphery of the body and can take up proteins and cell debris for presentation of peptides derived from them on major histocompatibility complex molecules. These cells respond to recognition of a pathogen by synthesizing cytokines such as interferons. *(Chapters 2, 3, and 4)*

Sepsis Uncontrolled, systemic inflammation induced by infection with a pathogen. *(Chapter 5)*

Shedding The release of virus particles from an infected host. *(Chapter 2)*

Signal transduction cascade A chain of sequential physical interactions among, and biochemical modifications of, membrane-bound and cytoplasmic proteins. *(Chapter 6)*

Sinusoids Small blood vessels characterized by a discontinuous basal lamina, with no significant barrier between the blood plasma and the membranes of surrounding cells. *(Chapter 2)*

Slow infection An extreme variant of the persistent pattern of infection; has a long incubation period (years) from the time of initial infection until the appearance of recognizable symptoms. *(Chapter 5)*

Slow viruses Viruses characterized by long incubation periods, typical for the genus lentivirus in the family *Retroviridae*. *(Chapter 7)*

S phase The phase of the cell cycle in which the DNA genome is replicated. *(Chapter 6)*

Structural plasticity The ability of virus particles to tolerate large numbers of amino acid substitutions in surface proteins without losing infectivity. *(Chapter 5)*

Subunit vaccine A vaccine formulated with purified components of virus particles, rather than intact virus particles. *(Chapter 8)*

Superantigen Extremely powerful membrane bound T cell proteins that nonspecifically activate many subsets of T cells. *(Chapter 5)*

Susceptibility The property of a cell that enables it to be infected by a particular virus (e.g., the presence of a viral receptor[s] on the cell surface). *(Chapter 2)*

Susceptible Producing the receptor(s) required for virus entry; refers to cells. *(Chapter 2)*

Systemic infection An infection that results in spread to many organs of the body. *(Chapter 2)*

Systemic inflammatory response syndrome A disproportionate host response that leads to large-scale release of inflammatory cytokines and stress mediators, resulting in severe pathogenesis or death. Also known as a cytokine storm. *(Chapter 5)*

Systems biology An approach in which experimental methods to identify all the components of a biological system and their interactions are linked to properties and functions of the system using computational models. *(Chapter 1)*

Transcytosis A mechanism of transport in which material in the intestinal lumen is endocytosed by M cells, transported to the basolateral surface, and released to the underlying tissues. *(Chapters 2 and 4)*

Transducing oncogenic retroviruses Retroviruses that include oncogenic, cell-derived sequences in their genomes and carry these sequences to each newly infected cell; such viruses are highly oncogenic. *(Chapter 6)*

Transformed Having changed growth properties and morphology as a consequence of infection with certain oncogenic viruses, introduction of oncogenes, or exposure to chemical carcinogens. *(Chapter 6)*

Transforming infection A class of persistent infection in which cells infected by certain DNA viruses or retroviruses may exhibit altered growth properties and proliferate faster than uninfected cells. *(Chapter 5)*

Tropism The predilection of a virus to invade, and reproduce, in a particular cell type. *(Chapter 2)*

Tumor A mass of cells originating from abnormal growth. *(Chapter 6)*

Tumor suppressor gene A cellular gene encoding a protein that negatively regulates cell proliferation; mutational inactivation of both copies of the genes is associated with tumor development. *(Chapter 6)*

Vaccination Inoculation of individuals with attenuated or related microorganisms, or their antigenic products, in order to elicit an immune response that will protect against later infection by the corresponding pathogen. *(Chapter 8)*

Variolation Inoculation of healthy individuals with material from a smallpox pustule, or in modern times from a related or attenuated cowpox (vaccinia) virus preparation, through a scratch on the skin (called scarification). *(Chapter 8)*

Vector A carrier, often an arthropod, that transmits a virus or other infectious agent from one host to another. *(Chapter 1)*

Viral pathogenesis The processes by which viral infections cause disease. *(Chapters 1, 2, and 5)*

Viremia The presence of infectious virus particles in the blood. *(Chapter 2)*

Viroceptor A viral protein that modulates cytokine signaling or cytokine production by mimicking host cytokine receptors. *(Chapters 3 and 5)*

Viroids Unencapsidated, small, circular, single-stranded RNA molecules that replicate autonomously when introduced mechanically into host plants. *(Chapter 12)*

Virokine A secreted viral protein that mimics cytokines, growth factors, or similar extracellular immune regulators. *(Chapters 3 and 5)*

Viroporin Hydrophobic viral protein that forms pores in cellular membranes; many facilitate release of progeny virus particles. *(Chapter 5)*

Virtual screening Computational methods for iterative docking of chemical compounds into a chosen site in a protein target to identify drug leads. *(Chapter 9)*

Virulence The relative capacity of a viral infection to cause disease. *(Chapter 5)*

Viruria The presence of viruses in the urine. *(Chapter 2)*

Virus evolution The constant change of a viral population in the face of selection pressures. *(Chapter 10)*

v-Oncogene An oncogene that is encoded in a viral genome. *(Chapter 6)*

Xenophagy The capture of virus particles for degradation in lysosomes. *(Chapter 3)*

Zoonoses (zoonotic infections) Diseases that are transferred from other animals to humans. *(Chapters 1 and 11)*

Index

A

A20 protein, 61
A46R protein, vaccinia virus, 58, 61
A52R protein, vaccinia virus, 58, 61
Abi system, 73
Abl tyrosine kinase, 203, 207
Abortive infections, 137, 157
Acanthamoeba castellanii, 331
Acanthamoeba polyphaga, 331
Acanthamoeba polyphaga mimivirus, 369
Accessibility of viral receptors, 36
Accessory proteins, human immunodeficiency
     virus type 1, 220, 223, 225–230
Accidental infections, 351
ACE2 (angiotensin-converting enzyme 2), 353
Acquired immunodeficiency syndrome (AIDS),
     403. *See also* Human immunodeficiency
     virus
  cancer and HIV, 246–248
  dynamics of HIV-1 reproduction in patients,
     241–242
  epidemic, 13
  historical aspects, 219–221
  HIV discovery and characterization, 219–221
  humoral response to HIV, 239
  immune dysfunction, 237–238
  lives saved by anti-HIV therapy, 312–313
  pandemic, 360
  prevention, multifaceted approach to, 252
  recognition as new disease, 15
  secondary infections that accompany, 360
  symptomatic phase of HIV infection, 236
  vaccine, quest for, 277–279
  variability of response to HIV infection,
     236–237
  worldwide impact of, 219
  zoonotic transmission of human
     immunodeficiency virus to humans, 11
Actin, 42
Activator protein 1 (AP-1), 80, 192, 195, 198–199,
     202
Active immunization, 260–261
Acute infection, 137–143, 172
  antigenic variation facilitating repeated,
     141–142
  course of typical, 140
  defined, 137, 139
  incubation period, 138
  limited by intrinsic and innate responses, 138
  MHC class I protein production, 144
  public health problems, 142–143
  r-replication strategy, 139
Acyclovir (Zovirax), 291, 293–294, 295
Adapter proteins, 198–199, 201, 225
Adaptive immunity, 98–130
  antigen presentation, 114–119
  B and T cell receptor diversity, 107–108
  cell-mediated response, 101, 103, 119–124
  events at site of infection, 108–114
  host response attributes, 99–100
    diversity and specificity, 100
    memory, 100
    self-control, 100
    speed, 99–100
  humoral (antibody) response, 101, 103, 122,
     125–128
  immunological memory, 100, 127, 128–130
  integration of intrinsic defense and innate
     immune response, 55
  lymphocyte development, diversity, and
     activation, 100–108
  overview, 99
  persistent infections, 144–147
  receptors and antigen specificity, 107–108
  self-limitation, 127
  specificity, 127
Adaptive response, 27
Adaptor protein 1 (AP-1), 229
Adar (adenosine deaminase acting on RNA), 71
ADCC (antibody-dependent cellular cytotoxicity),
     127, 239
Adefovir, 296, 297
Adematous polyposis coli *(apc)* gene, 176
Adeno-associated virus
    classification of, 370
    genome incorporation into host, 337
Adeno-associated virus vector, 251
Adenosine, 295
Adenosine arabinoside, 295
Adenosine deaminase, hepatitis delta satellite
     virus, 370
Adenosine deaminase acting on RNA (Adar), 71

Adenovirus
  antiviral drugs, 297
  apoptosis and, 66
  defensin promotion of infection, 34
  diseases, 380
  E1A protein, 66, 189–190, 192–193, 204–205, 210
  E1B protein, 66, 189–190, 209
  E3 protein, 61
  E4 Orf3 protein, 69
  E4 Orf6 protein, 209
  epidemiology, 380
  escape from immune surveillance by
     transformed cells, 210
  human infections, 380
  incubation period, 138
  integration, 187
  interferon response modulators, 82
  p53 inactivation, 209
  pathogenesis, 380
  persistent infection, 143
  spread in host, 114
  transformation of cultured mammalian cells,
     185
  transforming gene products, 190
  transforming genes, 185
  VA-RNA I, 84
Adenovirus type 2
  transforming gene products, 190
Adenovirus type 5
  RNA interference, suppression of, 71
Adenovirus type 12, 210
Adenovirus vaccine, 267, 269
Adenovirus vector, 158
Adjuvant, 110, 275–276
Adoptive transfer, 122, 261
Adult T-cell leukemia (ATL), 210–213, 402
Aedes aegypti, 16
Aedes albopictus, 16
Aerosols, 48
African swine fever virus, 360
Age, spread of infection influenced by, 350
Agricultural organisms
  economic toll of viral epidemics, 9–10
  movement of livestock, 360
  national programs for eradication of important
     viral diseases, 258–260

AIDS. *See* Acquired immunodeficiency syndrome
Air gun injection, 276
Akt, 179–180, 192, 195–196, 198–201, 203, 206, 208
Alimentary tract
  cellular organization of small intestine, 34
  commensal bacteria, 33
  skin compared, 33
  transcytosis in, 34–35
  viral entry, 30, 33–35
ALLINIs, 290
Allosteric antiviral, 290
Alpha-defensins, 34
Alphaherpesvirus
  diseases, 388–389
  epidemiology, 388–389
  evolution, 333
  human infections, 388–389
  latency, 43
  neural spread, 42–43
  pathogenesis, 388–389
  reactivation, 43
  ribonucleotide reductase, 162
  spread in host, 37
  thymidine kinase, 162
  use as self-amplifying tracers of synaptically connected neurons, 45
Alphavirus
  diseases, 405
  epidemiology, 405
  human infections, 405
  pathogenesis, 405
  persistent infection, 143
  vector, 405
Alternative hypothesis, 14
Alternative pathway, complement, 87–88
Alum, 276
Alveoli, 31–32
Amantadine (Symmetrel), 283–284, 298
Amprenavir, resistance to, 311
Anal sex, HIV entry with, 35
Anchorage independence, transformed cells, 178
Andrewes, Christopher, 6
Angiotensin-converting enzyme 2 (ACE2), 353
Animal models of disease, 135–136
  dead-end interactions, 345
Anogenital carcinoma, human immunodeficiency virus type 1, 248
Anterograde spread, 44
Antibody
  antibody-dependent cell-mediated cytotoxicity, 127
  autoantibodies, 238
  human immunodeficiency virus infection, 238, 239–240, 250
    broadly neutralizing antibody, 240, 250–251
    enhancing antibodies, 239–240
    immune system-based therapies, 250, 251
    interfering antibodies, 239
    neutralizing antibodies, 239–240
  monoclonal antibody-resistant mutants, 127
  natural, 89
  passive immunization, 260–261
  poliovirus escape from, 142
  primary antibody response, 125
  production by plasma cells, 122, 125
  structure, 126
  in tears, 35
  types and functions, 125–126, 128
  virus neutralization, 125–127

Antibody escape mutations, 127
Antibody response, 122, 125–128. *See also* Humoral immune response
Antibody-dependent cellular cytotoxicity (ADCC), 127, 239
Antigen, definition, 101
Antigen presentation, 114–119
  endogenous, 115–117
  exogenous, 116, 118–119
  major histocompatibility complex (MHC) class I, 115–119
  major histocompatibility complex (MHC) class II, 102, 115–119
Antigenic drift, 127, 141, 143, 268
Antigenic shift, 141–142, 268
Antigenic variation, repeated acute infections and, 141–142
Antigen-presenting cell
  acquisition of viral proteins, 108–111
  activation induced via costimulation, 114–115
  B cells, 102
  inflammasome, 109–110
  measles virus infection of, 170
  migration to lymph nodes, 111–114
Antiviral drugs, 282–313. *See also specific drugs*
  allosteric, 290
  bioavailability, 284, 293
  combination therapy, 310, 312
  cytotoxic index, 291
  delivery, 293
  direct-acting, 288, 309
  discovery, 284–293
    clinical trials, 292
    computational approaches, 287–288, 290
    descending staircase of, 291
    drug formulation and delivery, 293
    expense, 288, 291
    knowledge of viral life cycles, 284
    lexicon, 284
    path of, 285
    safety, 291, 293
    screening of compounds, 285–287
  dose effect on viral load, 310
  drug design, 287–288
    in silico, 288, 290
    structure-assisted, 287
    using genome sequencing information, 288
  efficacy, 283–284, 293, 305–306, 309
  examples, 293–299
  expense, 288, 291, 305
  formulation, 293
  genetics and, 285
  hepatitis C virus, 286
  historical perspective, 283–284
  human immunodeficiency virus, 285–286, 294, 297
    dynamics of HIV-1 reproduction in AIDS patients, 241–242
    prophylaxis, 250
    transmission rate, affect on, 233
    treatment, 248–249
  marketing, 283, 291
  overview, 283
  pharmokinetics, 284
  potency, 283
  research and development, 288, 291–293
  resistance, 309–312
  safety, 291, 293

  screening, 285–287
    blind, 283
    cell-based screens, 286
    combinatorial chemistry, 287
    high-content screens, 286–287
    high-throughput screens, 286–287
    mechanism-based screens, 285
    minireplicon, 286
    sources of chemical compounds used in, 287
    virtual, 288
  side effects, 291–293, 310
  success stories, 301–309
  targets, 300–301
    entry and uncoating inhibitors, 300–301
    genetics and drug discovery, 285
    nucleic acid processing enzymes, 301
    proteases, 285, 301
    regulatory RNA molecules, 300–301
    viral nucleic acid synthesis, 293–297, 301
    viral regulatory proteins, 300
  terminology, 294
  therapeutic index, 291
Antiviral state, interferon and, 79, 83–84
AP-1 (activator protein 1), 80, 192, 195, 198–199, 202
Apaf-1 protein, 208
apc (adematous polyposis coli) gene, 176
Apobec (apolipoprotein B mRNA-editing enzyme, catalytic polypeptide-like), 72
  Apobec3, 225–227
  Apobec3F, 227
  Apobec3G, 72, 226
Apoptosis, 27, 60, 62, 64–67
  blocked
    in Epstein-Barr virus infection, 156
    by herpes simplex virus type 1 latency-associated transcripts (LATs), 153
    by human T-lymphotropic virus type 1, 212
    in persistent infections, 144
  cytotoxic T lymphocyte induction of, 120–121
  as defense against viral infection, 66
  extrinsic pathway, 64–66
  HIV Vpr protein and, 227
  integration of inhibition of apoptosis with stimulation of cell proliferation, 206
  intrinsic pathway, 64–66
  monitoring by sentinel cells, 67
  process, 64
  programmed necrosis (necroptosis), 67
  pronunciation, 64
  spontaneous T cell apoptosis in HIV infection, 147
  viral gene products that modulate, 66–67
  viral inhibitors of, 206
Apoptosome, 65
Apple scar skin viroid, 365
Arachnoid, 44
Arbovirus vectors, 16
Archaea
  proteomic diversity, 332
  on tree of life, 332
Arenavirus
  attenuated reassortment viruses, 269
  disease, 381
  epidemiology, 381
  genetic map, 334
  human infections, 381
  pathogenesis, 381
  persistent infection, 143, 148

shedding in urine, 49
zoonotic infection, 344
Argentine hemorrhagic fever, 381
Arms race, host-virus, 337–339
Arthropod
  dead-end host-virus interactions, 345
  vectors, 4–6, 29, 49 (*see also specific arthropods*)
    breeding in used tires, 360
Asian tiger mosquito (*Aedes albopictus*), 16
Astrocytes, 124, 245
Ataxia telangiectasia mutated (*Atm*) gene, 208
Atf family, 80
ATL (adult T-cell leukemia), 210–213, 402
Atm (ataxia telangiectasia mutated) gene, 208
Atripla, 312
Attachment to host cells, human immunodefi-
    ciency virus, 230–231
Attenuated vaccine, 262, 264, 266–267, 269–271
  immune response, 269
  passage in nonhuman cell lines, 270
Attenuating mutations, 347
Australia, poxvirus release, 318, 345, 347
Autism, 12, 278
Autocrine growth stimulation, 178
Autoimmunity, 170–171, 238
Autolysosome, 68
Autophagy, 67–68
Autophosphorylation, 178–179
Auxiliary proteins, lentivirus, 220, 222
Avian influenza virus, 39, 325, 355–357
  H5N1, 48, 343, 355–356
  H7N9, 356, 361
  pandemic, 48
  reproduction in ferrets, 48
  scientific and societal implications of
      transmissibility experiments using animal
      models, 356–357
  transmission, 10
  tropism, 349
Avian leukemia, 182–183
Avian leukosis virus, 183
  insertional activation, 195–196
  properties, 184
Avian myelocytoma virus MH2, 194
Avocado sunblotch viroid, 365
Avsunviroid, 365–368
5-Azacytidine, 322
AZT (Retrovir, zidovudine), 294–297
  chain termination, 296
  resistance to, 311
  structure, 295–296
Aztecs, smallpox epidemic, 352
Azurophilic granules, 32

B

B cell lymphoma, 204, 210
  Epstein-Barr virus, 154–156
  human immunodeficiency virus type 1, 248
B cell receptor, 100–102, 104, 107–108, 115, 119,
    125
B cells
  activation to produce antibodies, 125
  antigen specificity, 107–108
  as antigen-presenting cell, 102
  Epstein-Barr virus latent infections in, 154–156
  in human immunodeficiency virus infection, 238
  humoral immune response, 101, 103, 122,
      125–128

immunopathology lesions caused by, 166–167
  memory cells, 53, 100–102, 128–130
  overview, 101–102
  plasma cell, 101–103
  positive and negative selection, 108
Bacteria
  commensal, 33
  defense systems, 73
  gastrointestinal tract, 33
  horizontal gene transfer, 326
  proteomic diversity, 332
  symbiotic, 33
  on tree of life, 332
Bacteriophage, recombination, 326
Bacteriophage 434, 326
Bacteriophage λ, 326
Bacteriophage Qβ, 321
Baculovirus, 272
Bak protein, 208
Bandicoot papillomatosis carcinomatosis virus
    type 1, 183
Bang, Olaf, 182
Barré-Sinoussi, Françoise, 221
Barriers to infection, 24–50
Basement membrane, 29, 31, 34, 39, 45–47
Bats
  life span, 349
  Nipah virus in, 350
  reservoir for SARS coronavirus, 353
  virus diversity in, 361
  zoonotic infections, 347, 349
Bax protein, 208
Bcl-2 protein, 65, 144
Benign, 176, 177
Betaherpesvirus
  diseases, 390
  epidemiology, 390
  evolution, 333
  human infections, 390
  pathogenesis, 390
  persistent infection, 144
β-propriolactone, 265
Bid protein, 65
Binary data, 14
Bioavailability, 284, 293
Biological weapon, 359
Biosafety, 356
Biosecurity, 356
Bioterrorism, 48, 359
Bird flu. *See* Avian influenza virus
Bishop, Michael, 188
BK virus, 186
  coevolution with host, 333
  diseases, 399
  epidemiology, 399
  human infections, 399
  pathogenesis, 399
  persistent infection, 143
  shedding in urine, 49
Blebbing, 62
Blind screening, 283
Blood
  hematogenous spread, 40–42
  hematopoietic stem cell, 100–102
  organ invasion by bloodstream, 45–47
  shedding of virions in, 49
  virus travel to tissues, 46
Blood supply, hepatitis viruses in the human,
    357–358

Blood transfusion, HIV infection via, 232
Blood-borne viruses, 40, 42
Blood-brain barrier, 46, 124
Blood-tissue junction, 45
Bone marrow, 101
Bone marrow stromal antigen 2 (Bst2), 83,
    227–228
Booster shots, 263
Bottleneck
  on the dissemination of a virus population,
      27–28
  genetic, 322–324, 327
Bovine immunodeficiency virus, 222
Bovine spongiform encephalopathy, 371, 374,
    376, 377
Bovine viral diarrhea virus
  persistent infection, 144
  recombination, 324
Bplf1, 61, 144
Brain
  attenuated herpes simplex virus used to clear
      human brain tumors, 163
  entry of virus by olfactory routes, 44
  herpes simplex virus infection, 39–40
  HIV infection, 41
  immune system and, 124
  measles virus infection, 148
  reduced immune surveillance in, 147
  tracing neuronal connections with viruses, 45
  virus entry, 46–47
Brand name, 294
Breakbone fever, 166
Breasts, human immunodeficiency virus patho-
    genesis and, 246
Brenner, Sydney, 317
Brown, Timothy Ray, 249
Brownian (random) motion, 42
Bru isolate, 221
BSE. *See* Bovine spongiform encephalopathy
Bst2 (bone marrow stromal antigen 2), 83,
    227–228
Bunyamwera virus, 382
Bunyavirus
  attenuated reassortment viruses, 269
  discovery of new, 358
  diseases, 382
  epidemiology, 382
  evolution, 319
  genetic maps, 334
  human infections, 382
  mutation rate, 319
  neuroinvasiveness, 164
  pathogenesis, 382
  transmission in blood, 49
  vector, 382
  zoonotic infection, 344
Burkitt's lymphoma, 146, 186, 210
Bystander vaccination, 270

C

C1q, 90
C3 convertase, 88–89
C5 convertase, 89
Cache Valley virus vector, 16
Cafeteria roenbergensis virus, 369
Calicivirus
  diseases, 383
  epidemiology, 383

Calicivirus *(continued)*
    human infections, 383
    lethal rabbit, 347
    pathogenesis, 383
California encephalitis virus, 382
cAMP response element binding protein
        (Creb), 212
cAMP response element binding protein
        (Creb)-binding protein (Cbp), 80, 209, 212
Campylobacter jejuni, 272
Canarypox, 275
Canarypox vector, 279
Cancer. *See also* Oncogenesis; Oncogenic virus;
        Transformed cells; Viral transformation;
        specific cancer types
    anticancer vaccine, 273–274
    cytokine stimulation of cell proliferation, 246–248
    defined, 176
    as genetic disease, 175–176
    genetic paradigm for, 182
    human immunodeficiency virus, 246–248
        anogenital carcinoma, 248
        B cell lymphoma, 248
        induction of cancers, 247
        Kaposi's sarcoma, 246–248
    simian virus 40 contribution to human cancer, 185
    terminology, 176
Canine distemper virus
    host range, 359
    neural spread, 43
Canine parvovirus, host-range shift in, 354
Capase activation and recruitment domain
        (CARD), 59
Capillary, blood-tissue junction, 45
Caprine arthritis-encephalitis virus, 222
Capsid
    satellite virus, 369
    virus-like particle vaccines, 272–273
Carcinogenesis, 176
Carcinogens, 175
Carcinoma
    colon, 176
    defined, 176
    hepatocellular, 176
CARD domain, 59
Cardif, 59
Cardinal data, 14
Carroll, James, 4, 5
Carrying capacity, 139
cas genes, 73
Case fatality ratio, 12–13
Case-controlled study, 11–12
Caspase(s), 62, 64–65, 67, 91
    activation, 60, 64
    extrinsic pathway, 64–66
    intrinsic pathway, 64–66
Caspase-3, 64–65
Caspase-8, 64–65, 67
Caspase-9, 65
Catalytic core domains (CCDs), 290
Cathelicidin, 233
Cause-and-effect relationship, 4
Cbp (cAMP response element binding protein
        [Creb]-binding protein), 80, 209, 212
CCl5, 121
CCr5, 20, 86, 223, 230–231, 234, 245, 248–250,
        285, 309, 312, 327
    CCr5Δ32 mutation, 248–249
CD (cluster-of-differentiation) markers, 104

CD1d, 93
CD2, 114
CD3, 230
CD4, 146, 227, 327
    autophagy enhancement of CD4+ cells, 68
    chemokine receptors, 230–231, 233
    cytotoxic T lymphocytes, CD4+ cell stimulation
        of development, 146
    degradation, 228
    human immunodeficiency virus infection, 225,
        227–231, 234–238, 241–243, 245, 247–250
    immunopathology lesions caused by CD4+cells,
        165–166
    infected CD4+ cells, 147
    soluble, 309
    structure, 104
    SU interaction with, 240, 309
    surface concentration downregulated by Nef
        protein, 229
CD4+ cells. *See* T helper cells
CD8, 92, 125, 236–238, 241, 243
    CD8+ cell proliferation in lymphocytic
        choriomeningitis virus infection, 100
    structure, 104
CD8+ cells. *See* Cytotoxic T lymphocyte (CTL)
CD19, 125
CD21, 90, 125
CD26, 114
CD28, 230
CD40 ligand, 125
CD46, 55, 77, 90
CD55, 90
CD59, 90
CD80, 114
CD81
CD86, 114
Cdc6, 204
Cdk (cyclin-dependent kinase), 180–182,
        201–204, 206
Cell culture
    benefits of system, 175
    establishment of, 177
    primary cells, 177, 188
    transformed, 175
Cell cycle
    disruption of control by viral transforming
        proteins, 201–205
        abrogation of restriction point control, 201–204
        inhibition of negative regulation, 204
        permanent activation, 201
    engine, 180–182
    phases, 179–182
    regulation, 179–180
    virus-specific cyclins, 204
Cell cycle arrest, HIV Vpr protein and, 227
Cell cycle checkpoint, 66
Cell lysis, 164
Cell proliferation
    control of, 178–182
    integration of inhibition of apoptosis with
        stimulation of cell proliferation, 206
Cell signaling, induced by receptor engagement, 55
Cell-autonomous protective programs. *See* Cellu-
        lar defense, intrinsic
Cell-based screens, 286
Cell-mediated immune response, 101, 103, 119–124.
        See also Adaptive immunity; T cells
    control of CTL proliferation, 122
    CTL lysis of virus-infected cells, 119–122

human immunodeficiency virus, 238–239
    measuring, 123
    noncytolytic control of infection by T cells, 122
    rashes and poxes, 122, 124
Cells. *See also* Transformed cells
    detection of infection, 54–62
    normal distinguished from transformed,
        177–178
    origin of, 329
Cellular defense, intrinsic, 54–74, 99–100
    acute infection, 138
    apoptosis, 62, 64–67
    autophagy, 67–68
    bacterial defense systems, 73
    cell signaling induced by receptor
        engagement, 55
    continuum between intrinsic and innate
        immunity, 74
    CRISPRs, 72–74
    cytosine deamination, 71–72
    detection of infection, 54–62
    epigenetic silencing, 68–69
    first critical moments of infection, 54–62
    genome editing, 71–72
    integration with innate and adaptive
        response, 55
    intracellular detectors of infection, 56
    Isgylation, 72
    phagocytes, 54, 60, 62, 67, 74, 87–89
    receptor-mediated recognition of microbe-
        associated molecular patterns (MAMPs),
        55–60
    RNA interference, 69–71
    summary, 63
Cellular FLICE-like inhibitory protein
        (cFLIP), 110
Cellular transformation, 175, 177
Central memory T cells, 129–130
Central nervous system
    entry of virus by olfactory routes, 43
    measles virus infection, 148
    reduced immune surveillance in, 147
    spread of virus, 39–40, 42–45
    terminology, 43
    virus entry, 46–47
Cerebral palsy, 390, 406
Cerebrospinal fluid, 46
Cervical cancer, 273–274
cFLIP (cellular FLICE-like inhibitory protein), 110
cGAS, 57, 60, 230
Checkpoint, 182
Checkpoint kinase 2 (Chk2), 207
Chemical name, 294
Chemokine(s), 74, 86, 195
    cytotoxic T lymphocytes secretion of, 121
    genetic differences in, 237
Chemokine receptors, 86, 195, 230–231, 237
Chermann, Jean-Claude, 221
Chickenpox, 47, 124, 269, 389
    acute infection, 140–141
    incubation period, 50
    parties, 50
Chikungunya virus
    diseases, 405
    epidemiology, 405
    human infections, 405
    pathogenesis, 405
    spread of, 16
    vector, 405

Chimpanzee, 220, 353–354, 358
Chk2 (checkpoint kinase 2), 207
Chloroplast, viroid replication in, 365, 367
Choroid plexus, 46
Chromatin, epigenetic silencing of DNA, 68–69
Chronic wasting disease, 376–377
Chrysanthemum chlorotic mottle viroid, 366, 368
Chrysanthemum stunt viroid, 366, 368
Cidofovir (Vistide), 296, 297
Cigarette smoking, influence on susceptibility to infection, 22
Cilia, respiratory tract, 31–32
Cip, 204
Circadian rhythms, 17
Circovirus
    error-prone replication, 319
    genome incorporation into host, 337
    in human blood supply, 357
    persistent infection, 143
Cirrhosis, 213–214, 273
cis-acting sequences, 321
CJD. *See* Creutzfeldt-Jakob disease
Clara cells. *See* Club cells
Classical pathway, complement, 87–88
Climate, influence on viral infections of populations, 17–19
Climate change, 361
Clinical latency, 236
Clinical trials, 292
Club cells, 38
Clustered regularly interspaced short palindromic repeats (CRISPRs), 72–74
Cluster-of-differentiation (CD) markers, 104
c-Myc protein, 193–195
Cobicistat, 312
Coconut cadang-cadang viroid, 365
Cohort study, 11–12
Colchicine, 43
Cold chain, 264
Cold-adapted mutants, 269
Collectins, 90
Colon carcinoma, 176
Colorado tick fever, 401
Colostrum, antibody in, 261
Coltivirus
    diseases, 401
    epidemiology, 401
    pathogenesis, 401
Combination therapy, 310, 312
Combinatorial chemistry, 287
Commensal bacteria, 33
Common cold, 395, 397
Complement, 87–90
    alternative pathway, 87–88
    cascade
        biological functions of, 87
        described, 88–89
        proteins and peptides released during, 89
        regulation, 89–90
    cell lysis, 87
    classical pathway, 87–88
    inflammation activation, 87
    inhibitors, viral, 163
    lectin pathway, 87–88
    opsonization, 87
    protection against infection, 89
    regulation of T cell response, 107
    solubilization of immune complexes, 87
Complement control protein, vaccinia virus, 163

Complement receptor type 1 (Cr1) protein, 90
Complement receptor type 2 (Cr2) protein, 90
Complement-mediated antibody enhancement, 239
Complete Freund's adjuvant, 276
Computational approaches to drug discovery, 287–288, 290
Computer analyses, 8
Concatemers
    satellite RNA, 369
    viroid RNA, 365, 367
c-oncogene, 183
Confidence interval, 14
Confucius, 99
Congenital brain infection, lymphocytic chorio-meningitis virus model of, 21
Conjunctivitis, 35
Contact inhibition, loss of, 178
Contagion (movie), 8
Copy choice, 324
Cormack, James, 64
Cornea
    herpes stromal keratitis, 165
    infection of, 35–36
Coronavirus, 3
    age-dependent susceptibility, 22
    M cell destruction by, 35
Cortez, Hernán, 352
Corticosteroid hormones, influence on suscepti-bility to infection, 22
Coughing, 32, 48
Cowpox virus, 256, 400
Coxsackievirus, 47, 138
CpG DNA, 56–58, 68, 78
Cpsf 6, 230
Cr1 (complement receptor type 1) protein, 90
Cr2 (complement receptor type 2) protein, 90
Creb (cAMP response element binding protein), 212
Cre-lox technology, 137
Creutzfeldt-Jakob disease (CJD), 375
    familial, 373
    iatrogenic, 371
    sporadic, 371
    variant, 374–376
Crimean-Congo hemorrhagic fever virus, 382
CRISPRs, 72–74
Cryptdins, 34
c-Src protein, 194–195, 199, 201
CTL. *See* Cytotoxic T lymphocyte
CTL escape mutations, 146
Cucumber mosaic virus, 368, 369
Cul 4-E3 ubiquitin ligase, 227
Culling, 260
Cutaneous immune system, 114
CXCr4, 86, 122, 223, 230–231, 236, 249–250, 309, 327
Cyclin(s), 179–182, 184, 195, 201–205
    virus-specific, 204
Cyclin A-Cdk2, 202–203
Cyclin D-dependent Cdks, 202–204
Cyclin E-dependent Cdks, 182, 202–203
Cyclin-Cdk complex, 180–182, 202–204
Cyclin-dependent kinase (Cdk), 180–182, 201–204, 206
Cyclin-dependent kinase inhibitor, 181, 193, 195–196, 204–206, 208
CypA, 230
Cytidine, 295

Cytokine(s). *See also* Interferon(s); Interleukin(s)
    of CD4$^+$ T$_h$1 cells, 165
    of CD4$^+$ T$_h$2 cells, 165
    cell proliferation stimulation, 246–248
    cytotoxic T lymphocytes secretion of, 121
    of early warning and action, 76–86
    in human immunodeficiency virus infection, 235, 237–238, 240–242, 245–246
    immunotherapy with, 276–277
    inflammatory, 74, 76, 105, 108–111, 165, 167, 242
    intracellular cytokine assay, 123
    overview of functions, 74, 76
    release in adaptive immune response, 109
    release in central nervous system infections, 147
    T helper cell synthesis of, 104–106
    therapies based on, 250
    virokines, 76
Cytokine storm, 21, 74, 167
Cytomegalovirus
    antiviral drugs, 294, 300
    apoptosis and, 66
    B cell function and, 238
    diseases, 390
    epidemiology, 390
    FasL-mediated CTL killing, 147
    Guillain-Barré syndrome, 272
    HIV infection and, 246
    human infections, 390
    IE1 proteins, 69
    IE2 protein, 147
    interference with MHC class I pathway, 122
    major histocompatibility complex (MHC) class I molecules, interference with production and function, 144–145
    pathogenesis, 390
    persistent infection, 143, 144–145
    pp71 protein, 68
    retinal infection, 35
    stable host-virus interaction, 345
    superoxide production in lungs, 168
    transmission in saliva, 49
    transplacental infection, 47
    UL89, 301
    US2 protein, 144
    US3 protein, 144
    US6 protein, 144
    US11 protein, 144
Cytomegalovirus vector, 279
Cytopathic viruses, 136, 143–144
Cytosine deamination, 71–72
Cytoskeleton, neuronal, 42
Cytotoxic index, 291
Cytotoxic T lymphocyte (CTL), 104
    adoptive transfer, 122
    CD4$^+$ cell stimulation of development, 146
    control of production, 122
    destruction of activated, 147
    in Epstein-Barr virus infection, 146
    escape mutants, 146
    FasL-mediated killing, 147
    in human immunodeficiency virus infections, 229–230, 235–236, 238–239
    immunological synapse, 119–120
    immunopathology lesions caused, 164–165
    interferon γ production, 104
    lysis of virus-infected cells, 119–122
    magnitude of response, 122
    measuring antiviral cellular immune response, 123

Cytotoxic T lymphocyte (CTL) *(continued)*
  mechanisms of killing, 120–121
  in persistent infections, 146–147
  recognizing infected cells by engaging MHC
    class I receptors, 115–116
  ritonavir effect on, 311
  tumor necrosis factor α (Tnf-α), 104
Cytovene (ganciclovir), 294, 295

D

Daf (decay-accelerating protein), 90
Damage-associated molecular patterns (DAMPs), 55
Darunavir, 307
Darwin, Charles, 352
Data mining, 8
Daxx, 68
DC-Sign, 230
Dead-end interaction, 344–345, 347–348
Decay-accelerating protein (Daf), 90
Defective particles, 28
Defensins, 34, 90, 233
Delavirdine, resistance to, 311
Delayed-type hypersensitivity, 122, 165
Deleted in colon carcinoma *(dcc)* gene,
  loss-of-function mutation in, 176
Delivery
  drug, 293
  vaccine, 264–265, 276
Delivery vehicles for vaccines, 265
Delta antigen, hepatitis delta satellite virus, 370
Deltaretrovirus, 211
  diseases, 402
  epidemiology, 402
  human infections, 402
  pathogenesis, 402
Dementia, AIDS, 244
Demyelination, 165
Dendritic cell, 54, 56, 58, 62, 67, 74, 76–77, 79,
  85, 87, 89, 91, 93–94, 99–100, 102–106,
  108–109, 111–116, 130. *See also* Antigen-presenting cell
  cytokine production, 105–106, 109, 111
  human immunodeficiency virus and, 233–235,
    238
  immature, 79, 93, 111, 114, 130
  infection, 114
  interferon synthesis, 77, 79
  lacking in brain, 124
  measles virus infection of, 169
  migration to lymphoid tissue, 111–112, 114
  nuclear factor κb (Nf-κb) activation, 108
  plasmacytoid, 77
Dengue virus, 266
  antibody-dependent enhancement of infection,
    166–167
  diseases, 385
  epidemiology, 385
  human infections, 385
  incubation period, 138
  innate immune response, 59
  pathogenesis, 385
  primary infection, 166
  vector, 385
  zoonotic infection, 344
Depression, Epstein-Barr virus and pregnancy, 155
Descriptive studies, 13, 15
Deubiquitinase, viral, 144
Diapedesis, 46

Diarrhea, rotavirus-induced, 164
Dictyostelium discoideum, 79
Dideoxycytidine, 295
Dideoxyinosine, 295
Direct-acting antivirals, 288, 309
Disease
  animal models of human diseases, 135–136
  detection in nasal brushings and urine, 375
  fungi, 4
  host-parasite interaction, classic theory of,
    317–318
Disseminated infection, 39
DNA
  epigenetic silencing of, 68–69
  as episome, 187
  integration of viral into cellular genome, 187
DNA chip, 288
DNA damage, p53 response to, 208
DNA damage checkpoint-DNA repair pathway, 349
DNA genome, as viral invention, 328–329
DNA ligase I, in viroid replication, 365
DNA methylation, 68–69, 73
DNA methyltransferase, 68, 73
DNA polymerase
  antiviral drugs, 296–297, 303
  error correction, 319
  phylogeny of, 330
  sputnik virophage, 370
DNA polymerase δ, 176
DNA synthesis phase (S), 180–182, 201–204, 206,
  208
DNA vaccines, 273–274
DNA virus
  endogenous sequences in vertebrate genomes,
    337–338
  error threshold, 322–323
  evolution, 319, 322–323, 326, 329, 333
  exchange of genetic information, 324
  giant viruses, 329–332
  host-virus relationships, 333
  mutation rate, 319
  nucleocytoplasmic large DNA viruses, 329–330,
    332
  oncogenic virus, 182, 185–186
  origins of eukaryotic nucleus and, 329
  sequence coherence of large DNA viruses, 330
  transforming infection, 157
DNA-dependent RNA polymerase, chloroplast, 365
DNA-dependent RNA polymerase II
  plant, 365
  viroid replication and, 365
Dolutegravir, 308
Dominant transforming agent, 192
Double blind study, 12
Doubled-stranded RNA virus, replication of, 328
DPT (diphtheria, pertussis, and tetanus) vaccine,
  263
Drosophila melanogaster
  Stat gene homologs, 79
  Toll signaling pathway, 56
Drug resistance, 309–312
dsRNA-activated protein kinase, 79, 81
Dynein, 42

E

E1 ligase, 59
E1A protein, adenovirus, 66, 189–190, 192–193,
  204–205, 210

E1B protein, adenovirus, 66, 189–190, 209
E2 protein, hepatitis C virus, 91
E2f family of transcriptional regulators, 202–205,
  208
E3 protein, adenovirus, 61
E3 ubiquitin ligase, 145, 205, 207, 209, 227–228
  herpesvirus, 145
  human herpesvirus 8, 145–146
  murine gammaherpesvirus 68, 145
  myxoma virus, 145–146
  poxvirus, 145
E3L protein, vaccinia virus, 84
E4 Orf3 protein, adenovirus, 69
E4 Orf6 protein, adenovirus, 209
E5 protein, papillomavirus, 199
E6 protein, human papillomavirus, 190, 193, 209
E7 protein, human papillomavirus, 190, 192,
  204–205
Early innate response, 27
Eastern equine encephalitis virus, 16, 405
EBER-1, Epstein-Barr virus, 155
EBER-2, Epstein-Barr virus, 155
EBNA-1 protein, Epstein-Barr virus, 146, 156, 210
EBNA3C protein, Epstein-Barr virus, 205
EBNA5 protein, Epstein-Barr virus, 69
Ebola virus, 3, 22, 359
  bat reservoir, 349
  Bundibugyo, 384
  cross-species infections, 343
  cytokine storm, 167
  dead-end virus-host interaction, 345
  diseases, 384
  epidemiology, 384
  human infections, 384
  incubation period, 138
  infectious period, 149
  outbreak (2014), 142, 265
  pathogenesis, 384
  public attention to, 359
  R_0 (R-naught) value, 318
  reservoir species for, 345
  Reston, 384
  RNA interference, suppression of, 71
  Sudan, 384
  Tai Forest, 384
  Zaire, 384
  ZMapp therapeutic, 125–126
  zoonotic infection, 344
Ecological parameters, in host-virus interaction,
  348–351
Ectromelia virus
  inadvertent creation of more virulent virus, 161
  pathogenesis of, 25–27
  recombinant containing IL-4 gene, 160–161
Effector memory T cells, 129
eIF2α, 68, 81
eIF2B, 71
Eigen, Manfred, 321
Ellerman, Vilhelm, 182
Elvitegravir, 308
Emerging viruses, 342–362
  climate change and, 361
  drivers of emergence, 344
  ecological parameters, 348–351
  expanding viral niches, 351–352
  host range expansion, 354–355
  host-virus interactions, 343–348
    accidental infections, 351
    dead-end, 344–345, 347–348

encounters of virus with new host, 348–351
evolving, 344–347, 352, 353
resistant host, 344–345, 348
sources of animal-to-human transmission, 347
stable, 344–345
immediate problems and issues illuminated
by, 360
new venues for infection, 361–362
parameters converging to promote infections,
344
predicting pandemics, 359–360
prevention, 361
public interest and concern, 358–359
uncovering unrecognized viruses, 355, 357–358
zoonoses, 343–344, 347–348, 352–354, 361
Emerson, Ralph Waldo, 175
Emtricitabine, 250
Encephalitis, 43
lymphocytic choriomeningitis virus, 150
measles virus, 148
Nipah virus, 350
West Nile virus, 346
Endemic disease, yellow fever as, 6
Enders, John, 269
Endogenous antigen presentation, 115–117
Endogenous retroviruses, 336–337
Endoplasmic reticulum, in antigen presentation
process, 116
Endothelioma, 176
Enfuvirtide, 309
Enhancer
IFN-β, 80
insertion, 198
viral tropism and, 36–37
Enhancing antibodies, human immunodeficiency
virus, 239–240
Enteric viruses, enhanced infection in the presence of intestinal bacteria, 33
Enterovirus
diseases, 398
epidemiology, 398
human infections, 398
pathogenesis, 398
Enterovirus 68, 398
Enterovirus 70, 35, 398
Enterovirus 71, 398
Entry into host, 29–36
alimentary tract, 30, 33–35
blood-borne viruses, 40
eyes, 30, 35–36
human immunodeficiency virus, 230–231
inhibitors of, 300, 309
respiratory tract, 30, 31–33
skin, 29–31
urogenital tract, 30, 35
Env protein
human immunodeficiency virus, 122, 220,
228, 240
neutralizing antibodies, 309
retrovirus, 220
Envelope, human immunodeficiency virus, 240
Environmental factors, infection and, 16–19
Epidemic, 3
accelerators of transmission, 10
acute infections, 142
artificial, 347
economic toll in agricultural organisms, 9–10
hantavirus, 352
influenza, 142, 354

poliomyelitis, 351
SARS, 353
stopping in agricultural animals by culling and
slaughter, 260
tracking by sequencing, 9
video game model, 12
West Nile virus, 9, 346
worldwide (*see* Pandemic)
yellow fever in Philadelphia (1793), 8–9
zoonotic infections, 10
Epidemiologist, 10–11
Epidemiology, 3, 10–16
case fatality ratio, 12–13
descriptive, 13, 15
fundamental concepts, 11–13
incidence of disease, 11, 13
morbidity, 12–13
mortality, 12–13
prevalence of disease, 11
prospective studies, 11–12
retrospective studies, 11–12
statistics, use of, 14
surveillance, 15–16
tools, 13, 15
Epidermal growth factor, 179, 192, 200
Epigenetic drugs, 248, 312
Epigenetic silencing, 68–69
Episome, 154, 187
Epithelial cell, polarized release of viruses from, 39
Epitope
definition, 100, 101
host-encoded, 108
immunodominant, 146, 168
Epivir (3TC, lamivudine), 295, 297, 321
Epsilonretrovirus
reproductive cycle, 184
rv-cyclin, 204
Epstein-Barr virus
B cell function and, 238
B cell infection, 105
B cell lymphoma, 154–156, 248
binding to complement receptor, 90
Bplf1-mediated Tlr evasion, 61
Burkitt's lymphoma, 146, 186, 210
depression, 155
diseases, 391
EBER-1, 155
EBER-2, 155
EBNA-1, 146, 156, 210
EBNA3C protein, 205
EBNA5 protein, 69
epidemiology, 391
episome, 187
heterologous T cell immunity, 167–168
Hodgkin's lymphoma, 146
human infections, 391
immune defenses, protection from, 210
incubation period, 48–49, 138
infection of immune system cells, 147
infectious mononucleosis, 48, 146, 154
interferon response modulators, 82
latent infection, 154–157
latent membrane protein 1 (LMP-1), 155–156,
198–200, 206
latent membrane protein 2A (LMP-2A), 155,
157
nasopharyngeal carcinoma, 146, 200, 210
oncogenic, 146, 182, 186
pathogenesis, 391

persistent infection, 143, 146, 155–156
primary infection, 154, 156
reactivation, 155–157
strains, 154
Tlr signaling pathway blockage, 61
transmission in saliva, 48–49
Zta protein, 157
Equine herpesvirus 2, 105
Equine infectious anemia virus, 222
Equine influenza virus, 265
Erk (extracellular signal-regulated kinase), 58
Error threshold, 322–323
Escape mutants, cytotoxic T lymphocytes, 146
Escherichia coli iap gene, 73
Etiology, 4
Eukaryotes
origin of nucleus, 329
proteomic diversity, 332
on tree of life, 332
European tick-borne flavivirus, 348
Evasion, definition, 26
Evolution, 316–340
bottleneck, 322–324, 327
coevolution, 333
constraints, 326–327
deliberate release of rabbitpox virus in
Australia, 347
DNA virus, 319, 322–323, 326, 329, 333
emerging viruses, 343
endogenous retroviruses, 336–337
error threshold, 322–323
exchange of genetic information, 324–325
fitness, 322–323
general pathways, 325–326
genetic shift and drift, 324
host range expansion, 354–355
host-parasite interaction, classic theory of,
317–318
host-virus relationship, 333–335
arms race, 337–339
DNA virus, 333
herpesvirus, 333
influenza virus, 334–335
papillomavirus, 333
polyomavirus, 333
retrovirus, 335
RNA virus, 333–335
(+) strand RNA virus, 333–334
(−) strand RNA virus, 333
human immunodeficiency virus-1 from
monkey and primate hosts, 11
large numbers of viral progeny, 319
mechanisms, 318–327
error-prone replication, 318–321
horizontal gene transfer, 326
incorporation of cellular sequences, 324
mutation, 318–323
nonhomologous recombination, 326
reassortment, 318, 324
recombination, 318, 324
myxoma virus in Australia, 318
origin of viruses, 327–332
paleovirology, 335–339
quasispecies concept, 321–322, 368
replication strategies and, 319, 326–328
RNA virus, 319, 322–323, 326, 333–335
selection, 318–319
for transmission and survival inside a
host, 327

Evolution *(continued)*
  sequence conservation, 321–322
  survival of the flattest, 366, 368
  viral transforming genes, origin and nature of, 189–192
  virulence, 318
Evolving host-virus interaction, 344–347, 352, 353
Exogenous antigen presentation, 116, 118–119
Exosome, Epstein-Barr virus LMP-1 containing, 200
Exponential growth, 140
Expression strategies, evolution of, 326–327
Extracellular signal-regulated kinase (Erk), 58
Extrinsic pathway to apoptosis, 64–66
Eye
  herpes stromal keratitis, 165
  viral entry, 35–36
Eyebrows, 36
Eyelashes, 36

F

Fab, 125–126
Fak (focal adhesion kinase), 193
Familial adematous polyposis, 176
Faroe Islands, 262
Fas ligand (FasL), 121–122, 147
Fatal familial insomnia, 373
Fc receptor, 125–128
FDA (Food and Drug Administration), 268, 273, 283, 305
Feces, shedding of virions in, 49
Feline immunodeficiency virus, 222
Feline panleukopenia virus, 354
Ferret model of influenza infection, 48, 356
Fetus, viral invasion of, 47
Fibroblast, 176, 177
Fibropapilloma, 176
Fifth disease, 47
Filovirus
  dead-end virus-host interaction, 345
  diseases, 384
  epidemiology, 384
  human infections, 384
  pathogenesis, 384
  transmission in blood, 49
  zoonotic infection, 344
Filterable agent, 4
Finlay, Carlos Juan, 4
Fitness, 322–323
  bottlenecks, 323
  survival of the fittest, 321
Flavivirus
  diseases, 385
  epidemiology, 385
  human infections, 385
  neural spread, 43
  NK modulators, 91
  oncogenic, 187, 213
  pathogenesis, 385
  persistent infection, 144
  resistance to, *FLV* gene and, 19
  stable host-virus interaction, 345
  vector, 385
  zoonotic infection, 344, 348
Flexner, Simon, 277
Flublok, 272
Fluorescence-activated cell sorting, 123
FLV gene, 19
Foamy virus, Fv1 protein targeting of, 72

Focal adhesion kinase (Fak), 193
Foci, 178
Fold superfamilies, 332
Fomites, 4
Fomivirsen, 300
Food and Drug Administration (FDA), 268, 273, 283, 305
Foot-and-mouth disease virus
  containment by culling/slaughter, 260
  epidemic (2001), 9–10
  interferon response modulators, 82
  national programs for eradication of, 258, 260
  vaccine, 10, 260
Formaldehyde, 265
Formulation, drug, 293
Foscarnet (Foscavir), 297
  resistance to, 311
Fossils, 317, 337
Fowlpox, 275
Franklin, Benjamin, 255
Free radical-mediated immunopathology, 168
Freund's adjuvant, 110
Furin proteases, 38
Fusion, inhibitors of, 309
Fv-1 gene, 337
Fv1 protein, 71–72

G

G protein
  signaling, 206
  vesicular stomatitis virus, 197
G protein-coupled receptors, 86
G_0 phase, 178, 180, 201, 204
G_1 phase, 180–182, 194, 201–206, 208
G_2 phase, 180–182, 227
G2/M arrest, HIV Vpr protein and, 227
Gag protein, human immunodeficiency virus, 220
Gag-Pol polyprotein, human immunodeficiency virus, 306
Gain-of-function studies, 356
Gajdusek, Carleton, 371
Galactose $\alpha(1,3)$-galactose (α-Gal), 89
Gallo, Robert, 221
GALT (gut-associated lymphoid tissue), 113, 234, 242
$\gamma\delta$ cells, 93
Gammaherpesvirus
  diseases, 391
  epidemiology, 391
  evolution, 333
  human infections, 391
  pathogenesis, 391
  recombination, 324
Gammaretrovirus, Fv1 protein targeting of, 72
Ganciclovir (Cytovene), 294, 295
Ganglia
  herpes simplex virus primary infection of, 151–152
  herpes simplex virus reactivation from, 153
GAP (Global AIDS Program), 219
Gap phases, 180
Gastrointestinal system
  human immunodeficiency virus pathogenesis, 245
  viral entry, 30, 33–35
Gates Foundation, 264
Gcn5 (general control nondepressible 5), 80
Gelsinger, Jesse, 158
Gene gun, 274

Gene knockouts, 137
Gene shuffling, 274
Gene therapy
  human immunodeficiency virus, 249–250
  retrovirus use, 197
General control nondepressible 5 (Gcn5), 80
Generic name, 294
Genes, nonessential, 160
Genetic bottleneck, 322–324
Genetic drift, 324
Genetic influences on infection, 19–21
Genetic shift, 324
Genetics, drug discovery and, 285
Genital herpes, 35
Genital tract, human immunodeficiency virus pathogenesis and, 246
Genital warts, 48
Genome, viral
  genome size as evolution constraint, 326
  metagenomic analysis, 330
  satellite virus, 368–370
  sequence coherence, 330
  sequence conservation, 321–322
  unknown, 330
Genome editing, 71–72
Genomic vaccine, 274
Geography, infection and, 16
German measles, 47
Gerstmann-Straussler-Scheinker disease, 373
Giant viruses, 329–332
  discovery, 331
  origins, 332
  phylogeny, 330
  sequence coherence, 330
Gilead Sciences, 305
Glioblastoma, 163
Glioma, 163
Global AIDS Program (GAP), 219
Global Vaccine Fund, 264
Global warming, 361
Glomerulonephritis, 166
Glucocorticoids, reactivation of latent herpesvirus, 154
Glucosaminoglycans, 86
Glycosylation, in prion strains, 374
Golgi apparatus, 227–229
Google-flu, 16, 359
Gould, Steven Jay, 319
Granzyme, 121
Growth
  control of cell proliferation, 178–182
  growth-promoting signals, 179–180
Growth factor
  autocrine growth stimulation, 178
  control of cell proliferation, 178, 195
  platelet-derived, 179, 192, 196, 199–200
  signaling and, 178, 199, 203
  T helper cell synthesis of, 104
  vascular endothelial (Vegf), 195–196
  viral transforming proteins similar to, 192, 195
Growth factor receptor, 178–179, 192, 200
Guanosine, 295
Guanosine triphosphatase (GTPase)
  interferon-inducible, 82
  Mx1 gene, 19
Guillain-Barré syndrome, 272
Guiteras, Juan, 4
Gut-associated lymphoid tissue (GALT), 113, 234, 242

H

HA (hemagglutinin) protein
    antigenic drift, 143
    cleavage, 37–39
    influenza virus, 37–39, 143, 268, 298, 300,
        324–325, 349, 360
    measles virus, 101
HA0 precursor, influenza virus, 37–39
Hairy-cell leukemia, 402
HAND (HIV-associated neurocognitive disorder),
    245
Hand-foot-and-mouth disease, 47, 398
Hantaan virus, 344, 382
Hantavirus
    antiviral drugs, 297
    diseases, 382
    epidemiology, 382
    evolution, 319
    human infections, 382
    mutation rate, 319
    pathogenesis, 382
    shedding in urine, 49
    zoonotic infection, 352
Hantavirus pulmonary syndrome, 352, 382
Hbz protein, human T-lymphotropic virus type 1,
    212
Health, spread of infection influenced by, 350
Heart muscle, human immunodeficiency virus
    pathogenesis and, 246
Hendra virus, immunosuppression and, 169–170
Helper virus, 188, 368–370
Hemagglutinin. *See* HA (hemagglutinin) protein
Hematogenous spread, 40–42
Hematopoietic stem cell, 100–102
Hemorrhagic fever, 49, 347, 352, 359
    dengue virus, 166–167
    R_0 (R-naught) value, 318
Hendra virus
    bat reservoir, 349
    zoonotic infection, 344, 347
Hepadnavirus
    diseases, 387
    epidemiology, 387
    genome incorporation into host, 337
    human infections, 387
    oncogenesis, 213
    pathogenesis, 387
Hepatitis A virus
    diseases, 396
    epidemiology, 396
    human infections, 396
    incubation period, 138
    passive immunization, 261
    pathogenesis, 396
    pregnant women, susceptibility of, 22
Hepatitis A virus vaccine, 265, 267
Hepatitis B virus
    accidental infection, 351
    antiviral drugs, 297
    concentration of particles in blood, 48
    cytotoxic T lymphocyte (CTL)-mediated
        immunopathology, 165
    difficulty of culturing, 5
    diseases, 213, 387
    epidemiology, 387
    as helper virus, 370–371
    hepatocellular carcinoma, 213, 273
    human infections, 387
    immune complex deposition, 166

inactivation in intestine, 49
incubation period, 138
interferon response modulators, 82, 84
interferon therapy, 276
noncytologic control of infection by T cells, 122
oncogenesis, 182, 213
pathogenesis, 387
persistent infection, 143
pregnant women, susceptibility of, 22
prevalence, 11
reproduction rate, 319
shedding in semen, 49
transmission, 48
viremia, 42
X protein, 213
Hepatitis B virus vaccine, 213, 267, 276
Hepatitis C virus
    antiviral drugs, 286, 297
        miR-122 antagonists, 301–309, 312
        monoclonal antibody, 300
    autophagy machinery co-opted by, 68
    discovery, 213, 357–358
    diseases, 213, 386
    DNA vaccine, 274
    E2 protein, 91
    epidemiology, 386
    hepatocellular carcinoma, 182, 187, 213
    in human blood supply, 357
    human infections, 386
    incubation period, 138
    interferon response modulators, 82
    interferon therapy, 276
    miRNA targeting genomic RNA, 69
    noncytologic control of infection by T cells, 122
    NS3/4A protein, 61, 144, 306–307
    NS4A protein, 214
    NS5A protein, 214, 308–309
    NS5B protein, 214, 303
    oncogenesis, 182, 187, 213–214
    p7 protein, 164
    pathogenesis, 386
    persistent infection, 143, 144
    polyprotein, 302
    protease, 302
    RNA polymerase, 303, 306
    viremia, 42
Hepatitis delta satellite virus, 370–371
Hepatitis E virus, 22
Hepatocellular carcinoma, 387
    defined, 176
    hepatitis B virus, 213, 273
    hepatitis C virus, 182, 187, 213
Herd immunity, 255, 262–264, 278
Herpes gladiatorium, 49
Herpes simplex virus
    animal model of disease, 345, 347
    antiviral drugs, 291, 293–294
    attenuated used to clear human brain tumors,
        163
    collectins and defensins binding glycoproteins
        of, 90
    genetic influence on infection, 20
    hygiene hypothesis, 151
    ICP0 protein, 60, 69, 153
    ICP34.5 protein, 68, 81, 162–163
    ICP47 protein, 153
    incubation period, 138
    keratitis, 165
    Langerhans cells and, 114

latent infection, 150–154
    establishment and maintenance of, 151–152
    latency-associated transcripts, 152–153
    multiple viral genomes in neurons, 154
    reactivation, 153–154
M cell infection, 114
neuroinvasiveness, 43
neurotropism, 36
neurovirulence, 44, 162
persistent infection, 143, 146
primary infection, 150–152
reactivation, 153–154
    from ganglia, 153
    signal pathways, 153–154
ribonucleotide reductase, 162
shedding from skin lesions, 49
spread in host, 39–40
stable host-virus interaction, 345
transplacental infection, 47
Herpes simplex virus type 1
    corneal infection, 35–36
    CTL response, 146
    diseases, 388
    epidemiology, 388
    human infections, 388
    ICP0, 60, 69
    ICP34.5 protein, 68, 81
    interferon inhibition of infection, 85
    interferon response modulators, 82
    latency-associated transcripts (LATs), 153
    pathogenesis, 388
    persistent infection, 143, 146
    US11 protein, 84
Herpes simplex virus type 2
    diseases, 388
    entry into host, 35
    epidemiology, 388
    HIV infection risk increase, 233
    human infections, 388
    latency-associated transcripts (LATs), 153
    pathogenesis, 388
    persistent infection, 143
Herpes stromal keratitis, 165
Herpesvirus
    antiviral drugs, 293–294, 297, 301
    corneal infection, 35–36
    E3 ubiquitin ligase, 145
    evolution, 333
    HIV infection risk increase, 233
    host-virus relationship, 333
    NK modulators, 91
    nucleocytoplasmic large DNA viruses, 329
    oncogenic, 186
    protease, 301
    shedding in semen, 49
    T cell receptor function reduction, 120
    transmission by sexual contact, 49
    viral transforming proteins, 192
Herpesvirus saimiri, 204
HERVs (human endogenous retroviruses),
    336–337
Heterologous T cell immunity, 167–168
Heterosexual contact, HIV infection via, 231–232
High-content screens, 286–287
High-throughput screens, 286–287
Hippocrates, 64
Histones
    acetylation, 68
    deacetylation, 68

History
  identification of first human viruses, 4–7
  microbes as infectious agents, 4
  technological developments, 7–8
  viral epidemics, 8–10
HIV. *See* Human immunodeficiency virus; Human
    immunodeficiency virus type 1
HIV-associated neurocognitive disorder
    (HAND), 245
HLA-E protein, 91–92
Hmg1(Y) protein, 80
Hmg(A1) protein, 80
Hodgkin's lymphoma, Epstein-Barr virus and, 146
Homosexual contact, HIV infection via, 231–232
Hop stunt viroid, 365
Horizontal gene transfer, 326
Hormones
  cytokines as, 74
  influence on susceptibility to infection, 22
Host defense, 52–94. *See also* Cellular defense,
    intrinsic; Immunity
  acute infection, 138
  adaptive immune response, 99–130
    diversity, 100
    memory, 100
    self-control, 100
    specificity, 100
    speed, 99–100
  bacterial defense systems, 73
  continuum between intrinsic and innate
    immunity, 74
  coordinated host response to infection, 27
  detection of infection, 54–62
  first critical moments of infection, 54–62
  host alterations as early signals of infection, 62
  innate immune response, 74–93
    chemokines, 74, 86
    complement, 87–90
    cytokines, 74–86
    γδ cells, 93
    natural killer (NK) cells, 90–92
    neutrophils, 92–93
    NKT cells, 93
    soluble mediators, 74–86
  intracellular detectors of infection, 56
  intrinsic immune response, 54–74
  intuition as, 36
  phagocytes, 54, 60, 62, 67, 74, 87–89
  viral offense against, 25
  virulence gene products that modify host
    defense mechanisms, 162–163
Host factors in viral infection, 19–22
  genetic and immune parameters, 19–21
  nongenetic risk factors, 21–22
Host range
  expansion, 354–355
  prions, 374
  viroid, 365
Host-parasite interaction, classic theory of,
    317–318
Host-virus interaction, 343–348
  accidental infections, 351
  dead-end, 344–345, 347–348
  encounters of virus with new host, 348–351
  evolution of viruses, 333–335
    arms race, 337–339
    DNA virus, 333
    herpesvirus, 333
    influenza virus, 334–335

papillomavirus, 333
polyomavirus, 333
retrovirus, 335
RNA virus, 333–335
  (+) strand RNA virus, 333–334
  (−) strand RNA virus, 333
evolving, 344–347, 352, 353
experimental analysis of, 350
human activity effects on, 360
resistant host, 344–345, 348
sources of animal-to-human transmission, 347
stable, 344–345
Hsc70, 204
Human cytomegalovirus. *See* Cytomegalovirus
Human endogenous retroviruses (HERVs),
    336–337
Human herpesvirus 4. *See* Epstein-Barr virus
Human herpesvirus 7, 143
Human herpesvirus 8
  B cell lymphoma in HIV-1-infected individuals,
    248
  diseases, 391
  E3 ubiquitin ligase, 145–146
  epidemiology, 391
  human infections, 391
  immune defenses, protection from, 210
  K5 protein, 145–146
  Kaposi's sarcoma, 66, 186, 194–196, 248
  oncogenesis, 182, 186, 194–196
  paracrine oncogenesis, 196
  pathogenesis, 391
  v-cyclin gene, 204
  vFLIP protein, 66
  v-IL-6, 210
Human immunodeficiency virus, 218–252. *See also*
    Acquired immunodeficiency syndrome;
    Human immunodeficiency virus type 1
  antibody response, 238, 239–240
    broadly neutralizing antibodies, 240, 250–251
    enhancing antibodies, 239–240
    interfering antibodies, 239
    neutralizing antibodies, 239
  antiviral drugs, 294, 297, 301–312
    approved drugs, 302
    combination therapy, 302, 310, 312
    dynamics of HIV-1 reproduction in AIDS
      patients, 241–242
    fusion and entry inhibitors, 309
    integrase inhibitors, 302, 306, 308
    lives saved by anti-HIV therapy, 312–313
    monoclonal antibody, 300
    nonnucleoside inhibitors, 302–303, 306
    nucleoside analogs, 302–303
    prophylaxis, 250
    protease inhibitors, 302, 306–307, 311
    resistance, 309–310
    steps in reproduction targeted by, 304
    transmission rate, affect on, 233
    treatment, 248–249
    unpredicted activities of, 311
  attachment to host cells, 230–231
  auxiliary proteins, 220, 222
  Berlin patient, 249
  Bru isolate, 221
  cancer, 246–248
    anogenital carcinoma, 248
    B cell lymphoma, 248
    induction of cancers, 247
    Kaposi's sarcoma, 246–248

CCr5 receptor, 223, 230–231, 234, 245, 248–250
CD4 cells, 225, 227–231, 234–238, 241–243,
    245, 247–250
  cellular targets, 230–231
  chemokine receptor, 309
  clinical latency, 236
  course of infections, 234–237
    acute phase, 234–235
    asymptomatic phase, 235–236
    pathological conditions associated with
      phases of infection, 236
    schematic diagram of events, 236
    symptomatic phase, 236
    variability of response to infection, 236–237
    virologic set point, 235
  cross-species infections, 343
  CXCr4, 223, 230–231, 236, 249–250
  cytotoxic lymphocytes (CTLs) and, 229–230,
    235–236, 238–239
  discovery and characterization, 13, 15, 219–221
  DNA vaccine, 274
  dynamics of HIV-1 reproduction in AIDS
    patients, 241–242
  entry into host, 35, 40, 230–231
  Env protein, 220, 228, 240
  FasL-mediated CTL killing, 147
  Gag-Pol polyprotein, 306
  immune dysfunction, 237–238
  immune responses, 238–241
    cell-mediated response, 238–239
    humoral response, 239–240
    innate response, 238
  immunosuppression, 168
  immunotherapy, 277
  incubation period, 138
  latent infection, 249
  long terminal repeat (LTR), 222, 224, 227, 250
  Nef protein, 68, 122, 147, 220, 223, 225,
    228–230, 237
  neurotropism, 236
  pandemic, 359
  pathogenesis
    breasts, 246
    gastrointestinal system, 245
    genital tract, 246
    heart muscle, 246
    kidneys, 246
    lungs, 246
    lymphoid organs, 242–244
    nervous system, 244–246
  perinatal infection, 47
  persistent infection, 143, 249
  phylogenetic relationships, 222
  prevention of infection, 248–250
    multifaceted approach to, 252
    postexposure prophylaxis (PEP), 250
    preexposure prophylaxis (PrEP), 250
  replication cycle, distinctive features,
    222–230
    accessory proteins, 225–230
    intrinsic defense mechanisms countered by
      viral capsid, 230
    regulatory proteins, 222–225
  reproduction in immune system cells, 41
  Rev protein, 220, 223–225
  salivary agglutinin interference with, 34
  slow infections, 157
  splice sites, 223
  spontaneous T cell apoptosis, 147

spread
    in host, 41–42, 47
    mechanics of, 233–234
structural plasticity, 141
SU protein, 147, 237–240, 245, 250, 309
syncytium formation, 227, 236
TAR (*trans*-activating response element), 222, 224
Tat protein, 122, 147, 220, 222, 224, 238, 245, 248
TM protein, 238–240, 245, 248, 250
transcription, 222
transmission, 231–234
    in blood, 49
    bottleneck, 327
    likelihood of infection, 232
    mechanics of spread, 233–234
    modes, 231–233
    mother to child, 47, 232–233
    by sexual contact, 49
treatment, 248–250
    antiviral drugs, 248–249
    gene therapy, 249–250
    immune system-based therapies, 250, 251
    problems of persistence and latency, 249
vaccine, 3, 277–279
Vif protein, 72, 220, 225–226
viremia, 42
Vpr protein, 220, 225–227
Vpu protein, 220, 225, 227–229, 237
worldwide impact of AIDS, 219
worldwide infections, 219–220
zoonotic transmission, 11, 344, 352–354
Human immunodeficiency virus type 1. *See also* Human immunodeficiency virus
accessory proteins, 220, 223, 225–230
Adar1 promotion of, 71
antibody response, 238, 239–240
    broadly neutralizing antibodies, 240, 250–251
    enhancing antibodies, 239–240
    interfering antibodies, 239
    neutralizing antibodies, 239
antiviral drugs, 285–286
Apobec, degradation of, 72
bottleneck for transmission, 327
cancer, 246–248
    anogenital carcinoma, 248
    B cell lymphoma, 248
    induction of cancers, 247
    Kaposi's sarcoma, 246–248
capsid protection of viral nucleic acids, 230
CCr5 coreceptor, 86, 327
clade A to K, 220
collectins and defensins binding glycoproteins of, 90
course of infections
    pathological conditions associated with phases of infection, 236
    schematic diagram of events, 236
    variability of response to infection, 236–237
CXCr4 coreceptor, 86, 327
cytotoxic T lymphocyte action, interference with, 122
cytotoxic T lymphocyte escape mutants, 146
defensin promotion of infection, 34
discovery and characterization, 220
diseases, 403
dynamics of HIV-1 reproduction in AIDS patients, 241–242

earliest record of infection, 223
Env protein, 114
epidemiology, 403
genetic influence on infection, 20
group M, 11, 220, 222, 223, 227
group N, 220, 222
group O, 220, 222
group P, 220
human infections, 403
immune dysfunction, 237–238
immune responses, 238–241
    cell-mediated response, 238–239
    humoral response, 239–240
infection of immune system cells, 147
infection risk increase with herpes simplex virus type 2, 233
interactome, 288–289
interferon response modulators, 82
isolation from body fluids, 232
Lai, 221
miRNA targeting of RNAs, 69
mutation, 320, 340
Nef protein, 68, 114, 220, 223, 225, 228–230, 237
oncogenic, 182
origin, 11
pathogenesis, 403
phylogenetic relationships, 222
preintegration complex, 227
protease, 286–288
quasispecies concept, 321
reproduction
    dynamics of HIV-1 reproduction in AIDS patients, 241–242
    rate of reproduction, 319
Rev protein, 220, 223–225
reverse transcriptase error rate, 320
RNA interference, suppression of, 71
sanctuary compartments, 241
shedding in semen, 49
spread in host, 114
Tat protein, 114, 220, 222, 224, 238, 245, 248
tetherin ubiquitination, 83
transmission, 231–234, 327
Trim proteins targeting of, 71
Vif protein, 72, 220, 225–226
Vpr protein, 220, 225–227
Vpu protein, 220, 225, 227–229, 237
zoonotic infection, 353–354
Human immunodeficiency virus type 2. *See also* Human immunodeficiency virus
discovery and characterization, 220
diseases, 403
epidemiology, 403
group A, 220, 222
group B, 220, 222
human infections, 403
pathogenesis, 403
phylogenetic relationships, 222
tetherin targeted by Env protein, 228–229
Vpx protein, 225, 227
zoonotic infection, 354
Human immunodeficiency virus vaccine, 250, 277–279
Human papillomavirus. *See also* Papillomavirus
cervical cancer, 273
E6 protein, 190, 193, 209
E7 protein, 190, 192, 204–205
entry into host, 29, 35
L1 protein, 273

oncogenic, 182
oropharyngeal squamous cell carcinoma, 34
persistent infection, 143
propagation, 286
vaccine, 272–273, 276
Human papillomavirus type 16
    anogenital carcinomas in HIV-1-infected individuals, 248
    coevolution with host, 333
    E6 protein, 193, 209
    E7 protein, 204–205
    interferon response modulators, 82
    transforming gene products, 190, 193
Human papillomavirus type 18
    anogenital carcinomas in HIV-1-infected individuals, 248
    coevolution with host, 333
    E6 protein, 193, 209
    transforming gene products, 190, 193
Human papillomavirus vaccine, 272–273, 276
Human rhinovirus 14, 129
Human T-lymphotropic virus
    FasL-mediated CTL killing, 147
    slow infections, 157
    Tax protein, 147
Human T-lymphotropic virus type 1
    adult T-cell leukemia (ATL), 210–213
    cancer and, 182
    diseases, 402
    epidemiology, 402
    Hbz protein, 212
    human infections, 402
    pathogenesis, 402
    persistent infection, 143
    properties, 184
    provirus, 211–212
    Tax transcriptional activator, 147, 211–212
    transcription map of proviral DNA, 211
    transmission, 211
    tumorigenesis with very long latency, 210–213
Human T-lymphotropic virus type 2
    diseases, 402
    epidemiology, 402
    human infections, 402
    pathogenesis, 402
    persistent infection, 143
Human T-lymphotropic virus type 5
    diseases, 402
    epidemiology, 402
    human infections, 402
    pathogenesis, 402
Human viruses, first identified, 4–7
Humanized mice, 137
Humans, facilitating emerging viruses, 350, 360–361
Humidity, influenza transmission and, 19
Humoral immune response, 101, 103, 122, 125–128. *See also* Adaptive immunity; B cells
antibody production by plasma cells, 122, 125
antibody-dependent cell-mediated cytotoxicity, 127
human immunodeficiency virus, 239–240
types and functions of antibodies, 125–126, 128
virus neutralization, 125–127
Huxley, Aldous, 135
Hybridization, subtractive, 189
Hygiene hypothesis, 151
Hypersensitivity, delayed-type, 122
Hypothesis testing, 14

I

Icam1 (intercellular adhesion molecule 1), 114
ICP0 protein, herpes simplex virus, 60, 69, 153
ICP34.5 protein, herpes simplex virus, 68, 81, 162–163
ICP47 protein, herpes simplex virus, 153
ID$_{50}$ (median infectious dose), 158
IE1 protein, human cytomeglovirus, 69
IE2 protein, human cytomegalovirus, 147
Ifi16, 57, 60
IgA, 114
Iκbα, 77
Iκk, 59
IL-1 receptor-associated kinase (Irak), 58
ILs. *See* Interleukin(s)
Immortality, of transformed cells, 177
Immune cell extravasation, 86
Immune complex
    deposition, 166
    solubilization of, 87
Immune defenses, inhibition of, 210
Immune memory, 100, 128–130
    natural experiment demonstrating, 262
    stimulated by vaccination, 261–264
Immune response
    complexity of, 53–54
    continuum between intrinsic and innate
        immunity, 74
    early host response, 53–94
    innate, 74–93
    intrinsic, 54–74
    overview, 53–54
    vaccination, 254–279
Immune surveillance, persistent infection in
    tissues with reduced, 147
Immune system
    brain and, 124
    complexity, 99
    cutaneous, 114
    dysfunctional immune modulation, 114
    immunotherapy, 250, 276–277
    inflammation integration and synergy with
        immune system components, 111
    persistent infections of immune system cells, 147
    short circuited by superantigens, 168
Immunity
    adaptive, 98–130
    autoimmunity, 170–171, 238
    cell-mediated immune response, 101, 103,
        119–124
    continuum between intrinsic and innate
        immunity, 74
    herd, 255, 262–264, 278
    host population susceptibility to disease and,
        20–21
    humoral immune response, 101, 103, 122,
        125–128
    mucosal, 113
Immunization. *See also* Vaccination
    active, 260–261
    passive, 260–261
Immunoadhesins, 250
Immunocontraception, 347
Immunodominant epitope, 146, 168
Immunoglobulin(s). *See also* Antibody
    in human immunodeficiency virus infection,
        238
    types and functions, 125–126, 128
Immunoglobulin A (IgA), 34, 125–128

Immunoglobulin D (IgD), 125–126
Immunoglobulin E (IgE), 125–126
Immunoglobulin G (IgG), 125–127
Immunoglobulin M (IgM), 125–126
Immunological memory, 100, 128–130
Immunological synapse, 119–120
Immunopathology, 164–168
    from adenovirus vector, 158
    free radical-mediated, 168
    heterologous T cell immunity, 167–168
    lesions caused by B cells, 166–167
    lesions caused by CD4$^+$cells, 165–166
    lesions caused by cytotoxic T lymphocytes,
        164–165
    superantigens, 168
    systemic inflammatory response syndrome
        (SIRS), 167
Immunosuppression, 168–170
Immunotherapy, 250, 276–277
In silico drug discovery, 288, 290
Inactivated ("killed") virus vaccine, 265–268, 269
Incas, smallpox and measles infection, 352
Incidence of disease, 4, 11, 13
Incubation periods, 137–138, 149
Index case, 9
Infection
    barriers to, 24–50
    cellular changes occurring following, 60, 62
    complement protection against, 89
    coordinated host response to, 27
    critical events in hypothetical acute virus
        infection, 94
    detection of, 54–62
    disseminated, 39
    early host response, 53–94
    ectromelia virus example, 25–27
    entry sites, viral, 29–36
        alimentary tract, 30, 33–35
        eyes, 30, 35–36
        respiratory tract, 30, 31–33
        skin, 29–31
        urogenital tract, 30, 35
    first critical moments of, 54–62
    host alterations as early signals of, 62
    host cell proteins that regulate the infectious
        cycle, 36–39
    host-parasite interaction, classic theory of,
        317–318
    imitation of, 27–29
    intracellular detectors of, 56
    organ invasion, 45–47
    patterns, 136–158
        abortive infection, 137, 157
        acute infection, 137–143, 172
        incubation periods, 137–138
        latent infection, 137–138, 150–157
        mathematics of growth correlating with,
            138–140
        persistent infection, 137–138, 143–150
        slow infection, 157
        transforming infection, 157–158
    as series of stochastic events, 27–28
    shedding of virus pariticles, 47–50
    spread throughout host, 39–45
    systemic, 39
    viral tropism, 36–39
Infections of populations
    epidemics (*See* Epidemic)
    epidemiology of (*See* Epidemiology)

parameters governing ability of virus to infect,
    16–22
    climate, 17–19
    environment, 16–19
    genetic and immune parameters, 19–21
    geography, 16
    host factors, 19–22
    nongenetic risk factors, 21–22
    population density, 17
    seasonal variations, 17–19
    urban legends concerning, 17, 20
Infectious hematopoietic necrosis virus, 334
Infectious mononucleosis, 146, 154, 391
Infectious period, 149
Infiltration, 76
Inflammasome, 109–110
Inflammation, 110–111
    activation by complement, 87
    classic signs, 110
    cytokines role in, 74, 76
    defined, 76
    free radical-mediated immunopathology, 168
    integration and synergy with immune system
        components, 111
    systemic inflammatory response syndrome
        (SIRS), 167
Influenza A virus. *See also* Influenza virus
    antiviral drugs, 283–323, 298–299
    discovery of, 7
    genomic and epidemiological dynamics, 336
    H1N1, 324–325
    H2N2, 324
    H3N2, 324–325
    hemagglutinin (HA), 298, 300
    M2 protein, 228, 298
    neuraminidase (NA), 298–299
    NS1A protein, 72
    pandemic, 324–325
    RNA interference, suppression of, 71
    seasonal variation in disease, 17
    stable host-virus interaction, 345
Influenza B virus, 298
    antiviral drugs, 299
    NS1B protein, 72
Influenza virus. *See also* Avian influenza virus;
        Influenza A virus
    1918 flu, 7, 21
    acute infection, 139
    antigenic variation, 141–143, 268
    cap-snatching, 82
    collectins and defensins binding glycoproteins
        of, 90
    cost of infection, 268
    cytokine storm, 167
    discovery of, 6, 7, 13, 15
    diseases, 392
    epidemics, 142, 354
    epidemiology, 392
    evolution, 334–335, 354–355
    in ferrets, 48, 356
    genetic map, 334
    genetic shift and drift, 324
    Guillain-Barré syndrome, 272
    H1N1, 13, 15, 354
    H3N2, 298
    H5N1, 38, 268, 355–356
    H7N9, 356, 361
    HA (hemagglutinin) protein, 37–39, 143, 268,
        298, 300, 324–325, 349, 360

HA0 precursor, 37–39
heterologous T cell immunity, 167–168
human infections, 392
inactivated vaccine, 265
incidence of, 11
incubation period, 138
interferon discovery, 76
interferon response modulators, 82
Isg15, countermeasures against, 72
M2 protein, 164, 228, 284, 298
monitoring tools, 359–360
monoclonal antibody inhibitors, 300
NA (neuraminidase) protein, 38, 268, 298–299, 324–325
NS1 protein, 84
pandemics, 3, 6, 7, 21, 48, 324–325, 354–355, 359–360
pathogenesis, 392
polarized release from epithelial cells, 39
R_0 (R-naught) value, 318
reassortment, 268, 355
reproduction blocked by Mx proteins, 82
resistance to, *Mx1* gene and, 19
salivary agglutinin interference with, 34
seasonal variation in disease, 17–19
shedding from host, 39
sialic acid, 136
structural plasticity, 141
superoxide production in lungs, 168
surveillance, 268
tropism, 37–39
urban legends concerning, 20
WSN/33 strain, 38
zoonotic infection, 344
Influenza virus vaccine, 20, 267–268
administration, 269
annual timeline for creating, 268
antigenic variation of virus, 141–142
subunit, 272
Information technology, 8
Ink4, 204, 208
INN (International Nonproprietary Name), 294
Innate immune response, 74–93, 99–100
acute infection, 138
cGAS/Sting axis, 60
complement, 87–90
defined, 27
early, 27
γδ cells, 93
human immunodeficiency virus, 238
integration with intrinsic defense and adaptive immune response, 55
natural killer (NK) cells, 90–92
neutrophils, 92–93
NKT cells, 93
sentinel cells, 67, 87
soluble mediators, 74–86
chemokines, 74, 86
cytokines, 74–86
Insertional activation, 195, 197–198
Institutional Review Board (IRB), 292
Integrase, 335
antiviral drugs, 302, 306, 308
human immunodeficiency virus type 1, 290
Integrase strand transfer inhibitors, 308
Integration
adenovirus, 187
polyomavirus, 187
provirus, 187, 189, 191, 195, 197, 200, 324

retrovirus, 187, 191, 197
into vertebrate genomes, 337–338
Integrins, 86
Interactome, human immunodeficiency virus (HIV) type 1 protein, 288–289
Intercellular adhesion molecule 1 (Icam1), 114
Interfering antibodies, human immunodeficiency virus, 239
Interferon(s), 20, 74–87
action blocked by v-IL-6, 210
antiviral state, 79, 83–84
CD4+ T$_h$1 cell production of, 165
cell signaling induced by receptor engagement, 55
crucial nature in antiviral defense, 77
discovery, 76
Hendra virus impairment of gene induction, 169
in human immunodeficiency virus infection, 238
IFN-induced gene products, 79, 81–83
2′-5′-oligo(A) synthetase, 81
dsRNA-activated protein kinase (Pkr), 79, 81
IFN regulatory proteins, 83
Mx proteins, 81–82
promyelocytic leukemia (Pml) proteins, 82
RNase L, 81
tetherin/Bst2, 83
ubiquitin-proteasome pathway components, 82–83
Mx2 and, 230
persistent infections, 144
promyelocytic leukemia (Pml) bodies, affect on, 68–69
regulators of IFN response, 83–84
synthesis of type I, 77–79
tetherin and, 228
transcription of *IFN-β gene*, switching on and off of, 80
viral gene products that counter IFN response, 84–86
viral modulators of response, 82
Interferon α/β (IFN-α/β), 56–57, 59, 61, 63, 69, 74, 76–82, 85–86
Interferon ϵ, 76
Interferon γ (IFN-γ), 76, 78, 82, 85, 90–91, 93, 104–105, 111, 113, 116, 119–122
Interferon genes, 60, 75, 77, 83
Interferon κ, 76
Interferon λ, 76
Interferon λ receptor 1, 78
Interferon receptors, 78–79
Interferon regulatory factors (Irfs), 56–60, 74, 81, 83
Irf1, 81, 83
Irf3, 57, 59–60, 78
Irf4, 83
Irf7, 57
Irf8, 83
Irf9, 81, 83
Interferon therapy, 276–277
Interferon ω, 76
Interferon-gamma activated sequence (GAS) elements, 81
Interferon-inducible genes/proteins, 57, 72, 79–83
Interferon-inducible myxovirus resistance gene, *Mx1*, 19
Interferon-inducible protein 16 (Ifi16), 57
Interferon-stimulated gene 15 (Isg15), 72

Interferon-stimulated response elements (ISREs), 81, 83
Interleukin(s), 76, 109
Interleukin-1 (IL-1), 58, 76, 79, 111, 113
Interleukin-1 receptor, 58, 76
Interleukin-1β (IL-1β), 74, 109–110
Interleukin-2 (IL-2), 91, 104–107, 116, 122, 165, 237–238, 249
Interleukin-4 (IL-4), 90, 105–106, 125
CD4+ T$_h$2 cell production of, 165
recombinant ectromelia virus containing IL-4 gene, 160–161
Interleukin-5 (IL-5), 105–106, 125, 165
Interleukin-6 (IL-6), 57, 74, 76, 79, 81, 105–106, 113
Interleukin-8 (IL-8), 61, 86, 111
Interleukin-10 (IL-10), 74, 105–106, 165
Interleukin-10 receptor 2, 78
Interleukin-12 (IL-12), 57, 74, 105–106, 109, 169–170
Interleukin-13 (IL-13), 90, 106
Interleukin-16 (IL-16), 121
Interleukin-17 (IL-17), 74, 106
Interleukin-18 (IL-18), 91, 109
Interleukin-21 (IL-21), 106
Interleukin-23 (IL-23), 106
Interleukin-28A (IL-28A), 76
Interleukin-28B (IL-28B), 76
Interleukin-29 (IL-29), 76
International Nonproprietary Name (INN), 294
International Society for Infectious Diseases, 15
Intestinal mucosa, HIV-1 effects on, 243
Intracellular cytokine assay, 123
Intracellular parasites, virus origin from, 328
Intracellular trafficking, of viroid RNA, 366–367
Intravenous drug injection, 232
Intrinsic immune response, 27, 54–74, 99–100
acute infection, 138
ancient mechanisms, 73
apoptosis, 62, 64–67
autophagy, 67–68
bacterial defense systems, 73
CRISPRs, 72–74
cytosine deamination, 71–72
detection of infection, 54–62
epigenetic silencing, 68–69
first critical moments of infection, 54–62
genome editing, 71–72
integration with innate and adaptive response, 55
intracellular detectors of infection, 56
Isgylation, 72
receptor-mediated recognition of microbe-associated molecular patterns (MAMPs), 55–60
RNA interference, 69–71
summary, 63
Intrinsic pathway to apoptosis, 64–66
Intuition, as host defense, 36
Iododeoxyuridine, 295
Ips1, 59, 61
Irak (IL-1 receptor-associated kinase), 58
IRB (Institutional Review Board), 292
Irfs. *See* Interferon regulatory factors
Iridovirus, 330, 360
Isgf3, 81–83
Isgylation, 72
ISREs (interferon-stimulated response elements), 81, 83

J

Jak kinases, 79, 84, 121
Jak/Stat pathway, 79, 81
Janeway, Charles, 275
Japanese encephalitis virus, 385
Japanese encephalitis virus vaccine, 267
JC virus, 186
    coevolution with host, 333
    diseases, 399
    enhancer, 37
    epidemiology, 399
    human infections, 399
    pathogenesis, 399
    persistent infection, 143
    progressive multifocal leukoencephalopathy
        (PML), 37
    shedding in urine, 49
    slow infections, 157
    tropism, 37
Jenner, Edward, 21, 255–256
Jnk kinase, 58, 59
Junin virus, 344, 347, 381

K

K5 protein, human herpesvirus 8, 145–146
Kaposi, Moritz, 246
Kaposi's sarcoma, 186, 194–196, 391
    human herpesvirus 8, 66, 248
    human immunodeficiency virus type 1,
        246–248
Kaposi's sarcoma-associated herpesvirus.
    See Human herpesvirus 8
Keratitis, herpes simplex virus, 165
Keystone virus, 16
Kidneys
    human immunodeficiency virus
        pathogenesis, 246
    immune complex deposition in, 166
Killer cell immunoglobulin-like inhibitory
        receptors (Kirs), 91
Kinase. *See* Tyrosine kinase
Kinesin, 42
Kip, 204
Kirs (killer cell immunoglobulin-like inhibitory
        receptors), 91
Knockout mice for studying viral pathogenesis, 137
Koch, Robert, 4, 5
Koch's postulates, 4, 5, 358
Koplik spots, 22, 47
K-replication strategy, 139
K-selection, 138–139
Kupffer cells, 46
Kuru, 371

L

L1 protein, human papillomavirus, 273
La Crosse virus, 382
Laidlaw, Patrick, 6
Lake Victoria Marburgvirus, 384
Lamin C, 311
Lamivudine (3TC, Epivir), 295, 297, 321
Langerhans cells, 114, 233–234
Lao Tzu, 219
Large delta antigen, hepatitis delta satellite virus,
        370
Large T antigen, simian virus 40 (SV40), 66, 190,
        192, 204–205

Lassa virus, 359, 381
    antiviral drugs, 297
    zoonotic infections, 347
Latency-associated transcripts (LATs), herpes
        simplex virus, 152–153
Latent infection, 137–138, 150–157
    clinical latency in HIV infections, 236
    described, 137, 150
    Epstein-Barr virus, 154–157
    general pattern of, 138
    herpes simplex virus, 150–154
    human immunodeficiency virus, 249
    K-replication strategy, 139
    long-latency oncogenic retrovirus, 183–184,
        210–213
    in peripheral nervous system, 43
    tumorigenesis with very long latency, 210–213
    varicella-zoster virus, 269
Latent membrane protein 1 (LMP-1), Epstein-Barr
        virus, 155–156, 198–200, 206
Latent membrane protein 2A (LMP-2A),
        Epstein-Barr virus, 155, 157
LAV (lymphadenopathy virus), 221
Lazear, Jesse, 4
Lck (lymphocyte-specific protein tyrosine kinase),
        230
LD$_{50}$ (median lethal dose), 158
L-dopa, 377
Leaky gut, 242
Lectin, 90
Lectin pathway, complement, 87–88
LEDGINs, 290
Lens epithelium-derived growth factor
        (Ledgf/p75), 290
Lentivirus
    auxiliary proteins, 220, 222
    diseases, 220, 222, 403
    epidemiology, 403
    Fv1 protein targeting of, 72
    genome organization, 220, 224
    human immunodeficiency virus as, 220
    human infections, 403
    pathogenesis, 403
    phylogenetic relationships, 222
    table of examples, 222
    vectors, 197
Lethal mutagenesis, 322
Leukemia, 176
Leukocyte
    definition, 102
    hematopoietic stem cell, 100–102
    types, 102
Leukocyte function antigen 1 (Lfa1), 114
Levovirin, 297
Life cycles, viral, 284
Limiting-dilution assay, 123
LINEs (long interspersed nuclear elements), 335, 337
Lipopolysaccharide, 33
Liver, routes of viral entry into, 46
Livestock
    movement of, 360
    national vaccination and disease control
        programs, 258–260
LJ-001, 297
LL-37 peptide, 233
LMP-1, Epstein-Barr virus, 155–156, 198–200, 206
Loeffler, Friedrich, 4
Logistic growth, 140
Logovirus, 383

Long interspersed nuclear elements (LINEs),
        335, 337
Long terminal repeat (LTR)
    human immunodeficiency virus, 222, 224,
        227, 250
    retroviral, 335–337
Longitudinal study, 11–12
Long-latency oncogenic retrovirus, 183–184,
        210–213
Long-latency retrovirus, 183–184
Loss-of-function mutation, 176
Lungs, human immunodeficiency virus
        pathogenesis, 246
Lymph node
    anatomy, 112
    antigen-presenting cell migration to, 111–114
    HIV pathogenesis, 242–244
    locations in human body, 100, 112
    role in rapid immune response, 100
Lymphadenopathy virus (LAV), 221
Lymphatic system
    anatomy, 112
    structure of, 41
    viral replication in lymphoid cells, 41
    viral spread, 41
Lymphocyte. *See also* B cells; T cells
    cell proliferation triggered by activation of, 119
    demise after activation, 100
    development, diversity, and activation, 100–108
    diapedesis, 46
    infected, 147
    viral spread by, 46–47
Lymphocyte receptor genes, somatic
        rearrangements of segments, 100
Lymphocyte-specific protein tyrosine kinase
        (Lck), 230
Lymphocytic choriomeningitis virus, 21, 381
    autoimmune disease, 171
    CD8$^+$ T cell proliferation, 100
    cytotoxic T lymphocyte (CTL)-mediated
        immunopathology, 164–165
    genetic map, 334
    immune complex deposition, 166
    persistent infection, 143, 148–150
    viremia, 42
    virulence, 159
    zoonoses, 149
Lymphoid tissue
    human immunodeficiency virus pathogenesis,
        242–244
    measles virus infection, 148
Lymphoma, 176
Lysosome, 55, 58
    autolysosome, 68
    autophagy, 67–68
    xenophagy, 68
Lysozymes, 34, 35
Lyssavirus
    diseases, 404
    epidemiology, 404
    genetic map, 334
    human infections, 404
    pathogenesis, 404

M

M cell, 34–35, 113–114
M (mitosis) phase, 180–181, 202
M2 protein, influenza virus, 164, 228, 284, 298

Maass, Clara Louise, 6
Machupo virus, 344
Macrophage, 54, 56, 58, 62, 64, 67, 74, 77, 83, 86–87, 89, 93–94
　antigen-presenting cell, 108, 114–115
　Fc receptors, 126–127
　HIV-1-infected, 41, 238, 245, 246
　in lymph node, 112
　M cells and, 114
　measles virus infection, 106, 169
　as portal for entry of viral particles into tissues, 45–46
　rashes and, 122
　recruitment to skin wound, 31
　removal of virus particles by, 46
Mad cow disease, 374
Maedi/visna, 59, 157, 222
Major histocompatibility complex (MHC)
　class I proteins
　　antigen presentation, 115–119
　　artificial MHC tetramer, 123
　　herpes simplex virus ICP47 protein blockage of presentation, 153
　　inhibition of EBNA-1 presentation, 210
　　inhibition of transcription by adenovirus, 210
　　receptors on NK cells, 91
　　self antigens, 91
　　surface concentration downregulated by Nef protein, 229–230
　　viral interference with production and function in persistent infections, 144
　class II proteins
　　antigen presentation, 102, 115–119
　　autophagy and, 68
　　cell types synthesizing, 119
　　in HIV, 237–238
　　modulation in persistent infections, 146
　diversity among genes, 21
　susceptibility to infection and, 21
Malignant, 176, 177
Malignant cutaneous lymphoma, 402
Malnutrition, influence on susceptibility to infection, 22
MALT (mucosa-associated lymphoid tissue), 113
Mammals, recognition of foreign nucleic acids in, 57
MAMPs (microbe-associated molecular patterns), 55–60
Mannan-binding lectin pathway, 88
Map (mitogen-activated protein) kinase, 176, 179–180, 193–195, 199–203, 206
Maraviroc, 285, 309
Marburg virus, 384
　cross-species infections, 343
　cytokine storm, 167
　dead-end virus-host interaction, 345
Martin, Malcolm, 221
Mathematical models, of host-virus interaction, 350
Mavs, 59
Maximum tolerated dose, 292
Mda5, 56, 59, 61, 68, 75, 81
Mdm-2 protein, 207–208
Measles virus, 47
　clinical course, 148–149
　deaths, 260
　diseases, 394
　epidemiology, 394
　eradication effort, 258, 264

evolving host-virus interaction, 345
　hemagglutinin protein, 101
　historical aspects, 256, 258, 260, 352
　host age influence on severity, 21
　human infections, 394
　immune memory, 262
　immune system cell infection, 41, 147
　immunosuppression, 168–169
　incidence, 147–148
　incubation period, 138
　infection by, 147–149
　infectious period, 149
　interferon response to, 77
　Koplik spots, 22, 47
　macrophage infection, 106
　neural spread, 43
　outbreak (2015), 278
　pathogenesis, 394
　persistent infection, 143, 147–149
　polarized release from epithelial cells, 39
　population density of hosts and, 17
　R_0 (R-naught) value, 318
　rash, 122, 124
　reproduction
　　blocked by Mx proteins, 82
　　in immune system cells, 41
　reproduction number, 258
　secondary infections, 148
　shedding from host, 39
　slow infections, 157
　spread in host, 47
　stable host-virus interaction, 345
　subacute sclerosing panencephalitis (SSPE), 148
　transplacental infection, 47
　vaccination campaign, 256–258, 260
Measles virus vaccine, 147, 267
　attenuated, 264, 269
　herd immunity threshold, 263
　interferon synthesis induced by, 85
　large-scale vaccination program, 257–258, 260
Measles-mumps-rubella (MMR) vaccine, 269, 278
Mechanism-based screens, 285
Median infectious dose (ID_{50}), 158
Median lethal dose (LD_{50}), 158
Median paralytic dose (PD_{50}), 158
Melanoma differentiation-associated protein 5 (Mda5), 56, 59, 61, 68, 75, 81
Membrane cofactor protein, 90
Memory cells
　B cells, 53, 100–102, 128–130
　natural killer (NK) cells, 91–92
　T cells, 53, 100, 104, 128–130
　　adoptive transfer, 261
　　central memory T cells, 129–130
　　effector memory T cells, 129
　　heterologous T cell immunity, 167–168
　　HIV infections, 234, 241–243, 249
Merkel cell carcinoma, 186
Merkel cell polyomavirus, 182, 186
MERS. *See* Middle East respiratory syndrome (MERS) coronavirus
Mesothelioma, 185
Metagenomic analysis, 330, 358
Metastasis, 177
Methylation, DNA, 68–69, 73
Miasma, 4
Mice
　ectromelia virus infection of, 25–27
　hantavirus zoonoses and, 352

humanized, 137
　models of human diseases, 136
　transgenic and knockout for studying pathogenesis, 137
Microbe-associated molecular patterns (MAMPs), 55–60
Microbicides, 300
Microglia, HIV infection and, 245–246
Microneedle patch, for vaccine delivery, 265
microRNA (miRNA), 186
　antagonists, 300–301
　herpesvirus latency, 153
　RNA interference (RNAi), 69
Microtubules, neuronal spread of viruses, 42–44
Microvilli, 34
Middle East Respiratory Syndrome, 3
Middle East respiratory syndrome (MERS) coronavirus
　identification of, 362
　reservoir, 361
　zoonotic infection, 353
Middle T protein (mT), mouse polyomavirus, 186, 198–199, 201
Milk, shedding of virions in, 49
Miller, Lois, 66
Mimicry, molecular, 165, 170–171, 272
Mimivirus
　cryo-electron micrograph, 331
　discovery, 331
　sputnik virophage and, 370
Minireplicon, 286
Misfolding of prion proteins, 375
Mismatch repair genes, defects in, 176
Mitochondrial pathway to apoptosis, 64–66
Mitogen-activated protein kinase (Mapk), 176, 179–180, 193–195, 199–203, 206
Mitosis, 180–181, 202
MK3 protein, murine gammaherpesvirus 68, 145
Mlkl, 67
Modulation, definition, 26
MOI (multiplicity of infection), 139
Molecular biology, developments in, 7
Molecular mimicry, 165, 170–171, 272
Molluscum contagiosum virus
　diseases, 400
　epidemiology, 400
　pathogenesis, 400
Moloney leukemia virus, gene therapy use of, 197
Monkeypox virus, 345, 400
Monoclonal antibody
　antiviral, 300
　for prion disease treatment, 377
Monoclonal antibody-resistant mutants, 127
Monoclonal tumor, 187
Monocyte
　diapedesis, 46
　HIV-1-infected, 238, 245
　infected, 147
　viral spread by, 46–47
Mononucleosis, 48, 146, 154, 391
Monotherapy, 303, 310
Morbidity, 12–13
Morbidity and Mortality Weekly Report (publication), 15
Morbillivirus
　diseases, 394
　epidemiology, 394
　human infections, 394
　pathogenesis, 394

Mortality, 12–13
mos gene, 188, 190, 192
Mosaic, 326
Mosquito vector
  breeding in used tires, 360
  Chikungunya virus, 16
  myxoma virus, 347
  West Nile virus, 346
  yellow fever, 4–6, 8
Mother to child transmission, of human immunodeficiency virus, 47, 232–233
Mouse hepatitis virus, 165
Mouse mammary tumor virus
  enhanced infection in the presence of intestinal bacteria, 33
  infectious cycle, 169
  superantigens, 168
Mouse polyomavirus, 186
  middle T protein (mT), 186, 198–199, 201
  viral adapter proteins, 198–199, 201
Mousepox virus, 26, 41, 106
Mucophagy, 32
Mucosa-associated lymphoid tissue (MALT), 113
Mucosal immunity, 113
Mucus
  alimentary tract, 33–34
  green, 32
  respiratory tract, 32
  small intestine, 34
  stomach, 34
  urogenital tract, 35
Mucus-secreting, respiratory tract, 31
Muller's ratchet, 323
Multiple sclerosis, 165
Multiplicity of infection (MOI), 139
Mumps virus
  central nervous system entry, 46
  host age influence on severity, 21
  incubation period, 138
  neuroinvasiveness, 43
  neurovirulence, 44
  transmission in saliva, 49
Mumps virus vaccine, 267
Murine gammaherpesvirus 68
  E3 ubiquitin ligase, 145
  MK3 protein, 145
Murine leukemia virus, 71
Murine norovirus, 33
Mutagenesis
  lethal, 322
  site-directed, 160
Mutation
  alteration of virulence, 159–160
  antibody escape, 127
  attenuated virus vaccines, 269–270
  attenuating, 347
  cancer and, 175–176
  drug resistance, 309–310
  escape mutants, cytotoxic T lymphocytes, 146
  immunodominant epitopes, 146
  lethal mutagenesis, 322
  loss-of-function, 176
  Muller's ratchet, 323
  quasispecies concept, 321–322
  structural plasticity of viruses and, 141
  in vaccine strains, 266
  viral evolution, 318–323
  virulence genes, 160–163

Mutation rate
  DNA virus, 319
  genome size relationship, 319, 321
  human immunodeficiency virus type 1, 320
  RNA virus, 319
  viroid, 368
MV-LAP protein, myxoma virus, 145–146
Mx proteins, 81–82
Mx2, 230
myc gene, 188, 190, 192–195, 197
Mycobacterium avium, 246
Mycobacterium tuberculosis, 246
Myeloid differentiation primary response protein 88 (Myd88), 57–58, 61
Myeloperoxidase, 32
Myxoma virus
  deliberate release in Australia, 347
  E3 ubiquitin ligase, 145–146
  entry into host, 29
  MV-LAP protein, 145–146
  release to control rabbits in Australia, 318

N

NA (neuraminidase) protein, influenza virus, 38, 268, 298–299, 324–325
Nairovirus, 382
Names, virus, 359
NANBH (non-A, non-B hepatitis) agents, 357–358
Nasal secretions, shedding of virions in, 48
Nasopharyngeal carcinoma, 146, 200, 210, 391
Natural killer (NK) cells, 87, 90–92, 105, 127
  in human immunodeficiency virus infection, 238
  memory, 91–92
  MHC class I receptors on, 91
  two-receptor mechanism, 91
  viral proteins that modulate, 91–92
Necroptosis, 67
Nef protein
  human immunodeficiency virus, 68, 122, 147, 220, 223, 225, 228–230, 237
  intracellular functions, 229–230
  posttranslational modification, 229
  simian immunodeficiency virus, 228
Negative selection, 108
Nelmes, Sarah, 256
Neoplasm, 176
Nervous system infection
  entry of virus by olfactory routes, 43
  human immunodeficiency virus pathogenesis, 244–246
  spread of virus, 39–40, 42–45
  terminology, 43
NETs (neutrophil extracellular networks), 92–93
Neural spread, 42–45
  antegrade, 44
  entry into central nervous system by olfactory routes, 44
  herpes simplex virus, 39–40
  pathways, 43
  retrograde, 44
  tracing neuronal connections with viruses, 45
Neuraminidase inhibitors, 298–299
Neuraminidase (NA) protein, 38, 268, 298–299, 324–325
Neuroinvasive virus, 43
Neuroinvasiveness, 164

Neurons
  herpes simplex virus infection, 150–154
    multiple genomes in cells, 154
  olfactory, 44
  structure, 42
Neurotropic virus, 36, 42–43, 45–46
Neurovirulent virus, 43
Neutralization assay, 123
Neutralization index, 123
Neutralizing antibodies
  broadly neutralizing antibody, 240
  human immunodeficiency virus, 239–240
Neutrophil extracellular networks (NETs), 92–93
Neutrophils, 62
  green mucus caused by, 32
  innate immune response, 92–93
  recruitment to skin wound, 31
Neutrotransmitters, cytokines as, 74
Newcastle disease virus, 85
Nf-κb. *See* Nuclear factor κb
Nipah virus
  bat reservoir, 349
  encephalitis, 350
  interferon response modulators, 82
  transmission, 10
  zoonotic infection, 344, 347, 349–350
Nitric oxide, 168
NK cells. *See* Natural killer (NK) cells
NKT cells, 93, 105
Nlrp3 inflammasome, 110
Non-A, non-B hepatitis (NANBH) agents, 357–358
Noncytopathic viruses, 137, 143–144
Nonessential genes, 160
Nongenetic risk factors for infection, 21–22
Nonhomologous recombination, 326, 337
Nonnucleoside inhibitors, 302–303, 306
Nonparametric test, 14
Nontransducing oncogenic retrovirus, 183–184, 187, 195–198, 200, 210–213
  insertional activation, 195–198
  long-latency, 183–184, 210–213
  properties, 184
Normally distributed data, 14
Norovirus
  acute infection, 139
  diseases, 383
  epidemics, 142
  epidemiology, 383
  human infections, 383
  pathogenesis, 383
  transmission, 10
Norwalk virus, 10, 383
NS1 protein, influenza virus, 84
NS1A protein, influenza A virus, 72
NS1B protein, influenza B virus, 72
NS3/4A protein, hepatitis C virus, 61, 144, 306–307
NS4A protein, hepatitis C virus, 214
NS5A protein, hepatitis C virus, 214, 308–309
NS5B protein, hepatitis C virus, 214, 303
Nuclear factor κb (Nf-κb), 55–61, 74–75, 77–78, 80, 90, 105, 108–109, 144, 195, 198–199, 206, 212–213, 222, 228–229, 249
Nuclear localization signal, hepatitis delta satellite virus, 370
Nucleocytoplasmic large DNA viruses, 329–330, 332
Nucleoside analogs, 295–296, 301–302, 302–303

Nucleosome, silencing of viral genomes, 151
Nucleotide analogs, 295
Nucleus
    origin of, 329
    viroid replication in, 365, 367
Null hypothesis, 14
Nüsslein-Volhard, Christiane, 56

O

Olfactory routes, viral spread by, 44
2′-5′-Oligo(A) synthetase, 81
Oncogene, 175, 182–184, 186–195, 197–198, 206, 208–209
    acquisition from host cell genome, 324
    capture, 188, 191, 198, 200
    c-oncogene, 183
    defined, 183
    dominant, 192
    functional classes, 192
    insertional activation, 195, 197–198
    probe, 189
    proto-oncogene, 182, 184, 186, 191–192, 194–195, 198, 213
    transduced, 192–194
    v-oncogene, 183–184, 187–188, 190, 192
Oncogene capture, 188, 191, 198, 200
Oncogenesis, 169–170, 175–214. *See also* Transformed cells
    cancer terminology, 176
    defined, 175
    human immunodeficiency virus type 1 and, 246–248
    by human tumor viruses, 210–213
        hepatitis B virus, 213
        hepatitis C virus, 213–214
        human T-lymphotropic virus type 1, 210–213
    immortality acquisition, 177
    paracrine, 196
    pathogenesis, 169–170
    tumorigenesis with very long latency, 210–213
    walleye dermal sarcoma virus transmission cycle, 184
Oncogenic virus, 169–170, 175, 176, 182–192
    bandicoot papillomatosis carcinomatosis virus type 1, 183
    contemporary identification, 186–187
    discovery of, 182–187
    DNA virus, 182, 185–186
    human immunodeficiency virus type 1, 246–248
    human tumor viruses, mechanisms of transformation and oncogenesis by, 210–214
    properties of, 187
    recombination, 324
    retrovirus, 182–184
        genome map, 190
        insertional activation, 195, 197–198
        long-latency, 183–184, 210–213
        nontransducing, 183–184, 187, 195–198, 200, 210–213
        oncogene capture, 190
        transducing, 183–184, 186–188, 190, 198, 200
    viral genetic information in transformed cells, 187–189
    viral transforming genes
        identification and properties, 187–189
        origin and nature of, 189–192

viral transforming proteins
    activation of cellular signal transduction pathways, 192–201
    disruption of cell cycle control, 201–205
    functions, 192
Opsonization, complement, 87
Orc1, 204
Orf A protein, walleye dermal sarcoma virus, 184
Orf B protein, walleye dermal sarcoma virus, 184
Orf C protein, walleye dermal sarcoma virus, 184
Organ invasion, 45–47
    central nervous system, 46–47
    fetus, 47
    liver, 46
    organs with dense basement membranes, 46–47
    organs with sinusoids, 45–46
    organs without sinusoids, 46
    skin, 47
Origin of replication, viroid, 365
Origin of viruses, 327–332
Original antigenic sin, 168
Oropharyngeal squamous cell carcinoma, human papillomaviruses, 34
Oropouche virus, 382
Orthomyxovirus
    attenuated reassortment viruses, 269
    diseases, 392
    epidemiology, 392
    genetic map, 334
    human infections, 392
    pathogenesis, 392
    reassortment, 324
    zoonotic infection, 344
Orthoreovirus
    diseases, 401
    epidemiology, 401
    human infections, 401
    pathogenesis, 401
Outbreak (movie), 8

P

P protein, paramyxovirus, 61
P value, 14
p7 protein, hepatitis C virus, 164
p38, 58
p53 gene, loss-of-function mutation in, 176
p53 protein, 192
    inactivation, 206–209
    regulation of stability and activity of, 207–208
    structure, 207
    viral proteins inactivating, 208–209
Paleovirology, 335–339
Palivizumab, 300
PAMPs (pathogen-associated molecular patterns), 55
Pan troglodytes troglodytes, 220, 222
Pandemic
    AIDS, 219, 360
    human immunodeficiency virus, 359
    influenza virus, 3, 6, 7, 21, 48, 324–325, 354–355, 359–360
    predicting next, 359–360
Pandoravirus, 329, 331
    discovery, 331
    electron micrograph, 331
Pandoravirus salinus, 331, 365
Paneth cells, 34
Panum, Peter, 262

Papillomavirus. *See also* Human papillomavirus
    antiviral drugs, 297
    bandicoot papillomatosis carcinomatosis virus type 1, 183
    coevolution with host, 333
    difficulty of culturing, 5
    diseases, 393
    E5 protein, 199
    entry into host, 34
    epidemiology, 393
    evolution, 333
    human infections, 393
    immunotherapy, 277
    incubation period, 138
    NK modulators, 91
    oncogenesis, 185
    p53 inactivation, 209
    particle-to-PFU ratio, 28
    pathogenesis, 393
    persistent infection, 143, 147
    shedding of virions, 48, 49
    transforming gene products, 190
    transmission by sexual contact, 49
Papillomavirus vaccine, 267
Paracrine oncogenesis, 196
Parametric test, 14
Paramyxovirus
    diseases, 394–395
    epidemiology, 394–395
    genetic maps, 334
    human infections, 394–395
    neural spread, 43
    P protein, 61
    pathogenesis, 394–395
    persistent infection, 147
    Tlr signaling pathway blockage, 61
    V proteins, 61
    zoonotic infection, 344
Particle-to-plaque forming unit (PFU) ratio, 28
Parvovirus, 47
    coevolution with host, 333
    error-prone replication, 319
    genome incorporation into host, 337
    host-range shift, 354
    TfR1 receptor for, 339
    transplacental infection, 47
Passive immunization, 260–261
Pasteur, Louis, 4, 256–257, 266
Pathogen, definition, 101
Pathogen-associated molecular patterns (PAMPs), 55
Pathogenesis
    animal models of human diseases, 135–136
    defined, 3
    ectromelia virus, 25–27
    history
        identification of first human viruses, 4–7
        microbes as infectious agents, 4
        technological developments, 7–8
        viral epidemics, 8–10
    immunopathology, 164–168
    immunosuppression induced by viral infection, 168–169
    infected cell lysis, 164
    mechanisms, 134–172
    molecular mimicry, 170–171
    oncogenesis, 169–170
    overview, 3

Pathogenesis *(continued)*
  patterns of infection, 136–158
    abortive infection, 137, 157
    acute infection, 137–143, 172
    incubation periods, 137–138, 149
    infectious period, 149
    latent infection, 137–138, 150–157
    mathematics of growth correlating with,
      138–140
    persistent infection, 137–138, 143–150
    slow infection, 157
    transforming infection, 157–158
   transgenic and knockout mice for studying, 137
   viral virulence, 158–164
Pattern recognition receptors, 55–60, 144
PCR, 7, 357–358
Pd-1 (programmed cell death protein 1), 250
Peach latent mosaic viroid, 366, 368
Penciclovir, 295
PEP (postexposure prophylaxis), human immuno-
  deficiency virus, 250
PEPFAR (President's Emergency Plan for AIDS
  Relief), 219
Peptide mimics (peptidomimetics), 306
Perameles bougainville, 183
Perforin, 120–121
Perinatal infections, 47
Peripheral nervous system
  herpes simplex virus infection confined to, 152
  infection of, 43
Permissive cell, 27–28, 36, 349
Peroxynitrite, 168
Persistent infection, 137–138, 143–150
  adaptive immune response modulation,
    144–147
    immunodominant epitopes, 146
    interference with MHC class I system,
      144–146
    interference with Toll-like receptor detection
      and signaling, 144
    MHC class II modulation, 146
    T cell destruction, 147
  bovine viral diarrhea virus, 144
  cellular mechanisms that promote, 143–144
  described, 137, 143
  Epstein-Barr virus, 143, 146, 155–156
  general pattern of, 138
  human examples, 143
  human immunodeficiency virus, 249
  in immune system cells, 147
  K-replication strategy, 139
  lymphocytic choriomeningitis virus, 143,
    148–150
  measles virus, 147–149
  Sindbis virus, 143–144
  in tissues with reduced immune surveillance, 147
Pestivirus, 144
Peyer's patches, 34, 113
PFU (particle-to-plaque forming unit) ratio, 28
Phagocytes, 54, 60, 62, 67, 74, 87–89
Phagocytosis, 62, 108
Phagosome, 108
Pharmacokinetics, 284
Phases, of clinical trials, 292
Phipps, James, 256
Phlebovirus, 334, 382
Phosphatidylinositol 3-kinase (Pi3k), 180,
  195–196, 198–199, 201, 203, 206, 208
Physarum polycephalum, 180

Picornavirus
  2B protein, 164
  central nervous system entry, 46
  detection within host cell, 59
  diseases, 396–398
  epidemiology, 396–398
  human infections, 396–398
  pathogenesis, 396–398
Pink eye, 35
Pithovirus
  discovery, 331
  electron micrograph, 331
Pithovirus sibericum, 331
Pizarro, Francisco, 352
Pkr, 71, 79, 81, 84, 162–163
Plague Inc. (app), 12
Planctomycetes, 329
Plants
  RNA silencing in, 368
  satellite viruses, 369–370
  viroid infection of, 365–368
Plaque reduction assay, 123
Plasma cell, 101–103, 122, 125–128
Plasma membrane, signal transduction pathway
  receptor activation, 199–200
Plasmid
  DNA vaccines, 273–274
  minireplicon system, 286
Plasmin, 38
Plasminogen, activation of, 38
Plasmodesmata, 366
Platelet-derived growth factor, 179, 192, 196,
  199–200
PML (progressive multifocal leukoencepha-
  lopathy), 37, 399
Pneumocystis jiroveci, 246, 250
Pneumovirus, 334
Pol protein, human immunodeficiency virus, 220
Poliomyelitis
  accidental infection, 351
  epidemic, 351
  eradication effort, 258–259
  globally reported incidence of, 259
  paralytic, 351
  seasonal variation in disease, 17–18
  unexpected consequences of modern sanitation,
    351
  vaccine-associated, 259
Poliovirus
  accidental infections, 351
  alteration of virulence, 159–160
  antibody stimulated by, 126
  autophagy machinery co-opted by, 68
  diseases, 396
  dropped foot consequence of infection, 164
  endemic, 351
  enhanced infection in the presence of intestinal
    bacteria, 33
  epidemic, 351
  epidemiology, 396
  eradication effort, 258–259
  escape from antibodies, 142
  host age influence on severity, 21
  human infections, 396
  inapparent infections, 351
  incubation period, 138
  interferon response to, 77
  outbreak (2010), 142
  particle-to-PFU ratio, 28

  pathogenesis, 396
  pregnant women, susceptibility of, 22
  quasispecies concept, 321
  R_0 (R-naught) value, 318
  reproduction rate, 319
  shedding from host, 39
  spread in host, 114
  vaccination campaign, 256–258, 260
Poliovirus vaccine, 126, 258–259, 267
  administration, 269
  attenuated, 270–271
  bystander vaccination, 270
  disease associated with, 259
  inactivated vaccine, 265
  replication-competent oral, 269
  reversion, 270–271
  Sabin, 126, 142, 259, 270–271
  Salk, 267
  simian virus 40 in, 270
  SV40-free, 185
Polyomavirus, 333
Polymerase chain reaction (PCR), 7, 357–358
Polymorphic, 116
Polyomavirus
  antiviral drugs, 297
  bandicoot papillomatosis carcinomatosis virus
    type 1, 183
  coevolution with host, 333
  diseases, 399
  epidemiology, 399
  host-virus relationship, 333
  human infections, 399
  integration, 187
  Merkel cell carcinoma and, 186
  oncogenic, 185–186
  p53 inactivation, 209
  pathogenesis, 399
  persistent infection, 143
  transforming gene products, 190
  transforming genes, 185
  viral adapter proteins, 198–199, 201
Polyps, 216
Population density
  as accelerator of viral transmission, 10
  spread of infection influenced by, 350
  sustainability of virus populations and, 17
Population dynamics, viral, 139
Population growth, world, 360
Populations, infections of
  epidemics (*See* Epidemic)
  epidemiology of (*See* Epidemiology)
  parameters governing ability of virus to infect,
    16–22
    climate, 17–19
    environment, 16–19
    genetic and immune parameters, 19–21
    geography, 16
    host factors, 19–22
    nongenetic risk factors, 21–22
    population density, 17
  seasonal variations, 17–19
  urban legends concerning, 17, 20
Porcine circovirus
  genome size, 365
  inactivated vaccine, 265
Portals of entry, 29–36
  alimentary tract, 30, 33–35
  eyes, 30, 35–36
  respiratory tract, 30, 31–33

skin, 29–31
    urogenital tract, 30, 35
Positive selection, 108
Pospivirus, 365–368
Postexposure prophylaxis (PEP), human immuno-
    deficiency virus, 250
Postherpetic neuralgia, 269
Potato spindle tuber viroid, 365–366
Potency, antiviral drug, 283
Potosi virus, 16
Potyvirus, 369
Powassan virus, 385
Power, 14
Poxes, 122, 124
Poxvirus, 47
    antiviral drugs, 297
    deliberate release in Australia, 345, 347
    diseases, 400
    E3 ubiquitin ligase, 145
    entry into host, 29
    epidemiology, 400
    human infections, 400
    immune-modulating proteins that affect viral
        virulence, 163
    inadvertent creation of more virulent poxvirus,
        161
    NK modulators, 91
    nucleocytoplasmic large DNA viruses, 329
    oncogenic, 186
    pathogenesis, 400
    recombination, 324
    shedding from skin lesions, 49
    T1-IFN binding protein, 163
    vaccine vectors, 274–275
    viral transforming proteins, 192
pp71 protein, human cytomeglovirus, 68
Preexposure prophylaxis (PrEP), human immuno-
    deficiency virus, 250
Pregnancy
    depression and Epstein-Barr virus, 155
    fetal infection, 47
Preintegration complex, human immunodefi-
    ciency virus type 1, 227
Prelamin A, 311
President's Emergency Plan for AIDS Relief
    (PEPFAR), 219
Prevalence of infection, 11
Prevention, of emerging virus infections, 361
Primary antibody response, 125
Primary cells, 177, 188
Primary viremia, 26, 40–42
Prions, 371–377
    bovine spongiform encephalopathy, 371, 374,
        376, 377
    chronic wasting disease, 376–377
    host range, 374
    human transmissible spongiform
        encephalopathies, 371, 373
    pathogenesis, 372
    prnp gene, 371, 372–374
    scrapie, 371–374
    strains, 374
    treatment of disease, 377
    variant Creutzfeldt-Jakob disease, 374–376
Prnp gene, 371, 372–374
Proapoptotic proteins, 65
Procaspase, 64–65
Processive transcription, 222
Prodrug, 293, 305

Programmed cell death. *See* Apoptosis
Programmed cell death protein 1 (Pd-1), 250
Programmed necrosis (necroptosis), 67
Progressive multifocal leukoencephalopathy
    (PML), 37, 399
ProMED (Program for Monitoring Emerging
    Diseases), 15
Promoter insertion, 195, 198
Promyelocytic leukemia (Pml) bodies, 68–69, 82
Prospective study, 11–12
Protease
    antiviral drugs, 302, 306–307, 311
    cellular, viral tropism and, 37–39
    viral
        antiviral drugs, 301–302
        human cytomegalovirus, 301
        human immunodeficiency virus type 1,
            286–288
        inhibitors of, 285–287, 301
        maturational, 301
Proteasome, 115–117
Protectin, 90
Protein interaction maps, 288
Protein kinase. *See also* Protein tyrosine kinase
    Akt, 179–180, 192, 195–196, 198–201, 203, 206,
        208
    mitogen-activated protein kinase (Mapk), 176,
        179–180, 193–195, 199–203, 206
    Src, 188–190
        c-Src protein, 194–195, 199, 201
        Src family, 193, 198
        v-Src protein, 188, 192–195, 206
Protein kinase R (Pkr), 71, 79, 81, 84, 162–163
Protein phophatase 2A
    inhibition of, 200–202
    viral adapter protein binding to, 199
Protein tyrosine kinase
    c-Src protein, 194–195, 199, 201
    focal adhesion kinase (Fak), 193
    receptor protein tyrosine kinase (Rptk),
        178–180, 192, 199, 206
    v-Src protein, 188, 192–195, 206
Proteinase K, 373–375
Proteins, cellular
    interactions with viral transforming proteins,
        195, 197–201
    regulation of infectious cycle, 36–39
    viroid-binding, 367–368
Proteins, viral
    accessory proteins, human immunodeficiency
        virus type 1, 220, 223, 225–230
    adapter proteins, 198–199, 201, 225
    alteration of production/activity of cellular
        proteins, 195, 197–201
    as antiviral target, 300
    apoptosis-modulating, 66–67
    cellular signal transduction pathway activation,
        192–201
    disruption of cell cycle control, 201–205
    interferon response modulators, 82
    NK-cell modulating, 91–92
    p53 protein inactivation, 208–209
    Rb-binding, 208
    superantigens, 168
    viral transforming proteins, 192–205
Proteome, 288–289
Proteomic diversity, 332
Proto-oncogene, 182, 184, 186, 191–192, 194–195,
    198, 213

Protovirus hypothesis for retrovirus, 335
Provirus
    antiviral therapies and, 312
    human T-lymphotropic virus type 1, 211–212
    integration, 187, 189, 191, 195, 197, 200, 324
    retrovirus, 187, 191, 324
    transformed cells, 184, 187, 189, 191, 195, 197,
        200, 211–212
Prusiner, Stanley, 372–373
Pten protein, 208

Q

Quarantine, 29, 30, 359
Quasispecies concept, 321–322
Quasispecies effect, 321
Quinacrine, 377

R

R_0 (R-naught), 317–318
Rabbitpox, 347
Rabbits, control in Australia, 318, 345, 347
Rabies virus
    bat reservoir, 349
    diseases, 404
    entry into host, 29
    epidemiology, 404
    human infections, 404
    inactivated vaccine, 265
    incubation period, 138
    neural spread, 42–43
    neuroinvasiveness, 43
    neurovirulence, 44
    passive immunization, 261
    pathogenesis, 404
    transmission, 10
Rabies virus vaccine, 267
    for wild animals, 275
Raccoonpox virus, 275
RAGs (recombinase-activating genes), 108
Raltegravir, 308
Ras protein, 179–180, 193–196, 199, 201–203, 206
Rash, 47, 49, 122, 124
Rats, Lice, and History (Zinsser), 3
Rb (retinoblastoma) protein, 192
    abrogation of restriction point control exerted
        by, 201–204
    inhibition of negative regulation by, 204
    stabilization of p53 by viral proteins that bind
        to Rb, 208
Reactivation, 150
    Epstein-Barr virus, 155–157
    herpes simplex virus, 153–154
    signal pathways, 153–154, 157
Reactive oxygen species, 349
Reassortment
    influenza virus, 268, 355
    viral evolution, 318, 324
Receptor protein tyrosine kinase (Rptk), 178–180,
    192, 199, 206
Receptor-mediated endocytosis, 121
Receptors
    accessibility of, 36
    constitutively active viral, 198–199
Recombinant DNA technology, 7
    hepatitis C virus identification, 357
    subunit vaccines, 266, 272–275
Recombinant mouse retrovirus (XMRV), 358

Recombinase-activating genes (RAGs), 108
Recombination
  B and T cell receptor, 108
  bacteriophage, 326
  genetic shift, 324
  nonhomologous, 326, 337
  viral evolution, 318, 324
Red Cross, 264
Reed, Walter, 4, 5
Regulation
  cell cycle, 179–180
  complement cascade, 89–90
Regulatory T cells (Treg), 106–107, 119
Release of virus particles
  directional, 39
  from polarized cells, 39
Relenaz (zanamivir), 298–299
Reovirus
  attenuated reassortment viruses, 269
  diseases, 401
  entry into host, 35
  epidemiology, 401
  human infections, 401
  interferon response modulators, 82
  pathogenesis, 401
  reassortment, 324
  σ3 protein, 84
  spread in host, 114, 164
  transcytosis of, 35
Replicase, 337
Replication
  errors and viral evolution, 318–321
  gene products that alter virus replication, 160–162
  noncoding sequences that affect virus
    replication, 160–162
Replication competent vaccines, 267, 269–271
Replication strategy
  evolution, 326–328
  K-replication strategy, 139, 319
  RNA virus, 327–328
  r-replication strategy, 139, 319
Reproduction number (R_0), 258, 317–318
Reservoir, 5, 9, 10, 361
Resistant host, 344–345, 348
Respiratory syncytial virus
  age-dependent susceptibility, 22
  antiviral drugs, 297
  diseases, 395
  epidemiology, 395
  genetic map, 334
  human infections, 395
  immunopathology, 165–166
  pathogenesis, 395
Respiratory tract
  mucus, 32
  shedding of virions in secretions of, 48
  viral entry, 30, 31–33
Resting state, 180
Restriction point control, 201–204
Restriction-modification (R-M) system, 73
Reticuloendothelial system, 45, 166, 272
Retina, infection of, 35–36
Retinoblastoma, 176
Retinoblastoma (Rb) protein, 192
  abrogation of restriction point control exerted
    by, 201–204
  inhibition of negative regulation by, 204
  stabilization of p53 by viral proteins that bind
    to Rb, 208

Retinoic acid-inducible protein I (Rig-I), 56–57,
    59, 68, 75, 230
Retrograde spread, 44
Retrospective study, 11–12
Retrotransposon, 335
Retrovir. *See* Zidovudine
Retrovirus
  Apobec, degradation of, 72
  diseases, 402–403
  endogenous, 336–337
  Env protein, 220
  epidemiology, 402–403
  evolution, 335
  Fv1 protein targeting of, 71–72
  Gag protein, 220
  gene therapy, 197
  genetic shift, 324
  human infections, 402–403
  integration, 187, 191, 197
  long terminal repeats (LTRs), 335–337
  mutation rate, 320
  NK modulators, 91
  oncogenic virus, 182–184, 324
    genome map, 190
    insertional activation, 195, 197–198
    long-latency, 183–184, 210–213
    nontransducing, 183–184, 187, 195–198, 200,
      210–213
    oncogene capture, 190
    transducing, 183–184, 186–188, 190, 198, 200
  pathogenesis, 402–403
  Pol protein, 220
  protovirus hypothesis, 335
  provirus, 187, 191, 324
  recombination, 324
  replication, 328
  shedding in semen, 49
  slow infections, 157
  T cell receptor function reduction, 120
  transforming infection, 157
  transmission in saliva, 49
  Trim proteins targeting of, 71
  zoonotic infection, 344
Rev protein, human immunodeficiency virus, 220
Reverse transcriptase
  error rate, 320
  human immunodeficiency virus type 1, 320
  inhibitors, 288, 294, 296–297, 302–303, 306,
    310, 312
    lipid metabolism affected by, 248
    nonnucleoside, 303, 306, 312
    nucleoside, 303, 312
    resistance, 310–311
  mutants, 310
Rev-responsive element (RRE), 224
Reye's syndrome, 392
Rhabdovirus
  diseases, 404
  epidemiology, 404
  genetic maps, 334
  human infections, 404
  neural spread, 43
  pathogenesis, 404
  use as self-amplifying tracers of synaptically
    connected neurons, 45
Rhinovirus, 17
  acute infection, 139
  diseases, 397
  epidemiology, 397

  human infections, 397
  incubation period, 138
  pathogenesis, 397
  shedding from host, 39
  structural plasticity, 141
Ribavirin (Virazole), 295, 297, 321
Ribonuclease L (RNase L), 19, 81
Ribonucleotide reductase, 329
  alphaherpesvirus, 162
  herpes simplex virus, 162
Ribozyme
  satellite RNA concatemer cleavage, 369
  viroid, 365, 366
Rift Valley fever virus, 360, 382
  zoonotic infection, 344
Rig-I (retinoic acid-inducible protein I), 56–57,
    59, 68, 75, 230
Ring-slaughter containment, 260
Rip1, 67
Rip3, 67
Risc (RNA-induced silencing complex), 69–70
Risk-taking, public's view of, 263
Ritonavir, 311
R-M (restriction-modification) system, 73
RNA
  circular, 369
  regulatory RNA molecules, 300–301
  satellite, 368–370
  viroid, 365–368
RNA editing
  hepatitis delta satellite virus, 370
  intrinsic immune response, 71–72
RNA genome
  DNA fossils derived from, 337
  genetic maps, 334
  organization, 335
  satellite, 368–370
RNA helicase, 56, 59, 230
RNA interference (RNAi), 69–71
RNA polymerase, 69
  antiviral drugs, 297, 301, 303
  error rate, 319
  helper virus, 369
  hepatitis C virus, 303, 306
  viroid replication, 365–367
RNA polymerase II
  satellite virus replication, 370
  viroid replication, 365–367
RNA sequencing methods (RNA-seq), 288
RNA silencing, in plants, 368
RNA virus
  antiviral drugs for infections, 297
  cis-acting sequences, 321
  error threshold, 322–323
  evolution, 319, 322–323, 326, 333–335
  exchange of genetic information, 324
  genetic maps, 334
  host-virus relationships, 333–335
  integration into vertebrate genomes, 337–338
  mutation rate, 319
  oncogenic virus, 182–184
  replication strategies, similarities in, 327–328
RNA-activated protein kinase, 71
RNA-binding proteins, 69
RNA-dependent RNA polymerase, 69
  antiviral drugs, 297, 301
RNA-induced silencing complex (Risc), 69–70
RNase H, 335
RNase L, 19, 81

Rodents, zoonotic infections and, 347
Rotavirus
    autophagy machinery co-opted by, 68
    diarrhea, 164
    diseases, 401
    epidemiology, 401
    human infections, 401
    M cell destruction by, 35
    pathogenesis, 401
Rotavirus vaccine, 267, 269
Rous, Peyton, 183, 195
Rous sarcoma virus, 193
    foci of transformed avian cells, 178
    insertional activation, 195
    mutants, 187–188
    oncogene acquisition, 324
    oncogene probe preparation, 189
    properties, 184
Rous-associated virus 1, 200
Rptk (receptor protein tyrosine kinase), 178–180,
        192, 199, 206
RRE (Rev-responsive element), 224
r-replication strategy, 319
r-selection, 138–139
Rubella virus, 47
    diseases, 406
    epidemiology, 406
    human infections, 406
    incubation period, 138
    pathogenesis, 406
    persistent infection, 143
    seasonal variation in disease, 17–18
    transplacental infection, 47
Rubella virus vaccine, 267
Rush, Benjamin, 8
Russell, Bertrand, 53
rv-cyclin, 184, 204

S

S (DNA synthesis) phase, 180–182, 201–204,
        206, 208
Sabin poliovirus vaccine, 126, 142, 259, 270–271
Saccharomyces cerevisiae cell cycle, 260
Safety, antiviral drug, 283, 291, 293
Saliva, 34
    shedding of virions in, 48–49
Salivary agglutinin, 34
Salk poliovirus vaccine, 267
SamHd1, 227
Sample size, 14
Sandfly fever, 382
Sandfly fever virus, 382
Sapovirus, 383
Sapporo virus, 383
Saquinavir, 307, 311
Sarcoma, defined, 176
SARS. *See* Severe acute respiratory syndrome
        coronavirus (SARS-CoV)
Satellite cells, 151
Satellite cucumber mosaic virus, 369
Satellite RNAs, 368–370
Satellite tobacco necrosis virus, 369
Satellites, 368–371
    genomes, 368–370
    hepatitis delta satellite virus, 370–371
    origin of, 368
    pathogenesis, 369
    replication, 369

viroids compared, 368
    virophages or satellites, 369–370
Scab, 31
Scarification, 30
Scf protein, 228
Schmallenburg virus, 358
SCID (severe combined immunodeficiency), 197
Sclera, 35
Scrapie
    physical nature of the scrapie agent, 371
    strains of scrapie prion, 374
Secondary structure, viroid RNA, 366
Secondary viremia, 26, 40–42
Secretory antibody (IgA), 34, 125–128
Segmented RNA virus
    genetic maps, 334
    reassortment, 324
Selection
    human immunodeficiency virus type 1, 340
    K-selection, 138–139
    negative, 108
    positive, 108
    r-selection, 138–139
    for transmission and survival inside a host, 327
    in viral evolution, 318–319
Self antigens, 91
Semen, shedding of virions in, 49
Semliki Forest virus, 28
Sendai virus
    polarized release from epithelial cells, 39
    shedding from host, 39
Sentinel cells, 67, 87
Sepsis, 167
Sequence coherence of large DNA viruses, 330
Sequence conservation, 321–322, 332
Sequencing, tracking epidemics by, 9
Serial passage, 323
Severe acute respiratory syndrome coronavirus
        (SARS-CoV), 357, 360
    bat reservoir, 349
    epidemic, 361
    identification of, 362
    zoonotic infection, 352–353
Severe combined immunodeficiency (SCID), 197,
        239
Sexual transmission, 35, 49, 231–234
Shakespeare, William, 25
Shedding of virions, 47–50
    blood, 40, 49
    chicken pox parties, 50
    feces, 49
    milk, 49
    respiratory secretions, 48
    saliva, 48–49
    semen, 49
    skin lesions, 49
    urine, 49
Shingles, 47, 269, 389
Shope, Richard, 185
Shope fibroma virus, 186
Short hairpin RNA (shRNA), 70
Sialic acid, influenza virus, 136
Sialyltransferase, 136
Side effects, antiviral drug, 291–293, 310
σ3 protein, reovirus, 84
Signal transduction pathway
    activation of cellular signal transduction
        pathways by viral transforming proteins,
        192–201

alteration of production/activity of cellular
        proteins, 195, 197–201
    cell signaling induced by receptor
        engagement, 55
    common for IFN-α/β and IL-6, 81
    herpes simplex virus reactivation, 153–154
    inhibition of protein phophatase 2A, 200–202
    insertional activation, 195, 197–198
    interferon γ (IFN-γ), 121
    Jak/Stat pathway, 79, 81
    mitogen-activated protein kinase (Mapk)
        pathway, 176, 179–180, 193–195,
        199–203, 206
    overview, 178–179
    plasma membrane receptor, activation of,
        199–200
    Toll-like receptors (Tlrs), 55–58, 144
    transformed cell survival, 206
    viral adapter proteins, 198–199, 201
    viral homologs of cellular genes, 194–195
    viral receptors, constitutively active, 198–199
    viral signaling molecules acquired from the cell,
        192–195
Significance level, 14
Sigurdsson, Bjorn, 157
Simian cytomegalovirus, stable host-virus
        interaction, 345
Simian immunodeficiency virus, 11
    chemokine receptors, 231
    Nef-defective mutant, 230
    phylogenetic relationships, 222
    SIV$_{cpz}$, 220, 222, 227
    SIV$_{sm}$, 220, 222
    stable host-virus interaction, 345
    as surrogate for studying HIV infection, 136
    tetherin targeted by Nef protein, 228
    vectors, 197
Simian virus 5, genetic map, 334
Simian virus 40 (SV40), 186
    apoptosis and, 66
    contribution to human cancer, 185
    large T antigen, 66, 190, 192, 204–205
    in poliovirus vaccines, 270
    small T antigen, 189–190, 201–202
    transformation of cultured mammalian
        cells, 185
    transforming gene products, 190
Sin Nombre virus, 352, 382
    zoonotic infections, 344, 347
Sindbis virus
    collectins and defensins binding glycoproteins
        of, 90
    persistent infection, 143–144
Sink-source model of viral ecology, 335
Sinusoids
    blood-tissue junction, 45
    virus entry into organs with, 45–46
siRNA. *See* Small interfering RNA
SIRS (systemic inflammatory response
        syndrome), 167
Skin
    alimentary canal compared, 33
    anatomy, 29
    reduced immune surveillance in, 147
    scarification, 30
    shedding of virions from, 49
    viral entry, 29–31, 47
Skin-associated lymphoid tissue, 114
Slaughter, 260

Slime mold, 180
Slow infection, 157
Slx 4 protein, 227
Small interfering RNA (siRNA)
  RNA interference (RNAi), 69–70
  vectors encoding, 70
Small intestine, cellular organization of, 34
Small T antigen, simian virus 40 (SV40), 189–190,
    201–202
Smallpox, 47, 400
  deaths from, 3
  epidemic, 352
  eradication, 255–258
  historical aspects, 255–257, 352
  incubation period, 138
  outbreak (New York City, 1947), 359
  R_0 (R-naught) value, 318
  variolation, 255–256
Smallpox inhibitor of complement, variola virus,
    163
Smallpox inhibitor of complement enzymes
    (Spice) protein, 90
Smallpox vaccine, 261, 264, 267
  administration, 30
  current U.S., 257
  historical aspects, 255–257, 279
  safety, 257, 262
  vaccine-associated infection, 262
Smallpox virus
  eradication of, 345
  herd immunity threshold, 263
  laboratory stocks of, 258
  lesions, 122, 124
  reproduction number, 258
  stable host-virus interaction, 345
Smith, Wilson, 6
Sneezing, 32–33, 48
Snowshoe hare virus, 334
Social parameters, facilitating transmission of
    infection to new hosts, 350
Socs (suppressor of cytokine signaling), 84
Sofosbuvir, 305
Somatic hypermutation, 129
Somatic rearrangements of lymphocyte receptor
    gene segments, 100
Sooty mangabey, 220, 354
Spice (smallpox inhibitor of complement
    enzymes) protein, 90
*Spillover: Animal Infections and the Next Human
    Pandemic* (Quammen), 17
Spread in host, 39–45
  hematogenous spread, 40–42
  neural spread, 42–45
  organ invasion, 45–47
  overview, 39–40
  virulence gene products, 164
Spread of virus
  host-parasite interaction, classic theory of,
    317–318
Sputnik virus, 369–370
Src, 188–190
  c-Src protein, 194–195, 199, 201
  Src family, 193, 198
  v-Src protein, 188, 192–195, 206
Src family tyrosine kinase, 125
SSPE (subacute sclerosing panencephalitis), 148
St. Louis encephalitis virus, 385
ST6Gal I, 136
Stable host-virus interactions, 344–345

Stat proteins, 79, 84
Stat1, 121, 169–170
Stat2, 169
Stat2 receptor, 210
Statistics, use in virology, 14
Stavudine (d4T), 295
Stem cell, hematopoietic, 100–102
Sting, 57, 60
Stomach acid, 34
(+) strand RNA virus
  evolution, 333–335
  genome organization, 335
  replication, 328
(−) strand RNA virus
  evolution, 333
  genetic maps, 334
  reproduction blocked by Mx proteins, 82
Stribild, 312
Structural plasticity, 141
Structure-assisted drug design, 287
SU protein, 147, 237–240, 245, 250, 309
Subacute sclerosing panencephalitis (SSPE), 148
Subtractive hybridization, 189
Subunit vaccines, 270–272
Subviral RNA Database, 365
Super transmitters, 318
Superantigens, 168
Supergroups, 333
Superoxide, 168
Suppressor of cytokine signaling (Socs), 84
Suppressor T cell. *See* Regulatory T cells (Treg)
Surveillance, 15–16
Survival of the fittest, 321
Survival of the flattest, 321, 366, 368
Survivin protein, 208
Susceptible cell, 27–28, 36, 349
SV40. *See* Simian virus 40
Swine influenza virus, 325, 357
Swi/Snf, 80
Symbiotic bacteria, 33
Symmetrel (amantadine), 283–284, 298
Synapse, immunological, 119–120
Systemic infection, 39
Systemic inflammatory response syndrome
    (SIRS), 167
Systems biology, 7

T

T cell receptor, 100–104, 107–109, 113–120, 123,
    125, 129, 168
T cells
  antigen specificity, 107–108
  autoreactive, 170–171
  CD4 (*see* CD4; T helper cells)
  CD8 (*see* CD8; Cytotoxic T lymphocyte)
  cell-mediated immune response, 101, 103,
    119–124
  coreceptors for HIV, 230–231
  cytotoxic (*see* Cytotoxic T lymphocyte)
  double-negative, 104
  double-positive, 104
  γδ cells, 93
  helper (*see* T helper cells)
  heterologous T cell immunity, 167–168
  memory cells, 53, 100, 104, 128–130
    adoptive transfer, 261
    central memory T cells, 129–130
    effector memory T cells, 129

  heterologous T cell immunity, 167–168
  HIV infection, 234, 241–243, 249
  naive, 100, 103–109, 114, 116, 119
  noncytolytic control of infection by, 122
  overview, 102, 104–107
  positive and negative selection, 108
  regulatory T cells (Treg), 106–107, 119
  single-positive, 104
  surface molecules and ligands, 114–115
  tropism, 114
T helper cells, 146
  antigen-presenting cell recognition, 116,
    118–119
  cytokine synthesis, 104–106
  differentiation of subsets, 106
  in HIV infection, 238
  hygiene hypothesis and, 151
  immunopathology lesions caused by, 165–166
  infected CD4+ cells, 147
  response to DNA vaccines, 274
  T$_h$1 cells, 104–106, 165–166
    balance of T$_h$1 and T$_h$2 cells, 166
    immunopathology lesions caused by,
      165–166
  T$_h$2 cells, 104–106, 165–166
    balance of T$_h$1 and T$_h$2 cells, 166
    immunopathology lesions caused by,
      165–166
  T$_h$17 cells, 106, 119
T1-IFN binding protein, poxvirus, 163
Tab2 protein, 59
Tak1 kinase, 59
Tamiflu, 298–299
Taps (transport-associated proteins), 116
TAR (*trans*-activating response element), 222, 224
Tat protein, human immunodeficiency virus, 122,
    147, 220, 222, 224, 238, 245, 248
Tax protein, human T-lymphotropic virus, 147,
    211–212
Tears, 35
Technological developments, 7–8
Telomerase, 178
Temin, Howard, 335
Temperature-sensitive mutants, 187–188, 269
Tenofovir, 250, 297
Tensaw virus, 16
Tetherin, 83, 227–228
TFR1 gene, 339
Tgf-β (transforming growth factor β), 106
Theiler's murine encephalomyelitis virus, 165
Therapeutic index, 291
Thomas, Lewis, 317, 343
3TC (Epivir, lamivudine), 295, 297, 321
Thymidine, 295
Thymidine kinase, 162, 293–294
Thymidylate synthetase, 329
Thymus gland, 102, 108
Tissue invasion. *See* Organ invasion
TM protein, human immunodeficiency virus,
    238–240, 245, 248, 250
Tnf-α (tumor necrosis factor-α), 57, 64, 67, 74, 79,
    90, 104, 109–111, 113, 122, 165, 238
Tobacco necrosis virus, 369
Togavirus
  central nervous system entry, 46
  diseases, 405–406
  epidemiology, 405–406
  human infections, 405–406
  pathogenesis, 405–406

stable host-virus interaction, 345
vector, 405–406
Toll-like receptor(s), 55–58, 105
  interference with detection and signaling, 144
Toll-like receptor 3 (Tlr3), 20
Toll-like receptor 9 (Tlr9), 276
Tolstoy, Leo, 283
TORCH, 47
Toxoplasma, transplacental infection, 47
Traf6, 59, 61
Transcription, processive, 222
Transcription signals, viral, 187
Transcytosis, 34–35, 46, 126
Transduced oncogene, 192–194
Transducing oncogenic retrovirus, 183–184, 186–188
  defective, 188
  genome maps, 190
  oncogene capture, 198, 200
  properties, 184
  transforming genes, 189–190, 192
Transferrin receptor, 339, 354
Transformation, 175–214
Transformed cells, 175–214
  anchorage independence, 178
  autocrine growth stimulation, 178
  cancer terminology, 176
  cellular transformation, 175, 177
  changes in properties of transformed cells
      required for tumorigenesis, 209–210
  contact inhibition, loss of, 178
  control of cell proliferation, 178–182
    cell cycle engine, 180–182
    cell cycle regulation, 179–180
    integration of mitogenic and growth-
        promoting signals, 179–180
    sensing the environment, 178–179
  defined, 175
  escape from immune surveillance, 210
  foci of, 178
  growth parameters and behavior, 177
  immortality, 177
  immune defenses, protection from, 210
  normal cells compared, 177–178
  properties of, 175–178
  survival, mechanisms permitting, 206–209
    apoptotic cascade, viral inhibitors of, 206
    p53 protein inactivation, 206–209
    signaling pathways, 206
  viral genetic information in, 187–189
    state of viral DNA, 187
    viral transforming genes, 187–189
  viral transforming proteins in
    activation of cellular signal transduction
        pathways, 192–201
    disruption of cell cycle control, 201–205
    functions, 192
Transforming growth factor β (Tgf-β), 106
Transforming infection, 157–158
Transgenic mice for studying viral pathogenesis, 137
Transmissible gastroenteritis virus, M cell
    destruction by, 35
Transmissible spongiform encephalopathies
    (TSEs), 371–377
  bovine spongiform encephalopathy, 9, 12, 14, 15
  chronic wasting disease, 376–377
  human, 371, 373
  pathogenesis, 372
  prnp gene, 371, 372–374
  scrapie, 371–374

treatment, 377
variant Creutzfeldt-Jakob disease, 374–376
Transmission of viruses, 47–50
  blood, 40, 49
  chicken pox parties, 50
  feces, 49
  host-parasite interaction, classic theory of,
      317–318
  human activity's effects on, 360–361
  human immunodeficiency virus, 231–234
  milk, 49
  parameters governing ability of virus to infect,
      16–22
    climate, 17–19
    environment, 16–19
    genetic and immune parameters, 19–21
    geography, 16
    host factors, 19–22
    nongenetic risk factors, 21–22
    population density, 17
  respiratory secretions, 48
  saliva, 48–49
  seasonal variation in, 17–19
  selection for transmission and survival inside a
      host, 327
  semen, 49
  skin lesions, 49
  urine, 49
Transplacental infections, 47
Transport-associated proteins (Taps), 116
Tree of life, 332
Treg (regulatory T cells), 106–107, 119
Tremblant du mouton, 371
Trif, 61
Trifluridine, 295
Trim proteins, 71–72, 338
Trim5α, 230
TrimCypA, 230
tRNA ligase, chloroplast, 366
Trojan Horse myth, 41
Tropical spastic paraparesis, 402
Tropism, 36–39
  defined, 36
  host cell proteins and, 36–39
  viral receptor accessibility, 36
Tryptase, 37
TSEs. *See* Transmissible spongiform
    encephalopathies
TT virus, 357
Tumor, 175, 183
  attenuated herpes simplex virus used to clear
      human brain tumors, 163
  benign, 177
  defined, 176, 177
  malignant, 177
  metastasis, 177
  monoclonal, 187
Tumor necrosis factor-α (Tnf-α), 57, 64, 67, 74,
    79, 90, 104, 109–111, 113, 122, 165, 238
Tumor suppressor gene, 176, 182, 186, 214
Tumorigenesis, 183, 200
  changes in properties of transformed cells
      required for, 209–210
  with very long latency, 183, 210–213
  walleye dermal sarcoma virus, 184
2B protein, picornavirus, 164
Tyrosine kinase
  Abl, 203, 207
  c-Src protein, 194–195, 199, 201

receptor protein tyrosine kinase (Rptk),
    178–180, 192, 199, 206
Src family, 193, 198
v-Src protein, 188, 192–195, 206

U

Ubc12, 59
Ubiquitination
  deubiquitinase, viral, 144
  inhibition in persistent infections, 145
  tetherin, 83
Ubiquitin-proteasome pathway components,
    82–83
UL89, human cytomegalovirus, 301
Unc-93B, 20
Uncoating inhibitors, 300
Ung2, 227
United Nations' AIDS program, 219
Urban legends about infections, 17, 20
Urine, shedding of virions in, 49
Urogenital tract, viral entry, 35
U.S. Centers for Disease Control and Prevention
    (CDC), 15
US2 protein, human cytomegalovirus, 144
US3 protein, human cytomegalovirus, 144
US6 protein, human cytomegalovirus, 144
US11 protein
  herpes simplex virus 1, 84
  human cytomegalovirus, 144
Uukuniemi virus, 334
Uve1A, 59

V

V proteins, paramyxovirus, 61
Vaccination. *See also* Vaccine
  active strategies to stimulate immune memory,
      261–264
  bystander, 270
  eradication of disease by, 257–260
  goal of, 255
  historical perspective, 255–257
  introduction of term, 256
  large-scale programs, 257–260, 278
  origins of, 255–260
  schedule, 267
Vaccine, 254–279
  active immunization, 260–261
  adjuvants, 275–276
  AIDS, quest for, 277–279
  anticancer, 273
  attenuated virus, 262, 264, 266–267, 269–271
  booster shot, 263
  cost, 264
  delivery/administration, 264–265, 276
  DNA, 264–274
  efficacy, 260, 262–264, 266
  foot-and-mouth disease virus, 10
  formulation, 276
  fundamental challenge, 264–265
  genomic, 274
  hepatitis B virus, 213
  historical perspective, 255–257
  immune memory stimulated by, 261–264
  immune therapy, 276–277
  immunization schedule, 267
  inactivated (killed) virus, 265–268, 269
  large-scale vaccination programs, 257–260, 278

Vaccine (*continued*)
   licensed in United States, 267
   mutations in vaccine strains, 266
   passive immunization, 260–261
   "poor take," 264
   practicality, 262–264
   protection from disease/infection, 261
   public's view of, 263, 278
   recombinant, 266, 272–275
   replication competent, 267, 269–271
   revertant viruses, 270
   safety, 262–264, 266, 269–270, 272, 275
   science and art of making, 265–275
   stability, 264
   subunit, 266, 270–272
   SV40-free, 185
   virus-like particles, 272–273
Vaccine immunoglobulin, 261
Vaccine thermoses, 264
Vaccine-preventable viruses, 3
Vaccinia virus
   A46R protein, 58, 61
   A52R protein, 58, 61
   complement control protein, 163
   diseases, 400
   E3L protein, 84
   entry into cells, 67
   epidemiology, 400
   immune-modulating proteins that affect viral
      virulence, 163
   interferon discovery, 76
   interferon response modulators, 82
   Isg15, countermeasures against, 72
   pattern receptor recognition function blockage, 61
   RNA interference, suppression of, 71
   scarification, 30
   smallpox vaccination, 261
   Toll-like signaling and, 58, 61
   vaccine vectors, 274–275
Vagina, 35
Valacyclovir (Valtrex), 293
Variant Creutzfeldt-Jakob disease (vCJD), 374–376
Varicella virus vaccine, 267
Varicella-zoster virus, 47
   acute infection, 140–141
   antiviral drugs, 293–294, 310
   chicken pox parties, 50
   diseases, 389
   epidemiology, 389
   human infections, 389
   incubation period, 138
   infectious period, 149
   lesions, 122
   model of infection and spread, 141
   pathogenesis, 389
   persistent infection, 143
   shedding from skin lesions, 49
   transmission, 47
   transplacental infection, 47
Varicella-zoster virus vaccine, 267
   administration, 269
   attenuated, 269
Variola virus
   diseases, 400
   epidemiology, 400
   human infections, 400
   immune-modulating proteins that affect viral
      virulence, 163
   pathogenesis, 400

   smallpox inhibitor of complement, 163
   Spice protein, 90
Variolation, 255–256
Varmus, Harold, 188
VA-RNA I, adenovirus, 84
Vascular endothelial growth factor (Vegf),
   195–196
vCJD (variant Creutzfeldt-Jakob disease), 374–376
v-cyclin, 196, 204
Vector
   adenovirus-associated virus, 251
   attenuated vaccine vectors, 274–275
   encoding siRNAs, 70
   gene therapy use, 197
   HIV immunoprophylaxis, 251
   lentivirus, 197
   lethal immunopathology, 158
   Moloney leukemia virus, 197
Vectors of disease, 4. *See also* Arthropod, vectors
Vegf (vascular endothelial growth factor), 195–196
Venezuelan equine encephalitis virus, 405
Venule, and blood-tissue junction, 45
v-ErbB, 193, 200
Vesicles, 47
Vesicular stomatitis virus
   diseases, 404
   epidemiology, 404
   G protein, 197
   genetic map, 334
   pathogenesis, 404
   polarized release from epithelial cells, 39
   reproduction blocked by Mx proteins, 82
   shedding from host, 39
Vesiculovirus, 334
vFLIP protein, 66, 110, 210
v-Fos, 193
v-gpcr gene, 195–196
Vibrio cholerae, 5
Video game, as epidemic model, 12
Vif protein, human immunodeficiency virus, 72,
   220, 225–226
v-IL-6, human herpesvirus 8, 210
Viral adapter proteins, 198–199, 201, 225
Viral evolution. *See* Evolution
Viral FLICE-like inhibitory protein (vFLIP), 66,
   110, 210
Viral load, antiviral drug dose and, 310
Viral pathogenesis. *See* Pathogenesis
Viral receptors, constitutively active, 198–199
Viral transformation, 175–214. *See also* Transformed cells
Viral transforming genes
   identification and properties, 187–189
   origin and nature of, 189–192
Viral transforming proteins
   activation of cellular signal transduction
      pathways, 192–201
   disruption of cell cycle control, 201–205
   functions, 192
Viral vector. *See* Vector
Viramidine, 297
Virazole (ribavirin), 295, 297, 321
Viremia, 37
   active, 41–42
   characteristics, 42
   defined, 41
   diagnostic value, 42
   passive, 41–42
   in pregnancy, 47

   primary, 26, 40–42
   secondary, 26, 40–42
   shedding of viruses, 49
Viroceptors, 76, 162–163
Virochip, 357, 362
Viroid(s), 365–368
   movement, 366–367
   mutation rate, 368
   as parasites of cellular transcription proteins,
      365
   pathogenesis, 368
   replication, 365–366, 367
   satellites compared, 368
   sequence diversity, 366
   structure, 366
Viroid small interfering RNAs, 368
Virokines, 76, 162–163
Virologic set point, 235
Virophages, 370
Viroporin, 164, 228–229
Virtual screening, 288
Virulence, viral, 158–164
   alteration of, 159–160
   gene products that alter virus replication,
      160–162
   gene products that enable spread in host, 164
   gene products that modify host defense
      mechanisms, 162–163
   inadvertent creation of more virulent poxvirus,
      161
   lymphocytic choriomeningitis virus, 159
   measuring, 158–159
   noncoding sequences that affect virus
      replication, 160–162
   revertant viruses, 270
   selection and evolution, 318
Virulence genes, 160–164
   classes, 160
   gene products that alter virus replication, 160–162
   gene products that enable spread in host, 164
   gene products that modify host defense
      mechanisms, 162–163
   mutations, 160–163
   noncoding sequences that affect virus
      replication, 160–162
Viruria, 49
Virus discovery, 355, 357–358
Virus neutralization, 125–127
Virus-like particles, as vaccine, 272–273
Visa, 59
Visna/maedi virus, 59, 157, 222
Vistide (cidofovir), 296, 297
v-Kit, 193
v-Myc, 193
v-oncogene, 183–184, 187–188, 190, 192
Vpr protein, human immunodeficiency virus, 220,
   225–227
Vpu protein, human immunodeficiency virus, 220,
   225, 227–229, 237
Vpx protein, human immunodeficiency virus
   type 2, 225, 227
v-Ras, 194
v-Sis, 193
v-Src protein, 188, 192–195, 206

W

Wain-Hobson, Simon, 221
Wakefield, Andrew, 278

Walleye dermal sarcoma virus, 184
Wart, 29, 35, 48, 49, 138, 147, 185, 273–274, 393
Washington, George, 8
Weiss, Robin, 221
West Nile virus, 266
   diseases, 385
   DNA vaccine, 274
   epidemiology, 385
   evolving host-virus interaction, 345–346
   genetic influence on infection, 20
   historical aspects, 346
   innate immune response, 59
   neural spread, 43
   pathogenesis, 385
   spread in Western Hemisphere, 9, 17, 346
   symptoms of infection, 346
   transmission, 10
   vector, 385
   zoonotic infection, 344
Western barred bandicoot, 183
Western equine encephalitis virus, 405
Whitehead, Alfred North, 365
Whooping cough, 263
Wieschaus, Eric, 56
Woodville, William, 256

World Health Organization (WHO), 15
World of Warcraft (video game), 12
World travel
   as accelerator of viral transmission, 10
   sustainability of virus populations and, 17

X

X protein, hepatitis B virus, 213
Xanthine oxidase, 168
Xenophagy, 68
Xenotransplantation, 348
Xist RNA, 153
XMRV (recombinant mouse retrovirus), 358
X-ray crystallography, 287, 290

Y

Yellow fever virus
   diseases, 385
   epidemic in Philadelphia (1793), 8–9
   epidemiology, 385
   historical aspects, 4–6
   human infections, 385
   pathogenesis, 385
   vector, 385

Yellow fever virus vaccine, 266, 267
Y-satellite RNA, 369

Z

Zanamivir (Relenza), 298–299
Zidovudine (AZT, Retrovir), 294–297
   chain termination, 296
   resistance to, 311
   structure, 295–296
Zinsser, Hans, 3
ZMapp, 125–126
Zmpste24 protease, 311
Zoonotic infection, 343–344, 347–348, 352–354
   common sources of animal-to-human transmission, 347
   defined, 10
   human behaviors and activities increasing the risk of, 17
   human immunodeficiency virus, 11, 344, 353–354
   lymphocytic choriomeningitis virus, 149
   preventing, 361
   viral epidemics and, 10
Zovirax (acyclovir), 291, 293–294, 295
Zta protein, Epstein-Barr virus, 157